T0325881

THE COSMIC MICROWAVE BACKGROUND

The cosmic microwave background (CMB), the radiation left over from the Big Bang, is arguably the most important topic in modern cosmology. Its theory and observation have revolutionized cosmology from an order-of-magnitude science to a precision science. This graduate textbook describes CMB physics from first principles in a detailed yet pedagogical way, assuming only that the reader has a working knowledge of general relativity. Among the changes in this second edition are new chapters on non-Gaussianities in the CMB and on large-scale structure, and extended discussions on lensing and baryon acoustic oscillations, topics that have developed significantly in the past decade. Discussions of CMB experiments have been updated from Wilkinson Microwave Anisotropy Probe (WMAP) data to the new Planck data. The CMB success story in estimating cosmological parameters is then treated in detail, conveying the beauty of the interplay of theoretical understanding and precise experimental measurements.

RUTH DURRER is a full professor at the Department of Theoretical Physics at Geneva University and was the department director for about 10 years. She previously held junior positions at Zürich, Princeton, and Cambridge Universities. Among her other duties, she was a member of the Swiss Research Council and of the evaluation board for European Research Council Advanced Grants. She has written more than 230 research papers and reviews on different topics in cosmology. Her main research subjects are the CMB and large-scale structure, topological defects, cosmological phase transitions, cosmic magnetic fields, and gravitational waves.

THE COSMIC MICROWAVE BACKGROUND

SECOND EDITION

RUTH DURRER

University of Geneva

CAMBRIDGE
UNIVERSITY PRESS

University Printing House, Cambridge CB2 8BS, United Kingdom

One Liberty Plaza, 20th Floor, New York, NY 10006, USA

477 Williamstown Road, Port Melbourne, VIC 3207, Australia

314–321, 3rd Floor, Plot 3, Splendor Forum, Jasola District Centre, New Delhi – 110025, India

79 Anson Road, #06–04/06, Singapore 079906

Cambridge University Press is part of the University of Cambridge.

It furthers the University's mission by disseminating knowledge in the pursuit of education, learning, and research at the highest international levels of excellence.

www.cambridge.org
Information on this title: www.cambridge.org/9781107135222
DOI: 10.1017/9781316471524

Cover illustration by Martin Zimmermann and Denis Jutzeler

First published 2021

A catalogue record for this publication is available from the British Library.

Library of Congress Cataloging-in-Publication Data
Names: Durrer, Ruth, author.
Title: The cosmic microwave background / Ruth Durrer, Université de Genève.
Description: Second edition. | New York : Cambridge University Press, 2021. |
Includes bibliographical references and index.
Identifiers: LCCN 2020034137 (print) | LCCN 2020034138 (ebook) |
ISBN 9781107135222 (hardback) | ISBN 9781316471524 (epub)
Subjects: LCSH: Cosmic background radiation.
Classification: LCC QB991.C64 D87 2021 (print) |
LCC QB991.C64 (ebook) | DDC 523.1–dc23
LC record available at https://lccn.loc.gov/2020034137
LC ebook record available at https://lccn.loc.gov/2020034138

ISBN 978-1-107-13522-2 Hardback

Additional resources for this publication at www.cambridge.org/durrer.

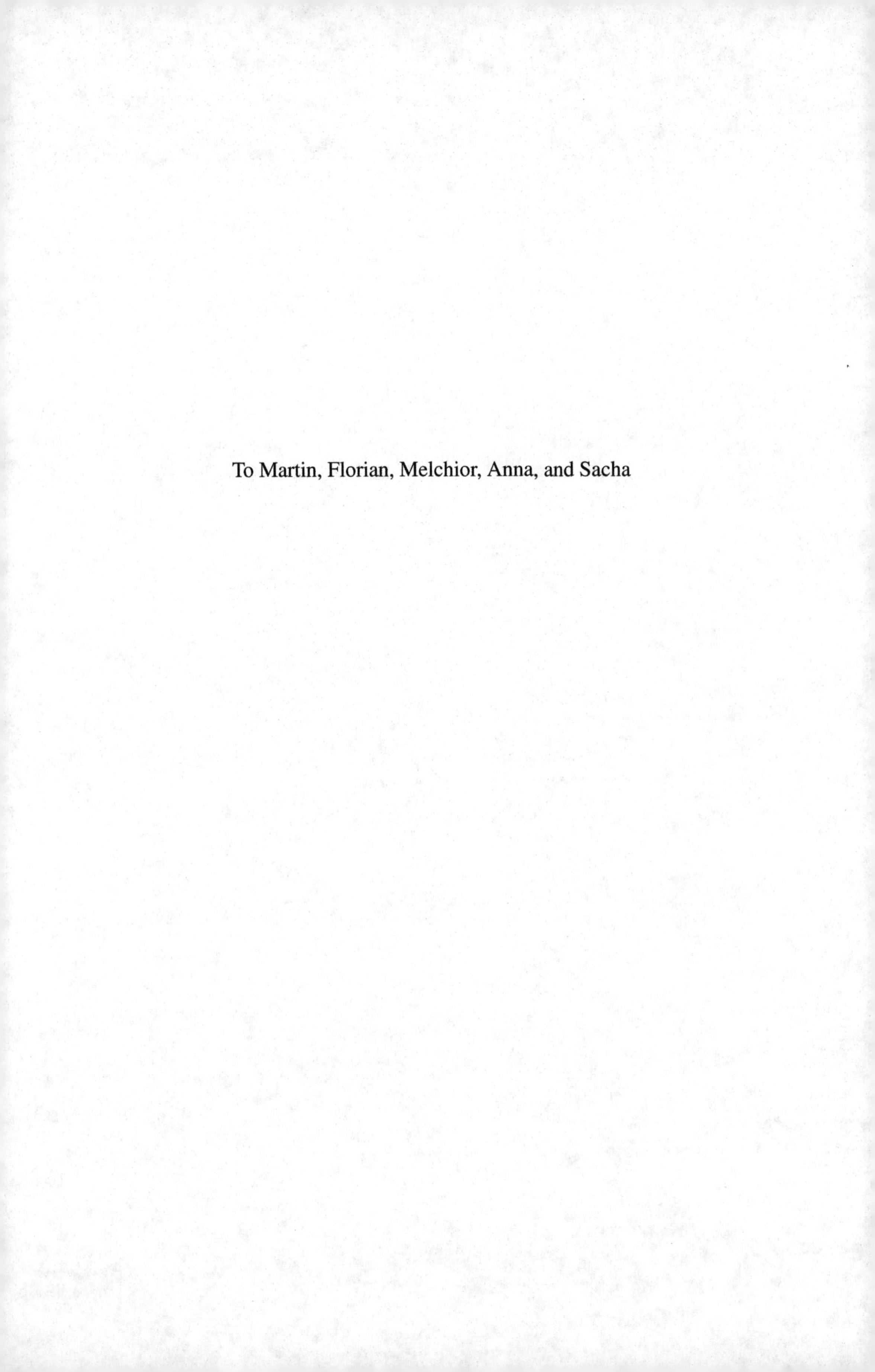

To Martin, Florian, Melchior, Anna, and Sacha

Contents

Preface

Cosmology, the quest concerning the Universe as a whole, has been a primary interest of study since the beginnings of mankind. For a long time our ideas about the Universe were dominated by religious beliefs – tales of creation. Only since the advent of general relativity in 1915 have we had a scientific theory at hand that might be capable of describing the Universe. Soon after Einstein's first attempt of assuming a static universe, Edwin Hubble and collaborators discovered that the observable Universe is expanding [Hubble, 1929; see, however, Nussbaumer & Bieri (2009) recounting the discovery of the expansion of the Universe]. This, together with the discovery of the cosmic microwave background (CMB) by Penzias and Wilson (Nobel Prize 1978), has established the theory of an expanding and cooling universe that started in a "big bang."

For a long time observations that have led to the determination of cosmological parameters, such as the rate of expansion, the so-called Hubble parameter, and the mean matter density of the Universe or its curvature, have been very sparse and we could only determine the order of magnitude of these parameters.

Roughly since the beginning of this century, this situation has changed drastically and cosmology has entered an era of precision measurements. This major breakthrough is to a large extent due to precise measurement and analysis of the CMB. In this book I develop the theory that is used to analyze and understand measurements of the CMB, especially of its anisotropies and polarization, but also its frequency spectrum. The 2006 Nobel Prize in Physics was awarded to George Smoot and John Mather, for the discovery of these anisotropies and for precise measurements of the CMB spectrum.

The book is directed mainly toward graduate students and researchers who want to obtain an overview of the main developments in CMB physics, and who want to understand the state-of-the-art techniques that are used to analyze CMB data. I believe that the theory of CMB physics is sufficiently mature for a book on this topic to be useful. I shall not enter into any details concerning CMB experiments.

This is by no means because I consider them less interesting, but rather that they are still in development and will hopefully make significant progress, especially in polarization measurements, in the near future. Of course, my background is also that of a theoretical physicist and my main interest lies in the theoretical aspects of CMB physics. I hope, however, that this book will also be useful to CMB experimentalists, or more precisely analysts, who want to know what happens inside their cosmic parameter estimation routines.

It is assumed that the reader is familiar with undergraduate physics including the basics of general relativity, and has an elementary knowledge of quantum field theory and particle physics. The beauty of cosmology lies in the fact that it employs more or less all fields of physics starting with general relativity over thermodynamics and statistical physics to electrodynamics, quantum mechanics, and particle physics. In this book I do not want to present an introduction to these topics as well, since, first of all, there exist wonderful textbooks on all of them, and second, you have learned them in your undergraduate physics courses.

Before we start, let me sketch the content of the different chapters and provide a guide on how to read this book.

The first chapter is an overview of the homogeneous and isotropic universe. We present and discuss the Friedmann equations, recombination, nucleosynthesis, and inflation. Readers familiar with cosmology may skip this chapter or just skim it to familiarize themselves with the notation used.

In Chapter 2 we develop cosmological perturbation theory. This is the basics of CMB physics. The main reason why the CMB allows such an accurate determination of cosmological parameters lies in the fact that its anisotropies are small and can be determined mainly within first-order perturbation theory. In Fourier space the linear perturbation equations become a series of ordinary linear differential equations, which can be solved numerically to high precision without any difficulty. We derive the perturbations of Einstein's equations and the energy–momentum conservation equations and solve them for some simple but relevant cases. We also discuss the perturbation equation for light-like geodesics. This is sufficient to calculate the CMB anisotropies in the so-called instant recombination approximation. The main physical effects that are missed in such a treatment are Silk damping on small scales and polarization. We then introduce the matter and CMB power spectrum and draw our first conclusions for its dependence on cosmological and primordial parameters. For example, we derive an approximate formula for the position of the acoustic peaks. Section 2.7 discusses fluctuations not laid down at some initial time but continuously sourced by some inhomogeneous component, a so-called source. This section lies somewhat outside the main scope of this book and can be skipped in a first reading. An experimentalist mainly interested in parameter estimation may jump, after Chapter 2, directly to Chapter 9 and skip the more theoretical parts between.

The third chapter is devoted to the initial conditions. Here we explain how the unavoidable quantum fluctuations are amplified during an inflationary phase and lead to a nearly scale-invariant spectrum of scalar and tensor perturbations. We also calculate the small non-Gaussianities generated during single-field inflation and discuss the initial conditions for mixed adiabatic and isocurvature perturbations.

In Chapter 4 we derive the perturbed Boltzmann equation for CMB photons. After a brief introduction to relativistic kinetic theory, we first derive the Liouville equation, that is, the Boltzmann equation without collision term. We also discuss the connection between the distribution function and the energy–momentum tensor. We then derive the collision term, that is, the right-hand side of the Boltzmann equation, due to Thomson scattering of photons and electrons. In this first attempt we neglect the polarization dependence of Thomson scattering. This treatment, however, includes the finite thickness of the last scattering surface and Silk damping. The chapter ends with a list of the full system of perturbation equations for a ΛCDM universe, including massless neutrinos.

In Chapter 5 we discuss polarization. Here we derive the total angular momentum method that is perfectly adapted to the problem of CMB anisotropies and polarization, taking into account its symmetry, which allows a decomposition into modes with fixed total angular momentum. The representation theory of the rotation group and the spin weighted spherical harmonics that are extensively used in this chapter are deferred to an appendix. We interpret some results using the flat sky approximation, which is valid on small angular scales. This is the most technical chapter of this book and may be glanced over by readers not interested in the gory details.

In Chapter 6 we present an introduction to the vast subject of non-Gaussian perturbations. We mainly concentrate on the bispectrum and the trispectrum. We define some standard shapes of the bispectrum in Fourier space and translate them to angular space. For a description of an arbitrary N-point function in the sky we introduce a basis of rotation-invariant functions on the sphere in Appendix 4, Section A4.2.5. This chapter has been added in the second edition.

In Chapter 7 we introduce weak lensing due to foreground structures with the aim of treating lensing of CMB anisotropies and polarization. This second-order effect is especially important on small scales but has to be taken into account for $\ell \gtrsim 400$ if we want to achieve an accuracy of better than 1%. We first derive the deflection angle and the lensing power spectrum. Then we discuss lensing of CMB fluctuations and polarization in the flat sky approximation, which is sufficiently accurate for angular harmonics with $\ell \gtrsim 50$ where lensing is relevant.

In Chapter 8 we present the analysis of the large scale matter distribution within linear perturbation theory in a fully relativistic way. We take into account that only directions and redshifts are observable while lengths scales are always inferred from a cosmological model. We first introduce the traditional density and redshift space distortion contribution to the observed fluctuations and then proceed to discuss the smaller lensing and large-scale relativistic terms. We express the clustering properties of matter in terms of directly observable quantities and study their scale and redshift dependence. We also discuss "intensity mapping," a new observational technique that will hopefully bear fruit in the near future. This chapter has been newly added in the second edition.

Chapter 9 is devoted to parameter estimation. We first discuss the physical dependence of CMB anisotropies on cosmological parameters. After a section on CMB data we then treat in some detail statistical methods for CMB data analysis. We discuss especially the Fisher matrix and explain Markov chain Monte Carlo methods. We also address degeneracies, combinations of cosmological parameters on which CMB anisotropies and polarization depend only weakly. Because of these degeneracies, cosmological parameter estimation also makes use of other, non-CMB related, observations, especially observations related to the large-scale matter distribution. We summarize them and other cosmological observations in two separate sections.

In the final chapter, spectral distortions of the CMB are discussed. We first introduce the three relevant collision processes in a universe with photons and nonrelativistic electrons: Compton scattering, Bremsstrahlung, and double Compton scattering. We derive the corresponding collision terms and Boltzmann equations. For Compton scattering this leads us to the Kompaneets equation, for which we present a detailed derivation. We introduce timescales corresponding to these three collision processes and determine at which redshift a given process freezes out, that is, becomes slower than cosmic expansion. We also discuss the generation of a chemical potential in the CMB spectrum by a hypothetical particle deçay and by Silk damping of small-scale fluctuations. Finally, we study the Sunyaev–Zel'dovich effect of CMB photons that pass through hot cluster gas.

All chapters are complemented with some exercises at the end.

In the appendices we collect useful constants and formulas, information on special functions, and some more technical derivations. The solutions to a selection of exercises can be found in Appendix 11 available online.

This book has grown out of a graduate course on CMB anisotropies that I have given on several occasions. Thanks are due to the students of these courses, who have motivated me to write it up in the form of a textbook. I am also indebted to many collaborators and colleagues with whom I have discussed various aspects of the book and who have helped me to clarify many issues. I especially want to mention Camille Bonvin, Chiara Caprini, Martin Kunz, Toni Riotto, Uros Seljak, Norbert Straumann, and Filippo Vernizzi. I am also immensely grateful to students and colleagues who have read parts of the draft and helped me correct numerous typographical errors and other mistakes: Giulia Cusin, Jean-Pierre Eckmann, Jérémie Francfort, Alice Gasparini, Basundhara Ghosh, Goran Jelic-Cizmek, Francesca Lepori, Ermis Mitsou, Sandro Scodeller, Vittorio Tansella, Szabolcs Zakany, and others. Of course any remaining mistakes are entirely my responsibility. Basundhara Ghosh, Francesca Lepori, Vittorio Tansella, Marcus Ruser, and Martin Kunz have also helped me with some of the figures. Finally, I wish to thank Susan Staggs, who provided me with a most useful dataset of the CMB frequency spectrum (which unfortunately had no need to be updated for the second edition of the book).

1

The Homogeneous and Isotropic Universe

Notation

In this book we denote the derivative with respect to physical time by a prime and the derivative with respect to conformal time by a dot,

$$\tau = \text{physical (cosmic) time} \qquad \frac{dX}{d\tau} \equiv X', \tag{1.1}$$

$$t = \text{conformal time} \qquad \frac{dX}{dt} \equiv \dot{X}. \tag{1.2}$$

Spatial 3-vectors are denoted by a boldface symbol such as $\mathbf{k}$ or $\mathbf{x}$ whereas four-dimensional spacetime vectors are denoted as $x = (x^\mu)$.

We use the metric signature $(-, +, +, +)$ throughout the book.

The Fourier transform is defined by

$$f(\mathbf{k}) = \int d^3x\, f(\mathbf{x})\, e^{i\mathbf{k}\cdot\mathbf{x}}, \tag{1.3}$$

so that

$$f(\mathbf{x}) = \frac{1}{(2\pi)^3} \int d^3k\, f(\mathbf{k})\, e^{-i\mathbf{k}\cdot\mathbf{x}}. \tag{1.4}$$

We use the same letter for $f(\mathbf{x})$ and for its Fourier transform $f(\mathbf{k})$. The spectrum $P_f(k)$ of a statistically homogeneous and isotropic random variable f is given by

$$\langle f(\mathbf{k}) f^*(\mathbf{k}') \rangle = (2\pi)^3\, \delta(\mathbf{k} - \mathbf{k}') P_f(k). \tag{1.5}$$

Since it is isotropic, $P_f(k)$ is a function only of the modulus $k = |\mathbf{k}|$.

Throughout this book we use units where the speed of light, c; Planck's constant, $\hbar$; and Boltzmann's constant, k_B, are unity: $c = \hbar = k_B = 1$. Length and time therefore have the same units and energy, mass, and momentum also have the same units, which are inverse to the unit of length. Temperature has the same units as energy. We may use cm^{-1} to measure energy, mass, and temperature, or eV^{-1} to measure distances or times. We shall use whatever unit is convenient to discuss a given problem. Conversion factors can be found in Appendix 1.

1.1 Homogeneity and Isotropy

Modern cosmology is based on the hypothesis that our Universe is to a good approximation homogeneous and isotropic on sufficiently large scales. This relatively bold assumption is often called the "cosmological principle." It is an extension of the Copernican principle stating that not only should our place in the Solar System not be a special one, but also that the position of the Milky Way in the Universe should be in no way statistically distinguishable from the position of other galaxies. Furthermore, no direction should be distinguished. The Universe looks statistically the same in all directions. This, together with the hypothesis that the matter density and geometry of the Universe are smooth functions of the position, implies homogeneity and isotropy on sufficiently large scales. Isotropy around each point together with analyticity actually already implies homogeneity of the Universe.[1] A formal proof of this quite intuitive result can be found in Straumann (1974).

But which scale is "sufficiently large"? Certainly not the Solar System or our Galaxy. But also not the size of galaxy clusters. [In cosmology, distances are usually measured in Mpc (Megaparsec). 1 Mpc $= 3.2615 \times 10^6$ light years $= 3.0856 \times 10^{24}$ cm is a typical distance between galaxies; the distance between our neighbor Andromeda and the Milky Way is about 0.7 Mpc. These and other connections between frequently used units can be found in Appendix 1.]

It turns out that the scale at which the *galaxy distribution* becomes homogeneous is difficult to determine. From the analysis of the Sloan Digital Sky Survey (SDSS) it has been concluded that the irregularities in the galaxy density are still on the level of a few percent on scales of 100 Mpc (Hogg *et al.*, 2005). Fortunately, we know that the *geometry* of the Universe shows only small deviations from the homogeneous and isotropic background, already on scales of a few Mpc. The geometry of the Universe can be tested with the peculiar motion of galaxies, with lensing, and in particular with the cosmic microwave background (CMB).

The small deviations from homogeneity and isotropy in the CMB are of utmost importance, since, most probably, they represent the "seeds," that, via gravitational instability, have led to the formation of large-scale structure, galaxies, and eventually solar systems with planets that support life in the Universe.

[1] If "analyticity" is not assumed, the matter distribution could also be fractal and still statistically isotropic around each point. For a detailed elaboration of this idea and its comparison with observations see Sylos Labini *et al.* (1998).

Furthermore, we suppose that the initial fluctuations needed to trigger the process of gravitational instability stem from tiny quantum fluctuations that have been amplified during a period of inflationary expansion of the Universe. I consider this connection of the microscopic quantum world with the largest scales of the Universe to be of breathtaking philosophical beauty.

In this chapter we investigate the background Universe. We shall first discuss the geometry of a homogeneous and isotropic spacetime. Then we investigate two important events in the thermal history of the Universe. Finally, we study the paradigm of inflation. This chapter lays the basis for the following ones where we shall investigate *fluctuations* on the background, most of which can be treated in first-order perturbation theory.

1.2 The Background Geometry of the Universe

1.2.1 The Friedmann Equations

In this section we assume a basic knowledge of general relativity. The notation and sign convention for the curvature tensor that we adopt are specified in Appendix 2, Section A2.1.

Our Universe is described by a four-dimensional spacetime $(\mathcal{M}, g)$ given by a pseudo-Riemannian manifold $\mathcal{M}$ with metric g. A homogeneous and isotropic spacetime is one that admits a slicing into homogeneous and isotropic, that is, maximally symmetric, 3-spaces. There is a preferred geodesic time coordinate τ, called "cosmic time," such that the 3-spaces of constant time, $\Sigma_\tau = \{\mathbf{x}|(\tau, \mathbf{x}) \in \mathcal{M}\}$, are maximally symmetric spaces, hence spaces of constant curvature. The metric g is therefore of the form

$$ds^2 = g_{\mu\nu}\, dx^\mu\, dx^\nu = -d\tau^2 + a^2(\tau)\gamma_{ij}\, dx^i\, dx^j. \tag{1.6}$$

The function $a(\tau)$ is called the scale factor and γ_{ij} is the metric of a 3-space of constant curvature K. Depending on the sign of K this space is locally isometric to a 3-sphere ($K > 0$); a three-dimensional pseudo-sphere ($K < 0$); or flat, Euclidean space ($K = 0$). In later chapters of this book we shall mainly use "conformal time" t defined by $a\, dt = d\tau$, so that

$$ds^2 = g_{\mu\nu}\, dx^\mu\, dx^\nu = a^2(t)\left(-dt^2 + \gamma_{ij}\, dx^i\, dx^j\right). \tag{1.7}$$

The geometry and physics of homogeneous and isotropic solutions to Einstein's equations were first investigated mathematically in the early 1920s by Friedmann (1922, 1924) and physically as a description of the observed expanding Universe in 1927 by Lemaître.[2] Later, Robertson (1936), Walker (1936), and others rediscovered the Friedmann metric and studied several additional aspects. However, since we consider the contributions by Friedmann and Lemaître to be far more fundamental than the subsequent work, we shall call a homogeneous and isotropic solution to Einstein's equations a "Friedmann–Lemaître universe" (FL universe) in this book.

It is interesting to note that the Friedmann solution breaks Lorentz invariance. Friedmann universes are not invariant under boosts; there is a preferred cosmic time τ, the proper time of an observer who sees a spatially homogeneous and isotropic universe. Like so often in physics, the Lagrangian and therefore also the field equations of general relativity are invariant under Lorentz transformations, but a specific solution in general is not. In that sense we are back to Newton's vision of an absolute time. But on small scales, for example, the scale of a laboratory, this violation of Lorentz symmetry is, of course, negligible.

The topology is not determined by the metric and hence by Einstein's equations. There are many compact spaces of negative or vanishing curvature (e.g., the torus), but there are no infinite spaces with positive curvature. A beautiful treatment of the fascinating, but difficult, subject of the topology of spaces with constant curvature and their classification is given in Wolf (1974). Its applications to cosmology are found in Lachieze-Rey and Luminet (1995).

Forms of the metric γ, which we shall often use, are

$$\gamma_{ij}\, dx^i\, dx^j = \frac{\delta_{ij}\, dx^i\, dx^j}{(1 + \frac{1}{4}K\rho^2)^2}, \tag{1.8}$$

$$\gamma_{ij}\, dx^i\, dx^j = dr^2 + \chi^2(r)\left(d\theta^2 + \sin^2(\theta)\, d\varphi^2\right), \tag{1.9}$$

$$\gamma_{ij}\, dx^i\, dx^j = \frac{dR^2}{1 - KR^2} + R^2\left(d\theta^2 + \sin^2(\theta)\, d\varphi^2\right), \tag{1.10}$$

where in Eq. (1.8)

$$\rho^2 = \sum_{i,j=1}^{3} \delta_{ij} x^i x^j, \quad \text{and} \quad \delta_{ij} = \begin{cases} 1 & \text{if } i = j, \\ 0 & \text{else}, \end{cases} \tag{1.11}$$

and in Eq. (1.9);

$$\chi(r) = \begin{cases} r & \text{in the Euclidean case,} \quad K = 0, \\ \frac{1}{\sqrt{K}} \sin(\sqrt{K}r) & \text{in the spherical case,} \quad K > 0, \\ \frac{1}{\sqrt{|K|}} \sinh(\sqrt{|K|}r) & \text{in the hyperbolic case,} \quad K < 0. \end{cases} \tag{1.12}$$

[2] In the English translation of (Lemaître, 1927) from 1931 Lemaître's somewhat premature but pioneering arguments that the observed Universe is actually expanding have been omitted.

Often one normalizes the scale factor such that $K = \pm 1$ whenever $K \neq 0$. One has, however, to keep in mind that in this case r and K become dimensionless and the scale factor a has the dimension of length. If $K = 0$ we can normalize a arbitrarily. We shall usually normalize the scale factor such that $a_0 = 1$ and the curvature is not dimensionless. The coordinate transformations that relate these coordinates are determined in Exercise 1.1.

Owing to the symmetry of spacetime, the energy–momentum tensor can only be of the form

$$(T_{\mu\nu}) = \begin{pmatrix} -\rho g_{00} & \mathbf{0} \\ \mathbf{0} & P g_{ij} \end{pmatrix}. \tag{1.13}$$

There is no additional assumption going into this ansatz, such as the matter content of the Universe being an ideal fluid. It is a simple consequence of homogeneity and isotropy and is also verified for scalar field matter, a viscous fluid, or free-streaming particles in a FL universe. As usual, the energy density ρ and the pressure P are defined as the time- and space-like eigenvalues of (T^{μ}_{ν}).

The Einstein tensor can be calculated from the definition (A2.12) and Eqs. (A2.32)–(A2.39),

$$G_{00} = 3\left[\left(\frac{a'}{a}\right)^2 + \frac{K}{a^2}\right] \qquad \text{(cosmic time)}, \tag{1.14}$$

$$G_{ij} = -\left(2a''a + a'^2 + K\right)\gamma_{ij} \qquad \text{(cosmic time)}, \tag{1.15}$$

$$G_{00} = 3\left[\left(\frac{\dot{a}}{a}\right)^2 + K\right] \qquad \text{(conformal time)}, \tag{1.16}$$

$$G_{ij} = -\left(2\left(\frac{\dot{a}}{a}\right)^{\bullet} + \left(\frac{\dot{a}}{a}\right)^2 + K\right)\gamma_{ij} \qquad \text{(conformal time)}. \tag{1.17}$$

The Einstein equations relate the Einstein tensor to the energy–momentum content of the Universe via $G_{\mu\nu} = 8\pi G T_{\mu\nu} - g_{\mu\nu}\Lambda$. Here Λ is the so-called cosmological constant. In an FL universe the Einstein equations become

$$\left(\frac{a'}{a}\right)^2 + \frac{K}{a^2} = \frac{8\pi G}{3}\rho + \frac{\Lambda}{3} \qquad \text{(cosmic time)}, \tag{1.18}$$

$$2\frac{a''}{a} + \frac{(a')^2}{a^2} + \frac{K}{a^2} = -8\pi G P + \Lambda \qquad \text{(cosmic time)}, \tag{1.19}$$

$$\left(\frac{\dot{a}}{a}\right)^2 + K = \frac{8\pi G}{3}a^2\rho + \frac{a^2\Lambda}{3} \qquad \text{(conformal time)}, \tag{1.20}$$

$$2\left(\frac{\dot{a}}{a}\right)^{\bullet} + \left(\frac{\dot{a}}{a}\right)^2 + K = -8\pi G a^2 P + a^2\Lambda \qquad \text{(conformal time)}. \tag{1.21}$$

Energy "conservation," $T^{\mu\nu}_{;\mu} = 0$, yields

$$\dot{\rho} = -3(\rho + P)\left(\frac{\dot{a}}{a}\right) \quad \text{or, equivalently} \quad \rho' = -3(\rho + P)\left(\frac{a'}{a}\right). \tag{1.22}$$

This equation can also be obtained by differentiating Eq. (1.18) or (1.20) and inserting (1.19) or (1.21); it is a consequence of the contracted Bianchi identities (see Appendix 2, Section A2.1). Equations (1.18)–(1.21) are the Friedmann equations. The quantity

$$H(\tau) \equiv \frac{a'}{a} = \frac{\dot{a}}{a^2} \equiv \mathcal{H}a^{-1}, \tag{1.23}$$

is called the Hubble rate or the Hubble parameter, where $\mathcal{H}$ is the comoving Hubble parameter. At present, the Universe is expanding, so that $H_0 > 0$. We parameterize it by

$$H_0 = 100\, h \text{ km s}^{-1} \text{Mpc}^{-1} \simeq 3.241 \times 10^{-18}\, h \text{ s}^{-1} \simeq 0.3336 \times 10^{-3} \text{ h Mpc}^{-1}.$$

Observations show (Freedman *et al.*, 2001) that $h \simeq 0.72 \pm 0.1$. Equation (1.22) is easily solved in the case $w = P/\rho =$ constant. Then one finds

$$\rho = \rho_0(a_0/a)^{3(1+w)}, \tag{1.24}$$

where ρ_0 and a_0 denote the value of the energy density and the scale factor at present time, τ_0. In this book cosmological quantities indexed by a "0" are evaluated today, $X_0 = X(\tau_0)$. For nonrelativistic matter, $P_m = 0$, we therefore have $\rho_m \propto a^{-3}$ while for radiation (or any kind of massless particles) $P_r = \rho_r/3$ and hence $\rho_r \propto a^{-4}$. A cosmological constant corresponds to $P_\Lambda = -\rho_\Lambda$ and we obtain, as expected, $\rho_\Lambda =$ constant. If the curvature K can be neglected and the energy density is dominated by one component with $w =$ constant, inserting Eq. (1.24) into the Friedmann equations yields the solutions

$$a \propto \tau^{2/3(1+w)} \propto t^{2/(1+3w)} \qquad w = \text{constant} \neq -1, \tag{1.25}$$

$$a \propto \tau^{2/3} \propto t^2 \qquad w = 0, \quad \text{(dust)}, \tag{1.26}$$

$$a \propto \tau^{1/2} \propto t \qquad w = 1/3, \quad \text{(radiation)}, \tag{1.27}$$

$$a \propto \exp(H\tau) \propto 1/|t| \qquad w = -1, \quad \text{(cosmol. const.)}. \tag{1.28}$$

It is interesting to note that if $w < -1$, so-called phantom matter, we have to choose $\tau < 0$ to obtain an expanding universe and the scale factor diverges in finite time, at $\tau = 0$. This is the so-called big rip. Phantom matter has many problems but it is discussed in connection with the supernova type 1a (SN1a) data, which are compatible with an equation of state with $w < -1$ or with an ordinary cosmological constant (Caldwell *et al.*, 2003). For $w < -\frac{1}{3}$ the time coordinate t

has to be chosen as negative for the Universe to expand and spacetime cannot be continued beyond $t = 0$. But $t = 0$ corresponds to a cosmic time, the proper time of a static observer, $\tau = \infty$; this is not singularity. (The geodesics can be continued until affine parameter ∞.)

We also introduce the adiabatic sound speed c_s determined by

$$c_s^2 = \frac{P'}{\rho'} = \frac{\dot{P}}{\dot{\rho}}. \tag{1.29}$$

From this definition and Eq. (1.22) it is easy to see that

$$\dot{w} = 3\mathcal{H}(1+w)(w - c_s^2). \tag{1.30}$$

Hence $w =$ constant if and only if $w = c_s^2$ or $w = -1$. Note that already in a simple mixture of matter and radiation $w \neq c_s^2 \neq$ constant (see Exercise 1.3).

Equation (1.18) implies that for a critical value of the energy density given by

$$\rho(\tau) = \rho_c(\tau) = \frac{3H^2}{8\pi G} \tag{1.31}$$

the curvature and the cosmological constant vanish. The value ρ_c is called the critical density. The ratio $\Omega_X = \rho_X/\rho_c$ is the "density parameter" of the component X. It indicates the fraction that the component X contributes to the expansion of the Universe. We shall make use especially of

$$\Omega_r \equiv \Omega_r(\tau_0) = \frac{\rho_r(\tau_0)}{\rho_c(\tau_0)}, \tag{1.32}$$

$$\Omega_m \equiv \Omega_m(\tau_0) = \frac{\rho_m(\tau_0)}{\rho_c(\tau_0)}, \tag{1.33}$$

$$\Omega_K \equiv \Omega_K(\tau_0) = \frac{-K}{a_0^2 H_0^2}, \tag{1.34}$$

$$\Omega_\Lambda \equiv \Omega_\Lambda(\tau_0) = \frac{\Lambda}{3H_0^2}. \tag{1.35}$$

1.2.2 The "Big Bang" and "Big Crunch" Singularities

We can absorb the cosmological constant into the energy density and pressure by redefining

$$\rho_{\text{eff}} = \rho + \frac{\Lambda}{8\pi G}, \qquad P_{\text{eff}} = P - \frac{\Lambda}{8\pi G}.$$

Since Λ is a constant and $\rho_{\text{eff}} + P_{\text{eff}} = \rho + P$, the conservation equation (1.22) still holds. A first interesting consequence of the Friedmann equations is obtained when subtracting Eq. (1.18) from (1.19). This yields

$$\frac{a''}{a} = -\frac{4\pi G}{3}(\rho_{\text{eff}} + 3P_{\text{eff}}). \tag{1.36}$$

Hence if $\rho_{\text{eff}} + 3P_{\text{eff}} > 0$, the Universe is decelerating. Furthermore, Eqs. (1.22) and (1.36) then imply that in an expanding and decelerating universe

$$\frac{\rho'_{\text{eff}}}{\rho_{\text{eff}}} < -2\frac{a'}{a},$$

so that ρ decays faster than $1/a^2$. If the curvature is positive, $K > 0$, this implies that at some time in the future, τ_{max}, the density has dropped down to the value of the curvature term, $K/a^2(\tau_{\text{max}}) = 8\pi G\rho_{\text{eff}}(\tau_{\text{max}})$. Then the Universe stops expanding and recollapses. Furthermore, this is independent of curvature; as a' decreases the curve $a(\tau)$ is concave and thus cuts the $a = 0$ line at some finite time in the past. This moment of time is called the "big bang." The spatial metric vanishes at this value of τ, which we usually choose to be $\tau = 0$; and spacetime cannot be continued to earlier times. This is not a coordinate singularity. From the Ricci tensor given in Eqs. (A2.32) and (A2.33) one obtains the Riemann scalar

$$R = 6\left[\frac{a''}{a} + \left(\frac{a'}{a}\right)^2 + \frac{K}{a^2}\right],$$

which also diverges if $a \to 0$. Also the energy density, which grows faster than $1/a^2$ as $a \to 0$, diverges at the big bang.

If the curvature K is positive, the Universe contracts after $\tau = \tau_{\text{max}}$ and, since the graph $a(\tau)$ is convex, reaches $a = 0$ at some finite time τ_c, the time of the "big crunch." The big crunch is also a physical singularity beyond which spacetime cannot be continued.

It is important to note that this behavior of the scale factor can be implied only if the so-called strong energy condition holds, $\rho_{\text{eff}} + 3P_{\text{eff}} > 0$. This is illustrated in Fig. 1.1.

1.2.3 Cosmological Distance Measures

It is notoriously difficult to measure distances in the Universe. The position of an object in the sky gives us its angular coordinates, but how far away is the object from us? This problem had plagued cosmology for centuries. It took until 1915–1920 when Hubble discovered that the "spiral nebulae" are actually not situated inside our own galaxy but much further away. This then led to the discovery of the expansion of the Universe.

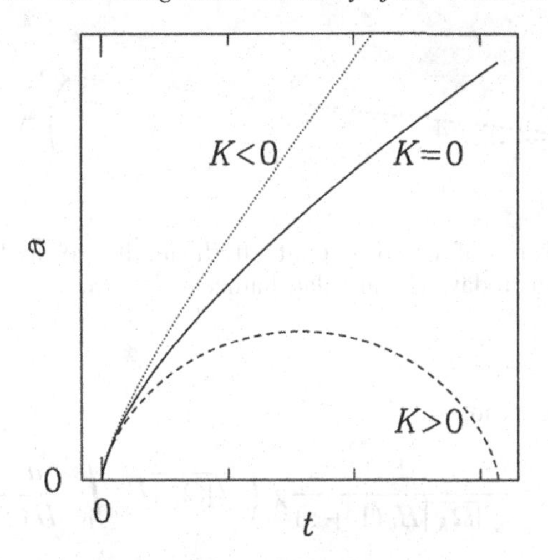

Fig. 1.1 The kinematics of the scale factor in a Friedmann–Lemaître universe that satisfies the strong energy condition, $\rho_{\rm eff} + 3P_{\rm eff} > 0$.

For cosmologically distant objects, a third coordinate, which is today relatively easy to obtain, is the redshift z experienced by the photons emitted from the object. A given spectral line with intrinsic wavelength λ is redshifted due to the expansion of the Universe. If it is emitted at some time τ, it reaches us today with wavelength $\lambda_0 = \lambda a_0/a(\tau) = (1+z)\lambda$. This leads to the definition of the cosmic redshift

$$z(\tau) + 1 = \frac{a_0}{a(\tau)}. \tag{1.37}$$

On the other hand, an object at physical distance $d = a_0 r$ away from us, at redshift $z \ll 1$, recedes with speed $v = H_0 d$. To the lowest order in z, we have $\tau_0 - \tau \approx d$ and $a_0 \approx a(\tau) + a'(\tau_0 - \tau)$, so that

$$1 + z \approx 1 + \frac{a'}{a}(\tau_0 - \tau) \approx 1 + H_0 d.$$

For objects that are sufficiently close, $z \ll 1$. We therefore have $v \approx z$ and hence $H_0 = z/d$. This is the method usually applied to measure the Hubble constant.

There are different ways to measure distances in cosmology, all of which give the same result in a Minkowski universe but differ in an expanding universe. They are, however, simply related, as we shall see.

One possibility is to define the distance d_A to a certain object of given physical size Δ seen at redshift z_1 such that the angle subtended by the object is given by

$$\vartheta = \Delta/d_A, \qquad d_A = \Delta/\vartheta. \tag{1.38}$$

This is the angular diameter distance; see Fig. 1.2.

Fig. 1.2 The two ends of the object emit a flash simultaneously from A and B at z_1 which reaches us today. The angular diameter distance to A (or B) is defined by $d_A = \Delta/\vartheta$.

We now derive the expression

$$d_A(z) = \frac{1}{\sqrt{|\Omega_K|}H_0(1+z)}\chi\left(\sqrt{|\Omega_K|}H_0\int_0^z \frac{dz'}{H(z')}\right), \tag{1.39}$$

for the angular diameter distance to redshift z. In a given cosmological model, this allows us to express the angular diameter distance for a given redshift as a function of the cosmological parameters.

To derive Eq. (1.39) we use the coordinates introduced in Eq. (1.9). Without loss of generality we set $r = 0$ at our position. We consider an object of physical size Δ at redshift z_1 simultaneously emitting a flash at both of its ends, A and B. Hence $r = r_1 = t_0 - t_1$ at the position of the flashes, A and B at redshift z_1. If Δ denotes the physical arc length between A and B we have $\Delta = a(t_1)\chi(r_1)\vartheta = a(t_1)\chi(t_0 - t_1)\vartheta$, that is,

$$\vartheta = \frac{\Delta}{a(t_1)\chi(t_0 - t_1)}. \tag{1.40}$$

According to Eq. (1.38) the angular diameter distance to t_1 or z_1 is therefore given by

$$a(t_1)\chi(t_0 - t_1) \equiv d_A(z_1). \tag{1.41}$$

To obtain an expression for $d_A(z)$ in terms of the cosmic density parameters and the redshift, we have to calculate $(t_0 - t_1)(z_1)$.

Note that in the case $K = 0$ we can normalize the scale factor a as we want, and it is convenient to choose $a_0 = 1$, so that comoving scales become physical scales today. However, for $K \neq 0$, we have already normalized a such that $K = \pm 1$ and $\chi(r) = \sin r$ or $\sinh r$. In this case, we have no normalization constant left and a_0 has the dimension of a length. The present spatial curvature of the Universe then is $\pm 1/a_0^2$.

The Friedmann equation Eq. (1.20) reads

$$\dot{a}^2 = \frac{8\pi G}{3}a^4\rho + \frac{1}{3}\Lambda a^4 - Ka^2, \tag{1.42}$$

where $\dot{a} = da/dt$. To be specific, we assume that ρ is a combination of dust, cold, nonrelativistic "matter" of $P_m = 0$ and radiation of $P_r = \rho_r/3$.

Since $\rho_r \propto a^{-4}$ and $\rho_m \propto a^{-3}$, we can express the terms on the right-hand side of Eq. (1.42) as

$$\frac{8\pi G}{3} a^4 \rho = H_0^2 \left(a_0^4 \Omega_r + \Omega_m a a_0^3\right), \tag{1.43}$$

$$\frac{1}{3} \Lambda a^4 = H_0^2 \Omega_\Lambda a^4, \tag{1.44}$$

$$-K a^2 = H_0^2 \Omega_K a^2 a_0^2. \tag{1.45}$$

The Friedmann equation then implies

$$\frac{da}{dt} = H_0 a_0^2 \left(\Omega_r + \frac{a}{a_0}\Omega_m + \frac{a^4}{a_0^4}\Omega_\Lambda + \frac{a^2}{a_0^2}\Omega_K\right)^{1/2}, \tag{1.46}$$

so that

$$\begin{aligned} r(z_1) &= t_0 - t_1 \\ &= \frac{1}{H_0 a_0} \int_0^{z_1} \frac{dz}{\left[\Omega_r (z+1)^4 + \Omega_m (z+1)^3 + \Omega_\Lambda + \Omega_K (z+1)^2\right]^{1/2}} \\ &= \frac{1}{a_0} \int_0^{z_1} \frac{dz}{H(z)}. \end{aligned} \tag{1.47}$$

Here we have used $z + 1 = a_0/a$ so that $da = -dz a_0/(1+z)^2$.

In principle, we could of course also add other matter components such as, for example, "quintessence" (Caldwell and Steinhardt, 1998), which would lead to a somewhat different form of the integral (1.47), but for definiteness, we remain with matter, radiation, and a cosmological constant.

From $-K/H_0^2 a_0^2 = \Omega_K$ we obtain $H_0 a_0 = 1/\sqrt{|\Omega_K|}$ for $\Omega_K \neq 0$. The expression for the angular diameter distance thus becomes

$$d_A(z) = \begin{cases} \frac{1}{\sqrt{|\Omega_K|} H_0 (z+1)} \chi\left(\sqrt{|\Omega_K|} \int_0^z \frac{dz'}{\left[\Omega_r (z'+1)^4 + \Omega_m (z'+1)^3 + \Omega_\Lambda + \Omega_K (z'+1)^2\right]^{1/2}}\right) \\ \quad \text{if } K \neq 0 \\ \frac{1}{H_0 (z+1)} \int_0^z \frac{dz'}{\left[\Omega_r (z'+1)^4 + \Omega_m (z'+1)^3 + \Omega_\Lambda\right]^{1/2}} \\ \quad \text{if } K = 0. \end{cases} \tag{1.48}$$

Using the Friedmann equation, this formula can also be written in the more general form of Eq. (1.39).

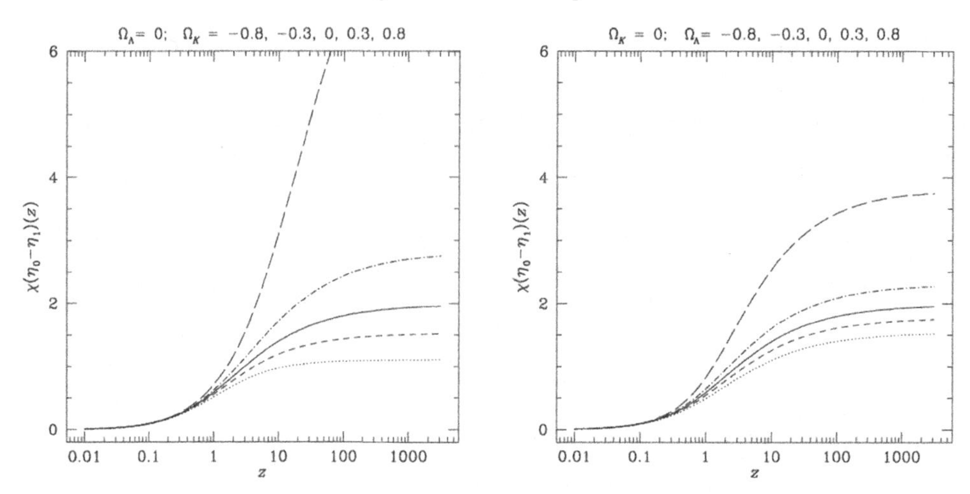

Fig. 1.3 The function $\chi(t_0 - t_1)$ as a function of the redshift z for different values of the cosmological parameters Ω_K (left, with $\Omega_\Lambda = 0$) and Ω_Λ (right, with $\Omega_K = 0$), namely -0.8 (dotted), -0.3 (short-dashed), 0 (solid), 0.3 (dot-dashed), 0.8 (long-dashed).

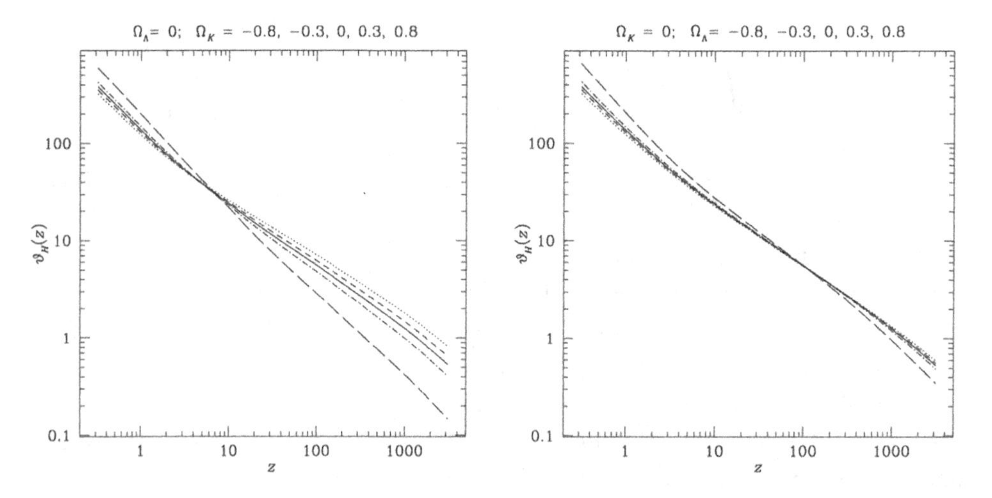

Fig. 1.4 $\vartheta_H(z_1)$ (in degrees) for different values of the cosmological parameters Ω_K and Ω_Λ. The line styles are as in Fig. 1.3.

In general, the integral in Eq. (1.48) has to be solved numerically. It determines the angle $\vartheta(\Delta, z) = \Delta/d_A(z)$ under which an object of size Δ placed at redshift z is seen (see Figs. 1.3 and 1.4).

If we are able to measure the redshifts and the angular extensions of a certain class of objects at different redshifts, of which we know the intrinsic size Δ, comparing with Eq. (1.48) allows us, in principle, to determine the parameters Ω_m, Ω_Λ, Ω_K, and H_0.

Observationally we know for certain that $10^{-5} < \Omega_r \le 10^{-4}$ as well as $0.1 \le \Omega_m \lesssim 1$, $|\Omega_\Lambda| \lesssim 1$, and $|\Omega_K| \lesssim 1$.

If we are interested in small redshifts, $z_1 \lesssim 10$, we may therefore safely neglect Ω_r. In this region, Eq. (1.48) is very sensitive to Ω_Λ and provides an excellent mean to constrain the cosmological constant.

At high redshift, $z_1 \gtrsim 1000$, neglecting radiation is no longer a good approximation.

We shall later also need the opening angle of the *horizon* distance,

$$\vartheta_H(z_1) = \frac{t_1}{\chi(t_0 - t_1)}, \tag{1.49}$$

$$t_1 = \frac{1}{H_0 a_0} \int_{z_1}^{\infty} \frac{dz}{\left[\Omega_r (z+1)^4 + \Omega_m (z+1)^3 + \Omega_\Lambda + \Omega_K (z+1)^2\right]^{1/2}}. \tag{1.50}$$

(Clearly this integral diverges if $\Omega_r = \Omega_m = 0$. This is exactly what happens during an inflationary period and leads there to the solution of the horizon problem; see Section 1.5.)

Neglecting Ω_r, for $\Omega_\Lambda = 0$ and small curvature, $0 < |\Omega_K| < \Omega_m z_1$ at high enough redshift, $z_1 \ge 10$, one has $t_0 - t_1 \simeq 2\sqrt{|\Omega_K|/\Omega_m} = 2/(H_0 a_0 \sqrt{\Omega_m})$. With $\chi(x) \simeq x$, which is valid for small curvature, this yields $\vartheta(\Delta, z_1) \simeq \sqrt{\Omega_m} H_0 a_0 \Delta/(2a_1) = \frac{1}{2}\sqrt{\Omega_m} H_0 \Delta/(z_1 + 1)$ (see also Exercise 1.10).

Another important distance measure in cosmology is the luminosity distance. It is defined as follows. Let L be the luminosity (energy emitted per second) of a source at redshift z_1 and F its flux (energy received per second per square centimeter) arriving at the observer position. We define the luminosity distance to the source by

$$d_L(z_1) \equiv \left(\frac{L}{4\pi F}\right)^{1/2}. \tag{1.51}$$

We now want to show that $d_L(z_1) = (1 + z_1)^2 d_A(z_1)$.

In a proper time interval of the emitter, $d\tau_1 = a(t_1)\, dt$, the source emits the energy $L a(t_1)\, dt$. This energy is redshifted by a factor of $(1 + z_1)^{-1} = a(t_1)/a(t_0)$. It is then distributed over a sphere with radius $a(t_0)\chi(t_0 - t_1)$. So that the flux per proper time of the observer $d\tau_0 = a(t_0)\, dt$ becomes

$$F = \frac{L a^2(t_1)}{4\pi a^4(t_0) \chi^2(t_0 - t_1)},$$

leading to

$$d_L(z_1) = \frac{a(t_0)^2}{a(t_1)}\chi(t_0 - t_1) = (1 + z_1)^2 d_A(z_1). \tag{1.52}$$

The luminosity distance hence contains two additional factors $(1 + z)$ compared to the angular diameter distance. One of them is due to the "redshift" of proper time and the other is due to the redshift of photon energy.

1.3 Recombination and Decoupling

We assume that, at sufficiently early times, reaction rates for particle interactions are much faster than the expansion rate, so that the cosmic fluid is in thermal equilibrium. During its expansion, the Universe then cools adiabatically. At early times, it is dominated by a relativistic radiation background with

$$\rho = C/a^4 = \frac{g_{\rm eff}}{2} a_{\rm SB} T^4. \tag{1.53}$$

This behavior implies that $T \propto a^{-1}$. Here $g_{\rm eff}$ is the effective number of degrees of freedom, which we define below and $a_{\rm SB}$ is the Stefan–Boltzmann constant, $a_{\rm SB} = \pi^2/15$ in our units. For massless (or extremely relativistic) fermions and bosons in thermal equilibrium at temperature T with N_b respectively N_f spin degrees of freedom we have (remember that we use units such that $\hbar = k_B = c = 1$)

$$\rho_b = \frac{N_b 4\pi}{(2\pi)^3}\int_0^\infty \frac{p^3\,dp}{\exp(p/T) - 1} = \frac{N_b T^4}{2\pi^2}\int_0^\infty \frac{x^3\,dx}{\exp(x) - 1}$$
$$= \frac{N_b T^4}{2\pi^2}\Gamma(4)\zeta(4) = \frac{N_b T^4 \pi^2}{30}, \tag{1.54}$$

$$\rho_f = \frac{N_f 4\pi}{(2\pi)^3}\int_0^\infty \frac{p^3\,dp}{\exp(p/T) + 1} = \frac{N_f T^4}{2\pi^2}\int_0^\infty \frac{x^3\,dx}{\exp(x) + 1}$$
$$= \frac{N_f T^4}{2\pi^2}\Gamma(4)\zeta(4)\frac{7}{8} = \frac{7}{8}\frac{N_f T^4 \pi^2}{30}, \tag{1.55}$$

where Γ denotes the Gamma-function and ζ is the Riemann zeta-function and we make use of the integrals (Gradshteyn and Ryzhik, 2000)

$$I_b(\alpha) = \int_0^\infty \frac{x^\alpha\,dx}{\exp(x) - 1} = \Gamma(\alpha + 1)\zeta(\alpha + 1), \tag{1.56}$$

$$I_f(\alpha) = \int_0^\infty \frac{x^\alpha\,dx}{\exp(x) + 1} = \left[1 - \left(\frac{1}{2}\right)^\alpha\right]\Gamma(\alpha + 1)\zeta(\alpha + 1). \tag{1.57}$$

Furthermore, $\zeta(2) = \pi^2/6$, $\zeta(4) = \pi^4/90$, and $\Gamma(n) = (n-1)!$ for $n \in \mathbb{N}$; see Abramowitz and Stegun (1970).

Hence $\rho = \rho_b + \rho_f = \frac{g_{\rm eff}}{2} a_{SB} T^4$ for $a_{SB} = \pi^2 k_B^4/(15\,\hbar^3 c^2) = \pi^2/15$ and $g_{\rm eff} = N_b + 7/8 N_f$, if all the particles are at the same temperature T. If the temperatures are different, such as, for example, the neutrino temperature after electron–positron annihilation, this has to be taken into account with a factor $(T_\nu/T_\gamma)^4$ multiplying N_ν in $g_{\rm eff}$.

At temperatures below the electron mass, at $T < m_e \simeq 0.511$ MeV, only neutrinos and photons are still relativistic. Very recently, $T \lesssim 0.06$ eV at least some of the neutrinos also become nonrelativistic so that the density parameter of relativistic particles today is probably given only by the photon density,[3]

$$\Omega_{\rm rel} = \Omega_\gamma = \frac{8\pi G}{3H_0^2} a_{\rm SB} T_0^4 = 2.49 \times 10^{-5} h^{-2}. \tag{1.58}$$

Here we have set $T_0 = 2.725$ K. The present CMB temperature is the most precisely measured number in cosmology. Its value is (Fixsen, 2009)

$$T_0 = 2.72548 \pm 0.00057\text{K}. \tag{1.59}$$

The pressure of relativistic particles is given by $P = T_i^i/3 = \rho/3$. The thermodynamic relation $dE = T\,dS - P\,dV$ therefore gives for the entropy density $s = dS/dV$

$$s = \frac{dS}{dV} = \frac{1}{T}\left(\frac{dE}{dV} + P\right) = \frac{\rho + P}{T} = \frac{4\rho}{3T}. \tag{1.60}$$

Using the expression for the energy density (1.54) and (1.55) this gives for each particle species X

$$s_X = \begin{cases} \frac{2\pi^2}{45} N_X T^3 & \text{for bosons,} \\ \frac{7\pi^2}{180} N_X T^3 & \text{for fermions.} \end{cases} \tag{1.61}$$

The particle density for relativistic particles is given by

$$n_X = \frac{N_X}{2\pi^2} \int \frac{p^2}{\exp(p/T) \pm 1} dp = \begin{cases} T^3 \frac{N_X}{\pi^2} \zeta(3) & \text{for bosons,} \\ T^3 \frac{N_X}{\pi^2} \zeta(3) \frac{3}{4} & \text{for fermions.} \end{cases} \tag{1.62}$$

[3] At present only neutrino mass differences are known from oscillation experiments. The lowest neutrino mass could still be zero, or at least lower than T_0. From oscillation experiments, however, we know that the heaviest neutrino mass is at least 0.05eV (see Olive *et al.*, 2014).

The particle and entropy densities both scale like T^3. Using $\zeta(3) \simeq 1.202\,057$ we obtain

$$s_X \simeq \begin{cases} 3.6 \cdot n_X & \text{for bosons,} \\ 4.2 \cdot n_X & \text{for fermions.} \end{cases} \tag{1.63}$$

The photons obey a Planck distribution ($\epsilon = ap =$ the photon energy),

$$f(\epsilon) = \frac{1}{e^{\epsilon/T} - 1}. \tag{1.64}$$

At a temperature of about $T \sim 4000\,\mathrm{K} \sim 0.4$ eV, the number density of photons with energies above the hydrogen ionization energy ($= \Delta = 1\,\mathrm{Ry} = 13.6$ eV) drops below the baryon density of the Universe, and the protons begin to (re)combine to neutral hydrogen. Even though electrons and protons were not combined to neutral hydrogen before, this process is called "recombination" rather than "combination."

Helium has already recombined earlier. The binding energy of the first electron to the He nucleus is $4\Delta = 54.4$eV. Using the Saha equation derived in the next section for the transition $\mathrm{He}^{+2} \to \mathrm{He}^{+}$, one finds that the recombination of the first electron transition takes place at $T_{2\to1} \simeq 1.4 \times 10^4$K. The binding energy of the second electron to the He nucleus is 24.6 eV and, again using the Saha equation, one finds that the transition $\mathrm{He}^{+} \to \mathrm{He}$ takes place at $T_{1\to0} \simeq 0.5 \times 10^4$K (see Exercise 1.5).

Before (re)combination photons and baryons are tightly coupled by Thomson scattering of electrons. During recombination the free electron density drops sharply and the mean free path of the photons grows larger than the Hubble scale. At the temperature $T_{\mathrm{dec}} \sim 3000$ K (corresponding to the redshift $z_{\mathrm{dec}} \simeq 1100$ and the physical time $\tau_{\mathrm{dec}} \simeq a_{\mathrm{dec}} t_{\mathrm{dec}} \simeq 10^5$ yr) photons decouple from the electrons and the Universe becomes transparent. We now want to study this process in somewhat more detail.

1.3.1 The Physics of Recombination

From Eq. (1.63) with $N_\gamma = 2$ we obtain that the photon entropy is given by

$$s_\gamma = \frac{4\pi^2}{45} T^3 \simeq 3.6 n_\gamma.$$

The conserved baryon number n_B satisfies $a^3 n_B =$ constant; hence $n_B \propto a^{-3} \propto T^3$. The entropy per baryon is therefore a constant,

$$\sigma = s_\gamma / n_B = \frac{\frac{4\pi^2}{45} T_0^3}{\Omega_B \rho_c(\tau_0)/m_p} = 1.4 \times 10^8 \frac{T_{2.7}^3}{\Omega_B h^2}. \tag{1.65}$$

Here we have used (see Appendix 1)

$$\rho_c(\tau_0) = 1.88h^2 \times 10^{-29}\,\mathrm{g\,cm^{-3}} = 8.1h^2 \times 10^{-11}\,(\mathrm{eV})^4,$$
$$m_p = 9.38 \times 10^8\ \mathrm{eV}, \qquad \text{(proton mass)},$$
$$T(\tau_0) = 2.3T_{2.7} \times 10^{-4}\ \mathrm{eV}, \quad T_{2.7} = T(\tau_0)/2.7\,\mathrm{K}.$$

As we shall see in the next section, the baryon density is approximately $\Omega_B h^2 \simeq 2.2 \times 10^{-2}$ so that $\sigma \simeq 10^{10}$. Correspondingly the ratio between the baryon and photon density is

$$\eta_B = n_B/n_\gamma = 2.7 \times 10^{-8}\left(\frac{\Omega_B h^2}{T_{2.7}^3}\right) \simeq 6 \times 10^{-10}. \tag{1.66}$$

As long as hydrogen is ionized, the timescale of interaction between photons and electrons (Thomson scattering) and between electrons and protons (Rutherford scattering) is much faster than expansion and we may therefore consider the latter as adiabatic. At every moment, the electron, proton, and photon plasma is in thermal equilibrium. As long as the temperature is above the ionization energy of neutral hydrogen, $T > 1\,\mathrm{Ry} = \Delta = \alpha^2 m_e/2 = 13.6$ eV, all hydrogen atoms that form are rapidly dissociated. Most electrons and protons are free and the neutral hydrogen density is very low. At some sufficiently low temperature, however, there will no longer be sufficiently many energetic photons around to disrupt neutral hydrogen and the latter becomes more and more abundant. To determine the temperature at which this transition, called "recombination,"[4] happens, we apply the standard rules of equilibrium statistical mechanics to the reaction

$$e^- + p \longleftrightarrow \mathrm{H} + \gamma\ (13.6\ \mathrm{eV}). \tag{1.67}$$

Supposing that pressure and temperature are fixed and only the number of free electrons, N_e; free protons, N_p; hydrogen atoms, N_H; and photons, N_γ, can change, the second law of thermodynamics implies that the Gibbs potential G is constant,

$$0 = dG = \mu_p\, dN_p + \mu_e\, dN_e + \mu_H\, dN_H + \mu_\gamma\, dN_\gamma,$$

Here μ_X denotes the chemical potential of species X. The different dN_X are not independent. Particle number conservation implies

$$dN_p + dN_H = dN_e + dN_H = 0. \tag{1.68}$$

As there is no conservation of photons, the chemical potential of photons vanishes, $\mu_\gamma = 0$. With this and Eq. (1.68) the Gibbs equation, $dG = 0$, implies

$$\mu_e + \mu_p - \mu_H = 0. \tag{1.69}$$

[4] The expression "combination" would be more adequate, since this is the first time that neutral hydrogen forms, but it is difficult to change historical misnamings. . . .

In principle, this result is valid only in full thermal equilibrium. But Thomson scattering between electrons and photons does not change the photon energy and the Rydberg photons are not readily thermalized. They actually have time to ionize another hydrogen atom before they lose energy. Therefore, neglecting recombination into excited states of the hydrogen atom is a bad approximation, but it leads roughly to the right recombination temperature.

In this discussion, where we are more interested in the basic concepts than in accuracy, we also neglect helium that has recombined earlier. We set $n_p + n_H = n_B$, which induces an error of about 25%. For an accurate calculation of the final ionization fraction, one would have to take into account both the recombination of helium and the recombination into excited states of hydrogen. We briefly discuss this in Section 1.3.3. Despite these complications, a discussion of recombination into the ground state gives the correct orders of magnitude for the recombination and decoupling redshifts which we now derive.

In thermal equilibrium, electrons, protons and hydrogen atoms obey a Maxwell–Boltzmann distribution. Their number densities are given by (see Exercise 1.7)

$$n_e = \frac{2}{(2\pi)^3}(2\pi m_e T)^{3/2} \exp\left(-\frac{m_e - \mu_e}{T}\right), \tag{1.70}$$

$$n_p = \frac{2}{(2\pi)^3}(2\pi m_p T)^{3/2} \exp\left(-\frac{m_p - \mu_p}{T}\right), \tag{1.71}$$

$$n_H = \frac{4}{(2\pi)^3}(2\pi m_H T)^{3/2} \exp\left(-\frac{m_H - \mu_H}{T}\right). \tag{1.72}$$

We now make use of the fact that the Universe is globally neutral, $n_e = n_p$. Furthermore, the binding energy of hydrogen $\Delta = \alpha^2 m_e/2$ (here $\alpha \simeq 1/137$ is the fine structure constant) is given by $\Delta = m_e + m_p - m_H$. With this we obtain

$$\frac{n_e^2}{n_H} = \frac{n_e n_p}{n_H} = \left(\frac{m_e T}{2\pi}\right)^{3/2} e^{-\Delta/T}. \tag{1.73}$$

Here we have neglected the small difference between the hydrogen and proton mass in the second factor of Eqs. (1.71) and (1.72) but not in the exponential. This is the *Saha equation*. The corresponding equation for helium, setting $n_{He^+} = n_{He^{2+}}$, yields the $He^{2+} \to He^+$ transition temperature and accordingly the $He^+ \to He$ transition temperature (see Exercice 1.5).

We now define the ionization fraction x_e by $x_e \equiv n_e/(n_e + n_H)$. In Section 1.4, we shall find that about 25% of all baryons in the Universe are bound in the form of He^4 so that $n_p + n_H = n_e + n_H = 0.75 n_B$. Equation (1.73) then leads to

$$\frac{x_e^2}{1-x_e} = \frac{n_e^2}{n_H(n_p+n_H)} = \frac{1}{0.75n_B}\left(\frac{m_e T}{2\pi}\right)^{3/2} e^{-\Delta/T}. \tag{1.74}$$

Inserting the entropy per baryon, $\sigma = (4\pi^2/45)T^3/n_{\rm B}$, in this equation yields

$$\frac{x_e^2}{1-x_e} = \frac{45\sigma}{0.75\times 4\pi^2}\left(\frac{m_e}{2\pi T}\right)^{3/2} e^{-\Delta/T}. \tag{1.75}$$

At very high temperatures, $T \gg \Delta$, the ionization fraction x_e is close to 1. Recombination happens roughly when $\sigma \exp(-\Delta/T)$ is of the order of unity. If $\sigma \sim 1$ this corresponds to $T \sim \Delta$. The fact that the entropy per baryon is very large, $\sigma = 1.4\times 10^8(\Omega_B h^2)^{-1} \sim 10^{10}$, delays recombination significantly. Since there are so many more photons than baryons in the Universe, even at a temperature much below $\Delta = 13.6$ eV there are still enough photons in the high-energy tail of the Planck distribution to keep the Universe ionized.

To be more specific, we define the recombination temperature $T_{\rm rec}$ as the temperature when $x_e = 0.5$ (as we shall see, the precise value is of little importance). Equation (1.75) then leads to

$$\left(\frac{T_{\rm rec}}{1\ {\rm eV}}\right)^{-3/2} e^{-\Delta/T_{\rm rec}} = 0.97\times 10^{-16}\,\Omega_B h^2. \tag{1.76}$$

For $\Omega_B h^2 \simeq 0.022$ we obtain

$$T_{\rm rec} = 3722\,{\rm K} = 0.321\,{\rm eV},\quad z_{\rm rec} = 1353.$$

The function $x_e(T)$ is shown in Fig. 1.5. Clearly, this function grows very steeply from $x_e \sim 0$ to $x_e \sim 1$ at $T \sim 3700$ K and $T_{\rm rec}$ depends only weakly on the value chosen for $x_e(T_{\rm rec})$.

Interestingly, at temperature $T_{\rm rec}$ the baryon and photon densities are of the same order, $\rho_\gamma(T_{\rm rec}) \simeq \rho_B(T_{\rm rec})$. This seems to be a complete coincidence. More precisely, the ratio of these two densities is given by

$$\frac{\rho_\gamma}{\rho_B} = \frac{(\pi^2/15)T^4}{n_B m_p} = \frac{\pi^2 T_0^4}{15 n_B(t_0) m_p}(z+1)$$
$$\simeq 2\times 10^{-5}\left(\Omega_B h^2\right)^{-1}(z+1). \tag{1.77}$$

This ratio is equal to 1 at redshift $z_{\gamma b}$ given by

$$(1+z_{\gamma b}) = 10^3\left(\frac{\Omega_B h^2}{2\times 10^{-2}}\right) \simeq 10^3 \sim 1 + z_{\rm rec}. \tag{1.78}$$

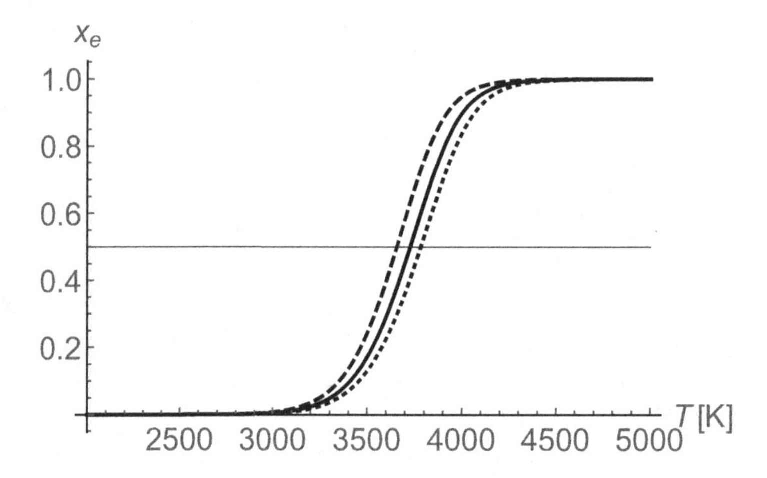

Fig. 1.5 The ionization fraction x_e as a function of the temperature is obtained via the Saha equation for $\Omega_B h^2 = 0.022$ (solid curve), for $\Omega_B h^2 = 0.01$ (dashed curve), and for $\Omega_B h^2 = 0.04$ (dotted curve). Our definition of recombination, $x_{\rm rec} = 0.5$, is indicated. Note that x decays from $x_e \simeq 1$ to $\simeq 0$ between $T = 4000$ and 3400 K. Below about $x_e \sim 0.9$ the shape of the true ionization fraction significantly differs from this Saha-equation result and levels off at the final ionization fraction computed in the text that follows.

1.3.2 Final Ionization and Photon Decoupling

We have determined the temperature at which electrons and protons recombine to neutral hydrogen. The Saha equation predicts an exponentially falling fraction of free electrons. But this is correct only as long as thermal equilibrium is established. As the free electron fraction drops, the interaction rate between electrons and protons decreases, and at some point the remaining free electrons and protons are too sparse to find each other, thermal equilibrium is lost, and the number of free electrons remains constant. But also the photon–electron interaction rate decreases. Whenever an interaction rate Γ drops below the expansion rate of the Universe,

$$\Gamma < H,$$

one considers the corresponding reaction as "frozen." It becomes negligible. The temperature at which $\Gamma = H$ is called the "freeze out" temperature of the reaction with rate Γ.

When the recombination rate drops below the expansion rate, recombination freezes out and the ionization fraction remains constant. When the scattering rate of photons on electrons falls below the expansion rate of the Universe, photons become free to propagate without further scattering. We want to calculate both the final ionization fraction, x_R, and the redshift, $z_{\rm dec}$, of the decoupling of photons.

Let us first determine the temperature T_g at which the process of recombination freezes out. The cross section of the reaction $p^+ + e^- \to \mathrm{H} + \gamma$ is (see, e.g., Rybicki and Lightman, 1979)

$$\langle \sigma_R v \rangle \simeq 4.7 \times 10^{-24} \left(\frac{T}{1\,\mathrm{eV}} \right)^{-1/2} \mathrm{cm}^2. \tag{1.79}$$

Here v is the thermal electron velocity and we have used the fact that $3T = m_e v^2$. The reaction rate is therefore

$$\begin{aligned} \Gamma_R = n_p \langle \sigma_R v \rangle &= x_e \left(\frac{0.75 n_B}{n_\gamma} \right) n_\gamma \langle \sigma_R v \rangle \\ &\simeq 2.1 \times 10^{-10}\,\mathrm{cm}^{-1} \left(\frac{T}{1\,\mathrm{eV}} \right)^{7/4} \exp(-\Delta/2T)(\Omega_B h^2)^{1/2}, \end{aligned}$$

where we have inserted the Saha equation, assuming that the ionization fraction is much smaller than 1, that is,

$$x_e \simeq (\sqrt{45\sigma/0.75}/2\pi)(m_e/2\pi T)^{3/4} \exp(-\Delta/2T) \ll 1.$$

We have also used Eq. (1.66).

To determine the expansion rate $H(T)$, we neglect curvature or a possible cosmological constant, which is certainly a good approximation for all redshifts larger than, say, 5. We also assume that the Universe is matter dominated at freeze-out, which induces an error of about 15% in H. The Friedmann equation (1.18) then gives

$$\begin{aligned} H^2 \simeq \frac{8\pi G}{3} \rho &\simeq \frac{8\pi G}{3} \rho_0 (a_0/a)^3 \\ &= \frac{8\pi G}{3} \Omega_m \rho_c(t_0)(T/T_0)^3, \end{aligned}$$

so that

$$H \simeq 3 \times 10^{-23}\,\mathrm{cm}^{-1} (\Omega_m h^2)^{1/2} \left(\frac{T}{1\,\mathrm{eV}} \right)^{3/2}. \tag{1.80}$$

Equation (1.80) is a very useful formula, valid whenever the Universe is dominated by nonrelativistic matter or dust, $P \ll \rho$, and curvature or a cosmological constant are negligible.

The temperature T_g is defined by $\Gamma_R(T_g) = H(T_g)$, which finally leads to

$$\left(\frac{T_g}{1\,\mathrm{eV}} \right)^{1/4} e^{-\Delta/2T_g} = 1.4 \times 10^{-13} \left(\frac{\Omega_m}{\Omega_B} \right)^{1/2}. \tag{1.81}$$

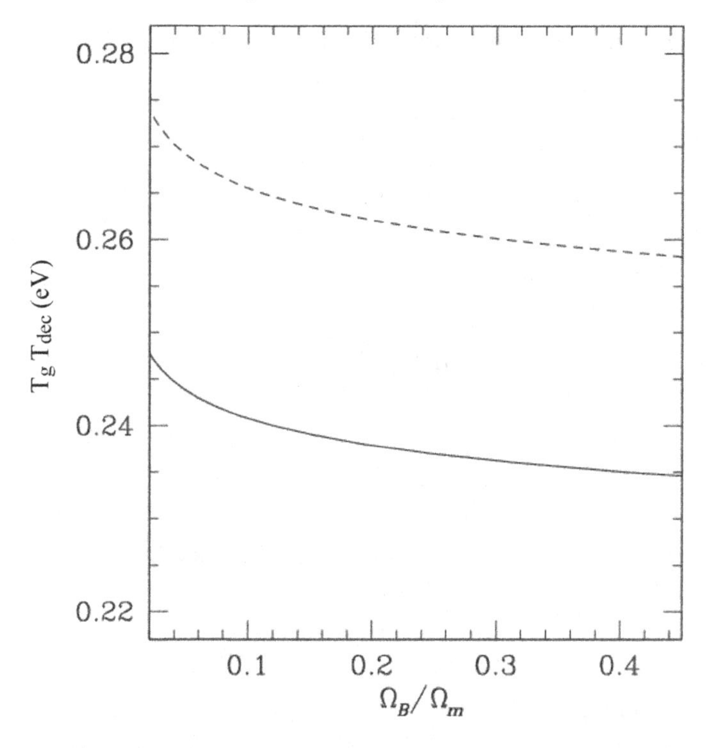

Fig. 1.6 The freeze-out temperatures of recombination, T_g (solid curve), and of Thomson scattering, $T_{\rm dec}$ (dashed curve), as functions of Ω_B/Ω_m.

This result is independent of h. For $\Omega_m \simeq 6.4\Omega_B$ (the value inferred from observations; see Planck Coll. XIII, 2016), we obtain $T_g \simeq 0.24$ eV and $z_g \simeq 1010$ (see Fig. 1.6). T_g depends only weakly on the ratio Ω_B/Ω_m.

The final ionization fraction is given by

$$x_R \simeq x_e(T_g) \simeq 7.3 \times 10^{-6} \left(\frac{T_g}{1\,\text{eV}}\right)^{-1} \Omega_m^{1/2}/(\Omega_B h) \simeq 3 \times 10^{-5}\Omega_m^{1/2}/(\Omega_B h). \tag{1.82}$$

A more detailed numerical analysis, taking into account the contribution of radiation to the expansion rate and, especially, the recombination into excited states of the hydrogen atoms and the presence of helium, gives $x_R \sim 1.2 \times 10^{-5}\Omega_m^{1/2}/(\Omega_B h)$ (Peebles, 1993; Mukhanov, 2005; Weinberg, 2008). We use this result to calculate the optical depth τ to Thomson scattering of photons by free electrons up to a redshift $z < z_g$ in a recombined universe. The optical depth to z is the scattering probability of a photon integrated from z until today. With the Thomson cross section

$$\sigma_T = \frac{8\pi}{3}\alpha^2 m_e^{-2} \simeq 6.65 \times 10^{-25}\,\text{cm}^2, \tag{1.83}$$

one finds

$$\tau(z) \equiv \int_{t(z)}^{t_0} \sigma_T n_e a dt \simeq 0.046 x_R (1+z)^{3/2} \Omega_B \Omega_m^{-1/2} h. \quad (1.84)$$

With the residual ionization computed in Eq. (1.82) we obtain $\tau(z = 800) \simeq 0.01$. As we shall see in Section 9.3, the Universe is reionized at low redshift $z \sim 7.5$, which increases the optical depth by about a factor of 6. This rescattering of CMB photons is relevant for the evolution of fluctuations, as we shall discuss in Section 9.3.

As long as the temperature is larger than T_g, the reaction $p + e \longleftrightarrow \mathrm{H} + \gamma$ is in thermal equilibrium. When the temperature drops below T_g, the recombination process freezes out and the degree of ionization remains nearly constant.

Let us also note that in deriving the Saha equation (1.73), we used the fact that the process of recombination is in thermal equilibrium, which we have verified only now since freeze-out happens after recombination, $T_g < T_{\rm rec}$.

We finally calculate the redshift of the decoupling of photons. The process that remains effective longest is elastic Thomson scattering. Its rate is given by

$$\Gamma_T = \sigma_T n_e = \sigma_T x_e \left(\frac{0.75 n_B}{n_\gamma} \right) n_\gamma$$
$$\simeq 2.6 \times 10^{-11}\,\mathrm{cm}^{-1} (\Omega_B h^2)^{1/2} \left(\frac{T}{1\,\mathrm{eV}} \right)^{9/4} \exp(-\Delta/2T). \quad (1.85)$$

Comparing it to the expansion rate, we find $T_{\rm dec}$, which is defined by $H(T_{\rm dec}) = \Gamma_T(T_{\rm dec})$. A rough estimate gives $T_{\rm dec} \sim 0.26$ eV (see Fig. 1.6), which corresponds to $z_{\rm dec} \sim 1100$. Again we have assumed $x_e \ll 1$ in Eq. (1.85), which is justified since $T_{\rm dec} \sim 3000$ K (see Fig. 1.5).

Even though after $z_{\rm dec}$ photons decouple from electrons, the latter are still coupled to photons. The scattering rate of electrons, given by $\Gamma_e = \sigma_T x_e n_\gamma = \sigma_T x_R n_\gamma$ at low redshifts, is sufficient to keep the remaining electrons and with them baryonic matter in thermal equilibrium with the photons until about $z \sim 100$. Therefore, even after recombination the matter temperature is equal to the temperature of the CMB and does not decay like $1/a^2$, as would be expected from a pure thermal gas of massive particles (see Section 1.3.4). This is an example of two species, electrons and photons, where the former is in thermal equilibrium with the latter but not vice versa.

1.3.3 An Accurate Treatment of Recombination

So far, we have given an approximate treatment of the process of recombination and photon decoupling. This yields the correct orders of magnitude, but to determine especially the anisotropies of the CMB and the polarization, in Chapters 4 and 5, with good precision, this is not sufficient. For precise results it is necessary to treat recombination of hydrogen into higher levels, especially 2S, but also Raman scattering by which the electrons of a hydrogen atoms are scattered into a higher energy level and then decay into a lower level by the emission of a photon of different energy.

It is actually interesting to note that recombination into the ground state (1S) is not efficient at all because the ionization cross section is very high for the resonant Rydberg photons so that most of these just ionize another hydrogen atom before being redshifted out of the resonance, leading to no net recombination. The same is true for recombination into the 2P excited state. The Lyα photons from the 2P$\rightarrow$1S transition are quickly absorbed and excite another hydrogen atom, which is then reionized via a 2P-ionization photon. The single most efficient channel is the capture of electrons into the 2S level, from which they can decay into the ground state via the emission of two photons. By angular momentum conservation, the emission of a single photon is not possible. The inverse process, excitation from 1S to 2S, is a three-body process and therefore highly unlikely. Even though the rate of the transition $(e, p) \rightarrow$ H(2S) $\rightarrow$ H(1S) is relatively low, it wins against direct recombination into the ground state and subsequent cosmological redshifting of the photon before the next ionization can take place. Since the binding energy of the 2S state is lower, this delays recombination somewhat. A semianalytic, dynamical treatment including recombination into the 2S and 2P states can be found in Peebles (1993), Mukhanov (2005), and Weinberg (2008).

For accurate results of helium and hydrogen recombination, as they are required to accurately study the CMB anisotropies discussed in the next chapters, a numerical computation is needed that takes into account many (up to 300) excited states and their decay. The most popular publicly available code for this is "RECFAST" (Seager *et al.*, 1999). The latest work (Shaw and Chluba, 2011), including even more details, has still found changes by up to 3% in the free electron fraction throughout the recombination process from $z \sim 2200$ (helium recombination) to $z \simeq 800$.

Interestingly, recombination also leads to lines and other distortions in the CMB frequency spectrum that might be observable with a future satellite mission measuring the CMB spectrum with high accuracy; see Rubino-Martin *et al.* (2006) and Wong *et al.* (2006).

1.3.4 Propagation of Free Photons and the CMB

After $t_{\rm dec}$, photons cease any interaction with the cosmic fluid and propagate freely. It is straightforward to estimate that the cross section for Rayleigh scattering with hydrogen atoms is much too weak to be relevant (see Exercise 1.6).

The free propagation of photons after decoupling is described with the Liouville equation for the photon distribution function, which we now develop. Since photons do not interact anymore, they simply move along geodesics. The Liouville equation translates this to a differential equation for the 1-particle distribution function f of the photons. The function f describes the particle density in the phase space P_0, the photon mass-shell, given by

$$P_0 = \{(x, p) \in T\mathcal{M} \mid g_{\mu\nu}(x)p^\mu p^\nu = 0\}, \quad f : P_0 \to \mathbb{R}.$$

The distribution function f gives the number of particles per phase space volume $|g|\, d^3x\, d^3p$ at fixed time t. In some general geometry a specific space-like hypersurface Σ has to be chosen and one then has to show that f does not depend on this choice [more details are found in Ehlers (1971) and Stewart (1971)]. In cosmology, due to the symmetries present, we simply use the hypersurfaces of constant time, $\Sigma = \Sigma_t$.

We choose the coordinates (x^μ, p^i) on the seven-dimensional mass-shell ($0 \le \mu \le 3$ and $1 \le i \le 3$). The energy p^0 is then determined by the mass-shell condition $g_{\mu\nu}(x)p^\mu p^\nu = 0$. Liouville's equation now says that the 1-particle distribution remains unchanged if we follow the geodesic motion of the particles, that is,

$$0 = \frac{df}{dt} = \dot{x}^\mu \partial_\mu f + \dot{p}^i \frac{\partial f}{\partial p^i},$$
$$0 = p^\mu \partial_\mu f - \Gamma^i_{\mu\nu} p^\mu p^\nu \frac{\partial f}{\partial p^i} \equiv L_{X_g} f\ . \tag{1.86}$$

A particle distribution obeying this equation is often also called a geodesic spray (see Abraham and Marsden, 1982). If the particles are not free, but collisions are so rare that an equilibrium description is not adequate, one uses the Boltzmann equation,

$$L_{X_g} f = C[f], \tag{1.87}$$

where $C[f]$ is the so-called collision integral, which depends on the details of the interactions.

It may be disturbing to some readers that we take over these concepts from non-relativistic physics so smoothly to the relativistic case. In cosmology, this does not cause any problems. But in general, it is true that the collision integral is not always well defined and certain conditions have to be posed to the nature of the spacetime and of the interaction. This problem has been studied in detail by Ehlers (1971).

Since the photons are massless, $|\mathbf{p}|^2 = \gamma_{ij} p^i p^j = (p^0)^2$. Here p^0 is the 0-component of the momentum 4-vector in *conformal* time so that $\epsilon = ap^0$ is the physical photon energy. Isotropy of the distribution implies that f depends on p^i only via $p \equiv |\mathbf{p}| = p^0$, and so

$$\frac{\partial f}{\partial p^i} = \frac{\partial p}{\partial p^i}\frac{\partial f}{\partial p} = \frac{p_i}{p}\frac{\partial f}{\partial p}. \tag{1.88}$$

Furthermore, f depends on x^i only through $p = \sqrt{\gamma_{ij} p^i p^i}$. Spatial derivatives are therefore given by

$$\begin{aligned} p^i \partial_i f &= \frac{1}{2} p^i \gamma_{lm,i} \frac{p^l p^m}{p}\frac{\partial f}{\partial p} = \frac{1}{2} p_j \gamma^{ij} \gamma_{lm,i} \frac{p^l p^m}{p}\frac{\partial f}{\partial p} \\ &= \frac{1}{2}\gamma^{ij}\left(\gamma_{li,m} + \gamma_{mi,l} - \gamma_{lm,i}\right)\frac{p_j p^l p^m}{p}\frac{\partial f}{\partial p} \\ &= \Gamma^j_{lm}\frac{p^l p^m p_j}{p}\frac{\partial f}{\partial p}. \end{aligned}$$

This leads to

$$p^i \partial_i f - \Gamma^i_{\mu\nu}\frac{p^\mu p^\nu p_i}{p}\frac{\partial f}{\partial p} = -\left(\Gamma^i_{j0} + \Gamma^i_{0j}\right)\frac{p^j p p_i}{p}\frac{\partial f}{\partial p} = -2p^2\frac{\dot a}{a}\frac{\partial f}{\partial p},$$

where we have used the expressions in Appendix 2, Section A2.3 for $\Gamma^i_{\mu\nu}$ and $p = p^0$. Inserting this result into (1.86) we obtain, with Eq. (1.88),

$$\partial_t f - 2p\frac{\dot a}{a}\frac{\partial f}{\partial p} = 0, \tag{1.89}$$

which is satisfied by an arbitrary function $f = f(pa^2) = f(a\epsilon)$. Hence the distribution of free-streaming photons changes only by redshifting the *physical energy* $\epsilon = ap^0$ or the *physical momentum* $a|\mathbf{p}| = \epsilon$. Therefore, setting $T \propto a^{-1}$ even after recombination, the blackbody shape of the photon distribution remains unchanged. This radiation of free photons with a perfect blackbody spectrum is the CMB. Its physics, especially its fluctuation and polarization, are the main topic of this book.

The same result is also obtained for massive particles,

$$\partial_t f - 2p\frac{\dot a}{a}\frac{\partial f}{\partial p} = 0, \tag{1.90}$$

where $p = |\mathbf{p}|$; hence the momentum is simply redshifted. Therefore, massive particles that decouple when they are still relativistic keep their extremely relativistic Fermi–Dirac (or Bose–Einstein) distribution, $f = (\exp(ap/T) \pm 1)^{-1}$, with a temperature that simply scales as $T \propto 1/a$. This is especially important for the cosmic neutrinos, which probably have masses in the range of a $0.1\text{eV} > m_\nu \gtrsim 0.01$ eV. But, as we shall see in the next section, they decouple at $T \sim 1.4$ MeV. We therefore expect them to be distributed according to an extremely relativistic Fermi–Dirac distribution, which is not a thermal distribution for nonrelativistic neutrinos. By the same argument, particles that decouple once they are nonrelativistic keep their Maxwell Boltzmann distribution, $f \propto \exp\left[(ap)^2/(mT)\right]$, if we assume the temperature to scale as $T \propto a^{-2}$, which is also the scaling in thermal equilibrium for massive particles [see discussion after Eq. (1.93)].

Note, however, that after decoupling the particles are no longer in thermal equilibrium and the T in their distribution function is not a temperature in the thermodynamical sense but merely a parameter, representing a measure of the mean kinetic energy.

The situation is different for the electron–proton–hydrogen plasma. As we have seen, the free electrons still scatter with photons and keep the same temperature as the latter. In other words: even though most photons are no longer interacting with the electrons, the latter are still interacting with the photons. (To have one collision with all the remaining electrons, only a fraction of about 10^{-14} of the photons have to be involved!)

Soon after recombination, the baryon energy density exceeds the photon energy density and one might expect that this would change the evolution of the temperature. To investigate this we use the energy conservation equation of the baryon–photon system. We neglect the tiny number of free electrons. The energy density and pressure are then given by

$$\rho = n_B m_B + (3/2) n_B T + \frac{\pi^2}{15} T^4, \tag{1.91}$$

$$p = n_B T + \frac{\pi^2}{45} T^4. \tag{1.92}$$

The energy conservation equation, $d\rho/da = -3(\rho + p)/a$, now gives

$$\frac{a}{T}\frac{dT}{da} = -\frac{3n_B + \frac{4\pi^2}{15}T^3}{(3/2)n_B + \frac{4\pi^2}{15}T^3} = -\frac{\sigma + 1}{\sigma + 1/2}. \tag{1.93}$$

Since $\sigma \gg 1$, the photons are so much more numerous than the baryons that the latter have no influence on the temperature, which keeps evolving as $1/a$. Note, however, that in the absence of photons, the temperature of a monoatomic gas would decrease like $1/a^2$ as mentioned earlier (just consider the limit $\sigma \to 0$).

The blackbody spectrum of the CMB photons is extremely well verified observationally (see Fig. 1.7 and Chapter 10). The limits on deviations are often parameterized in terms of three parameters: the chemical potential μ, the Compton-y parameter (which quantifies a well-defined change in the spectrum arising from interactions with a nonrelativistic electron gas at a different temperature; we

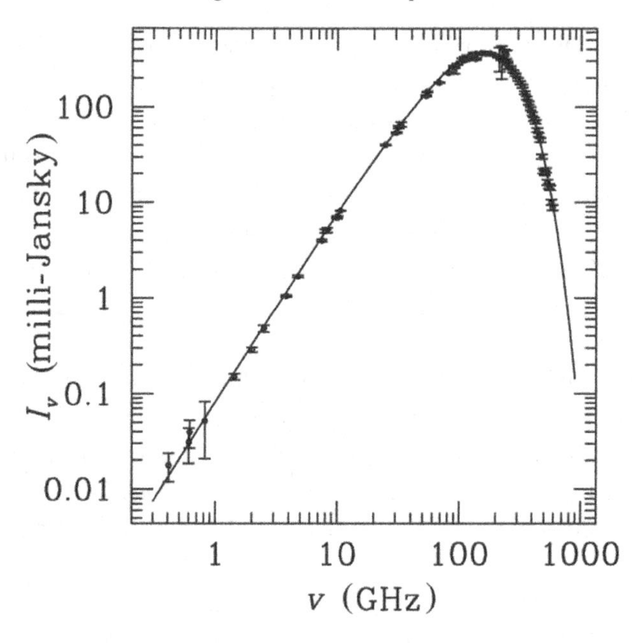

Fig. 1.7 The spectrum of the cosmic background radiation. I_ν is the energy flux per frequency. The data are from many different measurements that are all compiled in Kogut *et al.* (2007). The points around the top are the measurements from the FIRAS experiment on COBE (Fixsen *et al.*, 1996). The line traces a blackbody spectrum at a temperature of 2.728 K (the data are courtesy of Susan Staggs).

shall discuss this in detail in Chapter 10), and $Y_{\rm ff}$ (describing a contamination by free–free emission).

The present 95% confidence limits on these parameters are (Particle Data Group, 2006)

$$|\mu| < 9 \times 10^{-5}, \quad |y| < 1.2 \times 10^{-5}, \quad |Y_{\rm ff}| < 1.9 \times 10^{-5}. \tag{1.94}$$

These limits are mainly derived from the COBE satellite data, which had been taken more than 25 years ago. It would be very interesting to have newer data and better limits on these spectral distortions, as we will discuss in Chapter 10.

The CMB photons not only have a very thermal spectrum, but they are also distributed very isotropically, apart from a dipole that is (most probably) mainly due to our motion relative to the surface of last scattering.

Indeed, an observer moving with velocity $\mathbf{v}$ relative to a source in direction $\mathbf{n}$ emitting a photon with proper momentum $\mathbf{p} = -\epsilon\mathbf{n}$ sees this photon redshifted with frequency

$$\epsilon' = \gamma\epsilon\,(1 - \mathbf{n}\mathbf{v})\,, \tag{1.95}$$

where $\gamma = 1/\sqrt{1-v^2}$ is the relativistic γ-factor. For an isotropic emission of photons coming from all directions $\mathbf{n}$ this leads to a dipole anisotropy to first order in $\mathbf{v}$. This dipole anisotropy, which is of the order of

$$\left(\frac{\Delta T}{T}\right)_{\text{dipole}} \simeq 1.2 \times 10^{-3},$$

had already been discovered in the 1970s (Conklin, 1969; Henry, 1971). Interpreting it as due to our motion with respect to the last scattering surface implies a velocity for the Solar System barycenter of $v = 371 \pm 0.5\,\text{km}\,\text{s}^{-1}$ at 68% CL (Particle Data Group, 2006).

In addition to the dipole, the COBE[5] DMR experiment (differential microwave radiometer) has found fluctuations of order

$$\sqrt{\left\langle\left(\frac{\Delta T}{T}\right)^2\right\rangle} \sim (\text{a few}) \times 10^{-5}, \tag{1.96}$$

on all angular scales $\theta \gtrsim 7^\circ$ (Smoot *et al.*, 1992). On smaller angular scales many experiments found fluctuations (we shall describe the experimental results in more detail later) all of which satisfy $|\Delta T/T| \lesssim 10^{-4}$.

As we shall see in Chapter 2, the CMB fluctuations on large scales provide a measure for the deviation of the geometry from the Friedmann–Lemaître one. The geometry perturbations are thus small, and we may calculate their effects by *linear perturbation theory*. On smaller scales, $\Delta T/T$ reflects the fluctuations in the energy density in the baryon/radiation plasma prior to recombination. Their amplitude is just about right to allow the formation of the presently observed nonlinear structures (such as galaxies, clusters, etc.) by gravitational instability.

These findings strongly support our hypothesis that the large-scale structure (i.e., the galaxy distribution) observed in the Universe has been formed by gravitational instability from small ($\sim 10^{-4}$) initial fluctuations. As we shall see in Chapters 2, 4, and 5, such initial fluctuations leave an interesting "fingerprint" on the cosmic microwave background.

1.4 Nucleosynthesis

1.4.1 Expansion Dynamics at $T \sim$ a Few MeV

At high temperatures, $T > 30$ MeV, none of the light nuclei (deuterium, ^{2}H, helium-4, ^{4}He, helium-3, ^{3}He or lithium, ^{7}Li) are stable. At these temperatures, we expect the baryons to form a simple mixture of protons and neutrons in thermal

[5] Cosmic Background Explorer, NASA satellite launched 1990.

equilibrium with each other and with electrons, photons, and neutrinos. The highest binding energy is the one of ^{4}He, which is about 28 MeV. Nevertheless, ^{4}He cannot form at this temperature because the baryon density of the Universe is not high enough for three- or even four-body interactions to occur in thermal equilibrium. Therefore, before any nucleosynthesis can occur, the temperature has to drop below the binding energy of deuterium, which is about 2.2 MeV. But even at this temperature there are still far too many high-energy photons around for deuterium to be stable. This is due to the very low baryon to photon ratio, $\eta_B \simeq 10^{-10}$. Just as recombination is delayed from the naively expected temperature $T = 13.7$ eV to about $T_{\rm rec} \sim 0.3$ eV, nucleosynthesis does not happen at $T \sim 2.2$ MeV but around $T_{\rm nuc} \sim 0.1$ MeV. Most of the neutrons present at that temperature are converted into ^{4}He. Only small traces remain as deuterium or are burned into ^{3}He and ^{7}Li.

Let us study this in some more detail. At the time of recombination, the relativistic particle species are the photon and, probably, three types of neutrinos. As we shall see in the next paragraph, the neutrino temperature is actually a factor of $(4/11)^{1/3}$ lower than the temperature of the photons. With Eqs. (1.54) and (1.55), the energy density of these particles while they are relativistic is given by

$$\rho_{\rm rel}(t) = \left[\rho_\gamma(t) + \rho_\nu(t)\right] = \left[1 + 3\frac{7}{8}(4/11)^{4/3}\right]\frac{\pi^2}{15}T^4, \tag{1.97}$$

$$\simeq 10^{-33}\,{\rm g\,cm^{-3}}\left(\frac{T}{T_0}\right)^4, \tag{1.98}$$

$$\simeq \rho_c(t_0)\Omega_{\rm rel}h^2(1+z)^4, \text{ where}$$

$$\Omega_{\rm rel}h^2 \simeq 4.4 \times 10^{-5}\,. \tag{1.99}$$

Note that at temperatures below the highest neutrino mass, this is no longer the energy density of relativistic particles; therefore $\Omega_{\rm rel}$ is not the density parameter of relativistic particles today. Above the neutrino mass threshold and below the electron mass threshold we have

$$\frac{\rho_{\rm rel}}{\rho_m} = \frac{\Omega_{\rm rel}}{\Omega_m}(1+z) \simeq 4.4 \times 10^{-5}\left(\frac{1}{\Omega_m h^2}\right)(1+z), \tag{1.100}$$

Since $\Omega_m h^2 \simeq 0.14$, the redshift $z_{\rm eq}$ above which the Universe is dominated by relativistic particles is about

$$z_{\rm eq} \simeq 3.2 \times 10^3, \quad T_{\rm eq} \simeq 1\,{\rm eV}. \tag{1.101}$$

At temperatures significantly above $T_{\rm eq}$, we can also neglect a possible contribution from curvature or a cosmological constant to the expansion of the Universe, so that for

$$z \gg z_{\rm eq} \qquad P = \frac{1}{3}\rho, \quad a \propto \tau^{1/2} \propto t. \tag{1.102}$$

At these high temperatures the energy density of the Universe is given by

$$\rho = g_{\text{eff}} \frac{\pi^2}{30} T^4 \quad \text{where} \quad g_{\text{eff}} = N_B(T) + \frac{7}{8} N_F(T). \tag{1.103}$$

Here, N_B and N_F denote the number of bosonic and fermionic degrees of freedom of relativistic particles (i.e., particles with mass $m < T$) that are in thermal equilibrium at temperature T.

To discuss the physical processes at work at some temperature T, we need to know the spectrum of relativistic particles and their interactions at this temperature. Here, we shall study the Universe at $10\,\text{keV} < T < 100\,\text{MeV}$, where the physics is well known. The only relativistic particles present at these temperatures are electrons, positrons, photons, and three types of neutrinos. (The muons have a mass of $m_\mu \simeq 105.66$ MeV.) Even if the individual neutrino masses are not very well constrained, the oscillation experiments (Particle Data Group, 2004) imply that their masses are below 1 eV if there is no degeneracy. As we shall see later, also CMB data estimate masses below this value. Therefore, we may neglect the neutrino masses in our treatment. The baryon number is well conserved at these temperatures, so that we may set η_B equal to its present value, $\eta_B = n_B/n_\gamma \simeq 2.7 \times 10^{-8} \Omega_B h^2 =$ constant. We neglect the small contribution from muon/anti-muon pairs that decay exponentially $\propto \exp(-m_\mu/T)$ via the reaction

$$\mu + \bar{\nu}_\mu \rightarrow e + \bar{\nu}_e.$$

Thermal equilibrium between photons and electron/positrons is maintained mainly via the process $e^- + e^+ \longleftrightarrow 2\gamma$ (or $3\gamma \ldots$). The conservation of the chemical potential during this reaction implies

$$\mu_e + \bar{\mu}_e = 2\mu_\gamma = 0. \tag{1.104}$$

The last equals sign comes from the fact that photons can be generated and destroyed, their number is not conserved, and hence their chemical potential vanishes in thermal equilibrium. Here we use the notation $e^+ = \bar{e}$ and $\mu_{\bar{e}} = \bar{\mu}_e$. The difference in the density of electrons and positrons is therefore

$$n_e - \bar{n}_e = \frac{1}{\pi^2} \int p^2\, dp \left[\frac{1}{\exp\left(\frac{E-\mu_e}{T}\right) + 1} - \frac{1}{\exp\left(\frac{E+\mu_e}{T}\right) + 1} \right]. \tag{1.105}$$

At low temperatures this number is dictated by the neutrality of the Universe, and $n_e - \bar{n}_e \sim n_B$ is much smaller than $n_e + \bar{n}_e \sim n_\gamma$. Therefore, the chemical potential is much smaller than the electron mass, $\mu_e \ll m_e$. At high temperatures, $T \gg m_e$, we may therefore expand the electron number density in the small parameter μ_e/T. At first order this yields

$$n_e - \bar{n}_e \simeq \frac{2\mu_e}{\pi^2 T}\int p^2\,dp \frac{\exp(p/T)}{\left[\exp(p/T)+1\right]^2} = \frac{2\mu_e T^2}{\pi^2}\zeta(2). \qquad (1.106)$$

With $n_\gamma = 2T^3\zeta(3)/\pi^2$ this yields

$$\frac{n_e - \bar{n}_e}{n_\gamma} \simeq 1.4\frac{\mu_e}{T} \sim \frac{n_B}{n_\gamma} \simeq 2.7\times 10^{-8}\Omega_B h^2. \qquad (1.107)$$

We can therefore neglect the small chemical potential of the electrons and positrons. The interaction $e + \bar{e} \longleftrightarrow \nu + \bar{\nu}$ also implies that $\mu_\nu = -\bar{\mu}_\nu$. But unfortunately, the number $n_\nu - \bar{n}_\nu$ that determines, together with $n_e - \bar{n}_e$, the lepton number of the Universe, is not known from observations. We suppose that the lepton number, like the baryon number, is small and that we may also neglect the chemical potential of the neutrinos. Comparing our results with observations, we can check this hypothesis later.

At $T \lesssim 100$ MeV photons, electron/positrons, and neutrinos are still relativistic, so that $N_B = 2$ and $N_F = 4 + 6$; hence

$$g_{\rm eff}(T \sim 100\,{\rm MeV}) = \frac{43}{4} = 10.75. \qquad (1.108)$$

The Hubble parameter is given by

$$\left(\frac{a'}{a}\right)^2 = H^2 = \frac{1}{4\tau^2} = \frac{8\pi G}{3}\rho = \frac{8\pi^3 G}{90}g_{\rm eff}T^4.$$

With the Planck mass, m_P, defined by $G = 1/m_P^2 = 1/(1.22\times 10^{19}\,{\rm GeV})^2$, we find

$$H^2(T) \simeq 2.76 g_{\rm eff}(T)\left(\frac{T^2}{m_P}\right)^2, \qquad (1.109)$$

$$H \simeq 0.21\sqrt{g_{\rm eff}}\left(\frac{T}{1\,{\rm MeV}}\right)^2 {\rm s}^{-1}, \qquad (1.110)$$

$$\tau = \frac{1}{2H} \simeq 0.3 g_{\rm eff}(T)^{-1/2}\left(\frac{m_P}{T^2}\right) \simeq 2.4\,{\rm s}\left(\frac{1\,{\rm MeV}}{T}\right)^2 g_{\rm eff}^{-1/2}. \qquad (1.111)$$

Here we have used the formulas in Appendix 1 to convert MeV's to seconds, 1 MeV $= 1.5192\times 10^{21}{\rm s}^{-1}$. The temperature of $T \sim 100$ MeV corresponds thus to an age of $\tau \sim 7\times 10^{-5}$ s, and $T = 1$ MeV corresponds to $\tau \sim 0.7$ s. The relations (1.110) and (1.111) can be applied as long as the Universe is dominated by relativistic particles.

1.4.2 Neutrino Decoupling

Neutrinos are kept in thermal equilibrium via the exchange of a W-boson, $e + \bar{\nu} \longleftrightarrow e + \bar{\nu}$ and $\nu + \bar{e} \longleftrightarrow \nu + \bar{e}$, or a Z-boson, $e + \bar{e} \longleftrightarrow \nu + \bar{\nu}$. At low energies, $E \ll m_{Z,W} \sim 100$ GeV, we can determine the cross sections within the 4-fermion theory of weak interaction. Within this approximation, the effective interaction Langrangian is given by

$$L_{\rm int} = \frac{G_F}{\sqrt{2}} J_\mu^\dagger J^\mu + \text{hermitean conjugate}$$

$$= \frac{G_F}{\sqrt{2}} \left(u_e^* \gamma_\mu \frac{1}{2}(1 - \gamma^5) u_\nu \right) \left(u_\nu^* \gamma^\mu \frac{1}{2}(1 - \gamma^5) u_e \right) + \text{h.c.}, \qquad (1.112)$$

where the coupling parameter, G_F, is the Fermi constant, and γ^μ are Dirac's gamma-matrices, $\gamma_5 = i\gamma^0\gamma^1\gamma^2\gamma^3$.

$$G_F = 1.166 \times 10^{-5}\,\text{GeV}^{-2} = (293\,\text{GeV})^{-2}. \qquad (1.113)$$

The fermion $V - A$ current J_μ is expressed in terms of the electron and neutrino spinors $u_{e,\nu}$ and the Dirac γ-matrices.

The cross section of the W- and Z-boson exchange processes are identical within this approximation and they are given by

$$\sigma_F \simeq G_F^2 E^2 \sim G_F^2 T^2,$$

The involved particle density is $n_F(T) = g_F(T)\zeta(3)T^3/\pi^2 \sim 1.3T^3$, where we have set $g_F(T) = 3/4N_F(T) = 30/4$ for the three types of left-handed neutrinos and the $e^\pm$ s. Since the particles are relativistic, we can set $v \sim 1$ so that we obtain an interaction rate of

$$\Gamma_F = \langle \sigma_F v \rangle n_F \simeq 1.3 G_F^2 T^5.$$

Comparing this with the expansion rate H obtained in (1.109), we find

$$\frac{\Gamma_F}{H} \simeq 0.24 T^3 m_P G_F^2 \simeq \left(\frac{T}{1.4\,\text{MeV}} \right)^3. \qquad (1.114)$$

At temperatures below $T_F \sim 1.4$ MeV the mean number of interactions of a neutrino within one Hubble time, H^{-1}, becomes less than unity and the neutrinos effectively decouple. The plasma becomes transparent to neutrinos that are no longer in thermal equilibrium with electrons and positrons and hence photons and baryons.

As we have discussed in the previous section, even at temperatures far below their mass $m_\nu \gtrsim 0.05$ eV, their particle distribution remains an extremely relativistic Fermi–Dirac distribution with temperature

$$T_\nu = T_F \frac{a_F}{a},$$

since they are no longer in thermal equilibrium and their distribution is affected solely by redshifting of the momenta.

As long as the photon/electron/baryon temperature also scales like $1/a$, the neutrinos conserve the same temperature as the thermal plasma, but when the number of degrees of freedom, $g_{\rm eff}$, changes, the plasma temperature decays for a brief period of time less rapidly than $1/a$ and therefore remains higher than the neutrino temperature. This is exactly what happens at the electron–positron mass threshold, $T = m_e \simeq 0.5$ MeV. Below that temperature, only the process $e + \bar{e} \to 2\gamma$ remains in equilibrium while $2\gamma \to e + \bar{e}$ is exponentially suppressed. We calculate the reheating of the photons gas by electron–positron annihilation, assuming that the process takes place in thermal equilibrium and that the entropy remains unchanged. This is well justified because the cross section of this process is very high. Denoting the entropy inside a volume of size Va^3 before and after electron–positron annihilation by S_i and S_f, we therefore have $S_i = S_f$. With Eq. (1.60) this yields

$$S_i = \frac{2}{3} a_{\rm SB}\, g_{{\rm eff},i} (Ta)_i^3 V, \quad S_f = \frac{2}{3} a_{\rm SB}\, g_{{\rm eff},f} (Ta)_f^3 V.$$

The electron–positron degrees of freedom disappear in this process so that $g_{{\rm eff},f} = 2$ while $g_{{\rm eff},i} = 2 + 4(\frac{7}{8}) = 11/2$. From $S_i = S_f$ we therefore conclude

$$(Ta)_f = (Ta)_i \left(\frac{11}{4}\right)^{1/3}.$$

The neutrino temperature is not affected by $e^\pm$ annihilation, so that $(T_\nu a)_f = (T_\nu a)_i = (Ta)_i$. For the last equals sign we have used that the neutrino and photon temperatures are equal before $e^\pm$ annihilation. At temperatures $T \ll m_e$ we therefore have

$$T = \left(\frac{11}{4}\right)^{1/3} T_\nu. \tag{1.115}$$

Since there are no further annihilation processes, this relation remains valid until today and the present Universe not only contains a thermal distribution of photons, but also a background of cosmic neutrinos that have an extremely relativistic Fermi–Dirac distribution with temperature

$$T_\nu(\tau_0) = (4/11)^{1/3} T_0 = 1.95\,\text{K}. \tag{1.116}$$

We set

$$g_0 = 2 + \frac{7}{8} 6 \left(\frac{4}{11}\right)^{4/3} \simeq 3.36, \quad \text{and} \tag{1.117}$$

$$g_{0S} = 2 + \frac{7}{8} 6 \left(\frac{4}{11}\right) \simeq 3.91. \tag{1.118}$$

These are respectively the effective degrees of freedom of the energy and entropy densities as long as all the neutrinos are relativistic. Until then we therefore have

$$\rho_{\rm rel}(T) = \frac{\pi^2}{30} g_0 T^4 \simeq 8.1 \times 10^{-34}\,\text{g cm}^{-3} \left(\frac{T}{T_0}\right)^4, \tag{1.119}$$

$$s(T) = \frac{2\pi^2}{45} g_{0S} T^3 \simeq 3 \times 10^3\,\text{cm}^{-3} \left(\frac{T}{T_0}\right)^3. \tag{1.120}$$

The neutrino cross section at low energies is extremely weak, and so far the neutrino background has not been observed directly (see Exercise 1.9).

1.4.3 The Helium Abundance

The observed abundance of helium is universally about

$$\frac{n_{He} m_{He}}{n_H m_H} \equiv Y \simeq 0.24. \tag{1.121}$$

It is well known that this amount of helium cannot have been produced in stars. We now want to investigate how much helium is produced in the primordial Universe. At temperatures of a few MeV nuclei and baryons are non-relativistic and the equilibrium distribution for a nucleus with atomic mass (i.e., number of protons and neutrons) A and proton number Z is given by

$$n_A = N_A \left(\frac{m_A T}{2\pi}\right)^{3/2} \exp\left(-\frac{m_A - \mu_A}{T}\right). \tag{1.122}$$

The proton density is given in Eq. (1.71). The neutron density is correspondingly

$$n_n = 2 \left(\frac{m_B T}{2\pi}\right)^{3/2} \exp\left(-\frac{m_n - \mu_n}{T}\right). \tag{1.123}$$

Here, we neglect the small difference $Q = m_n - m_p = 1.293$ MeV in the prefactor, setting $m_n \sim m_p \sim m_B$. The conservation of the chemical potentials in nuclear reactions implies

$$\mu_A = Z\mu_p + (A - Z)\mu_n,$$

so that

$$\exp\left(-\frac{m_A - \mu_A}{T}\right) = \left(e^{\mu_p/T}\right)^Z \left(e^{\mu_n/T}\right)^{(A-Z)} e^{-m_A/T},$$
$$= \frac{1}{2^A}\left(\frac{2\pi}{m_B T}\right)^{3A/2} \exp(B_A/T) n_p^Z n_n^{A-Z}.$$

Here, $B_A = Zm_p + (A-Z)m_n - m_A$ is the binding energy of the nucleus (A, Z). In thermal equilibrium, the density of this ion is then given by

$$n_A = \frac{N_A}{2^A} A^{3/2}\left(\frac{2\pi}{m_B T}\right)^{3(A-1)/2} n_p^Z n_n^{A-Z} \exp(B_A/T). \qquad (1.124)$$

Here we have again neglected the nucleon mass difference Q and the binding energy B_A in the prefactor by setting $m_A \sim Am_B$, but not in the exponential.

We define the various mass abundances by

$$Y_A \equiv \frac{An_A}{n_B} = \frac{An_A}{\eta_B n_\gamma},$$
$$Y_p \equiv \frac{n_p}{n_B} = \frac{n_p}{\eta_B n_\gamma},$$
$$Y_n \equiv \frac{n_n}{n_B} = \frac{n_n}{\eta_B n_\gamma}.$$

Hence the thermal abundance of the nucleus (A, Z) is given by

$$Y_A = F(A)\left(\frac{T}{m_B}\right)^{3(A-1)/2} \eta_B^{A-1} Y_p^Z Y_n^{A-Z} e^{B_A/T}, \qquad (1.125)$$

$$\text{where} \quad F(A) = N_A A^{5/2} \zeta(3)^{A-1} \pi^{-(A-1)/2} 2^{(3A-5)/2}. \qquad (1.126)$$

This equation shows nicely the influence of the radiation entropy on nucleosynthesis. If we had $\eta_B \sim 1$, the nucleus (A, Z) would become stable and relatively abundant at $T \sim B_A$. At this temperature the formation of (A, Z) [controlled by the factor $\exp(B_A/T)$] is sufficiently important to counterbalance photodissociation (controlled by the factor η_B^{A-1}). In equilibrium, the exponential $\exp(B_A/T)$ is then of the order of $\eta_B^{1-A} \sim 1$ and the ratio Y_A then approaches the value $Y_A \sim Y_p^Z Y_n^{A-Z}$. However, if η_B is very small, the equilibrium between production of (A, Z) and photodissociation is delayed until $\exp(-B_A/T) \sim \eta_B^{A-1} \ll 1$, that is, to much lower temperatures. Neglecting the numerical factor $F(A)$, the temperature T_A, defined by $Y_A(T_A) \sim Y_p(T_A)^Z Y_n(T_A)^{A-Z}$, is

$$T_A \sim \frac{B_A}{(A-1)\left[\ln(\eta_B^{-1}) + 3/2 \ln(m_B/T_A)\right]}.$$

For the deuteron with binding energy $B_2 = 2.22\,\text{MeV}$ we find

$$T_2 \sim 0.085\,\text{MeV}. \tag{1.127}$$

The reaction rate Γ_{np} of the process $n + p \longleftrightarrow {}^2\text{H} + \gamma$ is given by

$$\Gamma_{np} = \langle \sigma_{np} v \rangle n_p \simeq 1.8 \times 10^{-17} (T/T_0)^3 \eta_B\,\text{s}^{-1} \simeq 10^{12} \eta_B \left(\frac{T}{\text{MeV}} \right)^3 \text{s}^{-1},$$

where we have used $\langle \sigma_{np} v \rangle = \text{constant} = 4.55 \times 10^{-20}\,\text{cm}^3\,\text{s}^{-1}$ at temperatures $1\,\text{keV} \le T \le 10\,\text{MeV}$, and $n_p = \eta_B n_\gamma \simeq 420 \eta_B (T/T_0)^3\,\text{cm}^{-3}$. Using $H \simeq 0.4\,(T/\text{MeV})^2\,\text{s}^{-1}$, we conclude that this interaction remains in thermal equilibrium as long as $T \gtrsim 0.004$ MeV. So the assumption of a thermal deuterium abundance is justified. As already mentioned, three-body interactions are not in thermal equilibrium; their reaction rate contains an additional factor $n_B/n_\gamma = \eta_B \ll 1$.

Therefore, at temperature T_2 only deuterium can form and subsequently virtually all the neutrons present are burned into ^{4}He. To determine the helium abundance, we have to determine the neutron density at this temperature. Let us first determine the temperature at which β and inverse β processes drop out of equilibrium,

$$\nu + n \longleftrightarrow p + e, \quad \bar{e} + n \longleftrightarrow p + \bar{\nu}, \quad n \rightarrow p + e + \bar{\nu}.$$

On one hand, particle conservation imposes

$$\mu_n - \mu_p = \mu_e - \mu_\nu.$$

On the other hand, the neutrality of the Universe requires $n_p = n_e$. Since $m_e \ll m_p$, Eqs. (1.70) and (1.71) imply $\mu_e \ll \mu_p$. Finally, setting $\mu_\nu \sim 0$, the chemical potentials of the neutron and the proton are approximately equal, that is, $\mu_n \simeq \mu_p$. The ratio of their densities is thus simply given by the mass difference $Q = m_n - m_p$,

$$\frac{n_n}{n_p} = \frac{Y_n}{Y_p} = \exp(-Q/T).$$

This ratio remains constant as long as the reactions $n \longleftrightarrow p$ are sufficiently rapid. At the decoupling temperature of these reactions,

$$\Gamma(T_D) = H(T_D) \simeq 3 \frac{T_D^2}{m_P},$$

the ratio (n_n/n_p) is hence given by

$$\left(\frac{n_n}{n_p} \right) (T_D) = \exp(-Q/T_D).$$

Afterwards, the neutron density decays exponentially by β-decay, $n \to p + e + \bar{\nu}$,

$$n_n(\tau) = n_n(\tau_D) \exp\left(-\frac{\tau - \tau_D}{\tau_n}\right) \quad \text{for } \tau > \tau_D, \tag{1.128}$$

where $\tau_n \simeq 886$ s is the neutron lifetime.

We now want to determine the temperature T_D. We can again use Fermi theory to determine the different cross sections. For nucleons, the pure $V - A$ current, $\psi^* \gamma_\mu (1 - \gamma_5) \psi$, is replaced by $\psi^* \gamma_\mu (g_V + g_A \gamma_5) \psi$, which takes into account the internal structure of the nucleons. In the Born approximation the cross section becomes (see, e.g., Maggiore, 2005)

$$\sigma(\nu + n \to p + e) = \frac{G_F^2}{\pi}(g_V^2 + 3g_A^2) v_e E_e^2.$$

The constants g_V and g_A are determined experimentally (e.g., by measuring the neutron lifetime), $g_V \simeq 1.00$ and $g_A \simeq 1.25$. The interaction rate per neutron is obtained by multiplying the preceding result with $v_\nu n_\nu$,

$$\Gamma(\nu + n \to p + e) = \langle \sigma v_\nu \rangle n_\nu = \frac{1}{2\pi^2} \int \frac{p_\nu^2 \, dp_\nu}{e^{p_\nu / T_\nu} + 1} v_\nu \sigma \left(1 - \frac{1}{e^{E_e/T} + 1}\right).$$

The factor $1 - 1/[\exp(E_e/T) + 1]$ is the probability that the electron state with energy E_e is free (it implements the Pauli principle). To simplify the integral we first use energy conservation, $E_\nu + E_n = E_p + E_e$. Since all the energies involved are of the order of MeV, we can set $E_n - E_p \sim m_n - m_p = Q = 1.293$ MeV and $E_e = p_\nu + Q$. Furthermore, $E_e = m_e \gamma = m_e/\sqrt{1 - v_e^2}$, which implies $v_e = \sqrt{(p_\nu + Q)^2 - m_e^2}/E_e$. Inserting these simplifications, we obtain finally

$$\Gamma(\nu + n \to p + e) = \frac{G_F^2 (g_V^2 + 3g_A^2) m_e^5}{2\pi^3} \times \int_0^\infty \frac{e^{\alpha(x+q)} x^2 (x+q) \sqrt{(x+q)^2 - 1}}{(1 + e^{\alpha(x+q)})(1 + e^{\beta x})} dx, \tag{1.129}$$

where we have set $x = p_\nu / m_e$, $\alpha = m_e / T_\gamma$, $\beta = m_e / T_\nu$, and $q = Q/m_e \simeq 2.5$. To compute the other processes we note that the matrix element $\mathcal{M}(p_\nu, p_n, p_p, p_e)$ that appears in the amplitude for $\nu + n \longleftrightarrow p + e$ is invariant under the transformations $(\mathbf{p}_\nu, p_n, p_p, \mathbf{p}_e) \to (-\mathbf{p}_\nu, p_n, p_p, -\mathbf{p}_e)$, and $(\mathbf{p}_\nu, p_n, p_p, \mathbf{p}_e) \to (-\mathbf{p}_\nu, p_n, p_p, \mathbf{p}_e)$, where p_ν, p_n, p_p and p_e are the momenta of the neutrino, neutron, proton and electron respectively,

$$\mathcal{M}(p_\nu, p_n, p_p, p_e) = \mathcal{M}(-p_\nu, p_n, p_p, -p_e),$$
$$\mathcal{M}(p_\nu, p_n, p_p, p_e) = \mathcal{M}(-p_\nu, p_n, p_p, p_e).$$

This observation allows us immediately to determine the reaction rates of the other processes. We simply have to take into account the different phase space constraints. With $x = E_e/m_e$ (the other parameters as earlier), we obtain

$$\Gamma(e + p \to n + \nu) = \frac{G_F^2(g_V^2 + 3g_A^2)m_e^5}{2\pi^3} \times \int_q^\infty \frac{e^{\beta(x-q)}x(x-q)^2\sqrt{x^2-1}\,dx}{(1+e^{\beta(x-q)})(1+e^{\alpha x})}, \quad (1.130)$$

and

$$\Gamma(n \to p + e + \bar{\nu}) \simeq \frac{G_F^2(g_V^2 + 3g_A^2)m_e^5}{2\pi^3} \times \int_1^q \frac{e^{\alpha x}e^{\beta(q-x)}(x-q)^2 x\sqrt{x^2-1}\,dx}{(1+e^{\beta(q-x)})(1+e^{\alpha x})}, \quad (1.131)$$

$$\Gamma(n \to p + e + \bar{\nu})|_{T \ll m_e} \simeq 1.6\frac{G_F^2}{2\pi^3}(g_V^2 + 3g_A^2)m_e^5 = \tau_n^{-1} \quad (1.132)$$

$$\tau_n^{-1} = \frac{1}{886\text{ s}}$$

for the β-decay of the neutron at low temperature.

The products $\tau_n\Gamma$ are functions of the temperature T. When $T \gg Q$, the kinetic energy in the system $e + \bar{\nu}$ is much higher than the electron mass. Hence $x \pm q \simeq x$ at the positions that contribute most to the foregoing integrals and the reaction rates go like

$$\left.\begin{array}{l}\tau_n\Gamma(n \to p)\\ \tau_n\Gamma(p \to n)\end{array}\right\} \propto T^5, \quad \text{for } T \gg Q.$$

In the regime $0.1\,\text{MeV} \leq T \leq 1\,\text{MeV}$, the product $\tau_n\Gamma(n \to p)$ is roughly proportional to $T^{4.4}$. The same is true for $\tau_n\Gamma(p \to n)$. But the phase space for β-decay is larger than for the reaction $p \to n$, so that $\tau_n\Gamma(n \to p) > \tau_n\Gamma(p \to n)$. Once the temperature drops below about 0.1 MeV, $\tau_n\Gamma(p \to n)$ decays exponentially while $\tau_n\Gamma(n \to p)$ converges to 1 [see Fig. 1.8, where $\tau_n\Gamma(n \to p)$, $\tau_n\Gamma(p \to n)$, and the expansion rate $\tau_n H$ are shown as functions of the temperature].

According to Fig. 1.8, the line $\tau_n H$ intersects the lines $\tau_n\Gamma(n \to p)$ and $\tau_n\Gamma(p \to n)$ around $T = 0.8$ MeV. A more detailed analysis gives a decoupling temperature of $T_D \simeq 0.7$ MeV, below which the three reactions are no longer in thermal equilibrium.

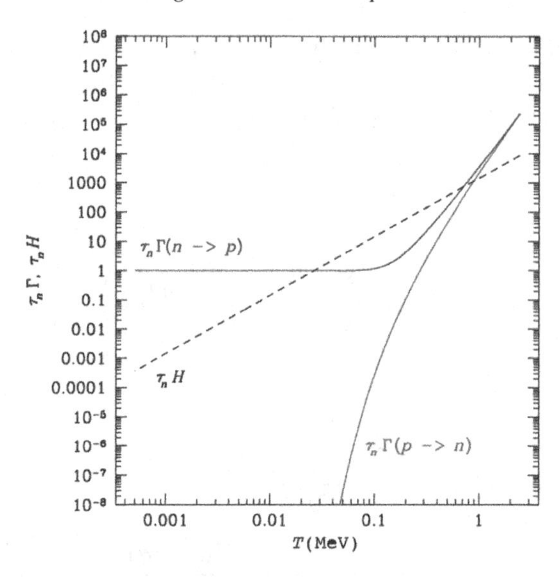

Fig. 1.8 The weak interaction rates, $\tau_n \Gamma(p \to n)$ and $\tau_n \Gamma(n \to p)$, are shown as functions of the temperature. The expansion rate, $\tau_n H$, is also indicated.

Another way to see this dropping out of the thermal equilibrium of weak interaction is to compare the true neutron abundance, Y_n, with the one obtained in thermal equilibrium. A semianalytical calculation gives (see Bernstein *et al.*, 1989) the behavior plotted in Fig. 1.9.

At decoupling, the ratio of the neutron to proton density is

$$\left(\frac{n_n}{n_p}\right)(T_D) = \exp(-Q/T_D) \simeq 1/6, \tag{1.133}$$

so that

$$Y_n = 1/7 \quad \text{and} \quad Y_p = 6/7. \tag{1.134}$$

Since T_2, the temperature of deuterium formation, is lower than T_D, in the interval $T_D > T > T_2$, neutrons simply β-decay. At τ_2 given by $T_2 = T(\tau_2) = 0.085$ MeV their density is

$$\left(\frac{n_n}{n_p}\right)(T_2) = e^{-Q/T_D} \exp(-\tau_2/\tau_n) \simeq 0.8/6 \simeq 1/7, \tag{1.135}$$

and therefore

$$Y_n = 1/8 \quad \text{and} \quad Y_p = 7/8. \tag{1.136}$$

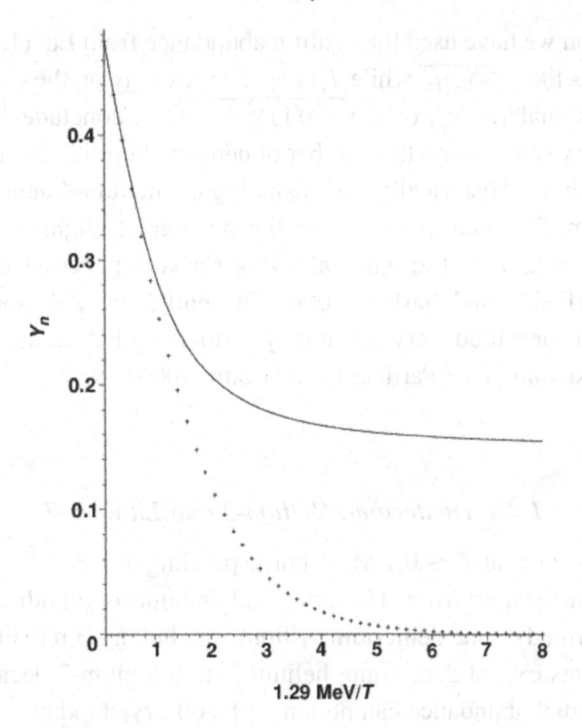

Fig. 1.9 The true neutron abundance as a function of $\Delta m/T$ (solid line) is compared with the equilibrium abundance (dotted line). Clearly, weak interaction freezes out around $T \sim 0.6 \times \Delta m \sim 0.7$ MeV.

For this we have used $\tau_2 \simeq 1.3\,\mathrm{s}\,(1/0.085)^2 \simeq 180$ s. Once deuterium is formed, helium-4 is very rapidly synthesized via the reactions

$$\begin{aligned} {}^2\mathrm{H} + {}^2\mathrm{H} &\longrightarrow n + {}^3\mathrm{He} \\ {}^3\mathrm{He} + {}^2\mathrm{H} &\longrightarrow p + {}^4\mathrm{He} \end{aligned}$$

$$\begin{aligned} {}^2\mathrm{H} + {}^2\mathrm{H} &\longrightarrow \mathrm{p} + {}^3\mathrm{H} \\ {}^3\mathrm{H} + {}^2\mathrm{H} &\longrightarrow n + {}^4\mathrm{He} \end{aligned}$$

$${}^2\mathrm{H} + {}^2\mathrm{H} \longrightarrow \gamma + {}^4\mathrm{He}$$

and essentially all deuterium is transformed in ^{4}He. The helium abundance is thus in good approximation, given by half the neutron abundance at temperature $T_2 \simeq 0.085$ MeV. With this approximation we obtain a helium-4 abundance of

$$Y_{^4He} = \frac{4(n_n/2)}{n_n + n_p} = \frac{2(n_n/n_p)}{n_n/n_p + 1} \simeq \frac{1}{4}. \tag{1.137}$$

In this expression we have used the neutron abundance from Eq. (1.136). Considering that τ_2 scales like $\sqrt{\log \eta_B}$ while T_D depends strongly on the expansion rate H, which is proportional to $\sqrt{g_{\rm eff}} \propto \sqrt{N_\nu (4/11)^{4/3} + 1}$, we conclude that the helium-4 abundance is very sensitive on the number of neutrino families, but does not change very rapidly with η_B. Historically, the cosmological helium-4 abundance has been the first experimental data to determine the number of (light) neutrino families in the range $N_\nu = 3.24 \pm 1.2$, when allowing for very generous error bars in the measurements (Fields and Sarkar, 2006). Presently, the Z-boson decay width, which has been measured very accurately with the LEP accelerator at CERN, gives the tightest value (see Particle Data Group, 2006), $N_\nu = 3.07 \pm 0.12$ at 95% confidence.

1.4.4 Deuterium, Helium-3 and Lithium-7

Nucleosynthesis starts at $T \sim 0.1$ MeV, corresponding to $\tau \sim 130$ s and terminates after a few minutes. Apart from ^{4}He very small amounts of all other elements up to lithium-7 are formed (some deuterium, tritium, and helium-3 remain unprocessed). All these elements except deuterium, helium-3, and lithium-7 decay radioactively and their primordial abundance can no longer be observed today.

The amount of deuterium and helium-3 that is not burned into helium-4 is a steep function of the baryon abundance in the Universe. The higher the baryon density, the more efficient is the conversion of deuterium and helium-3 into helium-4 (see Fig. 1.10). This can be used to determine the baryon density in the Universe very accurately. Measuring the primordial deuterium abundance is an art by itself on which we shall not dwell here. Most recent results are obtained by measuring it from the absorption lines in hydrogen (Ly-α) clouds intervening in the line of sight between us and quasars. Within generous error bars one obtains $2 \times 10^{-5} < Y_{^2H}/Y_p < 2 \times 10^{-4}$. This gives $4.7 \times 10^{-10} < \eta_B < 6.5 \times 10^{-10}$ [for more details see Olive *et al.* (2000), Burles *et al.* (2001), and Particle Data Group (2006)].

As one sees in Fig. 1.10, the lithium abundance is not a monotone function of η_B. This is so since, depending on the value of η_B, two different processes lead to lithium formation. If the baryon density is small, $\eta_B < 3 \times 10^{-10}$, lithium abundance is determined by the competition between the production process $^4\text{He} + {}^3\text{H} \to {}^7\text{Li} + \gamma$ and the destruction process $^7\text{Li} + p \to {}^4\text{He} + {}^4\text{He}$. In this regime, the abundance decays with growing η_B. For $\eta_B > 3 \times 10^{-10}$, the dominant channel goes over beryllium production $^4\text{He} + {}^3\text{He} \to {}^7\text{Be} + \gamma$, which is then converted into lithium-7 via the reaction $^7\text{Be} + e \to {}^7\text{Li} + \gamma$. The destruction process is the same as at low density. Since the conversion of beryllium into lithium increases with increasing baryon density, lithium abundance grows with η_B, for $\eta_B > 3 \times 10^{-10}$. The lithium abundance has a minimum around $\eta_B \simeq 3 \times 10^{-10}$. Inference of the

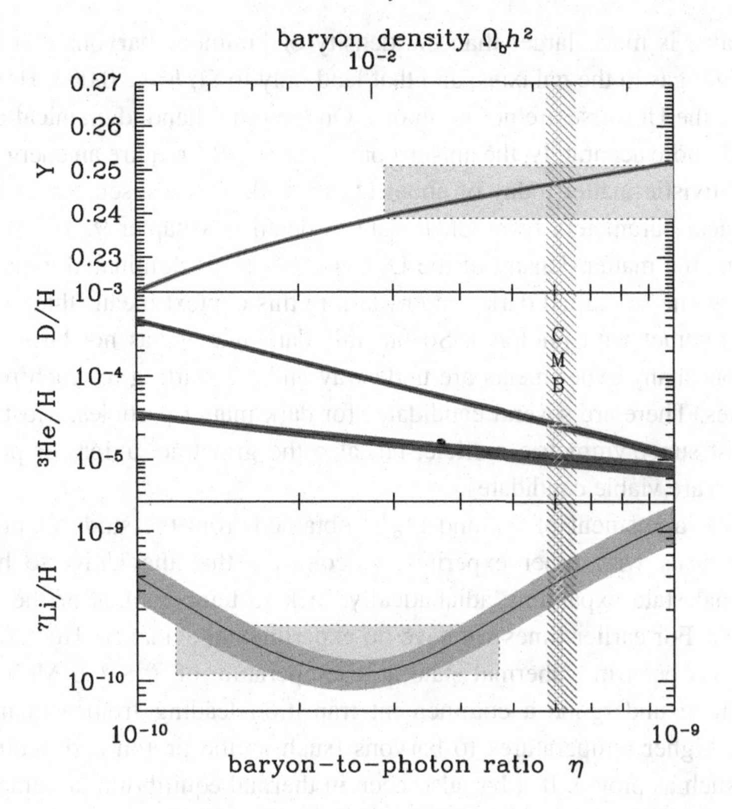

Fig. 1.10 The primordial element abundance as a function of the parameter $\eta_B = n_B/n_\gamma$. The bands compatible with the observations of the different nuclei are indicated. The horizontal band shows the range of η_B (or equivalently $\Omega_B h^2$) compatible with the nucleosynthesis data while the narrow vertical range is compatible with CMB anisotropies (see Chapter 9). It agrees very well with deuterium and helium abundances from nucleosynthesis but not so well with the lithium abundance. Figure from Tanabashi *et al.* (2019)

primordial lithium abundance is still a matter of considerable debate. It nevertheless allows us to constrain $10^{-10} < \eta_B < 10^{-9}$.

Finally, in the regime $10^{-10} < \eta_B < 10^{-9}$ the helium-4 abundance is well approximated by the formula

$$Y_{^4He} = 0.23 + 0.011 \ln(\eta_{10}) + 0.013(N_\nu - 3), \tag{1.138}$$

where we have introduced $\eta_{10} = \eta_B/10^{-10}$. All the present observations of light elements taken together limit $4.7 < \eta_{10} < 6.5$, leading to $0.017 < \Omega_B h^2 < 0.024$ [a constantly updated review can be found in Tanabashi *et al.* (2019)]. It is remarkable that this value is in very good agreement with the result obtained from measurements of the fluctuations in the CMB, which are based on completely different physics (see Chapter 9).

This value is much larger than the density of luminous baryons that make up the stars and gas in the galaxies, and that lead only to $\Omega_L h^2 \simeq 0.004$. Hence most baryons in the Universe are not luminous. On the other hand, dynamical measurements and, more accurately, the anisotropies in the CMB, require an energy density of nonrelativistic matter today of about $\Omega_m h^2 \simeq 0.13$. We discuss constraints on cosmological parameters from CMB data in detail in Chapter 9. To satisfy both constraints, the matter density of the Universe has to be dominated to about 80% by nonbaryonic, so-called dark matter (dark in this context means that this matter does not interact with photons). So far, this dark matter has not been observed directly, but many experiments are underway and are starting to reach promising sensitivities. There are several candidates for dark matter particles. Most notably, the lightest supersymmetric particle, but also the gravitino, axion, or primordial black holes are viable candidates.

The good agreement of N_ν and $\Omega_B h^2$ obtained from the study of primordial nucleosynthesis with other experiments, confirms that the Universe has been in a thermal state expanding adiabatically back to temperatures of the order of $T \sim 1$ MeV. For earlier times we have no experimental evidence. However, if the Universe has been in a thermal state at a temperature of $T \sim 200$ MeV, $\tau \sim 0.1$ s, it has then undergone a confinement transition leading from a quark gluon plasma at higher temperatures to baryons (such as the proton and neutron) and mesons (such as pions). If it has also been in thermal equilibrium at temperatures of up to $T \sim 200$ GeV, $\tau \sim 0.001$ s, it has then undergone the electroweak transition giving masses to the $W^\pm$ and Z bosons. At even higher temperatures we have no experimentally confirmed theory of fundamental interactions. Maybe, at $T \sim$ a few TeV the Universe becomes supersymmetric. Maybe, at $T \sim 10^{16}$ GeV a phase transition from a previous grand unified symmetry to the (supersymmetric) standard model symmetries took place. At this or higher energies the Universe may also have gone through (or emerged from) a superstring phase. To date such questions remain entirely speculative. Their quantitative investigation, especially possible observable signatures of a superstring phase, is an active field of research.

1.5 Inflation

1.5.1 Cosmological Problems

We first discuss the motivation for, and some consequences of a so-called inflationary phase. We then exemplify the idea with a cosmology dominated by a scalar field. It is, however, clear that this realization has to be regarded as a toy model because the actual physical degrees of freedom relevant in the very early Universe, where such a period has most probably to be situated (see Chapter 3), are not known. In that sense this section is on a different level from the previous ones. We do not have any direct evidence that an inflationary phase has taken place in our Universe. Such a period just addresses several otherwise mysterious initial conditions of the observed Universe. The most significant observed "prediction" of inflation is a nearly scale-invariant spectrum of initial fluctuations that we shall discuss in Chapter 3. What is more serious is that we have no "direct" experimental evidence of the existence of an "inflaton field."

We include a possible cosmological constant into the energy density and the pressure, so that Eqs. (1.20) and (1.21) reduce to

$$\mathcal{H}^2 = \frac{8\pi G}{3}a^2\rho - K, \tag{1.139}$$

$$\dot{\mathcal{H}} = -\frac{4\pi G}{3}a^2(\rho + 3P) = \left(\frac{\ddot{a}}{a}\right) - \mathcal{H}^2. \tag{1.140}$$

If $\rho + 3P > 0$ at all times, the homogeneous and isotropic cosmological model has several important problems.

First, as we have discussed in Section 1.2.2, there is the big bang singularity in the finite past, $t = 0$. At this time $a = 0$ and the curvature diverges.

Furthermore, the causal horizon at (conformal) time t, that is, the distance a photon has traveled from $t = \tau = 0$ until time t, is given by $a(t)t = a(t)\int_0^{\tau(t)} a^{-1}\,d\tau$. Since for $\rho + 3P > 0$, a grows slower than linear in τ, this integral converges, is finite. As we have seen (in Eq. (1.25)), $a(\tau) \propto \tau^{\frac{2}{3(1+w)}}$ if $w = P/\rho$ is constant.

For example, the size of the causal horizon at recombination is seen today under the angle of about 1°, if the Universe was radiation ($w = 1/3$) and matter ($w = 0$) dominated up to recombination see Exercise 1.10. It is therefore very mysterious that we see the same microwave background temperature on patches separated by much more than 1°, which had never been in causal contact before the microwave photons had been emitted. This is the "**horizon problem**."

Another problem is the following: the Friedmann equations, (1.139) and (1.140), allow us to derive an evolution equation for $\Omega(t) \equiv 8\pi G\rho a^4/3\dot{a}^2 \equiv 1 + K/\mathcal{H}^2$,

$$\frac{d}{dt}(\Omega(t) - 1) = (\Omega(t) - 1)\frac{8\pi Ga^2}{3}\left(\frac{\rho + 3P}{\mathcal{H}}\right). \tag{1.141}$$

This shows that, in an expanding universe with $\rho + 3P > 0$, $\Omega = 1$ is an unstable fixed point of evolution: if $\Omega(t) > 1$, the derivative is positive and $\Omega(t)$ increases while for $\Omega(t) < 1$, the derivative is negative and $\Omega(t)$ decreases. For a present value of $0.1 < \Omega_0 < 2$ we need $|\Omega(\eta_{\rm nuc}) - 1| \sim (z_{\rm eq}/z_{\rm nuc}^2)|\Omega_0 - 1| \lesssim 10^{-15}$ at nucleosynthesis, or $|\Omega(t_P) - 1| \lesssim 10^{-60}$ at the Planck time, $\tau_P = \sqrt{\hbar G/c^5} \simeq 5.4 \times 10^{-44}$ s. Why is $\Omega(t)$ still of order unity so long after the only timescale in the problem that is τ_P?

This "**flatness problem**" can also be formulated as an "**entropy problem**." The entropy inside the curvature radius is already of the order of $S_K \geq 10^{88}$ at the Planck time.

Another problem is the "**monopole problem**" or more generically the problem of unwanted "relics." Most particle physics models produce some stable "relics" at very high temperatures, which are not observed in the present Universe. A very rapid phase of expansion can help to dilute such relics.

To resolve these problems one introduces an "inflationary phase." Inflation is a phase during which the strong energy condition, $\rho + 3P > 0$, is violated and expansion can therefore be much more rapid than linear in τ.

1.5.2 Scalar Field Inflation

We now study the most common solution of the aforementioned problems, namely the introduction of a period in which the dynamics of the Universe is dominated by a scalar field, ϕ which is usually called the "inflaton." The scalar field Lagrangian is given by

$$\mathcal{L}_\phi = -\frac{1}{2}\,\partial_\mu\phi\,\partial^\mu\phi - W(\phi). \tag{1.142}$$

The sign of the kinetic term in the foregoing Lagrangian may differ from what you are used to from quantum field theory. This comes from the fact that we use the metric signature $(-, +, +, +)$.

The field ϕ can, in principle, interact with other fields such as fermions, gauge bosons, and so forth, but we assume that this interaction can be neglected during inflation, and that energy and pressure are dominated by the contribution from the inflaton. The energy–momentum tensor of ϕ is given by

$$T_{\mu\nu} = \frac{-2}{\sqrt{-g}}\frac{\partial}{\partial g^{\mu\nu}}(\sqrt{-g}\mathcal{L}_\phi),$$

where $g = \det(g_{\mu\nu})$. This yields

$$\begin{aligned} T_{\mu\nu} &= \partial_\mu\phi\,\partial_\nu\phi + g_{\mu\nu}\mathcal{L}_\phi \\ &= \partial_\mu\phi\,\partial_\nu\phi - \frac{1}{2}g_{\mu\nu}\,\partial_\lambda\phi\,\partial^\lambda\phi - g_{\mu\nu}W(\phi). \end{aligned}$$

Here we have used that the derivatives of the determinant A of an arbitrary matrix A_{ab} with respect to the elements of its inverse, A^{ab}, are given by $\partial A/\partial A^{ab} = AA_{ab}$.

For the energy density and pressure we thus obtain

$$\rho_\phi = -T^0_0 = \frac{1}{2a^2}\dot\phi^2 + \frac{1}{2a^2}(\nabla\phi)^2 + W(\phi), \tag{1.143}$$

and

$$P_\phi = \frac{1}{3}T^i_i = \frac{1}{2a^2}\dot{\phi}^2 - \frac{1}{6a^2}(\nabla\phi)^2 - W(\phi). \quad (1.144)$$

We now assume that there exists some region of space within which we may neglect the spatial derivatives of ϕ, at some initial time τ_i, and the temporal derivative is much smaller than the potential,

$$\nabla\phi(\mathbf{x}, \tau_i) \ll \dot{\phi}(\mathbf{x}, \tau_i) \ll W(\phi). \quad (1.145)$$

Furthermore, we assume that the potential is positive,

$$W(\phi(\mathbf{x}, \tau_i)) > 0. \quad (1.146)$$

We then have

$$\frac{3H^2}{8\pi G} = \rho = \rho_\phi = \frac{1}{2a^2}\dot{\phi}^2 + W(\phi) \simeq W(\phi), \quad (1.147)$$

$$P = P_\phi = \frac{1}{2a^2}\dot{\phi}^2 - W(\phi) \simeq -W(\phi), \quad (1.148)$$

so that $P_\phi \simeq -\rho_\phi$ and $\rho_\phi + 3P_\phi \simeq -2W(\phi) < 0$. (We have neglected a possible curvature term. Qualitatively nothing changes if we include it, since it soon becomes subdominant.)

This is the basic idea of inflation: at some early time, in some sufficiently large patch, the Universe is dominated by the potential of a slowly varying (slow rolling) scalar field, and hence it is in an inflationary phase. During inflation this patch expands rapidly and the causal horizon becomes very large and $\Omega(t)$ tends to 1, so that the curvature term is soon negligible. As time goes on, the scalar field starts evolving faster and inflation eventually comes to an end when the time derivative ϕ'^2 grows to the order of W. The scalar field then soon reaches the minimum of the potential and starts to oscillate. We suppose that at large values of $a^{-1}\dot{\phi}$, the coupling of the inflaton to other fields becomes significant so that it decays into a thermal mix of elementary particles, leading to a radiation-dominated universe. There are many detailed realizations of this basic picture that can be found in the literature; see, for example, Liddle and Lyth (2000). It is, however very difficult to deduce them from a serious high-energy physics theory such as string theory.

Let us study slow roll inflation in somewhat more detail. When neglecting spatial derivatives, the equation of motion of the scalar field becomes ($W_{,\phi} \equiv dW/d\phi$)

$$\ddot{\phi} + 2\left(\frac{\dot{a}}{a}\right)\dot{\phi} + a^2 W_{,\phi} = 0, \quad (1.149)$$

$$\phi'' + 3\left(\frac{a'}{a}\right)\phi' + W_{,\phi} = 0, \quad (1.150)$$

in conformal time, Eq. (1.149), and in cosmic time, Eq. (1.150). During slow rolling, the first term of this equation is negligible with respect to the two others, so that

$$3\left(\frac{a'}{a}\right)\phi' \simeq -W_{,\phi}. \tag{1.151}$$

The slow roll conditions are therefore

$$\frac{1}{2}\phi'^2 \ll W \quad \text{and} \quad |\phi''| \ll 3H|\phi'|. \tag{1.152}$$

With $H = a'/a$, slow rolling also implies that $H' \ll H^2$. Taking the time derivative of Eq. (1.147) and replacing ϕ' by (1.151) yields the slow roll conditions

$$\epsilon_1 \equiv -\frac{H'}{H^2} = \frac{\mathcal{H}^2 - \dot{\mathcal{H}}}{\mathcal{H}^2} \approx \frac{m_P^2}{16\pi}\left(\frac{W_{,\phi}}{W}\right)^2 \simeq \frac{3}{2}\frac{\phi'^2}{W} \ll 1. \tag{1.153}$$

The second condition of Eq. (1.152) gives

$$\left|\frac{\phi''}{3H\phi'}\right| \ll 1.$$

We now set

$$\epsilon_2 \equiv -\frac{m_P^2}{24\pi}\left(\frac{W_{,\phi\phi}}{W}\right) \quad \text{and require} \quad |\epsilon_2| \ll 1. \tag{1.154}$$

Note that ϵ_1 is always positive while ϵ_2 can have either sign. With $H^2 \simeq 8\pi W/(3m_P^2)$, and the derivative of $\phi' = -W_{,\phi}/(3H)$, one finds that the inequalities (1.153) and (1.154) are equivalent to the slow roll conditions (1.152). The parameters ϵ_1 and ϵ_2 are the slow roll parameters. Inflation terminates when ϵ_1 approaches unity. In the literature one often uses the notation $\epsilon \equiv \epsilon_1$ and $\delta \equiv -\epsilon_2/3$.

Taking the derivative (w.r.t. t) of Eq. (1.153) in the last equals sign, one obtains

$$\dot{\epsilon}_1 = 2\epsilon_1\left(3\epsilon_2 + 2\epsilon_1\right)\mathcal{H}, \qquad \frac{\dot{\epsilon}_1}{\mathcal{H}\epsilon_1} = 6\epsilon_2 + 4\epsilon_1 \equiv \eta. \tag{1.155}$$

The last equation can also be used as a definition of ϵ_2 (or, more consistently, η). The advantage of this definition is its independence of the realization of slow roll inflation by means of a scalar field. A more systematic procedure along these lines is to define $\tilde{\epsilon}_1 \equiv \epsilon_1$ and $\tilde{\epsilon}_2 = (\dot{\tilde{\epsilon}}_1/\tilde{\epsilon}_1)\mathcal{H}^{-1} = \eta$, $\tilde{\epsilon}_3 = (\dot{\tilde{\epsilon}}_2/\tilde{\epsilon}_2)\mathcal{H}^{-1}$, and so forth. Our parameter ϵ_2 is related to $\tilde{\epsilon}_2 \equiv \eta$ via

$$\epsilon_2 = -\frac{2}{3}\epsilon_1 + \frac{1}{6}\eta. \tag{1.156}$$

While ϵ_2 is usually of the same order of magnitude as ϵ_1, we expect η to be significantly smaller.

As an example we consider power law expansion, $a \propto t^q$. In this case we have

$$\mathcal{H} = \frac{q}{t}, \quad \epsilon_1 = 1 + \frac{1}{q}, \quad \epsilon_2 = -\frac{2}{3}\epsilon_1, \quad \tilde{\epsilon}_2 = \tilde{\epsilon}_n = 0. \tag{1.157}$$

During slow roll inflation, $q \sim -1$, the parameters ϵ_1 and ϵ_2 are small. Also note that $\epsilon_2 = -(2/3)\epsilon_1$ during power law expansion. The parameters $\tilde{\epsilon}_i$, $i > 1$ describe the deviation from power law expansion. They have been used in the literature to derive a systematic slow roll expansion to higher orders (Schwarz *et al.*, 2001). In this book we shall not go beyond the first order and we use the standard parameters ϵ_1 and ϵ_2 to make contact with the standard literature.

There are two principally different possibilities for slow roll inflation.

(1) We first consider a potential that is simply $\propto \phi^n$, so that $W_{,\phi\phi}/W \sim (W_{,\phi}/W)^2 \sim \phi^{-2}$. The slow roll conditions then require $\phi \gg m_P$ and inflation stops when the inflaton becomes of order the Planck mass. These models are termed **large-field inflation**. Setting $W = (\lambda/n)m_P^4(\phi/m_P)^n$, during the inflationary phase, Eq. (1.151) together with Eq. (1.147) implies

$$\sqrt{\frac{24\pi\lambda}{n}} m_P \left(\phi/m_P\right)^{n/2} \phi' = -\lambda m_P^3 \left(\frac{\phi}{m_P}\right)^{n-1}. \tag{1.158}$$

Dividing by ϕ^{n-1}, if $n \neq 4$ the left-hand side becomes the derivative of $(\phi/m_P)^{2-n/2}$, which hence is a constant. If $n = 4$, the left-hand side is $\propto 1/\phi$, that is, the derivative of $\log(\phi/m_P)$. The general solution is therefore given by

$$\phi(\tau)^{(4-n)/2} = \phi_i^{(4-n)/2} - \frac{4-n}{2}\sqrt{n\lambda/24\pi}\, m_P(\tau - \tau_i) \quad \text{if } n \neq 4, \tag{1.159}$$

$$\phi(\tau) = \phi_i \exp\left(-\sqrt{\frac{\lambda}{6\pi}} m_P(\tau - \tau_i)\right) \qquad \text{if } n = 4. \tag{1.160}$$

Inserting now $\phi' = -\sqrt{\lambda n/24\pi}\, m_P^2(\phi/m_P)^{n/2-1}$ in the Friedmann equation,

$$(\log(a))' = \sqrt{\frac{8\pi\lambda}{3n}} m_P(\phi/m_P)^{n/2},$$

we obtain

$$\frac{d\log(a)}{d\phi} = -\frac{8\pi}{n}\frac{\phi}{m_P^2},$$

with solution

$$a(\tau) = a_i \exp\left(\frac{4\pi}{nm_P^2}(\phi_i^2 - \phi^2)\right). \tag{1.161}$$

This case is illustrated in Fig. 1.11.

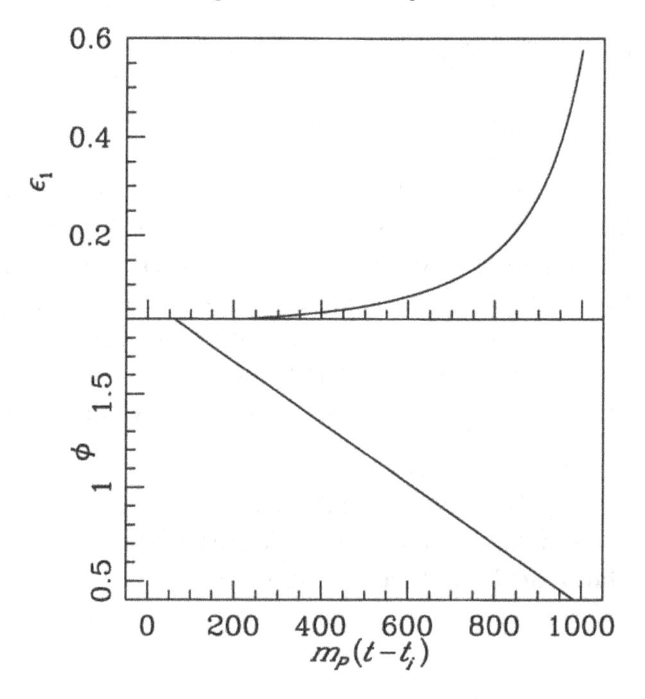

Fig. 1.11 Large-field inflation for $W = \lambda m_P^2 \phi^2/2$. The bottom panel shows the inflaton ϕ in units of m_P rolling linearly in time. In the upper panel the evolution of the slow roll parameter, $\epsilon_1(t)$, is indicated. As long as $\phi > m_P$, $\epsilon_1 = -2\epsilon_2$ stays small. At $\phi \sim m_P$, ϵ_1 starts to grow and inflation stops.

(2) If the potential is more complicated and has a very flat regime in the vicinity of its maximum $\phi = \sigma \ll m_P$, like, for example, the Coleman–Weinberg potential (Kolb and Turner, 1990),

$$W(\phi) = \frac{1}{2}\sigma^4 + \phi^4 \left[\ln\left(\frac{\phi^2}{\sigma^2}\right) - \frac{1}{2} \right],$$

we speak of **small-field inflation**. This potential passes through 0 at $\phi = \sigma$. In this case, the slow roll conditions are satisfied for field values $|\phi| \lesssim \sigma$, which are much smaller than the Planck mass.

During a potential dominated phase where $\rho \sim -P \sim W \sim$ constant, the solutions of the Friedmann equations are

$$a = a_0 \exp(\tau H) = \frac{1}{H|t|} \quad (-\infty < t < 0, \quad -\infty < \tau < \infty), \tag{1.162}$$

$$H^2 = \frac{8\pi G}{3} W = \text{constant}, \tag{1.163}$$

$$\mathcal{H} = aH = \frac{1}{|t|}. \tag{1.164}$$

The limit $\tau \to \infty$ corresponds to $t \to 0$. The foregoing solution is a portion of de Sitter spacetime.[6]

Denoting by indices i and f the beginning and the end of inflation, the number of e-foldings of expansion during inflation is given by

$$N(\phi_f, \phi_i) = \ln\left(\frac{a(\tau_f)}{a(\tau_i)}\right).$$

Using

$$N(\phi_f, \phi_i) = \ln a_f - \ln a_i = \int_{a_i}^{a_f} \frac{da}{a},$$

we obtain

$$N(\phi_f, \phi_i) = \int_{a_i}^{a_f} \frac{1}{a}\, da = \int_{\tau_i}^{\tau_f} \frac{a'}{a}\, d\tau = \int_{\tau_i}^{\tau_f} H\, d\tau. \tag{1.165}$$

With Eq. (1.151) we can write

$$H\, d\tau = H \frac{d\tau}{d\phi} d\phi = H \frac{d\phi}{\phi'} = -\frac{3H^2\, d\phi}{W_{,\phi}}.$$

The number of e-foldings is hence given by

$$N(\phi_f, \phi_i) = -3 \int_{\phi_i}^{\phi_f} \frac{H^2}{W_{,\phi}}\, d\phi \simeq -\frac{8\pi}{m_P^2} \int_{\phi_i}^{\phi_f} \frac{W}{W_{,\phi}}\, d\phi = -2\sqrt{\pi} \int_{\phi_i}^{\phi_f} \frac{1}{\sqrt{\epsilon_1}} \frac{d\phi}{m_P}$$
$$\sim \frac{8\pi}{n} \frac{\phi_i^2}{m_P^2}. \tag{1.166}$$

The last $\sim$ sign is valid only for large-field inflation, where $W \propto \phi^n$ and we suppose

$$\phi_f \sim m_P \ll \phi_i.$$

The slow roll conditions imply

$$N_{\rm tot} = N(\phi_f, \phi_i) \gg 1. \tag{1.167}$$

For $w = P/\rho = \text{constant}$ we have

$$|\Omega(\tau) - 1| = \frac{3|K|}{8\pi G a^2 \rho} \propto a^{1+3w}.$$

During an inflationary phase, $w = -1$, $|\Omega(\tau) - 1|$ decreases like $1/a^2$. To reduce it from a value of order unity down to $\sim 10^{-60}$ we therefore need about $30 \ln(10) \sim 70$ e-foldings of inflation.

[6] de Sitter spacetime is the solution to the Einstein equation $G_{\mu\nu} = \Lambda g_{\mu\nu}$ with $\Lambda > 0$. The solution with $\Lambda < 0$ is called anti-de Sitter; see Hawking and Ellis (1973).

1.5.3 Preheating and Reheating

When inflation ends, ϕ decays rapidly and starts oscillating about its minimum. The details of this process depend on the couplings of the inflaton to other degrees of freedom, which eventually decay into the degrees of freedom of the standard model. For this discussion we consider a simple toy model with $\mathcal{L}_\phi = -\frac{1}{2}\partial_\mu\phi\partial^\mu\phi - \frac{1}{2}m_\phi^2\phi^2$. At the end of inflation the inflaton oscillates as

$$\phi = \phi_0(\tau)\cos(m_\phi\tau)$$

with a slowly varying amplitude $\phi_0(\tau) \simeq m_P$. The inflatons have vanishing momentum and their number density is

$$n_\phi = \frac{\rho_\phi}{m_\phi} = \frac{1}{2m_\phi}((\phi')^2 + m_\phi^2\phi^2) \simeq \frac{1}{2}m_\phi m_P^2. \tag{1.168}$$

For example, for $m_\phi = 10^{15}$ GeV this amounts to the huge number density of $n_\phi \sim 10^{95}\ \mathrm{cm}^{-3}$.

Independent of the detailed form of the potential, to lowest order, ϕ is a harmonic oscillator with frequency $m_\phi^2 \simeq W_{,\phi\phi}(\phi_0)$ (as long as the quadratic term in the potential does not vanish). For a harmonic oscillator, when averaging over one period we have

$$\langle W \rangle = \frac{1}{2a^2}\langle\dot{\phi}^2\rangle,$$

so that

$$\langle p_\phi \rangle = \left\langle \frac{1}{2a^2}\dot{\phi}^2 - W \right\rangle = 0, \quad \text{and hence } \langle\rho_\phi\rangle \propto a^{-3}.$$

We assume that during these oscillations, the coupling of ϕ to other degrees of freedom becomes relevant and the inflaton finally decays into a mix of elementary particles. In a first approximation we can describe the coupling with the other degrees of freedom by means of a term of dissipation of the form $\Gamma\dot{\phi}$ in the equation of motion for ϕ,

$$\phi'' + 3H\phi' + \Gamma\phi' = -W_{,\phi}(\phi). \tag{1.169}$$

As long as $H \gg \Gamma$ (during inflation), particle production is negligible. When $H \simeq \Gamma$, reheating takes place and the inflaton energy is rapidly dissipated into other particles that couple to the inflaton.

In order to discuss the decay of the inflaton in somewhat more detail, we consider a toy model in which the interaction is dominated by the coupling of ϕ to a scalar field χ with Lagrangian

$$\mathcal{L}_\chi = -\frac{1}{2}\,\partial_\mu\chi\,\partial^\mu\chi - \frac{1}{2}m_\chi^2\chi^2. \tag{1.170}$$

The interaction between the inflaton ϕ and the matter field χ is supposed to be of the form

$$\mathcal{L}_{\rm int} = -\frac{1}{2} g \phi \chi^2, \tag{1.171}$$

where g is a coupling constant with the dimension of mass. The full Lagrangian is then given by

$$\mathcal{L} = \mathcal{L}_\phi + \mathcal{L}_{\rm int} + \mathcal{L}_\chi. \tag{1.172}$$

The decay rate of the ϕ particles in Born approximation is

$$\Gamma_\phi \sim \frac{g^2}{m_\phi}.$$

However, inserting this into Eq. (1.169) is a good approximation only as long as the mean number of χ particles already present in a given momentum mode k is small so that we may neglect stimulated emission. The effective mass of χ-particles is $m_{\rm eff} = \sqrt{m_\chi^2 + g\phi(t)}$ so that their momentum is

$$k = \left(\frac{m_\phi^2}{4} - m_{\rm eff}^2\right)^{1/2}.$$

Here we have taken into account that each inflaton decays into two χ-particles. Now, $\phi(t) \in [-m_P, m_P]$. Hence, if $m_\phi^2 \gg m_\chi^2 + g m_P$, the band of possible momenta is given by $k \in [k_0 - \Delta k, k_0 + \Delta k]$ with

$$k_0 = \sqrt{\frac{m_\phi^2}{4} - m_\chi^2} \simeq \frac{m_\phi}{2} \quad \text{and} \quad \Delta k \simeq \frac{g m_P}{m_\phi} \ll k_0.$$

Because $\Delta k \ll k_0$ this situation is called "narrow band preheating." As we shall see in the text that follows, this process leads to resonant amplification.

The number of χ-particles with momentum k is roughly given by the total number of χ-particles divided by the number of "elementary phase space volumes," $(2\pi)^3$, in the allowed volume of phase space, $4\pi k_0^2(2\Delta k)$. This yields

$$N_k \simeq \frac{4\pi^2 n_\chi}{g m_\phi m_P} \simeq \frac{2\pi^2 m_P n_\chi}{g n_\phi}.$$

For the second $\simeq$ sign we made use of Eq. (1.168). This occupation number exceeds unity as soon as a fraction g/m_P of ϕ-particles is converted into χ-particles. After that moment, stimulated emission can no longer be neglected and Eq. (1.169) becomes a bad approximation. Since g/m_P typically is very small, this is usually the case very soon. As we shall now see, when this happens, stimulated emission leads to resonant production of χ-particles in certain k-bands.

To calculate the generation of χ-particles in more detail we vary the Lagrangian with respect to χ to obtain the χ-equation of motion,

$$\chi'' + 3H\chi' - a^{-2}\nabla^2\chi + (m_\chi^2 + g\phi_0\cos(m_\phi\tau))\chi = 0.$$

To study qualitatively the decay of the ϕ-particles into χ, we neglect expansion by setting $H = 0$, $a = 1$ and $\phi_0 =$ constant. Fourier transforming the above equation, we then obtain for the mode χ_k

$$\chi_k'' + \left[\omega_k^2 + 2\mu\cos(m_\phi\tau)\right]\chi_k = 0, \quad \mu = \frac{g\phi_0}{2}, \quad \omega_k^2 = k^2 + m_\chi^2 .$$

This equation is known as the *Mathieu equation.* Its solutions are characterized by resonance bands of widths $\Delta\omega_k^{(n)}$ centered at the frequencies

$$\omega_k^{(n)} = \frac{n}{2}m_\phi.$$

The widths are of the order of

$$\frac{\Delta\omega_k^{(n)}}{\omega_k^{(n)}} \simeq \left(\frac{2\mu}{\omega_k^{(n)2}}\right)^n = \frac{\Delta k^{(n)}}{k^{(n)}} \simeq \left(\frac{4gm_P}{n^2m_\phi^2}\right)^n \propto \left(\frac{g}{n^2}\right)^n .$$

For frequencies within these bands, χ_k is amplified exponentially fast (for more details on the Mathieu equation and resonant amplification, see Arnold, 1978). Since the width of the nth resonance is proportional to g^n, it appears only at nth order in perturbation theory. For small couplings only the first resonance $\omega_k^{(1)} = m_\phi/2$ with $\Delta\omega_k^{(1)} = \Delta k$ is relevant.

When we take into account the expansion of the Universe, the frequency ω_k becomes time dependent. A given frequency therefore spends only a finite time in the resonance band and the energy transfer from ϕ into χ remains perfectly finite. Nevertheless, this parametric resonance is much more efficient than the decay obtained by some effective damping rate Γ.

After parametric resonance, χ is not yet in a thermal state. This period is therefore called "preheating." After preheating, the coupling of χ to other degrees of freedom leads to thermalization; this process is called *reheating*. The importance of preheating lies in its efficiency in transferring energy. If the χ-field couples strongly to the standard-model particles, reheating and thermalization can proceed much faster over resonant decay than over the necessarily weak average coupling of the inflaton to other particles.

If the condition $m_\phi^2 > m_\chi^2 + g\phi(t)$ is not satisfied, $\Delta\omega_k^{(1)} = \Delta k$ is not small and we have "broad-band" resonance. In this case, the effective mass of the χ-particles can be larger than the mass of the ϕ-particles and only the coherent decay of several inflatons can lead to χ-production. For a discussion of the main physical processes in this case see Mukhanov (2005). One of the most interesting consequences of broad-band resonance is that it can lead to the production of particles that are heavier than the inflaton.

Changing the coupling to $L_{\rm int} = -\frac{1}{2}\tilde{g}\phi^2\chi^2$ does not affect the generic behavior of preheating. We can again obtain a Mathieu equation but with resonant frequency $\omega^{(n)} = nm_\phi$ and width $\Delta\omega^{(n)}/\omega^{(n)} = (2\tilde{g}\phi_0^2/\omega^{(n)2})^n$; see Exercise 1.11. Due to the Pauli exclusion principle, couplings of the inflaton to fermions cannot give raise to parametric resonance. The details of the reheating process and the temperature at the end of reheating depend on the particle physics model describing the coupling of ϕ and χ to other particles, especially to the standard model particles. The reheating temperature can go from 1 TeV $< T <$ 10^{13} GeV.

1.5.4 Resolution of the "Cosmological Problems"

At the end of the reheating process, $\tau = \tau_{\rm rh}$, all the energy is supposed to be thermalized and the Universe is dominated by relativistic particles, satisfying $P = \rho/3$ such that

$$\rho \propto a^{-4}.$$

To determine the duration of inflation necessary in order to solve the horizon problem, we consider the entropy, S_H, contained in a volume that corresponds to one Hubble scale, H_i^{-1}, at the beginning of inflation. Since expansion is adiabatic after inflation, the entropy inside a given physical volume remains constant. The requirement that the present Hubble scale, H_0^{-1}, be smaller than the size of the causal horizon is therefore equivalent to $S_H > S_{H_0}$, where S_{H_0} denotes the entropy inside the volume H_0^{-3}. The entropy inside a causal volume, $H_i^{-3}(a/a_i)^3$, is given by its value

$$S_H \simeq H_i^{-3}\left(\frac{a_{\rm rh}}{a_i}\right)^3 T_{\rm rh}^3,$$

after reheating. The Hubble parameter at the beginning of inflation is

$$H_i^2 \simeq \frac{8\pi}{3m_P^2} W(\phi_i),$$

so that

$$S_H \simeq H_i^{-3}\left(\frac{a_f}{a_i}\right)^3\left(\frac{a_{\rm rh}}{a_f}\right)^3 T_{\rm rh}^3 \simeq \frac{m_P^3}{W_i^{3/2}} e^{3N_{\rm tot}} \frac{\rho_f}{T_{\rm rh}}.$$

For the last $\simeq$ sign we have assumed that the Universe was roughly matter dominated from the end of inflation until the end of reheating, $\rho \propto a^{-3}$ and $\rho_{\rm rh} \sim T_{\rm rh}^4$. With $\rho_f \sim W_f$, this yields

$$S_H \simeq \frac{m_P^3 W_f}{T_{\rm rh} W_i^{3/2}} e^{3N_{\rm tot}}.$$

In order to solve the entropy problem, we require that this entropy is at least as large at the entropy in the present Hubble horizon, $S_H > S_{H_0} \simeq T_0^3 H_0^{-3} \simeq 10^{88}$. This now results in

$$N_{\rm tot} \geq N_{\rm min} = \frac{88}{3}\ln(10) + \ln\left(\frac{T_{\rm rh}^{1/3} W_i^{1/2}}{m_P W_f^{1/3}}\right). \qquad (1.173)$$

For example, in a model with $W = \frac{1}{2}m_\phi^2\phi^2$, we have large-field inflation that stops roughly when $\phi = \phi_f \simeq m_P$ so that $W_f = \frac{1}{2}(m_\phi m_P)^2$ and $W_i = \frac{1}{2}(m_\phi \phi_i)^2$. Hence

$$N_{\rm min} = \frac{88}{3}\ln(10) + \frac{1}{3}\ln\left(\frac{T_{\rm rh} m_\phi}{m_P^2}\right) + \ln\left(\frac{\phi_i}{m_P}\right).$$

If $N_{\rm tot} \geq N_{\rm min}$ the horizon problem is also solved. Indeed, since the entropy inside a comoving volume is conserved after inflation, the present volume of radius H_0^{-1} has grown out of a radius that was smaller than H_i^{-1} at the beginning of inflation, and therefore was already in causal contact before the beginning of inflation.

To solve the flatness problem we must enlarge the curvature scale to $R_K(\tau_0) \geq H_0^{-1}$. This is equivalent to $S_K(\tau_0) \geq S_H(\tau_0) \simeq 10^{88}$. With

$$R_K^3(\tau_{\rm rh}) = R_K^3(\tau_i)\left(\frac{a_f}{a_i}\right)^3 = \frac{H_i^{-3}}{|\Omega_i - 1|^{3/2}}\left(\frac{a_f}{a_i}\right)^3,$$

this leads to

$$N_{\rm tot} \geq N_{\rm min} + \frac{1}{2}\log|\Omega_i - 1|. \qquad (1.174)$$

Comparing $N_{\rm min}$ with Eq. (1.166), we find that successful inflation with a simple $\frac{1}{2}m_\phi^2\phi^2$ potential requires $\phi_i \gtrsim$ a few times m_P. After an inflationary period that is sufficiently long, so that the conditions (1.173) and (1.174) are satisfied, both the horizon and flatness problems are resolved. During such an inflationary phase also all unwanted relics are diluted by a factor of $\exp(3N_{\rm tot})$.

Finally, it is important to note that we do not require a perfectly homogeneous and isotropic universe, or even thermal equilibrium prior to inflation. We just need a small "patch" in an otherwise arbitrary, chaotic, universe, within which the gradient and kinetic energy are much smaller than the potential energy, so that the slow roll conditions are satisfied. This patch then inflates to encompass the entire present Hubble volume. This idea of "chaotic inflation" goes back to Linde (1989) and it is of course much more satisfactory than a model in which the Universe has to start out with homogeneous and isotropic spatial sections before inflation.

When discussing inflation, one of the most mysterious problems of gravity becomes apparent: while adding a constant to the potential W of the scalar field does not affect any of the other interactions, it severely alters gravity. It modifies cosmic expansion in the same way as adding a cosmological constant. What determines the correct level of a potential? This question is equivalent to the problem of the cosmological constant. Why is the present cosmological constant so small, $\Lambda/(8\pi G) \simeq (2 \times 10^{-3}\,\text{eV})^4$, much smaller than any fundamental energy scale? The problem is even more serious when we remember that in quantum field theory we use the freedom to add or subtract a constant from the potential by absorbing the infinite zero-point energy into it. Furthermore, at each phase transition this zero-point energy changes by a finite, calculable amount. Before the discovery of the accelerated expansion of the Universe, which is most simply interpreted as a cosmological constant, $\Lambda/(8\pi G) \simeq (2 \times 10^{-3}\,\text{eV})^4 \neq 0$, it was justifiable to assume that the freedom of the cosmological constant has to be used in order to annihilate any vacuum energy contribution from quantum field theory, so that the effective cosmological constant would vanish, $\Lambda_{\text{eff}} = \Lambda + 8\pi G W_0 = 0$. Present observations, however, indicate that this compensation takes place only approximately, leaving a small but nonvanishing effective cosmological constant, $\Lambda_{\text{eff}} \neq 0$, which starts to dominate the expansion of the Universe just at present time, when there are sufficiently developed intelligent beings in the Universe that wonder about it. In all the cosmic past, this cosmological constant was completely negligible, and in all the cosmic future, it will be the only relevant contribution to expansion. Only at present it is comparable with the mean mass density of the Universe. Apart from the bizarre value of Λ_{eff}, we thus also have a strange coincidence problem.

This is presently one of the deepest problems of physics. Ordinary quantum field theory does not determine the vacuum energy of quantum fields, but only changes that may happen depending on the external conditions. We may hope that a quantum theory of gravity addresses the cosmological constant problem. The cosmological constant may even represent our first observational data related to quantum gravity.

Exercises

(The exercises marked with an asterisk are solved in Appendix 11 which is not in this printed book but can be found online.)

1.1 **Coordinates**

Find the coordinate transformation leading from the coordinates used in Eq. (1.9) to those of Eq. (1.10) and finally of Eq. (1.8).

1.2 **FL universes are conformally flat**

Show that FL universes are conformally flat (also when the curvature does not vanish) and find the coordinate transformation $(\tau, r) \to (\sigma, \rho)$ such that

$$-d\tau^2 + a^2(\tau)\gamma_{ij}\, dx^i\, dx^j = A^2(\sigma, \rho)\eta_{\mu\nu}\, dX^\mu\, dX^\nu, \quad (1.175)$$

with $\sigma = X^0$ and $\rho^2 = \sum_{i=1}^3 (X^i)^2$.

1.3 **Matter and radiation mixture**

Consider an FL universe containing a mixture of nonrelativistic matter (dust) and radiation with vanishing curvature. The respective densities and pressures are ρ_m, ρ_r, and $P_m = 0$, $P_r = \rho_r/3$. We denote the ratio of radiation to matter by $R = \rho_r/\rho_m$.

(a) Determine w and c_s^2 as functions of R. What is the time dependence of R?

(b) For a given redshift $z_{\rm eq} \gg 1$ of matter and radiation equality determine the scale factor as a function of conformal and of physical time; normalize the scale factor to 1 at equality, $a_{\rm eq} = 1$.

(c) Determine $t_{\rm eq}$ and $\tau_{\rm eq}$ as functions of $z_{\rm eq}$, and H_0.

1.4 **Cosmological constant***

Investigate the dynamics of an FL universe with matter ($P = 0$) and a cosmological constant Λ.

(1) Show for a sufficiently small cosmological constant and positive curvature that the Universe recollapses in a "big crunch," while for a larger cosmological constant or nonpositive curvature, the Universe expands forever.

(b) Show furthermore that for an even higher cosmological constant there are solutions that have no big bang in the past, but issue from a previous contracting phase. The transition from the contracting to an expanding phase is called the "bounce."

(c) Make a plot in the plane $(\Omega_m, \Omega_\Lambda)$ distinguishing the regimes determined earlier.

(d) For case (b), determine (numerically) the redshift of the bounce as a function of Ω_Λ for fixed $\Omega_m = 0.1$. Discuss.

1.5 **Helium recombination**

Write the Saha equation Eq. (1.73) for the two helium recombination processes and use it to determine the helium recombination temperatures, $T_{2\to 1}$ and $T_{1\to 0}$.

Hint: As a simplifying assumption neglect the fact that helium uses up some of the electrons and simply set $n_e \simeq n_B$.

1.6 **Rayleigh scattering**
The Rayleigh scattering cross section for an atom is $\sigma_{\rm Ra} \simeq \alpha^2/\lambda^4$, where α is the polarizability and λ the photon wavelength. For hydrogen atoms $\alpha \simeq 3.8\times10^{-24}{\rm cm}^3$. Show that after recombination the Rayleigh scattering rate of CMB photons on hydrogen atoms is much smaller than the expansion rate H.

1.7 **Distribution functions**
Show that in the nonrelativistic limit, $m \gg T$ both, the Fermi–Dirac and the Bose–Einstein distributions reduce to a Maxwell–Boltzmann distribution and the number and energy density are given by

$$n = \frac{2}{(2\pi)^3}\exp(-(m-\mu)/T)(2\pi mT)^{3/2}, \quad \rho = mn, \quad (1.176)$$

where μ is the chemical potential.

1.8 **Liouville equation**
Using that, in an FL universe the distribution function f depends only on (conformal) time t and $p = \sqrt{\gamma_{ij}p^i p^j}$, derive Eq. (1.88).

1.9 **The neutrino background**
Determine the neutrino cross section for the reaction $e^- + \bar\nu \to e^- + \bar\nu$ at energy $E_\nu = T_\nu(t_0)$. Compare it with the cross section of the neutrinos detected in the super-Kamiokande experiment. Keeping the efficiency of super-Kamiokande, how large a water tank would you need to detect neutrinos from the cosmic background?

1.10 **Angular diameter distance**
Determine the angular diameter distance to the last scattering surface under the assumptions $K = \Lambda = 0$. Under which angle do we presently see the causal horizon of this time, $a(t_{\rm rec})t_{\rm rec}$? How does this result change if one admits a cosmological constant so that $\Omega_m = 0.3$ and $\Omega_\Lambda = 0.7$?

1.11 **Resonant amplification**
Consider an inflaton coupled to a scalar field with $\mathcal{L}_\chi = -\frac{1}{2}\,\partial_\mu\chi\,\partial^\mu\chi - \frac{1}{2}$ $m_\chi^2\chi^2$ and $\mathcal{L}_{\rm int} = -\frac{1}{2}\tilde g\phi^2\chi^2$. Consider the equation of motion of χ in the classical background solution $\phi(t) = \phi_0\cos(m_\phi\tau)$. Neglect the cosmic expansion. Show that the equation for each Fourier mode can be written as a Mathieu equation and discuss the resonance frequencies and widths.

2

Perturbation Theory

2.1 Introduction

In this chapter we develop in detail the theory of linear perturbations of a Friedmann–Lemaître universe. This theory is of utmost importance, since we assume that the observed structure in the Universe (galaxies, cluster voids, etc.) have grown out of small initial fluctuations. Their entire evolution from the generation of the fluctuations until the time when they become of order unity can be studied within linear perturbation theory. This is especially relevant for the fluctuations in the CMB which are still very small today. It is also one of the main reasons why CMB anisotropies are so important for observational cosmology: they can be calculated to very good accuracy within linear perturbation theory, which is simple and lends itself to highly accurate and fast computations.

The idea that the large-scale structure of our Universe might have grown out of small initial fluctuations via gravitational instability goes back to Newton [letter to Bentley, 1692 (Newton, 1958)]. The first relativistic treatment of linear perturbations in a Friedmann–Lemaître universe was given by Lifshitz (1946). There he found that the gravitational potential cannot grow within linear perturbation theory and he concluded that galaxies have not been formed by gravitational instability.

Today we know that in order to form structures it is sufficient that matter density fluctuations can grow. Nevertheless, considerable initial fluctuations with amplitudes of the order of 10^{-5} are needed in order to reproduce the cosmic structures observed today. These are much larger than typical statistical fluctuations on scales of galaxies and we have to suggest a mechanism to generate them. Furthermore, the measurements of anisotropies in the CMB show that the amplitude of fluctuations in the gravitational potential is constant over a wide range of scales, that is, the fluctuation spectrum is scale independent.

As we shall see in Chapter 3, inflation generically produces such a spectrum of nearly scale-invariant fluctuations.

We begin this chapter by defining gauge-invariant perturbation variables. Then we present the basic perturbation equations. As examples for the matter equations we shall consider perfect fluids and scalar fields. Then we discuss light-like geodesics, which present a good approximation for CMB anisotropies on sufficiently large scales and are important for discussing the effect of lensing. The final section is devoted to the definition of the power spectrum and an elementary discussion of statistical issues. Owing to their complexity and importance for the goal of this book, we devote special chapters to the perturbed Boltzmann equation for CMB anisotropies, Chapter 4, and for polarization, Chapter 5.

2.2 Gauge-Invariant Perturbation Variables

The observed Universe is not perfectly homogeneous and isotropic. Matter is arranged in galaxies and clusters of galaxies and there are large voids in the distribution of galaxies. Let us assume, however, that these inhomogeneities grew out of small variations of the geometry and of the energy–momentum tensor, which we shall treat in first-order perturbation theory. For this we define the perturbed geometry by

$$g_{\mu\nu} = \bar{g}_{\mu\nu} + \varepsilon a^2 h_{\mu\nu}, \tag{2.1}$$

$\bar{g}_{\mu\nu}$ being the unperturbed Friedmann metric defined in the previous chapter. We conventionally set (absorbing the "smallness" parameter ε into $h_{\mu\nu}$)

$$\begin{aligned} g_{\mu\nu} &= \bar{g}_{\mu\nu} + a^2 h_{\mu\nu}, & \bar{g}_{00} &= -a^2, & \bar{g}_{ij} &= a^2\gamma_{ij}, & |h_{\mu\nu}| &\ll 1, \\ T^\mu_\nu &= \overline{T}^\mu_\nu + \theta^\mu_\nu, & \overline{T}^0_0 &= -\bar{\rho}, & \overline{T}^i_j &= \bar{P}\delta^i_j, & |\theta^\mu_\nu|/\bar{\rho} &\ll 1. \end{aligned} \tag{2.2}$$

2.2.1 Gauge Transformation, Gauge Invariance

The first fundamental problem we want to discuss is the choice of gauge in cosmological perturbation theory.

For linear perturbation theory to apply, the spacetime manifold $\mathcal{M}$ with metric g and the energy–momentum tensor T of the real, observable Universe must be in some sense close to an FL universe, that is, the manifold $\mathcal{M}$ with a Robertson–Walker metric $\bar{g}$ and a homogeneous and isotropic energy–momentum tensor $\overline{T}$. It is an interesting, nontrivial unsolved problem how to construct "the best" $\bar{g}$ and $\overline{T}$ from the physical fields g and T in practice. There are two main difficulties: first, spatial averaging procedures depend on the choice of a hypersurface of constant time and they do not commute with derivatives, so that averaged fields $\bar{g}$ and $\overline{T}$ will, in general, not satisfy Einstein's equations. Second, averaging is in practice impossible over superhorizon scales.

Even though we cannot give a constructive prescription of how to define the nearly homogeneous and isotropic spatial slices from the physical spacetime, or the spatially averaged metric and energy–momentum tensor, we now assume that there exists an averaging procedure that leads to an FL universe with spatially averaged tensor fields $\overline{S}$, such that the deviations are small,

$$\frac{|T_{\mu\nu} - \overline{T}_{\mu\nu}|}{\max_{\{\alpha\beta\}}\{|\overline{T}_{\alpha\beta}|\}} \ll 1 \quad \text{and} \quad \frac{|g_{\mu\nu} - \overline{g}_{\mu\nu}|}{\max_{\{\alpha\beta\}}\{\overline{g}_{\alpha\beta}\}} \ll 1,$$

and where $\bar{g}$ and $\overline{T}$ satisfy Friedmann's equations. The latter condition can be achieved, for example by defining $\overline{T}$ via the Friedmann equations. Let us call such an averaging procedure "admissible." There may be many different admissible averaging procedures (e.g., over different hypersurfaces) leading to slightly different FL backgrounds. But since $|g - \bar{g}|$ is small of order ε, the difference of the two FL backgrounds must also be small of order ε and we can interpret it as part of the perturbation.

We now consider a fixed admissible FL background $(\bar{g}, \overline{T})$ as chosen. Since the theory is invariant under diffeomorphisms (coordinate transformations), the perturbations are not unique. For an arbitrary diffeomorphism ϕ and its push forward ϕ_*, the two metrics g and $\phi_*(g)$ describe the same geometry. Since we have chosen the background metric $\bar{g}$ we only allow diffeomorphisms that leave $\bar{g}$ invariant, that is, that deviate only in first order from the identity. Such an "infinitesimal" diffeomorphism can be represented as the infinitesimal flow of a vector field X, $\phi = \phi^X_\varepsilon$. Remember the definition of the flow: for the integral curve, $\gamma_x(s)$, of X with starting point x, that is, $\gamma_x(s = 0) = x$ we have $\phi^X_s(x) = \gamma_x(s)$. In terms of the vector field X, to first order in ε, its push forward is then of the form

$$\phi_* = id + \varepsilon L_X + \mathcal{O}(\varepsilon^2),$$

where L_X denotes the Lie derivative in direction X (see Appendix 2, Section A2.2). The transformation $g \to \phi_*(g)$ is equivalent to $\bar{g} + \varepsilon a^2 h \to \bar{g} + \varepsilon(a^2 h + L_X\bar{g}) + \mathcal{O}(\varepsilon^2)$. Under an "infinitesimal coordinate transformation" the metric perturbation h therefore transforms as

$$h \to h + a^{-2} L_X \bar{g}. \tag{2.3}$$

In the context of cosmological perturbation theory, infinitesimal coordinate transformations are called "gauge transformations." The perturbation of an arbitrary tensor field $S = \bar{S} + \varepsilon S^{(1)}$ obeys the gauge transformation law

$$S^{(1)} \to S^{(1)} + L_X \bar{S}. \tag{2.4}$$

Since every vector field X generates a gauge transformation $\phi = \phi^X_\varepsilon$, we can conclude that only perturbations of tensor fields with $L_X\overline{S} = 0$ for all vector fields X, that is, with vanishing (or constant) "background contribution," are gauge invariant. This result is called the "Stewart–Walker lemma" (Stewart and Walker, 1974).

The gauge dependence of perturbations has caused many controversies in the literature, since it is often difficult to extract the physical meaning of gauge-dependent perturbations, especially on superhorizon scales. This problem is solved by gauge-invariant perturbation theory, which we are going to use throughout this book. The advantage of the gauge-invariant formalism is that the variables used have simple geometric and physical meanings and are not plagued by gauge modes. Although the derivation requires somewhat more work, the final system of perturbation equations is usually simple and well suited for numerical treatment. We shall also see that on subhorizon scales, the gauge-invariant matter perturbation variables approach the usual, gauge-dependent ones. Since one of the gauge-invariant geometrical perturbation variables corresponds to the Newtonian potential, the Newtonian limit can be performed easily.

First we note that all relativistic equations are covariant and can therefore be written in the form $S = 0$ for some tensor field S. The corresponding background equation is $\overline{S} = 0$; hence $S^{(1)}$ is gauge invariant. It is thus always possible to express the corresponding perturbation equations in terms of gauge-invariant variables.

The principal sources of this chapter are the following reviews on gauge-invariant cosmological perturbation theory: Bardeen (1980), Kodama and Sasaki (1984), Mukhanov *et al.* (1992), Durrer (1994).

2.2.2 Harmonic Decomposition of Perturbation Variables

Since the $\{t = \text{constant}\}$ hypersurfaces are homogeneous and isotropic, it is reasonable to perform a harmonic analysis: a (spatial) tensor field on these hypersurfaces can be decomposed into components that transform irreducibly under translations and rotations. All such components evolve independently. Decomposition into irreducible components of the translation symmetry corresponds to a harmonic analysis, that is, decomposition into eigenfunctions of the Laplacian. For a scalar quantity f in the case $K = 0$ this is nothing else than its Fourier decomposition:

$$f(\mathbf{x},t) = \frac{1}{(2\pi)^3}\int d^3k\, f(\mathbf{k},t)\, e^{i\mathbf{k}\mathbf{x}}. \tag{2.5}$$

(The exponentials $Q_{\mathbf{k}}(\mathbf{x}) = e^{i\mathbf{k}\mathbf{x}}$ are the unitary irreducible representations of the Euclidean translation group.) For $K = 1$ such a decomposition also exists, but the values k are the discrete eigenvalues of the Laplacian on the 3-sphere, $k^2 = \ell(\ell+2)$

and for $K = -1$, they are bounded from below, $k^2 > 1$. Of course, the functions $Q_{\mathbf{k}}$ depend on the curvature K.

They form the complete orthogonal set of eigenfunctions of the Laplacian,

$$\Delta Q_{\mathbf{k}}^{(S)} = -k^2 Q_{\mathbf{k}}^{(S)}. \tag{2.6}$$

In addition, a tensorial variable (at fixed position $\mathbf{x}$) can be decomposed into irreducible components under the rotation group $SO(3)$.

For a spatial vector field, this is its decomposition into a gradient and a curl,

$$V_i = \nabla_i \varphi + B_i, \tag{2.7}$$

where

$$B^i_{|i} = 0, \tag{2.8}$$

where we used $X_{|i}$ to denote the three-dimensional covariant derivative of X. Here φ is the spin-0 and $\mathbf{B}$ is the spin-1 component of the vector field V.

For a spatial symmetric tensor field we have

$$H_{ij} = H_L \gamma_{ij} + \left(\nabla_i \nabla_j - \frac{1}{3}\Delta\gamma_{ij}\right) H_T + \frac{1}{2}\left(H^{(V)}_{i|j} + H^{(V)}_{j|i}\right) + H^{(T)}_{ij}, \tag{2.9}$$

where

$$H_i^{(V)|i} = H_i^{(T)^i} = H_{i|j}^{(T)^j} = 0. \tag{2.10}$$

Here H_L and H_T are spin-0 components, $H_i^{(V)}$ is the spin-1 component, and $H_{ij}^{(T)}$ is the spin-2 component of the tensor field H.

We shall not need higher tensors (or spinors). As a basis for vector and tensor modes we use the vector- and tensor-type eigenfunctions of the Laplacian,

$$\Delta Q_j^{(V)} = -k^2 Q_j^{(V)} \quad \text{and} \tag{2.11}$$

$$\Delta Q_{ji}^{(T)} = -k^2 Q_{ji}^{(T)}, \tag{2.12}$$

where $Q_j^{(V)}$ is a transverse vector, $Q_j^{(V)|j} = 0$ and $Q_{ji}^{(T)}$ is a symmetric transverse traceless tensor, $Q_j^{(T)j} = Q_{ji}^{(T)|i} = 0$. Both $Q_j^{(V)}$ and $Q_{ji}^{(T)}$ have two degrees of freedom. In the case of vanishing curvature we can use an orthonormal basis $\mathbf{e}^{(1)}$, $\mathbf{e}^{(2)}$ in the plane normal to $\mathbf{k}$ and define helicity basis vectors,

$$\mathbf{e}^{\pm} = \frac{1}{\sqrt{2}}(\mathbf{e}^{(1)} \mp i\mathbf{e}^{(2)}). \tag{2.13}$$

In curved spaces the definition of the helicity basis is analogous, but somewhat more involved. Since we shall never need the explicit form of this basis, we shall not enter into this. Vector perturbations can be expanded in terms of this basis, while tensor perturbations are expanded either in terms of the standard tensor basis given by

$$e_{ij}^{d} = \frac{1}{2}\left[\mathbf{e}_i^{(1)}\mathbf{e}_j^{(1)} - \mathbf{e}_i^{(2)}\mathbf{e}_j^{(2)}\right], \tag{2.14}$$

$$e_{ij}^{\times} = \frac{1}{2}\left[\mathbf{e}_i^{(1)}\mathbf{e}_j^{(2)} + \mathbf{e}_i^{(2)}\mathbf{e}_j^{(1)}\right], \tag{2.15}$$

or also in terms of a helicity basis defined by

$$e_{ij}^{(+2)} = \mathbf{e}_i^{+}\mathbf{e}_j^{+} = e_{ij}^{d} + ie_{ij}^{\times}, \tag{2.16}$$

$$e_{ij}^{(-2)} = \mathbf{e}_i^{-}\mathbf{e}_j^{-} = e_{ij}^{d} - ie_{ij}^{\times}. \tag{2.17}$$

We can develop the vector and tensor basis functions as

$$Q_j^{(V)} = Q^{(1)}\mathbf{e}_j^{(1)} + Q^{(2)}\mathbf{e}_j^{(2)}, \tag{2.18}$$

$$= Q^{(+)}\mathbf{e}_j^{(+)} + Q^{(-)}\mathbf{e}_j^{(-)}, \tag{2.19}$$

$$Q_{ji}^{(T)} = Q^{(d)}e_{ij}^{(d)} + Q^{(\times)}e_{ij}^{(\times)}, \tag{2.20}$$

$$= Q^{(+2)}e_{ij}^{(+2)} + Q^{(-2)}e_{ij}^{(-2)}\,. \tag{2.21}$$

The components in the "helicity basis," $\mathbf{e}^{(\pm)}$ and $e_{ij}^{(\pm 2)}$ simply transform with a phase $e^{\pm i\varphi}$ and $e^{\pm 2i\varphi}$ respectively under rotations around $\mathbf{k}$ with angle φ. Hence vector perturbations are spin-1 fields while tensor perturbations are spin-2 fields. The functions $Q^{(+)}$ and $Q^{(+2)}$ have spin up, $m = +1$ and $m = +2$ respectively while $Q^{(-)}$ and $Q^{(-2)}$ have spin down. Scalar perturbations of course have spin 0. We shall make use of this spin structure especially in Chapters 4 and 5.

As in Eqs. (2.7) and (2.9), we can construct scalar-type vectors and symmetric, traceless tensors and vector-type symmetric tensors. To this goal we define

$$Q_j^{(S)} \equiv -k^{-1}Q_{|j}^{(S)}, \tag{2.22}$$

$$Q_{ij}^{(S)} \equiv k^{-2}Q_{|ij}^{(S)} + \frac{1}{3}\gamma_{ij}Q^{(S)} \quad \text{and} \tag{2.23}$$

$$Q_{ij}^{(V)} \equiv -\frac{1}{2k}\left(Q_{i|j}^{(V)} + Q_{j|i}^{(V)}\right). \tag{2.24}$$

In the following we shall extensively use this decomposition and write down the perturbation equations for a given mode k.

The decomposition of the k-mode of a vector field is then of the form

$$B_i = BQ_i^{(S)} + B^{(V)}Q_i^{(V)}. \tag{2.25}$$

The decomposition of a tensor field is given by compare [Eq. (2.9)]

$$H_{ij} = H_L Q^{(S)}\gamma_{ij} + H_T Q_{ij}^{(S)} + H^{(V)} Q_{ij}^{(V)} + H^{(T)} Q_{ij}^{(T)}. \tag{2.26}$$

Here B, $B^{(V)}$, H_L, H_T, $H^{(V)}$, and $H^{(T)}$ are functions of t and $\mathbf{k}$.

This decomposition is very useful because scalar vector and tensor amplitudes of each mode $\mathbf{k}$ evolve independently, obeying ordinary differential equations in time.

2.2.3 Metric Perturbations

Perturbations of the metric are of the form

$$g_{\mu\nu} = \bar{g}_{\mu\nu} + a^2 h_{\mu\nu}. \tag{2.27}$$

We parameterize them as

$$h_{\mu\nu}\, dx^\mu\, dx^\nu = -2A\, dt^2 - 2B_i\, dt\, dx^i + 2H_{ij}\, dx^i\, dx^j, \tag{2.28}$$

and we decompose the perturbation variables B_i and H_{ij} according to (2.25) and (2.26). In matrix form we have

$$\left(g_{\mu\nu}\right) = a^2 \begin{pmatrix} -(1+2A) & -B_i \\ -B_i & \gamma_{ij} + 2H_{ij} \end{pmatrix}. \tag{2.29}$$

The perturbations of the inverse metric are then given by

$$g^{\mu\nu} = \bar{g}^{\mu\nu} - a^2 h^{\mu\nu}, \quad \text{where} \quad h^{\mu\nu} = \bar{g}^{\mu\alpha}\bar{g}^{\nu\beta} h_{\alpha\beta}. \tag{2.30}$$

Let us consider the behavior of $h_{\mu\nu}$ under gauge transformations. We set the vector field defining the gauge transformation to

$$X = T\partial_t + L^i \partial_i. \tag{2.31}$$

Using the definition of the Lie derivative, we obtain (for details see Exercises)

$$\begin{aligned} L_{\mathbf{X}}\bar{g} = a^2 \big[&-2\left(\mathcal{H}T + \dot{T}\right) dt^2 + 2\left(\dot{L}_i - T_{,i}\right) dt\, dx^i \\ &+ \left(2\mathcal{H}T\gamma_{ij} + L_{i|j} + L_{j|i}\right) dx^i\, dx^j \big]. \end{aligned} \tag{2.32}$$

Comparing this with (2.28) and using (2.4), we obtain

$$\begin{aligned} A &\to A + \mathcal{H}T + \dot{T}, \\ B_i &\to B_i - \dot{L}_i + T_{,i}, \\ H_{ij} &\to H_{ij} + \frac{1}{2}\left(L_{i|j} + L_{j|i}\right) + \mathcal{H}T\gamma_{ij}. \end{aligned}$$

Using the decompositions (2.25) and (2.26) for B_i and H_{ij} this implies the following behavior of the perturbation variables under gauge transformations (we also decompose the vector $L_i = LQ_i^{(S)} + L^{(V)}Q_i^{(V)}$):

$$A \to A + \mathcal{H}T + \dot{T}, \tag{2.33}$$

$$B \to B - \dot{L} - kT, \tag{2.34}$$

$$B^{(V)} \to B^{(V)} - \dot{L}^{(V)}, \tag{2.35}$$

$$H_L \to H_L + \mathcal{H}T + \frac{k}{3}L, \tag{2.36}$$

$$H_T \to H_T - kL, \tag{2.37}$$

$$H^{(V)} \to H^{(V)} - kL^{(V)}, \tag{2.38}$$

$$H^{(T)} \to H^{(T)}. \tag{2.39}$$

Two scalar and one vector variable can be set to zero by gauge transformations. We shall use this in the text that follows to choose the longitudinal gauge for scalar perturbations, $B = H_T = 0$.

An interesting variable is also the "shear" on the t = constant hypersurfaces. We first introduce the normal to this hypersurface, which is given by $n_\mu dx^\mu = -a(1+A)dt$, so that to first order

$$(n^\mu) = (-g^{\mu\nu}a(1+A)\delta_\nu^0) = a^{-1}(1-A, B^i). \tag{2.40}$$

This vector field is normalized, $n^\mu n_\mu = -1$, and its scalar product with any vector field tangent to the t = constant hypersurfaces and hence of the form $X = X^i\partial_i$ vanishes. We now introduce its covariant derivative, setting

$$n_{\mu;\nu} = \frac{1}{3}P_{\mu\nu}\theta - a_\mu n_\nu + \sigma_{\mu\nu} + \omega_{\mu\nu}, \tag{2.41}$$

where

$$P_{\mu\nu} = n_\mu n_\nu + g_{\mu\nu} \tag{2.42}$$

is the projection tensor onto the subspace of tangent space normal to n, $\theta \equiv n^\mu_{\ ;\mu}$ is called the "expansion," $a_\mu = n^\nu n_{\mu;\nu}$ is the acceleration, and $\sigma_{\mu\nu}$ and $\omega_{\mu\nu}$ are the shear and vorticity of the vector field n respectively. They are defined as

$$\sigma_{\mu\nu} = \frac{1}{2}P_\mu^\lambda P_\nu^\rho(n_{\lambda;\rho} + n_{\rho;\lambda}) - \frac{1}{3}P_{\mu\nu}\theta \quad \text{and} \tag{2.43}$$

$$\omega_{\mu\nu} = \frac{1}{2}P_\mu^\lambda P_\nu^\rho(n_{\lambda;\rho} - n_{\rho;\lambda}). \tag{2.44}$$

This split of the covariant derivative of a vector field onto expansion, acceleration, shear, and vorticity is standard and sometimes very convenient. For example, Frobenius' theorem from differential geometry [see, e.g., Wald (1984), Appendix B] implies, that there exists a hypersurface that is normal to a given vector field if and only if its vorticity vanishes. The direction of the theorem which we use here, namely that the vorticity vanishes if n is (locally) orthogonal to a hypersurface follows from a simple calculation (see Exercise 2.2). The other direction is more involved. In three dimensions it boils down to the well known result that each vector field with vanishing curl can (locally) be written as the gradient of some function.

In the background FL universe, without perturbations, only the expansion, $\theta = 3\mathcal{H}/a = 3H$, is nonzero. In the presence of scalar perturbations we obtain

$$\theta = \frac{3}{a}\mathcal{H}[1 + \mathcal{K}Q], \qquad \mathcal{K} = -A + \frac{1}{3}\mathcal{H}^{-1}kB + \mathcal{H}^{-1}\dot{H}_L, \tag{2.45}$$

$$\sigma_{00} = \sigma_{0i} = \sigma_{i0} = 0, \tag{2.46}$$

$$\sigma_{ij} = ak\left(k^{-1}\dot{H}_T - B\right)Q_{ij} = ak\sigma Q_{ij}, \tag{2.47}$$

$$a_i = -kAQ_i, \qquad a_0 = 0, \tag{2.48}$$

$$\omega_{\mu\nu} = 0. \tag{2.49}$$

Another interesting variable is the spatial curvature on the hypersurface of constant time. It is easily calculated to first order and one finds

$$(\delta R)^s = 4\frac{k^2 - 3K}{a^2}\left(H_L + \frac{1}{3}H_T\right) = 4\frac{k^2 - 3K}{a^2}\mathcal{R}. \tag{2.50}$$

Since these variables depend only on the time coordinate, they transform only with T under coordinate transformation. Inserting the transformation laws found previously, Eqs. (2.33)–(2.37), we obtain

$$\mathcal{K} \to \mathcal{K} - \left(\mathcal{H} - \frac{\dot{\mathcal{H}}}{\mathcal{H}} + \frac{k^2}{3\mathcal{H}}\right)T, \tag{2.51}$$

$$\sigma \to \sigma + kT, \tag{2.52}$$

$$\mathcal{R} \to \mathcal{R} + \mathcal{H}T. \tag{2.53}$$

For vector and tensor perturbations only the perturbation of the shear does not vanish and we have

$$\sigma_{ij}^{(V)} = ak\left(k^{-1}\dot{H}^{(V)} - B^{(V)}\right)Q_{ij}^{(V)} = ak\sigma^{(V)}Q_{ij}^{(V)}, \tag{2.54}$$

$$\sigma_{ij}^{(T)} = a\dot{H}^{(T)}Q_{ij}^{(T)}. \tag{2.55}$$

Since there are no vector-type gauge transformations of the constant time hypersurfaces and no tensor-type gauge transformations at all, the quantities $\sigma^{(V)}$ and $H^{(T)}$ are gauge invariant.

One often chooses the gauge transformation $kL = H_T$ and $kT = B - \dot{L}$, so that the transformed variables H_T and B vanish. In this gauge (longitudinal gauge), scalar perturbations of the metric are of the form ($H_T|_{\rm long} = B|_{\rm long} = 0$)

$$h^{(S)}_{\mu\nu}dx^\mu dx^\nu = -2\Psi\, dt^2 - 2\Phi\gamma_{ij}\, dx^i dx^j, \tag{2.56}$$

where Ψ and Φ are the so-called *Bardeen* potentials. In a generic gauge the Bardeen potentials are given by

$$\Psi = A - \mathcal{H}k^{-1}\sigma - k^{-1}\dot{\sigma}, \tag{2.57}$$

$$\Phi = -H_L - \frac{1}{3}H_T + \mathcal{H}k^{-1}\sigma = -\mathcal{R} + \mathcal{H}k^{-1}\sigma, \tag{2.58}$$

where $\sigma = k^{-1}\dot{H}_T - B$, is the scalar potential for the shear of the hypersurface of constant time defined in Eq. (2.47). A short calculation using Eqs. (2.33), (2.52), and (2.53) shows that Ψ and Φ are indeed invariant under gauge transformations.

In an FL universe the Weyl tensor (see Appendix 2, Section A2.1) vanishes. It therefore is a gauge-invariant perturbation. For scalar perturbations one finds

$$E_{ij} \equiv C^{\mu}{}_{i\nu j}u_\mu u^\nu = -C^{0}{}_{i0j} = -\frac{1}{2}\left[(\Psi + \Phi)_{|ij} - \frac{1}{3}\Delta(\Psi + \Phi)\gamma_{ij}\right]. \tag{2.59}$$

All other components of the Weyl tensor are also given by E_{ij}; see Appendix 3, Section A3.1.

For vector perturbations it is convenient to set $kL^{(V)} = H^{(V)}$ so that $H^{(V)}$ vanishes and we have

$$h^{(V)}_{\mu\nu}\, dx^\mu\, dx^\nu = 2\sigma^{(V)}Q^{(V)}_i\, dt\, dx^i. \tag{2.60}$$

We shall call this gauge the "vector gauge."

The Weyl tensor from vector perturbation is given by

$$E_{ij} = -C^{0}{}_{i0j} = \frac{-k}{2}\dot{\sigma}^{(V)}Q^{(V)}_{ij}, \tag{2.61}$$

$$B_{ij} \equiv \frac{1}{2}\epsilon_{i\nu}{}^{\rho\sigma}C_{\rho\sigma}{}^{j\alpha}u^\nu u_\alpha = \epsilon_{ilm}C^{0}{}_{jlm},$$

$$= \frac{1}{2}\sigma^{(V)}\epsilon_{ilm}\left[Q^{(V)}_{l|jm} - Q^{(V)}_{m|jl} - \frac{k^2}{2}\gamma_{jl}Q^{(V)}_m + \frac{k^2}{2}\gamma_{jm}Q^{(V)}_l\right]. \tag{2.62}$$

Note that from their definition E_{ij} and B_{ij} are symmetric and since $u = (u^0, \mathbf{0})$ to lowest order, only $C^{0}{}_{i0j}$ and C^{0}_{ilm} respectively contribute. The tensors E_{ij} and B_{ij}, constructed as given earlier from the Weyl tensor for an arbitrary 4-velocity field u^μ, are normal to u^μ and they determine the Weyl tensor fully (see Appendix 3).

Clearly there are no tensorial (spin-2) gauge transformations and hence $H_{ij}^{(T)}$ is gauge invariant. The expression for the B-part of the Weyl tensor from tensor perturbation is

$$B_{ij} = -\dot{H}^{(T)}\epsilon_{ilm}\left[Q^{(T)}_{jl|m} - Q^{(T)}_{jm|l}\right]. \tag{2.63}$$

2.2.4 Perturbations of the Energy–Momentum Tensor

Let $T^\mu_\nu = \overline{T}^\mu_\nu + \theta^\mu_\nu$ be the full energy–momentum tensor. We define its energy density ρ and its energy flux 4-vector u as the time-like eigenvalue and eigenvector of T^μ_ν:

$$T^\mu_\nu u^\nu = -\rho u^\mu, \quad u^2 = -1. \tag{2.64}$$

We then parameterize their perturbations by

$$\rho = \bar{\rho}\,(1+\delta), \quad u = u^0\partial_t + u^i\partial_i. \tag{2.65}$$

The component u^0 is fixed by the normalization condition,

$$u^0 = \frac{1}{a}(1-A). \tag{2.66}$$

We further set

$$u^i = \frac{1}{a}v^i = \frac{1}{a}\left(vQ^{(S)i} + v^{(V)}Q^{(V)i}\right). \tag{2.67}$$

$P^\mu_\nu \equiv u^\mu u_\nu + \delta^\mu_\nu$ is the projection tensor onto the subspace of tangent space normal to u. We define the stress tensor

$$\tau^{\mu\nu} = P^\mu_\alpha P^\nu_\beta T^{\alpha\beta}. \tag{2.68}$$

With this we can write

$$T^\mu_\nu = \rho u^\mu u_\nu + \tau^\mu_\nu. \tag{2.69}$$

In the unperturbed case we have $\tau^0_\mu = \tau^\mu_0 = 0$ and $\tau^i_j = \bar{P}\delta^i_j$. Including first-order perturbations, the components $\tau_{0\mu}$ are determined by the perturbation variables that we have already introduced. We obtain

$$\tau^0_0 = 0, \quad \text{and} \quad \tau^j_0 = -\bar{P}v^j, \qquad \tau^0_j = \bar{P}(v_j - B_j). \tag{2.70}$$

But τ^i_j contains in general new perturbations. We define

$$\tau^i_j = \bar{P}\left[(1+\pi_L)\,\delta^i_j + \Pi^i_j\right], \quad \text{with} \quad \Pi^i_i = 0. \tag{2.71}$$

From our definitions we can determine the perturbations of the energy–momentum tensor. A short calculation gives

$$T_0^0 = -\bar{\rho}(1+\delta), \tag{2.72}$$

$$T^0{}_j = (\bar{\rho}+\bar{P})(v_j - B_j), \tag{2.73}$$

$$T^j{}_0 = -(\bar{\rho}+\bar{P})v^j, \tag{2.74}$$

$$T^i{}_j = \bar{P}\left[(1+\pi_L)\delta^i_j + \Pi^i{}_j\right]. \tag{2.75}$$

The traceless part of the stress tensor, Π^i_j, is called the anisotropic stress tensor. We decompose it as

$$\Pi^i_j = \Pi Q_j^{(S)\,i} + \Pi^{(V)} Q_j^{(V)\,i} + \Pi^{(T)} Q_j^{(T)\,i}. \tag{2.76}$$

We now study the gauge transformation properties of these perturbation variables. First we note that ρ is a scalar and $L_X\bar{\rho} = \dot{\bar{\rho}}T = -3(1+w)\mathcal{H}\bar{\rho}T$. Here we made use of Eq. (1.22). The same is true for $\bar{P}(1+\pi_L)$, which is one-third of the trace of $\tau^\mu{}_\nu$. With Eq. (1.29), we obtain $L_X\bar{P} = \dot{\bar{P}}T = -3\frac{c_s^2}{w}(1+w)\mathcal{H}\bar{P}T$. The background contribution to the anisotropic stress tensor, $\Pi^\mu_\nu = \tau^\mu_\nu - \frac{1}{3}\tau^\alpha_\alpha\delta^\mu_\nu$, vanishes; hence Π^μ_ν is gauge invariant (the Stewart–Walker lemma). For perfect fluids $\Pi^\mu_\nu = 0$. For the velocity we use $L_X\bar{u} = [X,\bar{u}] = (-T\dot{a}a^{-2} - a^{-1}\dot{T})\partial_t - a^{-1}\dot{L}^i\partial_i$. Inserting our decomposition into scalar, vector, and tensor perturbation variables for a fixed mode **k**, we obtain finally the following transformation behavior:

$$\delta \to \delta - 3(1+w)\mathcal{H}T, \tag{2.77}$$

$$\pi_L \to \pi_L - 3\frac{c_s^2}{w}(1+w)\mathcal{H}T, \tag{2.78}$$

$$v \to v - \dot{L}, \tag{2.79}$$

$$\Pi \to \Pi, \tag{2.80}$$

$$v^{(V)} \to v^{(V)} - \dot{L}^{(V)}, \tag{2.81}$$

$$\Pi^{(V)} \to \Pi^{(V)}, \tag{2.82}$$

$$\Pi^{(T)} \to \Pi^{(T)}. \tag{2.83}$$

Apart from the anisotropic stress perturbations, there is only one gauge-invariant variable that can be obtained from the energy–momentum tensor alone, namely

$$\Gamma = \pi_L - \frac{c_s^2}{w}\delta. \tag{2.84}$$

One can show (see Appendix 5) that Γ is proportional to the divergence of the entropy flux of the perturbations. Adiabatic perturbations are characterized by $\Gamma = 0$.

Gauge-invariant density and velocity perturbations can be found by combining δ, v and $v_i^{(V)}$ with metric perturbations. We shall use

$$V \equiv v - \frac{1}{k}\dot{H}_T = v^{\text{long}}, \tag{2.85}$$

$$D_s \equiv \delta + 3(1+w)\mathcal{H}(k^{-2}\dot{H}_T - k^{-1}B) \equiv \delta^{\text{long}}, \tag{2.86}$$

$$\begin{aligned} D &\equiv \delta^{\text{long}} + 3(1+w)\frac{\mathcal{H}}{k}V = \delta + 3(1+w)\frac{\mathcal{H}}{k}(v - B) \\ &= D_s + 3(1+w)\frac{\mathcal{H}}{k}V, \end{aligned} \tag{2.87}$$

$$\begin{aligned} D_g &\equiv \delta + 3(1+w)\left(H_L + \frac{1}{3}H_T\right) = \delta^{\text{long}} - 3(1+w)\Phi \\ &= D_s - 3(1+w)\Phi, \end{aligned} \tag{2.88}$$

$$V^{(V)} \equiv v^{(V)} - \frac{1}{k}\dot{H}^{(V)} = v^{(\text{vec})}, \tag{2.89}$$

$$\Omega \equiv v^{(V)} - B^{(V)} = v^{(\text{vec})} + \sigma^{(V)}, \tag{2.90}$$

$$\Omega - V^{(V)} = \sigma^{(V)}. \tag{2.91}$$

Here v^{long}, δ^{long}, and $v^{(\text{vec})}$ are the velocity (and density) perturbations in the longitudinal and vector gauge respectively, and $\sigma^{(V)}$ is the metric perturbation in vector gauge and the shear of the t = constant hypersurfaces [see Eqs. (2.54) and (2.60)].

These variables can be interpreted nicely in terms of gradients of the energy density and the shear and vorticity of the velocity field (Ellis and Bruni, 1989). Here we just calculate the covariant derivative of the velocity field u^μ and decompose it like the normal field n^μ. In a nonperturbed FL universe these two vector fields coincide. With our definition of variables, a short calculation using $u_{\mu;\nu} = u_{\mu,\nu} - \Gamma^\beta_{\mu\nu}u_\beta$ gives

$$u_{\mu;\nu} = \frac{1}{3}P^{(f)}_{\mu\nu}\theta^{(f)} - a^{(f)}_\mu u_\nu + \sigma^{(f)}_{\mu\nu} + \omega^{(f)}_{\mu\nu}, \tag{2.92}$$

where the projection, $P^{(f)}$; expansion, $\theta^{(f)}$; acceleration, $a^{(f)}$; shear, $\sigma^{(f)}$; and vorticity, $\omega^{(f)}$, are defined as in Eqs. (2.42)–(2.44); just the normal field n^μ is replaced by u^μ, the energy flux of the fluid. We indicate this by the superscript $^{(f)}$. For scalar perturbations one finds

$$\theta^{(f)} = \frac{3}{a}\mathcal{H}[1+\mathcal{K}^{(f)}Q], \qquad \mathcal{K}^{(f)} = -A + \mathcal{H}^{-1}\left(\dot{H}_L + \frac{k}{3}v\right), \tag{2.93}$$

$$\sigma^{(f)}_{00} = \sigma^{(f)}_{0i} = \sigma^{(f)}_{i0} = 0, \tag{2.94}$$

$$\sigma^{(f)}_{ij} = ak(k^{-1}\dot{H}_T - v)Q_{ij} \; = \; ak\sigma^{(f)}Q_{ij}, \tag{2.95}$$

$$a^{(f)}_i = -A^{(f)}Q_i, A^{(f)} = kA - \mathcal{H}(v-B) - (\dot{v} - \dot{B}), \tag{2.96}$$

$$a^{(f)}_0 = 0,$$

$$\omega_{\mu\nu} = 0. \tag{2.97}$$

Contrary to n^μ, the vector field u^μ is defined independently of the coordinate system. Therefore, and since $a^{(f)}_\mu$ and $\sigma^{(f)}_{\mu\nu}$ vanish in the background FL universe, the variables $A^{(f)}$ and $\sigma^{(f)} = V$ are gauge invariant. For V we have already noticed this before. Furthermore, it is easy to check that

$$A^{(f)} = k\Psi - \mathcal{H}V + \dot{V},$$

which is a gauge-invariant variable called the "peculiar acceleration."

For vector perturbations we obtain

$$\sigma^{(f)}_{00} = \sigma^{(f)}_{0i} = \sigma^{(f)}_{i0} = 0, \tag{2.98}$$

$$\sigma^{(f)}_{ij} = ak\left(k^{-1}\dot{H}^{(V)} - v^{(V)}\right)Q^{(V)}_{ij} = -akV^{(V)}Q_{ij}, \tag{2.99}$$

$$\omega^{(f)}_{i0} = \omega^{(f)}_{0i} = 0, \tag{2.100}$$

$$\omega^{(f)}_{ij} = \frac{a}{2}\left(v^{(V)} - B^{(V)}\right)\left[Q^{(V)}_{i|j} - Q^{(V)}_{j|i}\right] = \frac{a}{2}\Omega\left[Q^{(V)}_{i|j} - Q^{(V)}_{j|i}\right], \tag{2.101}$$

$$a^{(f)}_i = \left(\dot{\Omega} + \mathcal{H}\Omega\right)Q^{(V)}_i. \tag{2.102}$$

Note that the energy flux of scalar perturbations is hypersurface orthogonal, $\omega^{(S)} = 0$, while vector perturbations do have nonvanishing curl if $v^{(V)} \neq B^{(V)}$. A coordinate system with $v = B$ is called "comoving."

Tensor perturbations do not admit a perturbed energy flux so that for them the foregoing perturbation variables vanish.

We now want to show that on scales much smaller than the Hubble scale, $k \gg \mathcal{H} \sim t^{-1}$, the metric perturbations are much smaller than δ and v and we can thus neglect the difference between different gauges and/or gauge-invariant variables. This is especially important when comparing experimental results with gauge-invariant calculations. Let us neglect spatial curvature in the following order of magnitude argument. Then, the perturbations of the Einstein tensor are a combination of the second derivatives of the metric perturbations, $\mathcal{H}$ times the first derivatives, and $\mathcal{H}^2$ or $\dot{\mathcal{H}}$ times metric perturbations. The first-order perturbations of Einstein's equations therefore generically yield the following order of magnitude estimate $8\pi G\delta T_{\mu\nu} = \delta G_{\mu\nu}$:

$$\mathcal{O}\left(\frac{\delta T_{\mu\nu}}{a^2\rho}\right) \underbrace{\mathcal{O}\left(8\pi G a^2\rho\right)}_{\mathcal{O}(\dot{a}/a)^2=\mathcal{O}(1/t^2)} = \mathcal{O}\left(\frac{1}{t^2}h + \frac{k}{t}h + k^2 h\right), \tag{2.103}$$

$$\mathcal{O}\left(\frac{\delta T_{\mu\nu}}{a^2\rho}\right) = \mathcal{O}\left(h + kth + (kt)^2 h\right). \tag{2.104}$$

For $kt \gg 1$ this gives $\mathcal{O}(\delta, v) = \mathcal{O}\left(\delta T_{\mu\nu}/a^2\rho\right) \gg \mathcal{O}(h)$. Therefore, on subhorizon scales the differences between δ, $\delta^{\rm long}$, D_g, and D are negligible, as are the differences between v and V or $v^{(V)}$, $V^{(V)}$ and $\Omega^{(V)}$. For measurements of density and velocity perturbations that are made on deeply subhorizon scales, we may therefore use any of the gauge-invariant perturbation variables to compare with measurements. The issue is more subtle when fluctuations are measured at scales close to the horizon. Then, a detailed study of what is exactly measured reveals the gauge invariant quantity we have to consider. We shall come back to this issue for measurements of density fluctuations in Chapter 8.

2.3 The Perturbation Equations

We do not derive the first-order perturbations of Einstein's equations. By elementary algebraic methods, this is quite lengthy and cumbersome. However, we recommend that the student simply determines $\delta G_{\mu\nu}$ in longitudinal (vector) gauge using some algebraic package such as Maple or Mathematica and then writes down the resulting Einstein equations using gauge-invariant variables. Since we know that these variables do not depend on the coordinates chosen, the equations obtained in this way are valid in any gauge. Here, we just present the resulting equations in gauge-invariant form. A rapid derivation by hand is possible using the $3+1$ formalism of general relativity and working with Cartan's formalism for the Riemann curvature (see Durrer and Straumann, 1988). In order to simplify the notation, we suppress the overbar on background quantities whenever this does not lead to confusion.

2.3.1 Einstein's Equations

2.3.1.1 The Constraints

The Einstein equations $G_{0\mu} = 8\pi G T_{0\mu}$ lead to two scalar and one vector constraint equations,

$$\left.\begin{array}{rcll} 4\pi Ga^2\rho D &=& -(k^2-3K)\Phi & (00) \\ 4\pi Ga^2(\rho+P)V &=& k\left(\mathcal{H}\Psi+\dot{\Phi}\right) & (0i) \end{array}\right\} \quad \text{(scalar)}, \tag{2.105}$$

$$8\pi Ga^2(\rho+P)\Omega = \frac{1}{2}\left(2K-k^2\right)\sigma^{(V)} \quad (0i) \qquad \text{(vector)}. \tag{2.106}$$

2.3.1.2 The Dynamical Equations

The Einstein equations $G_{ij} = 8\pi GT_{ij}$ provide two scalar, one vector, and one tensor perturbation equations,

scalar:

$$k^2\left(\Phi-\Psi\right) = 8\pi Ga^2P\Pi^{(S)} \qquad (i \neq j), \tag{2.107}$$

$$\ddot{\Phi} + 2\mathcal{H}\dot{\Phi} + \mathcal{H}\dot{\Psi} + \left[2\dot{\mathcal{H}} + \mathcal{H}^2 - \frac{k^2}{3}\right]\Psi$$
$$= 4\pi Ga^2\rho\left[\frac{1}{3}D + c_s^2 D_s + w\Gamma\right] \qquad (ii), \tag{2.108}$$

vector:

$$k\left(\dot{\sigma}^{(V)} + 2\mathcal{H}\sigma^{(V)}\right) = 8\pi Ga^2P\Pi^{(V)}, \tag{2.109}$$

tensor:

$$\ddot{H}^{(T)} + 2\mathcal{H}\dot{H}^{(T)} + \left(2K+k^2\right)H^{(T)} = 8\pi Ga^2P\Pi^{(T)}. \tag{2.110}$$

The second dynamical scalar equation is somewhat cumbersome and not often used, since we may use one of the conservation equations given in the text that follows instead. For the derivation of the perturbed Einstein equation the following relations are useful. They can be derived from the Friedmann equations (1.20)–(1.22); a possible cosmological constant is included in ρ and P.

$$4\pi Ga^2\rho(1+w) = \mathcal{H}^2 - \dot{\mathcal{H}} + K, \tag{2.111}$$

$$\dot{\mathcal{H}} = -\frac{1+3w}{2}\left(\mathcal{H}^2+K\right), \tag{2.112}$$

$$4\pi Ga^2\rho(1+w)3c_s^2 = \frac{\ddot{\mathcal{H}}}{\mathcal{H}} - \dot{\mathcal{H}} - \mathcal{H}^2 - K, \tag{2.113}$$

$$c_s^2 = \frac{\frac{\ddot{\mathcal{H}}}{\mathcal{H}} - \dot{\mathcal{H}} - \mathcal{H}^2 - K}{3[\mathcal{H}^2 - \dot{\mathcal{H}} + K]}. \tag{2.114}$$

For the calculations that follow we shall also make use of

$$\dot{w} = 3(w-c_s^2)(1+w)\mathcal{H}. \tag{2.115}$$

Note that for perfect fluids, where $\Pi^i_j \equiv 0$, we have $\Phi = \Psi$. As we shall see in the text that follows behavior, for perfect fluids with $\Gamma = \Pi = 0$, the behavior of scalar perturbations is given by Ψ, which describes a damped wave propagating with speed c_s^2.

Tensor perturbations are given by $H^{(T)}$, which for perfect fluids also obeys a damped wave equation propagating with the speed of light. On small scales (over short time periods) when $t^{-2} \lesssim 2K + k^2$, the damping term can be neglected and H_{ij} represents propagating gravitational waves. For vanishing curvature or $k^2 \gg K$, small scales are just the sub-Hubble scales, $kt \gtrsim 1$. For $K < 0$, waves oscillate with a somewhat smaller frequency, $\omega = \sqrt{2K + k^2} < k$, while for $K > 0$ the frequency is somewhat higher than k.

Vector perturbations of a perfect fluid are determined by the $\sigma^{(V)}$ equation, Eq. (2.109), which implies $\sigma^{(V)} \propto 1/a^2$. Hence vector perturbations do not oscillate; they simply decay.

2.3.2 Energy–Momentum Conservation

The conservation equations, $T^{\mu\nu}_{;\nu} = 0$, lead to the following perturbation equations:

$$\left.\begin{aligned} &\dot{D}_g + 3\left(c_s^2 - w\right)\mathcal{H}D_g + (1+w)kV + 3w\mathcal{H}\Gamma = 0 \\ &\dot{V} + \mathcal{H}\left(1 - 3c_s^2\right)V = k\left(\Psi + 3c_s^2\Phi\right) + \tfrac{c_s^2 k}{1+w}D_g \\ &\qquad + \tfrac{wk}{1+w}\left[\Gamma - \tfrac{2}{3}\left(1 - \tfrac{3K}{k^2}\right)\Pi\right] \end{aligned}\right\} \quad \text{(scalar)}, \tag{2.116}$$

$$\dot{\Omega} + \left(1 - 3c_s^2\right)\mathcal{H}\Omega = -\frac{w}{2(1+w)}\left(k - \frac{2K}{k}\right)\Pi^{(V)} \quad \text{(vector)}. \tag{2.117}$$

It is sometimes also useful to express the scalar conservation equations in terms of the variable pair (D, V). Using $D = D_g + 3(1+w)\left[\mathcal{H}k^{-1}V + \Phi\right]$ in (2.116) one obtains after some algebra and making use of the $(0i)$ constraint equation (2.105)

$$\dot{D} - 3w\mathcal{H}D = -\left(1 - \frac{3K}{k^2}\right)\left[(1+w)kV + 2\mathcal{H}w\Pi\right], \tag{2.118}$$

$$\dot{V} + \mathcal{H}V = k\left[\Psi + \frac{c_s^2}{1+w}D + \frac{w}{1+w}\Gamma - \frac{2}{3}\left(1 - \frac{3K}{k^2}\right)\frac{w}{1+w}\Pi\right]. \tag{2.119}$$

Replacing Ψ in Eq. (2.119) via the (00) and (*ij*) Einstein equations, (2.105) and (2.107), and replacing V via Eq. (2.118), we can derive a second-order equation for D. A lengthy but straightforward calculation gives

$$\ddot{D} + (1 + 3c_s^2 - 6w)\mathcal{H}\dot{D} + \left[\left(\frac{9}{2}w^2 - 12w + 9c_s^2\right)\mathcal{H}^2 + \frac{9}{2}w^2 K + (k^2 - 3K)c_s^2 - 4\pi G\rho a^2\right]D = -(k^2 - 3K)w\Gamma - 2\left(1 - \frac{3K}{k^2}\right)\mathcal{H}w\dot{\Pi} + 2\left[(3w^2 + 3c_s^2 - 2w)\mathcal{H}^2 + w(3w + 2)K + \frac{k^2 - 3K}{3}w\right]\left(1 - \frac{3K}{k^2}\right)\Pi. \tag{2.120}$$

The conservation equations can, of course, also be obtained from the Einstein equations because they are equivalent to the contracted Bianchi identities (see Appendix 2, Section A2.1). For scalar perturbations we have four independent equations and six variables. For vector perturbations we have two equations and three variables, while for tensor perturbations we have one equation and two variables. To close the system we must add some matter equations. The simplest prescription is to set $\Gamma = \Pi_{ij} = 0$. This matter equation, which describes adiabatic perturbations of a perfect fluid, gives us exactly two additional equations for scalar perturbations and one each for vector and tensor perturbations. In this simple case, the tensor equation simply describes free gravitational waves propagating in an FL background. If $c_s^2 \neq 0$ also the scalar equation (2.120) is a wave equation. It describes what we shall call "acoustic oscillations" of the fluid where the fluid pressure counter-acts gravitational collapse. The vector perturbation equation, however, is of first order. $\Pi^{(V)} = 0$ implies $\sigma^{(V)} \propto 1/a^2$ and $\Omega \propto a^{-1+3c_s^2}$. Hence vector perturbations of the metric simply decay if there are no anisotropic stresses to source them.

Another simple example is a universe with matter content given by a scalar field. We shall discuss this case in the next chapter. More complicated are several interacting particle species, some of which have to be described by a Boltzmann equation. This is the actual universe at late times, $z \lesssim 10^7$.

2.3.3 Mixtures of Several Fluids

Here we only consider fluid components that are noninteracting, so that their energy–momentum tensor is separately conserved; that is, Eqs. (2.116) and (2.117) hold for each α component separately. The Einstein equations, however, determine the metric perturbations induced by the full perturbations,

$$\rho D_g = \sum_\alpha \rho_\alpha D_{g\alpha}, \tag{2.121}$$

$$(\rho + P)V = \sum_\alpha (\rho_\alpha + P_\alpha)V_\alpha, \tag{2.122}$$

$$P\Pi = \sum_\alpha P_\alpha \Pi_\alpha, \tag{2.123}$$

$$\begin{aligned} P\Gamma = P\pi_L - c_s^2\delta\rho &= \sum_\alpha P_\alpha\Gamma_\alpha + \sum_\alpha (c_\alpha^2 - c_s^2)\delta\rho_\alpha \\ &= \sum_\alpha P_\alpha\Gamma_\alpha + P\Gamma_{\rm rel}. \end{aligned} \tag{2.124}$$

In order to see that $\Gamma_{\rm rel}$ is gauge invariant we use Eq. (1.29):

$$c_s^2 = \frac{\dot P}{\dot\rho} = \sum_\beta \frac{(1 + w_\beta)\rho_\beta c_\beta^2}{(1+w)\rho}.$$

Also using $\sum_\beta (1 + w_\beta)\rho_\beta = (1 + w)\rho$ we find

$$\begin{aligned} P\Gamma_{\rm rel} &= \sum_\alpha (c_\alpha^2 - c_s^2)\delta\rho_\alpha = \sum_{\alpha\beta} \frac{(1 + w_\beta)\rho_\beta(1 + w_\alpha)\rho_\alpha}{\rho + P}(c_\alpha^2 - c_\beta^2)\frac{\delta_\alpha}{1 + w_\alpha} \\ &= \frac{1}{2}\sum_{\alpha\beta} \frac{(1 + w_\beta)(1 + w_\alpha)\rho_\beta\rho_\alpha}{\rho + P}(c_\alpha^2 - c_\beta^2)\left[\frac{\delta_\alpha}{1 + w_\alpha} - \frac{\delta_\beta}{1 + w_\beta}\right] \\ &= \frac{1}{2}\sum_{\alpha\beta} \frac{(1 + w_\beta)(1 + w_\alpha)\rho_\beta\rho_\alpha}{\rho + P}(c_\alpha^2 - c_\beta^2)\left[\frac{D_{g\alpha}}{1 + w_\alpha} - \frac{D_{g\beta}}{1 + w_\beta}\right] \\ &= \frac{1}{2}\sum_{\alpha\beta} \frac{(1 + w_\beta)\rho_\beta(1 + w_\alpha)\rho_\alpha}{\rho + P}(c_\alpha^2 - c_\beta^2)S_{\alpha\beta}, \end{aligned} \tag{2.125}$$

where we define

$$S_{\alpha\beta} = \left[\frac{D_{g\alpha}}{1 + w_\alpha} - \frac{D_{g\beta}}{1 + w_\beta}\right]. \tag{2.126}$$

For the third equals sign in Eq. (2.125) we have used the fact that the expression $[(1 + w_\beta)(1 + w_\alpha)\rho_\beta\rho_\alpha/(\rho + P)](c_\alpha^2 - c_\beta^2)$ is antisymmetric in α and β and we therefore may also antisymmetrize the remaining factor.

The individual components of the gauge-invariant velocity and density perturbations are defined via their energy–momentum tensors. Note that

$$V_\alpha = v_\alpha - k^{-1}\dot H_T, \qquad \text{and} \tag{2.127}$$

$$D_{g\alpha} = \delta_\alpha + 3(1 + w_\alpha)\mathcal{R}, \tag{2.128}$$

$$D_\alpha = \delta_\alpha + 3(1 + w_\alpha)\mathcal{H}k^{-1}(v_\alpha - B), \tag{2.129}$$

$$= D_{g\alpha} + 3(1 + w_\alpha)\left[\mathcal{H}k^{-1}V_\alpha + \Phi\right]. \tag{2.130}$$

It is easy to check that the conservation equations (2.116) are also valid for a mixture of conserved components, so that we have

$$\dot{D}_{g\alpha} + 3\left(c_\alpha^2 - w_\alpha\right)\mathcal{H}D_{g\alpha} = -(1+w_\alpha)kV_\alpha - 3w_\alpha\mathcal{H}\Gamma_\alpha, \tag{2.131}$$

$$\dot{V}_\alpha + \mathcal{H}\left(1-3c_\alpha^2\right)V_\alpha = k\left(\Psi + 3c_\alpha^2\Phi\right) + \frac{c_\alpha^2 k}{1+w_\alpha}D_{g\alpha} + \frac{w_\alpha k}{1+w_\alpha}\left[\Gamma_\alpha - \frac{2}{3}\left(1-\frac{3K}{k^2}\right)\Pi_\alpha\right]. \tag{2.132}$$

However, if we rewrite the conservation equations in terms of the variables (D_α, V_α) new terms appear, since we have to use the Einstein equations in the derivation. A somewhat tedious but straightforward calculation, replacing $D_{g\alpha}$ with the help of Eq. (2.130) and then eliminating Φ with the Einstein equation $(0i)$, gives

$$\dot{D}_\alpha - 3w_\alpha\mathcal{H}D_\alpha = \frac{9}{2}(\mathcal{H}^2+K)k^{-1}(1+w)(1+w_\alpha)[V-V_\alpha] - \left(1-\frac{3K}{k^2}\right)[(1+w_\alpha)kV_\alpha + 2\mathcal{H}w\Pi_\alpha], \tag{2.133}$$

$$\dot{V}_\alpha + \mathcal{H}V_\alpha = k\left[\Psi + \frac{c_\alpha^2}{1+w_\alpha}D_\alpha + \frac{w_\alpha}{1+w_\alpha}\Gamma_\alpha - \frac{2}{3}\left(1-\frac{3K}{k^2}\right)\frac{w_\alpha}{1+w_\alpha}\Pi_\alpha\right]. \tag{2.134}$$

It is sometimes more useful to describe mixed systems in terms of variables related to differences of individual components. With $S_{\alpha\beta}$ given in Eq. (2.126) and defining

$$V_{\alpha\beta} = V_\alpha - V_\beta, \tag{2.135}$$

$$\Gamma_{\alpha\beta} = \frac{w_\alpha}{1+w_\alpha}\Gamma_\alpha - \frac{w_\beta}{1+w_\beta}\Gamma_\beta, \tag{2.136}$$

$$\Pi_{\alpha\beta} = \frac{w_\alpha}{1+w_\alpha}\Pi_\alpha - \frac{w_\beta}{1+w_\beta}\Pi_\beta, \tag{2.137}$$

one can derive the following system of equations from Eqs. (2.131) and (2.132):

$$\dot{S}_{\alpha\beta} = -kV_{\alpha\beta} - 3\mathcal{H}\Gamma_{\alpha\beta}, \tag{2.138}$$

$$\dot{V}_{\alpha\beta} + \mathcal{H}V_{\alpha\beta} - \frac{3}{2}\mathcal{H}(c_\alpha^2+c_\beta^2)V_{\alpha\beta} - \frac{3}{2}\mathcal{H}(c_\alpha^2-c_\beta^2)\sum_\gamma\frac{\rho_\gamma+P_\gamma}{\rho+P}\left(V_{\alpha\gamma}+V_{\beta\gamma}\right)$$
$$= k\left[\frac{c_\alpha^2-c_\beta^2}{1+w}D + \frac{c_\alpha^2+c_\beta^2}{2}S_{\alpha\beta} + \frac{c_\alpha^2-c_\beta^2}{2}\sum_\gamma\frac{\rho_\gamma+P_\gamma}{\rho+P}\left(S_{\alpha\gamma}+S_{\beta\gamma}\right) + \Gamma_{\alpha\beta} - \frac{3}{2}\left(1-\frac{3K}{k^2}\right)\Pi_{\alpha\beta}\right]. \tag{2.139}$$

We present a detailed derivation of these equations in Appendix 6.

We shall use these equations when we discuss mixtures of cold dark matter and radiation. More details on mixed systems that also include interactions can be found in Kodama and Sasaki (1984). In Exercise 2.4, we discuss a simple example of a mixed system. Interacting mixed systems are not very relevant for us, since we shall describe them with a Boltzmann equation approach that we develop in Chapter 4.

2.3.4 The Bardeen Equation

The systems of equations that we have presented here are, of course, not closed. To close them one needs to add evolution equations for the matter variables, such as $\Pi^{(T)}$ for tensor perturbations, a relation between $\Pi^{(V)}$ and $\Omega^{(V)}$ for vector perturbations, and expressions for Γ and $\Pi = \Pi^{(S)}$ for scalar perturbations.

For scalar perturbations we can actually derive an evolution equation for Φ, where Γ and Π enter only as source terms. Replacing D and D_s in (2.108) by use of (2.87) and (2.105) and replacing Ψ by Π and Φ via Eq. (2.107) leads to

$$\ddot{\Phi} + 3\mathcal{H}(1 + c_s^2)\dot{\Phi} + \left[3(c_s^2 - w)\mathcal{H}^2 - (2 + 3w + 3c_s^2)K + c_s^2 k^2\right]\Phi$$
$$= \frac{8\pi G a^2 P}{k^2}\left[\mathcal{H}\dot{\Pi} + [2\dot{\mathcal{H}} + 3\mathcal{H}^2(1 - c_s^2/w)]\Pi - \frac{1}{3}k^2\Pi + \frac{k^2}{2}\Gamma\right]. \quad (2.140)$$

This is the Bardeen equation. To derive it we also made use of (2.112) to replace $\dot{\mathcal{H}}$.

This equation is especially useful in terms of another gauge-invariant variable that we now introduce: the scalar curvature on the comoving hypersurface. The comoving hypersurface is defined by having the normal n on the constant time hypersurface equal to the particle 4-velocity u. Using $(n^\nu) = a^{-1}(1 - A, B^j)$ and $(u^\nu) = a^{-1}(1 - A, v^j)$ this implies $v = B$. From the definitions of σ and V we thus have $V = -\sigma_{\text{co}}$ in this coordinate system. In comoving gauge we therefore have [see Eqs. (2.86)–(2.88)]

$$D_s = \delta_{\text{co}} - 3(1 + w)k^{-1}\mathcal{H}V, \quad (2.141)$$
$$D = \delta_{\text{co}}, \quad (2.142)$$
$$D_g = \delta_{\text{co}} + 3(1 + w)\mathcal{R}_{\text{co}} = D - 3(1 + w)[k^{-1}\mathcal{H}V + \Phi], \quad (2.143)$$

so that

$$-\mathcal{R}_{\text{co}} = \frac{1}{3(1 + w)}[D - D_g] = k^{-1}\mathcal{H}V + \Phi. \quad (2.144)$$

Here the index "co" indicates comoving coordinates. Using the $(0i)$ Einstein equation, Eq. (2.105), we obtain

$$-\mathcal{R}_{co} = \frac{2}{3(1+w)}\left[\Psi + \mathcal{H}^{-1}\dot{\Phi}\right] + \Phi \equiv \zeta. \tag{2.145}$$

We are especially interested in the evolution of the curvature perturbation variable ζ in situations in which we can neglect anisotropic stresses. Then the right-hand side of Eq. (2.140) simply becomes $4\pi G a^2 P\Gamma$ and $\Phi = \Psi$. The definition (2.145) together with the Bardeen equation then yields in the spatially flat case, $K = 0$,

$$\dot{\zeta} = \frac{\mathcal{H}}{\mathcal{H}^2 - \dot{\mathcal{H}}}\left[\frac{3w}{2}\mathcal{H}^2\Gamma - c_s^2 k^2 \Psi\right] \tag{2.146}$$

$$= \frac{w\mathcal{H}}{w+1}\Gamma - \frac{2c_s^2 k^2}{3(w+1)\mathcal{H}}\Psi. \tag{2.147}$$

For adiabatic perturbations, $\Gamma = 0$, the curvature perturbation ζ is therefore conserved on super-Hubble scales, $k/\mathcal{H} \ll 1$ at early times when curvature is certainly negligible. This will be very useful when we want to specify initial conditions in Chapter 3.

Also note that for constant w and constant Bardeen potential $\Psi = \Phi$, the curvature perturbation ζ differs from the Bardeen potential only by a multiplicative constant.

2.3.5 A Special Case

Here we want to discuss the scalar perturbation equations for a simple, but important, special case. We consider adiabatic perturbations of a perfect fluid. In this case there are no anisotropic stresses, $\Pi = 0$. Furthermore, the pressure fluctuation $\delta P = \pi_L P$ is related to the density fluctuation $\delta\rho$ by $\delta P = c_s^2\delta\rho$, hence $\Gamma = 0$. Equation (2.140) then becomes simply a second-order equation for the Bardeen potential $\Psi = \Phi$, which is in this case the only dynamical degree of freedom,

$$\ddot{\Psi} + 3\mathcal{H}(1+c_s^2)\dot{\Psi} + \left[(1+3c_s^2)(\mathcal{H}^2 - K) - (1+3w)(\mathcal{H}^2 + K) + c_s^2 k^2\right]\Psi = 0. \tag{2.148}$$

This is a damped wave equation. When we may neglect curvature, and if $w =$ constant so that $c_s^2 = w$, the time-dependent mass term $m^2(t) = -(1+3c_s^2) \times (\mathcal{H}^2 - K) + (1+3w)(\mathcal{H}^2 + K)$ vanishes. Equation (2.148) then reduces to

$$\ddot{\Psi} + 6\frac{1+w}{(1+3w)t}\dot{\Psi} + wk^2\Psi = 0, \tag{2.149}$$

where we have used that

$$a \propto t^{2/(1+3w)} = t^q \quad \text{and} \quad \mathcal{H} = \frac{2}{1+3w}\frac{1}{t} = \frac{q}{t} \quad q = \frac{2}{1+3w}.$$

Equation (2.149) has an exact solution of the form

$$\Psi = \frac{1}{a}\left(Aj_q(\sqrt{w}kt) + By_q(\sqrt{w}kt)\right), \tag{2.150}$$

where j_q and y_q denote the spherical Bessel functions of order q. Using $j_q(x) \propto x^q$ and $y_q(x) \propto x^{-q-1}$ for $x \ll 1$, we find that the A-mode is constant while the B-mode decays like $1/(a^2t)$ on super-Hubble scales. If both modes are generated with similar amplitudes, the B-mode is therefore negligible after a few expansion times. On sub-Hubble scales, $\sqrt{w}kt \gg 1$, the solution oscillates with frequency $\sqrt{w}k$ and decay like $1/(at)$. The only exception is the case of cosmic dust (CDM) with $w = 0$. In this case the oscillatory term drops and the solution is of the form

$$\Psi = A + \frac{B}{(kt)^5}. \tag{2.151}$$

For later use we collect the main results in the following equation: for power law expansion $a \propto t^q$ we find

$$\Psi = \begin{cases} \text{constant} & \text{for} \quad \sqrt{w}kt \ll 1 \\ \frac{A}{a\sqrt{w}kt}\sin(\sqrt{w}kt - \frac{q}{2}\pi) & \text{for} \quad \sqrt{w}kt \gg 1, \quad w \neq 0. \end{cases} \tag{2.152}$$

We now consider a universe that starts out in a radiation-dominated era with a spectrum (see Section 2.6) $\langle|\Psi|^2\rangle k^3 = A_\Psi(k/H_0)^{n_s-1}$ and that becomes matter dominated at some time $t_{\rm eq}$. Late in the matter-dominated era the spectrum of Ψ is therefore approximately given by (see Fig. 2.1)

$$\langle|\Psi|^2\rangle k^3 = A_\Psi(k/H_0)^{n_s-1}\begin{cases} 1 & \text{for} \quad kt_{\rm eq} < 1 \\ (kt_{\rm eq})^{-4}\cos^2(kt_{\rm eq}) & \text{for} \quad kt_{\rm eq} > 1. \end{cases} \tag{2.153}$$

As we shall see in Chapter 3, inflation generically leads to a spectrum that is close to scale invariant[1] $n_s \simeq 1$. A formal definition of the spectrum, interpreting Ψ, or more precisely the amplitude A as a random variable, is given in Section 2.6. Here we may just consider it as the square of the Fourier transform of Ψ and ignore the expectation value $\langle\cdots\rangle$.

Another interesting case (especially when discussing inflation) is the scalar field. There, as we shall see in Chapter 3, $\Pi = 0$, but in general $\Gamma \neq 0$ since $\delta p/\delta\rho \neq \dot{p}/\dot{\rho}$. Nevertheless, since this case again has only one dynamical degree of freedom, we can express the perturbation equations in terms of one single second-order

[1] The reason for the definition of n, such that $\langle|\Psi|^2\rangle k^3 \propto k^{n-1}$, is purely historical and not very logical, but as always, it is difficult to change conventions without leading to confusion. For compatibility with the literature we therefore keep this convention.

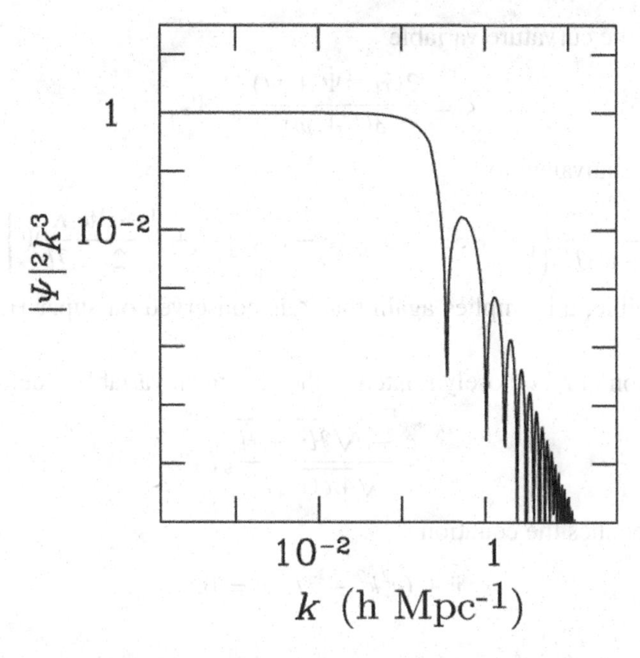

Fig. 2.1 The approximate form of the power spectrum $\langle|\Psi|^2\rangle k^3$ for a scale-invariant initial spectrum, $n = 1$, is plotted.

equation for Ψ. In Chapter 3 we shall find the following equation for a perturbed scalar field cosmology:

$$\ddot{\Psi} + 3\mathcal{H}(1 + c_s^2)\dot{\Psi} + [(1 + 3c_s^2)(\mathcal{H}^2 - K) - (1 + 3w)(\mathcal{H}^2 + K) + k^2]\Psi = 0. \tag{2.154}$$

The only difference between the perfect fluid and scalar field perturbation equation is that the latter is missing the factor c_s^2 in front of the oscillatory k^2 term. It is useful to define also the variable (Mukhanov *et al.*, 1992)

$$u = a[4\pi G(\mathcal{H}^2 - \dot{\mathcal{H}} + K)]^{-1/2}\Psi, \tag{2.155}$$

which satisfies the equation

$$\ddot{u} + (k^2 - \ddot{\theta}/\theta)u = 0, \tag{2.156}$$

where

$$\theta = \frac{3\mathcal{H}}{2a\sqrt{\mathcal{H}^2 - \dot{\mathcal{H}} + K}}. \tag{2.157}$$

A second-order linear differential equation of the form (2.154) can always be transformed into one of the form of Eq. (2.156) by a suitable transformation of variables. We show this in Exercise 2.6.

In terms of the curvature variable

$$\zeta \equiv \frac{2(\mathcal{H}^{-1}\dot{\Psi} + \Psi)}{3(1+w)} + \Psi, \tag{2.158}$$

Eq. (2.147) is equivalent to

$$\dot{\zeta} = \frac{2}{3(1+w)\mathcal{H}} \left\{ \left[(2 + 3w + 3c_s^2)K - c_s^2 k^2 \right] \Psi + \frac{1+3w}{2} \frac{K}{\mathcal{H}} \dot{\Psi} \right\}. \tag{2.159}$$

If K is negligible, this implies again that ζ is conserved on super-Hubble scales, $k/\mathcal{H} \ll 1$.

The evolution of ζ is closely related to the canonical variable v defined by

$$v = \frac{-a\sqrt{\mathcal{H}^2 - \dot{\mathcal{H}}}}{\sqrt{4\pi G} c_s \mathcal{H}} \zeta, \tag{2.160}$$

if $K = 0$. It satisfies the equation

$$\ddot{v} + (c_s^2 k^2 - \ddot{z}/z)v = 0, \tag{2.161}$$

for

$$z = \frac{a\sqrt{\mathcal{H}^2 - \dot{\mathcal{H}}}}{c_s \mathcal{H}}. \tag{2.162}$$

The significance of the canonical variable v that has been introduced in Mukhanov *et al.* (1992) will be discussed in Chapter 3.

2.4 Simple Examples

We first discuss two simple applications that are important to understand the CMB anisotropy spectrum.

2.4.1 The Pure Dust Fluid

We assume the dust to have $w = c_s^2 = p = 0$ and $\Pi = \Gamma = 0$. We first consider the case $K = 0, \Lambda = 0$. Equation (2.148) then reduces to

$$\ddot{\Psi} + \frac{6}{t}\dot{\Psi} = 0, \tag{2.163}$$

with the general solution,

$$\Psi = \Psi_0 + \Psi_1 \frac{1}{t^5}, \tag{2.164}$$

with arbitrary constants Ψ_0 and Ψ_1. Since the perturbations are supposed to be small initially, they cannot diverge for $t \to 0$, and we have therefore to choose the decaying mode, $\Psi_1 = 0$. Another way to argue is as follows: if the mode Ψ_1 has to be small already at some early initial time $t_{\rm in}$, it will be much smaller later and may hence be neglected at late times. But also the Ψ_0 mode is only constant and not growing. This fact led Lifshitz, who was the first to analyze relativistic cosmological perturbations, to the conclusions that linear perturbations do not grow in an FL universe and cosmic structure cannot have evolved by gravitational instability (Lifshitz, 1946). However, the important point to note here is that, even if the gravitational potential remains constant, matter density fluctuations do grow on subhorizon scales and therefore inhomogeneities can evolve on scales that are smaller than the Hubble scale. To see this we consider the conservation equations (2.116), (2.107), and the Poisson equation (2.105). For the pure dust case, $w = c_s^2 = \Pi = \Gamma = 0$, they reduce to

$$\dot{D}_g = -kV \quad \text{(energy conservation)}, \tag{2.165}$$

$$\dot{V} + \mathcal{H}V = k\Psi \quad \text{(gravitational acceleration)}, \tag{2.166}$$

$$-\frac{2k^2}{3\mathcal{H}^2}\Psi = \left(D_g + 3\left(\Psi + \mathcal{H}k^{-1}V\right)\right) \quad \text{(Poisson)}, \tag{2.167}$$

where we have used the relation

$$D = D_g + 3(1+w)\left(\Phi + \mathcal{H}k^{-1}V\right). \tag{2.168}$$

The Friedmann equation for dust gives $\mathcal{H} = 2/t$. Setting $kt = x$ and a prime $= d/dx$, the system (2.165)–(2.167) becomes

$$D_g' = -V, \tag{2.169}$$

$$V' + \frac{2}{x}V = \Psi, \tag{2.170}$$

$$\frac{6}{x^2}\left(D_g + 3\left(\Psi + \frac{2}{x}V\right)\right) = -\Psi. \tag{2.171}$$

We use (2.171) to eliminate Ψ and (2.169) to eliminate D_g, leading to

$$\left(18 + x^2\right)V'' + \left(\frac{72}{x} + 4x\right)V' - \left(\frac{72}{x^2} + 4\right)V = 0. \tag{2.172}$$

The general solution of Eq. (2.172) is

$$V = V_0 x + \frac{V_1}{x^4}. \tag{2.173}$$

The V_1 mode is the decaying mode (corresponding to Ψ_1), which we neglect. The perturbation variables are then given by

$$V = V_0 x, \tag{2.174}$$

$$D_g = -15V_0 - \frac{1}{2}V_0 x^2, \tag{2.175}$$

$$V_0 = \Psi_0/3. \tag{2.176}$$

We distinguish two regimes.

(1) Superhorizon, $x \ll 1$, where we have

$$V = \frac{1}{3}\Psi_0 x, \tag{2.177}$$

$$D_g = -5\Psi_0, \tag{2.178}$$

$$\Psi = \Psi_0. \tag{2.179}$$

Note that even though V is growing, it always remains much smaller than Ψ or D_g on superhorizon scales. Hence the largest fluctuations are of order Ψ, which is constant.

(2) Subhorizon, $x \gg 1$, where the solution is dominated by the terms

$$V = \frac{1}{3}\Psi_0 x, \tag{2.180}$$

$$D_g = -\frac{1}{6}\Psi_0 x^2, \tag{2.181}$$

$$\Psi = \Psi_0 = \text{constant}. \tag{2.182}$$

Note that for dust

$$D = D_g + 3\Psi + \frac{6}{x}V = -\frac{1}{6}\Psi_0 x^2.$$

In the variable D the constant term has disappeared and we have $D \ll \Psi$ on superhorizon scales, $x \ll 1$.

On subhorizon scales, the density fluctuations grow like the scale factor,

$$D \simeq D_g \simeq D_s \propto x^2 \propto a, \qquad x \gg 1. \tag{2.183}$$

Presently, the Universe seems to be dominated by a cosmological constant. Let us therefore also briefly discuss dust perturbations for $\Lambda \neq 0$. The case $K \neq 0$ is equivalent. In a ΛCDM universe the growth of density fluctuations is slowed down at late time due to the rapid expansion. More precisely, Eq. (2.120) for dust perturbations in a ΛCDM universe, or in a dust universe with curvature becomes

$$\ddot{D} + \mathcal{H}\dot{D} = 4\pi G a^2 \rho D = \frac{3}{2}\mathcal{H}^2\Omega_m(a)D. \tag{2.184}$$

Here, the wave number k no longer enters and the differental equation can be written as an equation in $y = \ln a$. Denoting the derivative w.r.t. y by a prime we find

$$D'' + \left(1 - \frac{1}{2}\Omega_m(a)\right) D' = \frac{3}{2}\Omega_m(a) D. \tag{2.185}$$

For $\Omega_m \equiv 1$ we find again a growing mode $D \propto a$ while for $\Omega_m = 0$ the "growing mode" is constant. For a redshift-dependent Ω_m from a ΛCDM or an open universe ($\Omega(a) < 1$) one obtains a numerical solution that can be approximated as

$$D(a) \equiv D_1(a) \simeq a^{\Omega_m^{0.56}(a)}. \tag{2.186}$$

Normalizing $a(t_0) = a_0 = 1$, this is the growing mode solution normalized to today.

Despite growing solutions for the density contrast, Lifshitz' conclusion (Lifshitz, 1946) that pure gravitational instability cannot be the cause of structure formation has some truth. If we start from tiny thermal fluctuations of the order of 10^{-35}, they can grow to only about 10^{-30} due to this mild, power law instability during the matter-dominated regime. Or, to put it differently, if we want to form structure by gravitational instability, we need initial fluctuations of the order of at least 10^{-5}, much larger than thermal fluctuations. One possibility for creating such fluctuations is quantum particle production in the classical gravitational field during inflation. The rapid expansion of the Universe during inflation quickly expands microscopic scales at which quantum fluctuations are important to cosmological scales where these fluctuations are then "frozen in" as classical perturbations in the energy density and the geometry. We will discuss the induced spectrum of fluctuations in Chapter 3.

2.4.2 The Pure Radiation Fluid, $K = 0, \Lambda = 0$

In this limit we set $w = c_s^2 = \frac{1}{3}$ and $\Pi = \Gamma = 0$ so that $\Phi = \Psi$. We conclude from $\rho \propto a^{-4}$ that $a \propto t$. For radiation, the general solution (2.150) becomes

$$\Psi(x) = \frac{1}{x}\left[A j_1(x) + B y_1(x)\right], \tag{2.187}$$

where we have set $x = kt/\sqrt{3} = c_s kt$ and used the fact that $a \propto x$. On superhorizon scales, $x \ll 1$, we have (see Appendix 4, Section A4.3)

$$\Psi(x) \simeq \frac{A}{3} + \frac{B}{x^3}. \tag{2.188}$$

We assume that the perturbations have been initialized at some early time $x_{\rm in} \ll 1$ and that at this time the two modes have been comparable. If this is the case then

$B \ll A$ and we may neglect the B-mode at later times, so that (see Abramowitz and Stegun, 1970)

$$\Psi(x) = \frac{A}{x} j_1(x) = A\left(\frac{\sin(x)}{x^3} - \frac{\cos(x)}{x^2}\right). \tag{2.189}$$

To determine the density and velocity perturbations, we use the energy conservation and Poisson equations for radiation, with a prime denoting d/dx these become, for radiation,

$$D_g' = -\frac{4}{\sqrt{3}} V, \tag{2.190}$$

$$-2x^2\Psi = D_g + 4\Psi + \frac{4}{\sqrt{3}x} V\,. \tag{2.191}$$

Inserting the solution (2.189) for Ψ, we obtain

$$D_g = 2A\left[\cos(x) - \frac{2}{x}\sin(x)\right], \tag{2.192}$$

$$V = -\frac{\sqrt{3}}{4} D_g', \tag{2.193}$$

$$\Psi = -\frac{D_g + \frac{4}{\sqrt{3}x} V}{4 + 2x^2}. \tag{2.194}$$

In the **superhorizon regime**, $x \ll 1$, we obtain

$$\Psi = \frac{A}{3}, \quad D_g = -2A\left(1 + \frac{1}{6}x^2\right), \quad V = \frac{A}{2\sqrt{3}} x. \tag{2.195}$$

On **subhorizon scales**, $x \gg 1$, we find oscillating solutions with constant amplitude and with frequency $k/\sqrt{3}$:

$$V = \frac{\sqrt{3}A}{2}\sin(x), \tag{2.196}$$

$$D_g = 2A\cos(x), \quad \Psi = -A\cos(x)/x^2. \tag{2.197}$$

The radiation fluid cannot simply "collapse" under gravity. As in acoustic waves, the restoring force provided by the pressure leads to oscillations with constant amplitude. These are called the "acoustic oscillations" of the radiation fluid. As we shall see in the next section, they are responsible for the acoustic peaks in the CMB fluctuation spectrum.

Also for radiation perturbations

$$D = -\frac{2A}{3}x^2 \ll \Psi$$

is small on superhorizon scales, $x \ll 1$.

The perturbation amplitude is given by the largest gauge-invariant perturbation variable. We conclude therefore that perturbations outside the Hubble horizon are frozen to first order. Once they enter the horizon they start to collapse, but pressure resists the gravitational force and the radiation fluid fluctuations oscillate at constant amplitude. The perturbations of the gravitational potential oscillate and decay like $1/a^2$ inside the horizon.

2.4.3 The Mixed Dust and Radiation Fluid for K = 0, Λ = 0

We now consider a mixed matter (also called "dust" since we neglect its pressure) and radiation fluid with comparable perturbation amplitudes in the fluid variables. At early times we are in the radiation-dominated era, and radiation perturbations will not be affected at all by the subdominant gravitational potential from matter fluctuations. As before, the radiation variables and the gravitational potential perform acoustic oscillations,

$$\Psi = \frac{A}{x} j_1(x) = A\left[\frac{\sin(x)}{x^3} - \frac{\cos(x)}{x^2}\right], \tag{2.198}$$

$$D_{gr} = 2A\left[\cos(x) - \frac{2}{x}\sin(x)\right], \tag{2.199}$$

$$V_r = \frac{\sqrt{3}A}{2}\left[\left(1 - \frac{2}{x^2}\right)\sin(x) + \frac{2}{x}\cos(x)\right]. \tag{2.200}$$

In the radiation era the matter equations become ($x = \frac{kt}{\sqrt{3}}$)

$$D'_{gm} + \sqrt{3}V_m = 0, \tag{2.201}$$

$$(aV_m)' = a\sqrt{3}\Psi = \frac{a\sqrt{3}A}{x} j_1(x). \tag{2.202}$$

Noting that a/x is constant in the radiation era, these equations can be solved simply by integration, leading to

$$V_m = \frac{-\sqrt{3}A}{x} j_0(x) + V_1/x = -\sqrt{3}A\frac{\sin(x)}{x^2} + V_1/x, \tag{2.203}$$

$$D_{gm} = -3A\left[\frac{\sin(x)}{x} + \mathrm{Ci}(x) - \ln(x) + z_0\right] - \sqrt{3}V_1\ln(x). \tag{2.204}$$

Here Ci is the integral cosine function defined by $\mathrm{Ci}(x) = \int_0^x \frac{1-\cos(z)}{z}\,dz$ (see Abramowitz and Stegun, 1970). The condition that V be small at very early times, $x \ll 1$, requires $V_1 = \sqrt{3}A$. The constant z_0 is an arbitrary integration constant. With this the foregoing solutions become

$$V_m = \frac{\sqrt{3}A}{x}\left[1 - \frac{\sin(x)}{x}\right], \tag{2.205}$$

$$D_{gm} = -3A\left[\frac{\sin(x)}{x} + \mathrm{Ci}(x) + z_0\right]. \tag{2.206}$$

On large scales, $x \ll 1$, we obtain the behavior

$$\Psi = \frac{A}{3}, \tag{2.207}$$

$$D_{gr} = -2A, \tag{2.208}$$

$$V_r = \frac{A}{2\sqrt{3}}x, \tag{2.209}$$

$$V_m = \frac{A}{2\sqrt{3}}x, \tag{2.210}$$

$$D_{gm} = -3A(1 + z_0). \tag{2.211}$$

The most natural condition to fix the constant z_0 is the requirement that at very early times perturbations are adiabatic, $\Gamma_{\mathrm{tot}} = \pi_L - (c_s^2/w)\delta = 0$. We use $\pi_L = \delta P_r/P_r = \delta\rho_r/\rho_r$ and

$$c_s^2/w = \frac{4}{R+3}, \qquad \text{where} \qquad R \equiv \frac{\rho_r}{\rho_m + \rho_r}.$$

Here we have used the fact that $P = P_r = \rho_r/3$ and $\rho_r \propto a^{-4}$, while $\rho_m \propto a^{-3}$. For the entropy production we then obtain

$$\Gamma_{\mathrm{tot}} = 4\frac{1-R}{R+3}\left(\frac{3}{4}\delta_r - \delta_m\right), \tag{2.212}$$

so that $\Gamma_{\mathrm{tot}} = 0$ implies $\delta_m = \frac{3}{4}\delta_r$. According to the definition of D_g, Eq. (2.87) this is equivalent to $D_{gm} = (3/4)D_{gr}$. To achieve this we have to set $z_0 = -\frac{1}{2}$ so that

$$D_{gm} = -\frac{3}{2}A. \tag{2.213}$$

With this choice, perturbations are adiabatic on super-Hubble scales. But since D_{gm} and D_{gr} evolve differently on sub-Hubble scales, there clearly $\Gamma_{\mathrm{tot}} \neq 0$. We shall use the notion "adiabatic" in the sense that the *initial conditions* are such that $\Gamma_{\rm tot}(t_{\rm in}) = 0$ for some early initial time $t_{\rm in}$ such that $kt_{\rm in} \ll 1$.

On sub-Hubble scales, $x \gg 1$, the radiation perturbations oscillate as in the ordinary radiation universe, but the matter perturbations grow logarithmically, $D_{gm} \simeq -3A\mathrm{Ci}(x) \simeq -3A\ln(x)$ for $x \gg 1$. This severe suppression of growth of matter perturbations during the radiation-dominated era is called the "Mészáros effect" (Mészáros, 1974). Physically, the reason for this suppression is that matter self-gravity $\propto 4\pi G\rho_m$ is too weak during the radiation-dominated regime to overcome damping, which (in the same units) is $\propto \mathcal{H}^2 \propto G\rho_r$. Neglecting self-gravity in the matter equation would yield D_{gm} = constant, which is nearly correct.

We now go over to the matter-dominated regime. There, the matter perturbations are not affected by radiation and behave as given in Eqs. (2.174)–(2.176),

$$\Psi = \Psi_0, \tag{2.214}$$

$$V_m = \frac{1}{\sqrt{3}}\Psi_0 x, \tag{2.215}$$

$$D_{gm} = -5\Psi_0\left(1 + \frac{1}{10}x^2\right). \tag{2.216}$$

Keeping in mind that $x = kt/\sqrt{3}$, these solutions correspond exactly to Eqs. (2.174)–(2.176). We now assume that the Bardeen potentials are those from the dominant matter perturbations, $\Phi = \Psi = \Psi_0$. The radiation perturbation equations then reduce to

$$D'_{gr} = -\frac{4}{\sqrt{3}}V_r, \tag{2.217}$$

$$D''_{gr} + D_{gr} = -8\Psi_0, \tag{2.218}$$

with the general solution

$$D_{gr} = B\sin(x) + C\cos(x) - 8\Psi_0,$$

$$V_r = -\frac{\sqrt{3}}{4}\left(B\cos(x) - C\sin(x)\right).$$

Requiring that these solutions be connected smoothly to the radiation dominated solutions fixes the constants B and C. Therefore, on large scales, $x \ll 1$, V has to grow like x, which implies $B \equiv 0$. The constant C is then determined by the condition that the perturbations be adiabatic for $x \ll 1$. This implies

$$D_{gr} \simeq C - 8\Psi_0 = \frac{4}{3}D_{gm} = -\frac{20}{3}\Psi_0 \quad \text{so that} \quad C = \frac{4}{3}\Psi_0. \tag{2.219}$$

This leads to the following solution for the radiation perturbations in the matter dominated era:

$$D_{gr} = 4\Psi_0\left(\frac{1}{3}\cos(x) - 2\right), \qquad (2.220)$$

$$V_r = \frac{1}{\sqrt{3}}\Psi_0 \sin(x). \qquad (2.221)$$

These are the exact solutions for decoupled but adiabatic matter and radiation fluctuations in the matter-dominated era. To connect them to the solutions in the radiation-dominated era, we require that D_{gm} be continuous at the transition, $x = x_{\rm eq} = kt_{\rm eq}/\sqrt{3}$. This implies

$$\Psi_0 = A\begin{cases} \frac{3}{10} & \text{for} \quad x_{\rm eq} \ll 1 \\ 6\frac{\ln(x_{\rm eq})}{x_{\rm eq}^2} & \text{for} \quad x_{\rm eq} \gg 1. \end{cases} \qquad (2.222)$$

This approximation is of course relatively crude, since the radiation to matter transition is very gradual and not as abrupt as it is implemented here. It is also easy to see that we would not obtain exactly the same condition when requiring Ψ to be continuous at the transition. The main difference is that we do not obtain the logarithmic growth of the potential in the radiation-dominated era from the continuity of Ψ. But this is clearly a failure, since the log growth of D_{gm} leads to a larger gravitational potential in the matter era. For $x_{\rm eq} \simeq 1$ both approximations are bad and should be taken simply as order of magnitude estimates. More details on the coupled matter radiation system are found in Section 3.5. In Fig. 2.2 the exact solutions are plotted.

Instead of requiring adiabatic initial conditions one sometimes also requires $\Psi = \Phi = 0$ on super-Hubble scales. This is the so-called *isocurvature* initial condition. We shall discuss it in Section 3.5.

2.5 Light-Like Geodesics and CMB Anisotropies

After decoupling, $t > t_{\rm dec}$, photons follow to a good approximation light-like geodesics. The temperature shift of a Planck distribution of photons is equal to the energy shift of any given photon. The relative energy shift, red or blue shift, is independent of the photon energy (gravity is "achromatic").

The unperturbed photon trajectory follows

$$(x^\mu(t)) \equiv \left(t, \int_t^{t_0} \mathbf{n}(t')\,dt' + \mathbf{x}_0\right),$$

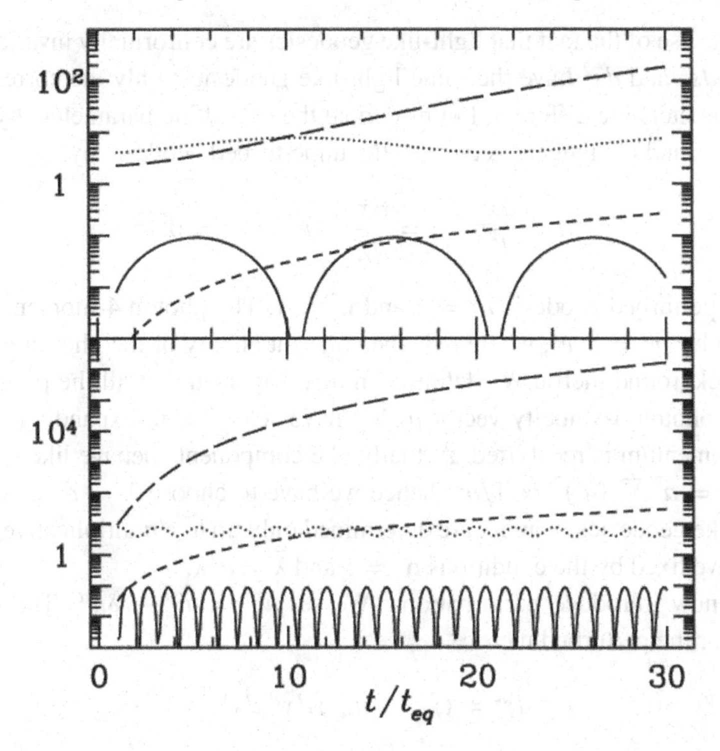

Fig. 2.2 The time evolution for $|D_{gm}|^2$ (long-dashed), $|D_{gr}|^2$ (dotted), $|V_m|^2$ (dashed), and $|V_r|^2$ (solid) is indicated as a function of $t/t_{\rm eq}$. The wave number in the top panel is $k_1 \ll 1/t_{\rm eq}$, while in the bottom panel $k_2 \gg 1/t_{\rm eq}$. Note that for a large wave number D_{gm} immediately starts growing and rapidly becomes much larger than D_{gr}, while for the small wave number (top panel) D_{gm} stays of the same order as D_{gr} until horizon entry, which is roughly at $t/t_{\rm eq} \sim 10$. After horizon entry D_{gm} starts growing while D_{gr} starts oscillating.

where $\mathbf{x}_0$ is the photon position at time t_0 and $\mathbf{n}$ is the (parallel transported) photon direction. We determine the components of the photon momentum with respect to a geodesic basis $(\mathbf{e}_i)_{i=1}^3$ on the constant time hypersurfaces. We choose

$$\mathbf{e}_i = \begin{cases} \frac{\partial}{\partial x^i}, & \text{if } K = 0, \\ \epsilon_i, & \text{with } \gamma(\epsilon_i, \epsilon_j) = \delta_{ij} \text{ if } K \neq 0. \end{cases} \tag{2.223}$$

In other words, the vector fields ϵ_i form an orthonormal basis for the spatial metric γ_{ij}.

Our metric is of the form

$$d\tilde{s}^2 = a^2 ds^2\,, \quad \text{with} \tag{2.224}$$

$$ds^2 = \left(\gamma_{\mu\nu} + h_{\mu\nu}\right) dx^\mu\, dx^\nu, \quad \gamma_{00} = -1,\ \gamma_{i0} = 0,\ \gamma_{ij} = \gamma_{ji}, \tag{2.225}$$

as before.

We make use of the fact that light-like geodesics are conformally invariant. More precisely, ds^2 and $d\tilde{s}^2$ have the same light-like geodesics; only the corresponding affine parameters are different. Let us denote the two affine parameters by λ and $\tilde{\lambda}$ respectively, and the tangent vectors to the unperturbed geodesic by

$$n = \frac{dx}{d\lambda}, \quad \tilde{n} = \frac{dx}{d\tilde{\lambda}}, \quad n^2 = \tilde{n}^2 = 0. \tag{2.226}$$

For the unperturbed geodesic $n^0 = 1$ and $\mathbf{n}^2 = 1$. The photon 4-momentum p^μ is then given by $p^\mu = \omega n^\mu$, where ω is the constant energy of the photon moving in the flat background metric.We define ω in this way such that all the perturbations are in the photon 4-velocity vector n. We have seen that in expanding space the photon momentum is redshifted. Actually, the components behave like $\tilde{n}^i \propto 1/a^2$ so that $\tilde{\mathbf{n}}^2 = a^2 \sum_i (\tilde{n}^i)^2 \propto 1/a^2$; hence we have to choose $\tilde{\lambda} = a^2\lambda$. As always for light-like geodesics, $\tilde{\lambda}$ and λ are determined only up to a multiplicative constant that we have fixed by the conditions $\mathbf{n}^2 = 1$ and $\tilde{\lambda} = a^2\lambda$.

Let us now introduce perturbations. We set $n^\mu = \bar{n}^\mu + \delta n^\mu$. The geodesic equation for the perturbed metric,

$$ds^2 = (\gamma_{\mu\nu} + h_{\mu\nu})\, dx^\mu\, dx^\nu, \tag{2.227}$$

yields, to first order,

$$\frac{d}{d\lambda}\delta n^\mu = -\delta\Gamma^\mu_{\alpha\beta} n^\alpha n^\beta. \tag{2.228}$$

For the energy shift, we have to determine δn^0. Since $g^{0\mu} = -\delta_{0\mu} +$ first order, we obtain $\delta\Gamma^0_{\alpha\beta} = -\frac{1}{2}(h_{\alpha 0|\beta} + h_{\beta 0|\alpha} - \dot{h}_{\alpha\beta})$, so that

$$\frac{d}{d\lambda}\delta n^0 = h_{\alpha 0|\beta} n^\beta n^\alpha - \frac{1}{2}\dot{h}_{\alpha\beta} n^\alpha n^\beta. \tag{2.229}$$

Integrating this equation we use $h_{\alpha 0|\beta} n^\beta n^\alpha = \frac{d}{d\lambda}(h_{\alpha 0} n^\alpha)$, so that the change of n^0 between some initial time t_i and some final time t_f is given by

$$\delta n^0|_i^f = \left[h_{00} + h_{0j} n^j\right]_i^f - \frac{1}{2}\int_i^f \dot{h}_{\mu\nu} n^\mu n^\nu d\lambda\,. \tag{2.230}$$

The energy of a photon with 4-momentum $\tilde{p}^\mu$ as seen by an observer moving with 4-velocity $\tilde{u}$ is given by $E = -(\tilde{u}\cdot\tilde{p})$. Hence, the ratio of the energy of a photon measured by some observer at t_f to the energy emitted at t_i is

$$\frac{E_f}{E_i} = \frac{(\tilde{n}\cdot\tilde{u})_f}{(\tilde{n}\cdot\tilde{u})_i} = \frac{a_i}{a_f}\frac{(n\cdot u)_f}{(n\cdot u)_i}, \tag{2.231}$$

where here $\tilde{\cdot}$ denotes the scalar product in an expanding universe, containing the factor a^2, and $\tilde{u}$ is the emitter and receiver 4-velocity in an expanding universe, $\tilde{u} = a^{-1}u$, while u_f and u_i are the 4-velocities of the observer and emitter respectively in the nonexpanding conformally related geometry given by

$$u = (1 - A)\partial_t + v^i \mathbf{e}_i = a\tilde{u} \,. \tag{2.232}$$

Together with $\tilde{n} = a^{-2}n$ this implies the result (2.231). The ratio $a_i/a_f = T_i/T_f$ is the usual (unperturbed) redshift that relates n and $\tilde{n}$.

Before continuing with the fully gauge-independent treatment, let us derive a useful formula for scalar perturbations in longitudinal gauge. In this gauge $h_{00} = -2\Psi$ and $\dot{h}_{\mu\nu}n^\mu n^\nu = -2(\dot{\Psi} + \dot{\Phi})$. The first-order geodesic perturbations then are

$$\delta n^0\big|_i^f = -\,2\Psi\big|_i^f + \int_i^f (\dot{\Psi} + \dot{\Phi})\, d\lambda, \tag{2.233}$$

$$\delta n^j\big|_i^f = 2n^j\Phi\big|_i^f - \int_i^f \partial^j(\Psi + \Phi)\, d\lambda. \tag{2.234}$$

Using also $v^i = V^i$ we obtain

$$\frac{E_f}{E_i} = \frac{1}{1+\bar{z}}\left(1 - \frac{\delta z}{1+\bar{z}}\right) \tag{2.235}$$

$$\frac{\delta z}{1+\bar{z}} = \left[V_j^{(b)} n^j + \Psi\right]_i^f - \int_i^f (\dot{\Psi} + \dot{\Phi})\, d\lambda \,. \tag{2.236}$$

Here $\mathbf{V}^{(b)}$ is the velocity of baryons (the emitters and observers of radiation). Expression (2.236) is valid only in longitudinal gauge. The redshift perturbation is not gauge invariant. For example, in the constant redshift time slicing, the redshift perturbation vanishes by definition.

Let us now continue with the generic analysis valid in an arbitrary gauge. An observer measuring a temperature T_0 receives photons that were emitted at the time $t_{\rm dec}$ of decoupling of matter and radiation, at the fixed temperature $T_{\rm dec}$. In first-order perturbation theory, we find the following relation between the unperturbed temperatures T_f, T_i, the true temperatures $T_0 = T_f + \delta T_f$, $T_{\rm dec} = T_i + \delta T_i$, and the photon density perturbation:

$$\frac{a_i}{a_f} = \frac{T_f}{T_i} = \frac{T_0}{T_{\rm dec}}\left(1 - \frac{\delta T_f}{T_f} + \frac{\delta T_i}{T_i}\right) = \frac{T_0}{T_{\rm dec}}\left(1 - \frac{1}{4}\delta_\gamma\big|_i^f\right), \tag{2.237}$$

where δ_r is the intrinsic density perturbation in the radiation and we have used $\rho_\gamma \propto T^4$ in the last equality. Inserting Eq. (2.237) and Eq. (2.230) into Eq. (2.231), and using Eq. (2.28) for the definition of $h_{\mu\nu}$, as well as Eqs. (2.57), (2.58), (2.88), and (2.85), one finds, after integration by parts, the following result for scalar perturbations:

$$\frac{E_f}{E_i} = \frac{T_0}{T_{\rm dec}} \left\{ 1 - \left[\frac{1}{4} D_g^{(r)} + V_j^{(b)} n^j + \Psi + \Phi \right]_i^f + \int_i^f (\dot{\Psi} + \dot{\Phi})\, d\lambda \right\} . \quad (2.238)$$

Here $D_g^{(r)}$ denotes the density perturbation in the radiation fluid.

Evaluating Eq. (2.238) at final time t_0 (today) and initial time $t_{\rm dec}$, we obtain the temperature difference of photons coming from different directions $\mathbf{n}_1$ and $\mathbf{n}_2$

$$\frac{\Delta T}{T} \equiv \frac{\Delta T(\mathbf{n}_1)}{T} - \frac{\Delta T(\mathbf{n}_2)}{T} \equiv \frac{E_f}{E_i}(\mathbf{n}_1) - \frac{E_f}{E_i}(\mathbf{n}_2). \quad (2.239)$$

Direction-independent contributions to E_f/E_i do not enter in this difference.

The largest contribution to $\Delta T/T$ is the dipole term, $V_j^{(b)}(t_0)n^j$, which simply describes our motion with respect to the emission surface. Its amplitude is about 1.2×10^{-3} and it has been measured so accurately that even the yearly variation due to the motion of the Earth around the Sun has been detected.

For the higher multipoles (polynomials in n^j of degree 2 and higher) we can set

$$\frac{\Delta T(\mathbf{n})}{T} = \left[\frac{1}{4} D_g^{(r)} + V_j^{(b)} n^j + \Psi + \Phi \right] (t_{\rm dec}, \mathbf{x}_{\rm dec}) + \int_{t_{\rm dec}}^{t_0} (\dot{\Psi} + \dot{\Phi})(t, \mathbf{x}(t))\, dt, \quad (2.240)$$

where $\mathbf{x}(t)$ is the unperturbed photon position at time t for an observer at $\mathbf{x}_0$, and $\mathbf{x}_{\rm dec} = \mathbf{x}(t_{\rm dec})$ (if $K = 0$ we simply have $\mathbf{x}(t) = \mathbf{x}_0 - (t_0 - t)\mathbf{n}$). The first term in Eq. (2.240) is the one we have discussed in the previous section. It describes the intrinsic inhomogeneities of the radiation density on the surface of last scattering, due to acoustic oscillations prior to decoupling; see Eq. (2.199). Depending on the initial conditions, it can also contribute significantly on superhorizon scales. This is especially important in the case of adiabatic initial conditions. As we have seen in Eq. (2.219), in a dust + radiation universe with $\Omega = 1$, adiabatic initial conditions imply $D_g^{(r)}(k,t) = -\frac{20}{3}\Psi(k,t)$ and $V^{(b)} = V^{(r)} \ll D_g^{(r)}$ when $kt \ll 1$. With $\Phi = \Psi$ the square bracket of Eq. (2.240) therefore gives for adiabatic perturbations

$$\left(\frac{\Delta T(\mathbf{n})}{T} \right)_{\rm adiabatic}^{\rm (OSW)} = \frac{1}{3} \Psi(t_{\rm dec}, \mathbf{x}_{\rm dec}),$$

on superhorizon scales. The contribution to $\Delta T/T$ from the last scattering surface on very large scales is called the "ordinary Sachs–Wolfe effect" (OSW). It was derived for the first time by Sachs and Wolfe (1967). For isocurvature perturbations, the initial conditions require $D_g^{(r)}(k,t) \to 0$ for $t \to 0$ so that the contribution of $D_g^{(r)}$ to the ordinary Sachs–Wolfe effect can be neglected,

$$\left(\frac{\Delta T(\mathbf{n})}{T}\right)^{(\mathrm{OSW})}_{\mathrm{isocurvature}} = 2\Psi(t_{\mathrm{dec}}, \mathbf{x}_{\mathrm{dec}}).$$

The second term in (2.240) describes the relative motion of emitter and observer. This is the Doppler contribution to the CMB anisotropies. It appears on the same angular scales as the acoustic term; we call the sum of the acoustic and Doppler contributions acoustic peaks.

The integral in Eq. (2.240) accounts for the red or blue shifts caused by the time dependence of the gravitational potential along the path of the photon, and represents the so-called integrated Sachs–Wolfe (ISW) effect. In a $\Omega = 1$, pure dust universe, as we have seen, the Bardeen potentials are constant and there is no integrated Sachs–Wolfe effect; the blue shift that the photons acquire by falling into a gravitational potential is exactly cancelled by the redshift induced by climbing out of it. This is no longer true in a universe with substantial radiation contribution, curvature, or a cosmological constant. The sum of the ordinary Sachs–Wolfe term and the integral is the full Sachs–Wolfe contribution.

For **vector** perturbations $\delta^{(r)}$ and A vanish and Eq. (2.231) leads to

$$\left(\frac{E_f}{E_i}\right)^{(V)} = \frac{a_i}{a_f}\left[1 - V_j^{(b)} n^j \Big|_i^f + \int_i^f \dot{\sigma}_j n^j \, d\lambda\right]. \tag{2.241}$$

We obtain a Doppler term and a gravitational contribution. For **tensor** perturbations, that is, gravitational waves, only the gravitational part remains:

$$\left(\frac{E_f}{E_i}\right)^{(T)} = \frac{a_i}{a_f}\left[1 - \int_i^f \dot{H}_{lj} n^l n^j d\lambda\right]. \tag{2.242}$$

Equations (2.238), (2.241), and (2.242) are the manifestly gauge-invariant results for the energy shift of photons due to scalar, vector, and tensor perturbations. Disregarding again the dipole contribution due to our proper motion, Eqs. (2.241) and (2.242) imply the vector and tensor temperature fluctuations

$$\left(\frac{\Delta T(\mathbf{n})}{T}\right)^{(V)} = V_j^{(b)}(t_{\mathrm{dec}}, \mathbf{x}_{\mathrm{dec}}) n^j + \int_i^f \dot{\sigma}_j(t, \mathbf{x}(t)) n^j \, d\lambda, \tag{2.243}$$

$$\left(\frac{\Delta T(\mathbf{n})}{T}\right)^{(T)} = -\int_i^f \dot{H}_{lj}(t, \mathbf{x}(t)) n^l n^j \, d\lambda. \tag{2.244}$$

Note that for models where initial fluctuations have been laid down in the very early universe, vector perturbations are irrelevant, as we have already pointed out. In this sense Eq. (2.243) is here mainly for completeness. However, in models where perturbations are sourced by some inherently inhomogeneous component [e.g., topological defects; see Durrer *et al.* (2002)] vector perturbations can be important.

2.6 Power Spectra

2.6.1 Generics

The quantities that we can determine from a given model are usually not the precise values of perturbation variables as $\Psi(\mathbf{k},t)$, but only expectation values like $\langle \Psi(\mathbf{k},t) \cdot \Psi^*(\mathbf{k}',t)\rangle$. In different realizations, for example, of the same inflationary model, the "phases" $\theta(\mathbf{k},t)$ given by $\Psi(\mathbf{k},t) = \exp(i\theta(\mathbf{k}))|\Psi(\mathbf{k},t)|$ are different. They are random variables. If we assume that the random process that generates the fluctuations Ψ is stochastically homogeneous and isotropic, these phases have a vanishing 2-point correlator for different values of $\mathbf{k}$ and $|\Psi|$ depends only on the modulus k.

The quantity that we can calculate for a given model and which then has to be compared with observations is the power spectrum, defined below. Power spectra are the "harmonic transforms" of the 2-point correlation functions.[2] If the perturbations of the model under consideration are Gaussian, a relatively generic prediction from inflationary models as we shall see, then the 2-point functions and therefore the power spectra contain all the statistical information of the model.

For inflationary models, the "randomness" is fully characterized by the initial conditions after inflation. Within linear perturbation theory, there exists a transfer function, $\mathcal{T}_X(k,t)$, that determines the solution at late time from the initial conditions. In the simplest models, for example, single-field inflation which is a good fit to present data, adiabatic scalar perturbations have only one degree of freedom, and within linear perturbation theory, every variable X is determined via a deterministic transfer function by the initial condition for the curvature perturbation ζ,

$$X(k,t) = \mathcal{T}_X(k,t)\zeta(k,t_{\text{in}}) \tag{2.245}$$

Naively one might think that two initial conditions are needed because the linear perturbation equations are of second order, but after inflation only the growing mode that satisfies ζ = constant on superhorizon scales is relevant. As $\mathcal{T}_X$ is deterministic, with ζ also X is a Gaussian random variable. This is no longer true when nonlinearities become relevant since products of Gaussian variables are not Gaussian.

[2] The "harmonic transform" in usual flat space is simply the Fourier transform. In curved space it is the expansion in terms of eigenfunctions of the Laplacian on that space, for example, on the sphere it corresponds to the expansion in terms of spherical harmonics.

Eq. (2.245) is easily generalized into a sum if several scalar modes are excited initially, for example, isocurvature modes; see Chapter 3. The situation is very different if perturbations are seeded by some inherently inhomogeneous and random component (such as, e.g., cosmic strings). Then the variables inherit the statistics of the typically non-Gaussian seeds and cannot be related in a simple way to initial conditions but need to be computed with a Green's function method; see Section 2.7.

There is one additional problem to consider: one can never "measure" expectation values. We have only one Universe, that is, one realization of the stochastic process that generates the fluctuations at our disposal for observations. The best we can do when we want to determine the mean square fluctuation on a given scale λ is to average over many disjoint patches of size λ, assuming that this spatial averaging corresponds to an ensemble averaging; a type of ergodic hypothesis. This works well as long as the scale λ is much smaller than the Hubble horizon, the size of the observable Universe. For $\lambda \sim \mathcal{O}(H_0^{-1})$ we can no longer average over many independent volumes and the value measured could be quite far from the ensemble average. This problem is known under the name "cosmic variance" and we shall come back to it in Chapter 9, where we shall quantify cosmic variance. More details about the formal aspects of power spectra can be found in Appendix 7.

For an arbitrary scalar variable X in position space, we define the power spectrum in Fourier space by

$$\left\langle X(\mathbf{k}, t_0)\, X^*(\mathbf{k}', t_0)\right\rangle = (2\pi)^3 \delta(\mathbf{k} - \mathbf{k}') P_X(k). \tag{2.246}$$

In flat space, $K = 0$, the function $X(\mathbf{k})$ is the ordinary Fourier transform of $X(\mathbf{x})$. If $K \neq 0$ the situation is more complicated. Then $X(\mathbf{k})$ represents an expansion of $X(\mathbf{x})$ in terms of eigenfunctions of the Laplacian and in the case $K > 0$ the Dirac δ-function has to be replaced by a discrete Kronecker δ.

The $\langle\ \rangle$ indicates a statistical average, ensemble average, over "random initial conditions" in a given model. We assume that no point in space is preferred, in other words that $X(\mathbf{x})$ and any other stochastic field that we consider has the same distribution in every point $\mathbf{x}$. Such random fields are called "statistically homogeneous" (or stationary). We further assume that the distribution of $X(\mathbf{x})$ has no preferred direction. This means that the random field X is statistically isotropic. These properties imply that the Fourier transform of the 2-point function is diagonal; that is, they explain the factor $\delta(\mathbf{k} - \mathbf{k}')$ in Eq. (2.246) (see Exercise 2.5).

A related and physically more intuitive quantity is the so-called dimensionless power spectrum defined by

$$\Delta_X(k) = \frac{1}{2\pi^2} k^3 P_X(k) = \frac{4\pi}{(2\pi)^3} k^3 P_X(k). \tag{2.247}$$

Contrary to P_X, this quantity has the same dimension as X. It also relates very simply to the correlation function. Suppressing the time variable we have

$$\xi(\mathbf{x},\mathbf{y}) = \langle X(\mathbf{x})X(\mathbf{y})\rangle = \int \frac{d^3k d^3k'}{(2\pi)^6} e^{i(\mathbf{k}\cdot\mathbf{x}-\mathbf{k}'\cdot\mathbf{y})} \left\langle X(\mathbf{k})\, X^*\left(\mathbf{k}'\right)\right\rangle$$
$$= \frac{1}{4\pi}\int \frac{d^3k}{k^3} e^{i(\mathbf{k}\cdot(\mathbf{x}-\mathbf{y}))}\Delta_X(k) = \frac{1}{2}\int_{-1}^{1} d\mu \int_0^\infty \frac{dk}{k} e^{i\mu kr}\Delta_X(k)$$
$$= \int_0^\infty \frac{dk}{k} j_0(kr)\Delta_X(k) = \xi(r). \qquad (2.248)$$

Here $r = |\mathbf{x}-\mathbf{y}|$ and having requested statistical homogeneity and isotropy for the power spectrum implies it therefore also for the correlation function (and vice versa). Here j_0 is the spherical Bessel function; see Appendix 4, Section A4.3. If Δ_X = constant, that is, is independent of k, we call the spectrum of X "scale invariant." In this case, the integral (2.248) is independent of r. However, it diverges logarithmically for $k \to 0$. This infrared divergence is not physical, since modes with $k < H_0$ are not observable. Therefore, once we compute directly observable quantities (like an angular correlation function), this divergence is no longer present.

2.6.2 The Matter Power Spectrum

Let us first consider the power spectrum of dark matter, $P_D(k)$, which is defined by

$$\left\langle D_{gm}(\mathbf{k},t_0)\, D^*_{gm}\left(\mathbf{k}',t_0\right)\right\rangle = P_D(k)(2\pi)^3\delta(\mathbf{k}-\mathbf{k}'). \qquad (2.249)$$

$P_D(k)$ is usually compared with the observed power spectrum of the galaxy distribution. This is clearly problematic, as it is by no means evident what the relation between these two spectra should be. This problem is known under the name of "bias" and it is very often simply assumed that the dark matter and galaxy power spectra differ only by a multiplicative factor. The hope is also that on sufficiently large scales, since the evolution of both galaxies and dark matter is governed by gravity, their power spectra should not differ much. This hope seems to be reasonably well justified. In Tegmark *et al.* (2004) it is found that the observed galaxy power spectrum and the matter power spectrum inferred from the observation of CMB anisotropies differ only by about 10% on very large scales. In Chapter 8, where we discuss the matter distribution and its fluctuations in more detail, we shall assume bias to be linear and scale independent, so that $P_g(k,z) = b^2(z)P_D(k,z)$.

The power spectrum of velocity perturbations satisfies the relation

$$\left\langle V_j(\mathbf{k},t_0)\, V_i^*\left(\mathbf{k}',t_0\right)\right\rangle = Q_j^{(S)}(\mathbf{k})Q_i^{(S)*}(\mathbf{k}')P_V(k)(2\pi)^3\delta(\mathbf{k}-\mathbf{k}'), \qquad (2.250)$$
$$P_V(k) \simeq H_0^2\Omega_m^{1.2}P_D(k)k^{-2}. \qquad (2.251)$$

For $\simeq$ we have used that $|kV(t_0)| = \dot{D}_g^{(m)}(t_0) \sim H_0\Omega_m^{0.6} D_g$ on subhorizon scales (see, e.g., Peebles, 1993); more precisely, one introduces the growth function,

$$f(z) = \mathcal{H}^{-1}(z)\dot{D}_g^{(m)}(z)/D_g^{(m)}(z).$$

In a matter-dominated universe $f(z) \simeq \Omega_m^{0.6}(z)$. This will be relevant for our discussion in Section 8.2

2.6.3 The CMB Power Spectrum

2.6.3.1 Definition

The spectrum that we are most interested in and that can be both measured and calculated to the best accuracy is the CMB anisotropy power spectrum. It is defined as follows: $\Delta T/T$ is a function of position $\mathbf{x}$, time t, and photon direction $\mathbf{n}$. Here, $\mathbf{x} = \mathbf{x}_0$ and now, $t = t_0$, $\Delta T/T$ is a function on the sphere, $\mathbf{n} \in \mathbb{S}^2$. We develop it in terms of spherical harmonics, $Y_{\ell m}$s. We will often suppress the arguments t_0 and $\mathbf{x}_0$ in the following calculations. Since our fields are statistically homogeneous, averages over an ensemble of realizations (expectation values) are independent of position. Furthermore, we assume that the process generating the initial perturbations is statistically isotropic. This means that the distribution of $\Delta T/T(\mathbf{n})$ is the same for all directions $\mathbf{n}$. As for the Fourier transforms of random fields in space, this implies that the harmonic transform of $\Delta T/T$ is diagonal. In other words, the off-diagonal correlators of the expansion coefficients $a_{\ell m}$ vanish and we have

$$\frac{\Delta T}{T}(\mathbf{x}_0, \mathbf{n}, t_0) = \sum_{\ell,m} a_{\ell m}(\mathbf{x}_0) Y_{\ell m}(\mathbf{n}), \quad \left\langle a_{\ell m} \cdot a^*_{\ell' m'}\right\rangle = \delta_{\ell\ell'}\delta_{mm'}C_\ell. \tag{2.252}$$

The C_ℓs are the CMB power spectrum.

The 2-point correlation function, $\mathcal{C}(\mu)$, $\mu = \mathbf{n}\cdot\mathbf{n}'$, is related to the C_ℓs by

$$\begin{aligned}
\mathcal{C}(\mu) &\equiv \left\langle \frac{\Delta T}{T}(\mathbf{n})\frac{\Delta T}{T}(\mathbf{n}')\right\rangle_{\mathbf{n}\cdot\mathbf{n}'=\mu} \\
&= \sum_{\ell,\ell',m,m'} \left\langle a_{\ell m} \cdot a^*_{\ell' m'}\right\rangle Y_{\ell m}(\mathbf{n})Y^*_{\ell' m'}(\mathbf{n}') \\
&= \sum_\ell C_\ell \underbrace{\sum_{m=-\ell}^{\ell} Y_{\ell m}(\mathbf{n})Y^*_{\ell m}(\mathbf{n}')}_{\frac{2\ell+1}{4\pi}P_\ell(\mathbf{n}\cdot\mathbf{n}')} \\
&= \frac{1}{4\pi}\sum_\ell (2\ell+1)C_\ell P_\ell(\mu),
\end{aligned} \tag{2.253}$$

where we have used the addition theorem of spherical harmonics for the last equality; the P_ℓs are the Legendre polynomials (see Appendix 4, Sections A4.2.3 and A4.1).

Clearly the a_{lm}s from scalar, vector, and tensor perturbations are uncorrelated,

$$\left\langle a^{(S)}_{\ell m} a^{(V)}_{\ell' m'} \right\rangle = \left\langle a^{(S)}_{\ell m} a^{(T)}_{\ell' m'} \right\rangle = \left\langle a^{(V)}_{\ell m} a^{(T)}_{\ell' m'} \right\rangle = 0. \tag{2.254}$$

Since vector perturbations decay, their contributions, the $C^{(V)}_\ell$, are negligible in models in which initial perturbations have been laid down very early, for example, after an inflationary period. Tensor perturbations are constant on superhorizon scales and perform damped oscillations once they enter the horizon.

2.6.3.2 Scalar Perturbations: The Sachs–Wolfe Term

Let us first discuss in somewhat more detail scalar perturbations. We specialize to the case $K = 0$ for simplicity. We suppose the initial perturbations to be given by a spectrum of the form

$$\langle \Psi(\mathbf{k}) \Psi^*(\mathbf{k}') \rangle k^3 = \frac{(2\pi)^6}{4\pi} \Delta_\Psi(k) \delta(\mathbf{k} - \mathbf{k}') = \frac{(2\pi)^6}{4\pi} A_\Psi (k t_0)^{n_s - 1} \delta(\mathbf{k} - \mathbf{k}'). \tag{2.255}$$

We multiply by the constant $t_0^{n_s-1}$, the actual comoving size of the horizon, in order to keep A_Ψ dimensionless for all values of n_s. The number n_s is called the scalar spectral index. A_Ψ then represents the amplitude of metric perturbations at horizon scale today, $k = 1/t_0$. More generally one sets

$$\Delta_\Psi(k) = A_\Psi \left(\frac{k}{k_*} \right)^{n_s - 1}, \tag{2.256}$$

with an arbitrary "pivot scale" k_*. If $n_s \neq 1$, the amplitude A_Ψ of course depends on the chosen pivot scale. In the scale invariant case, $n_s = 1$, the amplitude A_Ψ is independent of the pivot scale.

As we have seen in the previous section, the dominant contribution on *superhorizon scales* (neglecting the integrated Sachs–Wolfe effect $\int \dot\Phi + \dot\Psi$) is the ordinary Sachs–Wolfe effect, OSW, which for adiabatic perturbations is given by

$$\frac{\Delta T}{T}(\mathbf{x}_0, \mathbf{n}, t_0) \simeq \frac{1}{3} \Psi(x_{\text{dec}}, t_{\text{dec}}). \tag{2.257}$$

Since $\mathbf{x}_{\text{dec}} = \mathbf{x}_0 + \mathbf{n}(t_0 - t_{\text{dec}})$, the Fourier transform of (2.257) gives

$$\frac{\Delta T}{T}(\mathbf{k}, \mathbf{n}, t_0) = \frac{1}{3} \Psi(\mathbf{k}, t_{\text{dec}}) \cdot e^{i\mathbf{k}\mathbf{n}(t_0 - t_{\text{dec}})}. \tag{2.258}$$

Using the decomposition (see Appendix 4, Section A4.3)

$$e^{i\mathbf{kn}(t_0-t_{\rm dec})} = \sum_{\ell=0}^{\infty}(2\ell+1)i^\ell j_\ell(k(t_0-t_{\rm dec}))P_\ell(\hat{\mathbf{k}}\cdot\mathbf{n}), \tag{2.259}$$

where j_ℓ are the spherical Bessel functions, we obtain ($k = |\mathbf{k}|, \hat{\mathbf{k}} = \mathbf{k}/k$)

$$\begin{aligned}
&\left\langle \frac{\Delta T}{T}(\mathbf{x}_0,\mathbf{n},t_0)\frac{\Delta T}{T}(\mathbf{x}_0,\mathbf{n}',t_0)\right\rangle \qquad (2.260)\\
&= \frac{1}{(2\pi)^6}\int d^3k\, d^3k'\, e^{i\mathbf{x}_0\cdot(\mathbf{k}-\mathbf{k}')}\left\langle \frac{\Delta T}{T}(\mathbf{k},\mathbf{n},t_0)\left(\frac{\Delta T}{T}\right)^*(\mathbf{k}',\mathbf{n}',t_0)\right\rangle\\
&\simeq \frac{1}{(2\pi)^6 9}\int d^3kd^3k' e^{i\mathbf{x}_0\cdot(\mathbf{k}-\mathbf{k}')}\left\langle \Psi(\mathbf{k})\Psi^*(\mathbf{k}')\right\rangle \sum_{\ell,\ell'=0}^{\infty}(2\ell+1)(2\ell'+1)i^{\ell-\ell'}\\
&\cdot j_\ell(k(t_0-t_{\rm dec}))j_{\ell'}(k'(t_0-t_{\rm dec}))P_\ell(\hat{\mathbf{k}}\cdot\mathbf{n})\cdot P_{\ell'}(\hat{\mathbf{k}}'\cdot\mathbf{n}')\\
&= \frac{1}{(2\pi)^3 9}\int d^3k P_\Psi(k)\sum_{\ell,\ell'=0}^{\infty}(2\ell+1)(2\ell'+1)i^{\ell-\ell'}\\
&\times\ j_\ell(k(t_0-t_{\rm dec}))j_{\ell'}(k(t_0-t_{\rm dec}))P_\ell(\hat{\mathbf{k}}\cdot\mathbf{n})\cdot P_{\ell'}(\hat{\mathbf{k}}\cdot\mathbf{n}'). \qquad (2.261)
\end{aligned}$$

In the first equals sign we have used the unitarity of the Fourier transformation. Inserting $P_\ell(\hat{\mathbf{k}}\mathbf{n}) = \frac{4\pi}{2\ell+1}\sum_m Y^*_{\ell m}(\hat{\mathbf{k}})Y_{\ell m}(\mathbf{n})$ and $P_{\ell'}(\hat{\mathbf{k}}\mathbf{n}') = \frac{4\pi}{2\ell'+1}\sum_{m'} Y^*_{\ell' m'}(\hat{\mathbf{k}})$ $Y_{\ell' m'}(\mathbf{n}')$, integration over the directions $d\Omega_{\hat{k}}$ gives $\delta_{\ell\ell'}\delta_{mm'}\sum_m Y^*_{\ell m}(\mathbf{n})Y_{\ell m}(\mathbf{n}')$. Using also $\sum_m Y^*_{\ell m}(\mathbf{n})Y_{\ell m}(\mathbf{n}') = \frac{2\ell+1}{4\pi}P_\ell(\mu)$, where $\mu = \mathbf{n}\cdot\mathbf{n}'$, we find

$$\begin{aligned}
&\left\langle \frac{\Delta T}{T}(\mathbf{x}_0,\mathbf{n},t_0)\frac{\Delta T}{T}(\mathbf{x}_0,\mathbf{n}',t_0)\right\rangle_{\mathbf{nn}'=\mu}\\
&\simeq \sum_\ell \frac{2\ell+1}{4\pi}P_\ell(\mu)\frac{2}{\pi}\int\frac{dk}{k}\frac{1}{9}P_\Psi(k)k^3 j_\ell^2(k(t_0-t_{\rm dec})). \qquad (2.262)
\end{aligned}$$

Comparing this equation with Eq. (2.253) we obtain for *adiabatic perturbations* on scales $2 \le \ell \ll \chi(t_0-t_{\rm dec})/t_{\rm dec} \sim 100$:

$$C_\ell^{(\rm SW)} \simeq C_\ell^{(\rm OSW)} \simeq \frac{2}{9\pi}\int_0^\infty \frac{dk}{k}P_\Psi(k)k^3 j_\ell^2\left(k\left(t_0-t_{\rm dec}\right)\right). \tag{2.263}$$

The function $j_\ell^2\ (k(t_0-t_{\rm dec}))$ peaks roughly at $k\ (t_0-t_{\rm dec}) \simeq kt_0 \simeq \ell$. If Ψ is a pure power law on large scales, $kt_{\rm dec} \lesssim 1$ as in Eq. (2.255), and we set $k(t_0-t_{\rm dec}) \sim kt_0$, the integral (2.263) can be performed analytically. For the ansatz (2.255), using the integral (A4.150) one finds

$$C_\ell^{(\rm SW)} = \frac{2\pi^2 A_\Psi}{9}\frac{\Gamma(3-n_s)\Gamma(\ell-\frac{1}{2}+\frac{n_s}{2})}{2^{3-n_s}\Gamma^2(2-\frac{n_s}{2})\Gamma(\ell+\frac{5}{2}-\frac{n_s}{2})}\quad \text{for } -3 < n_s < 3\,. \tag{2.264}$$

Of special interest is the *scale-invariant* or Harrison–Zel'dovich (HZ) spectrum, $n_s = 1$. We shall see in Chapter 3 that inflationary initial conditions naturally generate a nearly scale invariant spectrum of scalar fluctuations. An HZ spectrum leads to

$$\ell(\ell+1)C_\ell^{(\mathrm{SW})} = \frac{2\pi A_\Psi}{9} \simeq \left\langle \left(\frac{\Delta T}{T}(\vartheta_\ell)\right)^2 \right\rangle, \quad \vartheta_\ell \equiv \pi/\ell. \tag{2.265}$$

This is precisely (within the accuracy of the experiment) the behavior observed by the DMR (differential microwave radiometer) experiment aboard the satellite COBE (Smoot *et al.*, 1992) and much more precisely with the Planck satellite experiment (Planck Coll. VI, 2018), which measured a scalar spectral index $n_s = 0.9652 \pm 0.0042$ (see Table 9.1).

As we shall see in Chapter 3, inflationary models predict very generically an HZ spectrum (up to small corrections). The DMR discovery has therefore been regarded as a great success, if not a proof, of inflation. There are, however, other models such as topological defects (see Section 2.7, or for more details Durrer *et al.*, 2002), or certain string cosmology models (Durrer *et al.*, 1999) that also predict scale-invariant, that is, Harrison–Zel'dovich spectra of fluctuations. These generically lead to isocurvature perturbations which are severely constrained by present data. In the case of string cosmology, the isocurvature perturbations can be transformed into adiabatic ones during reheating; see Enqvist and Sloth (2002). After that, they can be distinguished from standard inflationary models by their significant non-Gaussianity and by the absence of tensor modes. Models with topological defects are outside the class investigated here, since their perturbations are induced by "seeds" that evolve nonlinearly in time. They are not simply laid down as initial conditions for the fluid perturbations but typically affect the perturbations of a given wavelength until it crosses the Hubble scale. We investigate such models only briefly in Section 2.7.

2.6.3.3 Scalar Perturbations: The Integrated Sachs–Wolfe Term

For isocurvature perturbations, the main contribution on large scales comes from the integrated Sachs–Wolfe effect (ISW) and (2.263) is replaced by

$$C_\ell^{(\mathrm{ISW})} \simeq \frac{8}{\pi}\int \frac{dk}{k} k^3 \left\langle \left| \int_{t_{\mathrm{dec}}}^{t_0} \dot{\Psi}(k,t)\, j_\ell(k(t_0-t))\, dt \right|^2 \right\rangle. \tag{2.266}$$

Inside the horizon Ψ is roughly constant (matter dominated). Using the ansatz (2.255) for Ψ inside the horizon and setting the integral in (2.266) $\sim 2\Psi(k,t = 1/k)j_\ell^2(kt_0)$, we obtain again (2.264), but with $A_S/9$ replaced by $4A_S$. For a fixed

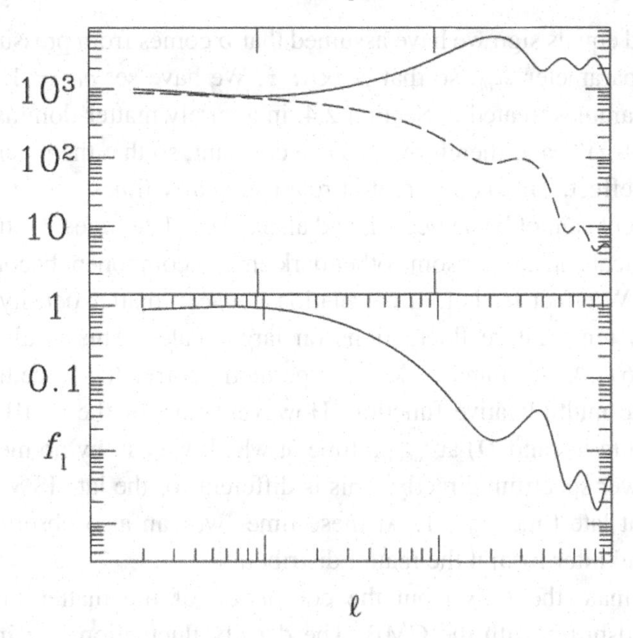

Fig. 2.3 Examples of COBE normalized adiabatic (solid line) and isocurvature (dashed line) CMB anisotropy spectra, $\ell(\ell+1)C_\ell/(2\pi)$ in units of $(\mu K)^2$ are shown on the top panel. In the bottom panel the ratio of the isocurvature to adiabatic temperature fluctuations is plotted.

amplitude A_S of perturbations, the Sachs–Wolfe temperature anisotropies coming from isocurvature perturbations are therefore about six times larger than those coming from adiabatic perturbations (see Fig. 2.3).

But also adiabatic perturbations from inflation generate an integrated Sachs–Wolfe term if the Bardeen potentials are not constant in time. This is the case, for example, at early times, right after decoupling, when the Universe is not yet fully matter dominated but also at late times, when a cosmological constant or some other form of dark energy becomes relevant. The early integrated Sachs–Wolfe term is well measured (see, e.g., Cabass *et al.*, 2015) and contributes to the first acoustic peak. The late ISW effect is relevant only on very large scales. A promising way to measure it is to correlate CMB fluctuations with density fluctuations at late times. This allows us to isolate it from the much larger ordinary SW term that originates from the last scattering surface and is not correlated with density fluctuations in the late Universe. Let us calculate the expected signal.

The Bardeen potential is determined by the Einstein equation (2.105). Neglecting curvature this is

$$\Psi = -\frac{4\pi G}{k^2}a^2\rho D = -\frac{3H_0^2\Omega_m(1+z)}{2k^2}D. \tag{2.267}$$

For the second equals sign we have assumed that ρ comes from pressureless matter with density parameter Ω_m, so that $\rho \propto a^{-3}$. We have set $a_0 = 1$. As we have seen in the examples treated in Section 2.4, in a purely matter-dominated universe $D \propto a = (1+z)^{-1}$ and therefore $a^2\rho D =$ constant, so that there is no integrated Sachs–Wolfe effect. This is different at relatively early times, $t \sim t_{\rm dec}$, where the radiation content cannot be neglected, and also at very late times if either curvature or a cosmological constant or some other dark energy component becomes relevant.

The late ISW effect leads to a correlation between matter density fluctuations and the CMB temperature fluctuations on large scales. This is already evident from Eq. (2.267); $\Psi(\mathbf{k},t)$ and $D(\mathbf{k},t)$ are perfectly correlated since they differ by a deterministic multiplicative function. However, most of the CMB anisotropies actually measure Ψ (and D) at $t_{\rm dec}$, a time at which we can by no means measure the matter power spectrum directly. This is different for the late ISW effect which measures Ψ at late times, $z \lesssim 1$. At these times we can also observe the galaxy distribution and infer from it the matter distribution.

Let us estimate the C_ℓ's from the correlation of the matter fluctuations at some fixed redshift z with the CMB. The density fluctuation at z in a direction $\mathbf{n}$ from us is

$$D(\mathbf{x}_0, \mathbf{n}, z) = D(x_0 - \mathbf{n}(t(z) - t_0), t(z)).$$

Here $\mathbf{x}_0$ is our position, t_0 denotes today and $t(z)$ is the conformal time at redshift z. Expressed in terms of the Fourier transform $D(\mathbf{k}, t(z))$ we obtain

$$\begin{aligned} D(\mathbf{x}_0, \mathbf{n}, z) &= \frac{1}{(2\pi)^3}\int d^3k\, e^{-i\mathbf{k}\cdot(\mathbf{x}_0 - \mathbf{n}(t(z)-t_0))} D(\mathbf{k}, t(z)) \\ &= -\frac{2}{3H_0^2\Omega_m(1+z)(2\pi)^3}\int d^3k\, k^2 e^{-i\mathbf{k}\cdot(\mathbf{x}_0 - \mathbf{n}(t(z)-t_0))}\Psi(\mathbf{k}, t(z)). \end{aligned} \tag{2.268}$$

We want to correlate these density fluctuations with the ISW effect,

$$\left\langle D(\mathbf{x}_0, \mathbf{n}', z)\left(\frac{\Delta T}{T}\right)_{\rm ISW}(\mathbf{x}_0, \mathbf{n})\right\rangle = \frac{1}{4\pi}\sum_\ell (2\ell+1) C_\ell^{(\rm xISW)}(z) P_\ell(\mathbf{n}\cdot\mathbf{n}'). \tag{2.269}$$

For the second term we insert the Fourier representation of the integrated term on the right-hand side of Eq. (2.240), setting $\Psi = \Phi$,

$$\left(\frac{\Delta T}{T}\right)_{\rm ISW}(\mathbf{x}_0, \mathbf{n}) = \frac{2}{(2\pi)^3}\int d^3k \int_{t_{\rm dec}}^{t_0} dt\, e^{-i\mathbf{k}\cdot(\mathbf{x}_0 - \mathbf{n}(t-t_0))}\dot{\Psi}(\mathbf{k}, t).$$

We now set

$$\Psi(\mathbf{k}, t) = g(t, k)\Psi_{\rm in}(\mathbf{k}). \tag{2.270}$$

We shall consider matter perturbations only and $t_{\rm in} \gtrsim t_{\rm dec}$. In this case the evolution of Ψ no longer depends on k; see Eq. (2.140) with $w = c_s^2 = P = 0$, and we can approximate $g(t,k) = g(t)$. Later we shall also introduce a growth function f defined by $f = d\log D/d\log a$. One easily checks using Eq. (2.105), Eqs. (2.118), and (2.119) that in a universe where perturbations are due to pressureless matter only, these growth functions are related by

$$\frac{\dot{g}(t)}{g(t)} = \mathcal{H}(t)\,(f(t) - 1)\,. \tag{2.271}$$

In a pure matter universe (where not only the perturbations but also the background is dominated by pressureless matter only) we have $f \equiv g \equiv 1$. Therefore during matter domination $\dot{g} = 0$. Once dark energy becomes relevant, perturbations grow slower than the scale factor and the Bardeen potential starts decaying, $\dot{g} < 0$.

In Fig. 2.4 we plot $g(z)$ and $\dot{g}(z)$ for a universe with $(\Omega_\Lambda, \Omega_m) = (0.7, 0.3)$.

We make use of $\dot{\Psi} = \dot{g}\Psi_{\rm in}$ and $\langle \Psi_{\rm in}(\mathbf{k})\Psi^*_{\rm in}(\mathbf{k}')\rangle = (2\pi)^3\delta(\mathbf{k}-\mathbf{k}')P_\Psi(k) = (2\pi)^6\delta(\mathbf{k}-\mathbf{k}')A_\Psi(k/H_0)^{n_s-1}k^{-3}/4\pi$; see Eqs. (2.255) and (2.256), where we have chosen the pivot scale H_0. Furthermore, we rewrite $e^{i\mathbf{k}\mathbf{n}(t_0-t)}$ in terms of spherical

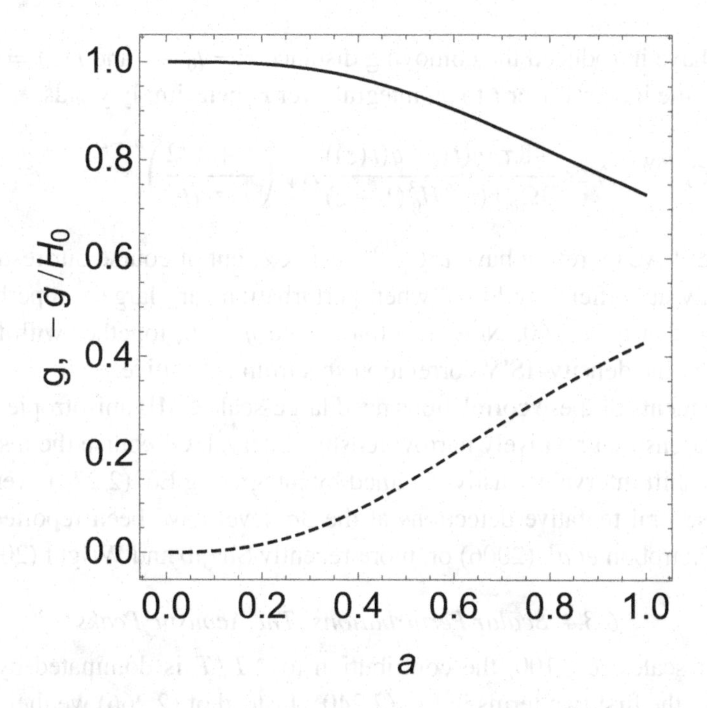

Fig. 2.4 The growth function g (solid) and its derivative, $-\dot{g}/H_0$ (dashed) are shown as functions of the scale factor $a = (1+z)^{-1}$ in a universe with $\Omega_\Lambda = 0.7$ and $\Omega_m = 0.3$. Note that only for $z \lesssim 2$, $a \gtrsim 0.3$, the growth function g starts to deviate significantly from 1.

Bessel functions and Legendre polynomials [see Eq. (A4.146)]. Applying the addition theorem for spherical harmonics, Eq. (A4.45), the integration over directions of **k** yields

$$C_\ell^{(\mathrm{xISW})}(z) = \frac{-8g(t(z))}{3\pi\Omega_m H_0^2(1+z)} \int dk\, k^4 P_\Psi(k) j_\ell(k(t_0 - t(z))) \times \int_{t_{\mathrm{in}}}^{t_0} dt\, \dot{g}(t) j_\ell(k(t_0 - t)). \tag{2.272}$$

Equation (2.272) is still exact. For a given initial power spectrum $k^3 P_\Psi(k) \propto (k/H_0)^{n_s-1}$, the transfer function $g(t)$ is determined by the cosmological parameters Ω_m and Ω_Λ. If $\Omega_m = 1$ and $\Omega_\Lambda = 0$, $g = $ constant and the effect vanishes.

To perform the integration over k we now use Limber's approximation (A4.152), which is not very accurate at low ℓ but provides the right order of magnitude. It yields

$$\int dk\, k^4 P_\Psi(k) j_\ell(k(t_0 - t(z))) j_\ell(k(t_0 - t)) \simeq \frac{\pi}{2} \frac{\delta(r - r(z))}{r^2} \left(k^2 P_\Psi(k)\right)\Big|_{k=(\ell+1/2)/r}, \tag{2.273}$$

where we have introduced the comoving distance $r = t_0 - t$ and $r(z) = t_0 - t(z)$. Converting the integral over t to an integral over r then simply yields

$$C_\ell^{(\mathrm{xISW})}(z) \simeq \frac{-8\pi^2 g(t(z))\dot{g}(t(z))}{3\Omega_m r(z)^2 H_0^3(1+z)} A_\Psi \left(\frac{\ell + 1/2}{r(z)H_0}\right)^{n_s-2}. \tag{2.274}$$

For $n_s \simeq 1$ we therefore have $\ell^2 C_\ell^{(\mathrm{xISW})}(z) \propto \ell$, but of course our result (2.274) is valid only at sufficiently low ℓ when perturbations are largely superhorizon at decoupling, that is, $\ell \lesssim 60$. Note also that, since $\dot{g} < 0$, together with the minus sign in front, the density–ISW correlation spectrum is positive.

Measurements of these correlations need large-scale CMB anisotropies and density fluctuations in a relatively narrow redshift interval. Of course the result over a broader redshift interval is easily obtained by integrating Eq. (2.274) over redshift.

So far, several tentative detections at the 3σ level have been reported; see for example, Pietrobon *et al.* (2006) or, more recently Shajib and Wright (2016).

2.6.3.4 Scalar Perturbations: The Acoustic Peaks

On smaller scales, $\ell \gtrsim 100$, the contribution to $\Delta T/T$ is dominated by acoustic oscillations, the first two terms in Eq. (2.240). Instead of (2.266) we then obtain

$$C_\ell^{(AC)} \simeq \frac{2}{\pi} \int_0^\infty \frac{dk}{k} k^3 \left\langle \left| \frac{1}{4} D_\gamma(k, t_{\mathrm{dec}}) j_\ell(kt_0) + V^{(b)}(k, t_{\mathrm{dec}}) j'_\ell(kt_0) \right|^2 \right\rangle. \tag{2.275}$$

To remove the SW contribution from $D_g^{(r)}$ we have simply replaced it by D_γ, which is much smaller than Ψ on superhorizon scales and therefore does not contribute to the SW terms. On subhorizon scales $D_\gamma \simeq D_g^{(\gamma)}$ and V_γ are oscillating like sine or cosine waves depending on the initial conditions. Correspondingly the $C_\ell^{(AC)}$ will show peaks and minima. For adiabatic initial conditions $D_g^{(\gamma)}$ and therefore D_γ also oscillates like a cosine. Its minima and maxima are at $k_n t_{\rm dec}/\sqrt{3} = n\pi$. Odd values of n correspond to maxima, "contraction peaks," while even numbers are minima, "expansion peaks."

These are the "acoustic peaks" of the CMB anisotropies. Sometimes they are misleadingly called "Doppler peaks" referring to an old misconception that the peaks are due to the velocity term in Eq. (2.275). Actually the contrary is true. At maxima and minima of the density contrast, the velocity (being proportional to the derivative of the density) nearly vanishes. We shall therefore consistently call the CMB peak structure "acoustic peaks."

The angle θ_n, which subtends the scale $\lambda_n = \pi/k_n$ at the last scattering surface, is determined by the angular diameter distance to the last scattering surface, $d_A(t_{\rm dec})$, via the relation $\theta_n = \lambda_n/d_A(t_{\rm dec})$. Expanding the temperature anisotropies in spherical harmonics, the angular scale θ_n corresponds (roughly) to the harmonic number

$$\ell_n \simeq \pi/\theta_n = \pi d_A(t_{\rm dec})/\lambda_n = d_A(t_{\rm dec})k_n = n\sqrt{3}\pi d_A(t_{\rm dec})/t_{\rm dec}. \qquad (2.276)$$

For a flat, matter-dominated universe $d_A(t_{\rm dec}) \simeq t_0$ leading to $\ell_n \simeq 180n$ (see Ex. 2.8). This crude approximation deviates by about 15% from the precise numerical value, which depends, with d_A, strongly not only on curvature but also on the Hubble parameter and on the cosmological constant. Furthermore, the peak positions depend on the sound speed of the radiation–baryon plasma, which we have simply set to $c_s = 1/\sqrt{3}$ in this approximation. We shall discuss this parameter dependence of the peak positions in detail in Chapter 9. Note, however, that the position of the first peak differs significantly for the isocurvature mode, for which $D_g^{(r)}$ oscillates like a sine. For generic initial conditions, we would expect a mixture of the sine and cosine modes that leads to a displacement of the first peak. The observed CMB anisotropies are consistent with a purely adiabatic mode and require, at least, that the adiabatic mode dominates (Bucher *et al.*, 2001; Trotta, 2006).

For a flat universe, $\Omega = 1$, the nth peak therefore is placed at

$$\ell_n \simeq k_n t_0 \cong n\pi\sqrt{3}\frac{t_0}{t_{\rm dec}}. \qquad (2.277)$$

For a flat matter dominated universe we have $\frac{t_0}{t_{\rm dec}} \sim \sqrt{z_{\rm dec}} \sim 33.2$, which yields $\ell_1 \sim 180$. Here we have used $z_{\rm dec} \sim 1100$ (see Section 1.3). This approximation is not very good, since the Universe is not very well matter dominated at $t_{\rm dec}$. A somewhat more accurate estimate (Exercise 2.8) gives $\ell_1 \sim 220$, in good agreement with the numerical value. Subsequent peaks are then given by $\ell_n = n\ell_1$.

Our discussion is valid only in flat space. In curved space the exponentials $\exp(ik(t_0 - t_{\rm dec}))$ have to be replaced with the harmonics of the curved spaces. For the positions of the peaks, this corresponds to replacing $k_n t_0$ by $k_n \chi(t_0)$, hence replacing t_0 by the comoving angular diameter distance to the last scattering surface. Instead of Eq. (2.277) we then obtain the following approximate relation for the peak positions:

$$\ell_n \sim n\pi\sqrt{3}\frac{\chi(t_0)}{t_{\rm dec}}. \tag{2.278}$$

For values of Ω close to unity this scales like $1/\sqrt{\Omega}$ (see Section 1.2).

On very small scales the acoustic peaks are damped by the photon diffusion that takes place during the recombination process. This effect will be discussed with the Boltzmann equation approach in Chapter 4.

2.6.3.5 *Tensor Perturbations*

For gravitational waves (which are tensor fluctuations), a formula analogous to (2.264) can be derived (see Appendix 8),

$$C_\ell^{(T)} = \frac{2}{\pi}\int dk\, k^2 \left|\int_{t_{\rm dec}}^{t_0} dt\, \dot{H}(t,k)\frac{j_\ell(k(t_0-t))}{(k(t_0-t))^2}\right|^2 \frac{(\ell+2)!}{(\ell-2)!}. \tag{2.279}$$

To a very crude approximation we may assume $\dot{H}^{(T)} = 0$ on superhorizon scales and $\int dt\, \dot{H}^{(T)} j_\ell(k(t_0-t)) \sim H^{(T)}(t = 1/k) j_\ell(kt_0)$. For a pure power law,

$$k^3\Delta_H(k) = A_t(kt_0)^{n_t}, \tag{2.280}$$

one obtains

$$\begin{aligned} C_\ell^{(T)} &\simeq 4\pi\frac{(\ell+2)!}{(\ell-2)!}A_t\int\frac{dx}{x}x^{n_t}\frac{j_\ell^2(x)}{x^4} \\ &= \frac{(\ell+2)!}{(\ell-2)!}A_t\frac{\pi^2\Gamma(6-n_t)\Gamma(\ell-2+\frac{n_t}{2})}{2^{5-n_t}\Gamma^2(\frac{7-n_t}{2})\Gamma(\ell+4-\frac{n_t}{2})}. \end{aligned} \tag{2.281}$$

For a scale-invariant spectrum ($n_t = 0$) this results in

$$\ell(\ell+1)C_\ell^{(T)} \simeq \frac{16\pi}{15}\frac{\ell(\ell+1)}{(\ell+3)(\ell-2)}A_t\,. \tag{2.282}$$

The singularity at $\ell = 2$ in this crude approximation is not real, but there is some enhancement of $\ell(\ell+1)C_\ell^{(T)}$ at $\ell \sim 2$ (see Fig. 2.5).

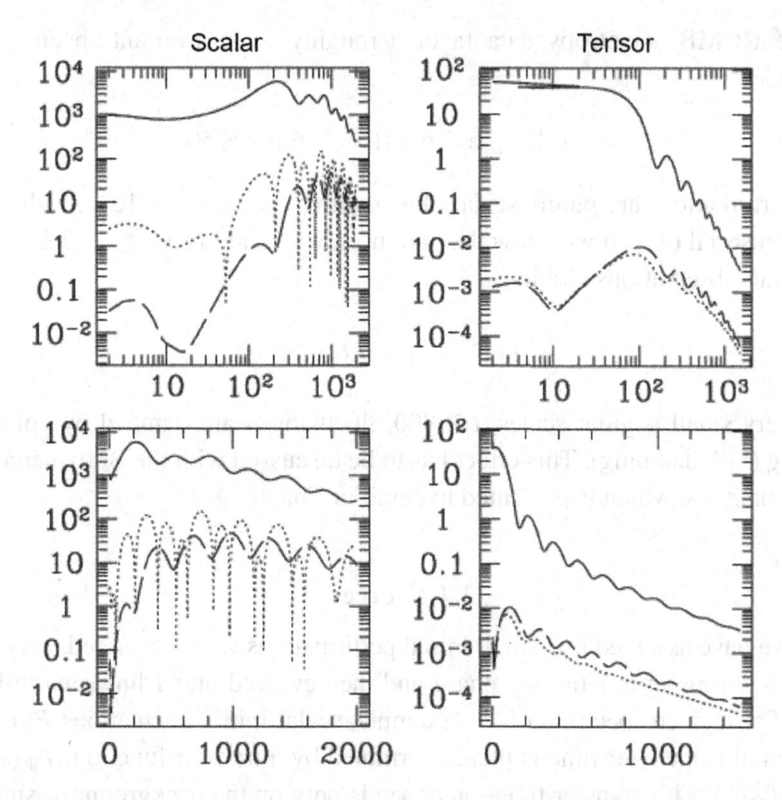

Fig. 2.5 Adiabatic scalar and tensor CMB anisotropy spectra are plotted, $\ell(\ell + 1)C_\ell/(2\pi)$ in units of $(\mu K)^2$ as functions of ℓ in log-scale (top panels), where the Sachs–Wolfe plateau is clearly visible and in linear scale (bottom panels), which shows the equal spacing of the acoustic peaks. The solid line shows the temperature spectrum, the dashed line is the polarization, and the dotted line shows the temperature–polarization cross correlation. We shall discuss the polarization of CMB radiation in Chapter 5. The temperature–polarization cross correlation can become negative, and the deep spikes in the dotted curves in the left-hand panels are actually sign changes (we show $|C_\ell^{TP}|$ in this log-plot). The left-hand side shows scalar fluctuation spectra, while the right-hand side shows tensor spectra. The observational data are well fitted by a purely scalar spectrum. A comparison of data and a model scalar spectrum are shown in Figs. 9.5 and 9.6.

Since tensor perturbations decay on subhorizon scales, $\ell \gtrsim 60$, they are not very sensitive to cosmological parameters.

Again, inflationary models (and topological defects) predict a scale-invariant spectrum of tensor fluctuations ($n_t \sim 0$).

Comparing the tensor and scalar result for scale-invariant perturbations we obtain for large scales, $\ell < 50$,

$$\frac{C_\ell^{(T)}}{C_\ell^{(S)}} \simeq \frac{72}{15}\frac{A_t}{A_\Psi} \equiv r. \tag{2.283}$$

Present CMB anisotropy data favor a roughly scale-invariant spectrum with amplitude

$$\ell(\ell+1)C_\ell \simeq 7.6 \times 10^{-10} \quad \text{for } \ell \lesssim 50.$$

If the perturbations are purely scalar, this requires $A_\Psi \simeq 1.1 \times 10^{-9}$; if they were purely tensorial (which we know they are not), we would need $A_t \simeq 2.2 \times 10^{-10}$. In general, observations require

$$\frac{2\pi}{9} A_\Psi (1+r) \simeq 7.6 \times 10^{-10}. \tag{2.284}$$

On very small angular scales, $\ell \gtrsim 800$, fluctuations are damped by collisional damping (Silk damping). This effect has to be discussed with the Boltzmann equation for photons, which is presented in detail in Chapter 4.

2.7 Sources

So far we have assumed that small initial perturbations were generated early in the Universe during an inflationary phase and then evolved under linear perturbation theory. For a given spectrum of, for example, scalar initial fluctuations $P_\Psi(k)$, the spectrum at some later time is then determined by a transfer function, $P_\Psi(k,t) = g^2(k,t)P_\Psi(k)$. This transfer function depends only on the background cosmology, that is, on the cosmological parameters.

There is, however, yet another possibility: an intrinsically inhomogeneous and anisotropic matter distribution, which makes up only a small perturbation, and which interacts with the cosmological matter and radiation only gravitationally. We consider the energy–momentum tensor of this component as a first-order perturbation. Within linear perturbation theory, it then evolves with the equations of motion determined by the background geometry.

Such a component is termed a **source** or "seed." The source's energy–momentum tensor seeds first-order perturbations in the geometry, which in turn affect the evolution of matter and radiation, generating fluctuations in the matter density and in the CMB.

2.7.1 Topological Defects

Topological defects that can form during symmetry breaking phase transitions are physically well-motivated seeds. If the vacuum manifold (i.e., the manifold of minima of the Higgs field or order parameter) that is responsible for the symmetry breaking is topologically nontrivial, regions where the field cannot relax to the minimum generically occur. The simplest examples are cosmic strings that form,

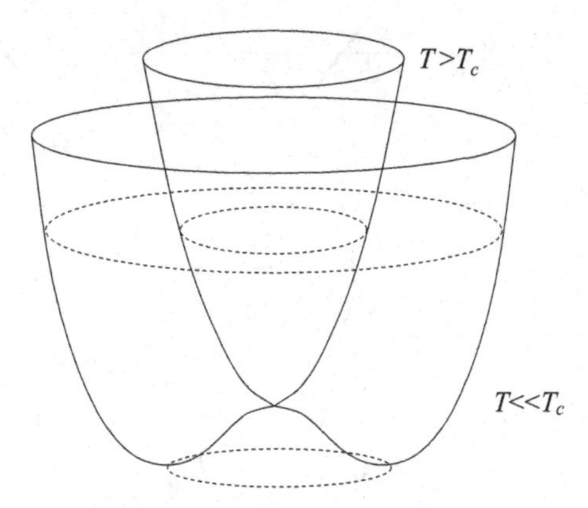

Fig. 2.6 The effective potential of a complex Higgs field for two values of the temperature, $T > T_c$ and $T < T_c$, is shown. The circle at the bottom is the vacuum manifold $\mathcal{S}$ of the low-temperature phase.

for example, when a $U(1)$ symmetry is broken. Below a critical temperature T_c, the temperature-dependent effective potential $V(\phi, T)$ of the complex Higgs field ϕ changes from a form with a single minimum at $\phi = 0$ to a Mexican hat shape with an entire circle $\mathcal{S}$ of minima; see Fig. 2.6.

When the temperature drops below T_c, the field at a given position $\mathbf{x}$ assumes some value in the new vacuum manifold $\mathcal{S}$. The field values at positions that are further apart than the Hubble horizon are uncorrelated. Therefore the configuration $\phi(s) = \phi(\mathbf{x}(s))$ along some large closed curve $\mathbf{x}(s)$ in a plane of physical space may well make one (or several) full turns in $\mathcal{S}$. If this happens, in order to remain continuous, ϕ has to leave the vacuum manifold and assume a value with higher potential energy somewhere in the interior of this curve. Continuing this argument in the third dimension, one obtains a line of higher energy. These lines, which are either closed or infinite, are cosmic strings; see Fig. 2.7.

As the Universe expands, the Higgs field straightens out. Strings that intersect exchange partners and can thereby chop off loops from the network of long strings. In this way the long string network loses energy by shortening the total length of strings. The strings from a broken gauge symmetry interact with other matter components only gravitationally. They shed energy only into a background of gravitational waves that they produce. This process is slow but sufficiently effective to lead to a mean energy density in cosmic strings that scales like the background energy density $\rho_S \propto 1/\tau^2$. If $M \simeq T_c$ is the energy scale of the phase transition, we expect $\rho_S \simeq M^2/\tau^2$ so that

$$\frac{\rho_S}{\rho} \simeq 4\pi G M^2 = 4\pi \left(\frac{M}{m_P}\right)^2 \equiv \epsilon. \tag{2.285}$$

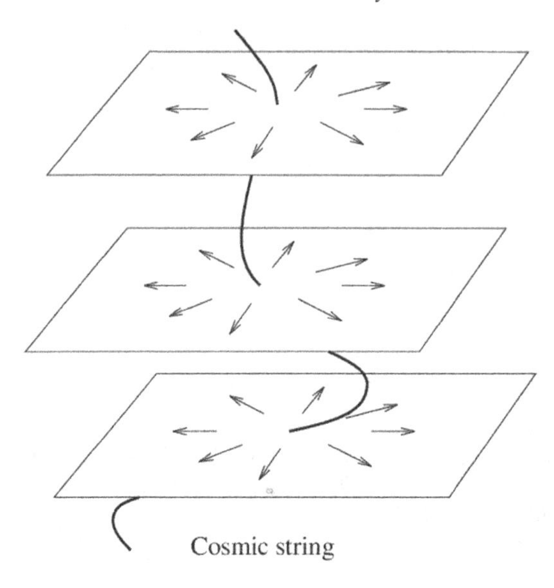

Fig. 2.7 A cosmic string in space is shown with the corresponding configuration of the complex Higgs field, indicated as arrows.

The amplitude of the induced perturbations will be of the order of ϵ. Recalling that the gravitational potential responsible for the CMB anisotropies is roughly 10^{-5}, we infer that the symmetry breaking scale cannot be much smaller than $M \sim 10^{-3} m_P \sim 10^{16}$ GeV, if such a component is to play a role for CMB anisotropies. Interestingly, this is a grand unified (GUT) scale where some drastic changes of physical interactions, for example, a phase transition, are expected to occur from the running of the coupling constants of gauge interactions. If cosmic strings were generated, for example, at the electroweak transition (which is not the case in the standard model), they would have far too low energy to play a role for structure formation or CMB anisotropies.

Another type of topological defects, called monopoles, occur at symmetry breaking phase transitions if the vacuum manifold of the broken phase, $\mathcal{S}$, has the topology of a sphere. More generically, monopoles form if the second homotopy group, $\pi_2(\mathcal{S})$ is nontrivial. Monopoles are points of higher potential energy. If the broken symmetry is gauged, such massive monopoles cease to interact soon after the phase transition. Their energy density then scales like ordinary matter $\rho \propto a^{-3}$ and soon dominates over the radiation density of the Universe. Therefore, local monopoles are ruled out by observations.

However, if the symmetry is not gauged, the gradients of the scalar field cannot be compensated by the presence of a gauge field. In this case, long-range interactions lead to very efficient annihilation of monopole–anti-monopole pairs and the remaining energy density has the correct scaling, $\rho_M \propto 1/\tau^2$.

This is true not just for a symmetry breaking Higgs field: an arbitrary, unordered multicomponent scalar field with a potential minimum at some scale $M \neq 0$ evolves in an expanding universe such that $\rho_S/\rho \simeq GM^2 = \text{constant}$. The field orders on the Hubble scale, so that its gradient and kinetic energy are of the order M^2/τ^2. These findings have been confirmed by numerical simulations and they become very accurate for fields with three or more components (Durrer *et al.*, 2002). For fields with only two components (global strings) this scaling law seems to obtain logarithmic corrections. One-component, or real scalar fields, do not scale at all. They generically lead to domain walls that soon come to dominate the energy density of the Universe and are therefore ruled out. (Their vacuum manifold consists of isolated points, so that they have negligible gradient and kinetic energy. Their energy is dominated by potential energy.)

2.7.2 Causal Scaling Seeds

If the initial conditions of the scalar field are uncorrelated on scales larger than the Hubble scale, correlations evolve causally and will always vanish on scales larger than the Hubble scale during noninflationary expansion. The correlation functions of arbitrary components of the energy–momentum tensor are therefore functions with compact support. An important mathematical theorem (Reed and Simon, 1980) states that the Fourier transform of a function with compact support is analytic. On scales that are much smaller than the Hubble scale, the field has already had sufficient time to order and will therefore not contribute much. The only scale in the problem is the Hubble scale $\mathcal{H} \simeq 1/t$. We therefore expect the power spectra to depend on scale only via the dimensionless variable $y \equiv kt$. Seeds that have this behavior are called "causal scaling seeds." They are most interesting since, as we shall see in the text that follows, they generically predict a scale-invariant spectrum of CMB fluctuations like inflation that we shall study in the next chapter.

We now consider an arbitrary seed energy–momentum tensor that may or may not come from a scalar field or even from cosmic strings, but that has the aforementioned properties of scaling and of causality and therefore analyticity. Let us parameterize the correlations of its energy–momentum tensor, $\Theta_{\mu\nu}$ in the form

$$\Theta_{\mu\nu}(\mathbf{k},t) = M^2\theta_{\mu\nu}(\mathbf{k},t), \tag{2.286}$$

$$\langle\theta_{\mu\nu}(\mathbf{k},t)\theta^*_{\rho\lambda}(\mathbf{k}',t)\rangle = (2\pi)^3 C_{\mu\nu\rho\lambda}(\mathbf{k},t)\delta(\mathbf{k}-\mathbf{k}'). \tag{2.287}$$

The correlators $C_{\mu\nu\rho\lambda}$ are analytic functions of $\mathbf{k}$. Scaling requires that they depend only on $t\mathbf{k}$ and on t, where the t dependence is a simple power law with the power required for dimensional reasons and the dependence on $t\mathbf{k}$ is analytic. We also require $C_{\mu\nu\rho\lambda} \to 0$ for $kt \to \infty$. The dimension of $\theta_{\mu\nu}(\mathbf{x})$ is $1/(\text{length})^2$ so that

$\theta_{\mu\nu}(\mathbf{k})$ has the dimension of a length. Hence $C_{\mu\nu\rho\lambda}$ must have the dimension of an inverse length and therefore be of the form $t^{-1} \times$ (an analytical function of $t\mathbf{k}$). For example,

$$C_{0000} = \frac{1}{t} F_1(kt), \quad \text{or} \tag{2.288}$$

$$C_{0i0j} = \frac{1}{t}\left[t^2 k_i k_j F_2(kt) + (kt)^2 \delta_{ij} F_3(kt)\right], \tag{2.289}$$

where the functions $F_n(y)$ are analytic in y^2 with the asymptotic behavior

$$\lim_{y\to\infty} F_n(y) = 0.$$

It can be shown that if the energy–momentum of the source is conserved, the correlators (2.287) can all be expressed in terms of five free functions with this asymptotic behavior.

In a given specific model, numerical simulations are usually employed to determine these functions; see Durrer *et al.* (2002).

Let us now show that, on large scales, the CMB anisotropy spectrum from causal scaling seeds is always scale invariant. Since the Laplacians of the Bardeen potentials are of the form

$$k^2\Phi, \quad k^2\Psi \sim \epsilon\theta,$$

where θ denotes some components of $\theta_{\mu\nu}$, their power spectrum must be of the form

$$\left\langle (\Psi+\Phi)(\mathbf{k},t)(\Psi+\Phi)^*(\mathbf{k}',t)\right\rangle = \epsilon^2(2\pi)^3\delta(\mathbf{k}-\mathbf{k}')\frac{F(y^2)}{k^4 t}, \tag{2.290}$$

$$\equiv (2\pi)^3\delta(\mathbf{k}-\mathbf{k}')P(k,t), \tag{2.291}$$

where again F is an analytic function of y^2 that tends to 0 for large y. We have written the power spectrum for $\Phi+\Psi$, since this is the quantity that determines the large-scale CMB anisotropy spectrum.

$$\frac{\Delta T}{T}(\mathbf{x}_0,\mathbf{n}) = (\Psi+\Phi)(\mathbf{x}(t_{\text{dec}}),t_{\text{dec}}) + \int_{t_{\text{dec}}}^{t_0} \partial_t(\Psi+\Phi)(\mathbf{x}(t),t)\,dt; \tag{2.292}$$

see Eq. (2.240). The Fourier transform of this equation yields

$$\frac{\Delta T}{T}(\mathbf{k},\mathbf{n}) = e^{i\mathbf{k}\cdot\mathbf{n}(t_0-t_{\text{dec}})}(\Psi+\Phi)(\mathbf{k},t_{\text{dec}}) + \int_{t_{\text{dec}}}^{t_0} e^{i\mathbf{k}\cdot\mathbf{n}(t_0-t)}\,\partial_t(\Psi+\Phi)(\mathbf{k},t)\,dt. \tag{2.293}$$

We have $\partial_t\left[e^{i\mathbf{k}\cdot\mathbf{n}(t_0-t)}(\Psi+\Phi)\right] = e^{i\mathbf{k}\cdot\mathbf{n}(t_0-t)}[-i\mathbf{k}\mathbf{n}(\Psi+\Phi)+\partial_t(\Psi+\Phi)]$. As long as $kt < 1$, the second term in this expression dominates and we may therefore approximate the derivative in Eq. (2.293) of $\Phi + \Psi$ by the time derivative of the total integrand. Since $\Psi + \Phi$ decays rapidly inside the horizon, it suffices to integrate until $t = 1/k$. The integral can now be performed and the value at the lower boundary simply cancels the "ordinary Sachs–Wolfe" term. We obtain

$$\frac{\Delta T}{T}(\mathbf{k},\mathbf{n}) \simeq e^{i\mathbf{k}\cdot\mathbf{n}(t_0-1/k)}(\Psi+\Phi)(\mathbf{k},1/k), \tag{2.294}$$

and

$$\left\langle \frac{\Delta T}{T}(\mathbf{k},\mathbf{n})\frac{\Delta T^*}{T}(\mathbf{k}',\mathbf{n}')\right\rangle \simeq e^{i\mathbf{k}\cdot(\mathbf{n}-\mathbf{n}')(t_0-1/k)}\frac{\epsilon^2}{k^3}F(1)(2\pi)^3\delta(\mathbf{k}-\mathbf{k}'). \tag{2.295}$$

Expanding $e^{i\mathbf{k}\cdot\mathbf{n}(t_0-1/k)}$ and $e^{-i\mathbf{k}\cdot\mathbf{n}'(t_0-1/k)}$ in Legendre polynomials and spherical Bessel functions, along the same steps as in Section 2.6, we arrive at

$$C_\ell \simeq \epsilon^2 F(1)\frac{2}{\pi}\int_0^\infty \frac{dk}{k}\, j_\ell^2(kt_0) = \frac{\epsilon^2 F(1)}{\pi}\frac{1}{\ell(\ell+1)}. \tag{2.296}$$

We have approximated $kt_0 - 1 \sim kt_0$ in the argument of the spherical Bessel function and used the integral (A4.150). As promised, we obtain a scale-invariant spectrum, $\ell(\ell+1)C_\ell = \text{constant}$. The numerical value obtained in this way is not accurate, but the scaling is correct. Note that the main ingredient for the scaling was that the power spectrum of $\Psi + \Phi$ does not contain any other dimensionful parameter other than t and k and that it decays inside the horizon. For dimensional reasons, the spectrum P then is such that $P(k, t = 1/k) \propto 1/k^3$, and the CMB anisotropies become scale invariant.

We expect $F(1)$ to be of order unity, so that ϵ determines the amplitude of the fluctuations.

In the next subsection, we explain how to go beyond such a rough approximation and calculate the CMB anisotropies and polarization from scaling causal seeds in more detail.

2.7.3 Calculating CMB Anisotropies from Sources

The linear perturbation equations in the presence of sources take the form (in $\mathbf{k}$-space)

$$\mathcal{D}X(\mathbf{k},t) = \epsilon\mathcal{S}(\mathbf{k},t), \tag{2.297}$$

where $\mathcal{D}$ is a first-order linear differential operator in time and X is a long vector containing as its components all the perturbation variables, for example, D_m, V_m; all the temperature fluctuation variables, $\mathcal{M}_\ell^{(m)}$; and the polarizations, $\mathcal{E}_\ell^{(m)}$, $\mathcal{B}_\ell^{(m)}$, which we shall discuss in detail in Chapters 4 and 5. For the discussion here the detailed definition of these variables is not relevant. We just need to know that they generically satisfy a linear differential equation of the form (2.297) where the source vector $\mathcal{S}$ consists of linear combinations of the source energy momentum tensor. For purely scalar perturbations, $\mathcal{S}$ can be described by two functions, for example, the Bardeen potentials generated by the source. It describes the gravitational interaction of the source with the cosmic fluid, and $\epsilon = 4\pi GM^2$ determines the gravitational coupling strength of the source.

In principle, one can simulate the source, Fourier transform it, and insert the random variable $\mathcal{S}(\mathbf{k},t)$ in Eq. (2.297). Averaging over directions in $\mathbf{k}$-space one can then obtain the correlation matrix $P_{nm}(k,t)(2\pi)^3\delta(\mathbf{k}-\mathbf{k}') = \langle X_n(\mathbf{k})X_m^*(\mathbf{k}')\rangle$. For this one needs as input $\mathcal{S}(\mathbf{k},t)$ depending on four variables, the three-dimensional $\mathbf{k}$-vector and time. This way has proved to be very tedious, requiring a huge $(3+1)$-dimensional numerical simulation with a dynamical range of several hundred only to determine the C_ℓs for $\ell \lesssim 100$. The following observation allows one to reduce the numerical complexity of the problem considerably.

To solve Eq. (2.297), we use the Green function method. If $\mathcal{G}(\mathbf{k},t,t')$ is the Green function for the operator $\mathcal{D}$ with initial condition $\mathcal{G}(t_1,t_1) = 0$ and $\mathcal{D}\mathcal{G}(t,t_1) = \delta(t-t_1)$, the general solution of Eq. (2.297) is

$$X(\mathbf{k},t_0) = \epsilon \int_{t_{\text{in}}}^{t_0} dt\, \mathcal{G}(\mathbf{k},t_0,t)\mathcal{S}(\mathbf{k},t) + X_0(\mathbf{k},t_0), \tag{2.298}$$

where t_{in} denotes the time at which the source first appears, for example, the phase transition, and $X_0(\mathbf{k},t_0)$ is an arbitrary homogeneous solution of Eq. (2.297). A specific example of a Green function is given in Exercise 2.9.

If perturbations are a mixture of two components, one coming from inflation and one from topological defects, X_0 denotes the component from inflation. These two components can be considered as uncorrelated and the resulting perturbation spectra can just be added. We shall discuss the computation of the perturbation spectra from inflation in detail in Chapters 3, 4, and 5. Here we concentrate on the part induced by the sources. We therefore neglect X_0, so that we obtain for the correlation matrix,

$$\langle X_i(\mathbf{k},t_0)X_j^*(\mathbf{k}',t_0)\rangle = \epsilon^2 \int_{t_{\text{in}}}^{t_0} dt'\, dt\, \mathcal{G}_{im}(\mathbf{k},t_0,t)\mathcal{G}_{nj}^*(\mathbf{k}',t_0,t') \times \langle \mathcal{S}_m(\mathbf{k},t)\mathcal{S}_n^*(\mathbf{k}',t')\rangle. \tag{2.299}$$

To calculate it, we need to determine the unequal time correlators of the source,

$$\left\langle \mathcal{S}_i(\mathbf{k},t)\mathcal{S}_j^*(\mathbf{k}',t')\right\rangle = (2\pi)^3 t^p \mathcal{F}_{ij}\left(\sqrt{tt'}\mathbf{k},r\right)\delta(\mathbf{k}-\mathbf{k}'). \tag{2.300}$$

Here we have introduced the ratio $r = t/t'$. The details of the correlation functions $\mathcal{F}_{ij}$ have to be determined case by case, via numerical simulations. But they are

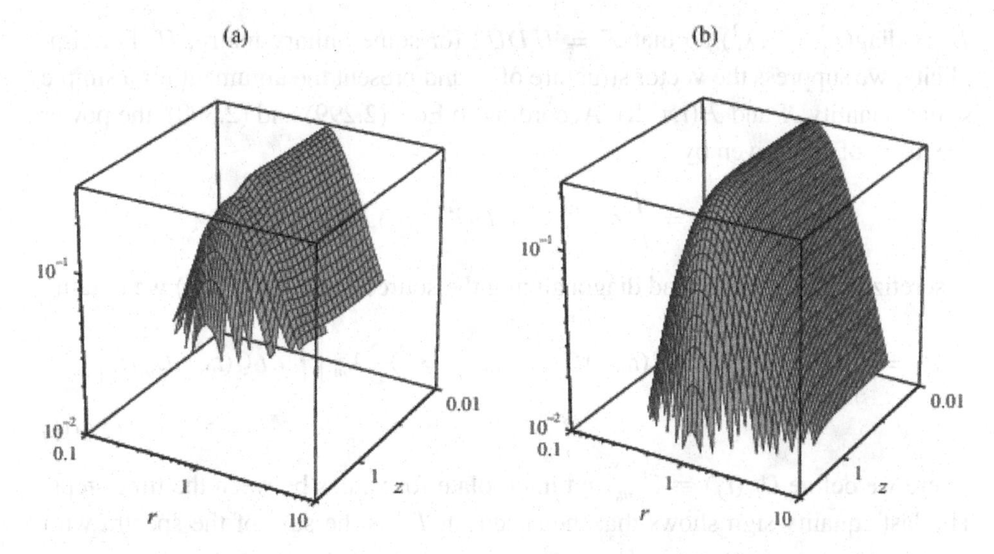

Fig. 2.8 The source correlation function $F(r,z)$ for vector perturbations from numerical simulations for a 4-component scalar field (texture) in panel (a) and for the semianalytical result for the large N limit, panel (b). From Durrer *et al.* (2002)

much easier to obtain on a large dynamical range than the full random variable $\mathcal{S}$. The source just consists of linear combinations of the energy–momentum tensor of the seed, which is determined by five functions, $F_n(r,y)$, $n \in \{1,2,3,4,5\}$ where now $y = \sqrt{tt'}k$. These functions are analytic in $\mathbf{k}$ and therefore in y^2. They go to zero for either $y \to \infty$ or $r \to \infty$, $r \to 0$; furthermore, $F_n(r,y) = F_n^*(1/r,y)$. We now only have to determine the amplitudes of the functions at $r \sim 1$, $y \sim 0{-}1$ and their behavior around these values. This is numerically very feasible and has been performed with good accuracy. The source functions for vector perturbations of a self-ordering scalar field are shown in Fig. 2.8; see also Lizarraga *et al.* (2016) for cosmic string simulations.

2.7.4 Decoherence

From its definition (2.300) it is clear that the source correlation function $\mathcal{F}(t,t',\mathbf{k})$ can be interpreted as a positive symmetric operator. For a given function $V(t)$ setting $(\mathcal{F}V)(t) = \int dt'\, \mathcal{F}(t,t')V(t')$, we find (suppressing the argument $\mathbf{k}$ and vector indices for simplicity)

$$\langle V, \mathcal{F}V \rangle \equiv \int dt\, dt'\, V^*(t)\mathcal{F}(t,t')V(t') \geq 0.$$

Discretizing it in time, $\mathcal{F}$ becomes a positive semidefinite hermitian matrix, which we can diagonalize. Let us denote its nonnegative eigenvalues by $\lambda_1^2, \ldots, \lambda_n^2$, and

$D = \mathrm{diag}(\lambda_1^2, \ldots, \lambda_n^2)$ so that $\mathcal{F} = UDU^*$ for some unitary matrix U. For simplicity, we suppress the vector structure of X and present the argument for a simple scalar quantity X and $\mathcal{F}(t, t', \mathbf{k})$. According to Eqs. (2.299) and (2.300), the power spectrum of X is given by

$$P_X = \int dt\, dt'\, \mathcal{G}(t_0, t)\mathcal{G}^*(t_0, t')\mathcal{F}(t, t').$$

Discretizing this integral and diagonalizing the source function $\mathcal{F}(t, t')$ we obtain

$$P_X = \sum_{ijm} (\Delta t)^2 \mathcal{G}(t_0, t_j)\mathcal{G}^*(t_0, t_i) U_{jm} U^*{}_{im} \lambda_m^2 = \sum_m \lambda_m^2 \left| \int dt\, \mathcal{G}(t_0, t) U_m(t) \right|^2, \tag{2.301}$$

where we define $U_m(t_j) \equiv U_{jm}$ and interpolate for values between the time steps. The last equality sign shows that the spectrum P_X is the sum of the spectra with deterministic source terms $\lambda_m U_m(t)$.

In this way, the problem of a stochastic source term is reduced to the problem of many deterministic source terms. In practice, one orders the eigenvalues according to size, $\lambda_1 > \lambda_2 \cdots$, and sums the contributions of about the 20–100 largest eigenvalues to achieve an accuracy of about 1%, which is also typically the accuracy of the source term from numerical simulations.

With this procedure in mind, let us discuss the acoustic peak structure of the CMB generated by sources. The acoustic peaks from inflationary perturbations reflect the maxima and minima in the radiation/baryon density at the moment of decoupling. The radiation density perturbation oscillates like a cosine wave, since it starts at maximum amplitude, $D(kt) \simeq A\cos(c_s kt)$. In the case of sourced perturbations, however, sources generate perturbations at different moments in time, so that $D \simeq \sum_n D_n(kt) = \sum_n A_n \cos(c_s kt - \delta_n)$, with different phases δ_n that are determined by the time at which the perturbation D_n is generated. At the time of decoupling, t_{dec}, many different wavelengths can have their maximum or minimum in one of the contributions $D_n(kt_{\mathrm{dec}})$. Instead of a distinct peak structure we therefore rather expect a broad hump in the acoustic peak region of the CMB spectrum. This phenomenon, which is rather generic for seeds, is called "decoherence." It was first pointed out by Albrecht *et al.* (1996). If only very few eigenvalues dominate, that is, if the aforementioned sum contains only a few terms, decoherence is not very effective and a peak structure can still be seen.

2.7.5 Results

In addition to decoherence, an important characteristic of scaling seeds is that they typically generate vector perturbations with an amplitude comparable to that of scalar perturbations; see Durrer *et al.* (2002) and Bevis *et al.* (2007). Tensor perturbations are usually somewhat smaller, $C_2^{(T)}/C_2^{(S)} \sim \frac{1}{4}$. Sources are probably the only way to obtain significant vector perturbations in the CMB. Scaling seeds always generate vector perturbations at the horizon scale while vector perturbations that are generated early in the Universe simply decay and leave no traces in the CMB.

Furthermore, the amplitude of the Sachs–Wolfe part of CMB anisotropies is roughly 2Ψ as compared to $\Psi/3$ for adiabatic inflationary perturbations. Therefore, we expect the acoustic peak structure, or the acoustic hump, to be not much higher than the Sachs–Wolfe plateau.

These are the main results for CMB anisotropies seeded by topological defects. They have a scale-invariant Sachs–Wolfe plateau that determines the normalization and that contains important contributions from all, scalar, tensor and especially vector perturbations. This is followed by a very low acoustic hump or peak structure. Since the perturbations are rather of isocurvature than of adiabatic nature (even though this classification does not strictly apply for sources), this "hump" is around $\ell \sim 300$–500 in a flat universe. This wide range stems from the uncertainty of the scale at which perturbations are induced. This may be the horizon scale, as for global defects, or somewhat less, as for cosmic strings.

In Fig. 2.9 we show the scalar, vector, and tensor CMB spectra from a 4-component global scalar field (cosmic texture). The contributions from the largest eigenvalues as well as their sum (bold solid line) are shown. Even though single eigenvalue contributions do show acoustic oscillations, these are washed out in the sum. Similar results have recently been obtained for cosmic strings (Lizarraga *et al.*, 2016).

From these results it is clear that topological defects or similar sources cannot generate the observed CMB anisotropy spectrum. However, they might make up a small contribution in models in which inflation ends with a symmetry breaking phase transition that leads to cosmic strings. It has been argued that the formation of cosmic strings is quite generic for GUTs and can actually be used to constrain them with the CMB (Rocher and Sakellariadou, 2005).

Another question of interest is the following: as the entire class of scaling causal seeds allows for five nearly free functions of two variables, is it possible to "manufacture" these functions such that they reproduce the observed CMB anisotropies and polarization or can this be excluded? It has been argued that the small hump at low $\ell \cong 100$ that is generated in the E-polarization spectrum from inflation during decoupling (see Chapter 5) cannot be reproduced by causal seeds (Spergel and Zaldarriaga, 1997). At decoupling, scales corresponding to $\ell \lesssim 100$ are still super-Hubble and causal seeds have no power on these scales. This question has been studied by "manufacturing" a model with causal scaling seeds that consists in very

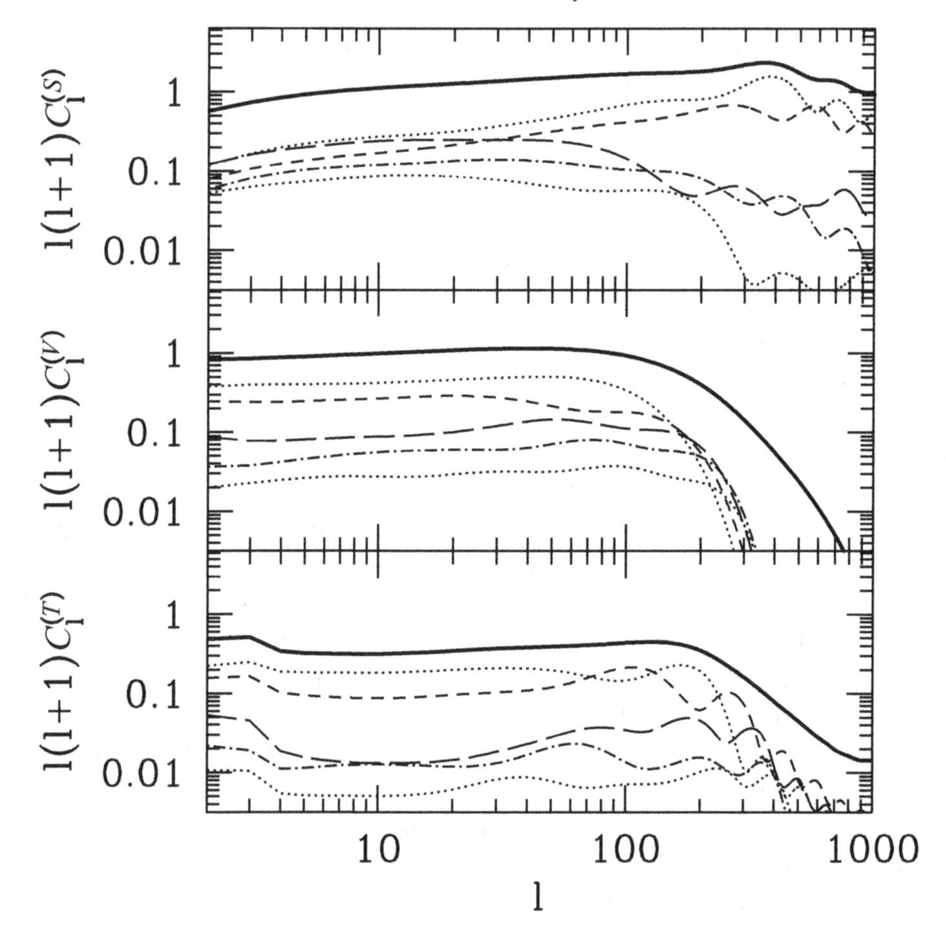

Fig. 2.9 The scalar, vector, and tensor contributions for the texture model of structure formation are shown. The dashed lines show the contributions from the first few single eigenfunctions while the solid line represents the sum (over 100 eigenfunctions). Note that the single contributions to the scalar and tensor spectra do show oscillations that are, however, washed out in the sum (vector perturbations do not obey a wave equation and thus do not show oscillations). Data courtesy of N. Bevis and M. Kunz (see Bevis *et al.*, 2004)

rapid spherical explosions and avoids decoherence. It has been shown numerically that this model fails to reproduce the first polarization peak as long as the explosion remain subliminal, that is, do not violate causality (Scodeller *et al.*, 2009).

2.8 Final Remarks

In this chapter we have developed the basics of cosmological perturbation theory. Perturbation theory is an important tool especially to calculate CMB anisotropies and polarization, since these are very small and can be determined reliably within linear cosmological perturbation theory. Here we have discussed the Einstein equations and the propagation of light-like geodesics. Linear perturbations of the somewhat more involved Boltzmann equation, which is more adequate to study CMB anisotropies and polarization, will be developed in Chapters 4 and 5. To determine the evolution of the matter density fluctuations at late times, linear perturbation theory has to be complemented with the theory of weakly nonlinear Newtonian gravity and with N-body simulations. Finally, to understand the formation of galaxies non-gravitational highly nonlinear physics, such as heating and cooling mechanisms, dissipation, and nuclear reactions have to be taken into account. These topics go beyond the scope of this book.

Exercises

(The exercises marked with an asterisk are solved in Appendix 11 which is not in this printed book but can be found online.)

2.1 Gauge transformations*

Using the general formulae of Appendix 2, Section A2.2 derive Eq. (2.32),

$$L_X\bar{g} = a^2[-2(\mathcal{H}T + \dot{T})dt^2 + 2(\dot{L}_i - T_{,i})dt\,dx^i + \left(2\mathcal{H}T\gamma_{ij} + L_{i|j} + L_{j|i}\right)dx^i dx^j]. \tag{2.302}$$

2.2 Normals to hypersurfaces

We consider the normal n^μ to some hypersurface of the form $t =$ constant for a suitably chosen time coordinate t. Then the corresponding 1-form is of the form $n_\mu\, dx^\mu = f\, dt$. The function $f = \pm\sqrt{-g^{00}}$ is determined by the normalization condition. Show that in this case the vorticity

$$\omega_{\mu\nu} = \frac{1}{2}P^\lambda_\mu P^\rho_\nu(n_{\lambda;\rho} - n_{\rho;\lambda}), \tag{2.303}$$

vanishes. Here P^ρ_ν is the projection tensor defined in Eq. (2.42).

Hint: If you are familiar with forms, you find from Wald (1984) that a vector field is hypersurface orthogonal if and only if the corresponding 1-form $\alpha = n_\mu\, dx^\mu$ satisfies $\alpha \wedge d\alpha = 0$. Show that for a form given by $\alpha = f(x)\, dt$ this is always the case. Here f is an arbitrary function on the spacetime manifold.

2.3 Synchronous and comoving gauge*

Show that in synchronous gauge (i.e., the gauge with $A = B = 0$) in a $K = 0$ pure dust universe the growing mode is also comoving, i.e., $v = 0$ for the growing mode.

2.4 **Adiabaticity***
Consider a mixture of two noninteracting fluids with sound speeds c_1, c_2, and enthalpies w_1, w_2. Determine Γ from the single fluid perturbation variables Γ_α, $D_{g\alpha}$, V_α, Π_α, $\alpha = 1, 2$. If the intrinsic perturbations of each fluid are adiabatic, $\Gamma_\alpha = 0$, what is the condition that the total perturbation be adiabatic, $\Gamma = 0$? Derive an evolution equation for Γ under the condition $\Gamma_\alpha = 0$.

2.5 **Power spectrum**

1) For a spatial, statistically homogeneous, and isotropic random variable $X(\mathbf{x})$ with vanishing mean the 2-point correlation function ξ_X can only depend on the distance, $r = |\mathbf{x} - \mathbf{y}|$,

$$\xi(r) = \langle X(\mathbf{x})X(\mathbf{y})\rangle \qquad r = |\mathbf{x} - \mathbf{y}|.$$

Homogeneity requires that ξ does not depend on the position of the first point, $\mathbf{x}$, and isotropy means that ξ is independent of the direction of $\mathbf{x} - \mathbf{y}$. Hence ξ depends only on the distance r. Show that this fact implies that the 2-point correlator in Fourier space is proportional to $\delta(\mathbf{k} - \mathbf{k}')$, that is, of the form of Eq. (2.246).

2) Consider a normalized, spherically symmetric smoothing function (or window function) $W_R(|\mathbf{x}|)$ that smoothes out fluctuations on scales smaller than R (e.g., a Gaussian or a top hat). We denote its Fourier transform by $\widehat{W}_R(\mathbf{k})$. The smoothed field is defined by

$$X_R(\mathbf{y}) = \int d^3x\, W_R(\mathbf{y} - \mathbf{x})X(\mathbf{x}). \tag{2.304}$$

Show that

$$X_R(\mathbf{y}) = \frac{1}{(2\pi)^3}\int d^3k\, \widehat{W}_R(k)X(\mathbf{k})$$

and

$$\langle X_R^2\rangle = \int_0^\infty \frac{dk}{k}\widehat{W}_R^2(k)\Delta_X(k). \tag{2.305}$$

For this reason, $\Delta_X(k)$ is a good measure for the amplitude of fluctuations of X smoothed over the scale $\lambda/2 = \pi/k$.

2.6 **Variable transformation**

We consider an ordinary linear second-order differential equation for ϕ, which is of the form

$$\ddot{\phi} + f(t)\dot{\phi} + \omega^2(t)\phi(t) = 0. \tag{2.306}$$

Show that the variable $\psi \equiv h(t)\phi$, with $h(t) = \exp\left(\frac{1}{2}\int^t f(t')\,dt'\right)$, satisfies the equation

$$\ddot{\psi} + \tilde{\omega}^2(t)\psi(t) = 0, \tag{2.307}$$

where

$$\tilde{\omega}^2 = \omega^2 + \ddot{h}/h.$$

Use these findings to derive Eq. (2.156).

Hint: Use

$$\frac{d}{dt}[\rho(1+w)]^{-1/2} = \frac{3}{2}\mathcal{H}(1+c_s^2)[\rho(1+w)]^{-1/2}.$$

To derive this equation make use of Eq. (2.115) and the energy conservation equation. Finally, use Eq. (2.111).

This shows that a linear second-order differential equation can always be brought into the form of the equation for a harmonic oscillator with a time-dependent frequency. This is useful to know, not only because we can quantize this system easily, but also since we know that an instability sets in, when $\tilde{\omega}^2 < 0$. By studying the time dependence of the frequency, we can therefore infer the qualitative behavior of the solution.

2.7 **Perturbations in universes with nonflat spatial section**

Consider a universe filled with dust, $c_s^2 = w = 0$. In this case, the Bardeen equation reduces to

$$\ddot{\Psi} + 3\mathcal{H}\dot{\Psi} - 2K\Psi = 0. \tag{2.308}$$

Solve this equation for both cases, $K > 0$ and $K < 0$, and discuss the results. What happens for $K < 0$ at late times? How do perturbations evolve during a collapsing universe?

2.8 **Acoustic peaks**

Consider a universe with matter and radiation but no curvature or cosmological constant. Solve the Friedmann equation exactly and determine $t_0/t_{\rm dec}$ given that $z_{\rm dec} \simeq 1100$. Insert a realistic value for Ω_r and keep h in the expression. Use the result to approximate the positions of the acoustic peaks in this Universe. Discuss qualitatively the change in the peak position in the following cases:

(a) addition of a cosmological constant (at fixed Ω_r and for vanishing curvature).
(b) addition of curvature (at fixed Ω_r and $\Lambda = 0$).

2.9 **The Green function**

Show that for

$$\mathcal{D}X = \ddot{X} + \alpha\dot{X} + \beta X$$

the Green function with initial condition $\mathcal{G}(t,t) = 0$ and $\dot{\mathcal{G}}(t,t) = 1$ is given by

$$\mathcal{G}(t_1,t) = \frac{D_1(t_1)D_2(t) - D_1(t)D_2(t_1)}{\dot{D}_1(t)D_2(t) - D_1(t)\dot{D}_2(t)},$$

where D_1 and D_2 are two linearly independent solutions of the homogeneous equation $\mathcal{D}X = 0$. Show that $\mathcal{G}$ is independent of the choice of D_1 and D_2. $\dot{\mathcal{G}}(t_1,t) = \frac{\partial}{\partial t_1}\mathcal{G}(t_1,t)$.

Consider the case $\alpha = 0$ and $\beta = c_s^2 k^2 =$ constant. Introduce a source of the form $\mathcal{S}_1(t) = A_1\delta(t - t_1)$. Discuss decoherence by adding the signal of several sources of this kind.

3
Initial Conditions

So far we have only studied the evolution of perturbations assuming that the initial conditions are fixed and given once and for all. Now we want to study how classical perturbations are generated out of quantum fluctuations during a simple inflationary phase. The fact that inflation generates a nearly scale-invariant spectrum of scalar perturbations in good agreement with the observations of the cosmic microwave background is to be considered as its greatest success. The solutions of the flatness and entropy problems with an inflationary phase are actually "post-dictions" while the scale-invariant spectrum of scalar perturbations was first predicted in Mukhanov and Chibishov (1982) [see Mukhanov *et al.* (1992) for a review] long before its discovery by the COBE satellite by Smoot *et al.* (1992). It represents therefore a real prediction of inflation. There are also other models for structure formation that predict a scale-invariant spectrum of fluctuations but that disagree with the detailed observed spectrum of CMB fluctuations such as topological defects (Durrer *et al.*, 2002).

In this chapter we first study perturbations in an FL universe filled with a scalar field. Next we discuss the generation of fluctuations during inflation. We especially determine the spectral index of scalar and tensor perturbations and the ratio of their amplitudes in the slow roll approximation. This will lead us to the well-known consistency relation for slow roll inflation. We study in detail the simple case of one scalar field, the "inflaton."

We then go beyond Gaussian fluctuations and study how nonlinearities in the evolution of the scalar field(s) induce non-Gaussianities, in particular an nonvanishing 3-point function and bispectrum.

Finally, we investigate more general initial conditions that are relevant if more than one scalar field plays a role during inflation, so-called mixed adiabatic and isocurvature fluctuations.

As we have discussed in Chapter 1, an inflationary phase supresses curvature and in order for curvature to be of order unity or smaller today, it must have been very small in the early Universe. In this chapter, which deals mainly with the early Universe, we therefore neglect curvature; $K = 0$. Note, however, that homogeneity and isotropy are not a simple consequence of inflation. We still have to assume a sufficiently large patch where inhomogeneities can be neglected and that then inflates to become the entire visible Universe. Inflation does not predict a homogeneous and isotropic Universe, but it renders it causally possible. A different point of view on this has been presented recently (Creminelli *et al.*, 2019).

3.1 Scalar Field Perturbations

We consider here the special case of an FL universe filled with self-interacting minimally coupled scalar field matter. The action is given by

$$S = \frac{1}{16\pi G}\int d^4x\,\sqrt{|g|}R - \int d^4x\,\sqrt{|g|}\left(\frac{1}{2}\partial_\mu\varphi\,\partial^\mu\varphi + W(\varphi)\right), \tag{3.1}$$

where φ denotes the scalar field and W is its potential. The energy–momentum tensor is obtained by varying the scalar-field action w.r.t. the metric $g^{\mu\nu}$,

$$T_{\mu\nu} = \partial_\mu\varphi\partial_\nu\varphi - \left[\frac{1}{2}\partial_\lambda\varphi\,\partial^\lambda\varphi + W\right]g_{\mu\nu}. \tag{3.2}$$

The energy density ρ and the energy flux u are defined as the timelike eigenvalue and eigenvector of $T^\mu{}_\nu$,

$$T^\mu{}_\nu u^\nu = -\rho u^\mu. \tag{3.3}$$

For the homogeneous and isotropic FL background we obtain (see also Chapter 1)

$$\rho = \frac{1}{2a^2}\dot\varphi^2 + W, \qquad (u^\mu) = \frac{1}{a}(1,\vec{0}). \tag{3.4}$$

The pressure is given by

$$T^i_j = P\delta^i_j, \qquad P = \frac{1}{2a^2}\dot\varphi^2 - W. \tag{3.5}$$

We want to derive the linear perturbation equations for the evolution of scalar field and metric perturbations. We define the scalar field perturbation modes,

$$\varphi = \bar\varphi + \delta\varphi Q^{(S)}. \tag{3.6}$$

Let us determine the first-order perturbations in the energy momentum tensor of the scalar field. With the definition (3.2) we obtain

$$\begin{aligned}\delta T_{\mu\nu} &= \partial_\mu\bar\varphi\,\partial_\nu\,\delta\varphi + \partial_\nu\,\bar\varphi\partial_\mu\,\delta\varphi + a^{-2}\dot{\bar\varphi}\,\delta\dot\varphi\,\bar g_{\mu\nu}\\ &\quad + \left[\frac{1}{2a^2}(\dot{\bar\varphi})^2 - \bar W\right]\delta g_{\mu\nu} - \dot{\bar\varphi}^2 g_{\mu\nu}A - \bar W'\delta\varphi\,\bar g_{\mu\nu}.\end{aligned} \tag{3.7}$$

Here A denotes the perturbation of g_{00}, $g_{00} = -a^2(1+2A)$; we consider a generic gauge as given in Eq. (2.29). Inserting Eq. (3.7) for a fixed Fourier mode $Q^{(S)}$ in the definition (3.3) of the energy density and energy flux, and setting $\rho = \bar{\rho} + \delta\rho Q^{(S)}$ and

$$(u^\mu) = \frac{1}{a}\left(1 - AQ^{(S)}, vQ_i^{(S)}\right), \tag{3.8}$$

we find

$$\delta\rho = \frac{1}{a^2}\dot{\varphi}\,\delta\dot{\varphi} - \frac{1}{a^2}\dot{\varphi}^2 A + W,_\varphi \delta\varphi, \tag{3.9}$$

and

$$-v = \frac{k}{\dot{\bar{\varphi}}}\left(\delta\varphi + \dot{\bar{\varphi}}k^{-1}B\right). \tag{3.10}$$

The stress tensor, $T_{ij} = \varphi,_i\varphi,_j - \left[\frac{1}{2}\partial_\lambda\varphi\partial^\lambda\varphi + W\right]g_{ij}$, yields

$$P\pi_L = \frac{1}{a^2}\dot{\bar{\varphi}}\delta\dot{\varphi} - \frac{1}{a^2}\dot{\bar{\varphi}}^2 A - W,_\varphi\delta\varphi \quad \text{and} \quad \Pi = 0. \tag{3.11}$$

We now define a gauge-invariant scalar field perturbation that corresponds to the value of $\delta\varphi$ in longitudinal gauge.

$$\delta\varphi^{(gi)} = \delta\varphi + \dot{\bar{\varphi}}k^{-1}(B - k^{-1}\dot{H}_T) = \delta\varphi - \dot{\bar{\varphi}}k^{-1}\sigma = \delta\varphi^{(\text{long})}. \tag{3.12}$$

The second and third expressions give $\delta\varphi^{(gi)}$ in a generic gauge. Under a gauge transformation the scalar field perturbation simply changes by $\delta\varphi \to \delta\varphi + \dot{\bar{\varphi}}T$. Since $\sigma \to \sigma + kT$, it is clear that the combination $\delta\varphi^{(gi)}$ is gauge invariant. On the other hand, in longitudinal gauge $B = H_T = 0$, so that $\delta\varphi^{(gi)} = \delta\varphi^{(\text{long})}$. This variable is very simply related to the other gauge-invariant scalar variables. Short calculations using Eq. (1.149) give

$$-V = k\delta\varphi^{(gi)}/\dot{\bar{\varphi}}, \tag{3.13}$$

$$D_g = -(1+w)\left[4\Psi + 2\mathcal{H}k^{-1}V - k^{-1}\dot{V}\right], \tag{3.14}$$

$$D_s = -(1+w)\left[\Psi + 2\mathcal{H}k^{-1}V - k^{-1}\dot{V}\right], \tag{3.15}$$

$$D = -(1+w)\left[\Psi - \mathcal{H}k^{-1}V - k^{-1}\dot{V}\right], \tag{3.16}$$

$$\Gamma = \frac{2W,_\varphi\dot{\bar{\varphi}}}{P\dot{\rho}}\left[\rho D_s - \dot{\rho}k^{-1}V\right] = -\frac{W,_\varphi\dot{\bar{\varphi}}}{kP}\left[V + \frac{1}{\mathcal{H}}\dot{V} - \frac{k}{\mathcal{H}}\Psi\right], \tag{3.17}$$

$$\Pi = 0. \tag{3.18}$$

Note that $\Gamma \neq 0$. However, this single scalar field is not in a (nearly) thermal state and hence Γ cannot be interpreted as the divergence of an entropy flux; see Appendix 5. Here Γ, like all other perturbation variables, is entirely fixed by $\delta\varphi^{(gi)}$ and Ψ. The last equation implies that the two Bardeen potentials are equal for scalar field perturbations, $\Phi = \Psi$. Using this we can write the perturbed Einstein equations fully in terms of the Bardeen potential Ψ and V. We actually only need (2.105). Since we need them mainly to discuss inflation where curvature is negligible, we write them down here only for the case $K = 0$:

$$k^2\Psi = 4\pi G\dot{\varphi}^2\left[\Psi - \mathcal{H}k^{-1}V - k^{-1}\dot{V}\right], \tag{3.19}$$

$$\dot{\Psi} + \mathcal{H}\Psi = 4\pi G\dot{\varphi}^2 k^{-1}V, \tag{3.20}$$

where we have used $a^2\rho(1+w) = \dot{\varphi}^2$. To simplify the notation, we have dropped the overbar on the background quantities. With the help of Eqs. (2.106) and (2.108) one can easily generalize these equations to the case with curvature. Using (3.20) to eliminate V and $\dot{V}$ from Eq. (3.19) leads to the following second-order equation for the Bardeen potential:

$$\ddot{\Psi} + 2(\mathcal{H} - \ddot{\varphi}/\dot{\varphi})\dot{\Psi} + (2\dot{\mathcal{H}} - 2\mathcal{H}\ddot{\varphi}/\dot{\varphi} + k^2)\Psi = 0. \tag{3.21}$$

Here we have also used the fact that $4\pi G\dot{\varphi}^2 = 4\pi G a^2\rho(1+w) = \mathcal{H}^2 - \dot{\mathcal{H}}$. Inserting the definition $c_s^2 = \dot{P}/\dot{\rho} = -\frac{1}{3\mathcal{H}}(2\frac{\ddot{\varphi}}{\dot{\varphi}} + \mathcal{H})$, we can also write (3.21) in the form

$$\ddot{\Psi} + 3\mathcal{H}(1 + c_s^2)\dot{\Psi} + (2\dot{\mathcal{H}} + (1 + 3c_s^2)\mathcal{H}^2 + k^2)\Psi = 0. \tag{3.22}$$

This equation differs from the Ψ equation for a perfect fluid only in the term proportional to k^2 that is not multiplied with the adiabatic sound speed c_s^2. Indeed the scalar field is not in a thermal state with fixed entropy, $\Gamma \neq 0$, but rather in a fully coherent state so that field fluctuations propagate with the speed of light and not with some adiabatic sound speed. On large scales, $|kt| \ll 1$ this difference is not relevant, but on sub-Hubble scales it does play a certain role.[1]

During slow roll inflation we can express the background variables in terms of $\mathcal{H}$ and the slow roll parameters ϵ_1 and ϵ_2 defined in Chapter 1. With the definition (1.155) we obtain

[1] Often, the terms "Hubble scale" and "horizon scale" are used interchangeably. For inflation, however, they can differ by many orders of magnitude. During inflation the (comoving) Hubble scale, $\mathcal{H}^{-1} \simeq |t| = -t$ is decreasing and much smaller than the comoving horizon scale, $\int_{t_i}^{t} dt \simeq -t_i \gg \mathcal{H} \simeq -t$. The scale relevant for the behavior of perturbations is, however, always the Hubble scale, since $\mathcal{H}$ enters into the perturbation equations. The horizon is a global quantity, an integral; it does not determine whether perturbations are oscillating or whether they behave like a power law. It just happens that in a decelerating universe the two scales are often of the same order. In this chapter we shall be careful not to mix them up. We shall use the terms "Hubble scale" for the Hubble scale and "Hubble exit" for a scale growing larger than the Hubble scale during inflation.

$$\frac{\ddot{\varphi}}{\mathcal{H}\dot{\varphi}} = 1 - \epsilon_1 + \frac{1}{2\mathcal{H}}\frac{\dot{\epsilon}_1}{\epsilon_1} = 1 + 3\epsilon_2 + \epsilon_1, \tag{3.23}$$

so that

$$2(\dot{\mathcal{H}} - \mathcal{H}\ddot{\varphi}/\dot{\varphi}) = -2\mathcal{H}^2(3\epsilon_2 + 2\epsilon_1). \tag{3.24}$$

Inserting these results in Eq. (3.21) we find

$$\ddot{\Psi} - 2(3\epsilon_2 + \epsilon_1)\mathcal{H}\dot{\Psi} - \left[2\mathcal{H}^2(3\epsilon_2 + 2\epsilon_1) - k^2\right]\Psi = 0. \tag{3.25}$$

Hence on small scales, $(3\epsilon_2 + 2\epsilon_1)\mathcal{H}^2 \ll k^2$, and Ψ oscillates, while on super-Hubble scales, $k/\mathcal{H} \ll 1$, it varies slowly as long as the slow roll parameters are small. During the transition from inflation to the radiation-dominated era, where the slow roll parameters reach order unity, the Bardeen potential can, however, vary substantially. It is therefore not very well suited to determine the amplitude of perturbations which have been induced during inflation in the radiation-dominated era. We now show that, on super-Hubble scales, the curvature variable ζ remains constant also during the transition from inflation to the radiation-dominated era. To study the evolution of super-Hubble perturbations from inflation into the radiation era, we shall therefore use the variable ζ.

As for the case of fluids [see Eqs. (2.156)–(2.158)] we introduce the variable u given by

$$u = a[4\pi G(\mathcal{H}^2 - \dot{\mathcal{H}})]^{-1/2}\Psi, \tag{3.26}$$

which now satisfies the equation

$$\ddot{u} + (k^2 - \ddot{\theta}/\theta)u = 0, \tag{3.27}$$

where

$$\theta = \frac{3\mathcal{H}}{2a\sqrt{\mathcal{H}^2 - \dot{\mathcal{H}}}}. \tag{3.28}$$

The difference to the fluid equations is just the factor c_s^2 in front of k^2, which for the scalar fields is replaced by 1 as already noted earlier. The curvature variable ζ in a scalar field background is given by (2.145)

$$\zeta \equiv \frac{2(\mathcal{H}^{-1}\dot{\Psi} + \Psi)}{3(1+w)} + \Psi. \tag{3.29}$$

Note that we need $w > -1$ so that ζ is well defined. In a pure de Sitter space, we cannot work with the ζ-variable. This becomes even more evident when expressing Eq. (3.29) in terms of the slow roll parameter ϵ_1. From (1.153) and

$$(1+w) = \frac{\dot{\varphi}^2}{\frac{1}{2}\dot{\varphi}^2 + a^2 W} = \frac{\frac{2}{3}\epsilon_1}{\frac{1}{3}\epsilon_1 + 1} \simeq \frac{2}{3}\epsilon_1, \tag{3.30}$$

we obtain

$$\zeta \simeq \frac{\mathcal{H}^{-1}\dot{\Psi} + \Psi}{\epsilon_1} + \Psi. \tag{3.31}$$

For ζ to be well defined we therefore need $\epsilon_1 \neq 0$ (or the perturbations have to decay obeying $\dot{\Psi} = -\mathcal{H}\Psi$). From Eq. (3.25), using Eq. (1.155), one finds

$$\dot{\zeta} = -\frac{2k^2}{3(1+w)\mathcal{H}}\Psi = -\frac{(1+\frac{1}{3}\epsilon_1)k^2}{\epsilon_1\mathcal{H}}\Psi \simeq \frac{-k^2}{\epsilon_1\mathcal{H}}\Psi. \tag{3.32}$$

As in the case of fluids, this implies that the curvature perturbation ζ is conserved on super-Hubble scales, $k/\mathcal{H} \ll 1$. Using Eqs. (3.20) and (3.13) and $\dot{\phi}^2 = a^2\rho(1+w)$, we can express ζ also as

$$\zeta = \frac{\mathcal{H}}{\dot{\varphi}}\delta\varphi^{(gi)} + \Psi. \tag{3.33}$$

As we have seen in Chapter 2, Eqs. (2.161)–(2.163), the evolution of ζ is closely related to the one of v defined by

$$v = \frac{a\sqrt{\mathcal{H}^2 - \dot{\mathcal{H}}}}{\sqrt{4\pi G}\mathcal{H}}\zeta = a\,\delta\varphi^{(gi)} + \frac{a\dot{\varphi}}{\mathcal{H}}\Psi. \tag{3.34}$$

The variable v satisfies the equation of motion

$$\ddot{v} + (k^2 - \ddot{z}/z)v = 0, \tag{3.35}$$

with

$$z = \frac{a\sqrt{\mathcal{H}^2 - \dot{\mathcal{H}}}}{\sqrt{4\pi G}\mathcal{H}} = a\sqrt{\frac{3(1+w)}{8\pi G}} = \frac{a\sqrt{a^2(\rho+P)}}{\mathcal{H}} = \frac{a\dot{\varphi}}{\mathcal{H}}. \tag{3.36}$$

Comparing Eqs. (3.33) and (3.34), this implies

$$v = z\zeta. \tag{3.37}$$

Also note that z is related to the slow roll parameter ϵ_1 by

$$\epsilon_1 = -\frac{dH/d\tau}{H^2} = \frac{\mathcal{H}^2 - \dot{\mathcal{H}}}{\mathcal{H}^2} = 4\pi G\frac{z^2}{a^2}. \tag{3.38}$$

The equation of motion (3.35) can be obtained from the Fourier decomposition of the action

$$S = -\frac{1}{2}\int d^4x\left(\partial_\mu v\,\partial^\mu v + m^2(t)v^2\right), \tag{3.39}$$

where $m^2 = -(\ddot{z}/z)$. This is the action of a simple free scalar field in Minkowski space with a time-dependent mass term. For a constant or slowly varying w we have $z \propto a$ and during inflation $\ddot{z}/z > 0$; hence $m^2(t) < 0$, which represents an instability and leads to the amplification of vacuum fluctuations (or particle creation). During ordinary expansion, $\ddot{z}/z < 0$ and the vacuum state is stable.

In Section 3.3 we will study quantum fluctuations of the variable v.

Remark: Note that also Eq. (3.27) can be written as a Euler–Lagrange equation for a canonical scalar field Lagrangian with time-dependent mass term $m_\theta^2 = -(\ddot{\theta}/\theta)$ for the variable u defined in Eq. (3.26). The problem there is, however, that we cannot "switch off" gravity for this variable, which diverges in the limit $\mathcal{H}, \dot{\mathcal{H}} \to 0$. Hence u does not have well-defined initial conditions when $k^2 \gg |\ddot{\theta}/\theta|$. Even though the perturbation equations take the form of canonical equations in this variable, it is therefore not the correct variable to quantize. In the next section we shall also see that the usual scalar field action leads to a quadratic action for ζ or v but not for u.

3.2 Perturbations of the Scalar Field Action in Unimodular Gauge

As we now show, the preceding action (3.39) can also be obtained by perturbing the original action of the system to second order,

$$S = \int dx^4 \sqrt{|g|} \left(\frac{R}{16\pi G} - \frac{1}{2}\partial_\mu \varphi \partial^\mu \varphi - W \right). \tag{3.40}$$

A long and cumbersome calculation, removing several total derivatives (Mukhanov *et al.*, 1992) shows that to second order, the perturbation of this action is given by (3.39). A much more elegant recent derivation of this result is given in Maldacena (2003). Here we present some elements of this paper. The first relevant point is the choice of the so-called unimodular gauge where, $\delta\varphi = 0$. From (3.12) is follows that in this gauge

$$\delta\varphi^{(gi)} = -\dot{\varphi} k^{-1} \sigma. \tag{3.41}$$

We use the $3+1$ formalism of General Relativity; see, for example, Padmanabhan (2010), with the general metric ansatz

$$ds^2 = -N^2 + h_{ij}(dx^i + N^i dt)(dx^j + N^j dt), \tag{3.42}$$

where N is the lapse function and N^i is the shift vector. In these variables the scalar field action (3.40) becomes (up to a total derivative)

$$S = \int dx^4 \sqrt{h} \Big(\frac{NR^{(3)} + N^{-1}(E_{ij}E^{ij} - E^2)}{16\pi G} + \frac{1}{2}\left(N^{-1}(\dot{\varphi} - N^i \partial_i \varphi)^2 - N h^{ij} \partial_i \varphi \partial_j \varphi\right) - NW \Big). \tag{3.43}$$

where we have introduced

$$E_{ij} = \frac{1}{2}\left(\dot{h}_{ij} - \nabla_i N_j - \nabla_j N_i\right). \tag{3.44}$$

In these expressions, spatial indices are lowered and raised with h_{ij} and its inverse, h^{ij}; h is the determinant of h_{ij}; $R^{(3)}$ is the scalar curvature of h_{ij}; and ∇_i denotes covariant derivatives w.r.t. h_{ij}. The constraint equations determine N and N_i, which are not dynamical variables. The dynamics can be described by a Hamiltonian for h_{ij} and its conjugate momenta. See Padmanabhan (2010) for details. For a scalar field in unimodular gauge, $\delta\varphi = 0$, the constraint equations become (see Ex. 3.3)

$$R^{(3)} - N^{-2}(E_{ij}E^{ij} - E^2) = 8\pi G\left(V + \frac{1}{2N^2}\dot{\varphi}^2\right) \tag{3.45}$$

$$\nabla_i\left(N^{-1}(E^i_j - \delta^i_j E)\right) = 0. \tag{3.46}$$

Let us first study scalar perturbations and choose as a second gauge condition $H_T = 0$ to all orders. The spatial metric is then of the form

$$h_{ij} = a^2 \exp(2\zeta)\delta_{ij}. \tag{3.47}$$

Here ζ is just a name but we shall see that in first-order perturbation theory, this is exactly our gauge invariant variable ζ. To first order

$$N = a(1 + \delta N), \qquad N_i = \partial_i \psi. \tag{3.48}$$

The constraint equations (3.45) and (3.46) then yield

$$\delta N = \frac{\dot{\zeta}}{\mathcal{H}}, \qquad \psi = -\frac{\zeta}{\mathcal{H}} + \frac{\dot{\varphi}^2}{2\mathcal{H}^2}\dot{\zeta}. \tag{3.49}$$

To find the quadratic action we can now simply insert these solutions and the ansatz (3.47) in the action and expand up to second order in ζ. After a little algebra we find (derive the details in Ex. 3.3)

$$S_s^{(2)} = \frac{1}{2}\int d^4x a^2 \frac{\dot{\varphi}^2}{\mathcal{H}^2}\left(\dot{\zeta}^2 - \partial_i\zeta\partial_i\zeta\right). \tag{3.50}$$

(Note that the factor a^{-2} to from raising one index μ of derivatives has already been included in the prefactor, $\sqrt{g} \to \sqrt{g}a^{-2} = a^2$.) It is interesting to see that in the absence of rolling, $\dot{\varphi} = 0$ and hence $\epsilon_1 = 0$; the second-order scalar action vanishes. The canoncially normalized field corresponding to this action is

$$v = \frac{a\dot{\varphi}}{\mathcal{H}}\zeta = z\zeta, \tag{3.51}$$

as we have found earlier in Eq. (3.37). Rewriting (3.50) in terms of the variable v yields (3.39).

To study tensor perturbations we simply set $\delta N = N^i = 0$ and make the ansatz

$$h_{ij} = \exp(2H_{ij}), \text{ with } \partial^i H_{ij} = H^i_i = 0. \tag{3.52}$$

Here, the exponential of a matrix is defined, as usual, via its series expansion. Inserting this in the action (3.43) we obtain after some algebra

$$S_T^{(2)} = \frac{1}{16\pi G}\int d^4x a^2 \left(\dot{H}_{ij}\dot{H}_{ij} - (\partial_k H_{ij})^2\right). \tag{3.53}$$

The canonically normalized variable, which we shall use later, is therefore h_{ij}, given by (see Ex. 3.4)

$$h_{ij} = \frac{a}{\sqrt{8\pi G}} H_{ij},\ h^i_i = 0,\ \partial^i h_{ij} = 0. \tag{3.54}$$

3.3 Generation of Perturbations during Inflation

So far we have simply assumed some initial fluctuation amplitude A_s, without investigating where it came from or what the k-dependence of A_s might be. In this section we discuss the most common idea about the generation of cosmological perturbations, namely their production from the quantum vacuum fluctuations during an inflationary phase.

The basic idea is simple: a time-dependent gravitational field generically leads to particle production, analogously to the electron–positron production in a classical, time-dependent electromagnetic field in quantum electrodynamics.

3.3.1 Scalar Perturbations

The main result of this section is the following: during inflation, the produced particles induce a perturbed gravitational potential with a (nearly) scale-invariant spectrum (see Section 2.6),

$$k^3\langle|\zeta(k,t)|^2\rangle \propto k^{n_s-1} \quad \text{with} \quad n_s \simeq 1. \tag{3.55}$$

The quantity $\Delta_\zeta(k) \propto k^3\langle|\zeta(k,t)|^2\rangle$ is the squared amplitude of the curvature perturbation at comoving scale $\lambda = \pi/k$. To make sure that this quantity is small on a broad range of scales, so that neither black holes are formed on small scales nor large deviations from homogeneity and isotropy on large scales appear, we must require $n_s \simeq 1$. These arguments were put forward for the first time by Harrison and Zel'dovich (Harrison, 1970; Zel'dovich, 1972) (still before the advent of inflation), leading to the name "Harrison–Zel'dovich spectrum" for a scale-invariant perturbation spectrum as discussed in Chapter 2.

To derive the foregoing result we consider a scalar field background with energy density that is dominated by the potential term; hence the slow roll parameters, ϵ_1 and ϵ_2 are small. Over a reasonably short period of time we can approximate them as constants leading to nearly power law expansion

$$a \propto |t|^q \quad \text{with} \quad q = -1 - \epsilon_1 + \mathcal{O}(\epsilon_1^2). \tag{3.56}$$

We want to determine the Bardeen potential during the subsequent radiation and matter-dominated era. For this we use the fact that the curvature perturbation ζ remains constant on super-Hubble scales. Hence if we calculate its amplitude at Hubble crossing, $k/\mathcal{H} = 1$, during inflation it will remain constant until it reenters the Hubble horizon in the radiation or matter-dominated era.

3.3.1.1 Quantization

To determine the initial conditions and the evolution of v we now quantize the variable v in the action (3.50) or equivalently (3.39)

$$S_s^{(2)} = -\frac{1}{2}\int dx^4 \left[(\partial_\mu v)^2 + m^2(t)v^2\right] = \int dx^4\, \mathcal{L}, \tag{3.57}$$

with canonical momentum $\pi = \partial\mathcal{L}/\partial\dot{v} = \dot{v}$.

We now interpret $\hat{v}$ and $\hat{\pi}$ as operators on a Hilbert space that satisfy the standard canonical commutation relations given by

$$[\hat{v}(t,\mathbf{x}), \hat{v}(t,\mathbf{x}')] = [\hat{\pi}(t,\mathbf{x}), \hat{\pi}(t,\mathbf{x}')] = 0, \quad \text{and} \tag{3.58}$$

$$[\hat{v}(t,\mathbf{x}), \hat{\pi}(t,\mathbf{x}')] = i\delta^3(\mathbf{x}-\mathbf{x}') \qquad (\hbar = 1). \tag{3.59}$$

We now expand the operator $\hat{v}$ in Fourier modes,

$$\hat{v}(t,\mathbf{x}) = \frac{1}{(2\pi)^{3/2}}\int d^3k[v_k(t)\hat{a}_{\mathbf{k}}e^{i\mathbf{k}\cdot\mathbf{x}} + v_k^*(t)\hat{a}_{\mathbf{k}}^*e^{-i\mathbf{k}\cdot\mathbf{x}}]. \tag{3.60}$$

The operators $\hat{a}_{\mathbf{k}}$ and their hermitean conjugates $\hat{a}_{\mathbf{k}}^*$ are the annihilation and creation operators. Since $\hat{v}$ describes a real field, v_k^* is the complex conjugate of v_k. Choosing the time-independent normalization

$$v_k^*\dot{v}_k - v_k\dot{v}_k^* = +i, \tag{3.61}$$

Eqs. (3.58) and (3.59) require that the operators $a_{\mathbf{k}}$ satisfy the usual commutation relations

$$[\hat{a}_{\mathbf{k}}, \hat{a}_{\mathbf{k}'}] = [\hat{a}_{\mathbf{k}}^*, \hat{a}_{\mathbf{k}'}^*] = 0 \quad \text{and} \quad [\hat{a}_{\mathbf{k}}, \hat{a}_{\mathbf{k}'}^*] = \delta^3(\mathbf{k}-\mathbf{k}'). \tag{3.62}$$

The time-dependent mode functions v_k obey the classical equation of motion (3.35).

At very early times, $k \gg \mathcal{H}$, we can neglect the mass term, $\ddot{z}/z$, in Eq. (3.35) and v_k is the mode function of a free massless scalar field. We assume that initially $\hat{v}$ is in the vacuum state so that

$$v_k(t) = \frac{1}{\sqrt{2k}} \exp(-ikt) \quad \text{for} \quad k \gg \mathcal{H}. \tag{3.63}$$

Note that we work (as is usual in quantum field theory) in the Heisenberg picture; the state, which we assume to be the vacuum state at very early times and that we denote by $|0\rangle$, does not evolve but the field operator does.

At late times, $k \ll \mathcal{H}$, we may neglect k^2 in Eq. (3.35) and the growing mode solution behaves like $v_k \propto z$, so that $\zeta_k = v_k/z$ remains constant, as expected.

We want to calculate the power spectrum of $\hat{\zeta} = \hat{v}/z$,

$$\begin{aligned}\hat{\zeta}(t,\mathbf{x}) &= \frac{1}{(2\pi)^{3/2}} \int d^3k[\zeta_k(t)\hat{a}_{\mathbf{k}} e^{i\mathbf{k}\cdot\mathbf{x}} + \zeta^*{}_k(t)\hat{a}^*_{\mathbf{k}} e^{-i\mathbf{k}\cdot\mathbf{x}}] \\ &= \frac{1}{(2\pi)^{3/2}} \int d^3k[\hat{\zeta}_{\mathbf{k}} e^{i\mathbf{k}\cdot\mathbf{x}} + \hat{\zeta}^*_{\mathbf{k}} e^{-i\mathbf{k}\cdot\mathbf{x}}],\end{aligned} \tag{3.64}$$

defined by

$$\langle 0|\, \hat{\zeta}_{\mathbf{k}} \hat{\zeta}^*_{\mathbf{k}'} \,|0\rangle \equiv P_\zeta(k)\delta^3(\mathbf{k}-\mathbf{k}'). \tag{3.65}$$

With $\zeta_k = v_k/z$ and using the properties of the vacuum, $\langle X|\, a_k \,|0\rangle = \langle 0|\, a^*_k \,|X\rangle = 0$, for an arbitrary state $|X\rangle$ as well as the commutation relations (3.62) one obtains (see Exercise 3.2)

$$P_\zeta(k) = \frac{|v_k(t)|^2}{z^2}. \tag{3.66}$$

3.3.1.2 Perturbation Spectrum from Power Law Inflation

As a first simple example we consider power law inflation, $a \propto |t|^q$, $q \lesssim -1$. In this case $\dot{\mathcal{H}} \propto \mathcal{H}^2$, so that $z \propto a$ The evolution equation (3.35) for v then reduces to (we suppress the index k again)

$$\ddot{v} + \left(k^2 - \frac{q(q-1)}{t^2}\right) v = 0. \tag{3.67}$$

The solutions to this equation are of the form $(k|t|)^{1/2} H^{(m)}_\mu(kt)$, where $\mu = \frac{1}{2} - q$ and $H^{(m)}_\mu$ is the Hankel function of the mth kind ($m = 1$ or 2) and of order μ. The initial condition (3.63) requires that solely $H^{(2)}_\mu$ appears. Fixing the constants we obtain

$$v = -\frac{i\sqrt{\pi}\exp(iq\pi/2)}{2\sqrt{k}} (k|t|)^{1/2} H^{(2)}_\mu(k|t|).$$

At late times, $k/\mathcal{H} \sim k|t| \ll 1$, we have $H_\mu^{(2)}(k|t|) \simeq (i/\pi)\Gamma(\mu)(k|t|/2)^{-\mu}$. Inserting this in the preceding equation we obtain

$$v \simeq C(\mu)e^{i\alpha}(k|t|)^{1/2-\mu}k^{-1/2}, \quad k|t| \ll 1, \tag{3.68}$$

with $C(\mu) = (2^{\mu-1}/\pi^{1/2})\Gamma(\mu)$. The phase $e^{i\alpha}$ is uninteresting, as it disappears in the power spectrum. The power spectrum of $\zeta = v/z$ is thus given by

$$P_\zeta(k,t) = \left|\frac{v}{z}\right|^2 = C(\mu)^2\frac{(k|t|)^{1-2\mu}}{z^2k} \simeq \frac{4\pi C(\mu)^2}{\epsilon_1 m_P^2}\frac{(k|t|)^{1-2\mu}}{a^2k}, \quad k|t| \ll 1, \tag{3.69}$$

where we have used Eq. (3.38) in the last equals sign. Recalling that $1 - 2\mu = 2q$, we see that P_ζ is time independent on super-Hubble scales, as expected. We now replace $|t|$ by $\mathcal{H} = -q/|t|$ and multiply the spectrum by k^3, which yields[2]

$$k^3 P_\zeta(k) = \frac{2\pi H^2}{\epsilon_1 m_P^2}\left(\frac{k}{\mathcal{H}}\right)^{3-2\mu} \tag{3.70}$$

$$\Delta_\zeta = \frac{k^3}{2\pi^2}P_\zeta(k) = \frac{H^2}{\pi\epsilon_1 m_P^2}\left(\frac{k}{\mathcal{H}}\right)^{3-2\mu}, \tag{3.71}$$

where we have used $H = \mathcal{H}/a$ and $C(\mu)^2 \simeq 1/2$. The latter approximation is obtained by setting $q = -1$ and $\mu = 3/2$ in the expression for $C(\mu)$ and $\mathcal{H}$ above. The amplitude of a given mode k at Hubble exit is given by

$$\Delta_\zeta(k)\big|_{k=\mathcal{H}} = \frac{k^3}{2\pi^2}P_\zeta(k)\bigg|_{k=\mathcal{H}} = \frac{H^2}{\pi\epsilon_1 m_P^2}. \tag{3.72}$$

The scalar spectral index n_s is defined by

$$n_s - 1 = \frac{d\log(\Delta_\zeta)}{d\log(k)}. \tag{3.73}$$

From Eq. (3.71), using $\epsilon_1 = 1 - \dot{\mathcal{H}}/\mathcal{H}^2$ [see Eq. (1.153)] we obtain [see also Eq. (3.56)]

$$n_s - 1 = 3 - 2\mu = 2 + 2q = -\frac{2\epsilon_1}{1-\epsilon_1} \simeq -2\epsilon_1. \tag{3.74}$$

While the equals signs are exact for power law inflation, the last approximate sign is valid only to first order in ϵ_1.

In Exercise 3.1 you show that power law inflation with a scalar field requires an exponential potential $W = W_0\exp(-\alpha\varphi/m_P)$. The slow roll parameters are readily

[2] Remember that the conformal time is negative during inflation, $t < 0$.

calculated, $\epsilon_1 = \alpha^2/16\pi = -(3/2)\epsilon_2$. We therefore can also write with the same accuracy as earlier

$$n_s - 1 = \frac{-6(\epsilon_1 + \epsilon_2)}{1 - \epsilon_1} \simeq -6(\epsilon_1 + \epsilon_2) = -2\epsilon_1 - \eta. \tag{3.75}$$

For the last equals sign we made use of (1.155). We shall see in the next paragraph that this is the general result for slow roll inflation.

3.3.1.3 Slow Roll Inflation

We now want to derive Eqs. (3.75) and (3.71) when expansion no longer follows a power law exactly, but the slow roll parameters ϵ_1 and ϵ_2 are small. Let us first calculate the mass term, $\ddot{z}/z$ in this case. From Eq. (3.36) we have $z = \dot{\varphi}a/\mathcal{H}$. Taking the derivative of this using Eq. (3.24) we obtain

$$\frac{\dot{z}}{z} = (1 + 2\epsilon_1 + 3\epsilon_2)\mathcal{H}, \tag{3.76}$$

leading to

$$\frac{\ddot{z}}{z} = \left(\frac{\dot{z}}{z}\right)^2 + \left(\frac{\dot{z}}{z}\right)^{\bullet}$$

$$= (1 + 2\epsilon_1 + 3\epsilon_2)\dot{\mathcal{H}} + (2\dot{\epsilon}_1 + 3\dot{\epsilon}_2)\mathcal{H} + (1 + 2\epsilon_1 + 3\epsilon_2)^2\mathcal{H}^2. \tag{3.77}$$

As we saw in Eq. (1.155), the derivatives $\dot{\epsilon}_1$ and $\dot{\epsilon}_2$ are second order and can be neglected. Neglecting also all the other second-order terms we obtain

$$\frac{\ddot{z}}{z} = (2 + 9\epsilon_1 + 9\epsilon_2)\,\frac{1}{t^2}. \tag{3.78}$$

Here we have used $\mathcal{H}^2 = q^2/t^2 = (1+\epsilon_1)^2/t^2 \simeq (1+2\epsilon_1)/t^2$ and $\dot{\mathcal{H}} = -q/t^2 \simeq (1+\epsilon_1)/t^2$. Neglecting the time dependence of ϵ_1 and ϵ_2 (which is second order), the v equation (3.35) is therefore again a Bessel equation with $\mu^2 - 1/4 = 2 + 9\epsilon_1 + 9\epsilon_2$, hence

$$\mu = \frac{3}{2}(1 + 2\epsilon_1 + 2\epsilon_2)\,. \tag{3.79}$$

Inserting this in the power spectrum for ζ given in Eq. (3.71) we find

$$\Delta_\zeta(k) = \frac{k^3}{2\pi^2}P_\zeta(k) = \frac{H^2}{\pi\epsilon_1 m_P^2}\left(\frac{k}{\mathcal{H}}\right)^{-6(\epsilon_1+\epsilon_2)}, \quad k/\mathcal{H} \ll 1. \tag{3.80}$$

The spectral index is therefore

$$n_s - 1 = -6(\epsilon_1 + \epsilon_2) = -2\epsilon_1 - \eta, \tag{3.81}$$

which corresponds exactly to the expression (3.75) for power law inflation.

From the curvature spectrum it is now easy to determine the spectrum for the Bardeen potential Ψ in the matter-dominated era, for example, at recombination. As we have seen in Eq. (2.153), during power law expansion the growing mode of the Bardeen potential is constant on super-Hubble scales. In a matter-dominated universe, $w = 0$, the Bardeen potential is even constant on all scales. The relation (3.29) then yields

$$\Delta_\Psi(k) = \frac{9}{25}\Delta_\zeta(k) = \frac{9H^2}{25\pi\epsilon_1 m_P^2}\left(\frac{k}{\mathcal{H}}\right)^{-6(\epsilon_1+\epsilon_2)}, \quad k/\mathcal{H} \ll 1. \tag{3.82}$$

The amplitude A_Ψ and the spectral index n of the Sachs–Wolfe contribution to the CMB power spectrum given in Eq. (2.256) are thus determined by the energy scale of inflation, H, and the slow roll parameters ϵ_1 and ϵ_2. The amplitude of Δ_ζ is called the amplitude of scalar perturbations and Δ_ζ is usually parameterized as

$$\Delta_\zeta = A_s\left(\frac{k}{k_*}\right)^{n_s-1} \tag{3.83}$$

and the amplitude A_s depends on the "pivot scale" k_* (except for $n_s = 1$).

It is possible to develop the slow roll approximation further, to second and third order, which has been done in the literature (Hoffman and Turner, 2001; Schwarz *et al.*, 2001; Martin and Schwarz, 2003). But also at first order in slow roll, it is possible to constrain inflationary models severely by using present CMB data. We shall discuss this in detail in Chapter 9.

3.3.2 Vector Perturbations

In simple models of inflation where the only degrees of freedom are one or several scalar fields and the metric, only scalar and tensor but no vector perturbations are generated. But even in more complicated inflationary models where vector perturbations are generated, these decay during the subsequent evolution after inflation. Indeed, in a perfect fluid background the anisotropic stress vanishes, $\Pi_{ij} = 0$. The evolution of vector perturbations given by Eq. (2.117) then implies for the fluid vorticity Ω

$$\Omega \propto a^{3c_s^2-1}. \tag{3.84}$$

For a radiation–matter fluid, $\dot{p}/\dot{\rho} = c_s^2 \leq 1/3$, this leads to a nongrowing vorticity. The dynamical Einstein equation (2.109) for a perfect fluid yields

$$\sigma^{(V)} \propto a^{-2}, \tag{3.85}$$

and the constraint (2.106) reads (at early times, so that we can neglect curvature)

$$\Omega \propto (kt)^2\sigma^{(V)}. \tag{3.86}$$

Therefore, even if they are created in the very early Universe on super-Hubble scales during an inflationary period, vector perturbations of the metric, which are given by $\sigma^{(V)}$, will decay and soon become entirely negligible. Furthermore, even if the vorticity remains constant in a radiation-dominated universe, it will be so small on relevant scales at formation ($kt_{\rm in} \ll 1$) that we may safely neglect it.

Vector perturbations are irrelevant if they have been created at some early time, for example, during inflation. This result changes completely when considering "active perturbations" such as topological defects where vector perturbations contribute significantly to the CMB anisotropies on large scales; see Durrer *et al.* (2002). It is interesting to note that, in a background without anisotropic stresses, vector perturbations do not satisfy a wave equation and therefore will not oscillate. Vorticity simply decays with time; see Eq. (3.85).

3.3.3 Tensor Perturbations

The situation is different for tensor perturbations. Again we consider the perfect fluid case, $\Pi_{ij}^{(T)} = 0$. Equation (2.110) implies, if K is negligible,

$$\ddot{H}_{ij} + 2\mathcal{H}\dot{H}_{ij} + k^2 H_{ij} = 0. \tag{3.87}$$

If the background has a power law evolution or is slowly rolling, $a \propto |t|^q$ with $q = -1-\epsilon_1, \mathcal{H} = q/t$, this equation can be solved in terms of Bessel functions (see Abramowitz and Stegun 1970, Eq. 9.1.52). For $q < 1/2$, the less decaying mode solution to Eq. (3.87) is $H_{ij} = e_{ij}x^{1/2-q}Y_{1/2-q}(x)$, where Y_ν denotes the Bessel function of order ν, $x = |kt|$, and e_{ij} is a transverse traceless polarization tensor. (Remember that $t < 0$ during inflation, hence $|t|$ is decreasing.). This leads to

$$H_{ij} = \text{constant} \quad \text{for } x \ll 1, \tag{3.88}$$

$$H_{ij} = \frac{1}{a} \quad \text{for } x \gtrsim 1. \tag{3.89}$$

One may also quantize the tensor fluctuations, which represent gravitons. Doing this, one obtains (up to small corrections) a scale-invariant spectrum of tensor fluctuations from inflation. For tensor perturbations the canonical variable is simply given by

$$h_{ij} = e_{ij}h = \frac{m_P a}{2\sqrt{8\pi}} H_{ij}. \tag{3.90}$$

Here e_{ij} is a normalized transverse traceless polarization tensor, $e_i^i = k^i e_{ij} = 0$ and $e_{ij}e^{ij} = 2$. In terms of h the quadratic action (3.53) takes the canonical form $\dot{h}^2/2 + \cdots$. The evolution equation for h is obtained by inserting the ansatz (3.90) in Eq. (3.87),

$$\ddot{h} + (k^2 + m^2(t))h = 0, \tag{3.91}$$

$$\text{with} \quad m^2(t) = -\frac{\ddot{a}}{a} = -(\dot{\mathcal{H}} + \mathcal{H}^2) = -(2 - \epsilon_1)\mathcal{H}^2$$

$$= -(2 - \epsilon_1)\frac{1 + 2\epsilon_1}{t^2} \simeq -\frac{2 + 3\epsilon_1}{t^2}. \tag{3.92}$$

As can be inferred from the action (3.53), the variable h is canonically normalized and can therefore be quantized with the usual commutation relation.

During inflation the mass term $m^2(t)$ in Eq. (3.91) is negative, leading to particle creation. As for scalar perturbations, the vacuum initial conditions are given on scales that are far inside the Hubble scale, $k^2 \gg |m^2|$, where expansion can be neglected and we may set

$$h_{\text{in}} = \frac{1}{\sqrt{2k}} \exp(-ikt) \quad \text{for } k|t| \gg 1.$$

The solution of Eq. (3.91) with this initial condition can be expressed in terms of a Hankel function (up to an uninteresting phase),

$$h = \frac{\sqrt{\pi}}{2\sqrt{k}}(k|t|)^{1/2} H_\nu^{(2)}(kt),$$

where $\nu^2 - 1/4 = 2 + 3\epsilon_1$ so that $\nu = 3/2 + \epsilon_1$. On super-Hubble scales, $k|t| \ll 1$, we obtain $H_\nu^{(2)}(kt) \simeq \sqrt{\frac{2}{\pi}}(k|t|)^{-\nu}$. This results in the tensor power spectrum

$$k^3 P_H = 4k^3|H_{ij}H^{ij}| = 8\frac{8\pi k^3|h|^2}{a^2 m_P^2} \simeq 32\pi\frac{H^2}{m_P^2}\left(\frac{k}{\mathcal{H}}\right)^{-2\epsilon_1}, \quad k/\mathcal{H} \ll 1. \tag{3.93}$$

The factor 4 after the first equals sign is due to our definition of tensor perturbations, as $2H_{ij}$ and the additional factor 2 after the second equals sign is due to the fact that there are two tensor modes. The prefactor after the $\simeq$ sign is then obtained by setting $\nu = 3/2$ and $q = -1$. In the exponent, however, we keep the slow roll parameter $\epsilon_1 \neq 0$.

From Eq. (3.93) we derive the tensor spectral index n_t defined by

$$n_t = \frac{d\log\left(k^3 P_H\right)}{d\log(k)} = -2\epsilon_1. \tag{3.94}$$

After inflation, H_{ij} is constant on super-Hubble scales. The gravity wave spectrum is therefore determined by the amplitude of the fluctuations at Hubble crossing,

$$\Delta_H = \frac{k^3}{2\pi^2} P_H = A_t \, (kt_0)^{n_t}, \qquad k/\mathcal{H} \ll 1, \qquad \text{with} \tag{3.95}$$

$$A_t = \left. \frac{16}{\pi} \frac{H^2}{m_P^2} \right|_{k=\mathcal{H}_0} \simeq \left. \frac{128}{3m_P^4} W \right|_{k=\mathcal{H}_0}, \tag{3.96}$$

where W is the inflaton potential. Note that here, as in Eq. (3.72), $H|_{k=\mathcal{H}} = k/a$ indicates the value of the Hubble parameter H at "Hubble exit," that is, during inflation when $1/k$ becomes larger than the comoving Hubble scale $1/\mathcal{H}$, corresponding to $a = a_1$ in Fig. 3.1. This is much larger than the value of $H = \mathcal{H}/a$ at reentry, long after inflation, when again $k = \mathcal{H}$. At this second Hubble crossing time the Hubble parameter $H = k/a_2$ is much smaller, since $a_2 \gg a_1$ is much larger; see Fig. 3.1.

The parameter A_t introduced in Eq. (3.95) is the amplitude of the tensor spectrum at the present Hubble scale $t_0 \simeq 1/\mathcal{H}_0 = 1/H_0$, if we normalize the scale factor such that $a_0 = 1$. Like for scalar perturbations, we could have chosen some arbitrary other pivot scale k_*. But for definiteness and because of its relevance for CMB anisotropies we choose the present Hubble scale. Equation (3.96) relates the tensor

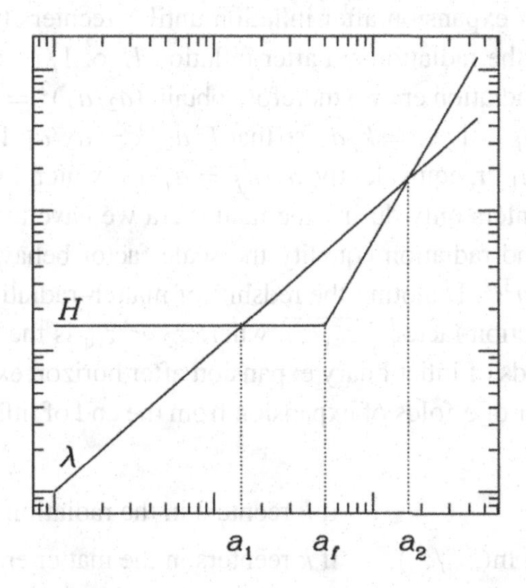

Fig. 3.1 We sketch the behavior of the Hubble scale $H^{-1} = a\mathcal{H}^{-1}$ and some wavelength $\lambda = a/k$ during and after inflation as functions of the scale factor. At $a = a_1$ during inflation, the scale λ exits the Hubble scale and after inflation, at $a = a_2$, it reenters. At $a = a_f$ inflation ends.

amplitude to the value of the inflaton potential at the moment when the comoving Hubble scale during inflation equals the present Hubble scale:

$$W_* \equiv W|_{\mathcal{H}=H_0} = \frac{3m_P^4}{128} A_t. \tag{3.97}$$

Measuring the amplitude of tensor perturbations therefore allows us to determine the energy scale of inflation. If we simply require that tensor fluctuations do not generate more than a third of the observed CMB anisotropies, according to Eq. (2.284), the present observational limits require $A_t \lesssim 10^{-10}$, so that

$$W_*^{1/4} \lesssim 0.001 \times m_P \simeq 10^{16}\,\mathrm{GeV}. \tag{3.98}$$

On the other hand, this energy scale must be larger than the reheating temperature after inflation. This limits the number of e-folds N_* of inflationary expansion after Hubble exit of the present Hubble scale H_0^{-1}. This is roughly the same number of e-folds it takes after inflation until the present.

Let us briefly derive this statement. We denote the Hubble parameter during inflation, which is nearly constant, by H_i and neglect its slight time dependence. We consider a comoving wave number k that exits the Hubble scale at the value a_1 of the scale factor during inflation, when its wavelength is $\lambda = \lambda_1 = a_1/k = H_i^{-1}$. It reenters after inflation at $a = a_2$, when its wavelength is $\lambda = \lambda_2 = a_2/k = H_2^{-1}$. The value of the scale factor at the end of inflation is denoted by a_f; see Fig. 3.1. The number of e-folds of inflation after a_1 is $N_1 = \ln(a_f/a_1)$, while the number of e-folds of expansion after inflation until λ reenters the Hubble scale is $N_2 = \ln(a_2/a_f)$. In the radiation era after inflation $H \propto 1/\tau \propto 1/a^2$. If the scale reenters during the radiation era we therefore obtain $(a_2/a_f)^2 = H_i/H_2 = H_i a_2/k$. On the other hand, $H_i = 1/\lambda_1 = k/a_1$, so that $H_i a_2/k = a_2/a_1$. Inserting this above yields $a_2^2/a_f^2 = a_2/a_1$ or, equivalently, $a_2/a_f = a_f/a_1$, which implies $N_1 = N_2$.

For a scale that enters only during the matter era we have to correct this result, since after matter and radiation equality the scale factor behaves as $a \propto \tau^{2/3}$ so that $H \propto 1/\tau \propto 1/a^{3/2}$. Denoting the redshift of matter–radiation equality by $z_{\rm eq}$, this leads to a correction factor $\sqrt{z_{\rm eq}/z_2}$, where $z_2 < z_{\rm eq}$ is the redshift at reentry. The number of e-folds of inflationary expansion after horizon exit, N_1, is therefore related to the number of e-folds of expansion from the end of inflation until reentry, N_2, by

$$N_1 \simeq \begin{cases} N_2 & \text{if } k \text{ reenters in the radiation era, } z_2 > z_{\rm eq} \\ N_2 - \frac{1}{2}\ln(z_{\rm eq}/z_2) & \text{if } k \text{ reenters in the matter era, } z_2 < z_{\rm eq}. \end{cases}$$

Neglecting the correction term $\frac{1}{2}\ln(z_{\rm eq}/z_2)$ which is never more than a few and denoting the reheating temperature by T_R, we therefore obtain the following limit for the number of e-folds of inflation after exit of the present Hubble scale,

$$N_* = \ln\left(\frac{T_R}{T_0}\right) \le \ln\left(\frac{W_*^{1/4}}{T_0}\right) \le \ln\left(\frac{2\times 10^{16}}{2.4\times 10^{-13}}\right) \simeq 66. \tag{3.99}$$

Note that this is a conservative estimate and the reheating temperature is most probably significantly lower.

3.3.3.1 The Consistency Relation

We have obtained the following results for the scalar and tensor power spectra induced during slow roll inflation:

$$n_s - 1 = -6(\epsilon_1 + \epsilon_2) = -2\epsilon_1 - \eta, \tag{3.100}$$

$$n_t = -2\epsilon_1, \tag{3.101}$$

$$\frac{\Delta_H}{\Delta_\zeta} = 16\epsilon_1 = -8n_t, \quad k/\mathcal{H} \ll 1. \tag{3.102}$$

Up to the usually small correction $-\eta = -\dot{\epsilon}_1/\epsilon_1/\mathcal{H}$, we obtain the same spectral index for scalar and tensor fluctuations. Using the relation (2.144) between ζ and $\Psi = \Phi$ in the radiation- and in the matter-dominated era, we find that on large scales, where $\dot{\Psi} = 0$, ζ and Ψ differ only by a constant factor,

$$\Delta_\Psi = \frac{4}{9}\Delta_\zeta, \quad k/\mathcal{H} \ll 1, \quad \text{(radiation-dominated era)}, \tag{3.103}$$

$$\Delta_\Psi = \frac{9}{25}\Delta_\zeta, \quad k/\mathcal{H} \ll 1, \quad \text{(matter-dominated era)}. \tag{3.104}$$

From $\Delta_\zeta = A_s(k/\mathcal{H}_0)^{n_s-1}$ for scalar perturbations and $\Delta_H = A_t(k/\mathcal{H}_0)^{n_t}$ for tensor perturbations on super-Hubble scales, the relation (3.102) implies

$$\frac{A_t}{A_s} \simeq 16\epsilon_1 = -8n_t. \tag{3.105}$$

Equation (3.102) or equivalently (3.105) is often also called the consistency relation of slow roll inflation. It is not surprising to find a relation, since A_s, A_t, n_s, and n_t are determined by the amplitude H_i/m_P and the two slow roll parameters ϵ_1 and ϵ_2. But the precise form of this relation is specific to slow roll inflation. It is one of the major goals of forthcoming CMB observations to measure tensor perturbations in order to test this relation, which holds for both slow roll and power law inflation (which is also slow roll for $q \simeq -1$) , but might be violated if inflation involved several scalar fields, occurred in several stages, or did not happen at all.

3.4 Non-Gaussianities from Inflation

It is well known that free quantum fields obey Gaussian statistics that are entirely described by the 2-point function or equivalently the power spectrum. All higher order moments are determined by Wick's theorem (see Appendix 7). This breaks down once interactions are considered and the equations of motion become non-linear. In quantum field theory, "small" interactions are treated using Feynman path integrals and in statistical mechanics small deviations from Gaussianity can be handled using the Feynman–Kac formula (see, e.g., Øksendal, 2007).

In inflationary cosmology, even if the potential W is of the form φ^2, so that the scalar field is essentially free, it is still subject to gravity, which is a nonlinear interaction and therefore will induce non-Gaussianities. In this section we discuss non-Gaussianities from inflation. A quadratic action only leads to free Gaussian fields. To determine the non-Gaussianities one has to compute the action to higher order and then compute, for example, $\langle\hat{\zeta}(\mathbf{k}_1)\hat{\zeta}(\mathbf{k}_2)\hat{\zeta}(\mathbf{k}_3)\rangle$ within the interacting quantum field theory. This is usually done using perturbation theory and the path integral. Denoting the interaction Hamiltonian by $H^{(3)}$, standard quantum field theory gives to lowest order in $H^{(3)}$ the following expression for the expectation value of an operator $\mathcal{O}$ [see, e.g., Weinberg (1995)]

$$\langle\mathcal{O}(t)\rangle = \left\langle 0\left|\left(1+i\int_{-\infty}^{t} dt' H^{(3)}(t')\right)\mathcal{O}\left(1-i\int_{-\infty}^{t} dt' H^{(3)}(t')\right)\right|0\right\rangle. \quad (3.106)$$

Here $1 \pm i\int_{-\infty}^{t} dt' H^{(3)}(t')$ is just the first term of the exponential that describes the unitary evolution of the vacuum state (or the operator). From this expression it is clear that setting $\mathcal{O} = \hat{\zeta}(\mathbf{k}_1)\hat{\zeta}(\mathbf{k}_2)\hat{\zeta}(\mathbf{k}_3)$ after multiplication with $H^{(3)}$, which is also third order in ζ, we obtain terms that are of sixth order in the free Gaussian field ζ, which we then can compute using Wick's theorem and that do not vanish. The detailed general calculation is lengthy and we just present the results and refer to the original literature (Acquaviva *et al.*, 2003; Creminelli, 2003; Maldacena, 2003; Weinberg, 2005). For the scalar part of the third order of the action (3.43) one finds

$$S^{(3)} = \int d^4x a^4\Big[e^{\zeta}\left(1+\frac{\dot{\zeta}}{\mathcal{H}}\right)\left(-2\partial^2\zeta-(\partial\zeta)^2\right)+\epsilon_1 e^{3\zeta}\dot{\zeta}^2\left(1-\frac{\dot{\zeta}}{\mathcal{H}}\right)$$
$$+\,e^{3\zeta}\left(\frac{1}{2}(\partial_i\partial_j\psi\partial^i\partial^j\psi-(\partial^2\psi)^2)\left(1-\frac{\dot{\zeta}}{\mathcal{H}}\right)-2\partial_i\psi\partial^i\zeta\partial^2\psi\right)\Big]. \quad (3.107)$$

Here indices are raised and lowered with the Minkowski metric and $\partial^2 = \partial_\mu\partial^\mu$. The variable ψ is defined in (3.49). This action can be simplified further; see Maldacena (2003). For the 3-point function one then finds to lowest order

$$\langle\hat{\zeta}(\mathbf{k}_1)\hat{\zeta}(\mathbf{k}_2)\hat{\zeta}(\mathbf{k}_3)\rangle = (2\pi)^3\delta(\mathbf{k}_1+\mathbf{k}_2+\mathbf{k}_3)\left(\frac{4\pi H^2}{\epsilon_1 m_P^2}\right)^2 \frac{1}{\Pi_{i=1}^3(2k_i)^3} \times\left[(3\epsilon_1+6\epsilon_2)\sum_i k_i^3+\epsilon_1\sum_{i\neq j}k_ik_j^2+\epsilon_1\frac{8}{k_t}\sum_{i>j}k_i^2k_j^2\right] \tag{3.108}$$

$$= (2\pi)^3\delta(\mathbf{k}_1+\mathbf{k}_2+\mathbf{k}_3)B(k_1,k_2,k_3). \tag{3.109}$$

Here $k_t = k_1+k_2+k_3$ is the sum of the moduli $k_i = |\mathbf{k}_i|$ and the Dirac-delta is a consequence of statistical homogeneity. It implies that the three vectors $(\mathbf{k}_1,\mathbf{k}_2,\mathbf{k}_3)$ form a closed triangle. The derivation of this formula is somewhat involved and we do not represent it here. We refer the reader to the papers cited earlier. Here we only want to make some remarks and give a derivation for a simple but important special case. First of all, as expected, for reasons of isotropy, the bispectrum $B(k_1,k_2,k_3)$ depends only on the moduli of the $\mathbf{k}_i$. Furthermore, for similar values of the $k_i \sim k$ it behaves as $B(k,k,k) \sim \epsilon P(k)^2 \propto k^{-6}$, where here ϵ denotes the larger of ϵ_1 and ϵ_2. Hence w.r.t. to the naively expected result of the order of the square of the power spectrum, there is an extra suppression by a slow roll parameter. This makes the non-Gaussianities from inflation very difficult to detect observationally, especially since the later nonlinear gravitational evolution tends to lead to a bispectrum of order $P(k)^2$ with no slow-roll suppression.

We now show that for $q = k_3 \to 0$ one obtains

$$\lim_{q\ll k} B(q,k_1,k_2) = -(n_s-1+\mathcal{O}(q^2/k^2))P(q)P(k), \qquad k_1\simeq k_2\simeq k. \tag{3.110}$$

This is a consequence of the so-called consistency relation that quantifies the following fact: consider a small density perturbation with wave number k on top of a much larger one with wave number $q \ll k$. The effect of the large-scale perturbation on the smaller one will then simply make it evolve as it would in a universe with density $\rho = \bar{\rho}(1+\delta_q)$. This idea can be formalized in so-called consistency relations, which show that the presence of a long mode, let us call it ζ_L, can be absorbed by a coordinate transformation on the short modes, which we shall specify in the text that follows; see Eq. (3.126). Denoting the transformed coordinates by $\tilde{x}_n = x_n + \delta x_n$ we then have

$$\langle\zeta(x_1)\zeta(x_2)|\zeta_L\rangle = \langle\zeta(\tilde{x}_1)\zeta(\tilde{x}_2)\rangle. \tag{3.111}$$

Averaging over ζ_L this yields

$$\langle\zeta_L(x)\zeta(x_1)\zeta(x_2)\rangle = \langle\zeta_L(x)\langle\zeta(x_1)\zeta(x_2)|\zeta_L\rangle\rangle = \langle\zeta_L(x)\langle\zeta(\tilde{x}_1)\zeta(\tilde{x}_2)\rangle\rangle. \tag{3.112}$$

To first order, the coordinate transformation gives

$$\langle \zeta(\tilde{x}_1)\zeta(\tilde{x}_2)\rangle = \langle \zeta(x_1)\zeta(x_2)\rangle + \sum_{n=1}^{2} \delta x_n^\mu \partial_{n\mu} \langle \zeta(x_1)\zeta(x_2)\rangle. \tag{3.113}$$

Here $\partial_{n\mu} \equiv \partial/\partial x_n^\mu$. The first term is simply the 2-point function and since $\langle \zeta_L(x)\rangle = 0$ it gives no contribution to (3.112). We therefore obtain for the 3-point function with one long mode

$$\langle \zeta_L(x)\zeta(x_1)\zeta(x_2)\rangle = \left\langle \zeta_L(x) \sum_{n=1}^{2} \delta x_n^\mu \partial_{n\mu} \langle \zeta(x_1)\zeta(x_2)\rangle \right\rangle. \tag{3.114}$$

As we shall show in the text that follows [see Eq. (3.126)] the presence of a long mode corresponds to a rescaling of the spatial coordinates

$$\delta x_n^i = \zeta_L(x_+) x_n^i, \qquad \delta t_n = 0. \tag{3.115}$$

Here x_+ is in principle arbitrary but it should be in the vicinity of x_1 and x_2, so we set it as $x_+ = (x_1 + x_2)/2$. We then have

$$\langle \zeta_L(x)\zeta(x_1)\zeta(x_2)\rangle = \left\langle \zeta_L(x)\zeta_L(x_+) \sum_{n=1}^{2} \delta x_n^i \partial_i \langle \zeta(x_1)\zeta(x_2)\rangle \right\rangle. \tag{3.116}$$

We now use that

$$\langle \zeta(x_1)\zeta(x_2)\rangle = \xi(|\mathbf{x}_1 - \mathbf{x}_2|, t) = \xi(|\mathbf{x}_-|, t) \tag{3.117}$$

depends only on the absolute value of the difference, $\mathbf{x}_1 - \mathbf{x}_2 = \mathbf{x}_-$. This implies

$$(\delta x_1^i \partial_{1i} + \delta x_2^i \partial_{2i} \langle \zeta(x_1)\zeta(x_2)\rangle = \zeta_L x_-^j \partial_{-j} \xi(x_-, t) \tag{3.118}$$

Let us write this in Fourier space using that ξ is the Fourier transform of the power spectrum, P_ζ

$$\xi(|\mathbf{x}_1 - \mathbf{x}_2|, t) = \frac{1}{(2\pi)^3} \int d^3k e^{-i\mathbf{k}\mathbf{x}_-} P_\zeta(k) \tag{3.119}$$

$$x_-^j \partial_j \xi(|\mathbf{x}_-|, t) = \frac{-i}{(2\pi)^3} \int d^3k \mathbf{k}\mathbf{x}_- e^{-i\mathbf{k}\mathbf{x}_-} P_\zeta(k) \tag{3.120}$$

$$= \frac{1}{(2\pi)^3} \int d^3k k^j \left(\partial_{k_j} e^{-i\mathbf{k}\mathbf{x}_-}\right) P_\zeta(k) \tag{3.121}$$

$$= \frac{-1}{(2\pi)^3} \int d^3k e^{-i\mathbf{k}\mathbf{x}_-} \partial_{k_j} \left(k^j P_\zeta(k)\right) \tag{3.122}$$

$$= \frac{-(n_s - 1)}{(2\pi)^3} \int d^3k e^{-i\mathbf{k}\mathbf{x}_-} P_\zeta(k). \tag{3.123}$$

For the last equals sign we used $\partial_{k_i} k^i = 3$ and

$$k^i \partial_{k_i} P_\zeta(x) = k\partial_k P_\zeta(k) = (-3 + n_s - 1)P_\zeta(k).$$

Putting everything together and Fourier transforming also $\langle \zeta_L(x)\zeta_L(x_+)\rangle$ we find

$$B(q,k,k) = -(n_s - 1)P(q)P(k), \qquad q \ll k. \tag{3.124}$$

This limit is often called the "squeezed limit" since the triangle $(\mathbf{k}_1, \mathbf{k}_2, \mathbf{k}_3)$ is nearly squeezed into a line, or the local limit since the points $\mathbf{x}_1$ and $\mathbf{x}_2$ are very close. It is easy to check that the general formula (3.108) reduces to (3.124) in the limit $q = k_1 \ll k_2, k_3 \simeq k$.

Of course such a non-Gaussianity is very small and can certainly not be measured within the near future. However, it also means that when measuring a primordial non-Gaussianity with squeezed bispectrum, finding $|B(q,k,k)| \gtrsim P(q)P(k)$ rules out single-field inflation on which this result is based.

Note that, other than single-field inflation, no ingredients have been used here; hence this result is valid for all single-field inflationary models beyond slow roll. It is also easily generalized beyond the 3-point function to a result for an arbitrary squeezed $(n+1)$-point function in terms of the n-point function and the power spectrum at low wave number; see Maldacena (2003) and Creminelli *et al.* (2012).

To understand the coordinate transformation (3.115), let us go back to the metric (3.47) and consider the case where $\zeta = \zeta_L + \zeta_S$ is the superposition of a long wave mode ζ_L and a short mode ζ_S,

$$h_{ij}dx^i dx^j = a^2 \exp(2(\zeta_L + \zeta_S))\delta_{ij}dx^i dx^j. \tag{3.125}$$

In a region where ζ_L is roughly constant we can remove the long mode by setting

$$\tilde{x}^i = x^i + \zeta_L x^i. \tag{3.126}$$

Up to first order in ζ_L this implies

$$h_{ij}dx^i dx^j = a^2 \exp(2\zeta_S)\delta_{ij}d\tilde{x}^i d\tilde{x}^j. \tag{3.127}$$

The constraint equations then determine N and N^i in the coordinates $\tilde{x}^i$ entirely in terms of ζ_S, and ζ_L has disappeared from the metric. As we have seen earlier, this coordinate transformation, however, leads to a nontrivial bispectrum as given in Eq. (3.124). In a region much smaller than the size of the long wavelength, this invariance is the relativistic analog to the fact that adding a constant to the gravitational potential does not change the physics. Generalizations to other coordinate transformations are discussed in Creminelli *et al.* (2012).

3.5 Mixture of Dust and Radiation Revisited

In this section we want to study the perturbation of a mixture of dust and radiation in more detail. We shall find that the system has two regular perturbation modes that we can identify with the adiabatic and an iso-curvature mode. We determine the solutions on super-Hubble scales for both modes explicitly and discuss the implications for CMB anisotropies, especially for the positions of the acoustic peaks.

After inflation and reheating the Universe is radiation dominated. Only very much later, at redshift $z < 10^4$, do dark matter and baryons start dominating. It may also be that a scalar field (called early quintessence) plays a certain subdominant role, but we neglect this possibility here. Curvature and a cosmological constant are certainly negligible at early times and we thus consider a mixture of radiation and matter only. As in Section 2.4.3 we define

$$R = \frac{\rho_r}{\rho} = \frac{\rho_r}{\rho_r + \rho_m}, \qquad a = \frac{\rho_m}{\rho_r} = \frac{1-R}{R}. \tag{3.128}$$

The scale factor is normalized to unity at equality, t_{eq}, defined by $\rho_m(t_{\mathrm{eq}}) = \rho_r(t_{\mathrm{eq}}) = \rho(t_{\mathrm{eq}})/2$. Also note that by definition $0 \le R \le 1$ and $R \simeq 1$ during the radiation era, while $R \simeq 0$ in the matter era. With $w_r = c_r^2 = \frac{1}{3}$ and $w_m = c_m^2 = 0$ we obtain the following useful relations (see also Section 2.4.3 and Exercise 1.3):

$$\frac{\rho_m}{\rho} = 1 - R, \tag{3.129}$$

$$w = \frac{\rho_r/3}{\rho} = \frac{1}{3}R, \tag{3.130}$$

$$c_s^2 = \frac{\dot\rho_r/3}{\dot\rho} = \frac{\frac{4}{3}R}{4R + 3(1-R)} = \frac{4R}{3(R+3)}. \tag{3.131}$$

Integrating the Friedmann equation,

$$\mathcal{H}^2 = \left(\frac{\dot a}{a}\right)^2 = \frac{4\pi G}{3}\rho_{\mathrm{eq}}\left(a^{-1} + a^{-2}\right), \tag{3.132}$$

we obtain the scale factor

$$a(t) = \left(\frac{t}{t_1}\right)^2 + 2\left(\frac{t}{t_1}\right) \quad \text{where} \quad t_1 \equiv \sqrt{\frac{3}{\pi G \rho_{\mathrm{eq}}}}. \tag{3.133}$$

The normalization $a(t_{\mathrm{eq}}) = 1$ yields $t_{\mathrm{eq}} = (\sqrt{2} - 1)t_1$. In terms of t_1, Eq. (3.132) leads to the following useful expression for the Hubble parameter:

$$\mathcal{H}^2 = \frac{4(1+a)}{t_1^2 a^2}. \tag{3.134}$$

The radiation/matter mixture has no anisotropic stresses and no intrinsic entropy perturbations. Equation (2.125) then leads to

$$\Gamma = \Gamma_{\rm rel} = \frac{(\rho_r + P_r)\rho_m}{3w(w+1)\rho^2} S_{rm} = \frac{4(1-R)}{3+R} S_{rm}. \tag{3.135}$$

Here we assume that both radiation and matter are themselves in thermal equilibrium.

The perturbation equations (2.138) and (2.139) for dust and radiation become, with $S \equiv S_{rm}$,

$$\dot{S} = -kV_{rm}, \tag{3.136}$$

$$k\dot{V}_{rm} + \frac{4R}{3+R}\mathcal{H}kV_{rm} = \frac{k^2}{3+R}D + \frac{k^2(1-R)}{3+R}S. \tag{3.137}$$

This is equivalent to the second-order equation

$$\ddot{S} + \frac{4R}{R+3}\mathcal{H}\dot{S} = -\frac{k^2}{3+R}\left[(1-R)S + D\right]. \tag{3.138}$$

In addition, we have the second-order equation (2.120) for D. Using our expressions for w, c_s^2, and Γ and $\Pi = K = 0$ we obtain

$$\begin{aligned}\ddot{D} + \frac{3-R-2R^2}{R+3}\mathcal{H}\dot{D} - \frac{9+3R+5R^2-R^3}{2(R+3)}\mathcal{H}^2 D + \frac{4R}{3(R+3)}k^2 D \\ = -\frac{4R(1-R)}{3(R+3)}k^2 S.\end{aligned} \tag{3.139}$$

We now want to transform this equation into a differential equation w.r.t. the variable R. For this we need $\dot{R} = -\dot{a}R^2 = \mathcal{H}R(R-1)$ and

$$\ddot{R} = \dot{\mathcal{H}}(R^2 - R) + \mathcal{H}(2R-1)\dot{R} = \frac{3}{2}\mathcal{H}^2(R-1)^2 R.$$

For the second equals sign we made use of $\dot{\mathcal{H}} = -(1+3w)\mathcal{H}^2/2$ [see Eq. (2.112)]. A lengthy but straightforward calculation gives

$$\begin{aligned}\frac{d^2D}{dR^2} + \left[\frac{1}{2R} - \frac{1}{R+3}\right]\frac{dD}{dR} - \frac{9+3R+5R^2-R^3}{2R^2(1-R)^2(R+3)}D \\ = \frac{-4}{3R(1-R)(R+3)}\left(\frac{k}{\mathcal{H}}\right)^2\left[\frac{1}{1-R}D + S\right].\end{aligned} \tag{3.140}$$

We also transform Eq. (3.138),

$$\frac{d^2S}{dR^2}+\left[\frac{3}{2R}-\frac{1}{1-R}-\frac{1}{R+3}\right]\frac{dS}{dR}$$
$$=-\frac{1}{R^2(1-R)^2(R+3)}\left(\frac{k}{\mathcal{H}}\right)^2[D+(1-R)S]. \qquad (3.141)$$

Equation (3.140) can be simplified by writing it as a differential equation for the variable $\Delta \equiv D(1-R)R^{-3/2}$,

$$\frac{d^2\Delta}{dR^2}+\left[\frac{7}{2R}-\frac{1}{R+3}+\frac{2}{1-R}\right]\frac{d\Delta}{dR}$$
$$=-\frac{4}{3R(R+3)}\left(\frac{k}{\mathcal{H}}\right)^2\left[\frac{1}{(1-R)^2}\Delta+R^{-3/2}S\right]. \qquad (3.142)$$

We want to study possible initial conditions after a generic inflationary phase and subsequent reheating (rh). We are interested in cosmological scales, $k^{-1}a_0 \sim \mathcal{O}(\text{Mpc})$. But from $\mathcal{H}_0^{-1}a_0 = H_0^{-1} \simeq 3000\,h^{-1}$ Mpc and our expressions for $\mathcal{H}$ and a one finds that $\mathcal{H}^{-1}(a=0.1)a_0 \sim \mathcal{O}\,(\text{Mpc})$. For the last estimate we have used $z_{\text{eq}} = a_0 \simeq 3300$ (see Appendix 1). The reheating temperature of the Universe is typically of the order $T_{\text{rh}} \sim 10^{10}$GeV, so that with $T_{\text{eq}} \sim 1$ eV, we obtain $a_{\text{rh}} \sim 10^{-19} \ll 0.1$. Therefore, to study the initial conditions it is sufficient to consider the limit of very long wave perturbations, $k/\mathcal{H} \to 0$. In this limit we may neglect the right-hand sides of Eqs. (3.141) and (3.142) and the equations decouple completely. They can then easily be solved by quadrature, leading to

$$\Delta = A_1R^{-5/2}\left[1-\frac{25}{9}R+\frac{5}{3}R^2-\frac{5}{3}R^3\right]+A_2, \qquad (3.143)$$

$$= A_1X(R)+A_2, \qquad (3.144)$$

$$S = B_1\left[3R^{-1/2}-2\log\left(\frac{1+\sqrt{R}}{1-\sqrt{R}}\right)\right]+B_2. \qquad (3.145)$$

We now transform Δ back into D and write the solutions in terms of the scale factor. If we just multiply the modes proportional to A_1 and A_2 by the factor $R^{3/2}/(1-R)$, both modes of D are singular. We would like to split the solution D into two modes $D = AU_R + BU_S$, where $U_S = R^{3/2}/(1-R)$ is decaying at late time, $R \to 0$, and we define $U_R = R^{3/2}/(1-R)X + bU_S$ with a constant b chosen such that U_R stays regular at early times, $R \to 1$. This can be achieved by choosing $b = \frac{16}{9}$. In terms of the scale factor, using

$$a = \frac{1-R}{R} \quad \text{and} \quad R = \frac{1}{a+1},$$

we get

$$D = AU_R + BU_S, \quad \text{with} \tag{3.146}$$

$$U_R = \frac{5}{3(1+a)} + \frac{(1+a)^2 - \frac{25}{9}(1+a) + \frac{16}{9}(1+a)^{-1/2}}{a}, \tag{3.147}$$

$$U_S = \frac{(1+a)^{-1/2}}{a}, \tag{3.148}$$

$$S = C_1 \left[2\log\left(\frac{\sqrt{a+1}+1}{\sqrt{a+1}-1}\right) - 3\sqrt{a+1} \right] + C_2, \tag{3.149}$$

$$= C_1 V_S + C_2. \tag{3.150}$$

Developing U_R we obtain

$$U_R(a) \simeq \frac{10}{9}a^2, \text{ if } \quad a \ll 1 \quad \text{ and } \quad U_R(a) \simeq a, \text{ if } \quad a \gg 1. \tag{3.151}$$

The singular modes behave like $U_S \simeq 1/a$ and $V_S \simeq 2\log(4/a)$ at early times, $a \ll 1$. In the most generic case, right after reheating, at $a = a_{\rm rh} \ll 1$, all the modes may have comparable amplitudes so that $|2\log(a_{\rm rh}/4)C_1| \sim |C_2| \simeq |Aa_{\rm rh}^2| \sim |B/a_{\rm rh}|$. Hence $C_1 \ll C_2$ and $B \ll A$. Therefore the U_S and V_S modes cannot be relevant at late times. We neglect them in what follows, setting $B = C_1 = 0$. On super-Hubble scales we hence end up with two possible modes, namely $A \neq 0$, $C_2 = 0$ and $A = 0$, $C_2 \neq 0$. The first is called the "adiabatic mode" and the second the "entropy mode." We shall also be interested in a linear combination of these modes, the so-called isocurvature mode in the text that follows.

3.5.1 Adiabatic Initial Conditions

Let us first consider the adiabatic mode, given by the initial condition

$$D = A\left[\frac{5}{3(1+a)} + \frac{(1+a)^2 - \frac{25}{9}(1+a) + \frac{16}{9}(1+a)^{-1/2}}{a}\right], \quad S = 0 \tag{3.152}$$

on super-Hubble scales, $k/\mathcal{H} \ll 1$. Here $A = A(k)$ is a function of the wave number that determines the spectrum. On sub-Hubble scales, radiation perturbations oscillate while matter perturbations that exert no pressure do not; therefore, adiabaticity, $S = 0$, cannot be maintained. The term "adiabatic perturbations" is, however, used for perturbations that have adiabatic initial conditions, that is, that

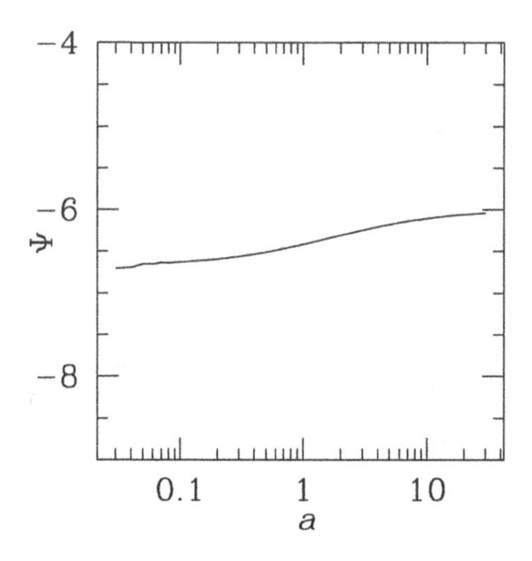

Fig. 3.2 The Bardeen potential (in units of A) for adiabatic perturbations of a mixed radiation and matter fluid on super-Hubble scales as a function of the scale factor normalized to equality, $a(t_{eq}) = 1$.

satisfy $S = 0$ at early times, when $k/\mathcal{H} \ll 0$. From the constraint Einstein equation we have $\Psi = -(3/2)(\mathcal{H}/k)^2 D$. With (3.134) this yields

$$\Psi = -\frac{6(1+a)}{(kt_1)^2 a^2} D. \tag{3.153}$$

This function is nearly constant in time; see Fig. 3.2. With Eq. (3.151) we find the following asymptotic behavior in the radiation- and matter-dominated eras:

$$\Psi = -6\frac{A}{(kt_1)^2} \times \begin{cases} \frac{10}{9}, & \text{if} \quad a \ll 1 \\ 1, & \text{if} \quad a \gg 1. \end{cases} \tag{3.154}$$

On super-Hubble scales we therefore have $\Psi \simeq \text{constant} \equiv \Psi_0$, where Ψ_0 denotes the value of the Bardeen potential in the late matter-dominated era. (In the Λ-dominated era the Bardeen potential starts to decay.) If the spectral index is defined as usual, $|\Psi|^2 k^3 \propto k^{n-1}$, we therefore have

$$|A|^2 \propto k^n, \text{ and } \quad |D|^2 \propto k^n \times \begin{cases} a^4, & \text{if} \quad a \ll 1 \\ a^2, & \text{if} \quad a \gg 1. \end{cases}$$

Let us now evolve forward to the matter-dominated era, $R \ll 1$ and $a \gg 1$, but no longer require $k/\mathcal{H} \ll 1$. Neglecting the terms that are subdominant in the matter era, Eq. (3.140) reduces to

$$\frac{d^2 D}{da^2} + \frac{3}{2a}\frac{dD}{da} - \frac{3}{2a^2} D = \frac{(kt_1)^2}{9a^2}[D + S]. \tag{3.155}$$

We first consider modes that enter the Hubble scale only in the matter-dominated era. For them $kt_1 \ll 1$ and we may always neglect the r.h.s. of Eq. (3.155). The growing mode solutions then also remain $D \simeq Aa$ and $\Psi \simeq \Psi_0$ on sub-Hubble scales.

Energy–momentum conservation for radiation (2.116) now becomes

$$D_g^{(r)\prime} = -\frac{4}{3}V_r, \tag{3.156}$$

$$V_r' = 2\Psi_0 + \frac{1}{4}D_g^{(r)}, \tag{3.157}$$

where here a prime denotes a derivative w.r.t. $x \equiv kt$. Now Ψ_0 acts like a constant source term. The general solution of this system is

$$D_g^{(r)} = A\cos\left(\frac{x}{\sqrt{3}}\right) - \frac{4}{\sqrt{3}}B\sin\left(\frac{x}{\sqrt{3}}\right) + 8\Psi_0\left[\cos\left(\frac{x}{\sqrt{3}}\right) - 1\right], \tag{3.158}$$

$$V_r = B\cos\left(\frac{x}{\sqrt{3}}\right) + \frac{\sqrt{3}}{4}A\sin\left(\frac{x}{\sqrt{3}}\right) + 2\sqrt{3}\Psi_0\sin\left(\frac{x}{\sqrt{3}}\right). \tag{3.159}$$

In Exercise 2.4 we show that adiabaticity requires $V_r = V_m$. But in the matter-dominated era $V_m \propto t \propto x$ so that

$$\lim_{x\to 0}\frac{V_r}{x} = \lim_{x\to 0}\frac{V_m}{x} = V_0 < \infty. \tag{3.160}$$

Therefore, we have to set $B = 0$ and $V_0 = A/4 + 2\Psi_0$. Using in addition $\Psi_0 = 3V_0$ [see (2.182)] we obtain

$$D_g^{(r)} = \frac{4}{3}\Psi_0\cos\left(\frac{x}{\sqrt{3}}\right) - 8\Psi_0, \tag{3.161}$$

$$V_r = \frac{1}{\sqrt{3}}\Psi_0\sin\left(\frac{x}{\sqrt{3}}\right), \tag{3.162}$$

$$D_{gm} = -\Psi_0\left(5 + \frac{1}{6}x^2\right), \tag{3.163}$$

$$V_m = \frac{1}{3}\Psi_0 x. \tag{3.164}$$

Here, we have neglected the influence of the radiation perturbations on the matter variable and simply used the pure dust solutions (2.176) and (2.175) for D_{gm} and V_m. On super-Hubble scales, $x \ll 1$, we have

$$D_g^{(r)} \simeq -\frac{20}{3}\Psi_0 \quad \text{and} \quad V_r \simeq \frac{1}{3}x\Psi_0. \tag{3.165}$$

This characterizes adiabatic initial conditions. Up to a constant, the density fluctuations oscillate like a cosine. At $x = 0$, $|D_g^{(r)}|$ has a minimum. The first maximum follows at $x = \sqrt{3}\pi$. This gives rise to the acoustic peak structure discussed in Section 2.6.

3.5.2 Isocurvature Initial Conditions

Let us now turn to the "entropy perturbations." We first recall that the curvature perturbation in the comoving gauge, ζ, is constant for adiabatic perturbations on super-Hubble scales. We want to calculate it for the entropy perturbation mode. From Eq. (2.147) with vanishing curvature we obtain

$$\dot{\zeta} = \frac{w}{w+1}\mathcal{H}\Gamma - \frac{2c_s^2}{3(w+1)}k^2\mathcal{H}^{-1}\Psi. \tag{3.166}$$

Inserting $\dot{\zeta} = \dot{R}\,d\zeta/dR = -R(1-R)\mathcal{H}\,d\zeta/dR$, $1+w = (3+R)/3$, and $c_s^2 = 4R/[3(3+R)]$, we obtain

$$\frac{d\zeta}{dR} = -\frac{4}{(R+3)^2}S + \frac{8}{3(R+3)^2(1-R)}\left(\frac{k}{\mathcal{H}}\right)^2\Psi, \tag{3.167}$$

$$= -\frac{4}{(R+3)^2}S - \frac{4}{(R+3)^2(1-R)}D, \tag{3.168}$$

where we have used the (00) Einstein equation in (2.105) for the second identity. For entropy perturbations one has $S = C =$ constant and $D = 0$ on large scales, $k \ll \mathcal{H}$, so that Eq. (3.168) can be integrated to

$$\zeta(R) = \frac{(1-R)}{(3+R)}C, \qquad (k/\mathcal{H}) \ll 1. \tag{3.169}$$

Here we have performed the definite integral from 1 to R in order not to add a constant to the result, because such a constant simply represents an adiabatic contribution. This mode therefore satisfies $\zeta \to 0$ and $\Psi \to 0$ for $R \to 1$, that is, in the radiation-dominated era on super-Hubble scales. Equations (2.144) and (2.145) imply

$$D_g = D - 3(1+w)\zeta. \tag{3.170}$$

For the isocurvature mode, $\zeta \to 0$ for $R \to 1$, $0 = D = RD_r + (1-R)D_m$ implies that $D_r \to 0$ and hence also $D_g^{(r)} \simeq D_r \to 0$ for $R \to 1$. Instead of the typical cosine oscillations of the adiabatic mode, we therefore obtain sine oscillations in $D_g^{(r)}$ when the scale $1/k$ enters the Hubble horizon.

As we have seen in Chapter 2, the CMB anisotropies contain a term

$$\frac{\Delta T}{T}(\mathbf{k}, t_0, \mathbf{n}) = \cdots + \frac{1}{4} D_g^{(\gamma)}(\mathbf{k}, t_{\rm dec})\, e^{i\mathbf{kn}(t_0 - t_{\rm dec})} + \cdots. \tag{3.171}$$

On scales where this term dominates, the peaks in $D_g^{(\gamma)}$ translate into peaks in the angular power spectrum of CMB anisotropies.

Since $D_g^{(\gamma)}$ oscillates like a sine for isocurvature perturbations, we find a first peak in $D_g^{(\gamma)} \propto \sin(c_s kt)$ at

$$x_i^{(0)} = k_i^{(0)} t_{\rm dec} = \frac{1}{c_s}\frac{\pi}{2}, \quad \lambda_i^{(0)} = \frac{\pi}{k_i^{(0)}} = 2c_s t_{\rm dec},$$

$$\vartheta_i^{(0)} \simeq \frac{2c_s t_{\rm dec}}{\chi\,(t_0 - t_{\rm dec})} \simeq \frac{2c_s t_{\rm dec}}{t_0}. \tag{3.172}$$

Here $\vartheta_i^{(0)}$ is the angle under which the comoving scale $\lambda_i^{(0)}$ at comoving distance $t_0 - t_{\rm dec}$ is seen; see Eqs. (1.38)–(1.48). Equation (3.172) shows clearly that $\vartheta_i^{(0)}$ strongly depends on the cosmological parameters, especially on curvature. The last $\simeq$ sign in Eq. (3/172) is true only if $K \simeq 0$.

The position of the acoustic peaks in the CMB anisotropy spectrum therefore presents an excellent means to determine the spatial curvature of the Universe. As we discussed in Chapter 2, when we expand the temperature fluctuations in terms of spherical harmonics, a fluctuation on angular scale ϑ shows up around the harmonic $\ell \sim \pi/\vartheta$. As an indication, we note that for $\Lambda = K = 0$, the harmonic of the first isocurvature peak is

$$\ell_i^{(0)} \sim \pi/\vartheta_i^{(0)} \sim 110.$$

In the adiabatic case the corresponding "first peak" would actually be at $k_a^{(0)} = 0$, but we have not discussed it because it is not visible at all. Since $k = 0$ is of course a super-Hubble scale at recombination, our discussion of the peak structure does not apply at this scale. This is also nearly true for the "first" peak of the isocurvature mode. Furthermore, $D_g^{(\gamma)}$ is negative for small x so that these "first" peaks are under-densities or "expansion peaks," and due to the gravitational attraction of the baryons (which we have neglected in this simple argument) they are less pronounced than the peaks due to over-densities, called "compression peaks."

These "second" peaks are usually called the first acoustic peaks. They are the first compression peaks. We shall also adopt the convention of calling them "first peak" for consistency with the literature. They correspond to wavelengths and angular scales

$$\left.\begin{aligned} \lambda_i^{(1)} &= \tfrac{2}{3}c_s t_{\rm dec}, \\ \vartheta_i^{(1)} &\simeq \tfrac{(2/3)c_s t_{\rm dec}}{\chi(t_0-t_{\rm dec})}, \\ \ell_i^{(1)} &\sim 330 \end{aligned}\right\} \quad \text{(isocurvature)}, \tag{3.173}$$

$$\left.\begin{aligned} \lambda_a^{(1)} &= c_s t_{\rm dec}, \\ \vartheta_a^{(1)} &\simeq \tfrac{c_s t_{\rm dec}}{\chi(t_0-t_{\rm dec})}, \\ \ell_a^{(1)} &\sim 220 \end{aligned}\right\} \quad \text{(adiabatic)}. \tag{3.174}$$

Here the indicated harmonic is the one obtained in the case $\Lambda = K = 0$, for a typical baryon density inferred from nucleosynthesis.

It is interesting to note that the distance between consecutive peaks is the same for adiabatic and isocurvature initial conditions. It is given by

$$\Delta k_i = k_i^{(1)} - k_i^{(0)} = \pi/(c_s t_{\rm dec}) = \Delta k_a, \quad \Delta\vartheta = \frac{c_s t_{\rm dec}}{\chi\,(t_0 - t_{\rm dec})}, \quad \Delta\ell \sim 200. \tag{3.175}$$

Again, the numerical value indicated for $\Delta\ell$ corresponds to a universe with $\Lambda = K = 0$. The result is strongly dependent, especially on K. This is the reason why the measurement of the peak position (or better of the interpeak distance) allows an accurate determination of curvature.

From our analysis we can draw the following important conclusions. For scales where the $D_g^{(\gamma)}$ term dominates, the CMB anisotropies show a series of acoustic oscillations with spacing Δk. The position of the first significant peaks is at $k = k_{a/i}^{(1)}$, depending on the initial condition; however, the spacing Δk is *independent* of initial conditions.

The angle $\Delta\vartheta$ onto which the scale Δk is projected in the sky is determined entirely by the matter content and the geometry of the Universe. According to our findings in Chapter 1 , ϑ will be larger if $\Omega_K < 0$ (positive curvature) and smaller if $\Omega_K > 0$ (see Fig. 1.4).

In our analysis we have neglected the presence of baryons, in order to obtain simple analytical results. Baryons have two effects: they lead to $(\rho - 3p)_{\rm rad+bar} > 0$, and therefore to an enhancement of the *compression* peaks (the first, third, etc. acoustic peak). In addition, the presence of baryons *decreases* the sound speed c_s of the baryon–photon plasma by about 10%, thereby increasing Δk and $\Delta\ell$ and decreasing $\Delta\vartheta$.

Another point that we have neglected is the fact that the Universe becomes matter dominated at $t_{\rm eq}$, only shortly before decoupling: $t_{\rm dec} \simeq 2.4 t_{\rm eq}$ for $\Omega_m \sim 0.3$. As we have seen, the gravitational potential on sub-Hubble scales is decaying in the radiation-dominated era. If the radiation-dominated era is not very long before decoupling, the gravitational potential is still decaying slightly and free-streaming photons fall into a deeper gravitational potential than they have to climb out of. This effect, called the "early integrated Sachs–Wolfe effect," adds to the photon temperature fluctuations at scales that are only slightly larger than the position of the first acoustic peak for adiabatic perturbations. It therefore "boosts" this peak and, at the same time, moves it to slightly larger scales (larger angles, lower spherical harmonics). Since $t_{eq} \propto h^{-1}$, the first acoustic peak is higher if h is smaller.

A small Hubble parameter therefore *increases* the amplitude of the first acoustic peak. A similar effect is observed if a cosmological constant or negative curvature is present, since t_{eq} is retarded in those cases. We shall discuss this dependence of the acoustic peak structure on cosmological parameters in detail in Chapter 9.

3.5.3 Mixed Adiabatic and isocurvature Perturbation

In general, inflation (from more than one scalar field) can lead to a mixture of adiabatic and isocurvature perturbations. At early time, $k/\mathcal{H} \ll 1$ and $R \to 1$, such a mixture is given by [see Eqs. (3.146) and (3.150)]

$$D = AU_R \quad \text{and} \quad S = C. \tag{3.176}$$

Here the "constants" A and C are random variables for each wave number $\mathbf{k}$. One usually assumes them to be Gaussian, so that all the expectation values are determined by $\langle A(\mathbf{k})A^*(\mathbf{k}')\rangle$, $\langle C(\mathbf{k})C^*(\mathbf{k}')\rangle$, and $\langle A(\mathbf{k})C^*(\mathbf{k}')\rangle$. Statistical homogeneity and isotropy require

$$\langle A(\mathbf{k})A^*(\mathbf{k}')\rangle = \delta(\mathbf{k}-\mathbf{k}')P_a(k), \tag{3.177}$$

$$\langle C(\mathbf{k})C^*(\mathbf{k}')\rangle = \delta(\mathbf{k}-\mathbf{k}')P_i(k), \tag{3.178}$$

$$\langle A(\mathbf{k})C^*(\mathbf{k}')\rangle = \delta(\mathbf{k}-\mathbf{k}')P_{ai}(k). \tag{3.179}$$

Clearly, $P_{ia}(k) = P_{ia}^*(k)$. Furthermore, Schwarz' inequality requires

$$\left|\langle A(\mathbf{k})C^*(\mathbf{k}')\rangle\right|^2 \le \langle A(\mathbf{k})A^*(\mathbf{k}')\rangle\langle C(\mathbf{k})C^*(\mathbf{k}')\rangle. \tag{3.180}$$

Hence the Hermitean 2×2 matrix (P_{mn}) is positive semidefinite. One calls A and C completely (anti-)correlated if

$$\langle A(\mathbf{k})C^*(\mathbf{k}')\rangle = \pm\sqrt{\langle A(\mathbf{k})A^*(\mathbf{k}')\rangle\langle C(\mathbf{k})C^*(\mathbf{k}')\rangle}.$$

To define such generic initial conditions one has, in principle, to specify four real functions, namely $P_a(k)$, $P_i(k)$, $\mathrm{Re}(P_{ai}(k))$, and $\mathrm{Im}(P_{ai}(k))$, which satisfy the inequality (3.180). The present data are fully compatible with purely adiabatic perturbations, $C = 0$. Nevertheless, a considerable iso-curvature contribution of a few % is still possible. (The precise percentage depends strongly on the definition of the ratio of isocurvature to adiabatic perturbations, e.g., on the scale at which it is defined, and whether it is the ratio of the CMB anisotropies or of some other perturbation variables.) It is interesting to note that the isocurvature contribution cannot be severely limited by CMB anisotropy data alone. The foregoing constraint comes mainly from the dark matter spectrum, to which isocurvature modes contribute very little on large scales. More details can be found in the literature (Trotta *et al.*, 2001/3; Moodley *et al.*, 2004). Another possibility for constraining the isocurvature mode are CMB polarization measurements with sufficient accuracy (Bucher *et al.*, 2001).

In reality, the situation is even more complicated. The real Universe contains not only photons and dark matter, but also neutrinos and baryons (and maybe quintessence). It has been found (Bucher *et al.*, 2000) that a mixture of photons, dark matter, baryons, and neutrinos allows five different modes that grow or stay constant, that is, that are "regular" in the sense that they do not grow very large into the past. These are the adiabatic mode and the dark matter isocurvature mode that we have just discussed, a similar baryon isocurvature mode (where only the baryon density is perturbed), and two neutrino modes (where only the neutrino density or velocity is perturbed). The acoustic peaks from the most generic initial conditions that allow for arbitrary correlations between the different modes are very unpredictable. For example, in a flat universe with a vanishing cosmological constant and fixed cosmic parameters we can obtain a first peak position in the range of $150 \leq \ell^{(1)} \leq 350$. However, combining CMB data and galaxy catalogs (LSS data) allows us to constrain the total contribution from all nonadiabatic modes to less than about 5%. This number will certainly still improve in the future, when even more accurate polarization data are available.

In the remainder of this book, we only discuss adiabatic perturbations, which are by far the most studied and that are in very good agreement with present data. However, one should keep in mind that all the results, especially those concerning the estimation of cosmological parameters, are not valid if we allow for more generic initial conditions (Bucher *et al.*, 2001; Trotta *et al.*, 2001/3; Moodley *et al.*, 2004).

Exercises

(The exercise marked with an asterisk is solved in Appendix 11 which is not in this printed book but can be found online.)

3.1 **Power law expansion***

Consider an FL universe filled with a (minimally coupled) scalar field with vanishing spatial curvature, $K = 0$. Show that the universe expands like a power law, $a \propto t^q$, so that $\mathcal{H} = q/t$ only if the scalar field potential is of the form

$$W(\varphi) = W_0 \exp\left(\alpha \frac{\varphi}{M_P}\right), M_P \equiv \frac{1}{\sqrt{8\pi G}}. \tag{3.181}$$

(M_P is the reduced Planck mass.)
Determine $\alpha(q)$ and $w(q) = P/\rho$. Determine also $p(q)$ such that $a \propto \tau^p$ and $\alpha(p)$. For which values of α do you obtain an inflationary universe?

Hint: Using the second Friedmann equation, show that power law expansion implies $w = P/\rho =$ constant and therefore both the quantities W and $\dot{\varphi}$ are power laws. Determine the corrsponding powers. Now solve for φ explicitely and insert the solutions into the expression (3.181) for W.

3.2 **The scalar power spectrum**
Using the canonical commutation relations for the generation and annihilation operators (3.62) show that the definition of the power spectrum(3.65) implies

$$P_\zeta(k) = \frac{\mid v_k(t) \mid^2}{z^2}.$$

3.3 **The second-order action for scalar perturbations**
Derive the constraint equations (3.45) and (3.46) by varying the action (3.43) w.r.t. N and N^j. Now derive Eq. (3.50) from (3.43), inserting the constraints in the form (3.49). Why do the second-order terms for N and N^j not appear?

3.4 **A canonical variable for tensor perturbations**
Consider a spatially flat FL universe with pure tensor perturbations,

$$ds^2 = a^2\left[-dt^2 + \exp(2H_{ij})\, dx^i\, dx^j\right]. \tag{3.182}$$

Consider only the gravitational part of the action (3.43) and show that up to second order in H_{ij} the perturbed action is given by

$$S_T^{(2)} = \frac{1}{2}\int d^4x \left[\dot{h}_{ij}\dot{h}^{ij} - h_{ij,l}h^{ij,l} + \frac{\ddot{a}}{a}h_{ij}h^{ij}\right]. \tag{3.183}$$

Indices in h_{ij} are raised and lowered with the flat metric δ_{ij} and

$$h_{ij} = \frac{m_p}{\sqrt{8\pi}} a H_{ij}.$$

Hint: Derive first (3.53) for H_{ij} and then transform to h_{ij}.

You can now Fourier transform h_{ij} and set $h_{ij}(\mathbf{k},t) = e_{ij}(\mathbf{k},+)h_+(\mathbf{k},t) + e_{ij}(\mathbf{k},\times)h_\times(\mathbf{k},t)$, where $e_{ij}(\mathbf{k},+)$ and $e_{ij}(\mathbf{k},\times)$ denote the two polarizations of the gravity wave that satisfy $e^i_i(\mathbf{k},\lambda) = k^i e_{ij}(\mathbf{k},\lambda) = 0$ and $e_{ij}(\mathbf{k},\lambda)e^{ij}(\mathbf{k},\lambda') = \delta_{\lambda\lambda'}$.

Calculate $e_{ij}(\mathbf{k},\lambda)$ for $\mathbf{k}$ along the z direction. Show that h satisfies Eq. (3.91) for each of the two polarizations λ.

3.5 **A mixture of matter and radiation**

Consider a mixture of a relativistic fluid, $P_r = (1/3)\rho_r$, and a nonrelativistic fluid, with energy density ρ_m and pressure $P_m = 0$ in a Friedmann universe with negligible curvature and cosmological constant. Assume that the fluids are noninteracting. As in Eq. (3.133),

$$R = \frac{\rho_r}{\rho} = \frac{\rho_r}{\rho_r + \rho_m}. \tag{3.184}$$

(1) Show that

$$a = \frac{\rho_m}{\rho_r} = R^{-1} - 1 = \frac{1-R}{R}$$

is the scale factor normalized to $a(t_{\rm eq}) = 1$, where $t_{\rm eq}$ is defined by $\rho_r(t_{\rm eq}) = \rho_m(t_{\rm eq}) \equiv \rho_{\rm eq}$.

(2) Show that

$$a(t) = \left(\frac{t}{t_1}\right)^2 + 2\left(\frac{t}{t_1}\right) \quad \text{where} \quad t_1 \equiv \sqrt{\frac{3}{\pi G \rho_{\rm eq}}}. \tag{3.185}$$

(3) Also derive the following useful relations that we have used throughout this chapter:

$$\mathcal{H}^2 = \frac{4(1+a)}{t_1^2 a^2},$$

$$w = \frac{R}{3},$$

$$c_s^2 = \frac{4R}{3(R+3)}.$$

Using $\tau_0 = 2/H_0$, $z_{\rm eq} = 2300h^2$, and $z_{\rm dec} = 1090$ calculate t_0, t_1, $t_{\rm eq}$, and $t_{\rm dec}$ in our flat model. Keep the explicit dependence on the Hubble parameter, h, in the expressions.

4
CMB Anisotropies

4.1 Introduction to Kinetic Theory

As we saw in Chapter 1, and as we know from statistical mechanics, the distribution function of photons in thermal equilibrium at temperature T is given by

$$f(\omega) = \frac{1}{e^{\omega/T} - 1}, \tag{4.1}$$

where $\omega = a|\tilde{\mathbf{p}}|$ is the physical photon energy. The comoving photon energy and momentum are denoted by $\tilde{p}^0$ and $\tilde{\mathbf{p}}$ and we have $\tilde{p} = |\tilde{\mathbf{p}}| = \tilde{p}^0 = a^{-1}\omega$. We shall denote the physical momenta by p^μ; hence $p^\mu = a\tilde{p}^\mu$. As long as interactions are sufficiently frequent to keep photons in thermal equilibrium, this distribution is maintained. Once the interaction rate drops below the Hubble rate, the distribution is affected only by redshifting photon momenta; this follows from Eq. (1.89) and was discussed in Chapter 1. As we saw there, if we define $T(a) = T_D a_D/a$ after decoupling, where $a(t_D) \equiv a_D$ is the scale factor at decoupling, the distribution retains its form as a Planck spectrum (4.1) even after decoupling. Of course, after decoupling $T(a)$ is no longer a temperature in the thermodynamical sense but merely a parameter of the distribution function. This point is especially interesting for neutrinos: even if they may have masses of the order of $m_\nu \sim 0.1\,\text{eV} \gg T_0$, their distribution is an extremely relativistic Fermi–Dirac distribution, since this is what it was at decoupling and it has changed since only by redshifting of neutrino momenta.

4.1.1 Generalities

We first present a brief introduction to relativistic kinetic theory. More details can be found in Ehlers (1971) and Stewart (1971).

In the context of general relativity on a spacetime $\mathcal{M}$, for a particle species with mass m we define the mass-shell, mass-bundle, or 1-particle phase space as the part of tangent space given by

$$P_m \equiv \{(x, p) \in T\mathcal{M} \mid g_{\mu\nu}(x)p^\mu p^\nu = -m^2\}. \tag{4.2}$$

This is a seven-dimensional subspace of the tangent space $T\mathcal{M}$. A (three-dimensional) "fiber" of the mass-bundle at a fixed event $x \in \mathcal{M}$ is defined by

$$P_m(x) \equiv \{p \in T_x\mathcal{M} \mid g_{\mu\nu}(x)p^\mu p^\nu = -m^2\}. \tag{4.3}$$

Here $T_x\mathcal{M}$ is the tangent space of $\mathcal{M}$ at point $x \in \mathcal{M}$. The 1-particle distribution function is defined on P_m,

$$f : P_m \to \mathbb{R} : (x, p) \mapsto f(x, p). \tag{4.4}$$

The distribution function is nonnegative and represents the phase-space density of particles with respect to the invariant measure $d\mu_M = 2\delta(p^2 + m^2)|g|\, d^4p\, d^4x$. Here g is the determinant of the metric and $p^2 = g_{\mu\nu}\tilde{p}^\mu\tilde{p}^\nu$. The factor 2 is a convention that we adopt here for convenience. We have chosen the coordinate basis $\partial_\mu = \partial/\partial x^\mu$ in tangent space, so that $p = \tilde{p}^\mu\partial_\mu$. We integrate over p^0 to get rid of the Dirac-δ. This yields the measure $d\mu_m$ on phase space in terms of the phase space coordinates $(\tilde{p}^i, x^\mu)$,

$$d\mu_m = \frac{|g(x)|}{|\tilde{p}_0(x, \tilde{\mathbf{p}})|}\, d^4x\, d^3\tilde{p} = \sqrt{|g(x)|}\, d\pi_m\, d^4x, \quad \text{where} \tag{4.5}$$

$$d\pi_m = \frac{\sqrt{|g(x)|}}{|\tilde{p}_0(x, \tilde{\mathbf{p}})|}\, d^3\tilde{p}. \tag{4.6}$$

Here $\tilde{\mathbf{p}} = (\tilde{p}^1, \tilde{p}^2, \tilde{p}^3)$ and $x = (x^0, x^1, x^2, x^3)$; $\tilde{p}_0 = g_{0\mu}\tilde{p}^\mu$ is determined as a function of $(x, \tilde{\mathbf{p}})$ via the mass-shell condition, $p^2 = -m^2$. The measure $\sqrt{|g(x)|}\, d^4x$ is the usual invariant measure on $\mathcal{M}$. Therefore densities on spacetime are obtained by integrating over the momenta with the measure $d\pi_m$. For example, the particle flux density 4-vector is given by

$$n^\mu(x) = \int_{P_m(x)} \frac{\sqrt{|g(x)|}}{|\tilde{p}_0(x, \tilde{\mathbf{p}})|} \frac{\tilde{p}^\mu}{\tilde{p}^0} f(x, \tilde{\mathbf{p}})\, d^3\tilde{p}. \tag{4.7}$$

More importantly, the energy–momentum tensor is given by

$$T^{\mu\nu}(x) = \int_{P_m(x)} \frac{\sqrt{|g(x)|}}{|\tilde{p}_0(x, \tilde{\mathbf{p}})|} \tilde{p}^\mu\tilde{p}^\nu f(x, \tilde{\mathbf{p}})\, d^3\tilde{p}. \tag{4.8}$$

If the particles are noninteracting, they move along geodesics,

$$\ddot{x}^\mu + \Gamma^\mu_{\nu\alpha}\dot{x}^\nu\dot{x}^\alpha = 0. \tag{4.9}$$

Here the dot denotes the derivative with respect to proper time s defined by the condition $g_{\mu\nu}(x)\dot{x}^\mu\dot{x}^\nu = \dot{x}^2 = -1$. In the case of massless (light-like) particles, the arc length cannot be defined. In this case the dot can be the derivative with respect to some arbitrary affine parameter. The geodesic equation (4.9) for massless particles, $\dot{x}^2 = 0$, is invariant under affine reparameterizations, $s \to As + B$, where A and B are constants.

Equation (4.9) is obtained as the Euler–Lagrange equation of the Lagrangian

$$\mathcal{L}(x, \dot{x}) = \frac{m}{2} g_{\mu\nu}(x)\dot{x}^\mu \dot{x}^\nu.$$

For massive particles m denotes the mass; for massless particles it is an arbitrary nonvanishing constant normally set to 1. The canonical momentum is then given by

$$\tilde{p}_\mu = \frac{\partial \mathcal{L}}{\partial \dot{x}^\mu} = m\dot{x}_\mu \quad \text{and} \quad \tilde{p}^\mu = m\dot{x}^\mu.$$

From the geodesic equation (4.9) we therefore have

$$m\dot{\tilde{p}}^\mu = -\Gamma^\mu_{\nu\alpha}\tilde{p}^\alpha\tilde{p}^\nu.$$

If there are no collisions, that is, no interactions other than gravity, the distribution function remains constant in a "comoving" volume element of phase space. Therefore

$$\frac{d}{ds} f \equiv \left[\dot{x}^\mu \partial_\mu + \dot{\tilde{p}}^i \frac{\partial}{\partial \tilde{p}^i}\right] f = 0, \tag{4.10}$$

$$\leftrightarrow \quad \left[\tilde{p}^\mu \partial_\mu - \Gamma^i_{\mu\nu}\tilde{p}^\mu\tilde{p}^\nu \frac{\partial}{\partial \tilde{p}^i}\right] f = 0. \tag{4.11}$$

This is the Liouville equation for collisionless particles. If collisions cannot be neglected, we have to replace the right-hand side by a collision term. Since collisions involve more than one particle, in principle the collision term depends on the 2- or even 3- and 4-particle distribution functions. To continue, one then has to derive an equation of motion for the 2-particle distribution function and so forth. This leads to the well known BBGKY (Bogoliubov–Born–Green–Kirkwood–Yvon) hierarchy of equations. Often, if the particles are sufficiently diluted, the 2-particle distribution function can be approximated by the product of the 1-particle distribution functions,

$$f_2(x, y, \mathbf{p}_x, \mathbf{p}_y) \simeq f(x, \mathbf{p}_x) f(y, \mathbf{p}_y). \tag{4.12}$$

This corresponds to the assumption that the particle positions in phase space are uncorrelated and is called "molecular chaos." In this case, the collision term becomes an integral over the momentum of the colliding particle and we obtain the Boltzmann equation,

$$\left[\tilde{p}^\mu \partial_\mu - \Gamma^i_{\mu\nu}\tilde{p}^\mu \tilde{p}^\nu \frac{\partial}{\partial \tilde{p}^i}\right] f = C[f]. \tag{4.13}$$

The collision integral $C[f]$ depends on the details of the interactions. We will calculate it for Thomson scattering of electrons and photons.

What we have discussed so far remains valid in the context of general relativity under some conditions on the number of collisions within a small volume that have to be satisfied in order for a coordinate-independent collision integral to exist (Ehlers, 1971).

In the kinetic approach it is often very useful to use a tetrad basis of vector fields, $e_\mu(x) = e^\nu_\mu \partial_\nu$, with $g(e_\mu, e_\nu) = g_{\alpha\beta} e^\alpha_\mu e^\beta_\nu = \eta_{\mu\nu}$. Here $\eta_{\mu\nu}$ denotes the flat Minkowski metric. With respect to such an orthonormal basis, $p = p^\mu e_\mu$, we have $|p_0| = |p^0| = \sqrt{m^2 - \mathbf{p}^2}$, where $\mathbf{p}^2 = \sum_{i=1}^3 (p^i)^2$ and $d\pi_m = d^3p/|p^0|$, as in flat Minkowski spacetime. This can also be written as

$$\eta_{\mu\nu} p^\mu p^\nu = g_{\mu\nu} \tilde{p}^\mu \tilde{p}^\nu.$$

4.1.2 Liouville's Equation in an FL Universe

We now want to discuss the Liouville equation in an FL universe. We choose the tetrad basis (orthonormal basis of four vector fields)

$$e_0 = a^{-1}\partial_t \quad \text{and} \quad e_i = a^{-1}\epsilon_i, \tag{4.14}$$

where (ϵ_i) is an orthonormal basis of vector fields for the metric of the 3-space of constant curvature γ_{ij}. If $K = 0$ we can choose $\epsilon_i = \partial_i$. But also if $K \neq 0$ we can always choose vector fields (ϵ_i) that form an orthonormal basis, that is, a basis that satisfies (see Exercise 4.1)

$$\gamma(\epsilon_i, \epsilon_j) = \delta_{ij}. \tag{4.15}$$

The expression for the energy–momentum tensor (with respect to the usual coordinate basis ∂_μ) in a Friedmann universe becomes

$$T^{\mu\nu}(x) = a^4\sqrt{\gamma(x)} \int_{P_m(x)} \frac{1}{|\tilde{p}_0|} \tilde{p}^\mu \tilde{p}^\nu f(x, \tilde{\mathbf{p}})\, d^3\tilde{p}, \tag{4.16}$$

where γ is the determinant of the three-dimensional metric (γ_{ij}) and we have used $|g| = a^8\gamma$.

The Liouville equation in a Friedmann universe in terms of the coordinates $(x^\mu, \tilde{p}^i)$ is given by

$$\tilde{p}^\mu \partial_\mu f|_{\tilde{p}} - \Gamma^i_{\mu\nu}\tilde{p}^\mu \tilde{p}^\nu \frac{\partial f}{\partial \tilde{p}^i} = 0. \tag{4.17}$$

Here we write $\partial_\mu f|_{\tilde{p}}$ in order to indicate that the components $\tilde{p}^i$ are fixed when the derivative w.r.t. x^μ is taken. Next we transform Eq. (4.17) into an equation for f with respect to the new coordinates (x^μ, p^i), that is, we consider f as a function of (x^μ, p^i). Since the FL universe is isotropic, f depends on the momentum only via[1] $p = \sqrt{\delta_{ij}p^i p^j} = \sqrt{a^2\gamma_{ij}\tilde{p}^i\tilde{p}^j} = a\tilde{p}$. The derivative of the distribution function with respect to t or $\mathbf{x}$ depends on the momentum variable, which we keep constant when performing this derivative. We denote by $\partial_\mu f|_p$ the derivative w.r.t. x^μ while keeping constant the momentum components p^i w.r.t. the orthonormal basis $\mathbf{e}_i$ while $\partial_\mu f|_{\tilde{p}}$ is the derivative w.r.t. x^μ keeping constant the momentum components $\tilde{p}^i$ w.r.t. the coordinate basis ∂_i. We then have

$$\partial_0 f|_{\tilde{p}} = \partial_0 f|_p + (\partial_0 p^j)|_{\tilde{p}}\frac{\partial f}{\partial p^j} = \partial_0 f|_p + \mathcal{H}p\frac{\partial f}{\partial p}, \tag{4.18}$$

$$\partial_i f|_{\tilde{p}} = \partial_i f|_p + (\partial_i p)|_{\tilde{p}}\frac{\partial f}{\partial p} = \partial_i f|_p + a^2\frac{\tilde{p}^k\tilde{p}^j\gamma_{kj},_i}{2p}\frac{\partial f}{\partial p},$$

$$\tilde{p}^i\partial_i f|_{\tilde{p}} = \tilde{p}^i\partial_i f|_p + a^2\frac{\tilde{p}^k\tilde{p}^j\tilde{p}^i\gamma_{kj},_i}{2p}\frac{\partial f}{\partial p}, \tag{4.19}$$

$$\frac{\partial f}{\partial \tilde{p}^i} = \frac{\partial p}{\partial \tilde{p}^i}\frac{\partial f}{\partial p} = a^2\frac{\gamma_{im}\tilde{p}^m}{p}\frac{\partial f}{\partial p}. \tag{4.20}$$

With Eq. (4.20) the terms with spatial Christoffel symbols in Eq. (4.17) become

$$\Gamma^i_{jm}\tilde{p}^m\tilde{p}^j\frac{\partial f}{\partial \tilde{p}^i} = a^2\Gamma^i_{jm}\gamma_{ik}\frac{\tilde{p}^m\tilde{p}^j\tilde{p}^k}{p}\frac{\partial f}{\partial p} = a^2\frac{1}{2}\gamma_{mj},_k\frac{\tilde{p}^m\tilde{p}^j\tilde{p}^k}{p}\frac{\partial f}{\partial p}. \tag{4.21}$$

In the last equals sign we have used the fact that $\tilde{p}^m\tilde{p}^j\tilde{p}^k$ is symmetrical in the indices m,k,j and therefore only the symmetrical part of the term $\Gamma^i_{jm}\gamma_{ik} = 1/2(\gamma_{kj,m} + \gamma_{km,j} - \gamma_{jm,k})$ contributes. With the help of Eq. (4.19) the terms $-\Gamma^i_{jk}\tilde{p}^j\tilde{p}^k(\partial f/\partial \tilde{p}^i)$ and $\tilde{p}^i(\partial_i p)|_{\tilde{p}}(\partial f/\partial p)$ in Eq. (4.17) cancel and we obtain

$$\tilde{p}^\mu\partial_\mu f|_p + \mathcal{H}\tilde{p}^0 p\frac{\partial f}{\partial p} - 2\Gamma^i_{0j}\tilde{p}^0\tilde{p}^j\frac{\partial f}{\partial \tilde{p}^i} = 0. \tag{4.22}$$

[1] Here we use p to denote the absolute value of the physical momentum while before we used it to denote the 4-vector p. Since these are very different objects we hope that there is no danger of confusion.

Inserting the Christoffel symbols of the FL universe (see Appendix 2, Section A2.3) $\Gamma^i_{0j} = \Gamma^i_{j0} = \mathcal{H}\delta^i_j$, we find

$$\tilde{p}^\mu \partial_\mu f|_p - \mathcal{H}\tilde{p}^0 p \frac{\partial f}{\partial p} = 0. \tag{4.23}$$

In an unperturbed FL universe we assume the distribution function to be homogeneous and isotropic, hence to depend on x^j and on p^i only via p. When using the coordinates p^i in momentum space we therefore expect f not to depend on the spatial coordinates x^i anymore. Therefore, the Liouville equation simplifies further to

$$\partial_0 f - \mathcal{H}p \frac{\partial f}{\partial p} = 0. \tag{4.24}$$

Or, setting $v = ap$ so that $\partial_0 f|_v = \partial_0 f|_p - \mathcal{H}p(\partial f/\partial p)$ and interpreting f as a function of (t, v), we obtain simply

$$\partial_0 f(t, v) = 0. \tag{4.25}$$

The Liouville equation in an FL universe therefore just requires that the distribution function of collisionless particles changes in time only by redshifting of the physical momentum p and therefore is simply a function of the redshift corrected momentum $v = ap$. Once this redshift correction is accounted for, it has no additional time dependence and is simply given by the initial condition. Note that this is true for spatially flat and spatially curved universes. We shall use the same letter f for $f(t, p)$ and $f(v)$.

4.2 The Liouville Equation in a Perturbed FL Universe

Let us consider small (first-order) deviations from an FL universe. This implies a small perturbation not only of the distribution function, but also of its domain of definition, the mass-shell (4.2), is modified due to the perturbations of the metric. We will keep track of this by modifying the tetrad fields (e_μ).

4.2.1 Scalar Perturbations

We derive the linear perturbation of Liouville's equation in longitudinal gauge. The perturbed metric is given by

$$ds^2 = -a^2(1 + 2\Psi)\, dt^2 + a^2(1 - 2\Phi)\gamma_{ij}\, dx^i\, dx^j. \tag{4.26}$$

The perturbed distribution function is $f = \bar{f}(v) + F^{(S)}(x^\mu, v, \theta, \phi)$, where (θ, ϕ) define the direction of the momentum $\mathbf{p}$. Liouville's equation now becomes, to first order, in the perturbations

$$\tilde{p}^\mu \partial_\mu f - \bar{\Gamma}^i_{\mu\nu} \tilde{p}^\mu \tilde{p}^\nu \frac{\partial f}{\partial \tilde{p}^i} - \delta\Gamma^i_{\mu\nu} \tilde{p}^\mu \tilde{p}^\nu \frac{\partial \bar{f}}{\partial \tilde{p}^i} = 0, \tag{4.27}$$

where the perturbations of the Christoffel symbols are given in Appendix 3, Eqs. (A3.2)–(A3.5). We have denoted background quantities by an over-bar. For simplicity, and also since this is the most relevant case, we restrict ourselves here to $K = 0$. The curved cases, $K \neq 0$ are treated in Appendix 9. However, in order to connect these results in a more straightforward manner to the $K \neq 0$ case, we do not yet make a Fourier decomposition of Φ, Ψ and $F^{(S)}$. We again use a tetrad basis, which is now given by

$$e_0 = a^{-1}(1 - \Psi)\partial_t \quad \text{and} \quad e_i = a^{-1}(1 + \Phi)\partial_i. \tag{4.28}$$

We want to transform Eq. (4.27) to the coordinates (x^μ, p^i) with $p^\mu e_\mu = \tilde{p}^\mu \partial_\mu$ so that

$$p^0 = a(1 + \Psi)\tilde{p}^0 \quad \text{and} \quad p^i = a(1 - \Phi)\tilde{p}^i. \tag{4.29}$$

For the transformation we use the derivatives

$$\partial_0 p^i|_{\tilde{p}} = \left[\mathcal{H}(1 - \Phi) - \dot{\Phi}\right] a\tilde{p}^i, \quad \text{so that}$$

$$\partial_0 f|_{\tilde{p}} = \partial_0 f|_p + [\mathcal{H}(1 - \Phi) - \dot{\Phi}]a\tilde{p}^i \frac{\partial f}{\partial p^i},$$

$$\partial_0 f|_{\tilde{p}} = \partial_0 f|_p + [\mathcal{H} - \dot{\Phi}]p\frac{\partial f}{\partial p}, \tag{4.30}$$

$$\tilde{p}^j \partial_j f|_{\tilde{p}} = \tilde{p}^j \partial_j f|_p - \tilde{p}^j \partial_j \Phi p \frac{\partial \bar{f}}{\partial p}. \tag{4.31}$$

As in the previous section, we indicate the momentum variable kept constant. With the help of the Liouville equation for $\bar{f}$, we then find

$$\tilde{p}^\mu \partial_\mu F^{(S)}\big|_p - \mathcal{H}\tilde{p}^0 p \frac{\partial F^{(S)}}{\partial p}$$
$$= a^{-1} v \frac{d\bar{f}}{dv}[p^i \partial_i \Phi + p^0 \dot{\Phi}] + a^{-1}\delta\Gamma^i_{\mu\nu} p^\mu p^\nu \frac{\partial \bar{f}}{\partial p^i}. \tag{4.32}$$

Inserting the perturbation of the Christoffel symbols (Eqs. (A3.2)–(A3.5) of Appendix 3), the right-hand side becomes

$$a^{-1} v \frac{d\bar{f}}{dv}\left[-p^0\dot{\Phi} + \frac{(\tilde{p}^0)^2}{\tilde{p}^2} p^k \partial_k \Psi\right],$$

where $\tilde{p}^2 = \sum_k (\tilde{p}^k)^2$ and we have used $p^i(\partial \bar{f}/\partial p^i) = v(d\bar{f}/dv)$.

We now rewrite the Liouville equation in terms of a new variable defined by $\mathcal{F} = F^{(S)} + \Phi v(d\bar{f}/dv)$. In most of the literature (Hu and Sugiayma, 1995;

Hu *et al.*, 1995, 1998; Hu and White, 1997a, 1997b) the variable $F^{(S)}$ is used directly. Note, however, that $\mathcal{F}$ and $F^{(S)}$ differ only by an isotropic (direction-independent) term. Hence, once we determine the CMB anisotropies this difference will only be present in the unmeasurable monopole term. The advantage of the variable $\mathcal{F}$ will become clear later.

Setting $\tilde{p}^j = \tilde{p}n^j$ with $1 = \delta_{ij}n^i n^j$ we have to lowest order, $\tilde{p} = p/a = v/a^2$. Defining also

$$q = a^2\tilde{p}^0 = a\omega = a\sqrt{p^2+m^2} = \sqrt{v^2+a^2m^2}, \tag{4.33}$$

we can rewrite the Liouville equation for the function $\mathcal{F}(t, \mathbf{x}, v, \mathbf{n})$ in the form

$$q\partial_0\mathcal{F} + vn^i\partial_j\mathcal{F} = n^i\partial_i\left[q^2\Psi + v^2\Phi\right]\frac{d\bar{f}}{dv}. \tag{4.34}$$

Here $v^j = ap^j$ are the redshift corrected physical momentum components and $\mathcal{F}$ is understood as a function of the variables x^μ and $v^j \equiv vn^j$. Since $\mathcal{F}$ and Φ, Ψ are already perturbations, we can use the background relations between $\mathbf{p}$ and $\mathbf{v}$ as well as q.

This is the Liouville equation for collisionless (massive) particles. If the right-hand side vanishes, it simply describes the free streaming of particles with momentum $(p^\mu) = (q, vn^i)/a$. If the particles are nonrelativistic, only the first term on the right-hand side contributes which simply describes the change of momenta by gravitational acceleration in the potential Ψ. If the particles are relativistic, the accelerating potential is $\Phi + \Psi$, which in the Newtonian case gives twice the Newtonian potential, as we also have it for light deflection; see Eq. (2.234). We shall see that the equation can be simplified in the massless case where $q = v$, which is relevant for the study of photons.

4.2.2 Vector Perturbations

Next we consider vector perturbations. For simplicity, here we do not use the vector gauge, but we set $B^i = 0$ so that

$$ds^2 = a^2\left(-dt^2 + (\gamma_{ij} + 2H_{ij})\,dx^i\,dx^j\right), \quad 2H_{ij} = H_{i|j} + H_{j|i}. \tag{4.35}$$

We use this gauge instead of the vector gauge, because it has a simpler perturbed orthonormal basis. The vector fields

$$e_0 = a^{-1}\partial_t \quad \text{and} \quad e_i = a^{-1}(\delta_i^j - H_i^j)\partial_j, \tag{4.36}$$

are orthonormal. If we used a gauge with $B^i \neq 0$ (nonvanishing "shift vector") this would lead to a mixing of time and space directions in the orthonormal basis and would complicate the calculations.

In the chosen basis the components of $p = \tilde{p}^\mu \partial_\mu = p^\mu e_\mu$ are related by

$$\begin{aligned}
\tilde{p}^0 &= a^{-1} p^0, \\
\tilde{p}^i &= a^{-1} p^j (\delta_j{}^i - H_j{}^i), \\
p^0 &= a \tilde{p}^0, \\
p^i &= a \tilde{p}^j (\delta_j{}^i + H_j{}^i).
\end{aligned}$$

The indices of H_{ij} are raised and lowered with the trivial metric δ_{ij}. In the gauge chosen in Eq. (4.35), the only nonvanishing perturbations of the Christoffel symbols are

$$\delta\Gamma^i_{j0} = \dot{H}^i{}_j, \quad \delta\Gamma^i_{jm} = H^i{}_{j|m} + H^i{}_{m|j} - H_{jm}{}^{|i}, \tag{4.37}$$

where in the spatially flat case, $_|$ is simply the ordinary partial derivative. Again, we want to write the Liouville equation $\tilde{p}^\mu \partial_\mu f - \Gamma^i_{\alpha\beta} \tilde{p}^\alpha \tilde{p}^\beta (\partial f / \partial \tilde{p}^i)$ for f as a function of (x^μ, p^i). The difference from the scalar case comes from the different basis and hence the difference in the relation between p^μ and $\tilde{p}^\mu$ and from the different Christoffel symbols. A short calculation gives for $f = \bar{f}(v) + F^{(V)}(t, \mathbf{x}, v, \mathbf{n})$

$$\begin{aligned}
\tilde{p}^i \partial_i f|_{\tilde{p}} &= a^{-1} p^i \left[\partial_i f|_p + \partial_i p^j|_{\tilde{p}} \frac{\partial f}{\partial p^j} \right] = a^{-1} p^i \left[\partial_i F^{(V)}|_p + p^k H^j_{k|i} \frac{\partial \bar{f}}{\partial p^j} \right] \\
&= a^{-1} p^i \left[\partial_i F^{(V)}|_p + \frac{p^j p^k}{p} H^j_{k|i} \frac{\partial \bar{f}}{\partial p} \right], \\
\Gamma^i_{jk} \tilde{p}^j \tilde{p}^k \frac{\partial f}{\partial \tilde{p}^i} &= a^{-1} \frac{p^i p^j}{p} p^k H^i_{k|j} \frac{\partial \bar{f}}{\partial p}.
\end{aligned}$$

Here we have used that the background contribution to f, $\bar{f}$, depends on momentum only via p so that $\partial \bar{f} / \partial p^j = (p^j / p)(\partial \bar{f} / \partial p)$. The other terms of the Liouville equation are

$$\begin{aligned}
\partial_0 f|_{\tilde{p}} &= \partial_0 f|_p + \partial_0 p^i|_{\tilde{p}} \frac{\partial f}{\partial p^i} \\
\partial_0 p^i &= \mathcal{H} a \tilde{p}^j (\delta^i_j + H_j{}^i) + a \tilde{p}^j \dot{H}^i_j = \mathcal{H} p^i + a \tilde{p}^j \dot{H}^i_j \\
\tilde{p}^0 \partial_0 f|_{\tilde{p}} &= a^{-1} \left[p^0 \partial_0 f|_p + p^0 \mathcal{H} p^i \frac{\partial f}{\partial p^i} + \dot{H}_{ij} \frac{p^i p^j}{p} \frac{\partial \bar{f}}{\partial p} \right].
\end{aligned}$$

Furthermore,

$$2\Gamma^i_{0j} \tilde{p}^j \tilde{p}^0 \frac{\partial f}{\partial \tilde{p}^i} = 2\mathcal{H} p^0 p^i \frac{\partial f}{\partial p^i} + 2\dot{H}_{ij} \frac{p^i p^j}{p} \frac{\partial \bar{f}}{\partial p}.$$

Together these results yield

$$\tilde{p}^{\mu}\partial_{\mu} f|_{\tilde{p}} - \Gamma^{i}_{\mu\nu}\tilde{p}^{\mu}\tilde{p}^{\nu}\frac{\partial f}{\partial \tilde{p}^{i}}$$
$$= a^{-1}\left[p^{0}\partial_{0} f|_{p} + p^{i}\partial_{i} f|_{p} - \mathcal{H}p^{0}p^{i}\frac{\partial f}{\partial p^{i}} - p^{0}\dot{H}^{i}_{m}\frac{p^{m}p^{i}}{p}\frac{\partial f}{\partial p}\right]. \tag{4.38}$$

Using the zeroth-order Liouville equation, and transforming to the redshift corrected momentum variable $v = ap$, all this finally leads to the following Liouville equation for $F^{(V)}(t, \mathbf{x}, v, \mathbf{n})$:

$$q\partial_{0} F^{(V)} + vn^{i}\partial_{i} F^{(V)} = qvn^{i}n^{j}\dot{H}_{ij}\frac{d\bar{f}}{dv}. \tag{4.39}$$

where for $B = 0$, $\sigma^{(V)}_{\ell m} = a\dot{H}^{(V)}_{\ell m}$ as defined in Chapter 2, Eq. (2.54).

The right-hand side of this equation has no Newtonian analog. For nonrelativistic particles it simply implies free streaming, $\partial_{0} F^{(V)} + \omega^{i}\partial_{i} F^{(V)} = 0$, where $\omega^{i} = v^{i}/(am)$ is the velocity. In the presence of a vector-type gravitational field, "frame dragging" a new gravitational term, the right-hand side of (4.39), has to be added that describes the evolution of the particle momenta due to frame dragging. Note that it was useful to choose the redshift-corrected momentum v and the directions n^{i} as our momentum variables. Otherwise the Liouville equation would be significantly more complicated.

4.2.3 Tensor Perturbations

For tensor perturbations the perturbed metric is given by

$$ds^{2} = a^{2}\left(-dt^{2} + (\gamma_{ij} + 2H_{ij})\, dx^{i}\, dx^{j}\right), \quad H^{i}_{i} = H^{j}_{i|j} = 0. \tag{4.40}$$

As before, we define the perturbation of the distribution function by

$$f = \bar{f}(v) + F^{(T)}(t, \mathbf{x}, v, \mathbf{n}).$$

The situation is exactly the same as for vector perturbations and we find the same Liouville equation,

$$q\partial_{0} F^{(T)} + vn^{i}\partial_{i} F^{(T)} = qvn^{i}n^{j}\dot{H}_{ij}\frac{d\bar{f}}{dv}. \tag{4.41}$$

This describes the evolution of the distribution function in the field of a gravitational wave H_{ij}. The equation is identical to (4.39) but of course now H_{ij} and $F^{(T)}$ are spin-2 perturbations.

4.3 The Energy–Momentum Tensor

From the perturbed distribution function and metric, we can determine the perturbed energy–momentum tensor. We start from the general expression

$$T^{\mu}{}_{\nu}(x) = \int_{P_m(x)} \frac{\sqrt{|g(x)|}}{|\tilde{p}_0(x,\tilde{\mathbf{p}})|} \tilde{p}^{\mu} \tilde{p}_{\nu} f(x,\tilde{\mathbf{p}})\, d^3\tilde{p}. \tag{4.42}$$

Observe that the components $\tilde{p}^{\mu}$ are the momentum components w.r.t. the coordinate basis ∂_{μ}. We now use

$$\tilde{p}^0 = a^{-2}(1-\Psi)q,$$
$$\tilde{p}_0 = -(1+\Psi)q,$$
$$\tilde{p}^i = a^{-2} v n^j (\delta_j{}^i - H_j{}^i),$$
$$\tilde{p}_i = v n_j (\delta^j{}_i + H^j{}_i).$$

Here we consider scalar, vector, and tensor perturbations together so that $H_{ij} = -\Phi\delta_{ij} + H_{ij}^{(V)} + H_{ij}^{(T)}$, where we have chosen longitudinal gauge for the scalar perturbations. In the following subsections we then isolate the contributions for the scalar, vector, and tensor perturbations of the energy–momentum tensor. We note that to first order $\det g = -a^8(1+2\Psi-6\Phi)$. To transform the integration $d^3\tilde{p}$ in Eq. (4.42) into an integration w.r.t. d^3p we use

$$\det\left(\frac{d\tilde{p}}{dp}\right) = a^{-3}\det\left(\delta_m{}^j - H_m{}^j\right) = a^{-3}[1+3\Phi]. \tag{4.43}$$

With this we find that the metric perturbations in T_0^0 and T_j^0 cancel and we obtain the following expressions for the energy–momentum tensor:

$$T_0^0 = -\int_{P_m(x)} p^0 p^2 f(x,\mathbf{p})\, dp\, d\Omega_{\mathbf{n}} = \frac{-1}{a^4}\int q v^2 f\, dv\, d\Omega_{\mathbf{n}}, \tag{4.44}$$

$$T_0{}^j = T^j{}_0 = -\int_{P_m(x)} n^j p^3 f(x,\mathbf{p})\, dp\, d\Omega_{\mathbf{n}} = \frac{-1}{a^4}\int n^j v^3 f\, dv\, d\Omega_{\mathbf{n}}, \tag{4.45}$$

$$T^i{}_j = (\delta^i{}_\ell - H^i{}_\ell)(\delta_{jk} + H_{jk})\frac{1}{a^4}\int n^\ell n^k \frac{v^4}{q} f\, dv\, d\Omega_{\mathbf{n}},$$
$$= \bar{T}^i{}_j + \frac{1}{a^4}\int n^i n_j \frac{v^4}{q} F\, dv\, d\Omega_{\mathbf{n}} = \bar{P}\delta^i{}_j + \delta T^i{}_j. \tag{4.46}$$

To find the last expression we use the fact that the background stress tensor is diagonal, $\int n^\ell n^k \bar{f}\, d\Omega_{\mathbf{n}} = \frac{4\pi}{3}\delta^{\ell k}\bar{f}$, and that to first order

$$(\delta^{ik} - H^{ik})(\delta_k{}^j + H_k^j) = \delta^{ij},$$

since $H^{km} = H^k{}_m = H_{km}$ is symmetric. In Eq. (4.45) we have neglected the term proportional to $H^\ell{}_m$ since in the direction integral $\int n^j f$ only the perturbation of the distribution function contributes, so that this term would be second order. The surface element $d\Omega_\mathbf{n}$ denotes the integral over the sphere of momentum directions, $p^i = pn^i$.

Before turning to the different modes, let us split the stress tensor into a trace and a traceless part,

$$T^i{}_j = P\delta^i{}_j + \bar{P}\Pi^i{}_j \qquad \text{with} \tag{4.47}$$

$$P = \frac{1}{3}T^i{}_i = \bar{P} + \frac{1}{3a^4}\int \frac{v^4}{q} F\, dv\, d\Omega_\mathbf{n} \qquad \text{and} \tag{4.48}$$

$$\bar{P}\Pi^i{}_j = T^i{}_j - P\delta^i{}_j = \frac{1}{a^4}\int \frac{v^4}{q}\left(n^i n_j - \frac{1}{3}\delta^i_j\right) F\, dv\, d\Omega_\mathbf{n}. \tag{4.49}$$

Here we have used the fact that

$$\bar{P} = \frac{1}{3}\bar{T}^i{}_i = \frac{4\pi}{3a^4}\int \frac{v^4}{q}\bar{f}\, dv.$$

4.3.1 Scalar Perturbations

We now use the foregoing general expressions to determine the variables defined in Chapter 2 that specify scalar perturbations of the energy–momentum tensor. We consider a Fourier mode $F^{(S)}(t, \mathbf{k}, \mathbf{n})e^{i\mathbf{k}\cdot\mathbf{x}}$. As before, we denote background quantities with an overbar.

• *Density:*
Equation (4.44) implies

$$\rho(t, \mathbf{k}) = \bar{\rho}(t) + \delta\rho^{(\text{long})}(t, \mathbf{k}) = \frac{4\pi}{a^4}\int q v^2 \bar{f}(v)\, dv + \frac{1}{a^4}\int q v^2 F^{(S)}(t, \mathbf{k}, \mathbf{n})\, dv\, d\Omega_\mathbf{n}. \tag{4.50}$$

Hence

$$D_s = \frac{\delta\rho^{(\text{long})}}{\bar{\rho}} = \frac{1}{\bar{\rho}a^4}\int q v^2 F^{(S)}\, dv\, d\Omega_\mathbf{n}. \tag{4.51}$$

To determine the integral of $\mathcal{F} = F^{(S)} + \Phi v(d\bar{f}/dv)$ we use

$$\frac{1}{a^4}\int q v^3 \frac{d\bar{f}}{dv} dv\, d\Omega_\mathbf{n} = -\frac{4\pi}{a^4}\int \left(3qv^2 + \frac{v^4}{q}\right)\bar{f}\, dv = -3(\bar{\rho} + \bar{P}).$$

For the first equals sign we have integrated by parts and used

$$\frac{dq}{dv} = \frac{d}{dv}\sqrt{a^2m^2 + v^2} = \frac{v}{q}.$$

There is no boundary term since f decays rapidly for large momenta. With this we obtain

$$\frac{1}{\bar{\rho}a^4}\int qv^2\mathcal{F}\,dv\,d\Omega_{\mathbf{n}} = D_s - 3(1+w)\Phi = D_g. \tag{4.52}$$

• *Velocity:*
The gauge-invariant velocity perturbation is given by T_i^0 in longitudinal gauge. Hence

$$T_i^{0(S)} = \frac{1}{a^4}\int n_i v^3 F^{(S)}\,dv\,d\Omega_{\mathbf{n}} = \frac{1}{a^4}\int n_i v^3\mathcal{F}\,dv\,d\Omega_{\mathbf{n}} = (\bar{\rho}+\bar{P})V_i. \tag{4.53}$$

Taking the divergence on both sides we obtain, with $V_j \equiv -i(k_j/k)V$,

$$kV = \frac{i}{a^4(\bar{\rho}+\bar{P})}\int n^i k_i v^3\mathcal{F}\,dv\,d\Omega_{\mathbf{n}},$$

$$V = \frac{i}{a^4(\bar{\rho}+\bar{P})}\int \mu v^3\mathcal{F}\,dv\,d\Omega_{\mathbf{n}}, \tag{4.54}$$

where we have introduced the direction cosine between $\mathbf{n}$ and $\mathbf{k}$, $\mu = n^i k_i/k = n^i\hat{k}_i$.

• *Entropy perturbation:*
To determine the variable Γ we first write

$$\pi_L = \frac{\delta P}{\bar{P}} = \frac{1}{3\bar{P}a^4}\int \frac{v^4}{q}F^{(S)}\,dv\,d\Omega_{\mathbf{n}},$$

and therefore

$$\frac{1}{3\bar{P}a^4}\int \frac{v^4}{q}\mathcal{F}\,dv\,d\Omega_{\mathbf{n}} = \pi_L + \Phi\frac{4\pi}{3\bar{P}a^4}\int \frac{v^5}{q}\frac{d\bar{f}}{dv}\,dv. \tag{4.55}$$

We use the background identity $\dot{\bar{\rho}} = -3\mathcal{H}\bar{\rho}(1+w) = -3\mathcal{H}\frac{(1+w)}{w}\bar{P}$ to replace $\bar{P}$ in the second term. After integration by parts we find

$$\Phi\frac{4\pi}{3\bar{P}a^4}\int \frac{v^5}{q}\frac{d\bar{f}}{dv}\,dv = \Phi\frac{\mathcal{H}(1+w)}{w\dot{\bar{\rho}}}\frac{4\pi}{a^4}\int\left(\frac{5v^4}{q} - \frac{v^6}{q^3}\right)\bar{f}\,dv.$$

On the other hand, using $\dot{q} = \mathcal{H}m^2a^2/q = \mathcal{H}\frac{q^2-v^2}{q}$, we obtain

$$\dot{\bar{P}} = -\mathcal{H}\frac{4\pi}{3a^4}\int\left(\frac{5v^4}{q} - \frac{v^6}{q^3}\right)\bar{f}dv.$$

With $\dot{P}/\dot{\bar\rho} = c_s^2$, these two equations together yield

$$\frac{1}{3\bar{P}a^4}\int \frac{v^4}{q}\mathcal{F}\,dv\,d\Omega_{\mathbf{n}} = \pi_L^{(\text{long})} - 3(1+w)\frac{c_s^2}{w}\Phi. \tag{4.56}$$

Furthermore,

$$\frac{c_s^2}{\bar{P}a^4}\int qv^2\mathcal{F}\,dv\,d\Omega_{\mathbf{n}} = \frac{c_s^2}{w}D_g = \frac{c_s^2}{w}\delta^{(\text{long})} - 3(1+w)\frac{c_s^2}{w}\Phi.$$

Combining this with Eq. (4.56) results in

$$\frac{1}{\bar{P}a^4}\int\left(\frac{v^4}{3q} - c_s^2 qv^2\right)\mathcal{F}\,dv\,d\Omega_{\mathbf{n}} = \pi_L^{(\text{long})} - \frac{c_s^2}{w}\delta^{(\text{long})} = \Gamma. \tag{4.57}$$

• *Anisotropic stress:*
The scalar anisotropic stress tensor is simply given by Eq. (4.49). It is related to its potential Π via $\Pi_{ij} = \left(-k^{-2}k_ik_j + \frac{1}{3}\delta_{ij}\right)\Pi$, so that $\Pi^{ij}{}_{|ij} = \frac{2}{3}k^2\Pi$. In Eq. (4.49) this leads to

$$\Pi = \frac{3}{2a^4\bar{P}}\int\frac{v^4}{q}\left(-(\mathbf{n}\cdot\mathbf{k})^2/k^2 + \frac{1}{3}\right)\mathcal{F}\,dv\,d\Omega_{\mathbf{n}} \tag{4.58}$$

$$= \frac{3}{2a^4\bar{P}}\int\frac{v^4}{q}\left(\frac{1}{3} - \mu^2\right)\mathcal{F}\,dv\,d\Omega_{\mathbf{n}}. \tag{4.59}$$

4.3.2 Vector Perturbations

Vector perturbations are given by divergence-free vector fields. For a fixed Fourier component $\mathbf{k}$, we expand them in the basis functions $Q_j^{(V)}$, which have two independent modes. Let us choose two basis vectors $\mathbf{e}^{(1)}$ and $\mathbf{e}^{(2)}$ so that $(\mathbf{e}^{(1)}, \mathbf{e}^{(2)}, \hat{\mathbf{k}})$ form a right-handed orthonormal basis. The $\mathbf{k}$-Fourier mode of an arbitrary vector perturbation is then of the form $A_j = (A^{(1)}\mathbf{e}_j^{(1)} + A^{(2)}\mathbf{e}_j^{(2)})\,e^{i\mathbf{kx}}$. We can also write it in terms of the helicity basis [see Eq. (2.13)]

$$\mathbf{e}^{(\pm)} = \frac{1}{\sqrt{2}}\left(\mathbf{e}^{(1)} \mp i\mathbf{e}^{(2)}\right), \tag{4.60}$$

$$A_j = \left(A^{(+)}\mathbf{e}_j^{(+)} + A^{(-)}\mathbf{e}_j^{(-)}\right)e^{i\mathbf{kx}},$$

where $A^{(\pm)} = \frac{1}{\sqrt{2}}\left(A^{(1)} \pm iA^{(2)}\right)$.

We write the vector perturbations of the distribution function for a given Fourier mode $\mathbf{k}$ in this form

$$F^{(V)}(t,\mathbf{x},\mathbf{n},v) = \left[F^{(V+)}(t,\mathbf{k},\mathbf{n},v)\mathbf{e}^{(+)}\cdot\mathbf{n} + F^{(V-)}(t,\mathbf{k},\mathbf{n},v)\mathbf{e}^{(-)}\cdot\mathbf{n}\right]e^{i\mathbf{k}\cdot\mathbf{x}}. \tag{4.61}$$

The functions $F^{(V+)}$ and $F^{(V-)}$ no longer depend on $\mathbf{e}^{(\pm)}$. Therefore, if the process that generated the fluctuations is isotropic, the components $F^{(V\pm)}$ depend on the direction $\mathbf{n}$ only via $\mu = \hat{\mathbf{k}} \cdot \mathbf{n}$. With respect to spherical coordinates chosen such that $\mathbf{k}$ points in the z-direction, the components of $\mathbf{n}$ are $\mathbf{n} = (\sqrt{1-\mu^2}\cos\varphi, \sqrt{1-\mu^2}\sin\varphi, \mu)$. With $\mathbf{e}^{(1)} = (1,0,0)$ and $\mathbf{e}^{(2)} = (0,1,0)$ we obtain

$$\mathbf{e}^{(\pm)} \cdot \mathbf{n} = n^{\mp} = \sqrt{\frac{1-\mu^2}{2}}e^{\mp i\varphi}. \tag{4.62}$$

The sign difference comes from the fact that $\mathbf{e}^{(\pm)} \cdot \mathbf{e}^{(\pm)} = 0$ while $\mathbf{e}^{(\pm)} \cdot \mathbf{e}^{(\mp)} = 1$; hence for $\mathbf{n} = n^+\mathbf{e}^{(+)} + n^-\mathbf{e}^{(-)} + \mu\hat{\mathbf{k}}$ we have $n^+ = \mathbf{n}\cdot\mathbf{e}^{(-)}$ and $n^- = \mathbf{n}\cdot\mathbf{e}^{(+)}$. With this

$$F^{(V)}(t,\mathbf{x},\mathbf{n},v) = \sqrt{\frac{1-\mu^2}{2}}\left[F^{(V+)}(t,k,\mu,v)e^{-i\varphi} + F^{(V-)}(t,k,\mu,v)e^{i\varphi}\right]e^{i\mathbf{k}\cdot\mathbf{x}}. \tag{4.63}$$

• *Vorticity:*

We now write the vorticity vector perturbation of the energy–momentum tensor in the helicity basis, $\Omega_i(t,\mathbf{k}) = \Omega^{(+)}(t,\mathbf{k})e_i^{(+)} + \Omega^{(-)}(t,\mathbf{k})e_i^{(-)}$.

$$\Omega^j(t,\mathbf{k}) = \frac{-1}{\bar\rho + \bar P}T_0^{(V)j},$$

$$\Omega^{(\pm)}(t,\mathbf{k}) = \frac{1}{(\bar\rho+\bar P)a^4}\int \mathbf{e}^{\mp}\cdot\mathbf{n}v^3F^{(V)}(t,\mathbf{k},\mathbf{n},v)\,dv\,d\Omega_{\mathbf{n}}$$

$$= \frac{\pi}{(\bar\rho+\bar P)a^4}\int v^3F^{(V\pm)}(t,k,\mu,v)(1-\mu^2)\,dv\,d\mu. \tag{4.64}$$

For the last equals sign we used that $\int_0^{2\pi}(\mathbf{e}^{\mp}\cdot\mathbf{n})e^{\pm i\varphi}d\varphi = 0$.

In the chosen gauge, $B^i = 0$, we obtain $T_0^i = (\bar\rho+\bar P)\Omega^i$; hence the first moment, $\int n^iF^{(V)}$, gives rise to the vorticity Ω^i and not to the shear $V^{(V)i}$ (for details see Section 2.2.4).

• *Anisotropic stress:*

We introduce the helicity decomposition of the vector potential for anisotropic stresses,

$$\Pi_j^{(V)} = \left(\Pi^{(V+)}\mathbf{e}_j^{(+)} + \Pi^{(V-)}\mathbf{e}_j^{(-)}\right)e^{i\mathbf{k}\mathbf{x}}. \tag{4.65}$$

The anisotropic stress tensor is defined by $\Pi^{(V)i}{}_j = \frac{-1}{2k}\left(\Pi^{(V)i}{}_{|j} + \Pi_j^{(V)|i}\right)$. But $\Pi^{(V)i}{}_j$ is also given by the integral of the distribution function over momentum space,

$$\Pi^{(V)i}{}_j = \frac{1}{a^4\bar{P}}\int \frac{v^4}{q}\left(n^i n_j - \frac{1}{3}\delta^i_j\right) F^{(V)}\, dv\, d\Omega_{\mathbf{n}}. \tag{4.66}$$

Taking the divergence of both expressions we obtain

$$\begin{aligned}\Pi^{(V)i}{}_{j|i} &= \frac{k}{2}\left(\Pi^{(V+)}\mathbf{e}^{(+)}_j + \Pi^{(V-)}\mathbf{e}^{(-)}_j\right) e^{i\mathbf{kx}} \\ &= \frac{ik}{a^4\bar{P}}\int \frac{v^4}{q}\left(n_j\mu - \frac{1}{3}\hat{k}_j\right) F^{(V)}\, dv\, d\Omega_{\mathbf{n}}.\end{aligned} \tag{4.67}$$

We multiply this vector with $\mathbf{e}^{\mp}$ to isolate the modes $\Pi^{\pm}$. We also make use of the helicity decomposition of the distribution function, Eq. (4.61):

$$\begin{aligned}\Pi^{(V\pm)} &= \frac{2i}{a^4\bar{P}}\int \frac{v^4}{q} n^{\pm}\mu F^{(V)}\, dv\, d\Omega_{\mathbf{n}} \\ &= \frac{2\pi i}{a^4\bar{P}}\int \frac{v^4}{q}\mu(1-\mu^2) F^{(V\pm)}(t,k,\mu,v)\, dv\, d\mu.\end{aligned} \tag{4.68}$$

For the second equals sign we made use of the decomposition $\mathbf{n} = n^+\mathbf{e}^{(+)} + n^-\mathbf{e}^{(-)} + \mu\hat{\mathbf{k}} = \sqrt{1-\mu^2}e^{-i\varphi}\mathbf{e}^{(+)} + \sqrt{1-\mu^2}e^{i\varphi}\mathbf{e}^{(-)} + \mu\hat{\mathbf{k}}$ introduced earlier.

4.3.3 Tensor Perturbations

For tensor perturbations only the anisotropic stresses survive. The ansatz for a tensor-type Fourier mode of the distribution function is

$$F^{(T)}(t,\mathbf{x},\mathbf{n},v) = \left[F^{(T\times)}(t,\mathbf{k},\mathbf{n},v)Q^{(T\times)}_{ij} + F^{(Td)}(t,\mathbf{k},\mathbf{n},v)Q^{(Td)}_{ij}\right] n^i n^j,$$

where

$$Q^{(T\times)}_{ij} = \frac{e^{i\mathbf{k}\cdot\mathbf{x}}}{\sqrt{2}}\left[e^{(1)}_i e^{(2)}_j + e^{(2)}_i e^{(1)}_j\right] \quad \text{and} \quad Q^{(Td)}_{ij} = \frac{e^{i\mathbf{k}\cdot\mathbf{x}}}{\sqrt{2}}\left[e^{(1)}_i e^{(1)}_j - e^{(2)}_i e^{(2)}_j\right].$$

These tensors form a basis of the symmetric traceless tensors normal to $\mathbf{k}$. Usually, when discussing gravity waves, the second mode function is denoted $Q^{(T+)}_{ij}$. Here we use $Q^{(Td)}_{ij}$ in order not to confuse this basis with the helicity basis, which we shall use later also for tensor perturbations. The superscript d indicates that this tensor is nonzero only on the diagonal with principal axes (eigenvectors) $\mathbf{e}^{(1)}$ and $\mathbf{e}^{(2)}$, while $Q^{(T\times)}_{ij}$ is purely off-diagonal. Its principal axes are rotated by 45° with respect to $\mathbf{e}^{(1)}$ and $\mathbf{e}^{(2)}$.

$$\Pi^{(T)i}{}_j = \frac{1}{a^4\bar{P}}\int \frac{v^4}{q}\left(n^i n_j - \frac{1}{3}\delta^i_j\right) F^{(T)}\, dv\, d\Omega_{\mathbf{n}}. \tag{4.69}$$

With the decomposition of the distribution function and $\Pi_{ij}^{(T)} = \Pi^{(T\times)}(t,\mathbf{k})Q_{ij}^{(T\times)} + \Pi^{(Td)}(t,\mathbf{k})Q_{ij}^{(Td)}$, we obtain for both modes

$$\Pi^{(T\bullet)} = \frac{\pi}{2a^4\bar{P}}\int \frac{v^4}{q}F^{(T\bullet)}(t,k,v,\mu)(1-\mu^2)^2\,dv\,d\mu. \tag{4.70}$$

As for vector perturbations, we assume that the process generating the perturbation is isotropic, so that $F^{(T\bullet)}$ depends on the direction $\mathbf{n}$ only via $\mu = \mathbf{n}\cdot\hat{\mathbf{k}}$.

In the massless case, which is most relevant for us, the energy–momentum tensor simplifies considerably. This is the subject of the next section.

4.4 The Ultrarelativistic Limit, the Liouville Equation for Massless Particles

The Liouville equation and the expressions for the perturbations of the energy–momentum tensor derived in the previous section are actually more important for massive collisionless particles, for example, massive neutrinos, than for massless particles. In the massless (or ultrarelativistic) case we have $q = v$ and the equations simplify significantly. Before discussing the different modes, let us introduce the "longitudinal temperature fluctuation" for a thermal bath of massless particles. "Longitudinal" indicates that we consider perturbations in longitudinal gauge. We integrate the perturbed distribution function over energy so that only the dependence on momentum directions, $\mathbf{n}$, remains,

$$\frac{4\pi}{a^4}\int v^3 f\,dv \equiv \bar{\rho}\,(1+4\Delta_L(\mathbf{n}))\,. \tag{4.71}$$

We call the variable $\Delta_L(\mathbf{n})$ the longitudinal temperature fluctuation in direction $\mathbf{n}$. $\Delta_L(\mathbf{n})$ depends also on $(t,\mathbf{x})$, which we suppress here for brevity. This definition is motivated by the following consideration: for a Planck distribution of photons that has a slightly direction-dependent temperature but is otherwise unperturbed (especially, it has a perfect blackbody spectrum, $f_B(p,T) = (\exp(p/T)+1)^{-1}$), the perturbed distribution function can be expanded to first order as

$$f(p,\mathbf{n}) = f_B(p,T(\mathbf{n})) = f_B(p,\bar{T}) - \frac{\delta T}{\bar{T}}p\partial_p f_B(p,\bar{T}). \tag{4.72}$$

Observe that f_B is purely a function of p/T so that $\partial_T f_B = -(p/T)\partial_p f_B$. The energy density of this photon distribution is given by

$$\rho_\gamma = \frac{1}{a^4}\int v^3 f(v,\mathbf{n})\,dv\,d\Omega_{\mathbf{n}} = \bar{\rho}_\gamma - \frac{1}{a^4}\int \frac{\delta T}{\bar{T}}v^4\partial_v f_B(v,\bar{T})\,dv\,d\Omega_{\mathbf{n}}$$

$$= \bar{\rho}_\gamma\left(1+\frac{4}{4\pi}\int\frac{\delta T}{\bar{T}}d\Omega_{\mathbf{n}}\right) = \bar{\rho}_\gamma\left(1+\frac{1}{\pi}\int\Delta_L(\mathbf{n})\,d\Omega_{\mathbf{n}}\right). \tag{4.73}$$

For the third equals sign we have performed an integration by parts to evaluate the integral over v and the last equals sign motivates our defintion of Δ_L as the temperature perturbation. We shall see that the Liouville equation for photons leads to a perturbation that can be described entirely by a direction-dependent temperature fluctuation. Of course f and also $\Delta_L = \delta T/\bar{T}$ also depend on position and time, arguments that we suppress here for brevity. The fact that the perturbation of the photon distribution can be described in such a simple way is not surprising. It is an expression of the "a-chromaticity" of gravity that is a consequence of the equivalence principle: the deflection and redshift of a photon in a gravitational field are independent of its energy.

4.4.1 Scalar Perturbations

For massless particles, $v = q$, the Liouville equation (4.34) reduces to

$$\partial_0 \mathcal{F} + n^i \partial_i \mathcal{F} = n^j \left[\Psi_{,j} + \Phi_{,j}\right] v \frac{d\bar{f}}{dv}. \tag{4.74}$$

We define the energy integrated fluctuation

$$\mathcal{M}^{(S)}(t, \mathbf{x}, \mathbf{n}) = \frac{\pi}{a^4 \bar{\rho}} \int v^3 \mathcal{F}\, dv. \tag{4.75}$$

In terms of the temperature fluctuation $\Delta_L(\mathbf{n})$ defined in Eq. (4.71) we get

$$\mathcal{M}^{(S)}(\mathbf{n}) = \Delta_L^{(S)}(\mathbf{n}) - \Phi. \tag{4.76}$$

Up to a (irrelevant) monopole contribution, the momentum integrated distribution function $\mathcal{M}^{(S)}$ is simply the temperature perturbation in longitudinal gauge. It is not surprising that the monopole terms of $\mathcal{M}^{(S)}(\mathbf{n})$ and $\Delta_L(\mathbf{n})$ do not agree because they are gauge dependent. Also the dipole terms might differ because they too are gauge dependent. (In a gauge with nonvanishing shear, the dipole contributions to Δ_L and $\mathcal{M}$ actually do differ.)

Integrating the Liouville equation (4.74) over momenta and performing an integration by parts on the right-hand side, we obtain the evolution equation for $\mathcal{M}^{(S)}$,

$$\partial_t \mathcal{M}^{(S)} + n^i \partial_i \mathcal{M}^{(S)} = -n^j \left[\Psi_{,j} + \Phi_{,j}\right]. \tag{4.77}$$

This equation can be solved formally for any given source term $\Phi + \Psi$. One easily checks that the solution with initial condition $\mathcal{M}^{(S)}(t_{\rm in}, \mathbf{x}, \mathbf{n})$ is

$$\begin{aligned}\mathcal{M}^{(S)}(t, \mathbf{x}, \mathbf{n}) &= \mathcal{M}^{(S)}\left(t_{\rm in}, \mathbf{x} - \mathbf{n}(t - t_{\rm in}), \mathbf{n}\right) \\ &\quad - \int_{t_{\rm in}}^{t} dt'\, n^i \partial_i (\Psi + \Phi)(t', \mathbf{x} - \mathbf{n}(t - t')).\end{aligned} \tag{4.78}$$

Using

$$\frac{d}{dt'}(\Psi+\Phi)(t',\mathbf{x}-\mathbf{n}(t-t')) = \partial_{t'}(\Psi+\Phi)(t',\mathbf{x}-\mathbf{n}(t-t')) + n^i\partial_i(\Psi+\Phi)(t',\mathbf{x}-\mathbf{n}(t-t')),$$

we can replace the second term on the right-hand side to obtain

$$\begin{aligned}\mathcal{M}^{(S)}(t,\mathbf{x},\mathbf{n}) &= \mathcal{M}^{(S)}(t_{\rm in},\mathbf{x}-\mathbf{n}(t-t_{\rm in}),\mathbf{n}) \\ &+ (\Psi+\Phi)(t_{\rm in},\mathbf{x}-\mathbf{n}(t-t_{\rm in})) \\ &+ \int dt'\,\partial_{t'}(\Psi+\Phi)(t',\mathbf{x}-\mathbf{n}(t-t')) + \text{monopole}.\end{aligned} \tag{4.79}$$

By "monopole" we denote an uninteresting **n**-independent contribution that does not affect the CMB anisotropy spectrum. The Bardeen potentials Ψ and Φ, however, are given via Einstein's equation in terms of the perturbations of the energy–momentum tensor that contain contributions from the photons, which are in turn the integrals over directions of $\mathcal{M}$ given below. Therefore, even though it might look like it, Eqs. (4.78) and (4.79) are not really a solution of the Liouville equation. The term on the right-hand side also depends on $\mathcal{M}^{(S)}$.

Let us compare Eq. (4.79) with the result from the integration of lightlike geodesics after decoupling in Eqs. (2.238) and (2.240). Here we have solved the Liouville equation, which also does not take into account the scattering of photons and is therefore equivalent to our approach in Chapter 2. They both correspond to the "sudden decoupling" approximation, where we assume that photons behave like a perfect fluid before decoupling and are entirely free after decoupling. This is a relatively good approximation for all scales that are much larger than the duration of the process of recombination, which we shall estimate in the next section. The comparison with Eqs. (2.238) and (2.240) yields

$$\mathcal{M}^{(S)}(t_{\rm dec},\mathbf{x}-\mathbf{n}(t-t_{\rm dec}),\mathbf{n}) = \left(\frac{1}{4}D_g + \mathbf{n}\cdot\mathbf{V}^{(b)}\right)(t_{\rm dec},\mathbf{x}-\mathbf{n}(t-t_{\rm dec})), \tag{4.80}$$

and

$$\mathcal{M}^{(S)}(t,\mathbf{x},\mathbf{n}) \equiv \frac{\Delta T}{T}(t,\mathbf{x},\mathbf{n}). \tag{4.81}$$

In other words, the temperature fluctuation defined via the energy shift of photons moving along geodesics corresponds to $\mathcal{M}^{(S)}$ while the temperature fluctuation defined via the energy density fluctuation in longitudinal gauge corresponds to $\Delta_L^{(S)} = \mathcal{M}^{(S)} + \Phi$. In addition to the energy shift, the latter includes a contribution from the perturbation of the volume element, $\sqrt{|\det(g_{ij})|}\,d^3x = a^3(1-3\Phi)\,d^3x$. The distinction is not very important because it is a monopole that does not show up in the angular power spectrum. However, the corresponding evolution equations are of course different. The variable $\Delta_L^{(S)}$ is used, for example, in Hu and Sugiayma (1995), Hu *et al.* (1995, 1998), and Seljak and Zaldarriaga (1996), while the variable $\mathcal{M}^{(S)}$ is used, for example, in Durrer and Straumann (1988), Durrer (1990, 1994), Durrer *et al.* (2002), Doran (2005), and Bashinsky (2006).

The initial condition in the sudden decoupling approximation is a distribution function that contains only a monopole and a dipole. Higher multipoles do not build up in a perfect fluid. In the next section we shall take into account the process of decoupling by studying the Boltzmann equation.

The scalar perturbations of the energy–momentum tensor of the radiation fluid for a given Fourier mode $\mathbf{k}$ can be found by integrating the right-hand sides of Eqs. (4.52), (4.54), (4.57), and (4.59) over energy, setting $q = v$,

$$D_g = 2\int_{-1}^{1} \mathcal{M}^{(S)}(\mu)\, d\mu = 4\mathcal{M}_0^{(S)}, \tag{4.82}$$

$$V = \frac{3i}{2}\int_{-1}^{1} \mu\mathcal{M}^{(S)}(\mu)\, d\mu = 3\mathcal{M}_1^{(S)}, \tag{4.83}$$

$$\Gamma = 0, \tag{4.84}$$

$$\Pi = 3\int_{-1}^{1} \left(1 - 3\mu^2\right)\mathcal{M}^{(S)}(\mu)\, d\mu = 12\mathcal{M}_2^{(S)}. \tag{4.85}$$

The general definition of $\mathcal{M}_\ell^{(S)}$ will be given later, Eq. (4.116). We have assumed that $\mathcal{M}^{(S)}(\mathbf{n})$ depends on the direction $\mathbf{n}$ only via $\mu = \hat{\mathbf{k}}\cdot\mathbf{n}$ and have performed the integration over φ, which simply gives a factor 2π. For isotropic perturbations there is no other vector that could single out a direction and therefore this assumption reflects statistical isotropy.

The exact equality $w = c_s^2 = \frac{1}{3}$ does not allow for any entropy perturbation in a pure radiation fluid.

4.4.2 Vector Perturbations

Vector perturbations of the distribution function are not gauge dependent. We have directly $\mathcal{M}^{(V)} \equiv \Delta_L^{(V)}$. The Liouville equation for vector perturbations of the radiation fluid is obtained by integrating Eq. (4.39) over energies,

$$\mathcal{M}^{(V)}(\mathbf{n}) = \frac{\pi}{a^4\bar{\rho}}\int v^3 F^{(V)}(\mathbf{n}, v)\, dv,$$

$$\mathcal{M}^{(V\pm)}(\mu) = \frac{\pi}{a^4\bar{\rho}}\int v^3 F^{(V\pm)}(\mu, v)\, dv, \tag{4.86}$$

$$\partial_t\mathcal{M}^{(V)} + n^i\partial_i\mathcal{M}^{(V)} = -n^i n^j \dot{H}_{ij}^{(V)}. \tag{4.87}$$

The formal solution to this equation is

$$\mathcal{M}^{(V)}(t,\mathbf{x},\mathbf{n}) = \mathcal{M}^{(V)}(t_{\rm in},\mathbf{x}-\mathbf{n}(t-t_{\rm in}),\mathbf{n}) - \int_{t_{\rm in}}^{t} dt'\, n^i n^j a(t')^{-1}\sigma_{ij}^{(V\pm)}(t',\mathbf{x}-\mathbf{n}(t-t')). \tag{4.88}$$

When setting $\mathcal{M}^{(V)}(t_{\rm in}) = \Omega_i(t_{\rm in})n^i$, this corresponds exactly to Eq. (2.243); see Exercise 4.2.

After Fourier transforming $\mathcal{M}^{(V)}$ and $\sigma_{ij}^{(V)}$ we can expand them in the helicity basis,

$$\dot{H}_j^{(V)} = \dot{H}^{(V+)}e_j^{(+)} + \dot{H}^{(V-)}e_j^{(-)} = \sigma^{(V+)}e_j^{(+)} + \sigma^{(V-)}e_j^{(-)},$$

so that

$$\begin{aligned}
\dot{H}_{ij}^{(V)} &\equiv \frac{-1}{2k}\left(\dot{H}_{i|j}^{(V)} + \dot{H}_{j|i}^{(V)}\right)\\
&= \frac{-i}{2}\left(\dot{H}^{(V+)}[e_i^{(+)}\hat{k}_j + e_j^{(+)}\hat{k}_i] + \dot{H}^{(V-)}[e_i^{(-)}\hat{k}_j + e_j^{(-)}\hat{k}_i]\right)\\
n^i n^j \dot{H}_{ij}^{(V)} &= \frac{-i}{\sqrt{2}}\mu\sqrt{1-\mu^2}\left(\dot{H}^{(V+)}e^{i\varphi} + \dot{H}^{(V-)}e^{-i\varphi}\right).
\end{aligned} \tag{4.89}$$

In the last equality we have introduced the representation of $\mathbf{n}$ in the helicity basis, Eq. (4.62). The φ dependence on the left- and right-hand sides of Eq. (4.87) shows that $\mathcal{M}^{(V\pm)}$ couples only to $\sigma^{(V\pm)}$ and both helicity components satisfy the equation

$$\partial_t\mathcal{M}^{(V\pm)} + ik\mu\mathcal{M}^{(V\pm)} = -i\mu\sigma^{(V\pm)}. \tag{4.90}$$

From Eqs. (4.64) and (4.68) we obtain the vector perturbations of the energy–momentum tensor in terms of $\mathcal{M}^{(V)}$,

$$\Omega^{(\pm)} = \frac{3}{4}\int_{-1}^{1}(1-\mu^2)\mathcal{M}^{(V\pm)}(\mu)\,d\mu, \tag{4.91}$$

$$\Pi^{(V\pm)} = 6i\int_{-1}^{1}(1-\mu^2)\mu\mathcal{M}^{(V\pm)}(\mu)\,d\mu. \tag{4.92}$$

4.4.3 Tensor Perturbations

For tensor fluctuations, the perturbed Liouville equation becomes

$$\partial_t\mathcal{M}^{(T)} + n^i\partial_i\mathcal{M}^{(T)} = -n^i n^j \dot{H}_{ij}^{(T)}. \tag{4.93}$$

For a given source term $H_{ij}^{(T)}$ this is solved by

$$\mathcal{M}^{(T)}(t,\mathbf{x},\mathbf{n}) = \mathcal{M}^{(T)}(t_{\rm in},\mathbf{x}-\mathbf{n}(t-t_{\rm in}),\mathbf{n}) - \int_{t_{\rm in}}^{t} dt'\, n^i n^j \dot{H}_{ij}^{(T)}(t',\mathbf{x}-\mathbf{n}(t-t')). \tag{4.94}$$

We decompose also $H^{(T)}$ and $\mathcal{M}^{(T)}$ in the basis $Q_{ij}^{(T\bullet)}$ defined in Eq. (2.20) (the bullet stands for the two modes d and $\times$),

$$H_{ij}^{(T)} = H^{(T\times)}Q_{ij}^{(T\times)} + H^{(Td)}Q_{ij}^{(Td)}, \tag{4.95}$$

$$\mathcal{M}^{(T\bullet)}(\mu) = \frac{\pi}{a^4\bar{\rho}}\int dv\, v^3 F^{(T\bullet)}(\mu,v), \tag{4.96}$$

$$\partial_t\mathcal{M}^{(T\bullet)} + ik\mu\mathcal{M}^{(T\bullet)} = -\dot{H}^{(T\bullet)}. \tag{4.97}$$

The tensor anisotropic stresses are

$$\Pi^{(T\bullet)} = \frac{3}{2}\int_{-1}^{1}(1-\mu^2)^2\mathcal{M}^{(T\bullet)}(\mathbf{n})d\Omega. \tag{4.98}$$

4.4.4 The Liouville Equation in Terms of the Weyl Tensor

(*This section is not required for the continuation of the book and can be left out in a first reading.*)

We know that the motion of photons in a gravitational field is conformally invariant. Therefore, the evolution of the photon distribution, once the redshift is taken out by using the conformally invariant momentum variable v, should depend only on the Weyl tensor. To find the Liouville equation in terms of the Weyl tensor, we first consider only scalar and tensor perturbations and assume that the vector perturbations vanish. Adding together the scalar and tensor parts of the Liouville equation $\mathcal{M} = \mathcal{M}^{(S)} + \mathcal{M}^{(T)}$ Liouville's equation becomes

$$(\partial_t + n^j\partial_j)\mathcal{M} = -n^i\partial_i(\Psi+\Phi) - n^j n^i \dot{H}_{ij}^{(T)} \equiv S_G. \tag{4.99}$$

S_G is the sum of the gravitational source terms for scalar and tensor perturbations. Now we apply the Laplacian on both sides. Using the expressions (A3.21) and (A3.57)–(A3.59) for the Weyl tensor, one finds that this corresponds to

$$(\partial_t + n^i\partial_i)\Delta\mathcal{M} = \Delta S_G \equiv -3n^i\,\partial^j E_{ij} - n^k n^j \epsilon_k{}^{i\ell}\,\partial_\ell B_{ij}\,, \tag{4.100}$$

where $\epsilon_{ki\ell}$ is the totally antisymmetric tensor in three dimensions, and E_{ij} and B_{ij} are the electric and magnetic parts of the Weyl tensor. We shall sometimes use S_G to denote the gravitational source term on the right-hand side of Liouville's equation.

As before, for a given source term ΔS_G this equation is simply solved by integration. With our variable $\mathcal{M}$ the Liouville equation can be directly written in terms of the Weyl tensor, while this is not possible with the variable Δ_L. The variable $\mathcal{M}$ manifests the conformal invariance of photon propagation. It remains zero if the Weyl curvature vanishes and therefore photon trajectories are not modified.

If we want to include also vector perturbations, a subtlety occurs. With the help of (A3.43)–(A3.45) one finds

$$-3n^i \partial^j E_{ij}^{(V)} - n^k n^j \epsilon_k{}^{i\ell} \partial_\ell B_{ij}^{(V)} = \frac{3}{4} n^j \Delta\dot{\sigma}_j + \frac{1}{4} n^i n^j \Delta\sigma_{i|j}, \tag{4.101}$$

which does not correspond to the right-hand side of Eq. (4.87). However, if we transform $\mathcal{M}^{(V)}$ by the addition of a simple dipole term that does not show up in the CMB multipoles for $\ell \geq 2$ to

$$\mathcal{M}^{(V2)} \equiv \mathcal{M}^{(V)} + \frac{3}{4} n^j \dot{H}_j, \tag{4.102}$$

one finds easily that

$$(\partial_t + n^i \partial_i) \Delta\mathcal{M}^{(V2)} = -3n^i \partial^j E_{ij}^{(V)} - n^k n^j \epsilon_k{}^{i\ell} \partial_\ell B_{ij}^{(V)}. \tag{4.103}$$

Hence with this redefinition, the variable

$$\mathcal{M} \equiv \mathcal{M}^{(S)} + \mathcal{M}^{(V2)} + \mathcal{M}^{(T)}$$

satisfies the Liouville equation

$$(\partial_t + n^i \partial_i) \Delta\mathcal{M} = -3n^i \partial^j E_{ij} - n^k n^j \epsilon_k{}^{i\ell} \partial_\ell B_{ij}. \tag{4.104}$$

It may be interesting to note that in a generic gauge this variable can be written as

$$\Delta\mathcal{M} = \Delta M + \Delta\mathcal{R} + \frac{3}{2} n^i \partial^j \sigma_{ij}. \tag{4.105}$$

The first term, M, is the momentum integration of the perturbation of the distribution function, F, while the second term is the perturbation of the spatial curvature given in Eq. (2.50). Only scalar perturbations contribute to it. The last term is related to the shear to which both scalar and vector perturbations contribute. Note that for scalar perturbation in longitudinal gauge the shear term vanishes while the vector part of the shear is gauge invariant. By construction, this variable is perfectly gauge invariant.

The right-hand side of Eq. (4.104) is written entirely in terms of tensor fields with vanishing background contribution and it is therefore manifestly gauge invariant. It would be interesting to attempt the same for the left-hand side, the variable $\Delta\mathcal{M}$.

4.4.5 The Liouville Equation in Fourier Space

A Fourier mode of $\mathcal{M}(t,\mathbf{x},\mathbf{n})$ is given by

$$\mathcal{M}(t,\mathbf{k},\mathbf{n}) \equiv \int d^3x\, e^{-i\mathbf{k}\cdot\mathbf{x}}\mathcal{M}(t,\mathbf{x},\mathbf{n}), \text{ and its inverse is}$$

$$\mathcal{M}(t,\mathbf{x},\mathbf{n}) = \frac{1}{(2\pi)^3}\int d^3k\, e^{i\mathbf{k}\cdot\mathbf{x}}\mathcal{M}(t,\mathbf{k},\mathbf{n}).$$

We have seen that the Liouville equation for a Fourier mode is given by

$$(\partial_t + ik\mu)\mathcal{M}(t,\mathbf{k},\mathbf{n}) = S_G(t,\mathbf{k},\mu), \tag{4.106}$$

where, as before, $\mu = \hat{\mathbf{k}}\cdot\mathbf{n}$ is the cosine between the unit vectors $\hat{\mathbf{k}} = \mathbf{k}/k$ and $\mathbf{n}$. The general solution to this equation for a given source term S_G can be written as

$$\mathcal{M}(t,\mathbf{k},\mathbf{n}) = e^{-ik\mu(t-t_{\text{in}})}\mathcal{M}(t_{\text{in}},\mathbf{k},\mathbf{n}) + \int_{t_{\text{in}}}^{t} dt' e^{-ik\mu(t-t')}S_G(t',\mathbf{k},\mathbf{n}). \tag{4.107}$$

The function S_G can be decomposed into scalar, vector, and tensor perturbations.

As already mentioned, the source term usually depends on $\mathcal{M}$ via Einstein's equations and Eq. (4.107) is not really a solution but simply corresponds to rewriting Eq. (4.106) as an integral equation. But as we shall see, this has serious advantages, especially for numerical computations.

From Eq. (4.107) using the decomposition (see Appendix 4, Section A4.2)

$$e^{i\mathbf{k}\cdot\mathbf{n}(t-t_{\text{in}})} = \sum_{\ell=0}^{\infty}(2\ell+1)i^{\ell} j_\ell(k(t-t_{\text{in}}))P_\ell(\mu),$$

one finds the CMB power spectrum, exactly as in Chapter 2, Eqs. (2.263)–(2.275) and (2.281). Before we do this, we want to include Thomson scattering, which is the relevant scattering process just before recombination. We will then derive the power spectrum taking into account this scattering process.

4.5 The Boltzmann Equation

At early times, long before recombination, scattering of photons with free electrons is very frequent. During recombination, however, the number density of free electrons, that is, of electrons not bound to an atom, drops drastically and soon the mean free path of photons is much larger than the Hubble scale so that, effectively, photons do not scatter any more. In the previous treatment we assumed this process of decoupling to be instantaneous; now we want to reconsider it in more detail.

The only scattering process that is relevant briefly before decoupling, that is, at temperatures of a few electron volts and less, is elastic Thomson scattering, where the photon energy is conserved and only its direction is modified. The Thomson scattering rate is

$$\Gamma_T = \sigma_T n_e,$$

where $\sigma_T = 6.6524 \times 10^{-25}\,\text{cm}^2$ is the Thomson scattering cross section and n_e is the number density of free electrons.

Before decoupling, in a matter dominated universe, we find

$$\Gamma_T \simeq 7 \times 10^{-30}\,\text{cm}^{-1}\Omega_b h^2 (1+z)^3 \quad \text{while}$$

$$H \simeq 10^{-28}\,\text{cm}^{-1} h (1+z)^{3/2}$$

$$\Gamma_T / H \simeq 0.07 \Omega_b h (z+1)^{3/2}.$$

Hence before recombination, which corresponds to redshifts $z > 1100$, say, Thomson scattering is much faster than expansion. During recombination, the free electron density drops and eventually the Thomson scattering rate drops below the expansion rate. To take scattering into account we add a so-called collision integral to the right-hand side of the Liouville equation, which leads us to the Boltzmann equation. To learn more about the Boltzmann equation and the approximations going into it see, for example, Lifshitz and Pitajewski (1983) or Diu et al. (1989). The collision integral $C[f]$ takes into account that the 1-particle distribution function can change due to collisions that scatter a particle into, f_+, or out of, f_-, a volume element $d^3x\, d^3p$ in phase space,

$$\left[\tilde{p}^\mu \partial_\mu - \Gamma^i_{\alpha\beta} \tilde{p}^\alpha \tilde{p}^\beta \frac{\partial}{\partial \tilde{p}^i}\right] f = C[f] = \frac{df_+}{dt} - \frac{df_-}{dt}. \tag{4.108}$$

Here f_+ and f_- denote the distribution of photons scattered into and out of the beam of photons at position $\mathbf{x}$ at time t with momentum $\mathbf{p}$ respectively.

In the baryon rest frame, which we denote by a prime, the photons scattered into the beam in direction $\mathbf{n}$ per unit of time are given by

$$\frac{df'_+}{dt'}(\mathbf{n}) = \frac{\sigma_T n_e}{4\pi} \int f'(p', \mathbf{n}') \omega(\mathbf{n}, \mathbf{n}')\, d\Omega',$$

where $\omega(\mathbf{n}, \mathbf{n}')$ denotes the normalized angular dependence of Thomson scattering after averaging over photon polarizations (Jackson, 1975):

$$\frac{d\sigma}{d\Omega} = \frac{\sigma_T}{4\pi} \omega(\mathbf{n}, \mathbf{n}') = \frac{3\sigma_T}{16\pi} \left[1 + (\mathbf{n} \cdot \mathbf{n}')^2\right], \tag{4.109}$$

$$= \frac{\sigma_T}{4\pi} \left[1 + \frac{3}{4} n_{ij} n'_{ij}\right] \quad \text{with} \quad n_{ij} = n_i n_j - \frac{1}{3}\delta_{ij}.$$

Here we have averaged over incoming polarizations and summed over final polarizations of the photons; see Jackson (1975). In this chapter we neglect the polarization dependence of Thomson scattering, which we discuss fully in Chapter 5.

In the baryon rest frame that moves with 4-velocity u^μ, the photon energy is

$$p' = -\tilde{p}_\mu u^\mu = p(1 - n_i v^i).$$

At first order, aberration does not contribute. Since Thomson scattering is energy independent, we may integrate f_+ over photon energies $p' = v'/a$ to obtain again an equation for $\mathcal{M}$. With $v'^3\,dv' = (1 - 4n_i v^i)v^3\,dv$ and Eq. (4.71), we obtain

$$\frac{4\pi}{a^4}\int v'^3 \frac{df'_+}{dt'}\,dv' = \bar{\rho}_\gamma \sigma_T n_e \left[1 - 4n_i v^i + \frac{1}{\pi}\int \Delta(\mathbf{n}')\omega(\mathbf{n},\mathbf{n}')\,d\Omega'\right]. \tag{4.110}$$

Here $\bar{\rho}_\gamma$ is the background photon density. The term $4\mathbf{n}\cdot\mathbf{v}$ is a Doppler term from the velocity of the electrons with respect to the longitudinal rest frame. The factor of 4 comes from the fact that we have to transform $p'^3\,dp' = p^3(1 + 4\mathbf{n}\cdot\mathbf{v})\,dp$ from the electron rest frame into the "laboratory" frame.

The distribution of photons scattered out of the beam per unit time is simply the scattering rate multiplied by the distribution function,

$$\frac{df_-}{dt'} = \sigma_T n_e f'(p', \mathbf{n}).$$

Integrating also this term over photon energies we obtain the collision term that enters the energy integrated Boltzmann equation for $\mathcal{M}$ in the baryon rest frame,

$$\begin{aligned} C'[\mathcal{M}] &= \frac{\pi}{a^4\bar{\rho}_\gamma}\int v^3\,dv\left(\frac{df_+}{dt'} - \frac{df_-}{dt'}\right) \\ &= \sigma_T n_e\left[\frac{1}{4}\delta_\gamma^{(\mathrm{long})} - \Delta(\mathbf{n}) - n_i V^{(b)i} + \frac{3n_{ij}}{16\pi}\int \Delta(\mathbf{n}')n'_{ij}\,d\Omega'\right]. \end{aligned} \tag{4.111}$$

Here $\delta_\gamma^{(\mathrm{long})}$ is the density perturbation in longitudinal gauge. To replace Δ in the collision term with $\mathcal{M}$ we use the relation Eq. (4.76) and $\delta_\gamma^{(\mathrm{long})} = D_g^{(\gamma)} + 4\Phi$ together with the fact that Δ and $\mathcal{M}$ differ only by a monopole term that does not contribute to the angular integral in Eq. (4.111). We introduce also

$$M_{ij} = \frac{3}{8\pi}\int n_{ij}\mathcal{M}(\mathbf{n})\,d\Omega$$

and observe that to lowest order $C = (dt'/dt)C' = aC'$. With all this the Boltzmann equation becomes

$$\left(\partial_t + n^i\partial_i\right)\mathcal{M} = S_G(\mathbf{n}) + a\sigma_T n_e\left[\frac{1}{4}D_g^{(\gamma)} - \mathcal{M} - n^i V_i^{(b)} + \frac{1}{2}n_{ij}M^{ij}\right], \tag{4.112}$$

where S_G is the gravitational term defined in Eq. (4.100). Note that the perturbation of the electron density, $n_e = \bar{n}_e + \delta n_e$, does not contribute to first order, since the isotropic background photon distribution $\bar{f}$ annihilates the collision term.

For the Fourier transform of $\mathcal{M}$ we obtain the equation

$$(\partial_t + ik\mu + a\sigma_T n_e)\,\mathcal{M}(\mathbf{k},\mathbf{n}) = S_G(\mathbf{k},\mathbf{n}) + a\sigma_T n_e\left[\frac{1}{4}D_g^{(\gamma)}(\mathbf{k}) - n^i V_i^{(b)}(\mathbf{k}) + \frac{1}{2}n_{ij}M^{ij}(\mathbf{k},\mathbf{n})\right]. \tag{4.113}$$

This can be converted to the integral equation

$$\mathcal{M}(t,\mathbf{k},\mathbf{n}) = e^{-ik\mu(t-t_{\rm in})-\kappa(t_{\rm in},t)}\mathcal{M}(t_{\rm in},\mathbf{k},\mathbf{n}) + \int_{t_{\rm in}}^{t} dt'\, e^{ik\mu(t'-t)-\kappa(t',t)}\left[S_G(\mathbf{k},\mathbf{n}) + \dot{\kappa}\left(\frac{1}{4}D_g^{(\gamma)}(\mathbf{k}) - n^i V_i^{(b)}(\mathbf{k}) + \frac{1}{2}n_{ij}M^{ij}(\mathbf{k},\mathbf{n})\right)\right]. \tag{4.114}$$

Here $\kappa(t_1,t_2) = \int_{t_1}^{t_2} a\sigma_T n_e\,dt$ is the optical depth and $\dot{\kappa}(t_1,t_2) = \partial_{t_2}\kappa(t_1,t_2) = a\sigma_T n_e(t_2)$ is independent of the initial value t_1.

We now decompose Eq. (4.114) into its scalar, vector, and tensor contributions.

4.5.1 Scalar Perturbation

We first consider scalar perturbations. Since the direction dependence enters the evolution equation only via the cosine $\mu = \hat{\mathbf{k}}\cdot\mathbf{n}$, we assume consistently that this is the only direction dependence of the Fourier transform $\mathcal{M}^{(S)}(t,\mathbf{k},\mathbf{n})$, so that $\mathcal{M}(t,\mathbf{k},\mathbf{n}) = \mathcal{M}(t,\mathbf{k},\mu)$. It therefore makes sense to expand $\mathcal{M}$ in Legendre polynomials,

$$\mathcal{M}^{(S)}(t,\mathbf{k},\mu) = \sum (2\ell+1)(-i)^\ell \mathcal{M}_\ell^{(S)}(t,\mathbf{k})P_\ell(\mu). \tag{4.115}$$

Using the orthogonality and normalization of Legendre polynomials, see Appendix 4, Section A4.1, we obtain the expansion coefficients,

$$\mathcal{M}_\ell^{(S)}(t,\mathbf{k}) = \frac{i^\ell}{2}\int_{-1}^{1} d\mu\,\mathcal{M}^{(S)}(t,\mathbf{k},\mu)P_\ell(\mu). \tag{4.116}$$

Statistical homogeneity and isotropy imply that the coefficients $\mathcal{M}_\ell$ for different values of ℓ and $\mathbf{k}$ are uncorrelated,

$$\left\langle \mathcal{M}_\ell^{(S)}(t,\mathbf{k})\mathcal{M}_{\ell'}^{(S)*}(t,\mathbf{k}')\right\rangle = M_\ell^{(S)}(t,k)(2\pi)^3\,\delta^3(\mathbf{k}-\mathbf{k}')\,\delta_{\ell\ell'}, \tag{4.117}$$

and that $M_\ell^{(S)}$ depends only on the norm $k = |\mathbf{k}|$.

We want to relate the spectrum $M_\ell^{(S)}(t,k)$ to the scalar CMB power spectrum $C_\ell^{(S)}$. We use the definition given in Eq. (2.253),

$$\left\langle \frac{\Delta T}{T}(t_0,\mathbf{x}_0,\mathbf{n})\frac{\Delta T}{T}(t_0,\mathbf{x}_0,\mathbf{n}')\right\rangle^{(S)} = \frac{1}{4\pi}\sum_\ell (2\ell+1)C_\ell^{(S)}P_\ell(\mathbf{n}\cdot\mathbf{n}')$$
$$= \frac{1}{(2\pi)^6}\int d^3k\, d^3k' \sum_{\ell_1\ell_2}(2\ell_1+1)(2\ell_2+1)(-i)^{\ell_1-\ell_2}e^{i\mathbf{x}_0\cdot(\mathbf{k}-\mathbf{k}')}$$
$$\times\left\langle \mathcal{M}_{\ell_1}^{(S)}(t_0,\mathbf{k})\mathcal{M}_{\ell_2}^{(S)*}(t_0,\mathbf{k}')\right\rangle P_{\ell_1}(\mu)P_{\ell_2}(\mu'),$$

where $\mu = \hat{\mathbf{k}}\cdot\mathbf{n}$ and $\mu' = \hat{\mathbf{k}}'\cdot\mathbf{n}'$. With Eq. (4.117) we obtain

$$\frac{1}{4\pi}\sum_\ell(2\ell+1)C_\ell^{(S)}P_\ell(\mathbf{n}\cdot\mathbf{n}')$$
$$= \frac{1}{(2\pi)^3}\sum_{\ell_1}\int d^3k\, M_{\ell_1}^{(S)}(t_0,k)(2\ell_1+1)^2 P_{\ell_1}(\mu)P_{\ell_1}(\mu')$$
$$= \frac{2}{\pi}\sum_{\ell_1}\int d^3k\, M_{\ell_1}^{(S)}(t_0,k)\sum_{m_1m_2}Y_{\ell_1m_1}(\mathbf{n})Y_{\ell_1m_1}^*(\hat{\mathbf{k}})Y_{\ell_1m_2}^*(\mathbf{n}')Y_{\ell_1m_2}(\hat{\mathbf{k}})$$
$$= \frac{2}{\pi}\sum_{\ell_1m_1}\int dk\, k^2 M_{\ell_1}^{(S)}(t_0,k)Y_{\ell_1m_1}(\mathbf{n})Y_{\ell_1m_1}^*(\mathbf{n}')$$
$$= \frac{1}{2\pi^2}\sum_{\ell_1}\left(\int dk\, k^2 M_{\ell_1}^{(S)}(t_0,k)\right)(2\ell_1+1)P_{\ell_1}(\mathbf{n}\cdot\mathbf{n}').$$

In several steps in this derivation we have applied the addition theorem of spherical harmonics derived in Appendix 4, Section A4.2.3. Comparing the first and the last expressions in the series of equalities above, we infer

$$C_\ell^{(S)} = \frac{2}{\pi}\int dk\, k^2 M_\ell^{(S)}(t_0,k). \tag{4.118}$$

To calculate the CMB power spectrum, we therefore have to determine the random variables $\mathcal{M}_\ell$. We now derive a hierarchical set of equations for them, the so-called Boltzmann hierarchy.

With Eqs. (4.82)–(4.85), Eq. (4.116), and the explicit expressions of the Legendre polynomials for $\ell \le 2$ given in Appendix 4, Section A4.1, one finds the relations of the scalar perturbations of the photon energy–momentum tensor to the expansion coefficients $\mathcal{M}_\ell(t,\mathbf{k})$, $\ell \le 2$,

$$D_g^{(\gamma)} = 4\mathcal{M}_0^{(S)}, \tag{4.119}$$

$$V_\gamma^{(S)} = 3\mathcal{M}_1^{(S)}, \tag{4.120}$$

$$\Pi_\gamma^{(S)} = 12\mathcal{M}_2^{(S)}. \tag{4.121}$$

Inserting Eq. (4.115) in the definition of M_{ij} and choosing the coordinate system such that $\mathbf{k}$ points in the z direction one can easily compute the integrals $M_{33} = -\mathcal{M}_2^{(S)}$ and $M_{11} = M_{22} = \mathcal{M}_2^{(S)}/2$ and all off-diagonal contributions vanish. With $n_1^2 + n_2^2 = 1 - \mu^2$ this yields

$$\frac{1}{2} n_{ij} M^{ij} = -\frac{1}{2} \mathcal{M}_2^{(S)} P_2(\mu).$$

Also using the fact that for scalar perturbations $\mathbf{V} = i\hat{\mathbf{k}} V$ we obtain the scalar Boltzmann equation

$$(\partial_t + ik\mu)\,\mathcal{M}^{(S)}(\mathbf{k},\mathbf{n}) = ik\mu(\Phi + \Psi) + \dot{\kappa}\left[\frac{1}{4} D_g^{(\gamma)}(\mathbf{k}) - \mathcal{M}^{(S)} - i\mu V^{(b)}(\mathbf{k}) - \frac{1}{2}\mathcal{M}_2(\mathbf{k}) P_2(\mu)\right]. \tag{4.122}$$

With the recurrence relation (see Appendix 4, Section A4.1)

$$\mu P_\ell(\mu) = \frac{\ell + 1}{2\ell + 1} P_{\ell+1}(\mu) + \frac{\ell}{2\ell + 1} P_{\ell-1}(\mu),$$

and the ansatz (4.115), we can convert Eq. (4.122) into the following hierarchy of equations:

$$\dot{\mathcal{M}}_\ell^{(S)} + k\frac{\ell + 1}{2\ell + 1}\mathcal{M}_{\ell+1}^{(S)} - k\frac{\ell}{2\ell + 1}\mathcal{M}_{\ell-1}^{(S)} + \dot{\kappa}\mathcal{M}_\ell^{(S)} = \delta_{\ell 0}\dot{\kappa}\mathcal{M}_0^{(S)} + \frac{1}{3}\delta_{\ell 1}\left[-k(\Phi + \Psi) + \dot{\kappa} V^{(b)}\right] + \dot{\kappa}\frac{1}{10}\delta_{\ell 2}\mathcal{M}_2^{(S)}. \tag{4.123}$$

Here the source terms on the right-hand side contribute only for $\ell = 0, 1$ and $\ell = 2$ respectively. In Eq. (4.123) each variable $\mathcal{M}_\ell^{(S)}$ couples to its neighbors, $\mathcal{M}_{\ell-1}^{(S)}$ and $\mathcal{M}_{\ell+1}^{(S)}$, via the left-hand side. The left-hand side of Eq. (4.122) and therefore also the first three terms of Eq. (4.123) just describe the free streaming of photons after decoupling.

If we want to determine the CMB power spectrum via the Boltzmann hierarchy, Eq. (4.123), in order to calculate, for example, C_{1000} we have to know all the other $\mathcal{M}_\ell^{(S)}$s that may influence $\mathcal{M}_{1000}^{(S)}$ via free streaming during a Hubble time, which is certainly more than 1000. Furthermore, at the beginning, when coupling is still relatively tight, we may simply take into account $\mathcal{M}_0^{(S)}$ and $\mathcal{M}_1^{(S)}$ given by the perfect fluid initial conditions and set all the other $\mathcal{M}_\ell^{(S)}$'s to zero. They then gradually build up mainly due to free streaming. But using the Boltzmann hierarchy (4.123), we cannot calculate $\mathcal{M}_{1000}^{(S)}$ with any accuracy if we have not determined all the $\mathcal{M}_\ell^{(S)}$'s with $\ell < 1000$ with the same (or rather better) accuracy.

On the other hand, if we knew the source term, the right-hand side of Eq. (4.123), we could simply write down the solution, Eq. (4.114). As the source term depends only on the first three moments of the hierarchy, it can usually be obtained with a precision of about 0.1% (see Seljak and Zaldarriaga, 1996) by solving the hierarchy only up to $\ell \simeq 10$. Inserting the corresponding moments into Eq. (4.114) one finds

$$\begin{aligned}\mathcal{M}^{(S)}(t_0, \mathbf{k}, \mu) &= e^{-ik\mu(t_0 - t_{\rm in}) - \kappa(t_{\rm in}, t_0)} \mathcal{M}^{(S)}(t_{\rm in}, \mathbf{k}, \mu) \\ &+ \int_{t_{\rm in}}^{t_0} dt\, e^{ik\mu(t - t_0) - \kappa(t, t_0)} \times \left[ik\mu(\Phi + \Psi)(\mathbf{k}) + \dot{\kappa} \left(\frac{1}{4} D_g^{(\gamma)}(\mathbf{k}) \right.\right. \\ &\left.\left. - i\mu V^{(b)}(\mathbf{k}) - \frac{1}{2} P_2(\mu) \mathcal{M}_2^{(S)}(\mathbf{k}, t) \right) \right]. \end{aligned} \tag{4.124}$$

If the only μ-dependent term was the exponential, we could use its representation in terms of spherical Bessel functions, using Eq. (A4.146), to isolate $\mathcal{M}_\ell^{(S)}$. With this in mind, we use

$$e^{ik\mu(t - t_0)} \mu f(t) = -ik^{-1} \frac{d}{dt} \left(e^{ik\mu(t - t_0)} \right) f(t)$$

to get rid of all the μ-dependence in the term in square brackets of Eq. (4.124). Furthermore, we move the derivative d/dt onto the function f via integration by parts. We want to choose the initial time $t_{\rm in}$ long before decoupling and t_0 denotes today. Therefore, $\kappa(t_{\rm in}, t_0)$ is huge and we can completely neglect the term from the initial condition. Since early times do not contribute, we can formally start the integral at $t_{\rm in} = 0$. We can also neglect the boundary terms in the partial integrations because terms from the upper boundary $t' = t_0$ contribute only to the uninteresting monopole and dipole terms.

Let us introduce the visibility function g, defined by

$$g(t) \equiv a\sigma_T n_e e^{-\kappa(t, t_0)} \equiv \dot{\kappa} e^{-\kappa(t)}. \tag{4.125}$$

This function is very small at early times, when the optical depth κ is very large. During decoupling, κ becomes smaller but also the prefactor, $a\sigma_T n_e = \dot{\kappa}$, then becomes small. Therefore, g is strongly peaked during decoupling and small both before and after; see Fig. 4.1. With the aforementioned integration by parts we then find

$$\mathcal{M}^{(S)}(t_0, \mathbf{k}, \mu) = \int_0^{t_0} dt\, e^{ik\mu(t - t_0)} S^{(S)}(t, \mathbf{k}), \tag{4.126}$$

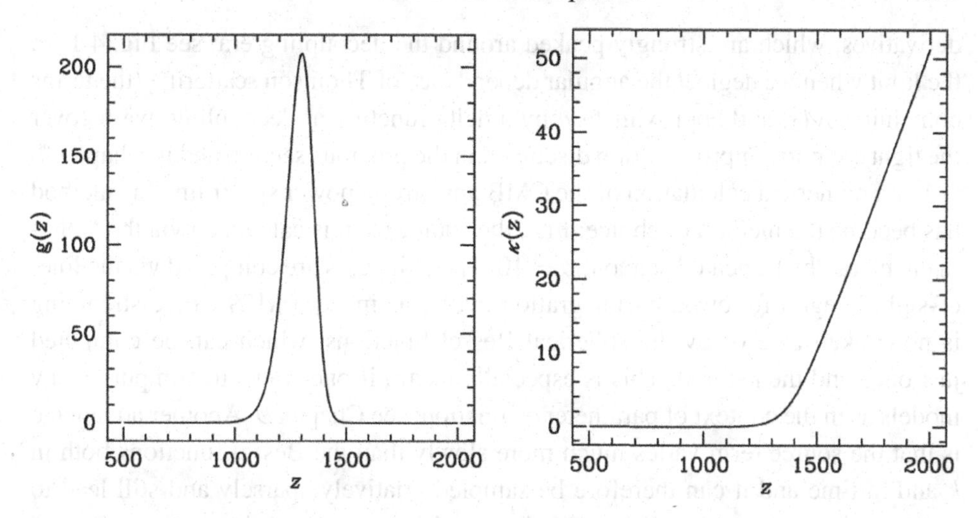

Fig. 4.1 The visibility function g (left) is plotted in units of H_0 as a function of redshift. For comparison we show also $\kappa(z)$ (right).

with

$$S^{(S)} = -e^{-\kappa}(\dot{\Phi} + \dot{\Psi}) + g\left(\Phi + \Psi + k^{-1}\dot{V}^{(b)} + \frac{1}{4}D_g^{(\gamma)} + \frac{1}{4}\mathcal{M}_2\right)$$
$$+ k^{-1}\dot{g}V^{(b)} - \frac{3}{4k^2}\frac{d^2}{dt^2}(g\mathcal{M}_2^{(S)}). \tag{4.127}$$

Rewriting the exponential in terms of spherical Bessel functions and Legendre polynomials and comparing the series with our ansatz (4.115), we find

$$\mathcal{M}_\ell^{(S)}(t_0, \mathbf{k}) = \int_0^{t_0} dt\, j_\ell(k(t_0 - t))S^{(S)}(t, \mathbf{k}). \tag{4.128}$$

Together with Eq. (4.118) this yields the scalar contributions to the CMB power spectrum, once the scalar source term is given:

$$C_\ell^{(S)} = 4\pi \int_0^\infty \frac{dk}{k}\Delta_{\mathcal{R}}(k)\left|\int_0^{t_0} dt\, j_\ell(k(t_0 - t))T_S(k,t)\right|^2. \tag{4.129}$$

Here T_S is the transfer function for the source term given in Eq. (4.127) and $\Delta_{\mathcal{R}} = A_s(k/k_*)^{n_s-1}$ is the primordial power spectrum.

In this chapter we still neglect the effect of polarization. As we shall see in the next chapter, including it simply leads to a slight modification of the source term $S^{(S)}$. Apart from the gravitational contribution $\dot{\Phi}+\dot{\Psi}$ that gives rise to the integrated Sachs–Wolfe effect, all the terms are multiplied with the visibility function g or its derivatives, which are strongly peaked around the decoupling era; see Fig. 4.1. In the limit when we neglect the angular dependence of Thomson scattering (the terms containing $\mathcal{M}_2$) and approximate g by a delta function at decoupling, we recover the tight coupling approximation discussed in the previous section and in Chapter 2.

For a numerical calculation of the CMB anisotropy power spectrum, this method has become the method of choice: first, the source term is calculated via the Boltzmann hierarchy truncated at about $\ell = 10$. Then, the C_ℓ's are computed via the line-of-sight integral followed by integration over k as in Eq. (4.129). Free streaming is now taken care of by the spherical Bessel functions, which can be computed just once and then stored. This is especially useful if one wants to compute many models as in the context of parameter estimation; see Chapter 9. Another advantage is that the source term varies much more slowly than the Bessel functions both in k and in time and it can therefore be sampled relatively sparsely and still lead to good accuracy. Also, not all the C_ℓ's have to be computed. It is usually sufficient to calculate every tenth ℓ and to interpolate smoothly between them. All these numerical advantages have are used in the publicly available codes CMBfast (Seljak and Zaldarriaga, 1996), CAMB (Lewis *et al.*, 2000), CMBeasy (Doran, 2005), and CLASS (Lesgourgues, 2011; Blas *et al.*, 2011). CMBfast is "the original" from which the others are drawn. CAMB and CLASS are presently the best maintained and updated of these codes. While CMBfast and CAMB are written in Fortran, CLASS is written in C++, which makes it more modern and user friendly. The agreement of CAMB and CLASS for a standard ΛCDM cosmology is on the level of 0.1%.

The time integral of the source term in Eq. (4.128) will smear out and damp fluctuations with wavelengths smaller than the width of the visibility function g. This phenomenon, called "Silk damping," will be discussed in more detail in Section 4.6.

But first we want to derive the Boltzmann hierarchy and its solution via line-of-sight integration also for vector and tensor perturbations.

4.5.2 Vector Perturbations

For vector perturbations, the Boltzmann equation (4.113) becomes

$$\dot{\mathcal{M}}^{(V)} + ik\mu\mathcal{M}^{(V)} = -n^i n^j a^{-1}\sigma_{ij}^{(V)} + a\sigma_T n_e\left[n^i\Omega_i^{(b)} - \mathcal{M}^{(V)} + \frac{1}{2}n^{ij}M_{ij}\right]. \tag{4.130}$$

As before we decompose $\mathbf{n}$ into

$$\begin{aligned}\mathbf{n} &= \mu\hat{\mathbf{k}} + \sqrt{1-\mu^2}\,(\cos\varphi\mathbf{e}_1 + \sin\varphi\mathbf{e}_2) = \mu\hat{\mathbf{k}} + n^{(+)}\mathbf{e}^{(+)} + n^{(-)}\mathbf{e}^{(-)} \\ &= \mu\hat{\mathbf{k}} + \sqrt{\frac{1-\mu^2}{2}}\left(e^{-i\varphi}\mathbf{e}^{(+)} + e^{i\varphi}\mathbf{e}^{(-)}\right).\end{aligned} \tag{4.131}$$

We split $\mathcal{M}^{(V)}$ and $-n^i n^j \sigma_{ij}^{(V)}$ into helicity modes as in Eqs. (4.86) and (4.89) and develop $\mathcal{M}^{(V\pm)}$ in Legendre polynomials:

$$\mathcal{M}^{(V)} = \sqrt{\frac{1-\mu^2}{2}}\left[\exp(i\varphi)\mathcal{M}^{(V+)} + \exp(-i\varphi)\mathcal{M}^{(V-)}\right], \tag{4.132}$$

$$\mathcal{M}^{(V+)} = \sum_\ell (-i)^\ell (2\ell+1)\mathcal{M}_\ell^{(V+)} P_\ell(\mu), \tag{4.133}$$

$$\mathcal{M}^{(V-)} = \sum_\ell (-i)^\ell (2\ell+1)\mathcal{M}_\ell^{(V-)} P_\ell(\mu), \tag{4.134}$$

$$a^{-1} n^i n^j \sigma_{ij}^{(V)} = \frac{-i}{\sqrt{2}}\mu\sqrt{1-\mu^2}\left(\sigma^{(V+)}e^{i\varphi} + \sigma^{(V-)}e^{-i\varphi}\right). \tag{4.135}$$

For the expansion in Legendre polynomials we have used the fact that the coefficients $\mathcal{M}^{(V\pm)}$ depend on $\mathbf{n}$ only via μ.

As for scalar perturbations, statistical homogeneity and isotropy require that the random variables $\mathcal{M}_\ell^{(V\pm)}(\mathbf{k})$ are uncorrelated for different values $\mathbf{k}$ and ℓ. Furthermore, we want to consider parity invariant perturbations; hence also $\mathcal{M}_\ell^{(V+)}(\mathbf{k})$ and $\mathcal{M}_\ell^{(V-)}(\mathbf{k})$ are uncorrelated and they have the same spectrum,

$$\begin{aligned}\left\langle \mathcal{M}_\ell^{(V+)}(\mathbf{k})\mathcal{M}_{\ell'}^{(V+)*}(\mathbf{k}')\right\rangle &= \left\langle \mathcal{M}_\ell^{(V-)}(\mathbf{k})\mathcal{M}_{\ell'}^{(V-)*}(\mathbf{k}')\right\rangle \\ &= (2\pi)^3\,\delta^3(\mathbf{k}-\mathbf{k}')\delta_{\ell\ell'}M_\ell^{(V)}(k).\end{aligned} \tag{4.136}$$

To relate this spectrum to the vector C_ℓ's we use, as for scalar perturbations,

$$\begin{aligned}&\left\langle \frac{\Delta T}{T}(t_0,\mathbf{n})\frac{\Delta T}{T}(t_0,\mathbf{n}')\right\rangle^{(V)} \\ &= \frac{1}{4\pi}\sum_\ell (2\ell+1)C_\ell^{(V)} P_\ell(\mathbf{n}\cdot\mathbf{n}') \\ &= \frac{1}{(2\pi)^6}\int d^3k\, d^3k'\,\langle \mathcal{M}^{(V)}(\mathbf{k},\mathbf{n})\mathcal{M}^{(V)*}(\mathbf{k}',\mathbf{n}')\rangle e^{i\mathbf{x}_0(\mathbf{k}-\mathbf{k}')} \\ &= \sum_\ell \frac{(2\ell+1)^2}{(2\pi)^3}\int d^3k\, P_\ell(\mu)P_\ell(\mu')\sqrt{(1-\mu^2)(1-\mu'^2)}\cos(\varphi-\varphi')M_\ell^{(V)}(k),\end{aligned} \tag{4.137}$$

where $\mu = \hat{\mathbf{k}}\cdot\mathbf{n}$ and $\mu' = \hat{\mathbf{k}}\cdot\mathbf{n}'$ and φ and φ' are the angles defined in the decomposition on $\mathbf{n}$ and $\mathbf{n}'$ respectively according to Eq. (4.131). The first equals sign is just the definition of the $C_\ell^{(V)}$'s and after the second equals sign we have inserted the Fourier representation of $\Delta T/T$. Using the decomposition (4.131) one finds

$$\mathbf{n}\cdot\mathbf{n}' = \mu\mu' + \sqrt{(1-\mu^2)(1-\mu'^2)}\cos(\varphi-\varphi'). \tag{4.138}$$

We therefore have $\sqrt{(1-\mu^2)(1-\mu'^2)}\cos(\varphi-\varphi') = \mathbf{n}\cdot\mathbf{n}' - \mu\mu'$. The term $\mathbf{n}\cdot\mathbf{n}'$ is independent of $\mathbf{k}$ and can be taken out of the integral. The terms containing additional factors μ and μ' respectively can be absorbed with the help of the recurrence relations of Legendre polynomials and again using the addition theorem of spherical harmonics. We finally arrive at

$$C_\ell^{(V)} = \frac{2\ell(\ell+1)}{\pi(2\ell+1)^2}\int dk\, k^2\left(M_{\ell+1}^{(V)} + M_{\ell-1}^{(V)}\right). \tag{4.139}$$

The details of the derivation are developed in Exercise 4.4.

A short calculation shows that the vector perturbations of the energy–momentum tensor are given in terms of the expansion coefficients $\mathcal{M}_\ell^{(V+)}$ and $\mathcal{M}_\ell^{(V-)}$ by

$$\Omega^\pm = \mathcal{M}_0^{(V\pm)} + \mathcal{M}_2^{(V\pm)}, \tag{4.140}$$

$$\Pi^{(V\pm)} = \frac{24}{5}\left[\mathcal{M}_1^{(V\pm)} + \mathcal{M}_3^{(V\pm)}\right]. \tag{4.141}$$

To write the Boltzmann equation with the help of moments of $\mathcal{M}^{(V\pm)}$ we still need $n^{ij}M_{ij}$. A short calculation shows that only $M_{13} = M_{31}$ and $M_{23} = M_{32}$ do not vanish. Using the basic properties of the Legendre polynomials (see Appendix 4, Section A4.1) we obtain ($\mathbf{k} = k\mathbf{e}_3$)

$$\begin{aligned} M_{\pm 3} \equiv M_{13} \mp iM_{23} &= \frac{3}{8}\int_{-1}^{1} d\mu\,\mu(1-\mu^2)\mathcal{M}^{(V\pm)} \\ &= -\frac{3i}{10}\left(\mathcal{M}_1^{(V\pm)} + \mathcal{M}_3^{(V\pm)}\right). \end{aligned}$$

With the definitions (4.132)–(4.135) and (4.140) the Boltzmann equation can then be written as

$$\begin{aligned} \dot{\mathcal{M}}^{(V\pm)} + ik\mu\mathcal{M}^{(V\pm)} + \dot\kappa\mathcal{M}^{(V\pm)} &= i\mu\sigma^{(V\pm)} \\ &\quad + \dot\kappa\left[\Omega^{(\pm)} - i\mu\frac{3}{10}\left(\mathcal{M}_1^{(V\pm)} + \mathcal{M}_3^{(V\pm)}\right)\right], \end{aligned} \tag{4.142}$$

where κ denotes the optical depth $\kappa(t) = \sigma_T \int_t^{t_0} a n_e \, dt'$. As in the case of scalar perturbations this yields a Boltzmann hierarchy equation for the $\mathcal{M}_\ell^{(V\epsilon)}$s,

$$\dot{\mathcal{M}}_\ell^{(V\pm)} + \frac{k}{2\ell+1}\left[(\ell+1)\mathcal{M}_{\ell+1}^{(V\pm)} - \ell\mathcal{M}_{\ell-1}^{(V\pm)}\right] = -\dot{\kappa}\mathcal{M}_\ell^{(V\pm)} + \delta_{\ell 0}\dot{\kappa}\Omega^{(\pm)} + \delta_{\ell 1}\left[\frac{-1}{3}\sigma^{(V\pm)} + \dot{\kappa}\frac{1}{10}\left(\mathcal{M}_1^{(V\pm)} + \mathcal{M}_3^{(V\pm)}\right)\right]. \tag{4.143}$$

Also for vector perturbations, the most rapid way of solving the equations numerically is to solve the above hierarchy only for the lowest few multipoles in order to determine the source term, the right-hand side of Eq. (4.143). For a given source term Eq. (4.142) is then easily solved by line-of-sight integration:

$$\mathcal{M}^{(V\pm)}(t_0, \mathbf{k}, \mu) = \int_0^{t_0} dt \, e^{ik\mu(t-t_0)-\kappa}\left[-i\mu\sigma^{(V\pm)} + \dot{\kappa}\left(\Omega^{(\pm)} - i\mu\frac{3}{10}\left(\mathcal{M}_1^{(V\pm)} + \mathcal{M}_3^{(V\pm)}\right)\right)\right]. \tag{4.144}$$

Absorbing the factors μ into time derivatives as in the scalar case, we find

$$\mathcal{M}^{(V\pm)}(t_0, \mathbf{k}, \mu) = \int_0^{t_0} dt \, e^{ik\mu(t-t_0)}\left[+k^{-1}e^{-\kappa}\dot{\sigma}^{(V\pm)} + g\left(\Omega^{(\pm)} - k^{-1}\sigma^{(V\pm)} + \frac{3}{10k}\left(\dot{\mathcal{M}}_1^{(V\pm)} + \dot{\mathcal{M}}_3^{(V\pm)}\right)\right) + \dot{g}\frac{3}{10k}\left(\mathcal{M}_1^{(V\pm)} + \mathcal{M}_3^{(V\pm)}\right)\right]. \tag{4.145}$$

We have again used the visibility function g defined in Eq. (4.125). The expansion of the exponential in terms of spherical Bessel functions and Legendre polynomials reproduces the $\mathcal{M}_\ell^{(V\pm)}$'s,

$$\mathcal{M}_\ell^{(V\pm)}(t_0, \mathbf{k}) = \int_0^{t_0} dt \, j_\ell(k\mu(t_0 - t))\left[+k^{-1}e^{-\kappa}\dot{\sigma}^{(V\pm)} + g\left(\Omega^{(\pm)} - k^{-1}\sigma^{(V\pm)} + \frac{3}{10k}\left(\dot{\mathcal{M}}_1^{(V\pm)} + \dot{\mathcal{M}}_3^{(V\pm)}\right)\right) + \dot{g}\frac{3}{10k}\left(\mathcal{M}_1^{(V\pm)} + \mathcal{M}_3^{(V\pm)}\right)\right]. \tag{4.146}$$

4.5.3 Tensor Perturbations

The Boltzmann equation (4.113) for tensor perturbations finally has the form

$$\dot{\mathcal{M}}^{(T)} + ik\mu\mathcal{M}^{(T)} = -n^i n^j \dot{H}_{ij}^{(T)} + a\sigma_T n_e\left[\frac{1}{2}n^{ij}M_{ij}^{(T)} - \mathcal{M}^{(T)}\right]. \tag{4.147}$$

Since $H_{ij}^{(T)}$ is entirely orthogonal to $\mathbf{k}$, it is of the form $H_{ij}^{(T)} = \hat{H}_{ab}^{(T)} e_i^a e_j^b$, where $\mathbf{e}^1$ and $\mathbf{e}^2$ denote the two polarization directions normal to $\mathbf{k}$. Using the decomposition Eq. (4.131) for $\mathbf{n}$, the gravitational source term in Eq. (4.147) is seen to be the time derivative of

$$\begin{aligned} n^i n^j H_{ij}^{(T)} &= (1-\mu^2)\left[\widehat{H}_{11}^{(T)}\cos^2\varphi + \widehat{H}_{22}^{(T)}\sin^2\varphi + 2\widehat{H}_{12}^{(T)}\cos\varphi\sin\varphi\right] \\ &= (1-\mu^2)\left[H_d\cos(2\varphi) + H_\times\sin(2\varphi)\right]. \end{aligned} \tag{4.148}$$

For the last equals sign we have used that $\hat{H}_{22}^{(T)} = -\widehat{H}_{11}^{(T)}$ and we have introduced the usual notation for the two polarizations of a gravity wave propagating in direction $\hat{\mathbf{k}}$, $H_d \equiv \widehat{H}_{11}^{(T)}$ and $\hat{H}_{12}^{(T)} \equiv H_\times$.

This motivates our ansatz for the tensor perturbations of the temperature anisotropy,

$$\begin{aligned} \mathcal{M}^{(T)}(\mathbf{k},\mathbf{n}) = (1-\mu^2)\big[&\mathcal{M}^{(Td)}(\mathbf{k},\mu)\cos(2\varphi) \\ &+ \mathcal{M}^{(T\times)}(\mathbf{k},\mu)\sin(2\varphi)\big]. \end{aligned} \tag{4.149}$$

The coefficients $\mathcal{M}^{(T\bullet)}$ only depend on μ and can again be expanded in Legendre polynomials,

$$\mathcal{M}^{(T\bullet)} = \sum_\ell (2\ell+1)(-i)^\ell \mathcal{M}_\ell^{(T\bullet)}(\mathbf{k}) P_\ell(\mu). \tag{4.150}$$

Here $^\bullet$ denotes either d or $\times$. Statistical homogeneity and isotropy again imply that expansion coefficients with different values of $\mathbf{k}$ or for different ℓ's are uncorrelated. Furthermore, also requiring invariance under parity implies that the two polarizations are uncorrelated and have the same spectra,

$$\left\langle \mathcal{M}_\ell^{(T\bullet)}(\mathbf{k})\mathcal{M}_{\ell'}^{(T\bullet)*}(\mathbf{k}')\right\rangle = (2\pi)^3\delta^3(\mathbf{k}-\mathbf{k}')\delta_{\ell\ell'}M_\ell^{(T)}(k). \tag{4.151}$$

The relation of this tensor spectrum to the C_ℓ's is obtained with the same reasoning as for the scalar and vector modes:

$$\begin{aligned} &\frac{1}{4\pi}\sum_\ell (2\ell+1)C_\ell^{(T)} P_\ell(\mathbf{n}\cdot\mathbf{n}') \\ &= \frac{1}{(2\pi)^6}\int d^3k d^3k' \langle \mathcal{M}^{(T)}(\mathbf{k},\mathbf{n})\mathcal{M}^{(T)*}(\mathbf{k}',\mathbf{n}')\rangle e^{-\mathbf{x}(\mathbf{k}-\mathbf{k}')} \\ &= \frac{1}{(2\pi)^3}\sum_\ell (2\ell+1)^2\int d^3k M_\ell^{(T)}(k)P_\ell(\mu)P_\ell(\mu') \\ &\quad\times (1-\mu^2)(1-\mu'^2)[\cos(2\varphi)\cos(2\varphi') + \sin(2\varphi)\sin(2\varphi')], \end{aligned} \tag{4.152}$$

where $\mu = \hat{\mathbf{k}} \cdot \mathbf{n}$ and $\mu' = \hat{\mathbf{k}} \cdot \mathbf{n}'$. The angles φ and φ' are those appearing in the decomposition of $\mathbf{n}$ and $\mathbf{n}'$ according to Eq. (4.131).

Again using Eq. (4.138) and $\cos(2\varphi)\cos(2\varphi') + \sin(2\varphi)\sin(2\varphi') = \cos(2(\varphi - \varphi'))$, the φ-dependence can be written as a function of $\mathbf{n} \cdot \mathbf{n}'$ and μ and μ'. The recurrence relations for the Legendre polynomials can then be applied to absorb the factors μ and μ', and with the addition theorem of spherical harmonics, we arrive after a somewhat lengthy calculation at

$$C_\ell^{(T)} = \frac{2}{\pi}\frac{(\ell+2)!}{(\ell-2)!}\int dk\, k^2 \frac{\Sigma^{(T)}(k)}{(2\ell+1)^2}, \quad \text{with} \tag{4.153}$$

$$\Sigma^{(T)}(k) = \frac{M_{\ell-2}^{(T)}}{(2\ell-1)^2} + \frac{4(2\ell+1)^2 M_\ell^{(T)}}{[(2\ell-1)(2\ell+3)]^2} + \frac{M_{\ell+2}^{(T)}}{(2\ell+3)^2}. \tag{4.154}$$

The details of this result are developed in Exercise 4.5.

We also express the tensor anisotropic stress in terms of the expansion coefficients $\mathcal{M}^{(T\bullet)}$; we use that it is transverse to $\mathbf{k}$:

$$\Pi_r^{(T\bullet)} = \frac{3}{2}\int_{-1}^{1} d\mu\,(1-\mu^2)^2 \mathcal{M}^{(T\bullet)} = \frac{24}{35}\mathcal{M}_4^{(T\bullet)} + \frac{16}{7}\mathcal{M}_2^{(T\bullet)} + \frac{8}{5}\mathcal{M}_0^{(T\bullet)}. \tag{4.155}$$

With the ansatz (4.149) we find that for tensor perturbations

$$n^{ij}M_{ij} = (1-\mu^2)\left\{\cos(2\varphi)\left[\frac{3}{35}\mathcal{M}_4^{(Td)} + \frac{2}{7}\mathcal{M}_2^{(Td)} + \frac{1}{5}\mathcal{M}_0^{(Td)}\right]\right.$$
$$\left. + \sin(2\varphi)\left[\frac{3}{35}\mathcal{M}_4^{(T\times)} + \frac{2}{7}\mathcal{M}_2^{(T\times)} + \frac{1}{5}\mathcal{M}_0^{(T\times)}\right]\right\}. \tag{4.156}$$

Inserting this in the Boltzmann equation we obtain

$$\dot{\mathcal{M}}^{(T\bullet)} + ik\mu\mathcal{M}^{(T\bullet)} + \dot{\kappa}\mathcal{M}^{(T\bullet)} = -\dot{H}_\bullet + \dot{\kappa}\left[\frac{3}{70}\mathcal{M}_4^{(T\bullet)} + \frac{1}{7}\mathcal{M}_2^{(T\bullet)} + \frac{1}{10}\mathcal{M}_0^{(T\bullet)}\right], \tag{4.157}$$

with the line-of-sight "solution"

$$\mathcal{M}^{(T\bullet)}(t_0, \mathbf{k}, \mu) = \int_0^{t_0} dt\, e^{ik\mu(t-t_0)-\kappa}\left[-\dot{H}_\bullet\right.$$
$$\left. + \dot{\kappa}\left(\frac{3}{70}\mathcal{M}_4^{(T\bullet)} + \frac{1}{7}\mathcal{M}_2^{(T\bullet)} + \frac{1}{10}\mathcal{M}_0^{(T\bullet)}\right)\right]. \tag{4.158}$$

Of course for this to solve the equation, the first moments, $\mathcal{M}_0^{(T\bullet)}$ to $\mathcal{M}_4^{(T\bullet)}$, which also determine $\dot{H}_\bullet$, have to be calculated via the Boltzmann hierarchy, which in this case is

$$\dot{\mathcal{M}}_\ell^{(T\bullet)} + \frac{k}{2\ell+1}\left[(\ell+1)\mathcal{M}_{\ell+1}^{(T\bullet)} - \ell\mathcal{M}_{\ell-1}^{(T\bullet)}\right]$$
$$= -\dot{\kappa}\mathcal{M}_\ell^{(T\bullet)} + \delta_{\ell 0}\left[-\dot{H}_\bullet + \dot{\kappa}\left(\frac{3}{70}\mathcal{M}_4^{(T\bullet)} + \frac{1}{7}\mathcal{M}_2^{(T\bullet)} + \frac{1}{10}\mathcal{M}_0^{(T\bullet)}\right)\right]. \tag{4.159}$$

The coefficients $\mathcal{M}_\ell^{(T\bullet)}$ are now given simply by

$$\mathcal{M}_\ell^{(T\bullet)}(t_0,\mathbf{k}) = \int_0^{t_0} dt\, j_\ell(k(t_0-t))S_\ell^{(T)}, \quad \text{with} \tag{4.160}$$

$$S_\ell^{(T)} = e^{-\kappa}\left[-\dot{H}_\bullet + \delta_{0\ell}\dot{\kappa}\left(\frac{3}{70}\mathcal{M}_4^{(T\bullet)} + \frac{1}{7}\mathcal{M}_2^{(T\bullet)} + \frac{1}{10}\mathcal{M}_0^{(T\bullet)}\right)\right]. \tag{4.161}$$

Also here, the only modification that this solution will experience once we include polarization is a change in the source term $S^{(T)}$, to which contributions from the polarization spectrum will have to be added (see Chapter 5).

The relations between the $\mathcal{M}_\ell^{(\bullet)}$ and the C_ℓ's for vector and tensor perturbations is not very straightforward, but rather somewhat "clumsy." In Chapter 5 we shall employ the much better adapted and more elegant "total angular momentum method" to fix this shortcoming. For this we shall make use of spin weighted spherical harmonics, which we have avoided in this chapter.

4.6 Silk Damping

In this section we want to discuss, in a more direct way, the damping on small scales that appears when the coupling between photons and the baryon/electron gas is still present but no longer perfect. We therefore do not want to describe the photon–baryon system as a perfect fluid, but want to take into account the force provided by Thomson scattering of electrons and photons. This leads to an additional force in the baryon equation of motion [Eq. (2.119) for $w = c_s^2 = 0$], the photon drag force due to Thomson scattering. For simplicity, and since this is the relevant case, we consider only scalar perturbations in this section. For them the photon drag force is given by

$$F_j = -\frac{\rho_\gamma}{\pi}\int C[\mathcal{M}]n_j\, d\Omega_{\mathbf{n}}, \tag{4.162}$$

$$\hat{\mathbf{k}}\cdot\mathbf{F} = \frac{-4i\sigma_T n_e a\rho_\gamma}{3}\left(3\mathcal{M}_1 - V^{(b)}\right). \tag{4.163}$$

For the second equals sign we have used Eq. (4.111) and integrated $\mathbf{n}C[\mathcal{M}] = \mathbf{n}C'[\mathcal{M}]a$ over angles. From the expansion of $\mathcal{M}$ in Legendre polynomials we know that

$$(-i)^\ell \mathcal{M}_\ell = \frac{1}{2}\int d\mu\, P_\ell(\mu)\mathcal{M}(\mu),$$

so that

$$\mathcal{M}_1 = \frac{i}{2}\int d\mu\, \mu \mathcal{M}(\mu) \quad \text{and} \quad \frac{1}{4}D_g^{(\gamma)} = \frac{1}{2}\int d\mu\, \mathcal{M}.$$

Adding the drag force to the baryon equation of motion yields (in Fourier space)

$$\dot{\mathbf{V}}^{(b)} + \mathcal{H}\mathbf{V}^{(b)} = -i\mathbf{k}\Psi + \rho_b^{-1}\mathbf{F}. \tag{4.164}$$

To discuss damping, we are only interested in small scales $kt \gg 1$ and therefore shall neglect the expansion of the Universe in our treatment. It then makes sense to model the time dependence of our variables with an exponential, $V, \mathcal{M} \propto e^{-i\omega t}$. Furthermore, we consider the epoch when there are still many collisions per Hubble expansion. Denoting the collision time by $t_c = 1/\dot{\kappa} = 1/(a\sigma_T n_e)$, this means $t \gg t_c$. For simplicity, we also neglect gravitational terms and the term $n^{ij}M_{ij}$ that is due to the direction dependence of Thomson scattering and is not important as long as scattering is sufficiently abundant. With the ansatz of a harmonic time dependence with frequency ω, we then obtain from Eqs. (4.122) and (4.164) with $\hat{\mathbf{k}} \cdot \mathbf{V} = +iV^{(b)}$,

$$-it_c(\omega - k\mu)\mathcal{M} = \frac{1}{4}D_g^{(\gamma)} - i\mu V^{(b)} - \mathcal{M}, \tag{4.165}$$

$$t_c\omega V^{(b)} = \frac{4i\rho_\gamma}{3\rho_b}[3\mathcal{M}_1 - V^{(b)}]. \tag{4.166}$$

Therefore, integrating the Boltzmann equation (4.165) over μ yields

$$\mathcal{M}_1 = \frac{i\omega}{4k}D_g^{(\gamma)}. \tag{4.167}$$

Inserting this in Eq. (4.166) we find, with $R \equiv 3\rho_b/4\rho_\gamma$,

$$V^{(b)} = \frac{3i\omega D_g^{(\gamma)}}{4k(1 - it_c\omega R)}. \tag{4.168}$$

Inserting this result for $V^{(b)}$ in Eq. (4.165) we obtain

$$\mathcal{M} = \frac{1 + \frac{3\mu\omega/k}{1 - it_c\omega R}}{1 - it_c(\omega - k\mu)}\frac{D_g^{(\gamma)}}{4}. \tag{4.169}$$

Integrating this equation over μ yields a dispersion relation for $\omega(k)$ in the form

$$1 = \frac{1}{2}\int_{-1}^{1} d\mu \, \frac{1 + \frac{3\mu\omega/k}{1-it_c\omega R}}{1 - it_c(\omega - k\mu)}$$
$$= \frac{3\omega}{it_c k^2 + t_c^2 k^2 \omega R} + \frac{1}{2}\left(\frac{1}{it_c k} + \frac{3\omega}{t_c^2 k^3}\frac{1 - it_c\omega}{1 - it_c\omega R}\right)$$
$$\times \left[\ln(1 + it_c(k - \omega)) - \ln(1 - it_c(\omega + k))\right]. \tag{4.170}$$

This equation cannot be solved analytically. If we expand it in $t_c k$ and $t_c\omega$ we find to lowest nonvanishing order

$$\omega(k) = k\left(\frac{1}{\sqrt{3(R+1)}} - i\frac{kt_c}{6}\frac{R^2 + \frac{4}{5}(R+1)}{(R+1)^2}\right). \tag{4.171}$$

The real part of $\omega(k)$ describes oscillations and $\text{Re}(\omega)/k$ is the group velocity of the oscillations. The imaginary term is a damping term. Over a time interval $(-\text{Im}(\omega))^{-1} = t_d = \frac{6}{k^2 t_c}\frac{(R+1)^2}{R^2 + \frac{4}{5}(R+1)}$, the amplitude is reduced by one e-fold. This is Silk damping (Silk, 1967), due to the imperfect coupling of electrons and photons. It vanishes in the limit $kt_c \to 0$. It is interesting to note that one has to expand Eq. (4.170) to third order in ωt_c and kt_c to find this relation (see Exercise 4.6). This indicates that damping is effective only for $t_c k$ relatively close to 1.

4.7 The Full System of Perturbation Equations

We end this chapter by writing down the full system of perturbation equations in a "standard" universe containing dark matter, baryons, photons, massless neutrinos, and a cosmological constant. The latter only influences the background evolution and does not appear in the perturbation equations. Even though we know that neutrinos are not truly massless, since their mass scale may be as low as 0.06 eV, it is nearly irrelevant for CMB anisotropies. We thus neglect it here. In standard inflationary models only scalar and tensor modes are generated; we therefore restrict this recapitulation to them. The fluid equations for dark matter and baryons and the Einstein equations have been derived in Chapter 2. The Boltzmann equation for photons and the evolution equation for neutrinos, which is simply the Liouville equation, have been derived in this chapter.

The evolution of cold dark matter perturbations, D_c and V_c, is determined by the energy–momentum conservation equations

$$\dot{D}_c = -kV_c + \frac{9}{2}\frac{\mathcal{H}^2}{k}(1 + w)(V - V_c), \tag{4.172}$$
$$\dot{V}_c + \mathcal{H}V_c = k\Psi. \tag{4.173}$$

For the evolution of baryons, we have also to take into account the photon drag force leading to

$$\dot{D}_b = -kV_b + \frac{9}{2}\frac{\mathcal{H}^2}{k}(1+w)(V - V_c), \tag{4.174}$$

$$\dot{V}_b + \mathcal{H}V_b = k\Psi + \frac{4\dot{\kappa}\rho_\gamma}{3\rho_b}(3\mathcal{M}_1 - V_b). \tag{4.175}$$

Here V is the total velocity perturbation,

$$(\rho + P)V = \rho_c V_c + \rho_b V_b + 4\rho_\gamma \mathcal{M}_1 + 3(\rho_\nu + P_\nu)\mathcal{N}_1, \tag{4.176}$$

where $\mathcal{N}_1$ is the first moment of the perturbation of the energy-integrated neutrino distribution function discussed in the text that follows.

For the low multipoles, $\ell < 10$, say, we have to solve the Boltzmann hierarchies

$$\dot{\mathcal{M}}_\ell^{(S)} + k\frac{\ell+1}{2\ell+1}\mathcal{M}_{\ell+1}^{(S)} - k\frac{\ell}{2\ell+1}\mathcal{M}_{\ell-1}^{(S)} = \frac{1}{3}\delta_{\ell 1}\left[-k(\Phi+\Psi) + \dot{\kappa}V^{(b)}\right] + \dot{\kappa}\left[\frac{1}{2}\delta_{\ell 2}\mathcal{M}_2^{(S)} + (\delta_{\ell 0} - 1)\mathcal{M}_\ell^{(S)}\right], \tag{4.177}$$

for scalar perturbations, and

$$\dot{\mathcal{M}}_\ell^{(T\bullet)} + \frac{k}{2\ell+1}\left[(\ell+1)\mathcal{M}_{\ell+1}^{(T\bullet)} - \ell\mathcal{M}_{\ell-1}^{(T\bullet)}\right] = -\dot{\kappa}\mathcal{M}_\ell^{(T\bullet)} + \delta_{\ell 0}\left[-\dot{H}_\bullet + \dot{\kappa}\left(\frac{3}{35}\mathcal{M}_4^{(T\bullet)} + \frac{2}{7}\mathcal{M}_2^{(T\bullet)} + \frac{1}{5}\mathcal{M}_0^{(T\bullet)}\right)\right] \tag{4.178}$$

for tensor perturbations. The higher multipoles, ℓ, can then be obtained via the integrals given in Eqs. (4.128) and (4.160).

Neutrino perturbations have to be treated via the collisionless Boltzmann equation. We neglect neutrino masses. Setting the collision term to zero and denoting the neutrino perturbation of the distribution function integrated over energies by $\mathcal{N}$, we obtain, by exactly the same steps as explained in the previous sections for photons,

$$\dot{\mathcal{N}}_\ell^{(S)} + \frac{k}{2\ell+1}\left[(\ell+1)\mathcal{N}_{\ell+1}^{(S)} - \ell\mathcal{N}_{\ell-1}^{(S)}\right] = \frac{-k}{3}\delta_{\ell 1}(\Phi+\Psi), \tag{4.179}$$

for scalar perturbations, and

$$\dot{\mathcal{N}}_\ell^{(T\bullet)} + \frac{k}{2\ell+1}\left[(\ell+1)\mathcal{N}_{\ell+1}^{(T\bullet)} - \ell\mathcal{N}_{\ell-1}^{(T\bullet)}\right] = -\delta_{\ell 0}\dot{H}_\bullet \tag{4.180}$$

for tensor perturbations.

The scalar and tensor metric perturbations are determined by Einstein's equations,

$$-k^2\Phi = 4\pi G a^2 \rho D, \tag{4.181}$$

$$k^2(\Phi - \Psi) = 4\pi G a^2 \left(P_\gamma \Pi_\gamma^{(S)} + P_\nu \Pi_\nu^{(S)}\right), \qquad \text{and} \tag{4.182}$$

$$\ddot{H}_\bullet + 2\mathcal{H}\dot{H}_\bullet + k^2 H_\bullet = 8\pi G a^2 \left(P_\gamma \Pi_{r\bullet}^{(T)} + P_\nu \Pi_{\nu\bullet}^{(T)}\right). \tag{4.183}$$

The scalar and tensor anisotropic stresses are given by

$$\Pi_\gamma^{(S)} = 12\mathcal{M}_2^{(S)}, \tag{4.184}$$

$$\Pi_\nu^{(S)} = 12\mathcal{N}_2^{(S)}, \tag{4.185}$$

$$\Pi_{r\bullet}^{(T)} = \frac{24}{35}\mathcal{M}_4^{(T)} + \frac{16}{7}\mathcal{M}_2^{(T)} + \frac{8}{5}\mathcal{M}_0^{(T)}, \tag{4.186}$$

$$\Pi_{\nu\bullet}^{(T)} = \frac{24}{35}\mathcal{N}_4^{(T)} + \frac{16}{7}\mathcal{N}_2^{(T)} + \frac{8}{5}\mathcal{N}_0^{(T)}. \tag{4.187}$$

The total density perturbation is

$$\rho D = \rho_c D_c + \rho_b D_b + \rho_\gamma D_\gamma + \rho_\nu D_\nu, \tag{4.188}$$

where

$$D_\gamma = D_g^{(\gamma)} + 4k^{-1}\mathcal{H}V_\gamma + 4\Phi,$$

$$= 4(\mathcal{M}_0 + 3k^{-1}\mathcal{H}\mathcal{M}_1 + \Phi), \tag{4.189}$$

$$D_\nu = D_{g\nu} + 4k^{-1}\mathcal{H}V_\nu + 4\Phi,$$

$$= 4(\mathcal{N}_0 + 3k^{-1}\mathcal{H}\mathcal{N}_1 + \Phi). \tag{4.190}$$

For a given background evolution, Eqs. (4.172)–(4.190) form a closed set of perturbation equations that can be solved. One obtains a good approximation by truncating the hierarchies for the photons and neutrinos at about $\ell = 10$ and determining the higher moments via the line-of-sight integrals. For photons these are given in Eqs. (4.128) and (4.160). For neutrinos one obtains the same equations just setting $\kappa = g = 0$.

The initial conditions are determined by inflation. In order not to miss any of the physical processes that can influence perturbations once they enter the Hubble horizon, we choose the initial time $t_{\rm in}$ so that $kt_{\rm in} \ll 1$ for the modes under study. Furthermore, we want to start deep in the radiation era, where we can use the results of Section 2.4.3. Requiring that perturbations remain regular for $t \to 0$ usually restricts us to the growing mode. Let us first consider scalar perturbations. On superhorizon scales the growing mode behaves as $\mathcal{M}_0 \propto \mathcal{N}_0 \propto$ constant.

$\mathcal{M}_1 \propto \mathcal{N}_1 \propto kt$ and $\Phi = \Psi =$ constant. For the nonrelativistic, subdominant species one can choose (for adiabatic perturbations) $V_c = V_b = V_\gamma = 3\mathcal{M}_1$. The initial condition for D_b and D_c is then determined by Eqs. (4.174) and (4.172) with the condition that $D \to 0$ for $t \to 0$. Adiabaticity also requires $\mathcal{N}_1 = \mathcal{M}_1$. For adiabatic scalar perturbations this leaves us with one initial condition, which is usually given as the initial power spectrum for Ψ determined during inflation,

$$\frac{k^3}{2\pi^2} P_\Psi = \Delta_\Psi = A_\Psi (k/k_*)^{n_s - 1}. \tag{4.191}$$

Here $2\pi^2 A_\Psi k_*^{-3}$ is the square of the perturbation amplitude of Ψ at scale k_* and n_s is the scalar spectral index. It is sufficient to start integration at $z \simeq 10^7 - 10^8$.

For tensor perturbations we can simply set $H_\bullet =$ constant and $\mathcal{N}_\ell \propto j_\ell(kt)$. At early times, the collision term suppresses the build up of higher moments in the photon distribution and imposes

$$\frac{4}{5}\mathcal{M}_0^{(T)} = \frac{2}{7}\mathcal{M}_2^{(T)} + \frac{3}{35}\mathcal{M}_4^{(T)}.$$

A simple possibility is $\mathcal{M}_0^{(T)} = \mathcal{N}_0^{(T)}$, $\mathcal{M}_2^{(T)} = \frac{14}{5}\mathcal{M}_0^{(T)}$ and $\mathcal{M}_4^{(T)} = 0$.

Of course one can also suggest some other initial conditions, for example, the neutrino isocurvature velocity mode, where $\mathcal{N}_1$ dominates.

From purely theoretical grounds one can define an initial perturbation that just induces a $\mathcal{N}_{13}^{(S)} \neq 0$ at some early time $t_{\rm in}$, while all other perturbation variables vanish. The system of equations presented here can then be solved given this initial condition. Such a condition is purely iso-curvature, since the energy–momentum perturbations vanish initially. But via Eq. (4.179), the perturbation will be induced in the lower moments of the neutrino distribution function and finally in the neutrino energy–momentum tensor. It then leads to perturbations of the gravitational field, which in turn induce perturbations in the dark matter, the baryons, and photons.

However, a physical mechanism leading to these kinds of initial perturbations has not been proposed so far.

Exercises

(The exercises marked with an asterisk are solved in Appendix 11 which is not in this printed book but can be found online.)

4.1 **An orthonormal basis for a curved universe**

We consider the usual coordinate basis ∂_i with respect to which the metric expressed in polar coordiates takes the following form:

$$dx^2 = \gamma_{ij}\, dx^i\, dx^j = dr^2 + \chi^2(r)\left(d\theta^2 + \sin^2\theta\, d\varphi\right),$$

with

$$\chi(r) = \begin{cases} \frac{1}{\sqrt{K}} \sin(\sqrt{K}r) & \text{for} \quad K > 0 \\ r & \text{for} \quad K = 0 \\ \frac{1}{\sqrt{-K}} \sinh(\sqrt{-K}r) & \text{for} \quad K < 0. \end{cases}$$

Find an orthonormal basis for this metric; that is, find vector fields $e_k = e_k{}^i \partial_i$ such that $\gamma_{ij} e_k{}^i e_n{}^j = \delta_{kn}$.

Hint: Start from polar coordinates.

4.2 **The Liouville equation for vector perturbations**
Perform an integration by parts to bring the Liouville equation (4.88) into the form of Eq. (2.243). What is the form of $\mathcal{M}^{(V)}(t_{\text{in}})$? Why?

4.3 **Vector perturbations of the CMB**
Consider a vector perturbation spectrum of the form

$$\langle \sigma_i(\mathbf{k}) \sigma_j^*(\mathbf{k}') \rangle = (\delta_{ij} - \hat{\mathbf{k}}_i \hat{\mathbf{k}}_j) A k^{n_V} \, \delta(\mathbf{k} - \mathbf{k}'). \qquad (4.192)$$

Using statistical isotropy (and symmetry under parity), explain why the k-space structure of the power spectrum has to be of this form.

Using the solution Eq. (4.146) in k-space, calculate the vector-type CMB anisotropies generated from σ_j. Which value of n_V leads to a scale-invariant spectrum? That is, for which n_V do you obtain $\ell(\ell+1)C_\ell \simeq$ constant for sufficiently large ℓ's?

4.4 **The vector C_ℓ's*** Derive Eq. (4.139) from Eqs. (4.136) and (4.137).

Hint: use

$$\sqrt{(1-\mu^2)(1-\mu'^2)} \cos(\varphi - \varphi') = \mathbf{n} \cdot \mathbf{n}' - \mu\mu', \qquad (4.193)$$

where $\mu = \hat{\mathbf{k}} \cdot \mathbf{n}$ and $\mu' = \hat{\mathbf{k}} \cdot \mathbf{n}'$. Replace now terms $\mu P_\ell(\mu)$ via the recursion relations in terms of $P_{\ell+1}$ and $P_{\ell-1}$. Show, using the addition theorem for spherical harmonics given in Appendix 4, Section A4.2.3 that

$$\int d\Omega_{\hat{\mathbf{k}}} \, P_\ell(\mu) P_{\ell'}(\mu') = \delta_{\ell\ell'} \frac{4\pi}{2\ell+1} P_\ell(\mathbf{n} \cdot \mathbf{n}'), \qquad (4.194)$$

and use this relation to perform angular integrations. Using the recursion relation for $(\mathbf{n} \cdot \mathbf{n}') P_\ell(\mathbf{n} \cdot \mathbf{n}')$ and finally collecting the terms that multiply $P_\ell(\mathbf{n} \cdot \mathbf{n}')$ one obtains Eq. (4.137).

4.5 The tensor C_ℓ's

From Eqs. (4.149) and (4.152) derive Eq. (4.153).

Indication: Follow exactly the same lines as for Ex. 4.4. Only this time the recursion formula has to be applied twice to reduce terms $\mu^2 P_\ell(\mu)$ and, at the end, $(\mathbf{n}\cdot\mathbf{n}')^2 P_\ell(\mathbf{n}\cdot\mathbf{n}')$. The calculation is lengthy but straightforward.

4.6 Silk damping

Derive the dispersion relation (4.171) from the integral (4.170).

Hint: Use an algebraic program, such as Maple or Mathematica, to expand (4.170) up to third order in $t_c k$ and $t_c \omega$.

5

CMB Polarization and the Total Angular Momentum Approach

The Thomson scattering cross section depends on the polarization of the outgoing photon. If its polarization vector lies in the scattering plane, the cross section is proportional to $\cos^2\beta$, where β denotes the scattering angle. If, however, the outgoing photon is polarized normal to the scattering plane, no such reduction by a factor $\cos^2\beta$ occurs [see Jackson (1975), Eq. (14.102), and note that for polarizations in the scattering plane, the angle between the incoming and outgoing polarization equals the scattering angle]. If photons come in isotropically from all directions, this does not lead to any net polarization of the outgoing radiation. However, if, for a fixed outgoing direction, the intensity of incoming photons from one direction is different from the intensity of photons coming in at a right angle with respect to the first direction, this anisotropy leads to a net polarization of the outgoing photon beam. In Fig. 5.1 we show the extremal case with $\beta = \pi/2$. In this case, the polarization in the scattering plane is entirely suppressed. It actually has to be, since photons can only carry transverse polarization. As it is clear from the figure, it is the quadrupole anisotropy in the reference frame of the scattering electron that is responsible for polarization.

In this chapter we discuss the induced polarization in detail. We derive the equations that govern the generation and propagation of polarization and we discuss their implications. This can be done by different methods, most of which are either rather involved or incomplete. Here we employ the so-called total angular momentum method that has been developed in Hu and White (1997b) and Hu *et al.* (1998), based on previous work mainly by Seljak (1996b), Kamionkowski *et al.* (1997), and Zaldarriaga and Seljak (1997). Even though the derivation of the results is quite involved, it is straightforward, in the sense that there are no "unexpected turns" in it. Nevertheless, readers who do not want to dwell on lengthy derivations may just read the first section and then go directly to the results, which are given in the form of integral solutions at the end of the chapter. Computationally, this is the most difficult chapter of this book.

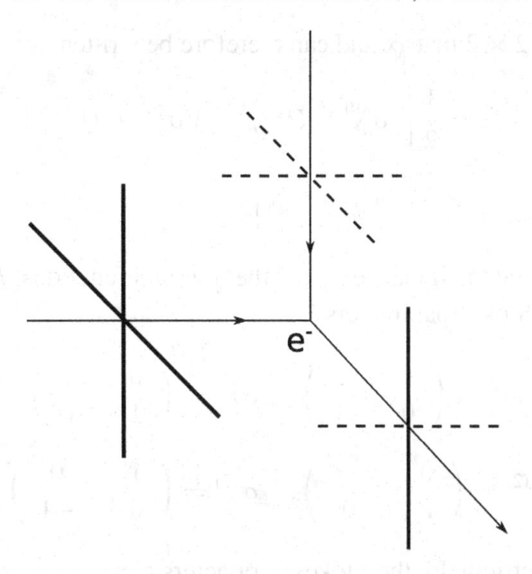

Fig. 5.1 More incoming photons from the left than from the top (indicated in the figure with longer polarization directions) lead to a net polarization of the outgoing photon beam. In the situation shown in the figure, where the scattering angle is $\pi/2$, the photons coming in from the left are scattered only if polarized vertically, while the photons coming in from the top are scattered only if polarized horizontally. In this way, an unpolarized photon distribution that exhibits a quadrupole anisotropy with respect to the scattering electron generates polarization on the surface of last scattering.

For our derivations we use spherical harmonics and spin weighted spherical harmonics. Also the basics of representation theory of the rotation group will be needed. All the notions on these topics that are employed here are presented in Appendix 4, especially in Section A4.2. Some detailed derivations are also deferred to that appendix or to the exercises.

5.1 The Stokes Parameters and the E-, B-Modes

We consider an electromagnetic wave propagating in direction $\mathbf{n}$. We define the polarization directions $\boldsymbol{\epsilon}^{(1)}$ and $\boldsymbol{\epsilon}^{(2)}$ such that $\left(\boldsymbol{\epsilon}^{(1)}, \boldsymbol{\epsilon}^{(2)}, \mathbf{n}\right)$ form a right-handed orthonormal system. The electric field of the wave is of the form $\mathbf{E} = E_1\boldsymbol{\epsilon}^{(1)} + E_2\boldsymbol{\epsilon}^{(2)}$. (The polarizations $\boldsymbol{\epsilon}^{(1)}$ and $\boldsymbol{\epsilon}^{(2)}$ are not to be confused with $\mathbf{e}^{(1)}$ and $\mathbf{e}^{(2)}$ that were introduced in Chapter 4 to form an orthonormal system with the wave vector $\mathbf{k}$.) The polarization tensor of an electromagnetic wave is defined as

$$P_{ij} = \widetilde{\mathcal{P}}_{ab}\epsilon_i^{(a)}\epsilon_j^{(b)}, \quad \text{with} \quad \widetilde{\mathcal{P}}_{ab} = E_a^* E_b. \tag{5.1}$$

$\widetilde{\mathcal{P}}_{ab}$ is a hermitian 2×2 matrix and can therefore be written as

$$\begin{aligned}\widetilde{\mathcal{P}}_{ab} &= \frac{1}{2}\left[I\sigma^{(0)}_{ab} + U\sigma^{(1)}_{ab} + V\sigma^{(2)}_{ab} + Q\sigma^{(3)}_{ab}\right] \\ &= \frac{1}{2}\left[I\sigma^{(0)}_{ab} + \mathcal{P}_{ab}\right],\end{aligned} \tag{5.2}$$

where $\sigma^{(\alpha)}$ denote the Pauli matrices, and the four real functions, $I(\mathbf{n})$, $U(\mathbf{n})$, $V(\mathbf{n})$, and $Q(\mathbf{n})$, are the Stokes parameters.

$$\begin{aligned}\sigma^{(0)} &= \begin{pmatrix} 1 & 0 \\ 0 & 1 \end{pmatrix}, \quad \sigma^{(1)} = \begin{pmatrix} 0 & 1 \\ 1 & 0 \end{pmatrix}, \\ \sigma^{(2)} &= \begin{pmatrix} 0 & -i \\ i & 0 \end{pmatrix}, \quad \sigma^{(3)} = \begin{pmatrix} 1 & 0 \\ 0 & -1 \end{pmatrix}.\end{aligned} \tag{5.3}$$

In terms of the electric field, the Stokes parameters are

$$\begin{aligned} I &= |E_1|^2 + |E_2|^2, \quad Q = |E_1|^2 - |E_2|^2, \\ U &= (E_1^* E_2 + E_2^* E_1) = 2\mathrm{Re}(E_1^* E_2), \quad V = 2\mathrm{Im}(E_1^* E_2). \end{aligned} \tag{5.4}$$

I is simply the intensity of the electromagnetic wave. Q represents the amount of linear polarization in directions $\boldsymbol{\epsilon}^{(1)}$ and $\boldsymbol{\epsilon}^{(2)}$; that is, Q is the difference between the intensity of radiation polarized along $\boldsymbol{\epsilon}^{(1)}$ and the intensity polarized in direction $\boldsymbol{\epsilon}^{(2)}$. The parameters Q and U describe the symmetric traceless part of the polarization tensor while V multiplies the antisymmetric Pauli matrix $\sigma^{(2)}$. This part describes a phase difference between E_1 and E_2 that results in circular polarization. This is best seen by expressing $\mathcal{P}_{ab}$ in terms of the helicity basis $\boldsymbol{\epsilon}^{(\pm)} = (1/\sqrt{2})\left(\boldsymbol{\epsilon}^{(1)} \pm i\boldsymbol{\epsilon}^{(2)}\right)$, where one finds that V is the difference between the left- and right-handed circular polarized intensities (see, e.g., Jackson, 1975). As we shall see in the text that follows, Thomson scattering does not introduce circular polarization. We therefore expect the V-Stokes parameter of the CMB radiation to vanish and we neglect it in the following. If $V = 0$, we have $\mathcal{P}_{ab} = \mathcal{P}^*_{ab} = \mathcal{P}_{ba}$. Hence $\mathcal{P}_{ab}$ is a real, symmetric, traceless matrix given by

$$\mathcal{P}_{ab} = \begin{pmatrix} Q & U \\ U & -Q \end{pmatrix}. \tag{5.5}$$

We often also use the quantities

$$P \equiv \mathcal{P}_{++} = 2\mathcal{P}^{ab}\epsilon_a^{(+)}\epsilon_b^{(+)} = Q + iU, \quad \text{and} \tag{5.6}$$

$$P^* \equiv \mathcal{P}_{--} = 2\mathcal{P}^{ab}\epsilon_a^{*(+)}\epsilon_b^{*(+)} = 2\mathcal{P}^{ab}\epsilon_a^{(-)}\epsilon_b^{(-)} = Q - iU. \tag{5.7}$$

Up to a factor of 2, these are the components of the polarization tensor expressed in the helicity basis. One easily verifies that the off-diagonal terms vanish since they are proportional to V, $\mathcal{P}_{+-} = \mathcal{P}_{-+} = \text{Tr}(\mathcal{P}) + iV = 0$.

The intensity is proportional to the energy density of the CMB, $\rho = \frac{1}{8\pi} I$, and therefore to our perturbation variable $\mathcal{M} = \delta T/T = \frac{1}{4}\delta\rho/\rho = \frac{1}{4}\delta I/I$. Correspondingly we define the dimensionless Stokes parameters

$$\mathcal{Q} \equiv \frac{Q}{4I} \quad \text{and} \quad \mathcal{U} \equiv \frac{U}{4I}. \tag{5.8}$$

Rotating the basis $(\boldsymbol{\epsilon}^{(1)}, \boldsymbol{\epsilon}^{(2)})$ by an angle ψ around the direction $\mathbf{n}$ we obtain $\boldsymbol{\epsilon}^{(1)'} = \cos\psi \boldsymbol{\epsilon}^{(1)} + \sin\psi \boldsymbol{\epsilon}^{(2)}$ and $\boldsymbol{\epsilon}^{(2)'} = \cos\psi \boldsymbol{\epsilon}^{(2)} - \sin\psi \boldsymbol{\epsilon}^{(1)}$ so that the coefficients with respect to the rotated basis are $E_1' = E_1 \cos\psi + E_2 \sin\psi$ and $E_2' = E_2 \cos\psi - E_1 \sin\psi$. For the Stokes parameters this implies

$$I' = I, \quad V' = V \quad \text{and}$$

$$Q' = Q\cos 2\psi - U \sin 2\psi, \quad U' = U \cos 2\psi + Q \sin 2\psi, \tag{5.9}$$

or more simply

$$Q' \pm iU' = e^{\pm 2i\psi}(Q \pm iU). \tag{5.10}$$

Hence $Q \pm iU$ transform like spin-2 variables with a magnetic quantum number ± 2 under rotations around the $\mathbf{n}$-axis. They depend not only on the direction $\mathbf{n}$, but also on the orientation of the polarization basis $(\boldsymbol{\epsilon}^{(1)}, \boldsymbol{\epsilon}^{(2)})$. For example, when rotating the polarization basis by $\pi/4$ we turn U into $-Q$ and Q into U. Hence U measures the linear polarization in the basis $(\boldsymbol{\epsilon}^{(1)'}, \boldsymbol{\epsilon}^{(2)'})$, which is rotated by $-\pi/4$ from the original basis.

The eigenvalues of $\mathcal{P} = \begin{pmatrix} Q & U \\ U & -Q \end{pmatrix}$ are

$$\lambda_{1,2} = \pm\sqrt{Q^2 + U^2}$$

with eigenvectors

$$\begin{pmatrix} x_1 \\ y_1 \end{pmatrix} = A \begin{pmatrix} Q + \sqrt{Q^2 + U^2} \\ U \end{pmatrix}, \quad \text{and} \quad \begin{pmatrix} x_2 \\ y_2 \end{pmatrix} = A \begin{pmatrix} Q - \sqrt{Q^2 + U^2} \\ U \end{pmatrix}.$$

Here $A \neq 0$ is an arbitrary constant. The first eigenvector encloses the angle ϕ_1 with the $\boldsymbol{\epsilon}^{(1)}$-axis, which is given by

$$\tan(2\phi_1) = \frac{2\sin\phi_1 \cos\phi_1}{\cos^2\phi_1 - \sin^2\phi_1} = \frac{2x_1 y_1}{x_1^2 - y_1^2} = \frac{U}{Q}.$$

The same equation is fulfilled for $\phi_2 = \phi_1 + \pi$. A polarizer oriented in the directions $\phi_{1,2}$ will detect a maximal signal, while when oriented at 90° to this polarization direction the signal is minimal.

It is not very convenient to work with the basis dependent amplitudes Q and U. The results will depend on the arbitrary choice of $\boldsymbol{\epsilon}^{(1)}$ and $\boldsymbol{\epsilon}^{(2)}$. For this reason, we shall not work directly with the Stokes parameters, but we make use of the spin weighted spherical harmonic functions ${}_sY_{\ell m}(\mathbf{n})$. These are defined for each integer s with $|s| \leq \ell$ and have the property that they transform under rotations about $\mathbf{n}$ by an angle ψ like ${}_sY_{\ell m}(\mathbf{n}) \to e^{is\psi}\ {}_sY_{\ell m}(\mathbf{n})$. The spin weighted spherical harmonics transform like the components of a symmetric, traceless rank $|s|$ tensor field defined on the tangent space of the sphere in the canonical basis ($\mathbf{e}_\vartheta \equiv \partial_\vartheta$, $\mathbf{e}_\varphi \equiv (1/\sin\vartheta)\partial_\varphi$). Note that $(\mathbf{e}_\vartheta,\ \mathbf{e}_\varphi)$ are not well defined at the north and south poles. Setting

$$\mathbf{e}^{\pm} = \frac{1}{\sqrt{2}}\left(\mathbf{e}_\vartheta \mp i\mathbf{e}_\varphi\right),$$

${}_sY_{\ell m}(\mathbf{n})$ transforms like the $+\cdots+$ component of a rank s tensor, if $s > 0$ and like the $-\cdots-$ component of a rank $|s|$ tensor, if $s < 0$. A traceless totally symmetric tensor in two dimensions has only two independent components, namely $T^{+\cdots+}$ and $T^{-\cdots-}$, which have helicities s and $-s$. All mixed components, $T^{\cdots+-\cdots}$, correspond to a partial trace and hence vanish; see Appendix 4, Section A4.2.6 for details.

With respect to the helicity basis $\mathbf{e}^{(\pm)}$, the dimensionless Stokes parameters $\mathcal{Q} \pm i\mathcal{U}$ can be expanded as

$$(\mathcal{Q} \pm i\mathcal{U})(\mathbf{n}) = \sum_{\ell=2}^{\infty}\sum_{m=-\ell}^{\ell} a_{\ell m}^{(\pm 2)}\ {}_{\pm 2}Y_{\ell m}(\mathbf{n}), \tag{5.11}$$

$$= \sum_{\ell=2}^{\infty}\sum_{m=-\ell}^{\ell} (e_{\ell m} \pm i b_{\ell m})\ {}_{\pm 2}Y_{\ell m}(\mathbf{n}). \tag{5.12}$$

Here we have introduced

$$e_{\ell m} = \frac{1}{2}\left(a_{\ell m}^{(2)} + a_{\ell m}^{(-2)}\right), \quad b_{\ell m} = \frac{-i}{2}\left(a_{\ell m}^{(2)} - a_{\ell m}^{(-2)}\right). \tag{5.13}$$

Under a "parity" transformation, $\mathbf{n} \mapsto -\mathbf{n}$, the basis vectors $\mathbf{e}^{(\pm)}$ transform as $\mathbf{e}^{(\pm)} \mapsto \mathbf{e}^{(\mp)}$. Hence helicities change sign, $s \mapsto -s$, and the coefficient $a_{\ell m}^{(2)}$ turns into $(-1)^\ell a_{\ell -m}^{(-2)}$ and $a_{\ell m}^{(-2)} \mapsto (-1)^\ell a_{\ell -m}^{(2)}$ (see Appendix 4, Section A4.2.6) so that $e_{\ell m} \mapsto (-1)^\ell e_{\ell -m}$ while $b_{\ell m} \mapsto (-1)^{\ell+1} b_{\ell -m}$.

The spin weighted spherical harmonics are defined in Appendix 4, Section A4.2.6, where also other useful properties are derived. Note that the sum over ℓ starts only at $\ell = 2$. As is clear from their definition, the spin weighted spherical harmonics ${}_sY_{\ell m}$ vanish for $|s| > \ell$. The coefficients $a_{\ell m}^{(\pm 2)}$ represent a decomposition of polarization into positive and negative helicity, while $e_{\ell m}$ and $b_{\ell m}$ provide a decomposition into the components with parity $(-1)^\ell$ and $(-1)^{\ell+1}$.

In Appendix 4, Section A4.2.6 we also define the spin raising and lowering operators $\eth$ and $\eth^*$. They are similar to the quantum mechanical angular momentum operators L_+ and L_- that raise and lower the magnetic quantum number m, but they raise and lower the helicity s. The operators $\eth^{(*)}$ have the properties $\eth\, {}_sY_{\ell m} \propto {}_{s+1}Y_{\ell m}$ and $\eth^*\, {}_sY_{\ell m} \propto {}_{s-1}Y_{\ell m}$. Hence, when acting twice with $\eth$ respectively $\eth^*$ on ${}_{-2}Y_{\ell m}$ respectively ${}_2Y_{\ell m}$, we reproduce ordinary spherical harmonics. More precisely (see Appendix 4, Section A4.2.6),

$$\eth^2\,({}_{-2}Y_{\ell m}) = \sqrt{\frac{(\ell+2)!}{(\ell-2)!}}\,Y_{\ell m}, \tag{5.14}$$

$$(\eth^*)^2\,({}_2Y_{\ell m}) = \sqrt{\frac{(\ell+2)!}{(\ell-2)!}}\,Y_{\ell m}. \tag{5.15}$$

Applying this to $\mathcal{Q} \pm i\mathcal{U}$ we find

$$(\eth^*)^2(\mathcal{Q}+i\mathcal{U})(\mathbf{n}) = \sum_{\ell=2}^{\infty}\sum_{m=-\ell}^{\ell} a_{\ell m}^{(2)}\sqrt{\frac{(\ell+2)!}{(\ell-2)!}}\,Y_{\ell m}(\mathbf{n}), \tag{5.16}$$

$$\eth^2(\mathcal{Q}-i\mathcal{U})(\mathbf{n}) = \sum_{\ell=2}^{\infty}\sum_{m=-\ell}^{\ell} a_{\ell m}^{(-2)}\sqrt{\frac{(\ell+2)!}{(\ell-2)!}}\,Y_{\ell m}(\mathbf{n}). \tag{5.17}$$

With this, we can define the scalar quantities

$$\mathcal{E}(\mathbf{n}) = \sum_{\ell=2}^{\infty}\sum_{m=-\ell}^{\ell} e_{\ell m}\sqrt{\frac{(\ell+2)!}{(\ell-2)!}}\,Y_{\ell m}(\mathbf{n}), \tag{5.18}$$

$$\mathcal{B}(\mathbf{n}) = \sum_{\ell=2}^{\infty}\sum_{m=-\ell}^{\ell} b_{\ell m}\sqrt{\frac{(\ell+2)!}{(\ell-2)!}}\,Y_{\ell m}(\mathbf{n}). \tag{5.19}$$

Like temperature fluctuations, $\mathcal{E}$ and $\mathcal{B}$ are invariant under rotation. Since the sign of $b_{\ell m}Y_{\ell m}(\mathbf{n})$ changes under parity, $\mathcal{B}$ has negative parity while $\mathcal{E}$ and $\mathcal{M}$ have positive parity. At the end of Section 5.5 we shall show that $\mathcal{E}$ measures gradient contributions while $\mathcal{B}$ measures curl contributions to the polarisation tensor.

Equations (5.16)–(5.19) imply

$$\mathcal{E}(\mathbf{x},\mathbf{n}) = \frac{1}{2}\left[(\eth^*)^2(\mathcal{Q}+i\mathcal{U})(\mathbf{x},\mathbf{n}) + \eth^2(\mathcal{Q}-i\mathcal{U})(\mathbf{x},\mathbf{n})\right] \tag{5.20}$$

$$\mathcal{B}(\mathbf{x},\mathbf{n}) = \frac{-i}{2}\left[(\eth^*)^2(\mathcal{Q}+i\mathcal{U})(\mathbf{x},\mathbf{n}) - \eth^2(\mathcal{Q}-i\mathcal{U})(\mathbf{x},\mathbf{n})\right]. \tag{5.21}$$

In Appendix 4, Section A4.2.6 we show that the operators $\eth$ and $\eth^*$ are the covariant derivatives on the sphere in the direction $\mathbf{e}_\pm = (1/\sqrt{2})(\mathbf{e}_1 \mp i\mathbf{e}_2)$. The quantities $Q \pm iU$ actually correspond to the $+,+$ and $-,-$ components of the polarization tensor $\mathcal{P}_{ab}$ defined in Eq. (5.1). Noting also that $P_{++} = P^{--}$ and $P_{--} = P^{++}$ (see Appendix 4, Section A4.2.6), we find that

$$(\eth^*)^2(Q + iU) = 2\nabla_+\nabla_+\mathcal{P}^{--}, \quad \eth^2(Q - iU) = 2\nabla_-\nabla_-\mathcal{P}^{++}. \tag{5.22}$$

Here we have used $\eth = -\sqrt{2}\nabla_-$ and $\eth^* = -\sqrt{2}\nabla_+$, which is derived in Appendix 4, Section A4.2.6. We also note that in two dimensions the curl of a vector, $\text{rot}V \equiv \epsilon_{ij}\nabla_i V_j$, is a (pseudo-)scalar; hence the double curl of a tensor, $\text{rot rot}T \equiv \epsilon_{lm}\epsilon_{ij}\nabla_l\nabla_i T_{jm}$, is a scalar. Here ϵ_{ij} denotes the totally antisymmetric tensor in two dimensions and we work in an orthonormal 2D basis, a dyad; hence rising and lowering indices has no effect. More precisely in two dimensions

$$\epsilon_{lm}\epsilon_{ij} = \delta_{li}\delta_{mj} - \delta_{lj}\delta_{mi} \qquad \text{(in 2D)}$$

so that (tr $\equiv$ trace)

$$\text{rot rot}T \equiv \epsilon_{lm}\epsilon_{ij}\nabla_l\nabla_i T_{jm} = \Delta\text{tr}\, T - \nabla_j\nabla_m T_{jm} = \Delta\text{tr}\, T - \text{div div}\, T. \tag{5.23}$$

A short calculation (see Appendix 4, Section A4.2.6) now shows that

$$\begin{aligned}\mathcal{E} = \nabla_+\nabla_+\mathcal{P}^{--} + \nabla_-\nabla_-\mathcal{P}^{++} &= \nabla_i\nabla_j\mathcal{P}_{ij} - \epsilon_{lm}\epsilon_{ij}\nabla_l\nabla_i\mathcal{P}_{jm} \\ &= 2\text{div div}\mathcal{P},\end{aligned} \tag{5.24}$$

$$\begin{aligned}i\mathcal{B} = \nabla_+\nabla_+\mathcal{P}^{--} - \nabla_-\nabla_-\mathcal{P}^{++} &= i\epsilon_{lm}\left(\nabla_i\nabla_l + \nabla_l\nabla_i\right)\mathcal{P}_{im} \\ &= i\left(\text{div rot} + \text{rot div}\right)\mathcal{P},\end{aligned} \tag{5.25}$$

so that

$$\mathcal{E} = 2\text{div div}\mathcal{P} \quad \text{and} \quad \mathcal{B} = (\text{div rot} + \text{rot div})\,\mathcal{P}. \tag{5.26}$$

$\mathcal{E}$ measures "gradient-type" polarization while $\mathcal{B}$ measures curl-type polarization. More precisely, if $\mathcal{B} = 0$, the vector rot$\mathcal{P}$ is a pure curl while the vector div$\mathcal{P}$ is a pure gradient. In full generality we can split the electric field tangent to the sphere of directions $\mathbf{n}$ into a gradient part and a curl part, $E_i = \nabla_i f + \epsilon_{ij}\nabla_j g$. Since there is no circular polarization, we can choose f and g real. Let us locally consider a flat 2d sky with coordinates (x, y). In the case where f and g are functions of $r = \sqrt{x^2 + y^2}$ only, if either f or g vanishes only E-polarization is generated. On the other hand, if $g(r) = \pm f(r)$ only B-polarization is generated. These are exactly the cases depicted in Fig. 5.2. A more general interpretation of $\mathcal{E}$ and $\mathcal{B}$ modes is studied in Exercise 5.4.

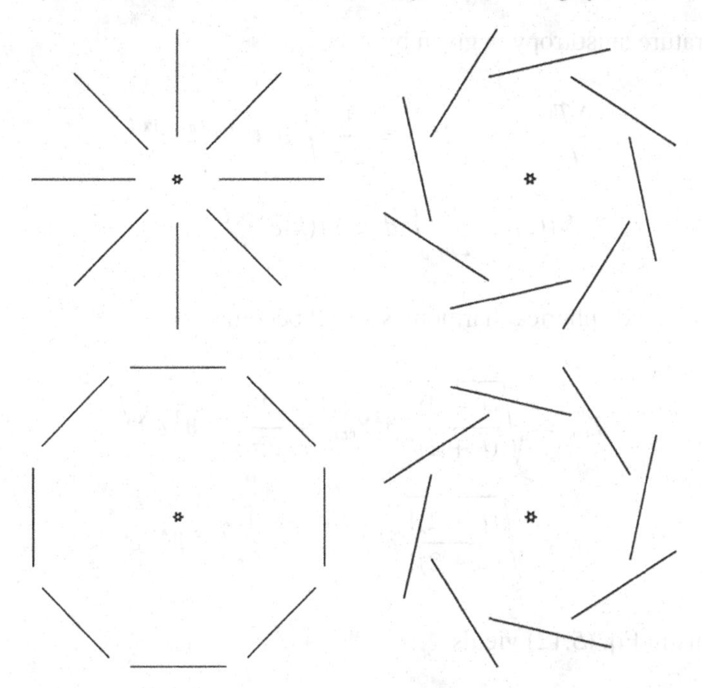

Fig. 5.2 E-polarization (left) and B-polarization (right) patterns are shown around the photon direction indicated as the center. E-polarization can be either radial or tangential, while B-polarization is clearly of curl type.

5.2 The Small-Scale Limit and the Physical Meaning of $\mathcal{E}$ and $\mathcal{B}$

The polarization variables $\mathcal{E}$ and $\mathcal{B}$ are easier to interpret in the small-scale limit or the so-called **flat sky approximation**. For $\ell \gtrsim 100$, which corresponds to angles of less than about 2°, we may neglect the curvature of the sphere of directions and consider it as a plane normal to $\mathbf{e}_z$. In this approximation, the spherical harmonics can be replaced by exponentials, the eigenfunctions of the Laplacian on the plane:

$$Y_{\ell m}(\mathbf{n}) \to \frac{1}{2\pi} \exp(i\boldsymbol{\ell} \cdot \mathbf{x}), \tag{5.27}$$

where $\mathbf{x}$ is a small vector in the plane normal to $\mathbf{e}_z$ and $\boldsymbol{\ell} = \ell(\cos\varphi_\ell, \sin\varphi_\ell)$ is a vector in the "Fourier plane." In this approximation the magnetic quantum number m is replaced by the continuous direction of the vector $\boldsymbol{\ell}$. The orthogonality relation now becomes

$$\frac{1}{(2\pi)^2} \int d^2x \, e^{i\mathbf{x}(\boldsymbol{\ell}-\boldsymbol{\ell}')} = \delta^2(\boldsymbol{\ell} - \boldsymbol{\ell}').$$

The temperature anisotropy is given by

$$\frac{\Delta T}{T}(\mathbf{x}) = \mathcal{M}(\mathbf{x}) = \frac{1}{2\pi}\int d^2\boldsymbol{\ell}\,\mathcal{M}(\boldsymbol{\ell})\,e^{i\mathbf{x}\cdot\boldsymbol{\ell}}, \tag{5.28}$$

$$\mathcal{M}(\boldsymbol{\ell}) = \frac{1}{2\pi}\int d^2\mathbf{x}\,\mathcal{M}(\mathbf{x})e^{-i\mathbf{x}\cdot\boldsymbol{\ell}}. \tag{5.29}$$

The spin weighted spherical harmonics $s = 2$ become

$${}_2Y_{\ell m} = \sqrt{\frac{(\ell-2)!}{(\ell+2)!}}\,\eth^2 Y_{\ell m} \to \frac{1}{2\pi}\ell^{-2}\,\eth^2\,e^{i\mathbf{x}\cdot\boldsymbol{\ell}}, \tag{5.30}$$

$${}_{-2}Y_{\ell m} = \sqrt{\frac{(\ell-2)!}{(\ell+2)!}}\,\eth^{*2} Y_{\ell m} \to \frac{1}{2\pi}\ell^{-2}\,\eth^{*2}\,e^{i\mathbf{x}\cdot\boldsymbol{\ell}}. \tag{5.31}$$

Inserting this in Eq. (5.12) yields

$$(\mathcal{Q}+i\mathcal{U})(\mathbf{x}) = \frac{1}{2\pi}\int d^2\ell\,(e(\boldsymbol{\ell})+ib(\boldsymbol{\ell}))\,\frac{1}{\ell^2}\,\eth^2\,e^{i\mathbf{x}\cdot\boldsymbol{\ell}}, \tag{5.32}$$

$$(\mathcal{Q}-i\mathcal{U})(\mathbf{x}) = \frac{1}{2\pi}\int d^2\ell\,(e(\boldsymbol{\ell})-ib(\boldsymbol{\ell}))\,\frac{1}{\ell^2}\,\eth^{*2}\,e^{i\mathbf{x}\cdot\boldsymbol{\ell}}. \tag{5.33}$$

We orient the coordinate system such that $\eth = -(\nabla_\vartheta + i\nabla_\varphi) = -(\nabla_x + i\nabla_y)$ at $\mathbf{e}_z$ and $\eth e^{i\mathbf{x}\cdot\boldsymbol{\ell}} = -i(\ell_x + i\ell_y)e^{i\mathbf{x}\cdot\boldsymbol{\ell}} = -i\ell e^{i\varphi_\ell}e^{i\mathbf{x}\cdot\boldsymbol{\ell}}$. With this we obtain

$$\eth^2 e^{i\mathbf{x}\cdot\boldsymbol{\ell}} = -\ell^2 e^{2i\varphi_\ell}e^{i\mathbf{x}\cdot\boldsymbol{\ell}}, \tag{5.34}$$

$$\eth^{*2} e^{i\mathbf{x}\cdot\boldsymbol{\ell}} = -\ell^2 e^{-2i\varphi_\ell}e^{i\mathbf{x}\cdot\boldsymbol{\ell}}. \tag{5.35}$$

In the small-scale limit, the Stokes parameters are therefore given in terms of $e(\boldsymbol{\ell})$ and $b(\boldsymbol{\ell})$ by

$$\mathcal{Q}(\mathbf{x}) = \frac{-1}{2\pi}\int d^2\ell\,[e(\boldsymbol{\ell})\cos(2\varphi_\ell) - b(\boldsymbol{\ell})\sin(2\varphi_\ell)]\,e^{i\mathbf{x}\cdot\boldsymbol{\ell}}, \tag{5.36}$$

$$\mathcal{U}(\mathbf{x}) = \frac{-1}{2\pi}\int d^2\ell\,[e(\boldsymbol{\ell})\sin(2\varphi_\ell) + b(\boldsymbol{\ell})\cos(2\varphi_\ell)]\,e^{i\mathbf{x}\cdot\boldsymbol{\ell}}. \tag{5.37}$$

These relations were introduced by Seljak (1996b), where E- and B-polarizations had been introduced for the first time. They can be inverted to

$$e(\boldsymbol{\ell}) = \frac{-1}{2\pi}\int d^2x\,[\mathcal{Q}(\mathbf{x})\cos(2\varphi) + \mathcal{U}(\mathbf{x})\sin(2\varphi)]\,e^{-i\mathbf{x}\cdot\boldsymbol{\ell}}, \tag{5.38}$$

$$b(\boldsymbol{\ell}) = \frac{-1}{2\pi}\int d^2x\,[\mathcal{U}(\mathbf{x})\cos(2\varphi) - \mathcal{Q}(\mathbf{x})\sin(2\varphi)]\,e^{-i\mathbf{x}\cdot\boldsymbol{\ell}}. \tag{5.39}$$

A short calculation (see Ex. 5.4) leads to the following relation of e, b and $\mathcal{Q},\mathcal{U}$ in real space:

$$e(\mathbf{x}) = -\nabla^{-2}\left[(\partial_x^2 - \partial_y^2)\mathcal{Q}(\mathbf{x}) + i2\partial_x\partial_y\mathcal{U}(\mathbf{x})\right], \tag{5.40}$$

$$ib(\mathbf{x}) = -\nabla^{-2}\left[(\partial_x^2 - \partial_y^2)\mathcal{U}(\mathbf{x}) - i2\partial_x\partial_y\mathcal{Q}(\mathbf{x})\right]. \tag{5.41}$$

Hence e and b, which are the inverse Laplacians of combinations of second derivatives of $\mathcal{Q}$ and $\mathcal{U}$, are more closely related to the latter than $\mathcal{E}$ and $\mathcal{B}$ as they have no additional factors of ℓ. However, the relation between (e, b) and (Q, U) is nonlocal due to the inverse Laplacians. We have to know $\mathcal{U}$ and $\mathcal{Q}$ globally to determine e and b, while $\mathcal{E} = \nabla^2 e$ and $\mathcal{B} = \nabla^2 b$ are locally related to $\mathcal{Q}$ and $\mathcal{U}$. Furthermore, since Q-polarization turns into U-polarization and vice versa by a rotation of 45°, a pure E-polarization configuration turns into pure B, if we turn all the polarization vectors by 45°.

Vanishing polarization corresponds to $\mathcal{B} = \mathcal{E} = 0$. Positive values of $\mathcal{E}$ around a zero indicate radial polarization patterns while negative values indicate tangential polarization. A B-polarization pattern can then be obtained by simply rotating the polarization vectors by 45°. Hence the B-polarization patterns rotate around their zeros; see Fig. 5.3.

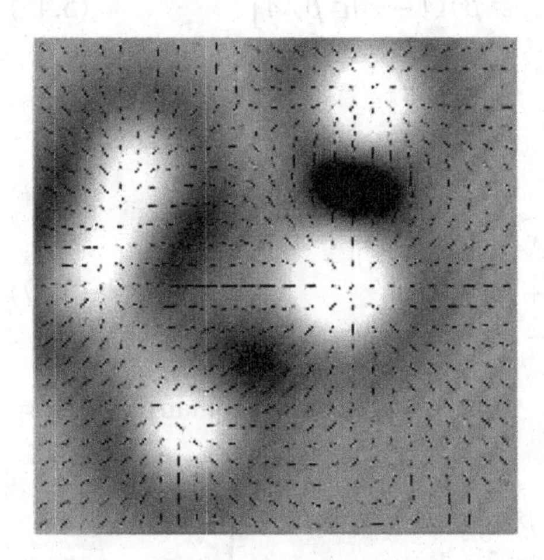
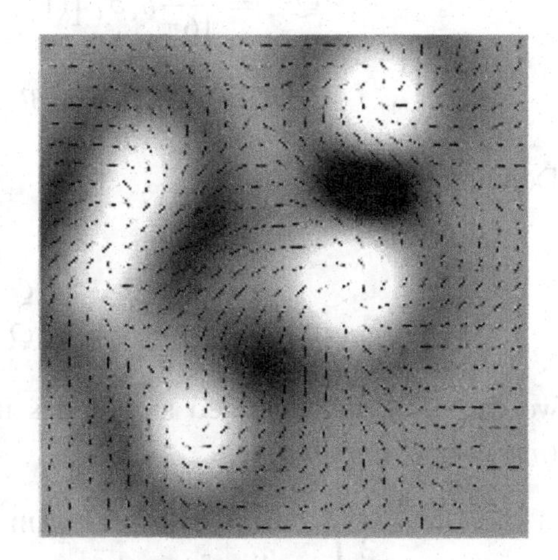

Fig. 5.3 An E-polarization pattern (left) is compared with B-polarization (right). The function $\mathcal{E}$ is indicated in grayscale, and the polarization directions are drawn. E-polarization is tangential along the dark negative regions while it is radial from the white positive regions. The B-polarization pattern is obtained by rotating the polarization directions by 45°.

5.3 Polarization-dependent Thomson Scattering

5.3.1 The Scattering Matrix and Collision Term

We now want to determine the change of each of the components, $\mathcal{M}$, $\mathcal{Q}+i\mathcal{U}$, and $\mathcal{Q}-i\mathcal{U}$ by Thomson scattering. We consider incoming radiation from direction $\mathbf{n}'$ that is then scattered into direction $\mathbf{n}$ with scattering angle β, $\mathbf{n}\cdot\mathbf{n}' = \cos\beta$. The cross section for scattering off a nonrelativistic electron depends on the polarization of the photon. For photons polarized in the scattering plane it is suppressed by a factor $\cos^2\beta$, while it is unsuppressed for photons polarized normal to the scattering plane. The scattered electric field generated per unit of time in a plasma with electron density n_e is proportional to $\sqrt{n_e\sigma_T}\,\mathbf{E}$, where σ_T is the scattering cross section. In the rest frame of the electron we have (Jackson, 1975)

$$E_{\parallel}^{(c)} = \frac{\sqrt{n_e}e^2}{m_e}\cos\beta E_{\parallel} = \sqrt{\frac{3}{8\pi}n_e\sigma_T}\cos\beta E_{\parallel}, \tag{5.42}$$

$$E_{\perp}^{(c)} = \frac{\sqrt{n_e}e^2}{m_e}E_{\perp} = \sqrt{\frac{3}{8\pi}n_e\sigma_T}E_{\perp}. \tag{5.43}$$

We now choose the polarization basis such that $\boldsymbol{\epsilon}^{(1)}(\mathbf{n})$ lies in the scattering plane and $\boldsymbol{\epsilon}^{(2)}(\mathbf{n})$ is normal to it. Using $I = |E_{\parallel}|^2 + |E_{\perp}|^2$, $Q = |E_{\parallel}|^2 - |E_{\perp}^2|$ and $U = Re(2E_{\parallel}E_{\perp}^*)$ we obtain

$$\mathcal{M}^{(c)} = \frac{3}{16\pi}n_e\sigma_T\left[(1+\cos^2\beta)\mathcal{M} - \sin^2\beta\mathcal{Q}\right], \tag{5.44}$$

$$\mathcal{Q}^{(c)} = \frac{3}{16\pi}n_e\sigma_T\left[(1+\cos^2\beta)\mathcal{Q} - \sin^2\beta\mathcal{M}\right], \tag{5.45}$$

$$\mathcal{U}^{(c)} = \frac{3}{8\pi}n_e\sigma_T\cos\beta\mathcal{U}. \tag{5.46}$$

Defining the vector

$$\mathcal{V} = \begin{pmatrix} \mathcal{M} \\ \mathcal{Q}+i\mathcal{U} \\ \mathcal{Q}-i\mathcal{U} \end{pmatrix}, \tag{5.47}$$

we can write the scattered amplitudes in terms of a scattering matrix, $\mathcal{V}^{(c)} = (n_e\sigma_T/4\pi)S\mathcal{V}$ with

$$S = \frac{3}{4}\begin{pmatrix} \cos^2\beta+1 & -\frac{1}{2}\sin^2\beta & -\frac{1}{2}\sin^2\beta \\ -\frac{1}{2}\sin^2\beta & \frac{1}{2}(\cos\beta+1)^2 & \frac{1}{2}(\cos\beta-1)^2 \\ -\frac{1}{2}\sin^2\beta & \frac{1}{2}(\cos\beta-1)^2 & \frac{1}{2}(\cos\beta+1)^2 \end{pmatrix}. \tag{5.48}$$

This is the scattering matrix expressed in the polarization basis $\left(\boldsymbol{\epsilon}^{(1)}(\mathbf{n}),\ \boldsymbol{\epsilon}^{(2)}(\mathbf{n})\right)$ which is chosen such that $\boldsymbol{\epsilon}^{(1)}(\mathbf{n})$ lies in the scattering plane. In the expansion (5.11), we express $\mathcal{Q} \pm i\mathcal{U}$ in the basis $(\mathbf{e}_\vartheta,\ \mathbf{e}_\varphi)$. To obtain the scattering matrix with respect to this basis, we first rotate $\mathcal{Q} \pm i\mathcal{U}$ by an angle γ' around $\mathbf{n}'$ to turn the basis $(\mathbf{e}_\vartheta(\mathbf{n}'),\ \mathbf{e}_\varphi(\mathbf{n}'))$ into $\left(\boldsymbol{\epsilon}^{(1)}(\mathbf{n}'),\ \boldsymbol{\epsilon}^{(2)}(\mathbf{n}')\right)$; only then can we apply the scattering matrix S on $\mathcal{V}$. Finally, we rotate the polarizations $\left(\boldsymbol{\epsilon}^{(1)}(\mathbf{n}),\ \boldsymbol{\epsilon}^{(2)}(\mathbf{n})\right)$ back into $(\mathbf{e}_\vartheta(\mathbf{n}),\ \mathbf{e}_\varphi(\mathbf{n}))$ by the rotation with angle $-\gamma$ around $\mathbf{n}$.

The rotation with angle γ' around direction $\mathbf{n}'$ multiplies $\mathcal{Q}(\mathbf{n}') \pm i\mathcal{U}(\mathbf{n}')$ by a factor $\exp(\pm 2i\gamma')$ and the rotation around $\mathbf{n}$ with angle $-\gamma$ multiplies $\mathcal{Q}^{(c)}(\mathbf{n}) \pm i\mathcal{U}^{(c)}(\mathbf{n})$ by $\exp(\mp 2i\gamma)$. The intensity perturbation is invariant under rotations. The scattering matrix that multiplies $\mathcal{V}$ with Stokes parameters oriented in the fixed polarization basis $(\mathbf{e}_\vartheta,\ \mathbf{e}_\varphi)$ is therefore simply $R(-\gamma)SR(\gamma')$ where we define the 3×3 matrix $R(\alpha) = \mathrm{diag}\left(1, e^{2i\alpha}, e^{-2i\alpha}\right)$.

Using the expressions for ${}_{\pm s}Y_{\ell m}(\vartheta, \varphi)$, $\ell \leq 2$ given in Appendix 4, Section A4.2.6, straightforward comparison gives

$$R(-\gamma)SR(\gamma') = \frac{1}{2}\sqrt{\frac{4\pi}{5}} \times$$
$$\begin{pmatrix} Y_{20}(\beta, -\gamma') + 2\sqrt{5}Y_{00}(\beta, -\gamma') & -\sqrt{\frac{3}{2}}Y_{2-2}(\beta, -\gamma') & -\sqrt{6}Y_{22}(\beta, -\gamma') \\ -\sqrt{6}\,{}_2Y_{20}(\beta, -\gamma')e^{-2i\gamma} & 3\,{}_2Y_{2-2}(\beta, -\gamma')e^{-2i\gamma} & 3\,{}_2Y_{22}(\beta, -\gamma')e^{-2i\gamma} \\ -\sqrt{\frac{3}{2}}\,{}_{-2}Y_{20}(\beta, -\gamma')e^{2i\gamma} & 3\,{}_{-2}Y_{2-2}(\beta, -\gamma')e^{2i\gamma} & 3\,{}_{-2}Y_{22}(\beta, -\gamma')e^{2i\gamma} \end{pmatrix}. \tag{5.49}$$

Note that the angle γ' that rotates $\left(\mathbf{e}_\vartheta(\mathbf{n}'),\ \mathbf{e}_\varphi(\mathbf{n}')\right)$ into $\left(\boldsymbol{\epsilon}^{(1)}(\mathbf{n}'), \boldsymbol{\epsilon}^{(2)}(\mathbf{n}')\right)$ actually corresponds to $-\varphi$ in the spherical harmonics of (5.49). We now use the addition theorem for spin weighted spherical harmonics; see Appendix 4, Section A4.2.6, Eq. (A4.101):

$${}_sY_{2s'}(\beta, -\gamma')e^{-is\gamma} = \sqrt{\frac{4\pi}{5}} \sum_m {}_{-s'}Y^*_{2m}(\mathbf{n}')\,{}_sY_{2m}(\mathbf{n}).$$

Note that the Euler angles of the rotation $R_1^{-1}R_2$ defined in Appendix 4, Section A4.2.6 are $(-\gamma', \beta, \gamma)$. With this we can write the matrix $R(-\gamma)SR(\gamma') = \frac{4\pi}{10}P(\mathbf{n}, \mathbf{n}') + \mathrm{diag}(1, 0, 0)$, where the matrix $P(\mathbf{n}, \mathbf{n}')$ is given by (${}_0Y_{\ell m} \equiv Y_{\ell m}$):

$$P(\mathbf{n}, \mathbf{n}') = \sum_{m=-2}^{2} P_m(\mathbf{n}, \mathbf{n}'),$$

where

$$P_m(\mathbf{n},\mathbf{n}') = \begin{pmatrix} Y_{2m}(\mathbf{n})Y^*_{2m}(\mathbf{n}') & -\sqrt{\frac{3}{2}}Y_{2m}(\mathbf{n})\,{}_2Y^*_{2m}(\mathbf{n}') & -\sqrt{\frac{3}{2}}Y_{2m}(\mathbf{n})\,{}_{-2}Y^*_{2m}(\mathbf{n}') \\ -\sqrt{6}\,{}_2Y_{2m}(\mathbf{n})Y^*_{2m}(\mathbf{n}') & 3\,{}_2Y_{2m}(\mathbf{n})\,{}_2Y^*_{2m}(\mathbf{n}') & 3\,{}_2Y_{2m}(\mathbf{n})\,{}_{-2}Y^*_{2m}(\mathbf{n}') \\ -\sqrt{6}\,{}_{-2}Y_{2m}(\mathbf{n})Y^*_{2m}(\mathbf{n}') & 3\,{}_{-2}Y_{2m}(\mathbf{n})\,{}_2Y^*_{2m}(\mathbf{n}') & 3\,{}_{-2}Y_{2m}(\mathbf{n})\,{}_{-2}Y^*_{2m}(\mathbf{n}') \end{pmatrix}. \tag{5.50}$$

The three-component collision term for $\mathcal{V}$ in the electron rest frame is now obtained by integrating over the incoming photon directions and subtracting the photons scattered out of the beam, as in Eq. (4.111),

$$C[\mathcal{V}]_{\text{rest}} = an_e\sigma_T\left[\frac{1}{10}\int d\Omega_{\mathbf{n}'}\sum_{m=-2}^{2}P_m(\mathbf{n},\mathbf{n}')\mathcal{V}(\mathbf{n}') - \mathcal{V}(\mathbf{n}) + \frac{1}{4\pi}\int d\Omega_{\mathbf{n}'}\mathcal{M}(\mathbf{n}')\begin{pmatrix}1\\0\\0\end{pmatrix}\right]. \tag{5.51}$$

The Y_{00} term in Eq. (5.49) results in the second integral in Eq. (5.51), which provokes isotropization in the electron rest frame. The other terms of $R(-\gamma)SR(\gamma')$ lead to $\sum_{m=-2}^{2} P_m(\mathbf{n},\mathbf{n}')$.

As we shall see, the contributions to the scattering term coming from the spin weighted spherical harmonics with $|m| = 0$, 1 and 2 correspond to the contributions for scalar, vector, and tensor perturbations respectively. To transform the scattering term from the electron (or baryon) rest frame to our coordinate frame, we simply add the Doppler term $\mathbf{n}\cdot\mathbf{V}^{(b)}$ as in Eq. (4.112). Also as there, we obtain an additional factor a, since we calculate the scattering per conformal time interval. The collision term per unit of conformal time in the coordinate frame then becomes

$$C[\mathcal{V}] = an_e\sigma_T\left[\frac{1}{10}\int d\Omega_{\mathbf{n}'}\sum_{m=-2}^{2}P_m(\mathbf{n},\mathbf{n}')\mathcal{V}(\mathbf{n}') - \mathcal{V}(\mathbf{n}) + \left[\frac{1}{4\pi}\int d\Omega_{\mathbf{n}'}\mathcal{M}(\mathbf{n}') + \mathbf{n}\cdot\mathbf{V}^{(b)}\right]\begin{pmatrix}1\\0\\0\end{pmatrix}\right]. \tag{5.52}$$

5.4 Total Angular Momentum Decomposition

In the previous section we calculated the scattering term of the vector $\mathcal{V}$ at some fixed position $\mathbf{x}$ as a function of the photon direction $\mathbf{n}$. Now we also want to consider the $\mathbf{x}$ dependence.

In Chapter 4, we Fourier transformed the $\mathbf{x}$ dependence of the temperature fluctuation $\mathcal{M}$, and then decomposed $\mathcal{M}(t,\mathbf{k},\mathbf{n})$ into its scalar, vector, and tensor contributions. We found that $\mathcal{M}^{(S)}$ depends on $\mathbf{n}$ only via $\mu = \hat{\mathbf{k}}\cdot\mathbf{n}$, while $\mathcal{M}^{(V)}$ and $\mathcal{M}^{(T)}$ are of the form

$$\mathcal{M}^{(V)} = \sqrt{1-\mu^2}\frac{1}{2}\left[\exp(i\phi)\mathcal{M}_+^{(V)}(\mu) + \exp(-i\phi)\mathcal{M}_-^{(V)}(\mu)\right], \tag{5.53}$$

$$\mathcal{M}^{(T)} = (1-\mu^2)\frac{1}{2}\left[\exp(i2\phi)\mathcal{M}_+^{(T)}(\mu) + \exp(-2i\phi)\mathcal{M}_-^{(T)}(\mu)\right]. \tag{5.54}$$

Here ϕ is the angle with respect to some fixed (but arbitrary) direction in the plane normal to $\mathbf{k}$. We then expanded the functions $\mathcal{M}_\pm^{(V,T)}$ in Legendre polynomials. But, according to Appendix 4, Section A4.2,

$$Y_{\ell,\pm1}(\mathbf{n}) \propto e^{\pm i\phi}\sqrt{1-\mu^2}\,P_\ell'(\mu),$$
$$Y_{\ell,\pm2}(\mathbf{n}) \propto e^{\pm 2i\phi}(1-\mu^2)P_\ell''(\mu).$$

For a fixed wave vector $\mathbf{k}$, we can therefore expand the $\mathbf{n}$ dependence of the vector contribution to $\mathcal{M}$ in terms of spherical harmonics of order $|m|=1$ and the tensor contributions in terms of spherical harmonics of order $|m|=2$. These are the spherical harmonics of the photon direction $\mathbf{n}$ in the coordinate system with $\mathbf{k}\parallel\mathbf{e}_z$.

For a fixed Fourier mode $\mathbf{k}$ we now introduce the basis functions

$${}_sG_{\ell m}(\mathbf{x},\mathbf{n}) = (-i)^\ell\sqrt{\frac{4\pi}{2\ell+1}}\,e^{i\mathbf{k}\cdot\mathbf{x}}\,{}_sY_{\ell m}(\mathbf{n}), \tag{5.55}$$

where the spin weighted spherical harmonics are evaluated in a coordinate system with $\mathbf{k}\parallel\mathbf{e}_z$. According to our findings in Chapter 4, the temperature fluctuation can now be expanded as

$$\mathcal{M}(t,\mathbf{x},\mathbf{n}) = \int\frac{d^3k}{(2\pi)^3}\sum_{\ell=0}^{\infty}\sum_{m=-2}^{2}\mathcal{M}_\ell^{(m)}(t,\mathbf{k})\,{}_0G_{\ell m}(\mathbf{x},\mathbf{n}). \tag{5.56}$$

As we have seen, the $m=0$ term represents scalar fluctuations while the $|m|=1$ terms are of vector-type and the $|m|=2$ terms are tensor fluctuations. The coefficients $\mathcal{M}_\ell^{(\pm2)}$ are easily related to the expansion coefficients $\mathcal{M}_\ell^{(T\pm)}$ defined in Eq. (4.149), and $\mathcal{M}_\ell^{(\pm1)}$ are related to $\mathcal{M}_\ell^{(V\pm)}$ given in Eqs. (4.133) and (4.134) (see Exercise 5.1).

Next, we use that the polarization can be expanded in terms of spin weighted spherical harmonics ${}_{\pm2}Y_{\ell m}$ [see Eqs. (5.11) and (5.12)]:

$$\mathcal{Q} \pm i\mathcal{U} = \int \frac{d^3k}{(2\pi)^3} \sum_{\ell=2}^{\infty} \sum_{m=-2}^{2} {}_{\pm 2}\mathcal{A}_\ell^{(m)}(t,\mathbf{k})\, {}_{\pm 2}G_{\ell m}(\mathbf{x},\mathbf{n}), \tag{5.57}$$

$$= \int \frac{d^3k}{(2\pi)^3} \sum_{\ell=2}^{\infty} \sum_{m=-2}^{2} \Big(\mathcal{E}_\ell^{(m)}(t,\mathbf{k}) \pm i\mathcal{B}_\ell^{(m)}(t,\mathbf{k})\Big)\, {}_{\pm 2}G_{\ell m}(\mathbf{x},\mathbf{n}). \tag{5.58}$$

Here, as in Eqs. (5.11) and (5.12), the coefficients $\mathcal{A}$ are related to $\mathcal{E}$ and $\mathcal{B}$ by

$${}_{\pm 2}\mathcal{A}_\ell^{(m)}(t,\mathbf{k}) = \mathcal{E}_\ell^{(m)}(t,\mathbf{k}) \pm i\mathcal{B}_\ell^{(m)}(t,\mathbf{k}).$$

As in the case of temperature fluctuations, $m = 0$ are scalar perturbations, while $|m| = 1$ and $|m| = 2$ are vector and tensor perturbations respectively. The above $\mathcal{Q}$ and $\mathcal{U}$ polarizations are defined with respect to some fixed coordinate system in real space, while the Fourier coefficients $\mathcal{E}_\ell$ and $\mathcal{B}_\ell$ correspond to the $\mathcal{Q}$ and $\mathcal{U}$ polarization with respect to the coordinate system where $\mathbf{k}$ points in the z-direction. Therefore, the inverse Fourier transform of $\mathcal{E}$ and $\mathcal{B}$ respectively will in general not simply give $\mathcal{Q}$ and $\mathcal{U}$ with respect to any fixed real space coordinate system.

The basis functions ${}_sG_{\ell m}$ have three different types of indices. Let us briefly recapitulate their meaning. As we have seen, m determines the tensor character of the perturbations. The index ℓ labels the expansion in an orthonormal set of functions of $\mu = \hat{\mathbf{k}} \cdot \mathbf{n} = \cos\vartheta$. Under rotations around the photon direction $\mathbf{n}$ temperature fluctuations are tensorial quantities of rank $s = 0$ that gives them the index 0, while the polarization variables, $\mathcal{Q} \pm i\mathcal{U}$, are tensorial quantities of rank $|s| = 2$ with helicity ± 2.

It is important to note that when expanding in ${}_sG_{\ell m}$ we express the spherical harmonics $Y_{\ell m}$ with respect to a coordinate system that depends on $\mathbf{k}$.

We now consider the situation in which the observer is placed at $\mathbf{x} = 0$ and the incoming photon is at a distance r from her, so that the photon position is $\mathbf{x} = -r\mathbf{n}$, where $\mathbf{n}$, as earlier, denotes the direction of propagation of the photon. This situation will be relevant for the line-of-sight integration that we shall use to solve the Boltzmann equation. We want to expand our basis functions ${}_sG_{\ell m}(-r\mathbf{n},\mathbf{n})$ for fixed $\mathbf{k}$ in their total angular momentum components. The functions ${}_sG_{\ell m}$ have "spin" ℓ but the "orbital" angular momentum of the exponential is a sum,

$$e^{i\mathbf{k}\cdot\mathbf{x}} = e^{-ikr\mu} = \sum_{L=0}^{\infty} \sqrt{4\pi(2L+1)}(-i)^{-L} j_L(kr) Y_{L0}(\mathbf{n}),$$

where we have used Eq. (A4.146) and $P_\ell(\mu) = \sqrt{4\pi/(2\ell+1)}\,Y_{\ell 0}(\mathbf{n})$. Hence

$${}_sG_{\ell m}(-r\mathbf{n},\mathbf{n}) = 4\pi \sum_{L=0}^{\infty} \sqrt{\frac{2L+1}{2\ell+1}}\, i^{-L-\ell} j_L(kr) Y_{L0}(\mathbf{n})\, {}_sY_{\ell m}(\mathbf{n}). \tag{5.59}$$

The spin weighted spherical harmonics are related to the matrix elements of the representations of the rotation group by (see Appendix 4, Section A4.2)

$$ {}_sY_{\ell m}(\theta,\phi) = \sqrt{\frac{2\ell+1}{4\pi}}D^{(\ell)}_{-sm}(\phi,\theta,0). \tag{5.60}$$

Here $\mathbf{n} = (\sin\theta\cos\phi,\, \sin\theta\sin\phi,\, \cos\theta)$ and $(\phi,\theta,0)$ denote the Euler angles of the rotation that first rotates around the y-axis with angle θ and then around the z-axis with angle ϕ. This is a rotation that turns the z-axis into $\mathbf{n}$. We also use the relation of Y_{L0} to $D^{(L)}_{00}$. The product $D^{(L)}_{SM}D^{(\ell)}_{sm}$ is the matrix element $(S,M;s,m)$ of the representation $D^{(L)}\otimes D^{(\ell)}$ in the basis $Y_{LM}\otimes Y_{\ell m}$. With the help of the Clebsch–Gordan series (see Appendix 4, Section A4.2) this representation can be decomposed as a sum of irreducible representations,

$$D^{(L)}\otimes D^{(\ell)} = \sum_{j=|L-\ell|}^{L+\ell} D^{(j)}.$$

The basis $(Y_{LM}\otimes Y_{\ell m})^{M=L,m=\ell}_{M=-L,m=-\ell}$ is transformed into the basis $((Y_{jr})^{j}_{r=-j})^{j=L+\ell}_{j=|L-\ell|}$ with the Clebsch–Gordan coefficients $\langle L,\ell;M,m|j,r\rangle$. Using the fact that Clebsch–Gordan coefficients are nonvanishing only if $r = M+m$, we can write the matrix elements

$$D^{(L)}_{00}D^{(\ell)}_{-sm} = \sum_{j=|L-\ell|}^{j=L+\ell}\langle L,\ell;0,m|j,m\rangle\langle L,\ell;0,-s|j,-s\rangle D^{(j)}_{-sm}. \tag{5.61}$$

(For more details see Appendix 4, Section A4.2.) Using Eq. (5.60) this yields

$$\begin{aligned} &4\pi\sqrt{\frac{2L+1}{2\ell+1}}Y_{L0}(\mathbf{n})\,{}_sY_{\ell m}(\mathbf{n}) \\ &= (2L+1)\sum_{j=|L-\ell|}^{j=L+\ell}\langle L,\ell;0,m|j,m\rangle\langle L,\ell;0,-s|j,-s\rangle\sqrt{\frac{4\pi}{(2j+1)}}\,{}_sY_{jm}(\mathbf{n}). \end{aligned} \tag{5.62}$$

When inserting this in the sum, Eq. (5.59) we can exchange the sums over L and j. Extending the sum over j from zero to infinity, we have to sum for each given j over all Ls for which this j contributes in Eq. (5.62). These are simply the values $|j-\ell|\le L\le j+\ell$. We define functions ${}_sf^{(\ell m)}_j$ that represent the sums over L,

$${}_sf^{(\ell m)}_j(x) \equiv \sum_{L=|j-\ell|}^{j+\ell}(-i)^{L+\ell-j}\frac{2L+1}{2j+1}\langle L,\ell;0,m|j,m\rangle\langle L,\ell;0,-s|j,-s\rangle j_L(x), \tag{5.63}$$

we can then write the sum (5.59) as

$$ {}_sG_{\ell m}(-r\mathbf{n},\mathbf{n}) = \sum_{j=0}^{\infty}(-i)^j\sqrt{4\pi(2j+1)}\,{}_sf_j^{(\ell m)}(kr)\,{}_sY_{jm}(\mathbf{n}). \tag{5.64} $$

We are only really interested in the cases $s=0$ and $s=\pm 2$. For these we define

$$ \alpha_j^{(\ell m)} \equiv {}_0f_j^{(\ell m)}, \tag{5.65} $$

$$ \epsilon_j^{(\ell m)} \pm i\beta_j^{(\ell m)} \equiv {}_{\pm 2}f_j^{(\ell m)}. \tag{5.66} $$

We repeat Eq. (5.64) for the relevant cases $s=0$ and $|s|=2$:

$$ {}_0G_{\ell m}(-r\mathbf{n},\mathbf{n}) = \sum_{j=0}^{\infty}\sqrt{4\pi(2j+1)}\,(-i)^j\alpha_j^{(\ell m)}(kr)Y_{jm}(\mathbf{n}), \tag{5.67} $$

$$ {}_{\pm 2}G_{\ell m}(-r\mathbf{n},\mathbf{n}) = \sum_{j=0}^{\infty}\sqrt{4\pi(2j+1)}\,(-i)^j\left(\epsilon_j^{(\ell m)}(kr) \pm i\beta_j^{(\ell m)}(kr)\right){}_{\pm 2}Y_{jm}(\mathbf{n}). \tag{5.68} $$

This is the total angular momentum expansion of ${}_sG_{\ell m}(-r\mathbf{n},\mathbf{n})$. We want to use it to find the integral solution of the Boltzmann equation. For this we shall need the functions $\alpha_j^{(\ell m)}$, $\epsilon_j^{(\ell m)}$ and $\beta_j^{(\ell m)}$ only for ℓ and $|m|\le 2$, since the "source terms" of the Boltzmann equation, which are the collision terms collected in $C[\mathcal{V}]$ in Eq. (5.52) and the gravitational contributions that we have determined in Chapter 4, all have $\ell \le 2$ and $|m|\le 2$.

Using the Clebsch–Gordan coefficients given in Appendix 4, Section A4.2 and the recurrence relations of spherical Bessel functions presented in Appendix 4, Section A4.3 one obtains

$$ \alpha_\ell^{(00)}(x) = j_\ell(x), \tag{5.69} $$

$$ \alpha_\ell^{(10)}(x) = j_\ell'(x), \quad \alpha_\ell^{(1\,\pm 1)}(x) = \sqrt{\frac{\ell(\ell+1)}{2}}\frac{j_\ell(x)}{x}, \tag{5.70} $$

$$ \alpha_\ell^{(20)}(x) = \frac{1}{2}[3j_\ell''(x)+j_\ell(x)], \quad \alpha_\ell^{(2\,\pm 1)}(x) = \sqrt{\frac{3\ell(\ell+1)}{2}}\left(\frac{j_\ell(x)}{x}\right)', \tag{5.71} $$

$$ \alpha_\ell^{(2\,\pm 2)}(x) = \sqrt{\frac{3(\ell+2)!}{8(\ell-2)!}}\frac{j_\ell(x)}{x^2}, \tag{5.72} $$

$$ \epsilon_\ell^{(20)}(x) = \sqrt{\frac{3(\ell+2)!}{8(\ell-2)!}}\frac{j_\ell(x)}{x^2} \equiv \alpha_\ell^{(2\,\pm 2)}(x), \tag{5.73} $$

$$\epsilon_\ell^{(2\,\pm1)}(x) = \frac{1}{2}\sqrt{(\ell-1)(\ell+2)}\left[\frac{j_\ell(x)}{x^2} + \frac{j'_\ell(x)}{x}\right], \tag{5.74}$$

$$\epsilon_\ell^{(2\,\pm2)}(x) = \frac{1}{4}\left[-j_\ell(x) + j''_\ell(x) + 2\frac{j_\ell(x)}{x^2} + 4\frac{j'_\ell(x)}{x}\right], \tag{5.75}$$

$$\beta_\ell^{(20)}(x) = 0, \tag{5.76}$$

$$\beta_\ell^{(2\,\pm1)}(x) = \pm\frac{1}{2}\sqrt{(\ell-1)(\ell+2)}\frac{j_\ell(x)}{x}, \tag{5.77}$$

$$\beta_\ell^{(2\,\pm2)}(x) = \pm\frac{1}{2}\left[j'_\ell(x) + 2\frac{j_\ell(x)}{x}\right]. \tag{5.78}$$

To (hopefully) avoid confusion we have used the letter ℓ here as the total angular momentum, since j is the name of the spherical Bessel functions.

The functions $\alpha_j^{(\ell m)}$, $\epsilon_j^{(\ell m)}$, and $\beta_j^{(\ell m)}$ will be investigated in more detail when we discuss the integral solution of the Boltzmann equation. They peak around $x \simeq j$, like spherical Bessel functions, and then oscillate and decay like $1/x$ or faster.

From the definition of ${}_sY_{\ell m}$ it follows that under the parity operation, $\mathbf{n} \to -\mathbf{n}$, $\mathbf{e}_\vartheta(\mathbf{n}) \to \mathbf{e}_\vartheta(-\mathbf{n}) = \mathbf{e}_\vartheta(\mathbf{n})$, $\mathbf{e}_\varphi(\mathbf{n}) \to \mathbf{e}_\varphi(-\mathbf{n}) = -\mathbf{e}_\varphi(\mathbf{n})$ one finds ${}_sY_{\ell m}(-\mathbf{n}) = (-1)^\ell\,{}_{-s}Y_{\ell m}(\mathbf{n})$. The first factor simply reflects the behavior of $Y_{\ell m}$ under parity, while the transformation $s \to -s$ comes from the fact that $\mathbf{e}_\varphi$ changes sign under parity, while $\mathbf{e}_\vartheta$ does not. This, together with the parity of the spherical Bessel functions, $j_\ell(-x) = (-1)^\ell j_\ell(x)$ explains that $\alpha_j^{(\ell\,-m)}(x) = \alpha_j^{(\ell m)}(x)$ and $\epsilon_j^{(\ell\,-m)}(x) = \epsilon_j^{(\ell m)}(x)$ while $\beta_j^{(\ell\,-m)}(x) = -\beta_j^{(\ell m)}(x)$. Furthermore, since $\mathcal{E}_\ell^{(m)}$ couples to the sum ${}_sY_{\ell m} + {}_{-s}Y_{\ell m}$ it has parity $(-1)^\ell$, while $\mathcal{B}_\ell^{(m)}$, which couples to the difference ${}_sY_{\ell m} - {}_{-s}Y_{\ell m}$, has parity $(-1)^{\ell+1}$. With $Y_{\ell m}$, the $\mathcal{M}_\ell^{(m)}$ have parity $(-1)^\ell$.

5.5 The Spectra

To find the power spectra in terms of the random variables $\mathcal{M}_\ell^{(m)}$, $\mathcal{E}_\ell^{(m)}$, and $\mathcal{B}_\ell^{(m)}$ in Fourier space, we use the definition of the temperature perturbation spectrum given in Chapter 2,

$$C_\ell^{(\mathcal{M})} = \langle |a_{\ell m}|^2\rangle, \quad \text{where} \tag{5.79}$$

$$\mathcal{M}(\mathbf{x},\mathbf{n}) = \sum_{\ell=0}^{\infty}\sum_{m=-\ell}^{\ell} a_{\ell m}(\mathbf{x}) Y_{\ell m}(\mathbf{n}). \tag{5.80}$$

From this and the addition theorem of spherical harmonics,

$$Y_{\ell 0}(\cos\vartheta = \mathbf{n}\cdot\mathbf{n}') = \sqrt{\frac{4\pi}{2\ell+1}}\sum_m Y_{\ell 0}(\mathbf{n}) Y^*_{\ell m}(\mathbf{n}'), \tag{5.81}$$

we have derived the expression for the correlation function,

$$\langle \mathcal{M}(t,\mathbf{x},\mathbf{n})\mathcal{M}(t,\mathbf{x},\mathbf{n}')\rangle = \frac{1}{4\pi}\sum_{\ell=0}^{\infty}(2\ell+1)P_\ell(\mathbf{n}\cdot\mathbf{n}')C_\ell^{(\mathcal{M})}, \tag{5.82}$$

where we have used $P_\ell(\mathbf{n}\cdot\mathbf{n}') = \sqrt{4\pi/(2\ell+1)}Y_{\ell 0}(\cos\vartheta = \mathbf{n}\cdot\mathbf{n}')$. In the same way we now define the rotationally invariant spectra

$$C_\ell^{(\mathcal{E})} = \langle |e_{\ell m}|^2\rangle, \tag{5.83}$$

$$C_\ell^{(\mathcal{B})} = \langle |b_{\ell m}|^2\rangle, \tag{5.84}$$

$$C_\ell^{(\mathcal{ME})} = \langle a_{\ell m}^* e_{\ell m}\rangle, \tag{5.85}$$

with the expansion coefficients $e_{\ell m}$ and $b_{\ell m}$ defined in Eq. (5.13). The coefficients $b_{\ell m}$ have parity $(-1)^{\ell+1}$ while $a_{\ell m}$ and $e_{\ell m}$ have parity $(-1)^\ell$. We shall always assume that the random process that generates the initial fluctuations is invariant under parity, so that expectation values with negative parity such as $C_\ell^{(\mathcal{MB})}$ and $C_\ell^{(\mathcal{EB})}$ vanish. But, in principle, this has to be tested experimentally. It is possible that parity violating processes, such as weak interactions, lead to effects in the CMB; see Caprini *et al.* (2004).

We now want to relate the spectra to the **k**-space expressions for the variables $\mathcal{M}$, $\mathcal{E}$, and $\mathcal{B}$. To do this we have to be careful about our use of spherical harmonics. In Eq. (5.80) we employ them with respect to some arbitrary but fixed z direction, let us call it **e**, while in the Fourier decomposition, Eq. (5.56), the spherical harmonics are to be taken in the coordinate system where $\hat{\mathbf{k}}$ denotes the z direction. To make this dependence clear, in this section we indicate the spherical harmonics with respect to a given z-axis, **e** by $Y_{\ell m}(\mathbf{n};\mathbf{e})$. To relate $Y_{\ell m}(\mathbf{n};\hat{\mathbf{k}})$ to $Y_{\ell m}(\mathbf{n};\mathbf{e})$ we use the fact that a basis with $\hat{\mathbf{k}}$ in the z direction can be obtained from a basis with **e** in the z direction by first rotating with the angle $-\phi_{\mathbf{k}}$ around the z-axis, **e** and then with $-\theta_k$ around the y-axis. Here, $(\theta_{\mathbf{k}},\phi_{\mathbf{k}})$ are the polar angles of **k** in the coordinate system with **e** in the z direction. We therefore rotate the basis with the rotation given by the Euler angles $(0,-\theta_{\mathbf{k}},-\phi_{\mathbf{k}})$. This is the inverse of the rotation with Euler angles $(\phi_{\mathbf{k}},\theta_{\mathbf{k}},0)$. Since the representation matrices are unitary,

$$D_{mm'}^{(\ell)}(0,-\theta_{\mathbf{k}},-\phi_{\mathbf{k}}) = D_{m'm}^{*(\ell)}(\phi_{\mathbf{k}},\theta_{\mathbf{k}},0).$$

Furthermore using the fact that the basis vectors $Y_{\ell m}$ transform with the transpose of the matrix with which the coefficients of vectors transform, we obtain [see also Appendix 4, Section A4.2.3, Eqs. (A4.40) and (A4.44)]:

$$\begin{aligned} Y_{\ell m}(\mathbf{n};\hat{\mathbf{k}}) &= \sum_{m'} Y_{\ell m'}(\mathbf{n};\mathbf{e}) D_{mm'}^{*(\ell)}(\theta_{\mathbf{k}},\phi_{\mathbf{k}},0) \\ &= \sqrt{\frac{4\pi}{2\ell+1}}\sum_{m'} Y_{\ell m'}(\mathbf{n};\mathbf{e})\ {}_{-m}Y_{\ell m'}^*(\hat{\mathbf{k}};\mathbf{e}). \end{aligned} \tag{5.86}$$

Note how the magnetic quantum number m in the **k**-basis becomes the spin weight in the **e**-basis. Equation (5.86) is a generalization of the addition theorem of spherical harmonics (see also Appendix 4, Section A4.2).

Inserting this in the Fourier decomposition, Eq. (5.56) we can isolate the coefficient $a_{\ell m}$ as the term proportional to $Y_{\ell m}(\mathbf{n};\mathbf{e})$. We use the orthogonality of spherical harmonics, which implies

$$a_{\ell m}(\mathbf{x}) = \int d\Omega_{\mathbf{n}} Y^*_{\ell m}(\mathbf{n};\mathbf{e})\mathcal{M}(\mathbf{x},\mathbf{n}).$$

Inserting $\mathcal{M}(\mathbf{x},\mathbf{n})$ from Eq. (5.56) and making use of the identity, Eq. (5.86) we obtain finally

$$a_{\ell m}(\mathbf{x}) = (-i)^\ell \frac{4\pi}{2\ell+1}\sum_{s=-2}^{2}\int \frac{d^3k}{(2\pi)^3}\mathcal{M}_\ell^{(s)}(\mathbf{k})\, {}_sY^*_{\ell m}(\hat{\mathbf{k}};\mathbf{e})e^{-i\mathbf{x}\cdot\mathbf{k}}. \tag{5.87}$$

Because of statistical homogeneity, coefficients $\mathcal{M}_\ell^{(s)}(\mathbf{k})$ with different values of $\mathbf{k}$ are uncorrelated. We introduce the power spectrum of $\mathcal{M}_\ell^{(s)}(\mathbf{k})$ which is of the form

$$\left\langle \mathcal{M}_\ell^{(s)}(\mathbf{k})\mathcal{M}_\ell^{(s)*}(\mathbf{k}')\right\rangle \equiv (2\pi)^3\delta^3(\mathbf{k}-\mathbf{k}')M_\ell^{(s)}(k). \tag{5.88}$$

Because of statistical isotropy, $M_\ell^{(s)}(k)$ is a function of the modulus $k = |\mathbf{k}|$ only, and the $\mathcal{M}_\ell^{(s)}(\mathbf{k})$s with different ℓ or s are uncorrelated. With this and Eq. (5.87) integration over angles leads to

$$(2\ell+1)^2 C_\ell = (2\ell+1)^2\langle|a_{\ell m}|^2\rangle = \frac{2}{\pi}\sum_{s=-2}^{2}\int dk\, k^2 M_\ell^{(s)}(k). \tag{5.89}$$

Here $s = 0$ is the contribution from scalar perturbations while $s = \pm 1$ and $s = \pm 2$ are vector and tensor modes respectively.

We now address the polarization spectra. Here, the situation is somewhat more complicated, since apart from the dependence of ${}_sY_{\ell m}(\mathbf{n};\hat{\mathbf{k}})$ on the chosen z-axis, the spin weighted spherical harmonics also depend on the polarization basis normal to $\mathbf{n}$. In addition to the rotation from the **k**-basis into the **e**-basis outlined earlier, we would also have to fix the polarization basis. To avoid this complication we use the spin raising and lowering operator $\eth$ and its hermitian conjugate $\eth^*$. With Eqs. (5.14) and (5.15) we can also relate $\mathcal{E}$ and $\mathcal{B}$ to the Fourier coefficients $\mathcal{E}_\ell^{(m)}(\mathbf{k})$ and $\mathcal{B}_\ell^{(m)}(\mathbf{k})$. Equations (5.20) and (5.21) imply

$$\mathcal{E}(\mathbf{x},\mathbf{n}) = \sum_{\ell=2}^{\infty}\sqrt{\frac{(\ell+2)!}{(\ell-2)!}}\sum_{m=-\ell}^{\ell} e_{\ell m}(\mathbf{x})Y_{\ell m}(\mathbf{n};\mathbf{e}), \tag{5.90}$$

and

$$\mathcal{B}(\mathbf{x},\mathbf{n}) = \sum_{\ell=2}^{\infty} \sqrt{\frac{(\ell+2)!}{(\ell-2)!}} \sum_{m=-\ell}^{\ell} b_{\ell m}(\mathbf{x}) Y_{\ell m}(\mathbf{n};\mathbf{e}). \tag{5.91}$$

Here we have inserted the original expansion of $\mathcal{Q} \pm i\mathcal{U}$ given in Eq. (5.12).

To relate $\mathcal{E}(\mathbf{x},\mathbf{n})$ and $\mathcal{B}(\mathbf{x},\mathbf{n})$ to their Fourier transforms, which can be obtained from Eq. (5.58), we first rotate ${}_{\pm 2}G_{\ell m}(\mathbf{x},\mathbf{n})$ into the $\mathbf{e}$-basis, using

$${}_sY_{\ell m}(\mathbf{n};\hat{\mathbf{k}}) = \sqrt{\frac{4\pi}{2\ell+1}} \sum_{m'} {}_sY_{\ell m'}(\mathbf{n};\mathbf{e})\, {}_{-m}Y^*_{\ell m'}(\hat{\mathbf{k}};\mathbf{e}), \tag{5.92}$$

which is derived exactly like for $s = 0$. But since we do not take notice of the orientation of the polarization basis, the latter is still oriented in a $\mathbf{k}$ dependent manner and ${}_sY_{\ell m'}(\mathbf{n};\mathbf{e})$ still depends on $\mathbf{k}$ over the orientation of the polarization basis. We now act with the operator $\eth^2$ for $s = -2$ and $(\eth^*)^2$ for $s = 2$ on ${}_sY_{\ell m'}(\mathbf{n};\mathbf{e})$ to obtain $Y_{\ell m'}(\mathbf{n};\mathbf{e})$. Using Eq. (5.86) we then find

$$\mathcal{E}(\mathbf{x},\mathbf{n}) = \int \frac{d^3k}{(2\pi)^3} e^{i\mathbf{k}\cdot\mathbf{x}} \sum_{\ell=2}^{\infty} \sqrt{\frac{(\ell+2)!}{(\ell-2)!}} \sum_{m=-2}^{2} \mathcal{E}_\ell^{(m)}(\mathbf{k}) Y_{\ell m}(\mathbf{n};\hat{\mathbf{k}}), \tag{5.93}$$

$$\mathcal{B}(\mathbf{x},\mathbf{n}) = \int \frac{d^3k}{(2\pi)^3} e^{i\mathbf{k}\cdot\mathbf{x}} \sum_{\ell=2}^{\infty} \sqrt{\frac{(\ell+2)!}{(\ell-2)!}} \sum_{m=-2}^{2} \mathcal{B}_\ell^{(m)}(\mathbf{k}) Y_{\ell m}(\mathbf{n};\hat{\mathbf{k}}). \tag{5.94}$$

This is exactly the same result as when acting directly with $\eth$ and $\eth^*$ on ${}_sY_{\ell m'}(\mathbf{n};\hat{\mathbf{k}})$, which is not entirely obvious since in general the operators $\eth$ and $\eth^*$ are basis dependent. However, as we have seen, Eqs. (5.14) and (5.15) are valid in every basis. Since both sides of these equations have spin-0, they are independent of the polarization basis.

Now that we have expressed polarization in terms of ordinary spherical harmonics, we can proceed as for the temperature anisotropies. We rotate $Y_{\ell m}(\mathbf{n};\hat{\mathbf{k}})$ into spin weighted harmonics ${}_mY_{\ell m'}(\mathbf{n};\mathbf{e})$, and obtain

$$\begin{aligned}(2\ell+1)^2 C_\ell^{(\mathcal{E})} &\equiv (2\ell+1)^2 \langle |e_{\ell m}(\mathbf{x})|^2 \rangle \\ &= \frac{2}{\pi} \sum_{s=-2}^{2} \int dk\, k^2 E_\ell^{(s)}(k),\end{aligned} \tag{5.95}$$

$$\begin{aligned}(2\ell+1)^2 C_\ell^{(\mathcal{B})} &\equiv (2\ell+1)^2 \langle |b_{\ell m}(\mathbf{x})|^2 \rangle \\ &= \frac{2}{\pi} \sum_{s=-2}^{2} \int dk\, k^2 B_\ell^{(s)}(k),\end{aligned} \tag{5.96}$$

and

$$(2\ell+1)^2 C_\ell^{(\mathcal{ME})} \equiv (2\ell+1)^2 \langle a_{\ell m}^* e_{\ell m}(\mathbf{x})\rangle = \frac{2}{\pi}\sum_{s=-2}^{2}\int dk\, k^2 F_\ell^{(s)}(k), \tag{5.97}$$

where we have introduced the power spectra

$$\left\langle \mathcal{E}_\ell^{(s)}(\mathbf{k})\mathcal{E}_\ell^{(s)*}(\mathbf{k}')\right\rangle \equiv (2\pi)^3\delta^3(\mathbf{k}-\mathbf{k}')E_\ell^{(s)}(k), \tag{5.98}$$

$$\left\langle \mathcal{B}_\ell^{(s)}(\mathbf{k})\mathcal{B}_\ell^{(s)*}(\mathbf{k}')\right\rangle \equiv (2\pi)^3\delta^3(\mathbf{k}-\mathbf{k}')B_\ell^{(s)}(k), \tag{5.99}$$

$$\left\langle \mathcal{E}_\ell^{(s)}(\mathbf{k})\mathcal{M}_\ell^{(s)*}(\mathbf{k}')\right\rangle \equiv (2\pi)^3\delta^3(\mathbf{k}-\mathbf{k}')F_\ell^{(s)}(k). \tag{5.100}$$

To relate these spectra to meaningful correlation functions, we correlate quantities that are scalars under rotations around $\mathbf{n}$ and $\mathbf{n}'$ respectively; hence quantities that can be expanded in ordinary, $s = 0$ spherical harmonics. For this we use our quantities $\mathcal{E}$ and $\mathcal{B}$. The same derivation that led to Eq. (5.82) now yields

$$\left\langle \mathcal{E}(t,\mathbf{x},\mathbf{n})\mathcal{E}(t,\mathbf{x},\mathbf{n}')\right\rangle = \frac{1}{4\pi}\sum_{\ell=0}^{\infty}\frac{(2\ell+2)!}{(2\ell-2)!}(2\ell+1)P_\ell(\mathbf{n}\cdot\mathbf{n}')C_\ell^{(\mathcal{E})}, \tag{5.101}$$

$$\left\langle \mathcal{B}(t,\mathbf{x},\mathbf{n})\mathcal{B}(t,\mathbf{x},\mathbf{n}')\right\rangle = \frac{1}{4\pi}\sum_{\ell=0}^{\infty}\frac{(2\ell+2)!}{(2\ell-2)!}(2\ell+1)P_\ell(\mathbf{n}\cdot\mathbf{n}')C_\ell^{(\mathcal{B})}, \tag{5.102}$$

$$\left\langle \mathcal{M}(t,\mathbf{x},\mathbf{n})\mathcal{E}(t,\mathbf{x},\mathbf{n}')\right\rangle = \frac{1}{4\pi}\sum_{\ell=0}^{\infty}\sqrt{\frac{(2\ell+2)!}{(2\ell-2)!}}(2\ell+1)P_\ell(\mathbf{n}\cdot\mathbf{n}')C_\ell^{(\mathcal{ME})}. \tag{5.103}$$

5.5.1 Correlation Functions in the Flat Sky Approximation

We finally want to derive expressions for the correlation functions in the small-scale limit, that is, the flat sky approximation. For the temperature anisotropies we use the fact that the correlation function is simply the Fourier transform of the power spectrum. In the small-scale limit [see Section 5.2 for other useful relations], the definition of the temperature anisotropy spectrum yields

$$\langle \mathcal{M}(\boldsymbol{\ell})\mathcal{M}^*(\boldsymbol{\ell}')\rangle = \delta(\boldsymbol{\ell}-\boldsymbol{\ell}')C_\ell^{(\mathcal{M})}.$$

Hence

$$\xi_{\mathcal{M}}(\mathbf{x}) \equiv \langle \mathcal{M}(\mathbf{y})\mathcal{M}(\mathbf{y}+\mathbf{x})\rangle = \frac{1}{(2\pi)^2}\int d^2\boldsymbol{\ell}\, e^{i\boldsymbol{\ell}\mathbf{x}}C_\ell^{(\mathcal{M})}$$

$$= \frac{1}{(2\pi)^2}\int d\ell\,\ell C_\ell^{(\mathcal{M})}\int_0^{2\pi} d\phi e^{i\ell r\cos\phi} = \frac{1}{2\pi}\int_0^\infty \ell\, d\ell\, J_0(r\ell)C_\ell. \tag{5.104}$$

For the integral over the angle ϕ between $\mathbf{x}$ and $\boldsymbol{\ell}$ we have set $r = |\mathbf{x}|$ and used that

$$\int_0^{2\pi} d\phi\, e^{ir\ell\cos\phi} = 2\pi\, J_0(r\ell),$$

where J_n is the Bessel function of order n. To see this we can use the formula given in Appendix 4, Section A4.3,

$$e^{iy\cos\phi} = \sum_{n=-\infty}^{\infty} i^n J_n(y)\, e^{in\phi} = J_0(y) + 2\sum_{n=1}^{\infty} i^n J_n(y)\cos(n\phi). \tag{5.105}$$

Integrating this expansion yields

$$\frac{1}{2\pi}\int_0^{2\pi} d\phi\, e^{iy\cos\phi}\, e^{-in\phi} = i^n J_n(y). \tag{5.106}$$

Equivalently, starting from the correlation function we can derive the expression for the power spectrum,

$$C_\ell = 2\pi \int_0^{\infty} r\, dr\; J_0(r\ell)\xi(r). \tag{5.107}$$

To derive the correlation functions for polarization, we introduce the variable $\mathcal{P} = \mathcal{Q} + i\mathcal{U}$ and correspondingly $\mathcal{P}^* = \mathcal{Q} - i\mathcal{U}$. According to Eqs. (5.36)–(5.39), their Fourier representations are

$$\mathcal{P} = \mathcal{Q} + i\mathcal{U} = -\int \frac{d^2\boldsymbol{\ell}}{2\pi}\,[\mathcal{E}(\boldsymbol{\ell}) + i\mathcal{B}(\boldsymbol{\ell})]\, e^{2i\phi}\, e^{i\boldsymbol{\ell}\cdot\mathbf{x}}, \tag{5.108}$$

$$\mathcal{P}^* = \mathcal{Q} - i\mathcal{U} = -\int \frac{d^2\boldsymbol{\ell}}{2\pi}\,[\mathcal{E}(\boldsymbol{\ell}) - i\mathcal{B}(\boldsymbol{\ell})]\, e^{-2i\phi}\, e^{i\boldsymbol{\ell}\cdot\mathbf{x}}, \tag{5.109}$$

or, inversely

$$\mathcal{E}(\boldsymbol{\ell}) + i\mathcal{B}(\boldsymbol{\ell}) = -\int \frac{d^2\mathbf{x}}{2\pi}\,\mathcal{P} e^{-2i\phi}\, e^{-i\boldsymbol{\ell}\cdot\mathbf{x}}, \tag{5.110}$$

$$\mathcal{E}(\boldsymbol{\ell}) - i\mathcal{B}(\boldsymbol{\ell}) = -\int \frac{d^2\mathbf{x}}{2\pi}\,\mathcal{P}^* e^{2i\phi}\, e^{-i\boldsymbol{\ell}\cdot\mathbf{x}}. \tag{5.111}$$

We want to define correlation functions of $\mathcal{P}$ and $\mathcal{P}^*$ in a coordinate independent way. For two given points $\mathbf{x} \neq \mathbf{x}'$, $\mathbf{r} \equiv \mathbf{x} - \mathbf{x}'$ we rotate the polarization basis by the angle ϕ_r that $\mathbf{r}$ encloses with the x-axis. The new polarization basis $\hat{\mathbf{r}}$, and the direction orthogonal to it, is uniquely defined by $\mathbf{r}$. The rotated polarization is given by

$$\mathcal{P}_r(\mathbf{x}) = e^{-2i\phi_r}\,\mathcal{P}(\mathbf{x}).$$

With respect to this intrinsic basis we can now define the coordinate independent correlation functions

$$\xi_+(r) = \langle \mathcal{P}_r^*(\mathbf{x})\mathcal{P}_r(\mathbf{x}')\rangle = \langle \mathcal{P}^*(\mathbf{x})\mathcal{P}(\mathbf{x}')\rangle$$
$$= \langle \mathcal{Q}(\mathbf{x})\mathcal{Q}(\mathbf{x}')\rangle + \langle \mathcal{U}(\mathbf{x})\mathcal{U}(\mathbf{x}')\rangle, \tag{5.112}$$

$$\xi_-(r) = \langle \mathcal{P}_r(\mathbf{x})\mathcal{P}_r(\mathbf{x}')\rangle = \langle e^{-4i\phi_r}\mathcal{P}(\mathbf{x})\mathcal{P}(\mathbf{x}')\rangle$$
$$= \langle \mathcal{Q}_r(\mathbf{x})\mathcal{Q}_r(\mathbf{x}')\rangle - \langle \mathcal{U}_r(\mathbf{x})\mathcal{U}_r(\mathbf{x}')\rangle + i\left(\langle \mathcal{Q}_r(\mathbf{x})\mathcal{U}_r(\mathbf{x}')\rangle - \langle \mathcal{U}_r(\mathbf{x})\mathcal{Q}_r(\mathbf{x}')\rangle\right), \tag{5.113}$$

$$\xi_\times(r) = \langle \mathcal{P}_r(\mathbf{x})\mathcal{M}(\mathbf{x}')\rangle = \langle e^{-2i\phi_r}\mathcal{P}(\mathbf{x})\mathcal{M}(\mathbf{x}')\rangle$$
$$= \langle \mathcal{Q}_r(\mathbf{x})\mathcal{M}(\mathbf{x}')\rangle + i\langle \mathcal{U}_r(\mathbf{x})\mathcal{M}(\mathbf{x}')\rangle. \tag{5.114}$$

Under parity, $\mathbf{r} \to -\mathbf{r}$, ϕ_r and with it the imaginary part of the terms $e^{ni\phi_r}$ change sign. If we assume statistical parity invariance, they therefore have to vanish,

$$\langle \mathcal{U}_r(\mathbf{x})\mathcal{Q}_r(\mathbf{x}')\rangle \equiv \langle \mathcal{U}_r(\mathbf{x})\mathcal{M}(\mathbf{x}')\rangle \equiv 0.$$

This expresses the fact that $\mathcal{B}$-polarization is uncorrelated with $\mathcal{E}$-polarization and the temperature anisotropies in terms of the correlation functions.

The calculation of the correlation function ξ_+ now is exactly analogous to that for the temperature anisotropy; one just has to replace C_ℓ by $C_\ell^{(\mathcal{E})} + C_\ell^{(\mathcal{B})}$,

$$\xi_+(r) = \langle \mathcal{P}^*(\mathbf{x})\mathcal{P}(\mathbf{x}')\rangle$$
$$= \frac{1}{2\pi}\int \ell\, d\ell \left[C_\ell^{(\mathcal{E})} + C_\ell^{(\mathcal{B})}\right] J_0(\ell r). \tag{5.115}$$

For ξ_- and $\xi_\times$, the situation is somewhat different because of the exponentials $e^{im\phi_r}$. Inserting the Fourier transform of $\mathcal{P}(\mathbf{x})$ given in Eq. (5.108) in the expression for ξ_-, we find

$$\xi_-(r) = \langle \mathcal{P}_r(\mathbf{x})\mathcal{P}_r(\mathbf{x}')\rangle$$
$$= \int \frac{d^2\ell}{2\pi}\frac{d^2\ell'}{2\pi}\langle[\mathcal{E}(\boldsymbol{\ell}) + i\mathcal{B}(\boldsymbol{\ell})][\mathcal{E}^*(\boldsymbol{\ell}') + i\mathcal{B}^*(\boldsymbol{\ell}')]\rangle\, e^{i(\boldsymbol{\ell}\cdot\mathbf{x} - \boldsymbol{\ell}'\cdot\mathbf{x}')} e^{i(2\phi_\ell + 2\phi_{\ell'} - 4\phi_r)}$$
$$= \int \frac{d^2\ell}{(2\pi)^2}\left[C_\ell^{(\mathcal{E})} - C_\ell^{(\mathcal{B})}\right] e^{ir\ell\cos(\phi_\ell - \phi_r)} e^{4i(\phi_\ell - \phi_r)}$$
$$= \frac{1}{2\pi}\int_0^\infty d\ell\, \ell J_4(r\ell)\left[C_\ell^{(\mathcal{E})} - C_\ell^{(\mathcal{B})}\right]. \tag{5.116}$$

For the last equals sign we have used Eq. (5.106) and the fact that Bessel functions with an even index are even. Similarly we obtain for the cross correlation function

$$\begin{aligned}
\xi_\times(r) &= \langle \mathcal{P}_r(\mathbf{x})\mathcal{M}_r(\mathbf{x}')\rangle \\
&= -\int \frac{d^2\ell\, d^2\ell'}{2\pi\ 2\pi}\langle[\mathcal{E}(\boldsymbol{\ell}) + i\mathcal{B}(\boldsymbol{\ell})]\mathcal{M}^*(\boldsymbol{\ell}')\rangle e^{i(\boldsymbol{\ell}\cdot\mathbf{x}-\boldsymbol{\ell}'\cdot\mathbf{x}')}e^{i(2\phi_\ell-2\phi_r)} \\
&= \frac{1}{2\pi}\int_0^\infty d\ell\,\ell J_2(r\ell)C_\ell^{(\mathcal{ME})}. \qquad (5.117)
\end{aligned}$$

As for the temperature anisotropy, the polarization power spectra and correlation functions are related via two-dimensional Fourier transforms. Taking into account the correct factors $e^{i(\phi_\ell-\phi_r)}$ coming from the definitions, Eqs. (5.113) and (5.114), and the expression (5.108), we find

$$C_\ell^{(\mathcal{E})} + C_\ell^{(\mathcal{B})} = 2\pi\int r\,dr\ J_0(\ell r)\xi_+(r), \qquad (5.118)$$

$$C_\ell^{(\mathcal{E})} - C_\ell^{(\mathcal{B})} = 2\pi\int r\,dr\ J_4(\ell r)\xi_-(r), \qquad (5.119)$$

$$C_\ell^{(\mathcal{EM})} = 2\pi\int r\,dr\ J_2(\ell r)\xi_\times(r). \qquad (5.120)$$

These small-scale expressions for the temperature and polarization power spectra and for the correlation functions will be especially useful when we discuss lensing in Chapter 7.

5.6 The Boltzmann Equation

In this section we write the Boltzmann equation for the mode functions $\mathcal{M}_\ell^{(m)}$, $\mathcal{E}_\ell^{(m)}$ and $\mathcal{B}_\ell^{(m)}$ in Fourier space introduced in Section 5.5. First we note that the usual free-streaming term is given by

$$i\mu k \equiv ik\sqrt{\frac{4\pi}{3}}Y_{10}(\mathbf{n}). \qquad (5.121)$$

Furthermore, representing the ${}_sY_{\ell m}$ as matrix elements of $D^{(\ell)}$ and using the Clebsch Gordan decomposition of $D^{(1)} \otimes D^{(\ell)}$ one finds (see Appendix 4, Section A4.2.6 and solve Exercise 5.2 for details),

$$\begin{aligned}
\sqrt{\frac{4\pi}{3}}Y_{10}\cdot {}_sY_{\ell m} &= \sqrt{\frac{\left[(\ell+1)^2-m^2\right]\left[(\ell+1)^2-s^2\right]}{(\ell+1)^2(2\ell+3)(2\ell+1)}}\,{}_sY_{\ell+1\,m} \\
&\quad - \frac{ms}{\ell(\ell+1)}\,{}_sY_{\ell m} + \sqrt{\frac{(\ell^2-m^2)(\ell^2-s^2)}{\ell^2(2\ell+1)(2\ell-1)}}\,{}_sY_{\ell-1\,m}. \qquad (5.122)
\end{aligned}$$

This determines the free streaming of the modes ${}_sG_{\ell m}$, which hence couple to ${}_sG_{\ell+1,m}$ and ${}_sG_{\ell-1,m}$. The $\mathcal{E}_\ell^{(m)}$-mode, which is proportional to ${}_2G_{\ell m} + {}_{-2}G_{\ell m}$, couples to the $\mathcal{E}_{\ell\pm1}^{(m)}$-modes and to the $\mathcal{B}_\ell^{(m)}$-mode, which multiplies ${}_2G_{\ell m} - {}_{-2}G_{\ell m}$. Correspondingly, free streaming couples $\mathcal{B}_\ell^{(m)}$ to $\mathcal{B}_{\ell\pm1}^{(m)}$ and $\mathcal{E}_\ell^{(m)}$. Therefore, even if B-modes are not generated by Thomson scattering, as we shall see in the text that follows, they are generated by free streaming from the $\mathcal{E}$-modes. Only the scalar $\mathcal{B}$- and $\mathcal{E}$-modes, for which $m = 0$, are not coupled. Therefore, if $\mathcal{B}_\ell^{(0)}$ vanishes initially it will remain zero. With Eq. (5.122), the left-hand side of the Boltzmann equation turns into the mode equations

$$
\begin{aligned}
&(\partial_t + \mathbf{n}\cdot\nabla)\left[\mathcal{M}_\ell^{(m)} Y_{\ell m}\right] \\
&\quad = \left[Y_{\ell m}\partial_t + ik\sqrt{\frac{\left[(\ell+1)^2 - m^2\right]}{(2\ell+3)(2\ell+1)}}Y_{\ell+1,m} + ik\sqrt{\frac{(\ell^2-m^2)}{(2\ell+1)(2\ell-1)}}Y_{\ell-1,m}\right]\mathcal{M}_\ell^{(m)},
\end{aligned}
\tag{5.123}
$$

$$
\begin{aligned}
&(\partial_t + \mathbf{n}\cdot\nabla)\left[(\mathcal{E}_\ell^{(m)} \pm i\mathcal{B}_\ell^{(m)})\,({}_{\pm2}Y_{\ell m})\right] \\
&\quad = \Bigg[{}_{\pm2}Y_{\ell m}\partial_t + ik\sqrt{\frac{\left[(\ell+1)^2 - m^2\right]\left[(\ell+1)^2-4\right]}{(\ell+1)^2(2\ell+3)(2\ell+1)}}\,({}_{\pm2}Y_{\ell+1,m}) \\
&\qquad \mp ik\frac{2m}{\ell(\ell+1)}\,({}_{\pm2}Y_{\ell m}) + ik\sqrt{\frac{(\ell^2-m^2)(\ell^2-4)}{\ell^2(2\ell+1)(2\ell-1)}}\,({}_{\pm2}Y_{\ell-1,m})\Bigg]\left(\mathcal{E}_\ell^{(m)} \pm i\mathcal{B}_\ell^{(m)}\right).
\end{aligned}
\tag{5.124}
$$

To obtain the scattering term we integrate $P_m(\mathbf{n},\mathbf{n}')\mathcal{V}(\mathbf{n}')$ given in Eq. (5.50) over the $\mathbf{n}'$-sphere. For this we use the mode expansion

$$
\begin{aligned}
\mathcal{V}(\mathbf{k},\mathbf{n}') &= \begin{pmatrix} \mathcal{M} \\ \mathcal{Q}+i\mathcal{U} \\ \mathcal{Q}-i\mathcal{U} \end{pmatrix} \\
&= \begin{pmatrix} \sum_{\ell=0}^\infty \sum_{m=-2}^2 \mathcal{M}_\ell^{(m)}(\mathbf{k})\,{}_0G_{\ell m}(\mathbf{n}') \\ \sum_{\ell=2}^\infty \sum_{m=-2}^2 \left(\mathcal{E}_\ell^{(m)}(\mathbf{k}) + i\mathcal{B}_\ell^{(m)}(\mathbf{k})\right){}_2G_{\ell m}(\mathbf{x},\mathbf{n}') \\ \sum_{\ell=2}^\infty \sum_{m=-2}^2 \left(\mathcal{E}_\ell^{(m)}(\mathbf{k}) - i\mathcal{B}_\ell^{(m)}(\mathbf{k})\right){}_{-2}G_{\ell m}(\mathbf{x},\mathbf{n}') \end{pmatrix}.
\end{aligned}
\tag{5.125}
$$

Using the orthogonality relation $\int d\Omega_{\mathbf{n}'}\,{}_sY_{\ell m}(\mathbf{n})\,{}_sY_{\ell' m'}(\mathbf{n}) = \delta_{\ell\ell'}\,\delta_{mm'}$ we obtain

$$
\int d\Omega_{\mathbf{n}'}\,P_m(\mathbf{n},\mathbf{n}')\mathcal{V}(\mathbf{n}') = \begin{pmatrix} \mathcal{M}_2^{(m)}(\mathbf{k})\,{}_0G_{2m}(\mathbf{n}) - \sqrt{6}\mathcal{E}_2^{(m)}(\mathbf{k})\,{}_0G_{2m}(\mathbf{n}) \\ -\sqrt{6}\mathcal{M}_2^{(m)}(\mathbf{k})\,{}_2G_{2m}(\mathbf{n}) + 6\mathcal{E}_2^{(m)}(\mathbf{k})\,{}_2G_{2m}(\mathbf{n}) \\ -\sqrt{6}\mathcal{M}_2^{(m)}(\mathbf{k})\,{}_{-2}G_{2m}(\mathbf{n}) + 6\mathcal{E}_2^{(m)}(\mathbf{k})\,{}_{-2}G_{2m}(\mathbf{n}) \end{pmatrix}.
\tag{5.126}
$$

Hence, Thomson scattering does not depend on B-mode polarization. Finally, we also need the gravitational scalar, vector, and tensor terms that enter the Boltzmann equation for the temperature anisotropy. They do not directly couple to the polarization because there is no zeroth-order polarization. We obtain exactly the same terms as in Chapter 4, which we now write in terms of the basis functions $Y_{\ell m}$. We get

$$i\mu k(\Psi+\Phi) = ik\sqrt{\frac{4\pi}{3}}(\Psi+\Phi)Y_{10} = -i\sqrt{\frac{4\pi}{3}}Y_{10}S_1^{(0)}, \tag{5.127}$$

$$-\frac{ik}{\sqrt{2}}\mu\left[(\mathbf{n}\cdot\mathbf{e}_+)\sigma_+ + (\mathbf{n}\cdot\mathbf{e}_-)\sigma_-\right] = ik\sqrt{\frac{4\pi}{15}}\left[\sigma_+Y_{21}+\sigma_-Y_{2-1}\right]$$
$$= -\sqrt{\frac{4\pi}{5}}\left[Y_{21}S_2^{(1)} + Y_{2-1}S_2^{(-1)}\right], \tag{5.128}$$

$$-(1-\mu^2)\left[\dot{H}_d\cos(2\varphi) + \dot{H}_\times\sin(2\varphi)\right]$$
$$= -\frac{1}{2}(1-\mu^2)\left[(\dot{H}_d - i\dot{H}_\times)e^{2i\varphi} + (\dot{H}_d + i\dot{H}_\times)e^{-2i\varphi}\right]$$
$$= -\sqrt{\frac{4\pi}{15}}\left[\dot{H}_2Y_{22} + \dot{H}_{-2}Y_{2-2}\right] = -\sqrt{\frac{4\pi}{5}}\left[Y_{22}S_2^{(2)} + Y_{2-2}S_2^{(-2)}\right], \tag{5.129}$$

where we have set $H_{\pm 2} = \sqrt{2}(H_d \pm iH_\times)$. The source terms $S_\ell^{(m)}$ are defined by these equations. In addition, we must take into account the Doppler term, which is of the form $i\mu kV^{(b)} = ik\sqrt{4\pi/3}\left[V_b^{(0)}Y_{10} + V_b^{(1)}Y_{11} + V_b^{(-1)}Y_{1\,-1}\right]$. Here, $V_b^{(0)}$ denotes the scalar part of the baryon velocity field and $V_b^{(\pm 1)}$ are the vector perturbations with helicity ± 1. With all this and taking care of the normalization of the mode function ${}_0G_{\ell m}$, the Boltzmann equation for temperature anisotropies, $(\partial_t + \mathbf{n}\cdot\nabla)\mathcal{M} = S + \dot{\kappa}C[\mathcal{M}]$, turns into the mode equations

$$\dot{\mathcal{M}}_\ell^{(m)} + k\left[\frac{\sqrt{(\ell+1)^2-m^2}}{(2\ell+3)}\mathcal{M}_{\ell+1}^{(m)} - \frac{\sqrt{\ell^2-m^2}}{(2\ell-1)}\mathcal{M}_{\ell-1}^{(m)}\right]$$
$$= S_\ell^{(m)} + \dot{\kappa}\left[P_\ell^{(m)} - \mathcal{M}_\ell^{(m)}\right]. \tag{5.130}$$

with

$$S_\ell^{(0)} = \delta_{\ell 1}k(\Psi+\Phi), \tag{5.131}$$

$$S_\ell^{(\pm 1)} = -\delta_{\ell 2}\frac{i}{\sqrt{3}}k\sigma_\pm, \tag{5.132}$$

$$S_\ell^{(\pm 2)} = \delta_{\ell 2}\frac{1}{\sqrt{3}}\dot{H}_{\pm 2}, \tag{5.133}$$

$$P_\ell^{(0)} = \delta_{\ell 0}\mathcal{M}_0^{(0)} + V_b^{(0)}\delta_{\ell 1} + \delta_{\ell 2}\frac{1}{10}\left[\mathcal{M}_2^{(0)} - \sqrt{6}\mathcal{E}_2^{(0)}\right], \tag{5.134}$$

$$P_\ell^{(\pm 1)} = V_b^{(\pm 1)}\delta_{\ell 1} + \delta_{\ell 2}\frac{1}{10}\left[\mathcal{M}_2^{(\pm 1)} - \sqrt{6}\mathcal{E}_2^{(\pm 1)}\right], \tag{5.135}$$

$$P_\ell^{(\pm 2)} = \delta_{\ell 2}\frac{1}{10}\left[\mathcal{M}_2^{(\pm 2)} - \sqrt{6}\mathcal{E}_2^{(\pm 2)}\right]. \tag{5.136}$$

For the left-hand side of Eq. (5.130) we used Eq. (5.123) and have isolated terms proportional to $Y_{\ell m}$ in the expansion (5.56) for fixed $\mathbf{k}$. The terms $P_\ell^{(m)}$ come from the collision integral (5.126). Apart from the new E-polarization contribution they agree with the result found in Chapter 4.

For the evolution of the polarizations, we only need to take into account free streaming and the collision term. As mentioned earlier, the coupling of polarization to gravity is a second-order effect that is, to some extent, taken into account when discussing lensing in Chapter 7. Isolating terms proportional to ${}_{\pm 2}Y_{\ell m}$ in Eq. (5.122), taking the sum and the difference, ${}_2Y_{\ell m} \pm {}_{-2}Y_{\ell m}$, leads to the left-hand side of the Boltzmann equation for E- and B-mode polarization. The right-hand side is obtained from (5.126). Putting it all together we find

$$\dot{\mathcal{E}}_\ell^{(m)} + k\left[\frac{\sqrt{[(\ell+1)^2-4][(\ell+1)^2-m^2]}}{(2\ell+1)(2\ell+3)}\mathcal{E}_{\ell+1}^{(m)} - \frac{2m}{\ell(\ell+1)}\mathcal{B}_\ell^{(m)} - \frac{\sqrt{(\ell^2-4)(\ell^2-m^2)}}{(2\ell+1)(2\ell+3)}\mathcal{E}_{\ell-1}^{(m)}\right] = -\dot{\kappa}\left[\mathcal{E}_\ell^{(m)} + \sqrt{6}P_\ell^{(m)}\right], \tag{5.137}$$

$$\dot{\mathcal{B}}_\ell^{(m)} + k\left[\frac{\sqrt{[(\ell+1)^2-4][(\ell+1)^2-m^2]}}{(2\ell+1)(2\ell+3)}\mathcal{B}_{\ell+1}^{(m)} + \frac{2m}{\ell(\ell+1)}\mathcal{E}_\ell^{(m)} - \frac{\sqrt{(\ell^2-4)(\ell^2-m^2)}}{(2\ell+1)(2\ell+3)}\mathcal{B}_{\ell-1}^{(m)}\right] = -\dot{\kappa}\mathcal{B}_\ell^{(m)}. \tag{5.138}$$

Equations (5.130)–(5.138) represent the full Boltzmann hierarchy, which has to be truncated at some value $\ell_{\max}$ and can then be solved, using the relations given in Section 4.7 that determine the gravitational source terms. As in Chapter 4, it is numerically very costly to solve the hierarchy until some large value $\ell_{\max} \sim 2000$, which determines the fluctuations on angular scales larger than about 5 arc minutes. One therefore solves it only up to $\ell \sim 10$ and uses this result to determine the source terms, which depend only on the multipoles $\ell = 0$, 1, and 2. The higher multipoles are then again calculated with the help of an integral solution, which for a given source term is obtained by simple quadrature. We now derive this integral solution.

5.6.1 Integral Solution

To find the integral solution, we consider, as in Chapter 4, the sums of the harmonic expansions. We define for $m = 0,\ 1$, and 2 (scalar, vector, and tensor perturbations)

$$\mathcal{M}^{(m)}(t,\mathbf{n},\mathbf{k}) = \sum_\ell \mathcal{M}_\ell^{(m)}(t,\mathbf{k})(-i)^\ell\sqrt{\frac{4\pi}{2\ell+1}}Y_{\ell m}, \tag{5.139}$$

$$\begin{aligned}&\mathcal{E}^{(m)}(t,\mathbf{n},\mathbf{k}) + i\mathcal{B}^{(m)}(t,\mathbf{n},\mathbf{k})\\ &\quad= \sum_\ell (\mathcal{E}_\ell^{(m)}(t,\mathbf{k}) + i\mathcal{B}_\ell^{(m)}(t,\mathbf{k}))(-i)^\ell\sqrt{\frac{4\pi}{2\ell+1}}\,{}_2Y_{\ell m},\end{aligned} \tag{5.140}$$

$$\begin{aligned}&\mathcal{E}^{(m)}(t,\mathbf{n},\mathbf{k}) - i\mathcal{B}^{(m)}(t,\mathbf{n},\mathbf{k})\\ &\quad= \sum_\ell (\mathcal{E}_\ell^{(m)}(t,\mathbf{k}) - i\mathcal{B}_\ell^{(m)}(t,\mathbf{k}))(-i)^\ell\sqrt{\frac{4\pi}{2\ell+1}}\,{}_{-2}Y_{\ell m}.\end{aligned} \tag{5.141}$$

For each of these variables, the Boltzmann equation is of the form

$$(\partial_t + i\mu k + \dot\kappa)X = S, \tag{5.142}$$

For $X = \mathcal{M}^{(m)}$ the source term is

$$S^{(m)}(t,\mathbf{n}) = \sum_{\ell=0}^{2}\left(S_\ell^{(m)} + \dot\kappa P_\ell^{(m)}\right)(-i)^\ell\sqrt{\frac{4\pi}{2\ell+1}}Y_{\ell m}, \tag{5.143}$$

while for $X = \mathcal{E}^{(m)} \pm \mathcal{B}^{(m)}$

$$S^{(m)}(t,\mathbf{n}) = -\sqrt{6}\frac{\dot\kappa}{10}\left(\mathcal{M}_2^{(m)} - \sqrt{6}E_2^{(m)}\right)\sqrt{\frac{4\pi}{5}}\,{}_{\pm 2}Y_{2m}. \tag{5.144}$$

The general solution to Eq. (5.142) with initial condition $X(t_{\rm in})$ is simply

$$X(t) = X(t_{\rm in})\,e^{-ik\mu(t-t_{\rm in})-\kappa(t,t_{\rm in})} + \int_{t_{\rm in}}^t dt' e^{-ik\mu(t-t')-\kappa(t,t')}S(t'), \tag{5.145}$$

where $\kappa(t,t_1) = \int_{t_1}^t dt'\,\dot\kappa(t')$ and $e^{\kappa(t_0,t')}\dot\kappa(t') = g(t')$ is the visibility function defined in Chapter 4, which is strongly peaked at the last scattering surface, where the collision terms induce the higher moments and polarization due to scattering. At much earlier times the photons behave like a perfect fluid and at much later times, collisions are very rare and all the evolution is determined by free streaming including the gravitational part of the source term.

We are interested in the solution at t_0 and choose the initial time early enough so that we can neglect the initial value for all modes we are interested in. We then obtain the integral solution

$$\mathcal{M}^{(m)}(t_0,\mathbf{n}) = \sum_{\ell=0}^{2}(-i)^\ell\sqrt{\frac{4\pi}{2\ell+1}}\int_{t_{\rm in}}^{t_0} dt\left(S_\ell^{(m)}(t)+\dot\kappa P_\ell^{(m)}(t)\right)Y_{\ell m}e^{-ik\mu(t_0-t)-\kappa(t_0,t)}, \tag{5.146}$$

$$\mathcal{E}^{(m)}(t_0,\mathbf{n}) \pm i\mathcal{B}^{(m)}(t_0,\mathbf{n}) = -\sqrt{\frac{24\pi}{5}}\int_{t_{\rm in}}^{t_0} dt\frac{\dot\kappa}{10}\left(\mathcal{M}_2^{(m)}-\sqrt{6}\mathcal{E}_2^{(m)}\right){}_{\pm 2}Y_{2m}\, e^{-ik\mu(t_0-t)-\kappa(t_0,t)}. \tag{5.147}$$

To expand this solution in spherical harmonics we use that

$$(-i)^\ell\sqrt{\frac{4\pi}{2\ell+1}}\,{}_sY_{\ell m}(\mathbf{n})e^{-ik\mu(t_0-t)} = {}_sG_{\ell m}(-\mathbf{n}(t_0-t),\mathbf{n}).$$

Furthermore, the total angular momentum expansion of ${}_sG_{\ell m}$ gives [see Eqs. (5.67) and (5.68)]

$${}_0G_{\ell m}(-\mathbf{n}r,\mathbf{n}) = \sum_L(-i)^L\sqrt{4\pi(2L+1)}\alpha_L^{(\ell m)}(kr)Y_{Lm}, \tag{5.148}$$

as well as

$${}_{\pm 2}G_{\ell m}(-\mathbf{n}r,\mathbf{n}) = \sum_L(-i)^L\sqrt{4\pi(2L+1)}\left(\epsilon_L^{(\ell m)}(kr)\pm i\beta_L^{(\ell m)}(kr)\right){}_{\pm 2}Y_{Lm}(\mathbf{n}). \tag{5.149}$$

Introducing Eq. (5.148) in Eq. (5.146) and making use of the explicit form of the source term given in Eqs. (5.131)–(5.136) yields with $x \equiv k(t_0-t)$

$$\begin{aligned}\mathcal{M}^{(0)}(t_0,\mathbf{n}) = \sum_\ell(-i)^\ell\sqrt{4\pi(2\ell+1)}Y_{\ell m}(\mathbf{n})\int_{t_{\rm in}}^{t_0} dt\, e^{-\kappa(t_0,t)}\\ \times\left[ik(\Psi+\Phi+\dot\kappa V^{(b)})\alpha_\ell^{(10)}(x)+\dot\kappa\left(\mathcal{M}_0^{(0)}\alpha_\ell^{(00)}(x)\right.\right.\\ \left.\left.+\frac{1}{10}\left[\mathcal{M}_2^{(0)}-\sqrt{6}\mathcal{E}_2^{(0)}\right]\alpha_\ell^{(20)}(x)\right)\right],\end{aligned} \tag{5.150}$$

$$\begin{aligned}\frac{\mathcal{M}_\ell^{(0)}(t_0)}{2\ell+1} = \int_{t_{\rm in}}^{t_0} dt\, e^{-\kappa(t_0,t)}\left[ik(\Psi+\Phi+\dot\kappa V^{(b)})\alpha_\ell^{(10)}(x)\right.\\ \left.+\dot\kappa\left(\mathcal{M}_0^{(0)}\alpha_\ell^{(00)}(x)+\frac{1}{10}\left[\mathcal{M}_2^{(0)}-\sqrt{6}\mathcal{E}_2^{(0)}\right]\alpha_\ell^{(20)}(x)\right)\right],\end{aligned} \tag{5.151}$$

$$\mathcal{M}^{(\pm1)}(t_0,\mathbf{n}) = \sum_\ell (-i)^\ell \sqrt{4\pi(2\ell+1)} Y_{\ell m}(\mathbf{n}) \int_{t_{\rm in}}^{t_0} dt\, e^{-\kappa(t_0,t)} \times \left[-\frac{ik}{\sqrt{3}} \sigma_\pm \alpha_\ell^{(2\,\pm1)}(x) + \dot\kappa V_\pm^{(b)} \alpha_\ell^{(1\,\pm1)}(x) + \frac{\dot\kappa}{10}\left[\mathcal{M}_2^{(\pm1)} - \sqrt{6}\mathcal{E}_2^{(\pm1)}\right] \alpha_\ell^{(2\,\pm1)}(x)\right], \tag{5.152}$$

$$\frac{\mathcal{M}_\ell^{(\pm1)}(t_0)}{2\ell+1} = \int_{t_{\rm in}}^{t_0} dt\, e^{-\kappa(t_0,t)} \left[-\frac{ik}{\sqrt{3}} \sigma_\pm \alpha_\ell^{(2\,\pm1)}(x) + \dot\kappa V_\pm^{(b)}\, \alpha_\ell^{(1\,\pm1)}(x) + \frac{\dot\kappa}{10}\left[\mathcal{M}_2^{(\pm1)} - \sqrt{6}\mathcal{E}_2^{(\pm1)}\right] \alpha_\ell^{(2\,\pm1)}(x)\right], \tag{5.153}$$

$$\mathcal{M}^{(\pm2)}(t_0,\mathbf{n}) = \sum_\ell (-i)^\ell \sqrt{4\pi(2\ell+1)} Y_{\ell m}(\mathbf{n}) \int_{t_{\rm in}}^{t_0} dt\, e^{-\kappa(t_0,t)} \times \left[\frac{1}{\sqrt{3}} k \dot H_{\pm2} + \frac{\dot\kappa}{10}\left[\mathcal{M}_2^{(\pm2)} - \sqrt{6}\mathcal{E}_2^{(\pm2)}\right]\right] \alpha_\ell^{(2\,\pm2)}(x), \tag{5.154}$$

$$\frac{\mathcal{M}_\ell^{(\pm2)}(t_0)}{2\ell+1} = \int_{t_{\rm in}}^{t_0} dt\, e^{-\kappa(t_0,t)} \left[\frac{1}{\sqrt{3}} k \dot H_{\pm2} + \frac{\dot\kappa}{10}\left[\mathcal{M}_2^{(\pm2)} - \sqrt{6}\mathcal{E}_2^{(\pm2)}\right]\right] \alpha_\ell^{2\,\pm2}(x). \tag{5.155}$$

Equivalently, introducing Eq. (5.149) in Eq. (5.147) we obtain

$$\mathcal{E}^{(m)}(t_0,\mathbf{n}) \pm i\mathcal{B}^{(m)}(t_0,\mathbf{n}) = -\sqrt{6} \sum_\ell (-i)^\ell \sqrt{4\pi(2\ell+1)}\, {}_{\pm2}Y_{\ell m}(\mathbf{n}) \times \int_{t_{\rm in}}^{t_0} dt\, e^{-\kappa(t_0,t)} \frac{\dot\kappa}{10} \left(\mathcal{M}_2^{(m)} - \sqrt{6}\mathcal{E}_2^{(m)}\right) \left(\epsilon_\ell^{(2m)}(x) \pm i\beta_\ell^{(2m)}(x)\right), \tag{5.156}$$

$$\frac{\mathcal{E}_\ell^{(m)}(t_0,\mathbf{n}) \pm i\mathcal{B}_\ell^{(m)}(t_0,\mathbf{n})}{2\ell+1} = -\sqrt{6} \int_{t_{\rm in}}^{t_0} dt\, e^{-\kappa(t_0,t)} \frac{\dot\kappa}{10} (\mathcal{M}_2^{(m)} - \sqrt{6}\mathcal{E}_2^{(m)}) \left(\epsilon_\ell^{(2m)}(x) \pm i\beta_\ell^{(2m)}(x)\right). \tag{5.157}$$

Taking the sum and the difference of the last equation we obtain

$$\frac{\mathcal{E}_\ell^{(m)}(t_0,\mathbf{n})}{2\ell+1} = -\sqrt{6} \int_{t_{\rm in}}^{t_0} dt\, e^{-\kappa(t_0,t)} \frac{\dot\kappa}{10} (\mathcal{M}_2^{(m)} - \sqrt{6}\mathcal{E}_2^{(m)}) \epsilon_\ell^{(2m)}(x), \tag{5.158}$$

$$\frac{\mathcal{B}_\ell^{(m)}(t_0,\mathbf{n})}{2\ell+1} = -\sqrt{6} \int_{t_{\rm in}}^{t_0} dt\, e^{-\kappa(t_0,t)} \frac{\dot\kappa}{10} (\mathcal{M}_2^{(m)} - \sqrt{6}\mathcal{E}_2^{(m)}) \beta_\ell^{(2m)}(x). \tag{5.159}$$

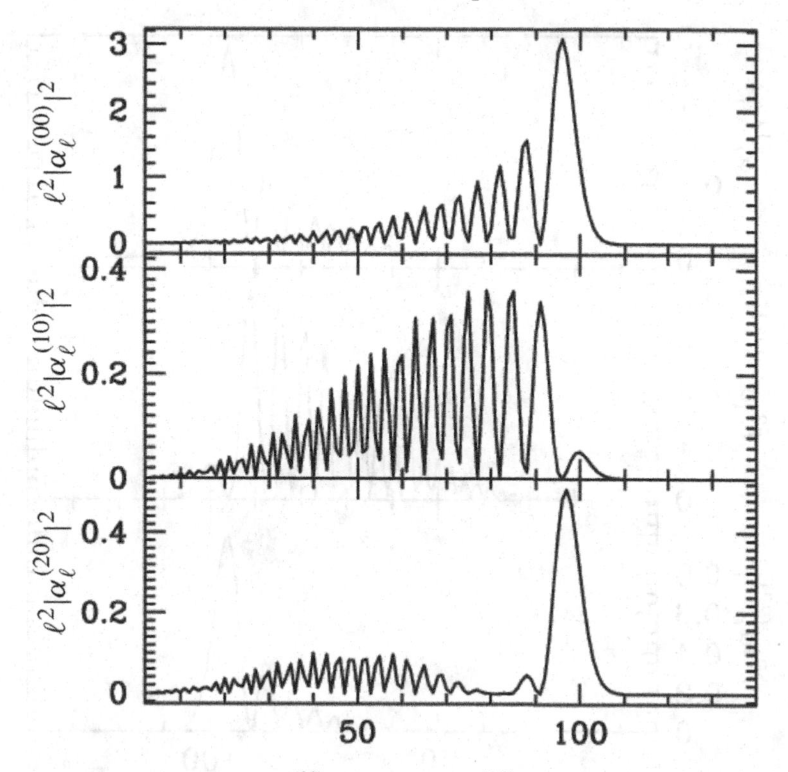

Fig. 5.4 The functions $\ell^2|\alpha_\ell^{(00)}|^2$ (top), $\ell^2|\alpha_\ell^{(10)}|^2$ (middle), and $\ell^2|\alpha_\ell^{(20)}|^2$ (bottom) are shown as function of ℓ for fixed $x = 100$. These are the kernels relevant for the scalar temperature anisotropies. Their amplitude and shape determine how strongly the corresponding source terms influence the final anisotropy spectrum.

The fact that the scalar B-mode, $\mathcal{B}_\ell^{(0)}$, vanishes is now a consequence of $\beta^{(20)} = 0$.

To have some insight into the kernels $\alpha_\ell^{(ij)}$, $\epsilon_\ell^{(ij)}$ and $\beta_\ell^{(ij)}$, we plot them in Figs. 5.4–5.7 as functions of ℓ for fixed $x = k(t_0 - t) = 100$. They are all peaked at $\ell \simeq x$. For temperature anisotropies this peak is strongest for the tensor kernel $\alpha^{(22)}$. The kernel that dominates scalar temperature anisotropies by a factor of nearly 10 is $\alpha^{(00)}$, which comes from the free streaming of density fluctuations on the last scattering surface and therefore is responsible for the acoustic peaks. The kernel $\alpha^{(10)}$ that multiplies the ordinary and integrated Sachs–Wolfe terms and the Doppler term is significantly lower and somewhat less strongly peaked. Finally, the kernel $\alpha^{(20)}$ that couples to polarization has a narrow peak at $\ell \simeq x$ and a lower, broader one around $\ell \simeq x/2$. The decay of all kernels for $\ell > x$ is very rapid.

The kernel $\alpha^{(21)}$ that couples vector perturbations to polarization and to the gravitational vector modes is smaller and less strongly peaked than $\alpha^{(11)}$, which

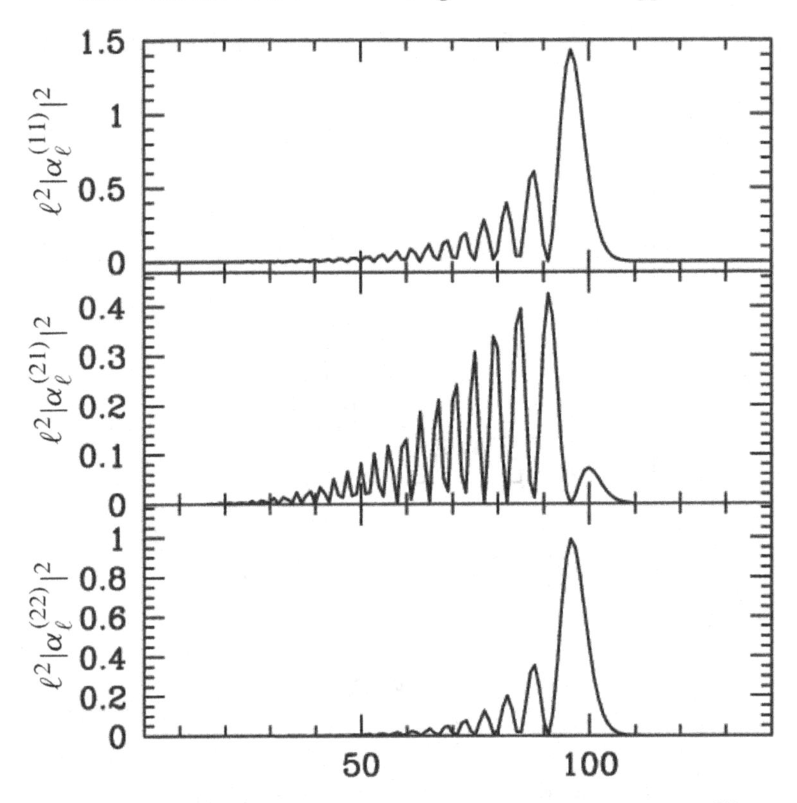

Fig. 5.5 The functions $\ell^2|\alpha_\ell^{(11)}|^2$ (top), $\ell^2|\alpha_\ell^{(21)}|^2$ (middle), and $\ell^2|\alpha_\ell^{(22)}|^2$ (bottom) are shown as functions of ℓ for fixed $x = 100$. These are the kernels relevant for vector, $\alpha_\ell^{(11)}$ and $\alpha_\ell^{(21)}$, and tensor, $\alpha_\ell^{(22)}$, temperature anisotropies.

couples to the vector-type Doppler term. Finally, tensor temperature anisotropies have only one kernel, $\alpha^{(22)}$, for their coupling to both the gravitational term and polarization.

Considering the polarization kernels $\epsilon_\ell^{(ij)}$ and $\beta_\ell^{(ij)}$ shown in Figs. 5.6 and 5.7, it is interesting to note that the vector B-kernel, $\beta_\ell^{(21)}$, is nearly eight times larger than the tensor one. For E-polarization, the situation is reversed. Hence, vector perturbations are very effective in generating B-polarization, while tensor perturbations generate somewhat more E- than B-polarization. Summing up the relevant contributions one finds for $x = k(t_0 - t) \gg 1$,

$$\frac{\sum_\ell \ell^2|\beta_\ell^{(2m)}|^2}{\sum_\ell \ell^2|\epsilon_\ell^{(2m)}|^2} \simeq \begin{cases} 6 & \text{for} \quad m = \pm 1 \\ \frac{8}{13} & \text{for} \quad m = \pm 2. \end{cases} \tag{5.160}$$

The scalar polarization kernel, $\epsilon^{(20)} = \alpha^{(22)}$, is the highest of all polarization kernels. As we have already mentioned, however, scalar perturbations generate no B-polarization at all.

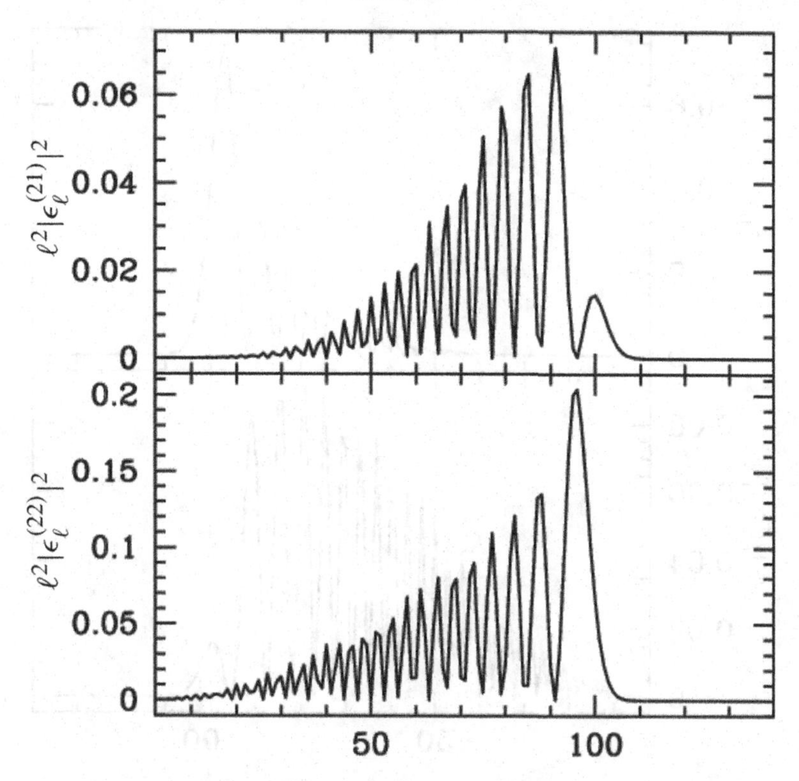

Fig. 5.6 The functions $\ell^2|\epsilon_\ell^{(21)}|^2$ (top) and $\ell^2|\epsilon_\ell^{(22)}|^2$ (bottom) are shown as functions of ℓ for fixed $x = 100$. These are the kernels relevant for E-polarization of vector and tensor modes respectively. Since $\ell^2|\epsilon_\ell^{(20)}|^2 = \ell^2|\alpha_\ell^{(22)}|^2$ this kernel for scalar E-polarization is not replotted. Note that the vector E-polarization kernel is very small and the scalar kernel is still about a factor of 5 larger than the tensor kernel.

Exercises

5.1 **Relation to Chapter 4**

Using the expressions for spherical harmonics from Appendix 4, Section A4.2.3 and our definitions of $\mathcal{M}_\ell^{(T)\pm}$ and $\mathcal{M}_\ell^{(V)\pm}$ given in Eqs. (4.149) and (4.133) and (4.134) in Chapter 4 show that

$$\mathcal{M}_\ell^{(0)} = (2\ell+1)\mathcal{M}_\ell^{(S)}$$

$$\mathcal{M}_\ell^{(\pm 1)} = i\sqrt{\ell(\ell+1)}\left[\mathcal{M}_{\ell+1}^{(V\pm)} + \mathcal{M}_{\ell-1}^{(V\pm)}\right], \tag{5.161}$$

$$\mathcal{M}_\ell^{(\pm 2)} = \ldots \mathcal{M}^{(T\pm)}. \tag{5.162}$$

Hint: For Eq. (5.161) use the recurrence relation (A4.20) to relate the Legendre functions $P_{\ell 1}$ to Legendre polynomials.

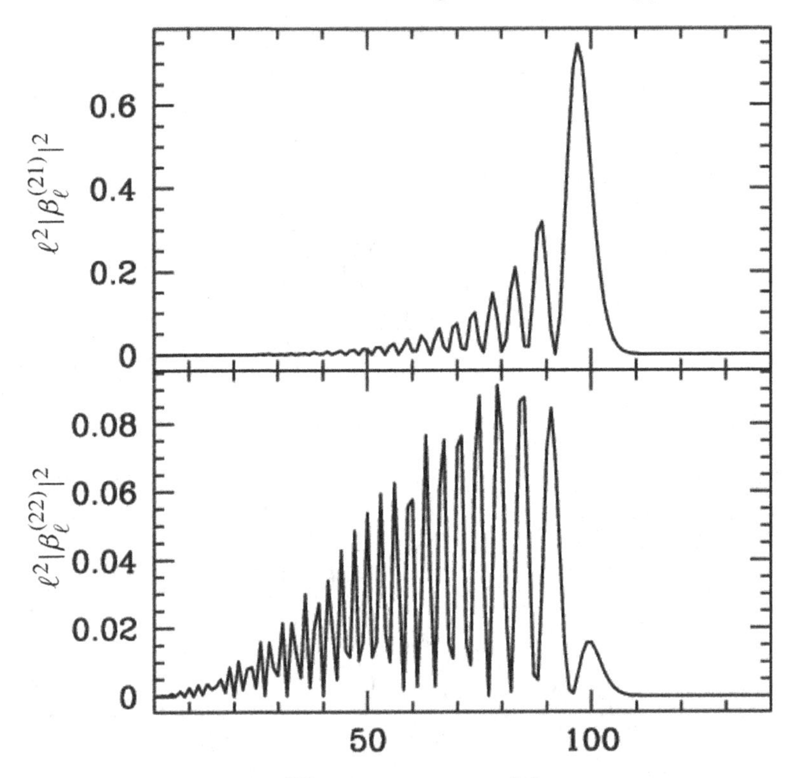

Fig. 5.7 The functions $\ell^2|\beta_\ell^{(21)}|^2$ (top) and $\ell^2|\beta_\ell^{(22)}|^2$ (bottom) are shown as functions of ℓ for fixed $x = 100$. These are the kernels relevant for B-polarization of vector and tensor modes respectively. Note that the vector B-polarization kernel is much larger than the tensor one. This is the opposite of what we find for E-polarization.

5.2 **Free streaming**
Show Eq. (5.122) by using the representation of spherical harmonics as matrix elements, ${}_sY_{\ell m} = \sqrt{(2\ell+1)/4\pi}\, D^{(\ell)*}_{m\,-s}$, the Clebsch–Gordan series, $D^{(1)} \otimes D^{(\ell)} = D^{(\ell-1)} \oplus D^{(\ell)} \oplus D^{(\ell+1)}$, and the corresponding Clebsch–Gordan coefficients for the basis transformation given in Table A4.1.

5.3 **Tensors in two dimensions**
Show that an arbitrary tensor field on the plane (flat 2d space) can be written in terms of four functions, α, γ, ε, β; two scalars; and two pseudo-scalars as follows:

$$T_{ab} = \alpha\delta_{ab} + \gamma\epsilon_{ab} + \left(\partial_a\partial_b - \frac{1}{2}\delta_{ab}\Delta\right)\varepsilon + \frac{1}{2}\left(\epsilon_{ca}\partial^c\partial_b + \epsilon_{cb}\partial^c\partial_a\right)\beta. \tag{5.163}$$

Here ϵ_{ab} is the totally antisymmetric tensor in two dimensions.

(1) Express the functions α to β in terms of the components of T_{ab}.
(2) Show that α and ε are scalars (i.e., even under parity) while γ and β are pseudo scalars (i.e., odd under parity).
(3) As the polarization P_{ab} is symmetric and traceless, it is of the form

$$\left(\partial_a\partial_b - \frac{1}{2}\delta_{ab}\Delta\right)\varepsilon + \frac{1}{2}\left(\epsilon_{ca}\partial^c\partial_b + \epsilon_{cb}\partial^c\partial_a\right)\beta. \qquad (5.164)$$

Show that in the flat sky approximation $\Delta^2\varepsilon = \mathcal{E}$ and $\Delta^2\beta = \mathcal{B}$.

Hint: Use Eqs. (5.24) and (5.25).

5.4 ***E*- and *B*-polarization in real space**
Using $\ell_x + i\ell_y = \ell e^{i\varphi_\ell}$ derive the following relation between e, b and $\mathcal{Q}, \mathcal{U}$ in real space:

$$e(\mathbf{x}) = \nabla^{-2}(\partial_x^2 - \partial_y^2)\mathcal{Q}(\mathbf{x}) + \nabla^{-2}2\partial_x\partial_y\mathcal{U}(\mathbf{x}), \qquad (5.165)$$

$$b(\mathbf{x}) = \nabla^{-2}(\partial_x^2 - \partial_y^2)\mathcal{U}(\mathbf{x}) - \nabla^{-2}2\partial_x\partial_y\mathcal{Q}(\mathbf{x}). \qquad (5.166)$$

6
Non-Gaussianities

6.1 Introduction

In Chapter 3 we have seen that inflationary models introduce not only Gaussian fluctuations that are fully described by the power spectrum, but also non-Gaussianities that lead to a bispectrum (or trispectrum, i.e., reduced 4-point function and higher N-point functions). Even though for single-field slow roll inflation these deviations from Gaussianity are usually very small, this is not the case for inflationary phases that deviate from slow roll [see, for example, Senatore *et al.* (2010); Renaux-Petel (2013)] or inflationary models that involve several scalar fields as discussed, for example, in Mazumdar and Wang (2012) and Achucarro *et al.* (2014).

Under linear evolution, a Gaussian field or a collection of Gaussian fields remains Gaussian. In other words, if $\zeta(\mathbf{k}, t_{\text{in}})$ is Gaussian then, for a deterministic transfer function $T_X(\mathbf{k}, t)$, the variable X given by

$$X(\mathbf{k}, t) = T_X(\mathbf{k}, t)\zeta(\mathbf{k}, t_{\text{in}}) \tag{6.1}$$

is also Gaussian. Therefore, within linear perturbation theory all variables remain (nearly) Gaussian if $\zeta(\mathbf{k}, t_{\text{in}}$ is (nearly) Gaussian. However, gravitational clustering is nonlinear, that is, the value of a given variable, for example, $\Psi(\mathbf{k}, t)$ depends not only on $\Psi(\mathbf{k}, t_{\text{in}})$ but also on $\Psi(\mathbf{k}_1, t_{\text{in}})\Psi(\mathbf{k}_2, t_{\text{in}})$ and higher powers. Therefore, once we go beyond linear perturbation theory, perturbations become non-Gaussian as the product of two Gaussian variables is non-Gaussian. In this chapter we give a brief introduction to non-Gaussianities in Fourier space and on the sphere. The prerequisite for this chapter is Appendix 7, of which we make full use here.

There are many ways to determine non-Gaussianity. Weak non-Gaussianities are often characterized by higher order correlators, the reduced N-point correlation functions, which are defined as the remainder after subtraction of the Gaussian piece given by Wick's theorem; see Appendix 7. But also other characteristics such as, for example, Minkowski functionals [see, for example, Schmalzing and Gorski (1998)] can be used.

We shall mainly concentrate on the bispectrum, that is, the 3-point function in Fourier space and on the sphere and mention further characteristics at the end of the chapter.

6.2 The Bispectrum in Fourier Space

As already introduced in Chapter 3, the 3-point function in Fourier space is of the form

$$\langle X(\mathbf{k}_1)X(\mathbf{k}_2)X(\mathbf{k}_2)\rangle = (2\pi)^3\delta(\mathbf{k}_1+\mathbf{k}_2+\mathbf{k}_3)B_X(k_1,k_2,k_3), \tag{6.2}$$

where $k_i = |\mathbf{k}_i|$. The function $B_X(k_1,k_2,k_3)$ is called the bispectrum of the variable X. The Dirac$-\delta$ is a consequence of statistical homogeneity. It implies that B is well defined only if the k_i can be connected into a closed triangle, for example,

$$|k_1 - k_2| \leq k_3 \leq k_1 + k_2 \tag{6.3}$$

It is easy to verify that if the triangle inequality holds for one combinations of the k_i's it holds for all. The fact that B depends only on the norm of the $\mathbf{k}_i$ is a consequence of statistical isotropy.

If X is a dimensionless function in real space, its Fourier transform has dimension L^3 so that B has dimension L^6. One often defines the dimensionless "shape function" $S_X(k_1,k_2,k_3)$ by [see Liguori *et al.* (2010) and Lazanu *et al.* (2016)]

$$S_X(k_1,k_2,k_3) = \alpha(k_1k_2k_3)^2 B_X(k_1,k_2,k_3) \tag{6.4}$$

The normalization constant α is chosen such that $S_X(k,k,k) = 1$, if S is scale invariant. Otherwise this normalization can be chosen for some characteristic scale k_c. Another possibility is to replace the factor $(k_1k_2k_3)^2$ by some $f(k_t)$ where $k_t = k_1 + k_2 + k_3$. Often S is separable or a sum of a small number of separable terms, that is, terms of the form $A_1(k_1)A_2(k_2)A_3(k_3) + 5$ perms. We have to add the five permutations of the first term. As we shall see in the next section, this simplifies the calculation of the bispectrum on the sphere considerably. Note, however, that even though S_X is dimensionless, it still usually depends not only on the shape of the triangle k_1,k_2,k_3 but also on its size. Typically there are scales in the bispectrum like, for example, $k_{\rm eq}$, the equality scale that enters in the power spectrum or some nonlinearity scale $k_{\rm nl}$.

Scale invariant bispectra can be classified by the shape of the triangle for which they are maximal. Let us describe some frequently obtained scale-invariant shapes. We present the trivial or constant shape that gives the same weight to every triangle,

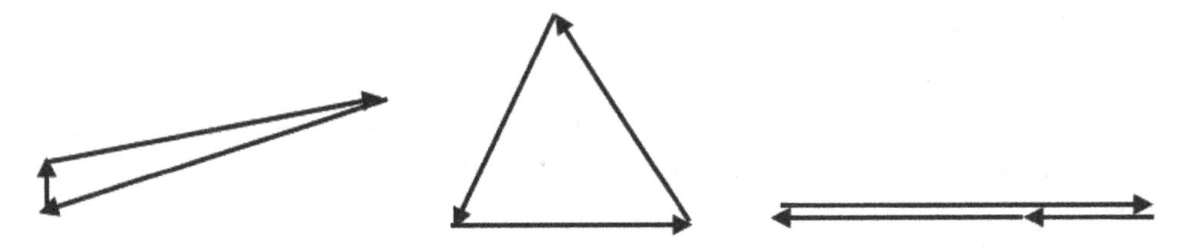

Fig. 6.1 We show a squeezed triangle (left), an equilateral triangle (middle), and a flattened triangle (right).

the squeezed or local shape, the equilateral shape, the orthogonal shape, and the flattened shape. They are defined as

$$S^{(\text{const})}(k_1,k_2,k_3) = 1 \tag{6.5}$$

$$S^{(\text{local})}(k_1,k_2,k_3) = \frac{1}{3}\left(\frac{k_1^2}{k_2k_3} + \frac{k_2^2}{k_1k_3} + \frac{k_3^2}{k_2k_1}\right) \tag{6.6}$$

$$S^{(\text{equi})}(k_1,k_2,k_3) = \frac{(k_1+k_2-k_3)(k_2+k_3-k_1)(k_3+k_1-k_2)}{k_1k_2k_3} \tag{6.7}$$

$$S^{(\text{ortho})}(k_1,k_2,k_3) = \frac{-3k_1^3 + 3(k_1^2-(k_2-k_3)^2)(k_3+k_2)}{k_1k_2k_3} + \frac{3k_3^2 + k_2(3k_2-8k_3)}{k_2k_3} \tag{6.8}$$

$$S^{(\text{flat})}(k_1,k_2,k_3) = \frac{(k_1^2+k_2^2+k_3^2)^3}{27(k_1+k_2-k_3)^2(k_2+k_3-k_1)^2(k_3+k_1-k_2)^2}. \tag{6.9}$$

Even though it is not evident at first sight, also $S^{(\text{ortho})}(k_1,k_2,k_3)$ is symmetric in its three arguments. The triangles for which the local, equilateral, and flat shape functions are maximal are shown in Fig. 6.1.

The normalization of the shape functions is chosen such that $S(k,k,k)=1$. An important point of many shape functions is that they all can be written as a sum of several functions of the form $X(k_1)Y(k_2)Z(k_3)$. In the above examples all except the flat shape are of this form. As we shall see in the next section, for shape functions of separable type, the calculation of the CMB bispectrum is numerically quite light. In the text that follows we shall discuss a so-called local non-Gaussianity and we shall find that its bispectrum is maximized for the squeezed shape, which we therefore also call the local shape.

In Fig. 6.2 we see that the local shape is maximal when $x = k_1/k_3 \to 0$ or $y = k_2/k_3 \to 0$ (as we have fixed k_3 to 1). The equilateral and the orthogonal shapes are maximal when $k_1 \simeq k_2 \simeq k_3$, which means $x \simeq y \simeq 1$ in our plot. But the orthogonal shape is negative close to the diagonal while the equilateral shape

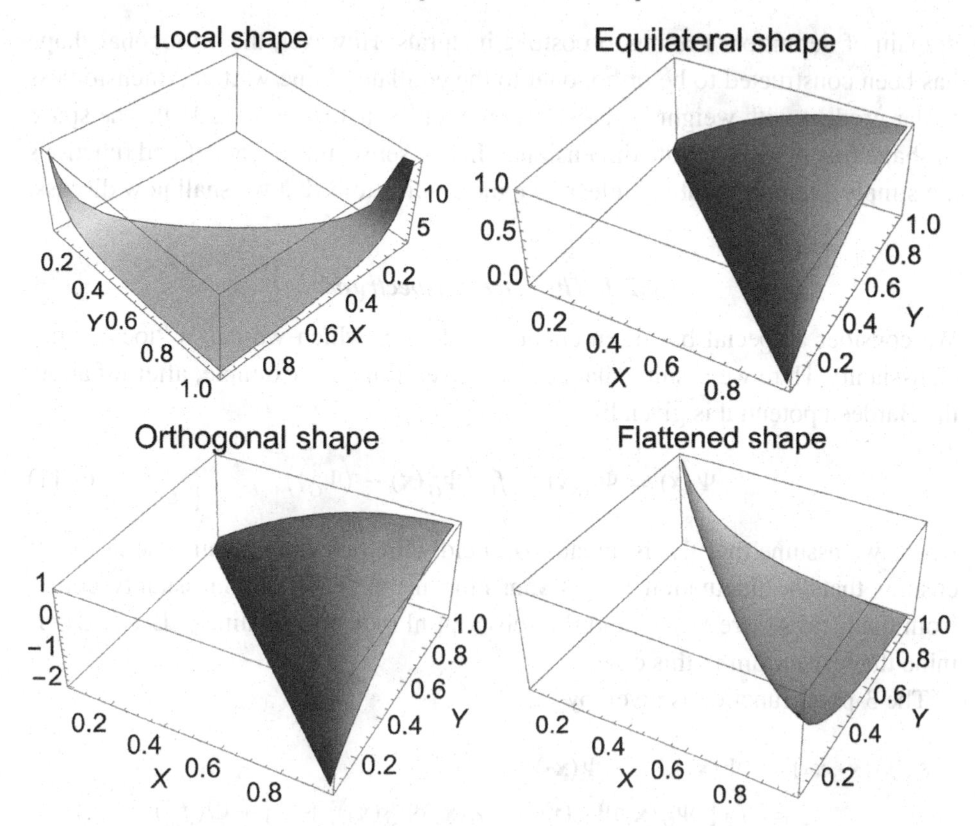

Fig. 6.2 We show the squeezed shape function (top left), the equilateral shape function (top right), the orthogonal shape function (bottom left), and the log of the flattened shape function (bottom right) as functions of $x = k_1/k_3$ and $y = k_2/k_3$.

is positive definite. The flattened shape is maximal on the diagonal $x + y \simeq 1$ and even larger at the edges when $x \to 0$ or $y \to 0$. The exact values $x, y \in \{0, 1\}$ and $y = 1 - x$ are removed in the plots, since our shapes diverge in some of these points (see Fig. 6.2).

The aforementioned shapes are not the only possible ones. Furthermore, they are not all orthogonal to each other in any sense. We can introduce a positive definite scalar product,

$$\langle S_1, S_2 \rangle = \int_V S_1(k_1, k_2, k_3) S_2(k_1, k_2, k_3) w(k_1, k_2, k_3) dk_1 dk_2 dk_3, \tag{6.10}$$

where V is a suitably chosen region within which $|k_1 - k_2| \leq k_3 \leq k_1 + k_2$ so that the shape functions S_i are well defined and w is some positive definite weight function. Clearly for all choices the above defined local, equilateral, and flat shapes, are not orthogonal to each other, since they are positive definite functions in their domain of definition, leading to positive integrals. However, the orthogonal shape has been constructed to be orthogonal to the equilateral one w.r.t. a dimensionless scalar product with weight $(k_1k_2k_3)^{-1}$ defined in Senatore *et al.* (2010). The space of shape functions is infinite dimensional. In this sense, the above defined functions are simply examples that are relevant in the applications that we shall now discuss.

6.2.1 The Local Bispectrum

We consider a special but often encountered case, which we call a "local" non-Gaussianity. Here we assume that at some given time, for example, after inflation, the Bardeen potential is given by

$$\Psi(\mathbf{x}) = \Psi_G(\mathbf{x}) + f_{\rm nl}\left(\Psi_G^2(\mathbf{x}) - \langle\Psi_G^2\rangle\right), \tag{6.11}$$

where we assume that Ψ_G is a Gaussian field with vanishing mean. The last term ensures that the fluctuation Ψ has vanishing mean. The non-Gaussianity comes from the local square $\Psi_G^2(\mathbf{x})$ and the subscript nl indicates nonlinear. Let us determine the bispectrum in this case.

The 3-point function is given by

$$\begin{aligned}\xi_3(\mathbf{x}_1,\mathbf{x}_2,\mathbf{x}_3) &= \langle\Psi(\mathbf{x}_1)\Psi(\mathbf{x}_2)\Psi(\mathbf{x}_3)\rangle \\ &= f_{\rm nl}\left[\langle\Psi_G(\mathbf{x}_1)\Psi_G(\mathbf{x}_2)\rangle\langle\Psi_G(\mathbf{x}_1)\Psi_G(\mathbf{x}_3)\rangle + \circlearrowleft\right] + \mathcal{O}(f_{\rm nl}^2) \\ &= f_{\rm nl}\left[\xi_2(r_{12})\xi_2(r_{13}) + \xi_2(r_{21})\xi_2(r_{23}) + \xi_2(r_{31})\xi_2(r_{32})\right] + \mathcal{O}(f_{\rm nl}^2).\end{aligned} \tag{6.12}$$

Here ξ_2 is the 2-point function of the Gaussian variable Ψ_G and we have used Wick's theorem (A7.8). The symbol $+\circlearrowleft$ indicates that the two cyclic permutations have to be added. We neglect terms higher order in $f_{\rm nl}$ since we assume Ψ to be small so that first-order perturbation theory is valid. As we know, this is a very good approximation, since the amplitude of Ψ measured in the CMB is of the order of 10^{-5}.

Fourier transforming Eq. (6.12) we find

$$B_\Psi(k_1,k_2,k_3) = f_{\rm nl}\left[P_{\Psi_G}(k_1)P_{\Psi_G}(k_2) + \circlearrowleft\right]. \tag{6.13}$$

Approximating $P_{\Psi_G}(k) \propto k^{-3}$ we obtain exactly the local shape function $S^{(\rm local)}(k_1,k_2,k_3)$ given in Eq. (6.6), as already anticipated. In the matter-dominated Universe, using $\zeta = (5/3)\Psi$ we can convert this into a relation for ζ,

$$\zeta(\mathbf{x}) = \zeta_G(\mathbf{x}) + \frac{3}{5}f_{\rm nl}\left(\zeta_G^2(\mathbf{x}) - \langle\zeta_G^2\rangle\right), \tag{6.14}$$

which implies

$$B_\zeta^{(\mathrm{local})}(k_1,k_2,k_3) = \frac{3}{5} f_{\mathrm{nl}} \left[P_{\zeta G}(k_1)P_{\zeta G}(k_2) + \circlearrowleft\right] \tag{6.15}$$

$$= \frac{9(2\pi^2 A_s)^2 f_{\mathrm{nl}}}{5} \frac{S^{(\mathrm{local})}(k_1,k_2,k_3)}{(k_1k_2k_2)^2}. \tag{6.16}$$

The small non-Gaussianity that we have found for single-field slow roll inflationary models, Eq. (3.108) is not of this form. However, we can write its shape function as

$$S^{(\mathrm{inf})}(k_1,k_2,k_3) = \alpha\left[-3(n_s-1)S^{(\mathrm{local})} + \epsilon(S^{(\mathrm{equi})} + 2S^{(\mathrm{const})} + 2T)\right]$$

where (6.17)

$$T(k_1,k_2,k_3) = \frac{4(k_1^2k_2^2 + k_1^2k_3^2 + k_2^2k_3^2) - k_t(k_1^3 + k_2^3 + k_3^3)}{k_1k_2k_3k_t} \tag{6.18}$$

(remember, $k_t = k_1 + k_2 + k_3$). The detailed form of $S^{(\mathrm{inf})}$ is not so relevant, but it is important that $S^{(\mathrm{equi})}$, $S^{(\mathrm{const})}$ and T all remain finite when one of the $k_i \to 0$ as one can check easily. If we go to this "squeezed" limit, $k_1 \sim k_2 = k \gg k_3 = q$, $S^{(\mathrm{local})}$, which diverges when $q \to 0$ dominates the other contributions and we find [see also Eq. (3.124)]

$$B_{\mathrm{inf}}(k,k,q) \simeq -(n_s-1)\left[P_{\zeta G}(k)P_{\zeta G}(q) + \circlearrowleft\right]. \tag{6.19}$$

On the other hand, a local non-Gaussianity of ζ becomes

$$B_\zeta(k,k,q) \simeq \frac{3}{5} f_{\mathrm{nl}} \left[P_{\zeta G}(k)P_{\zeta G}(q) + \circlearrowleft\right]. \tag{6.20}$$

Hence in the squeezed limit the single-field inflationary bispectrum looks like a local bispectrum with

$$f_{\mathrm{nl}}^{(\mathrm{inf})} = -\frac{5}{3}(n_s-1). \tag{6.21}$$

With the presently measured $n_s - 1 \sim -0.04$, this number is much too small to be measured in any near future CMB or large-scale structure experiment.

One can now introduce f_{nl} parameters also for other shapes, not only the local one. For this we write the bispectrum as

$$B_\zeta^{(\mathrm{shape})}(k,k,k) = \frac{9}{5} f_{\mathrm{nl}}^{(\mathrm{shape})} P_\zeta^2(k) \tag{6.22}$$

$$B_\zeta^{(\mathrm{shape})}(k_1,k_1,k_3) = \frac{1}{\alpha(k_1k_2k_3)^2} S_\zeta^{(\mathrm{shape})}(k_1,k_1,k_3). \tag{6.23}$$

The parameters $f_{\mathrm{nl}}^{(\mathrm{local})}$, $f_{\mathrm{nl}}^{(\mathrm{equi})}$, and $f_{\mathrm{nl}}^{(\mathrm{ortho})}$ have been constrained with CMB experiments (see Table 9.3).

6.2.2 Nonlinearities

Contrary to General Relativity, when considering Newtonian gravity only, it is quite simple to go to higher order in perturbation theory and one finds at second order for the density perturbations of a pressureless fluid [see, for example, Bernardeau *et al.* (2001)]

$$D^{(2)}(\mathbf{k},t) = \frac{1}{(2\pi)^3}\int d^3q\, F_2\left(\mathbf{q},\mathbf{k}-\mathbf{q}\right) D\left(\mathbf{q},t\right) D\left(\mathbf{k}-\mathbf{q},t\right), \tag{6.24}$$

where

$$F_2(\mathbf{q}_1,\mathbf{q}_2) = \frac{5}{7} + \frac{1}{2}\frac{\mathbf{q}_1\cdot\mathbf{q}_2}{q_1 q_2}\left(\frac{q_1}{q_2}+\frac{q_2}{q_1}\right) + \frac{2}{7}\left(\frac{\mathbf{q}_1\cdot\mathbf{q}_2}{q_1 q_2}\right)^2. \tag{6.25}$$

To convert Eq. (6.24) into an equation for the gravitational potential we use the Poisson equation,

$$\Psi(\mathbf{k},t) = -4\pi G a^2 \rho_m D(\mathbf{k},t)/k^2 = -\frac{3}{2}\frac{\Omega_m H_0^2}{k^2}(1+z)D(\mathbf{k},t). \tag{6.26}$$

Here Ω_m denotes the matter density parameter today and we have used $\rho_m(t) = \rho_m(t_0)(1+z)^3$. Replacing D and $D^{(2)}$ by Ψ and $\Psi^{(2)}$ in Eq. (6.24) using Eq. (6.26), we obtain

$$\Psi^{(2)}(\mathbf{k},t) = \frac{-2}{3\Omega_m H_0^2(1+z)}\frac{1}{(2\pi)^3 k^2}\int d^3q\, q^2(\mathbf{k}-\mathbf{q})^2 \times F_2\left(\mathbf{q},\mathbf{k}-\mathbf{q}\right)\Psi\left(\mathbf{q},t\right)\Psi\left(\mathbf{k}-\mathbf{q},t\right). \tag{6.27}$$

Inserting this expression in the bispectrum for

$$\Psi^{(\mathrm{tot})} = \Psi + \Psi^{(2)} \tag{6.28}$$

we obtain at tree level[1] (suppressing the time variable in Ψ for clarity)

$$\langle\Psi^{(\mathrm{tot})}(\mathbf{k}_1)\Psi^{(\mathrm{tot})}(\mathbf{k}_2)\Psi^{(\mathrm{tot})}(\mathbf{k}_3)\rangle = \langle\Psi^{(2)}(\mathbf{k}_1)\Psi(\mathbf{k}_2)\Psi(\mathbf{k}_3)\rangle + \circlearrowleft \tag{6.29}$$

$$= \frac{-2}{3\Omega_m H_0^2(1+z)}\frac{1}{(2\pi)^3 k_1^2}\int d^3q\, q^2(\mathbf{k}_1-\mathbf{q})^2 \times F_2\left(\mathbf{q},\mathbf{k}_1-\mathbf{q}\right)\langle\Psi(\mathbf{q})\Psi(\mathbf{k}_1-\mathbf{q})\Psi(\mathbf{k}_2)\Psi(\mathbf{k}_3)\rangle + \circlearrowleft. \tag{6.30}$$

Applying Wick's theorem on the 4-point correlator of the Gaussian field Ψ, we obtain

[1] This means when inserting only one $\Psi^{(\mathrm{tot})} = \Psi^{(1)} + \Psi^{(2)}$ and two $\Psi^{(1)} \equiv \Psi$ that we assume to be Gaussian.

$$\langle \Psi^{(\text{tot})}(\mathbf{k}_1)\Psi^{(\text{tot})}(\mathbf{k}_2)\Psi^{(\text{tot})}(\mathbf{k}_3)\rangle = \frac{-4}{3\Omega_m H_0^2(1+z)}(2\pi)^3\,\delta(\mathbf{k}_1+\mathbf{k}_2+\mathbf{k}_3)$$
$$\times\left[\frac{k_2^2k_3^2}{k_1^2}F_2(\mathbf{k}_2,\mathbf{k}_3)\,P_\Psi(k_2)P_\Psi(k_3)+\circlearrowleft\right]. \quad (6.31)$$

For the bispectrum this implies

$$B_\Psi^{(\text{nl})}(k_1,k_2,k_3) = \frac{-4}{3\Omega_m H_0^2(1+z)}\left[\frac{k_2^2k_3^2}{k_1^2}F_2(\mathbf{k}_2,\mathbf{k}_3)\,P_\Psi(k_2)P_\Psi(k_3)+\circlearrowleft\right]. \quad (6.32)$$

Note that $\mathbf{k}_2\cdot\mathbf{k}_3 = (k_1^2-k_2^2-k_3^2)/2$ so that the bispectrum truly only depends on the moduli k_i. More precisely,

$$\begin{aligned} F_2(\mathbf{k}_2,\mathbf{k}_3) &\equiv F_2(k_2,k_3;k_1) \\ &= \frac{5}{7}+\frac{1}{4}\frac{k_1^2-k_2^2-k_3^2}{k_2k_3}\left(\frac{k_2}{k_3}+\frac{k_3}{k_2}\right)+\frac{1}{14}\frac{(k_1^2-k_2^2-k_3^2)^2}{k_2^2k_3^2}. \end{aligned} \quad (6.33)$$

In this case, the term in brackets of Eq. (6.32) does not have dimension L^6 but only L^4, since the prefactor is dimensionful. Let us study also the squeezed limit for this bispectrum.

$$\text{For}\quad k_1 = q \ll k_2 \simeq k_3 = k \qquad F_2(k_2,k_2;q) \rightarrow \frac{3q^2}{14k^2}+\mathcal{O}(q^4). \quad (6.34)$$

Hence in the squeezed limit

$$B_\Psi^{(\text{nl})}(q,k,k) \simeq \frac{-2}{7\Omega_m H_0^2(1+z)}k^2P_\Psi^2(k). \quad (6.35)$$

Most importantly, nonlinearities are not enhanced in the squeezed limit, which makes the prospect for detecting primordial non-Gaussianities of the local type better.

6.3 The CMB Bispectrum

The bispectrum on the sphere is given by the 3-point correlator of the temperature fluctuations,

$$\frac{\Delta T}{T}(\mathbf{n}) = \sum_{\ell m} a_{\ell m}Y_{\ell m}(\mathbf{n}). \quad (6.36)$$

Like for the bispectrum in Fourier space, for perfectly Gaussian initial condition the bispectrum vanishes to first order in perturbation theory. We want to study the effects of a primordial non-Gaussianity and of nonlinear evolution on the bispectrum in the CMB.

Spherical isotropy implies that the 3-point function

$$\xi_3(\mathbf{n}_1, \mathbf{n}_2, \mathbf{n}_3) = \left\langle \frac{\Delta T}{T}(\mathbf{n}_1) \frac{\Delta T}{T}(\mathbf{n}_2) \frac{\Delta T}{T}(\mathbf{n}_3) \right\rangle \tag{6.37}$$

depends only on the direction cosines $\mu_{ij} = \mathbf{n}_i \cdot \mathbf{n}_j$. Expanding this dependence in Legendre polynomials we have

$$\xi_3(\mathbf{n}_1, \mathbf{n}_2, \mathbf{n}_3) = \sum_{\ell_1 \ell_2 \ell_3} b^{(2)}_{\ell_1 \ell_2 \ell_3} P_{\ell_1}(\mu_{12}) P_{\ell_2}(\mu_{23}) P_{\ell_3}(\mu_{31}). \tag{6.38}$$

On the other hand, using the expansion (6.36) we can determine the 3-point function in terms of the expectation values of three expansion coefficients,

$$\begin{aligned} \xi_3(\mathbf{n}_1, \mathbf{n}_2, \mathbf{n}_3) &= \left\langle \frac{\Delta T}{T}(\mathbf{n}_1) \frac{\Delta T}{T}(\mathbf{n}_2) \frac{\Delta T}{T}(\mathbf{n}_3) \right\rangle \\ &= \sum_{\substack{\ell_1 \ell_2 \ell_3 \\ m_1 m_2 m_3}} \langle a_{\ell_1 m_1} a_{\ell_2 m_2} a_{\ell_3 m_3} \rangle Y_{\ell_1 m_1}(\mathbf{n}_1) Y_{\ell_2 m_2}(\mathbf{n}_2) Y_{\ell_3 m_3}(\mathbf{n}_3). \end{aligned} \tag{6.39}$$

The product of three $a_{\ell m}$'s forms an invariant tensor and is therefore proportional to the Wigner 3J symbol; see Appendix 4, Section A4.2.5

$$\left\langle a_{\ell_1 m_1} a_{\ell_2 m_2} a_{\ell_3 m_3} \right\rangle = \mathcal{G}^{\ell_1 \ell_2 \ell_3}_{m_1 m_2 m_3} b_{\ell_1 \ell_2 \ell_3} \tag{6.40}$$

$$= \begin{pmatrix} \ell_1 & \ell_2 & \ell_3 \\ m_1 & m_2 & m_3 \end{pmatrix} B_{\ell_1 \ell_2 \ell_3}. \tag{6.41}$$

Equation A4.66 gives the relation between $b_{\ell_1 \ell_2 \ell_3}$ and $B_{\ell_1 \ell_2 \ell_3}$:

$$B_{\ell_1 \ell_2 \ell_3} = \sqrt{\frac{\prod_{i=1}^3 (2\ell_i + 1)}{4\pi}} \begin{pmatrix} \ell_1 & \ell_2 & \ell_3 \\ 0 & 0 & 0 \end{pmatrix} b_{\ell_1 \ell_2 \ell_3}. \tag{6.42}$$

The quantity $b_{\ell_1 \ell_2 \ell_3}$ is called the reduced bispectrum. It is related to $b^{(2)}_{\ell_1 \ell_2 \ell_3}$ via (see Exercise 6.2)

$$b_{\ell_1 \ell_2 \ell_3} = \sum_{L_i} Q^{L_1 L_2 L_3}_{\ell_1 \ell_2 \ell_3} b^{(2)}_{L_1 L_2 L_3} \quad \text{where} \tag{6.43}$$

$$\begin{aligned} Q^{L_1 L_2 L_3}_{\ell_1 \ell_2 \ell_3} &= (4\pi)^3 \begin{pmatrix} \ell_1 & \ell_2 & \ell_3 \\ 0 & 0 & 0 \end{pmatrix}^{-1} \sum_{m_i; M_i} \begin{pmatrix} \ell_1 & \ell_2 & \ell_3 \\ m_1 & m_2 & m_3 \end{pmatrix} \\ &\times \prod_{i=1}^{3} \begin{pmatrix} \ell_i & L_i & L_{[i-1]} \\ 0 & 0 & 0 \end{pmatrix} \begin{pmatrix} \ell_i & L_i & L_{[i-1]} \\ m_i & M_i & M_{[i-1]} \end{pmatrix}. \end{aligned} \tag{6.44}$$

Here $[i-1] = i-1$ if $i > 1$ and $[i-1] = i-1+3 = 3$ for $i = 1$. The reduced bispectrum $b_{\ell_1\ell_2\ell_3}$ contains all the relevant, (that is, coordinate independent) information about the non-Gaussianity at order ζ^3.

Let us consider an inflationary model that generates some arbitrary primordial bispectrum,

$$\langle \zeta(\mathbf{k}_1)\zeta(\mathbf{k}_2)\zeta(\mathbf{k}_3)\rangle = (2\pi)^3\delta(\mathbf{k}_1+\mathbf{k}_2+\mathbf{k}_3)B_\zeta(k_1,k_2,k_3). \tag{6.45}$$

We want to determine the bispectrum $b_{\ell_1\ell_2\ell_3}$ induced in the CMB by this non-Gaussianity within linear perturbation theory. For this we remember that in full generality we can write

$$\frac{\Delta T}{T}(\mathbf{n}) = \frac{1}{(2\pi)^3}\int d^3k\mathcal{T}(\mathbf{k},\mathbf{n})\zeta(\mathbf{k}),$$

$$a_{\ell m} = \frac{1}{(2\pi)^3}\int d^3k d\Omega_{\mathbf{n}}\mathcal{T}(\mathbf{k},\mathbf{n})Y^*_{\ell m}(\mathbf{n})\zeta(\mathbf{k}). \tag{6.46}$$

Here $\mathcal{T}(\mathbf{k},\mathbf{n})$ is the linear transfer function for CMB anisotropies. Considering, for example, expression (2.240), which neglects Silk damping, we find

$$\mathcal{T}(\mathbf{k},\mathbf{n}) = \left[\frac{1}{4}T_D + (\widehat{\mathbf{k}}\cdot\mathbf{n})T_V + T_\Psi + T_\Phi\right]\exp(i\mathbf{kn}(t_0-t_{\rm dec}))$$
$$+ \int_{t_{\rm dec}}^{t_0} dt(\dot{T}_\Psi(k,t) + \dot{T}_\Phi(k,t)\exp(i\mathbf{kn}(t_0-t)). \tag{6.47}$$

Here T_D is the transfer function of the radiation density fluctuation $D_g^{(r)}$, T_V is the transfer function of the velocity potential, and T_Ψ and T_Φ are the transfer functions of the Bardeen potentials, and all the transfer functions in the square bracket are to be evaluated at k and $t = t_{\rm dec}$. In a matter dominated universe, the transfer functions of the Bardeen potentials are simply $3/5$. Of course a much more accurate transfer function can be obtained from the numerical public codes CLASS (Lesgourgues, 2011; Blas *et al.*, 2011) or CAMB (Lewis *et al.*, 2000), which include also Silk damping and polarization. The equations that we derive here can also be derived for the bispectrum of polarization and mixtures of polarization and temperature. The only difference is that the transfer function has to be modified. Furthermore, since B-polarization has the opposite parity of E-polarization and of temperature, only an even number of $a_{\ell m}^{(\mathcal{B})}$ can lead to a nonvanishing result. For simplicity, we present the explicit derivations here only for the temperature anisotropy, $a_{\ell m}^{(\mathcal{M})} \equiv a_{\ell m}$.

Note that the transfer functions depend on $\mathbf{k}$ and $\mathbf{n}$ only via k and $\mu = \widehat{\mathbf{k}}\cdot\mathbf{n}$. This is simply a consequence of statistical isotropy. Writing the μ dependence of the transfer function in terms of Legendre polynomials we can write

$$\mathcal{T}(\mathbf{k},\mathbf{n}) = \sum_\ell (-i)^\ell (2\ell+1)\mathcal{T}(k,\ell)P_\ell(\mu). \tag{6.48}$$

The prefactor $(-i)^\ell(2\ell+1)$ is a pure convention; we have chosen it to simplify future formulae and to agree with the convention of Liguori *et al.* (2010). We use this and the addition theorem of spherical harmonics to perform the **n**-integral in Eq. (6.46) with the result

$$a_{\ell m} = 4\pi(-i)^\ell \int \frac{d^3k}{(2\pi)^3}\mathcal{T}(k,\ell)Y^*_{\ell m}(\widehat{\mathbf{k}})\zeta(\mathbf{k}). \tag{6.49}$$

In terms of the transfer function $\mathcal{T}(k,\ell)$ the CMB power spectrum becomes

$$C_\ell = \langle a_{\ell m}a^*_{\ell m}\rangle = \frac{2}{\pi}\int dk\, k^2\, |\mathcal{T}(k,\ell)|^2\, P_\zeta(k). \tag{6.50}$$

We now use Eq. (6.49) to determine the CMB bispectrum in terms of the one for ζ. For the expectation value of three expansion coefficients we obtain

$$\begin{aligned}
\langle a_{\ell_1 m_1}a_{\ell_2 m_2}a_{\ell_3 m_3}\rangle &= (4\pi)^3(-i)^{\ell_1+\ell_2+\ell_3}\int \frac{d^3k_1}{(2\pi)^3}\frac{d^3k_2}{(2\pi)^3}\frac{d^3k_3}{(2\pi)^3}\mathcal{T}(k_1,\ell_1)\mathcal{T}(k_2,\ell_2)\\
&\quad\times \mathcal{T}(k_3,\ell_3)Y^*_{\ell_1 m_1}(\widehat{\mathbf{k}}_1)Y^*_{\ell_2 m_2}(\widehat{\mathbf{k}}_2)Y^*_{\ell_3 m_3}(\widehat{\mathbf{k}}_3)\langle\zeta(\mathbf{k}_1)\zeta(\mathbf{k}_2)\zeta(\mathbf{k}_3)\rangle \qquad (6.51)\\
&= (4\pi)^3(-i)^{\ell_1+\ell_2+\ell_3}\int \frac{d^3k_1}{(2\pi)^3}\frac{d^3k_2}{(2\pi)^3}\frac{d^3k_3}{(2\pi)^3}\\
&\quad\times \mathcal{T}(k_1,\ell_1)\mathcal{T}(k_2,\ell_2)\mathcal{T}(k_3,\ell_3)Y^*_{\ell_1 m_1}(\widehat{\mathbf{k}}_1)Y^*_{\ell_2 m_2}(\widehat{\mathbf{k}}_2)Y^*_{\ell_3 m_3}(\widehat{\mathbf{k}}_3)\\
&\quad\times B_\zeta(k_1,k_2,k_3)(2\pi)^3\delta(\mathbf{k}_1+\mathbf{k}_2+\mathbf{k}_3).
\end{aligned}$$

To simplify this formula we use the fact that the Dirac delta is the Fourier transform of $(2\pi)^{-3}$. More precisely,

$$(2\pi)^3\delta(\mathbf{k}_1+\mathbf{k}_2+\mathbf{k}_3) = \int d^3x \exp\left(i(\mathbf{k}_1+\mathbf{k}_2+\mathbf{k}_3)\cdot\mathbf{x}\right).$$

Here $\mathbf{x}$ is just some dummy integration variable with units of length. Furthermore, we use Eq. (2.259),

$$e^{i\mathbf{k}_i\mathbf{x}} = \sum_{\ell=0}^{\infty}(2\ell+1)i^\ell j_\ell(k_i x)P_\ell(\widehat{\mathbf{k}}_i\cdot\widehat{\mathbf{x}}) = 4\pi\sum_{\ell m} i^\ell j_\ell(k_i x)Y_{\ell m}(\widehat{\mathbf{k}}_i)Y^*_{\ell m}(\widehat{\mathbf{x}}),$$

where $x = |\mathbf{x}|$ and $\widehat{\mathbf{x}} = \mathbf{x}/x$. In the second sum ℓ still goes from 0 to ∞ while for each ℓ, m goes from $-\ell$ to ℓ. The integral $d\Omega_{\widehat{\mathbf{k}}_i}$ then simply yields $\delta_{\ell\,\ell_i}\delta_{m\,m_i}$ and the integral over the directions of $\mathbf{x}$ is a Gaunt integral [see Eq. (A4.66)] so that we obtain

$$\left\langle a_{\ell_1 m_1} a_{\ell_2 m_2} a_{\ell_3 m_3} \right\rangle = \left(\frac{2}{\pi}\right)^3 \mathcal{G}^{m_1 m_2 m_3}_{\ell_1 \ell_1 \ell_3} \int_0^\infty dx x^2 \times \int_0^\infty dk_1 \int_0^\infty dk_2 \int_0^\infty dk_3 \left[\prod_{i=1}^3 k_i^2 \mathcal{T}(k_i, \ell_i) j_{\ell_i}(k_i x)\right] B_\zeta(k_1, k_2, k_3). \tag{6.52}$$

In terms of the dimensionless shape function

$$S_\zeta(k_1, k_2, k_3) = \alpha\, (k_1 k_2 k_3)^2 B_\zeta(k_1, k_2, k_3)$$

we can write the CMB bispectrum as

$$b_{\ell_1 \ell_2 \ell_3} = \frac{1}{\alpha}\left(\frac{2}{\pi}\right)^3 \int_0^\infty dx\, x^2 \int_0^\infty dk_1 \int_0^\infty dk_2 \int_0^\infty dk_3 \times \left[\prod_{i=1}^3 \mathcal{T}(k_i, \ell_i) j_{\ell_i}(k_i x)\right] S_\zeta(k_1, k_2, k_3). \tag{6.53}$$

This quantity is well defined when the ℓ_i's satisfy the triangle inequality, $|\ell_1 - \ell_2| \leq \ell_3 \leq \ell_1 + \ell_2$, and when the sum $\ell_1 + \ell_2 + \ell_3$ is even. Otherwise the Gaunt integral that multiplies $b_{\ell_1 \ell_2 \ell_3}$ vanishes. To see that also the k_i satisfy the triangle inequality one has to perform the x-integration. Let us, without loss of generality, assume that k_3 is the largest of the three wave numbers, $k_3 \geq k_1$ and $k_3 \geq k_2$. If the triangle inequality is violated, $k_3 > k_1 + k_2$, one then can show that the x-integral of the three spherical Bessel functions vanishes by using the first of the integrals 6.578 in Gradshteyn and Ryzhik (2000) (Mathematica cannot perform this integral.).

6.3.1 Separable Shapes

In most of the applications so far, the shape function has been a sum of a few separable terms of the form

$$S_\zeta(k_1, k_2, k_3) \propto X(k_1) Y(k_2) Z(k_3) + \text{ perms.}, \tag{6.54}$$

where we have in general to add all six permutations of the three wave numbers. In this case, Eq. (6.53) reduces to the x-integral of three one-dimensional integrals. Setting

$$X_\ell(x) = \frac{2}{\pi} \int_0^\infty dk \mathcal{T}(k, \ell) j_\ell(kx) X(k) \tag{6.55}$$

$$Y_\ell(x) = \frac{2}{\pi}\int_0^\infty dk\mathcal{T}(k,\ell)j_\ell(kx)Y(k) \tag{6.56}$$

$$Z_\ell(x) = \frac{2}{\pi}\int_0^\infty dk\mathcal{T}(k,\ell)j_\ell(kx)Z(k), \tag{6.57}$$

we have

$$b_{\ell_1\ell_2\ell_3} = \frac{1}{\alpha}\int_0^\infty dxx^2 X_{\ell_1}(x)Y_{\ell_2}(x)Z_{\ell_3}(x) \; + \; \text{perms.} \tag{6.58}$$

This is numerically a very tractable problem and is relevant for many cases of interest.

6.3.2 Simple Examples

Here we study some examples in the low ℓ regime where we consider adiabatic perturbations and concentrate on the Sachs–Wolfe contribution given in Eq. (2.257), without loss of generality we set $\mathbf{x}_0 = 0$,

$$\frac{\Delta T}{T}(\mathbf{n}) \simeq \frac{1}{3}\Psi(-(t_0 - t_{\text{dec}})\mathbf{n}) = \frac{1}{5}\zeta(-(t_0 - t_{\text{dec}})\mathbf{n}) \tag{6.59}$$

$$= \frac{1}{5}\int\frac{d^3k}{(2\pi)^3}\exp(-i\mathbf{nk}(t_0 - t_{\text{dec}}))\zeta(\mathbf{k}); \quad \text{hence} \tag{6.60}$$

$$\mathcal{T}(\mathbf{k},\mathbf{n}) = \frac{1}{5}\exp(-i\mathbf{nk}(t_0 - t_{\text{dec}})) \tag{6.61}$$

$$= \frac{1}{5}\sum_\ell(-i)^\ell(2\ell+1)j_\ell(k(t_0 - t_{\text{dec}})P_\ell(\mu) \quad \text{and} \tag{6.62}$$

$$\mathcal{T}(k,\ell) = \frac{1}{5}j_\ell(k(t_0 - t_{\text{dec}}). \tag{6.63}$$

For the last equals sign we make use of the definition (6.48).

The Constant Shape

Let us first consider a constant bispectrum, S_ζ =constant. We can choose $X(k) = Y(k) = Z(k) = 1$ so that

$$X_\ell(x) = \frac{2}{5\pi}\int_0^\infty dkj_\ell(k(t_0 - t_{\text{dec}})j_\ell(kx) \tag{6.64}$$

$$= \frac{\frac{1}{5}z^{\ell+1/2}}{\sqrt{x(t_0 - t_{\text{dec}})}}\frac{1}{2\ell+1} \quad \text{where } z = \min\left\{\frac{x}{t_0 - t_{\text{dec}}}, \frac{t_0 - t_{\text{dec}}}{x}\right\}.$$

The integral of $x^2 X_{\ell_1}(x) X_{\ell_2}(x) X_{\ell_3}(x)$ over x now becomes elementary with the result

$$b_{\ell_1\ell_2\ell_3} = \frac{S_\zeta}{125\alpha}\left(\frac{1}{\ell_1+\ell_2+\ell_3} + \frac{1}{\ell_1+\ell_2+\ell_3+3}\right)\prod_{i=1}^{3}\frac{1}{2\ell_i+1}. \tag{6.65}$$

This result has been derived in Fergusson and Shellard (2009). Note the scaling of $b_{\ell_1\ell_2\ell_3} \propto \ell^{-4}$. This bispectrum is maximized at small ℓ's, while the ratio

$$\frac{b_{\ell\ell\ell}}{C_\ell^2} \tag{6.66}$$

remains roughly constant. But of course our result is valid only at low $\ell \lesssim 100$.

The Local Shape

Next we consider the local shape with (6.6)

$$S(k_1,k_2,k_3) = \frac{1}{3}\left(\frac{k_1^2}{k_2k_3} + \frac{k_2^2}{k_1k_3} + \frac{k_3^2}{k_2k_1}\right). \tag{6.67}$$

To calculate $b_{\ell_1\ell_2\ell_3}$ we need the two integrals (A4.149) and (A4.151), which yield

$$\frac{2}{\pi}\int dk k^2 j_\ell(kx) j_\ell(k(t_0 - t_{\rm dec})) = \frac{\delta(x - t_0 + t_{\rm dec})}{(t_0 - t_{\rm dec})^2} \tag{6.68}$$

$$\int \frac{dk}{k} j_\ell(k(t_0 - t_{\rm dec})) j_\ell(k(t_0 - t_{\rm dec})) = \frac{1}{2}\frac{1}{\ell(\ell+1)} \tag{6.69}$$

From Eq. (6.15) we infer also that $\alpha = (9f_{\rm nl}/5)^{-1}$ so that finally

$$b_{\ell_1\ell_2\ell_3} = \frac{3f_{\rm nl}(2\pi^2 A_s)^2}{4\times 5^4}\left[\frac{1}{\ell_1(\ell_1+1)\ell_2(\ell_2+1)} + \circlearrowleft\right]. \tag{6.70}$$

Also this bispectrum behaves like ℓ^{-4}. Each of the terms is maximized when two ℓ's are small while the third is arbitrary, but it has to be chosen such that $(\ell_1, \ell_2\ell_3)$ satisfy the triangle inequality; otherwise the prefactor $\mathcal{G}^{m_1m_2m_3}_{\ell_1\ell_2,\ell_3}$ vanishes.

6.3.3 General Remarks and Nonseparable Shapes

The scaling $\ell^{-4} \propto C_\ell^2$ is a direct consequence of the scaling $k^{-6} \propto P_\zeta^2$ of the scale-invariant bispectrum in Fourier space and is model and shape independent. In Exercise 6.4 you will find the same scaling also for the equilateral shape. It also holds for nonseparable shapes at low ℓ's as long as $B(k_1,k_2,k_3)$ is scale invariant.

To calculate the full bispectrum one has to use one of the public codes CLASS or CAMB to determine the full transfer functions $\mathcal{T}(k,\ell)$, which then are numerically integrated to obtain $X_\ell(x)$, $Y_\ell(x)$, and $Z_\ell(x)$ for each term in the sum of separable shapes. The resulting product, $x^2 X_{\ell_1}(x) Y_{\ell_2}(x) Z_{\ell_3}(x)$, is then integrated over x to obtain the bispectrum. One is allowed to integrate over all values of the k_i because in principle, the integral over x takes care of the triangle inequality $|k_1 - k_2| \le k_3 \le k_1 + k_2$. However, depending on the form of $X(k)$, the k-integral from 0 to infinity may not exist and so for numerical integration one is advised to integrate only over the tetrahedron that satisfies the triangle inequality in k-space.

For a nonseparable shape like T of the inflationary bispectrum [see Eq. (6.18)] a full 4-dimensional integral $dx dk_1 dk_2 dk_3$ has to be performed, which is numerically quite challenging and is usually done employing Monte Carlo techniques. Another possibility is to expand the k-dependence into separable functions on the tetrahedron; see Liguori *et al.* (2010).

6.4 Beyond the Bispectrum

6.4.1 The 4-Point Function and Beyond

Non-Gaussian fluctuations are of course expected not only to have a nontrivial 3-point function or bispectrum but we typically expect nonvanishing N-point functions at all orders. For even N it is useful to subtract the Gaussian piece, which leads to the "connected" $2n$-point function[2] defined by

$$\xi_{2n}^{(c)}(\mathbf{x}_1, \cdots, \mathbf{x}_{2n}) = \xi_{2n}(\mathbf{x}_1, \cdots, \mathbf{x}_{2n}) - \frac{1}{2^n n!} \sum_{\sigma \in S_{2n}} \prod_{i=1}^{n} \xi(\mathbf{x}_{\sigma 2i-1}, \mathbf{x}_{\sigma 2i}). \qquad (6.71)$$

The second term is simply the $2n$-point function of a Gaussian field according to Wick's theorem see Appendix 7. The sum is over all permutations of the numbers $1, \dots 2n$ and the prefactor divides by the number of permutations leading to the same pairs (just in different order).

In Fourier space, statistical homogeneity again gives rise to a Dirac delta. The Fourier transform of the N-point function is

$$P_N(\mathbf{k}_1, \cdots \mathbf{k}_N) = B_N(\mathbf{k}_1, \cdots, \mathbf{k}_{N-1}) \, (2\pi)^3 \delta\left(\sum_{i=1}^{N} \mathbf{k}_i\right). \qquad (6.72)$$

But now, isotropy no longer determines all the possible angles between the $\mathbf{k}_i$ once their moduli are given. For an N-lateral figure the length of its sides determines the shape completely only if $N = 3$. Therefore, in addition to the moduli $|\mathbf{k}_i|$ we need to know either $N - 3$ spatial angles (θ, ϕ) or the length of $N - 3$ diagonals and

[2] Here we commit an abuse of language: the true "connected" part of an N-point function subtracts all possible lower point functions, while here we subtract only those that appear in the Gaussian case. This is "nearly" the connected part if our N-point function is nearly Gaussian.

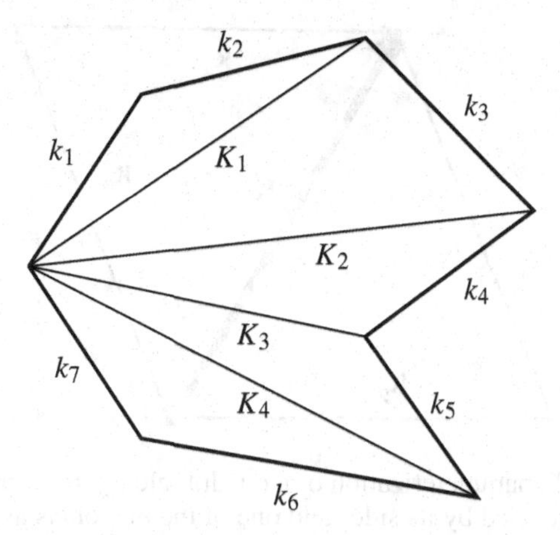

Fig. 6.3 A multilateral figure for the case $N = 7$.

$N - 3$ single angles. This is due to the fact that an N-lateral figure can always be decomposed into $N - 2$ triangles of which two have only one side in common with one of the others while $N - 4$ have two sides in common; see Fig. 6.3. Each triangle is determined by 3 data (of which at least one has to be a length); this means that we need $2N - 3$ data to determine the shape (and size) of all the triangles. In $3d$ space they can be connected to each other at an arbitrary angle that requires another $N - 3$ data so that in total we need $3N - 6$ data to fully determine the shape (and size) of an N-lateral figure in 3d space.

The most studied of the N-point correlations functions and spectra with $N > 3$ is the 4-point function and the corresponding trispectrum. This is especially important for the CMB, since the most significant nonlinearity in the CMB comes from lensing by foreground structures, the topic of Chapter 7. There we shall see that this nonlinearity does not generate a bispectrum but only a trispectrum. The CMB trispectrum due to lensing has been measured in the Planck data; see Planck Coll. VIII (2018).

The simplest "ansatz" leading to a primordial trispectrum is, like for the bispectrum, of the form

$$\zeta = \zeta_G + g_{\rm nl}\frac{9}{25}\zeta_G^3, \tag{6.73}$$

where we assume ζ_G to be Gaussian. In this case the bispectrum vanishes at tree level (the expectation value of a product of five Gaussian variables vanishes), and the trispectrum is simply given by

$$B_4(\mathbf{k}_1, \mathbf{k}_2, \mathbf{k}_3, \mathbf{k}_4) = \frac{54}{25} g_{\rm nl}\left[P_\zeta(k_1)P_\zeta(k_2)P_\zeta(k_3) + \circlearrowleft\right], \tag{6.74}$$

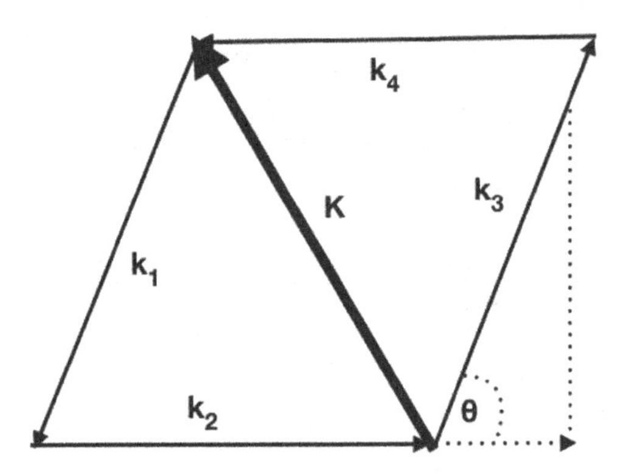

Fig. 6.4 A possible parameterization of a quadrangle by the length of the sides of the two triangles formed by its sides and one of the diagonals as well as the angle θ that defines the inclination of the second triangle w.r.t. to the first.

where ↻ indicates the three permutations where one of the $k_i \neq k_4$ is not included in the product of spectra. The Planck analysis Planck Coll. XVII (2016) has derived upper limits for $g_{\rm nl}$.

We can characterize a generic trispectrum $B_4(\mathbf{k}_1, \mathbf{k}_2, \mathbf{k}_3, \mathbf{k}_4)$ by six parameters. A possibility are the four lengths k_i, the length of the diagonal K, and the angle θ between the two triangles $(\mathbf{k}_1, \mathbf{k}_2, \mathbf{K})$ and $(\mathbf{k}_3, \mathbf{k}_4, -\mathbf{K})$; see Fig. 6.4.

With this parameterization we can write

$$B_4(\mathbf{k}_1, \cdots \mathbf{k}_4; \mathbf{K}) = \sum_{n=0}^{\infty} B_4^{(n)}(k_1, k_2, k_3, k_4; K) P_n(\cos\theta). \tag{6.75}$$

In the text that follows we shall also use the following form for the trispectrum of ζ:

$$\langle \zeta(\mathbf{k}_1)\zeta(\mathbf{k}_2)\zeta(\mathbf{k}_3)\zeta(\mathbf{k}_4)\rangle$$
$$= (2\pi)^3 \int d^3\mathbf{K}\delta(\mathbf{k}_1 + \mathbf{k}_2 + \mathbf{K})\delta(\mathbf{k}_3 + \mathbf{k}_4 - \mathbf{K}) B_4(\mathbf{k}_1, \cdots \mathbf{k}_4; \mathbf{K}). \tag{6.76}$$

Of course there are also other parameterizations possible, for example, in terms of the lengths k_i and the lengths of the two diagonals K and the one connecting the corners $(1, 2)$ and $(3, 4)$. We use the one given in Eq. (6.75) because it will show us that the CMB trispectrum is sensitive only to the $n = 0$ component of the primordial trispectrum $B_4(\mathbf{k}_1, \cdots \mathbf{k}_4; \mathbf{K})$, that is, to the flat, $\theta = 0$, component.

Let us now discuss the CMB trispectrum. Rotation invariance requires that the function $\left\langle \frac{T(\mathbf{n}_1)}{T}\frac{T(\mathbf{n}_2)}{T}\frac{T(\mathbf{n}_3)}{T}\frac{T(\mathbf{n}_4)}{T}\right\rangle$ is an invariant function on $(\mathbb{S}^2)^4$ According to Eq. (A4.91) we can therefore represent it in the form

$$\left\langle \frac{T(\mathbf{n}_1)}{T}\frac{T(\mathbf{n}_2)}{T}\frac{T(\mathbf{n}_3)}{T}\frac{T(\mathbf{n}_4)}{T}\right\rangle = \sum \mathcal{T}^{\ell_1\ell_2\ell_3\ell_4|L} Y_{\ell_1\ell_2\ell_3\ell_4|L}(\mathbf{n}_1, \mathbf{n}_2, \mathbf{n}_3, \mathbf{n}_4)$$

where

$$\mathcal{T}^{\ell_1\cdots\ell_4|L} = \sum W^{\ell_1\cdots\ell_4|L}_{m_1\ldots m_4}\langle a_{\ell_1 m_1}\cdots a_{\ell_4 m_4}\rangle. \tag{6.77}$$

The previous sum is over all ℓ_i and L, while the sum in (6.77) is over the m_i, $-\ell_i \le m_i \le \ell_i$. The generalized Wigner symbols $W^{\ell_1\ell_2\ell_3\ell_4|L}_{m_1\ldots m_4}$ are defined in Appendix 4, Section A4.2.5; see also Mitsou *et al.* (2019). For the case of four m_i they are given by

$$W^{\ell_1\cdots\ell_4|L}_{m_1\ldots m_4} = \sum_M (-1)^{L+M}\sqrt{2L+1}\begin{pmatrix}\ell_1 & \ell_2 & L\\ m_1 & m_2 & -M\end{pmatrix}\begin{pmatrix}\ell_3 & \ell_4 & L\\ m_3 & m_4 & M\end{pmatrix}. \tag{6.78}$$

As the correlators $\langle a_{\ell_1 m_1}\cdots a_{\ell_4 m_4}\rangle$ are totally symmetric under permutations of pairs (ℓ_i, m_i) , this is also true for permutations of the ℓ_i in $\mathcal{T}^{\ell_1\cdots\ell_4|L_1}$. Furthermore, since $Y_{\ell m}(-\mathbf{n}) = (-1)^\ell Y_{\ell m}(\mathbf{n})$, parity invariance requests that only even sums $\ell_1+\ell_2+\ell_3+\ell_4$ can contribute. In addition, since the 3J symbol $\begin{pmatrix}\ell_1 & \ell_2 & L\\ m_1 & m_2 & -M\end{pmatrix}$ changes by a factor $(-1)^{\ell_1+\ell_2+L}$ under the exchange of, for example, the first and second column, only terms where also $\ell_1+\ell_2+L$ is even can contribute. Since $\ell_1+\ell_2+\ell_3+\ell_4$ is even it follows that in this case also $\ell_3+\ell_4+L$ is even. We also note that the generalized Wigner symbols above vanish unless $m_1+m_2+m_2+m_4=0$.

Using the transfer functions (6.48) for the $a_{\ell m}$'s we can relate the trispectrum for ζ to the CMB trispectrum via

$$\langle a_{\ell_1 m_1}\cdots a_{\ell_4 m_4}\rangle_c = (4\pi)^4(-i)^{\sum\ell_i}\int d^3\mathbf{K}\prod_{i=1}^{4}\left[\frac{d^3k_i}{(2\pi)^3}\mathcal{T}(k_i,\ell_i)Y^*_{\ell_i m_i}(\widehat{k}_i)\right]$$
$$\times(2\pi)^3\delta(\mathbf{k}_1+\mathbf{k}_2+\mathbf{K})\delta(\mathbf{k}_3+\mathbf{k}_4-\mathbf{K})B_{4c}(\mathbf{k}_1,\cdots\mathbf{k}_4;\mathbf{K}). \tag{6.79}$$

Here the index c indicates that we consider the connected (or irreducible) part of the trispectra. This means we have subtracted the Gaussian contribution that is a consequence of Wick's theorem,

$$\begin{aligned}\langle a_{\ell_1 m_1}\cdots a_{\ell_4 m_4}\rangle_c = &\langle a_{\ell_1 m_1}\cdots a_{\ell_4 m_4}\rangle\\ &-(-1)^{m_1+m_3}C_{\ell_1}C_{\ell_3}\delta_{\ell_1\ell_2}\delta_{\ell_3\ell_4}\delta_{-m_1m_2}\delta_{-m_3m_4}-(-1)^{m_1+m_2}C_{\ell_1}C_{\ell_2}\\ &\times\left[\delta_{\ell_1\ell_3}\delta_{\ell_2\ell_4}\delta_{-m_1m_3}\delta_{-m_2m_4}+\delta_{\ell_1\ell_4}\delta_{\ell_2\ell_3}\delta_{-m_1m_4}\delta_{-m_2m_3}\right].\end{aligned} \tag{6.80}$$

Correspondingly we have subtracted the Gaussian part also from B_4 to obtain B_{4c}. Since the transformation (6.79) is linear, the connected part of B_4 generates only the connected part of the CMB trispectrum.

Writing the two Dirac deltas in (6.79) as Fourier transforms w.r.t. two dummy variables $\mathbf{x}_1$ and $\mathbf{x}_2$ and decomposing the exponentials as sums of products of spherical Bessel functions and spherical harmonics like for Eq. (6.52), we obtain

$$\begin{aligned}\langle a_{\ell_1 m_1}\cdots a_{\ell_4 m_4}\rangle_c &= \left(\frac{2}{\pi}\right)^5 (-i)^{\sum \ell_i}\int d^3\mathbf{K}\prod_i\left[d^3k_i\mathcal{T}(k_i,\ell_i)Y^*_{\ell_i m_i}(\widehat{\mathbf{k}}_i)\right]\\ &\times B_{4c}(\mathbf{k}_1,\cdots\mathbf{k}_4;\mathbf{K})\sum_{\ell_i' LL'}(i)^{\sum \ell_i'}\sum_{m_i' MM'}\int d^3x_1 d^3x_2\\ &\times\Big[j_{\ell_1'}(x_1k_1)Y_{\ell_1'm_1'}(\widehat{\mathbf{k}}_1)Y^*_{\ell_1'm_1'}(\widehat{\mathbf{x}}_1)\\ &\qquad j_{\ell_2'}(x_1k_2)Y_{\ell_2'm_2'}(\widehat{\mathbf{k}}_2)Y^*_{\ell_2'm_2'}(\widehat{\mathbf{x}}_1)j_L(x_1K)Y^*_{LM}(\widehat{\mathbf{K}})Y_{LM}(\widehat{\mathbf{x}}_1)\Big]\\ &\times\Big[j_{\ell_3'}(x_2k_3)Y_{\ell_3'm_3'}(\widehat{\mathbf{k}}_3)Y^*_{\ell_3'm_3'}(\widehat{\mathbf{x}}_2)j_{\ell_4'}(x_2k_4)\\ &\qquad Y_{\ell_4'm_4'}(\widehat{\mathbf{k}}_4)Y^*_{\ell_4'm_4'}(\widehat{\mathbf{x}}_2)j_{L'}(x_2K)Y^*_{L'M'}(\widehat{\mathbf{K}})Y_{L'M'}(\widehat{\mathbf{x}}_2)\Big].\end{aligned}\tag{6.81}$$

We can now perform the angular integration of $\mathbf{x}_1$ and $\mathbf{x}_2$ using (A4.66) for the integration of three spherical harmonics. We may assume that the angle θ on which the primordial trispectrum B_4 depends [see Eq. (6.75)] is the polar angle of $\mathbf{k}_4$, θ_4. Then we can also perform the angular integrations of $\mathbf{k}_1$ to $\mathbf{k}_3$ and $\mathbf{K}$, which just yield $\delta_{\ell_i\ell_i'}\delta_{m_im_i'}$ and $\delta_{LL'}\delta_{MM'}$. Inserting also the definition of $\mathcal{T}^{\ell_1\cdots\ell_4|L}$ and using Eq. (6.78) we find

$$\begin{aligned}\mathcal{T}^{\ell_1\cdots\ell_4|L} &= \frac{h_{\ell_1\ell_2L}h_{\ell_3\ell_4L}}{\sqrt{2L+1}}\left(\frac{2}{\pi}\right)^5\int K^2dKx_1^2dx_1x_2^2dx_2 j_L(Kx_1)j_L(Kx_2)\\ &\times\left[\prod_{i=1}^4 dk_ik_i^2\mathcal{T}(k_i,\ell_i)\right]j_{\ell_1}(k_1x_1)j_{\ell_2}(k_2x_1)j_{\ell_3}(k_3x_2)j_{\ell_4}(k_4x_4)\\ &\times\frac{1}{2\ell_4+1}\sum_{m_4}\int d\Omega_4 B_{4c}(k_1,k_2,k_3,k_4,K,\theta_4)Y_{\ell_4m_4}(\widehat{\mathbf{k}}_4)Y^*_{\ell_4m_4}(\widehat{\mathbf{k}}_4).\end{aligned}\tag{6.82}$$

Here we have introduced

$$h_{\ell_i\ell_jL} = \sqrt{\frac{(2\ell_i+1)(2\ell_j+1)(2L+1)}{4\pi}}\begin{pmatrix}\ell_i & \ell_j & L\\ 0 & 0 & 0\end{pmatrix}.\tag{6.83}$$

Now the sum

$$\sum_{m_4}Y_{\ell_4m_4}(\widehat{\mathbf{k}}_4)Y^*_{\ell_4m_4}(\widehat{\mathbf{k}}_4) = \frac{2\ell_4+1}{4\pi}P_{\ell_4}(1) = \frac{2\ell_4+1}{4\pi}.$$

Therefore in the expansion of $B_{4c}(k_1,k_2,k_3,k_4,K,\theta_4)$ in Legendre polynomials (6.75), only the first term, $n = 0$, contributes a factor 4π; all other terms vanish. This shows that only the planar projection, $n = 0$, of the primordial trispectrum contributes to the CMB trispectrum. If we want to probe the full 3D configuration of the trispectrum, we need a 3D dataset. The CMB that provides data on a 2D sphere cannot probe the 3D nature of the tri- and higher N-spectra. Interestingly, several of the most common non-Gaussianities, for example, the local non-Gaussianity, only induce a flat trispectrum. The Ω_4-integral in (6.82) therefore just cancels the factor $(2\ell_4 + 1)^{-1}$ and we obtain

$$\begin{aligned}\mathcal{T}^{\ell_1\cdots\ell_4|L} &= \frac{h_{\ell_1\ell_2L}h_{\ell_3\ell_4L}}{\sqrt{2L+1}}\left(\frac{2}{\pi}\right)^5\int K^2dKx_1^2dx_1x_2^2dx_2j_L(Kx_1)j_L(Kx_2)\\ &\times\left[\prod_{i=1}^{3}dk_ik_i^2\mathcal{T}(k_i,\ell_i)\right]j_{\ell_1}(k_1x_1)j_{\ell_2}(k_2x_1)j_{\ell_3}(k_3x_2)j_{\ell_4}(k_4x_4)\\ &\times B_{4c}^{(0)}(k_1,k_2,k_3,k_4;K).\end{aligned} \tag{6.84}$$

Like for the bispectrum, one usually defines the reduced trispectrum as

$$t^{\ell_1\cdots\ell_4|L} = \frac{\sqrt{2L+1}}{h_{\ell_1\ell_2L}h_{\ell_3\ell_4L}}\mathcal{T}^{\ell_1\cdots\ell_4|L}, \tag{6.85}$$

in order to avoid the cumbersome prefactors.

Using the machinery developed in this section and in the appendix, the formulas (6.77) and (6.84) can be generalized to arbitrary N-spectra. In a more elegant way, one can immediately expand the N-point functions in terms of the rotation-invariant functions $Y_{\ell_1\cdots\ell_N|L_1\cdots L_{N-3}}$ defined in Appendix 4, Section A4.2.5. The expansion coefficients are then exactly the N-spectra. More details about the trispectrum and how to measure it can be found in Regan *et al.* (2010).

6.4.2 Minkowski Functionals

For a Gaussian field, the 1- and 2-point statistics contain the full information, since all N-point statistics are determined by it. For a non-Gaussian field, measuring the N-point functions might not always be the best way to characterize its statistical properties. Imagine a CMB sky where the temperature fluctuations are correlated along arbitrarily oriented line segments. The N-point functions may still be isotropic. Such a lower dimensional arrangement, which certainly is non-Gaussian, cannot be discovered by simply considering N-point functions. In this and other cases different statistical tools that directly consider the map $\Delta T(\mathbf{n})$ on the sphere may be more adapted and informative than N-point functions. Here we briefly discuss the Minkowski functionals. They have first been used for the COBE data of CMB anisotropies by Schmalzing and Gorski (1998). Other interesting statistics are, for example, the line correlation function; see Obreschkow *et al.* (2013) or wavelets; see, for example, Vielva *et al.* (2004).

Definition 6.1 Minkowski functionals
Let $K \subset \mathbb{R}^d$ be an open subset with a smooth boundary ∂K and $\kappa_1, \cdots \kappa_{d-1}$ the $d-1$ principal curvatures of ∂K (i.e., the eigenvalues of the extrinsic curvature). The polynomial

$$\prod_{j=1}^{d-1}(x-\kappa_j) = \sum_{j=1}^{d} x^{d-j} M_j(\kappa, \cdots, \kappa_{d-1}) \tag{6.86}$$

defines d symmetrical multilinear functions in the κ_j. Each term in M_j has $j-1$ different factors κ_i. The Minkowski functionals of K are defined as follows:

$$V_0(K) = \int_K dv \tag{6.87}$$

$$V_j(K) = \frac{\Gamma(j/2)}{2\pi^{j/2}\binom{d}{j}} \int_{\partial K} ds M_j(\kappa, \cdots \kappa_{d-1}) \quad \text{for } 1 \le j \le d. \tag{6.88}$$

Here dv is the volume element on K and ds is the surface element on ∂K. Clearly, V_j has dimension $d-j$. Note also that $2\pi^{j/2}/\Gamma(j/2)$ is the surface of the $j-1$ dimensional sphere. The factor $\binom{d}{j}$ is the binomial coefficient.

We are interested in Minkowski functionals not on $\mathbb{R}^d$ but on the sphere $\mathbb{S}^2$. The main difference is that in Euclidean space V_d is simply the Euler characteristic while in spaces with curvature this is not the case. But the Minkowski functionals are still well defined. One can show that the "morphological properties" of a smooth subset are determined by its Minkowski functionals [see Schneider (1993) for more details about Minkowski functionals]. On the sphere we have

$$V_0(K) = \int_K d\Omega, \quad V_1(K) = \frac{1}{4}\int_{\partial K} ds, \quad V_2(K) = \frac{1}{2\pi}\int_{\partial K} \kappa(s) ds, \tag{6.89}$$

where ds is the line element along ∂K and κ denotes the geodesic curvature of the curve ∂K.

To characterize the temperature fluctuations in the sky one now considers the Minkowski functionals of excursion sets defined by

$$K(\nu) = \left\{ \mathbf{n} \in \mathbb{S}^2 \left| \frac{\Delta T}{T}(\mathbf{n}) > \nu \right. \right\}. \tag{6.90}$$

If fluctuations are very elongated, for example, we shall find that for sufficiently large values of ν, the Minkowski functional $V_0(\nu) \equiv V_0(K(\nu))$ is much smaller than $V_1(\nu)^2$. The value of $V_2(\nu)$ measures how strongly isotemperature lines are curved. Considering an arbitrary scalar field u on the sphere, one finds that

$$V_j(\nu) = \int_{\mathbb{S}^2} d\Omega A_j(\mathbf{n}) \quad \text{with} \tag{6.91}$$

$$A_0(\mathbf{n}) = \Theta(u(\mathbf{n}) - \nu) \tag{6.92}$$

$$A_1(\mathbf{n}) = \frac{1}{4}\delta(u(\mathbf{n}) - \nu)\sqrt{(\nabla u)^2} \tag{6.93}$$

$$A_2(\mathbf{n}) = \frac{1}{2\pi}\delta(u(\mathbf{n}) - \nu)\frac{\sum_{ij=1}^{2}(-1)^{j+i+1}\nabla_i u \nabla_j u \nabla_i \nabla_j u}{(\nabla u)^2}. \tag{6.94}$$

Here Θ denotes the Heaviside function. The expression for A_0 is evident. Equations 6.93 and (6.94) are derived in Exercise 6.7, which is solved in Appendix 11 which is found online.

For a given sky map we can in principle measure the $V_j(\nu)$ for different thresholds ν. For a Gaussian field their expectation values can be computed explicitly; see, for example, Tomita (1986). For Gaussian temperature fluctuations with vanishing mean one finds[3]

$$V_0(\nu) = 2\pi\left(1 - \mathrm{erf}\left(\frac{\nu}{\sqrt{2\sigma}}\right)\right) \tag{6.95}$$

$$V_1(\nu) = \frac{\pi\sqrt{\tau}}{2\sqrt{\sigma}}\exp\left(-\frac{\nu^2}{2\sigma}\right) \tag{6.96}$$

$$V_2(\nu) = \frac{2\tau\nu}{\sqrt{\pi\sigma^3}}\exp\left(-\frac{\nu^2}{2\sigma}\right), \tag{6.97}$$

where

$$\sigma = \left\langle\left(\frac{\Delta T}{T}\right)^2\right\rangle = \sum_\ell (2\ell+1)C_\ell \quad \text{and} \tag{6.98}$$

$$\tau = \frac{1}{2}\left\langle\left(\nabla\frac{\Delta T}{T}\right)^2\right\rangle = \frac{1}{2}\sum_\ell (2\ell+1)\ell(\ell+1)C_\ell. \tag{6.99}$$

The Planck temperature data has been analyzed using Minkowski functionals [see Section 10 of Planck Coll. XVII (2016)] and it is entirely compatible with a purely Gaussian field.

[3] Here $\mathrm{erf}(x) = \frac{2}{\sqrt{\pi}}\int_o^x dy \exp(-y^2)$ is the Gaussian error function.

Exercises

(The exercises marked with an asterisk are solved in Appendix 11 which is not in this printed book but can be found online.)

6.1 **Symmetries of the bispectrum***

Assuming statistical homogeneity and isotropy, show that the bispectrum B_X, defined as the Fourier transform of the 3-point function $\langle X(\mathbf{x}_1)X(\mathbf{x}_2)X(\mathbf{x}_3)\rangle$, is of the form (6.2), where $B(k_1,k_2,k_3)$ is symmetric in its arguments.

6.2 **The reduced CMB bispectrum***

Show that the terms $b^{(2)}_{\ell_1\ell_2\ell_3}$ are related to the reduced bispectrum $b_{\ell_1\ell_2\ell_3}$ via

$$b^{(2)}_{\ell_1\ell_2\ell_3} = \sum_{L_1L_2L_3} Q^{L_1L_2L_3}_{\ell_1\ell_2\ell_3} b_{L_1L_2L_3} \tag{6.100}$$

where

$$Q^{L_1L_2L_3}_{\ell_1\ell_2\ell_3} = \sum_{m_i;M_i} \begin{pmatrix} \ell_1 & \ell_2 & \ell_3 \\ m_1 & m_2 & m_3 \end{pmatrix} \prod_{i=1}^{3} \sqrt{\frac{2\ell_i+1}{4\pi}}$$
$$\times \begin{pmatrix} \ell_i & L_i & L_{[i-1]} \\ 0 & 0 & 0 \end{pmatrix} \begin{pmatrix} \ell_i & L_i & L_{[i-1]} \\ m_i & M_i & M_{[i-1]} \end{pmatrix}. \tag{6.101}$$

Here $[i-1] = i-1$ for $i = 2,3$ and $[i-1] = 3$ for $i = 1$.

Hint: Use the Clebsch–Gordan decomposition and the Wigner 3J symbols given in Appendix 4, Section A4.2.3.

6.3 **Integrals of spherical harmonics**

Using the derivation in Section 6.3, show that for an arbitrary function F such that the integral below exists we have

$$\int d^3k_1 d^3k_2 d^3k_3 F(k_1,k_2,k_3) \left(\prod_{i=1}^{3} Y_{\ell_i m_i}(\widehat{\mathbf{k}}_i)\right) \delta(\mathbf{k}_1+\mathbf{k}_2+\mathbf{k}_3)$$
$$= A\,\mathcal{G}^{\ell_1\ell_2\ell_3}_{m_1m_2m_3} \int_0^\infty x^2 dx dk_1 dk_2 dk_3 F(k_1,k_2,k_3) \left(\prod_{i=1}^{3} k_i^2 j_{\ell_i}(xk_i)\right). \tag{6.102}$$

Determine the ℓ_i dependent prefactor A.

Hint: As earlier this chapter, use the Fourier representation of the Dirac-delta and then write the exponential as a sum of products of spherical Bessel functions and spherical harmonics.

6.4 **Equilateral bispectrum**
Calculate the CMB bispectrum in the low ℓ regime for the equilateral shape,

$$S^{(\mathrm{equi})}(k_1,k_2,k_3) = \frac{(k_1+k_2-k_3)(k_2+k_3-k_1)(k_3+k_1-k_2)}{k_1k_2k_3}. \tag{6.103}$$

6.5 **The trispectrum**
Verify in detail Eq. (6.84) starting from Eq. (6.79).

6.6 **The local trispectrum**
Compute B_{4c} and the CMB trispectrum $t^{\ell_1\cdots\ell_4|L}$ for small ℓ_i in the local case, where

$$\zeta(\mathbf{x}) = \zeta_G(\mathbf{x}) + \frac{3}{5}f_{\mathrm{nl}}\left(\zeta_G^2(\mathbf{x}) - \langle\zeta_G^2\rangle\right). \tag{6.104}$$

Here $\zeta_G(\mathbf{x})$ is assumed to be a Gaussian field. Small ℓ_i means that you may use the transfer function

$$\mathcal{T}(k,\ell) = \frac{1}{5}j_\ell(k(t_o - t_{\mathrm{dec}})).$$

6.7 **The Minkowski functionals***
Derive in detail Eqs. (6.93) and (6.94) from the definitions of $V_1(\nu)$ and $V_2(\nu)$.

7
Lensing and the CMB

In this chapter we discuss the most important second-order effect on CMB anisotropies and polarization. Patches of higher or lower CMB temperature are modified and polarization patterns are distorted when they propagate through an inhomogeneous gravitational field. The content of this chapter is strongly inspired by the excellent review by Lewis and Challinor (2006) on the subject.

7.1 An Introduction to Lensing

On their path from the last scattering surface into our antennas, the CMB photons are deflected by the perturbed gravitational field. If the CMB were perfectly isotropic, the net effect of this deflection would vanish, since, by the conservation of photon number, as many photons would be deflected out of a small solid angle as into it. On the other hand, if there is no perturbation in the gravitational field, the latter is perfectly isotropic and the effect also vanishes. Hence, gravitational lensing of the CMB is a second-order effect and we have not discussed it within linear perturbation theory.

To estimate the effect let us consider the CMB temperature in a point $\mathbf{n}$ in the sky, $T(\mathbf{n})$. If the direction $\mathbf{n}$ is deflected by a small angle $\boldsymbol{\alpha}$, we receive the temperature $T(\mathbf{n})$ from the direction $\mathbf{n}' = \mathbf{n} + \boldsymbol{\alpha}$. Note that, since $\boldsymbol{\alpha}$ is a vector normal to $\mathbf{n}$, also $\mathbf{n}'$ is a unit vector to first order in $\boldsymbol{\alpha}$. To lowest order, this induces a change $\delta T = \boldsymbol{\alpha} \cdot \nabla_{\mathbf{n}} T(\mathbf{n})$; since the angular dependence of the temperature as well as $\boldsymbol{\alpha}$ are first-order quantities this effect is second order.

The deflection angle from a gravitational potential Ψ of an isolated mass distribution is roughly given by $4\Psi_m$, where Ψ_m is the maximum of the gravitational potential along the photon trajectory (see Exercise 7.1). The mean amplitude of the cosmic gravitational potential is about $\sqrt{\langle \Psi^2 \rangle} \simeq 2 \times 10^{-5}$ so that we have $\langle |\boldsymbol{\alpha}| \rangle \sim 10^{-4}$. The typical size of a primordial "potential well" is difficult to estimate, since the potential is scale invariant, but let us approximate it by the horizon size at equality, which is roughly 300 Mpc (comoving). The distance from the last scattering surface to us is about 14 000 Mpc, so that a light ray passes through of the order of 50 such potential wells. Assuming the direction of deviation to be random, this yields a total deviation of about $\sqrt{50}|\alpha| \sim 7 \times 10^{-4} \simeq 2$ arc minutes. This corresponds to a deviation of order unity for a patch with an angular size of 2 arc minutes, that is, for $\ell \sim 4000$. In fact, primary CMB anisotropies on these scales are severely damped by Silk damping so that lensing and other secondary effects such as the Sunyaev–Zel'dovich (SZ) effect already dominate on scales larger than $\ell \sim 3000$. In the acoustic peak region, which corresponds to about 1°, we expect lensing to change the size of the patches by roughly half a percent on average. Some patches are enlarged while others are reduced in size. In the C_ℓ spectrum this leads to a broadening of the peak. The peak position is somewhat less well defined.

Requiring better than 1% accuracy, we have to take into account lensing for $\ell \gtrsim 400$.

7.1.1 The Deflection Angle

We first want to compute the deflection of a light ray in a perturbed FL universe. We consider only scalar perturbations so that the metric is of the form

$$ds^2 = a^2(t)\left(-(1+2\Psi)\,dt^2 + (1-2\Phi)\gamma_{ij}\,dx^i\,dx^j\right), \tag{7.1}$$

with [see Eqs. (1.9) and (1.12)]

$$\gamma_{ij}\,dx^i\,dx^j = dr^2 + \chi^2(r)(d\vartheta^2 + \sin^2\vartheta\,d\varphi^2). \tag{7.2}$$

Since we are interested only in deflection, we may also consider the conformally related metric

$$d\tilde{s}^2 = -(1+4\Psi_W)\,dt^2 + \gamma_{ij}\,dx^i\,dx^j, \tag{7.3}$$

where

$$\Psi_W = \frac{1}{2}(\Psi+\Phi), \tag{7.4}$$

is the Weyl potential. According to Eq. (A3.21) the Weyl tensor from scalar perturbations is given by $(\nabla_i\nabla_j - \frac{1}{3}\gamma_{ij}\Delta)\Psi_W$. Without loss of generality we set the observer position to $\mathbf{x} = 0$ and we consider a photon with an unperturbed trajectory radially toward the observer, $(\bar{x}^\mu) = (s\bar{n}^\mu) = s(1, \mathbf{n})$, where $\mathbf{n}$ is the radially inward photon direction fixed by two angles, ϑ_0 and φ_0, and s is an affine parameter. With our choice for s we have $dt/ds = dx^0/ds = 1$; hence $s = t - t_0$ for the unperturbed trajectory. The perturbed photon velocity is given by $(n^\mu) = (1+\delta n^0(s), \mathbf{n}+\delta\mathbf{n}(s))$. The Christoffel symbols for $d\tilde{s}^2$ to first order in Ψ_W are easily determined as

$$\tilde{\Gamma}^0_{00} = 2\partial_t \Psi_W, \qquad \tilde{\Gamma}^0_{0i} = \tilde{\Gamma}^0_{i0} = 2\partial_i \Psi_W, \quad \tilde{\Gamma}^0_{ij} = 0,$$
$$\tilde{\Gamma}^i_{00} = 2\gamma^{ij}\partial_j \Psi_W, \quad \tilde{\Gamma}^i_{j0} = 0, \qquad \tilde{\Gamma}^i_{jm} = \bar{\Gamma}^i_{jm},$$

where $\bar{\Gamma}^i_{jm}$ are the Christoffel symbols of the unperturbed three-dimensional metric γ_{ij}.

With this, the geodesic equation of motion, $\frac{d^2x^\mu}{ds^2} + \Gamma^\mu_{\alpha\beta}\frac{dx^\alpha}{ds}\frac{dx^\beta}{ds} = 0$, leads to the following equations of motion for the perturbation of the photon 4-velocity δn^μ:

$$\frac{d}{ds}\delta n^0 = -2\partial_t \Psi_W - 4n^i\partial_i \Psi_W = -2\frac{d}{ds}\Psi_W - 2n^i\partial_i \Psi_W, \tag{7.5}$$
$$\frac{d}{ds}\delta n^i = -2\gamma^{ij}\partial_j \Psi_W - 2\delta n^j n^m \bar{\Gamma}^i_{jm}. \tag{7.6}$$

Here s denotes the affine parameter along the photon trajectory and we made use of $(d/ds)\Psi_W = \partial_t \Psi_W + n^i\partial_i \Psi_W$. For the rest of this section we set $\dot{} = d/ds$.

The deflection is given by the ϑ- and φ-components of $\delta\mathbf{n} = \epsilon\mathbf{n} + \dot{\vartheta}\partial_\vartheta + \dot{\varphi}\partial_\varphi$. In spherical coordinates (r, ϑ, φ) we have $\mathbf{n} = (-1, 0, 0) \equiv -\partial_r$. The unperturbed Christoffels in Eq. (7.6) are given by

$$\bar{\Gamma}^\vartheta_{r\vartheta} = \bar{\Gamma}^\varphi_{r\varphi} = \frac{\partial_r \chi}{\chi}$$

and all other $\bar{\Gamma}^j_{ri} = 0$. With this we obtain the following equations of motion for $\dot{\vartheta}$ and $\dot{\varphi}$:

$$\ddot{\vartheta} = \frac{-2}{\chi^2}\partial_\vartheta \Psi_W + 2\frac{\partial_r \chi}{\chi}\dot{\vartheta}, \tag{7.7}$$
$$\ddot{\varphi} = \frac{-2}{\chi^2 \sin^2\vartheta}\partial_\varphi \Psi_W + 2\frac{\partial_r \chi}{\chi}\dot{\varphi}, \quad \text{so that} \tag{7.8}$$
$$-\frac{d}{ds}\left(\chi^2\dot{\vartheta}\right) = 2\partial_\vartheta \Psi_W, \tag{7.9}$$
$$-\frac{d}{ds}\left(\chi^2\dot{\varphi}\right) = \frac{2}{\sin^2\vartheta}\partial_\varphi \Psi_W. \tag{7.10}$$

For the last two lines we have used the fact that to lowest order $\dot{\chi} = -\partial_r \chi$ for radial geodesics. Integrating these equations and using the fact that to lowest order $ds = dt$ we obtain at first order in the deflection angle

$$\chi^2(t_0 - t)\dot{\vartheta}(t) = 2\int_{t_0}^{t} dt'\, \partial_\vartheta \Psi_W(t', t_0 - t', \vartheta_0, \varphi_0), \tag{7.11}$$
$$\chi^2(t_0 - t)\dot{\varphi}(t) = 2\int_{t_0}^{t} \frac{dt'}{\sin^2\vartheta_0}\, \partial_\varphi \Psi_W(t', t_0 - t', \vartheta_0, \varphi_0). \tag{7.12}$$

Here t_0 is the time at which we receive the photon at $\mathbf{x} = 0$ and the constant of integration has been fixed by requiring $\chi^2|_{t=t_0} = 0$. Integrating this equation once more leads to

$$\vartheta(t_*) = \vartheta_0 - 2\int_{t_*}^{t_0} dt\, \frac{\chi(t-t_*)\partial_\vartheta \Psi_W(t, t_0 - t, \vartheta_0, \varphi_0)}{\chi(t_0 - t_*)\chi(t_0 - t)}, \tag{7.13}$$

$$\varphi(t_*) = \varphi_0 - \frac{2}{\sin^2\vartheta_0}\int_{t_*}^{t_0} dt\, \frac{\chi(t-t_*)\partial_\varphi \Psi_W(t, t_0 - t, \vartheta_0, \varphi_0)}{\chi(t_0 - t_*)\chi(t_0 - t)}. \tag{7.14}$$

The easiest way to see that these are the integrals of Eqs. (7.11) and (7.12) is to take the derivative of Eqs. (7.13) and (7.14) with respect to t_* and use $\chi'(t-t_*)\chi(t_0-t_*) - \chi(t-t_*)\chi'(t_0-t_*) = \chi(t_0-t)$ for all three functions χ given in Eq. (1.12).

The deflection angle $\boldsymbol{\alpha} = (\vartheta - \vartheta_0, \sin\vartheta_0(\varphi - \varphi_0))$ is therefore given by

$$\boldsymbol{\alpha} = -2\int_{t_*}^{t_0} dt\, \frac{\chi(t-t_*)}{\chi(t_0 - t_*)\chi(t_0 - t)} \nabla_\perp \Psi_W(t, t_0 - t, \vartheta_0, \varphi_0), \tag{7.15}$$

where $\nabla_\perp = (\partial_\vartheta, (\sin\vartheta)^{-1}\partial_\varphi)$ is the gradient on the sphere and Eq. (7.15) gives the components of the deflection angle $\boldsymbol{\alpha}$ in this basis. The application relating the observed direction (ϑ_0, φ_0) to the direction of emission (ϑ_*, φ_*),

$$(\vartheta_0, \varphi_0) \to (\vartheta_*, \varphi_*) = (\vartheta_0, \varphi_0) + \boldsymbol{\alpha}(\vartheta_0, \varphi_0), \tag{7.16}$$

is called the lens map. When investigating lensing of the CMB, we choose the time of emission to be the decoupling time, $t_* = t_{\rm dec}$.

Einstein's equation relates Ψ_W to the energy–momentum tensor. Equations (2.105) and (2.107) together with the definition (7.4) yield

$$(\Delta + 3K)\Psi_W = 4\pi G a^2 \rho \left[D + w(1 + 3K\Delta^{-1})\Pi\right]. \tag{7.17}$$

Using the canonical basis $\mathbf{e}_1 \equiv \mathbf{e}_\vartheta = \partial_\vartheta$ and $\mathbf{e}_2 \equiv \mathbf{e}_\varphi = (\sin\vartheta)^{-1}\partial_\varphi$ we introduce the gradient of the lens map,

$$A_{ab}(\vartheta, \varphi) = \delta_{ab} - 2\int_{t_*}^{t_0} dt\, \frac{\chi(t-t_*)\nabla_a\nabla_b \Psi_W(t, t_0 - t, \vartheta, \varphi)}{\chi(t_0 - t_*)\chi(t_0 - t)}, \tag{7.18}$$

$$\equiv \begin{pmatrix} 1-\kappa-\gamma_1 & -\gamma_2 \\ -\gamma_2 & 1-\kappa+\gamma_1 \end{pmatrix}. \tag{7.19}$$

The matrix A describes the deformation of a bundle of light rays from direction (ϑ, φ). Its trace, ${\rm tr}A = 2(1-\kappa)$, is a measure for the amount of focusing while its traceless part is often represented as the complex number $\gamma = \gamma_1 + i\gamma_2$ represents the shear. As a double gradient of a scalar, A_{ab} is symmetric. To first order in perturbation theory, lensing from scalar perturbations does not induce rotation.

The flux from a source $\iota(\mathbf{n}')$ becomes, after passing through the lensing potential, $\iota(\mathbf{n}) = \det(A^{-1})\iota(\mathbf{n}')$. With $\det(A^{-1}) = \left[(1-\kappa)^2 - |\gamma|^2\right]^{-1} \simeq 1 + 2\kappa$, we obtain the magnification $\mu = \det(A^{-1})$. To first order in the gravitational potential,

$$\mu = 1 + 2\kappa. \tag{7.20}$$

Focusing not only increases the number of photons that reach us from a source (or a patch in the CMB sky), but it also enhances the solid angle under which we see this patch exactly by the factor det A, so that the number of photons per unit solid angle is conserved. Lensing conserves surface brightness. Photons are neither absorbed nor created by lensing; they are just deflected and redshifted.

The shear is very important for the weak lensing of galaxy surveys, as it renders spherical sources elliptical. For CMB lensing both the focusing κ and the shear γ are relevant.

7.2 The Lensing Power Spectrum

Let us introduce the lensing potential

$$\psi(\mathbf{n}) = -2\int_{t_*}^{t_0} dt \frac{\chi(t-t_*)}{\chi(t_0-t_*)\chi(t_0-t)} \Psi_W(t, \mathbf{n}(t_0-t)). \tag{7.21}$$

This is a function on the sphere and the deflection angle is its gradient. The deflection potential seems to be divergent because $\chi(t_0 - t) \to 0$ for $t \to t_0$. But this divergence affects only the monopole term, which we may set to zero since it does not affect the lens map, which is given by

$$A_{ab}(\vartheta,\varphi) = \delta_{ab} + \nabla_a\nabla_b\psi \tag{7.22}$$

$$\kappa = -\frac{1}{2}\Delta_\Omega\psi \tag{7.23}$$

$$\gamma_1 = \frac{1}{2}(\nabla_1^2 - \nabla_2^2)\psi \tag{7.24}$$

$$\gamma_2 = -\nabla_1\nabla_2\psi. \tag{7.25}$$

Here Δ_Ω denotes the Laplacian on the sphere. Note that while κ is a scalar, $\gamma_\pm = \gamma_1 \pm i\gamma_2$ is a spin-2 tensor on the sphere with $s = \pm 2$; see Exercise 7.2.

We consider the CMB as a single source at fixed $t_* = t_{\rm dec}$. We expand the lensing potential in spherical harmonics,

$$\psi(\mathbf{n}) = \sum_{\ell m} \psi_{\ell m} Y_{\ell m}(\mathbf{n}), \tag{7.26}$$

$$\langle \psi_{\ell m}\psi^*_{\ell' m'}\rangle \equiv \delta_{\ell\ell'}\delta_{mm'}C_\ell^\psi. \tag{7.27}$$

The expectation values C_ℓ^ψ are the lensing power spectrum, and the Kronecker deltas are a consequence of statistical isotropy like for the CMB temperature and polarization spectra. The same manipulations as in Chapter 2, Eq. (2.253), now give the lensing correlation function in terms of the power spectrum,

$$\langle \psi(\mathbf{n})\psi(\mathbf{n}')\rangle = \frac{1}{4\pi}\sum_\ell (2\ell+1)C_\ell^\psi P_\ell(\mathbf{n}\cdot\mathbf{n}'). \tag{7.28}$$

We want to relate the lensing power spectrum to the primordial power spectrum of the Weyl potential. For simplicity we restrict ourselves to the case $K = 0$, with $\chi_0(r) = r$. In this case, the power spectrum of the Weyl potential is given by the Fourier transform,

$$\Psi_W(t,\mathbf{x}) = \frac{1}{(2\pi)^3}\int d^3k\, \Psi_W(t,\mathbf{k})\, e^{-i\mathbf{k}\cdot\mathbf{x}}, \tag{7.29}$$

$$\langle \Psi_W(t,\mathbf{k})\Psi_W^*(t',\mathbf{k}')\rangle = (2\pi)^3 T(k,t)T^*(k,t')P_\Psi(k)\,\delta(\mathbf{k}-\mathbf{k}'). \tag{7.30}$$

Here we have introduced the primordial power spectrum P_Ψ and the transfer function $T(k,t)$. For a fixed wave number k the transfer function is the solution of the evolution equation for Ψ with initial condition $T(k,t) \to 1$ for $kt \to 0$. The transfer function for a matter/radiation universe, neglecting the cosmological constant, is constant during the matter era and given by Eq. (2.222), setting $A = \frac{10}{3}$.

For simplicity, we neglect the difference between Ψ and Ψ_W, which is given by the anisotropic stresses and never contributes more than a few percent. This is easily corrected for in a numerical treatment.

Inserting Eqs. (7.30) and (7.21) in Eq. (7.28) and expanding

$$e^{i\mathbf{k}\cdot\mathbf{n}(t_0-t)} = 4\pi\sum_{\ell m} i^\ell j_\ell(k(t_0-t))Y_{\ell m}(\mathbf{n})Y_{\ell m}^*(\hat{\mathbf{k}}),$$

we obtain

$$C_\ell^\psi = \frac{8}{\pi}\int_0^\infty dk\, k^2 P_\Psi(k)\left|\int_{t_*}^{t_0} dt\, T(k,t)\, j_\ell(k(t_0-t))\frac{t-t_*}{(t_0-t_*)(t_0-t)}\right|^2. \tag{7.31}$$

The relevant quantity for us is the spectrum of the deflection angle $\boldsymbol{\alpha}(\mathbf{n}) = \nabla_\perp\psi(\mathbf{n})$. The correlation function of ψ depends only on the angle between $\mathbf{n}$ and $\mathbf{n}'$. It is invariant under simultaneous infinitesimal variations $\mathbf{n} \to \mathbf{n}+\boldsymbol{\epsilon}$ and $\mathbf{n}' \to \mathbf{n}'+\boldsymbol{\epsilon}$ so that $\langle\psi(\mathbf{n})\psi(\mathbf{n}'+\boldsymbol{\epsilon})\rangle = \langle\psi(\mathbf{n}-\boldsymbol{\epsilon})\psi(\mathbf{n}')\rangle$. Therefore $\langle\nabla_\perp\psi(\mathbf{n})\, \nabla_\perp\psi(\mathbf{n}')\rangle = -\langle\Delta\psi(\mathbf{n})\psi(\mathbf{n}')\rangle$. Since $\Delta Y_{\ell m} = -\ell(\ell+1)Y_{\ell m}$, the power spectrum of the deflection angle is simply given by $\ell(\ell+1)C_\ell^\psi$. This power spectrum, multiplied by the usual factor $\ell(\ell+1)/2\pi$, is shown in Fig. 7.1.

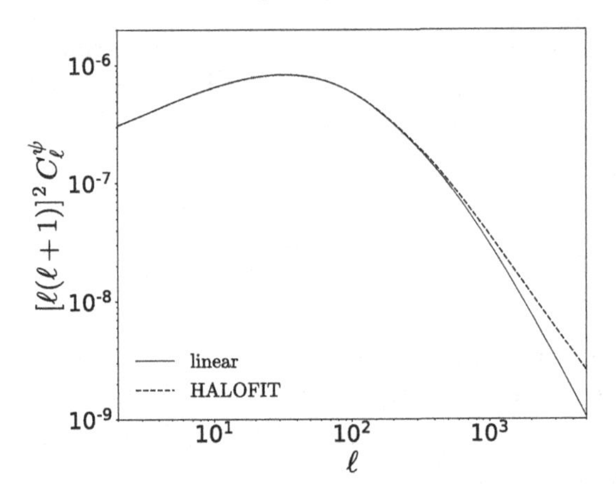

Fig. 7.1 The lensing power spectrum for a ΛCDM concordance model. The solid line is the linear approximation while in the dashed line nonlinear corrections are included using an analytic approximation to the nonlinear matter power spectrum called "halofit."

7.2.1 The Limber Approximation

We can also consider the correlation for lensing potentials at different redshifts, z and z'. Denoting the corresponding comoving distances by $r(z)$ and $r(z')$, transforming the integrals w.r.t. t to integrals w.r.t. r and exchanging the k and r integrals we can write the analog of Eq. (7.31) for two different redshifts as

$$C_\ell^\psi(z,z') = \frac{8}{\pi}\int_0^{r(z)} \frac{dr(r(z)-r)}{r(z)r}\int_0^{r(z')}\frac{dr'(r(z')-r')}{r(z')r'} \times \int_0^\infty dkk^2 T(k,t_0-r)T(k,t_0-r')j_\ell(kr)j_\ell(kr')P_\Psi(k). \quad (7.32)$$

Here we have dropped the complex-conjugation of the second transfer function, since the transfer function of the Bardeen potential is in fact real. This equation can be simplified using the so called Limber approximation; see Eq. (A4.152),

$$\frac{2}{\pi}\int dkk^2 f(k)j_\ell(kr)j_\ell(kr') \simeq \frac{\delta(r-r')}{r^2} f\left(\frac{\ell+1/2}{r}\right). \quad (7.33)$$

For a rather slowly varying function f (it is exact if f is constant) this is a very good approximation for large enough values of ℓ. Here ℓ must be sufficiently large so that $f(k)$ has no significant peak at $k > \ell/r$. For the Bardeen potential, $T^2(k,t)P_\Psi(k)$, this turns out to be a very good approximation for $\ell \gtrsim 20$.

Performing the integral over k in (7.32) with the Limber approximation we obtain

$$C_\ell^\psi(z,z') \simeq 4\int_0^{r_*} dr \frac{(r(z)-r)(r(z')-r)}{r(z)r(z')r^4} T^2\left(\frac{\ell+1/2}{r}, t_0 - r\right) P_\Psi\left(\frac{\ell+1/2}{r}\right), \tag{7.34}$$

where $r_* = \min\{r(z), r(z')\}$.

7.3 Lensing of the CMB Temperature Anisotropies

We now want to determine how lensing affects the CMB. On small scales, the lensing potential is nearly completely uncorrelated with the CMB anisotropies, since most of the lensing potential comes from low redshifts. At $\ell > 60$ the correlation of the lensing and CMB spectra is less than 10% of its maximum value and at $\ell \gtrsim 600$ it drops below 0.1%. Most of the lensing power was generated relatively recently, at $z \lesssim 20$, and it therefore does not correlate with the CMB anisotropies that were generated at $z_{\rm dec} \simeq 1100$. The lensing signal correlates significantly only with the late integrated Sachs–Wolfe effect, which is relevant on very large scales. But the latter has very little structure, so that lensing on large scales is negligible.

Since we are mostly interested in small scales, we approximate the sky by a flat plane, as in Section 5.2. The temperature anisotropy is given by Eq. (5.28). The correlation function between two points $\mathbf{x}$ and $\mathbf{x}'$ in the sky,

$$\langle \mathcal{M}(\mathbf{x})\mathcal{M}(\mathbf{x}')\rangle = \xi(|\mathbf{x}-\mathbf{x}'|) = \xi(|\mathbf{r}|), \quad \mathbf{r} = \mathbf{x}-\mathbf{x}',$$

is related to the power spectrum by Eq. (5.107),

$$\langle \mathcal{M}(\boldsymbol{\ell})\mathcal{M}^*(\boldsymbol{\ell}')\rangle = \delta^2(\boldsymbol{\ell}-\boldsymbol{\ell}')C_\ell^{(\mathcal{M})}$$
$$C_\ell^{(\mathcal{M})} = 2\pi\int_0^\infty dr\, r J_0(r\ell)\xi(|\mathbf{r}|).$$

The inverse relation is given in Eq. (5.104).

The same equations also relate the lensing potential correlation function to its power spectrum, C_ℓ^ψ.

7.3.1 Approximation for Small Deflection Angles

We now expand the lensed temperature fluctuation in the deflection angle $\boldsymbol{\alpha} = \nabla\psi$,

$$\begin{aligned}\widetilde{\mathcal{M}}(\mathbf{x}) &= \mathcal{M}(\mathbf{x}+\nabla\psi)\\ &\simeq \mathcal{M}(\mathbf{x}) + \nabla^a\psi(\mathbf{x})\nabla_a\mathcal{M}(\mathbf{x}) + \frac{1}{2}\nabla^a\psi(\mathbf{x})\nabla^b\psi(\mathbf{x})\nabla_b\nabla_a\mathcal{M}(\mathbf{x}) + \cdots.\end{aligned}$$

This is a good approximation only if the deflection angle is much smaller than the scales of interest to us. If not, we cannot truncate this expansion at second order.

Using

$$\nabla_a\psi(\mathbf{x}) = \frac{-i}{2\pi}\int d^2\ell\,\ell_a\psi(\ell)\,e^{-i\ell\cdot x}, \qquad \text{and}$$

$$\nabla_a\mathcal{M}(\mathbf{x}) = \frac{-i}{2\pi}\int d^2\ell\,\ell_a\mathcal{M}(\ell)\,e^{-i\ell\cdot x},$$

we can obtain the Fourier components for $\widetilde{\mathcal{M}}(\ell)$. For this we use that the Fourier transform of a product is equal to the convolution of the Fourier transforms. For example, for the functions $\nabla_a\psi$ and $\nabla_a\mathcal{M}$ we obtain

$$\begin{aligned}
&\frac{1}{2\pi}\int d^2x\,\nabla^a\psi(\mathbf{x})\nabla_a\mathcal{M}(\mathbf{x})e^{i\mathbf{x}\cdot\ell} \\
&\quad= \frac{-1}{(2\pi)^3}\int d^2x\int d^2\ell_1\int d^2\ell_2\,(\ell_1\cdot\ell_2)\,e^{i\mathbf{x}\cdot(\ell-\ell_1-\ell_2)}\psi(\ell_1)\mathcal{M}(\ell_2) \\
&\quad= \frac{-1}{2\pi}\int d^2\ell_2\,((\ell-\ell_2)\cdot\ell_2)\,\psi(\ell-\ell_2)\mathcal{M}(\ell_2) \\
&\quad= \frac{-1}{2\pi}\,(\ell\psi\star\ell\mathcal{M})\,(\ell).
\end{aligned}$$

Here $\star$ indicates convolution and for the second equals sign we have used that

$$\frac{1}{(2\pi)^2}\int d^2x\,e^{i\mathbf{x}\cdot(\ell-\ell_1-\ell_2)} = \delta^2(\ell-\ell_1-\ell_2).$$

Using the above for the second term in the Fourier transform of $\widetilde{\mathcal{M}}$ and the equivalent identity for the third term, we find

$$\begin{aligned}
\widetilde{\mathcal{M}}(\ell) \simeq \mathcal{M}(\ell) &- \int\frac{d^2\ell'}{2\pi}\ell'\cdot(\ell-\ell')\psi(\ell-\ell')\mathcal{M}(\ell') \\
&- \frac{1}{2}\int\frac{d^2\ell_1}{2\pi}\int\frac{d^2\ell_2}{2\pi}\ell_1\cdot(\ell_1+\ell_2-\ell)\ell_1\cdot\ell_2\mathcal{M}(\ell_1)\psi(\ell_2)\psi^*(\ell-\ell_1-\ell_2).
\end{aligned} \tag{7.35}$$

To work out the lensed power spectrum we neglect correlations of $\mathcal{M}$ with ψ and use $\psi^*(\ell) = \psi(-\ell)$. Up to first order in the lensing power spectrum C_ℓ^ψ the lensed temperature anisotropy spectrum becomes

$$\tilde{C}_\ell \simeq C_\ell + \int\frac{d^2\ell'}{(2\pi)^2}\left[\ell'\cdot(\ell-\ell')\right]^2 C^\psi_{|\ell-\ell'|}C_{\ell'} - C_\ell\int\frac{d^2\ell'}{(2\pi)^2}(\ell'\cdot\ell)^2 C^\psi_{\ell'}. \tag{7.36}$$

Integrating the second term over the angle gives

$$\tilde{C}_\ell \simeq (1-\ell^2 R^\psi)C_\ell + \int \frac{d^2\ell'}{(2\pi)^2}\left[\boldsymbol{\ell}'\cdot(\boldsymbol{\ell}-\boldsymbol{\ell}')\right]^2 C^\psi_{|\boldsymbol{\ell}-\boldsymbol{\ell}'|}C_{\ell'}, \tag{7.37}$$

where we have introduced the mean square of the deflection angle

$$R^\psi \equiv \frac{1}{2}\langle\alpha^2\rangle = \frac{1}{4\pi}\int_0^\infty d\ell\,\ell^3 C^\psi_\ell. \tag{7.38}$$

The deflection power spectrum peaks at relatively large scales, $\ell \simeq 50$ (see Fig. 7.1) and the bulk of the contribution of the convolution integral in Eq. (7.37) comes from $\boldsymbol{\ell}\sim\boldsymbol{\ell}'$.

We first investigate the result for a scale-invariant CMB power spectrum, that is, $\ell^2 C_\ell =$ constant. For such a scale-invariant spectrum the above integral becomes

$$\begin{aligned}\tilde{C}^{\rm si}_\ell &\simeq (1-\ell^2R^\psi)C_\ell + \ell^2 C_\ell\int\frac{d^2\ell'}{(2\pi)^2}\frac{\left[\boldsymbol{\ell}'\cdot(\boldsymbol{\ell}-\boldsymbol{\ell}')\right]^2}{\ell'^2}C^\psi_{|\boldsymbol{\ell}-\boldsymbol{\ell}'|}\\ &= (1-\ell^2R^\psi)C_\ell + \ell^2 C_\ell\int\frac{d^2\ell_1}{(2\pi)^2}\frac{[(\boldsymbol{\ell}_1-\boldsymbol{\ell})\cdot\boldsymbol{\ell}_1]^2}{(\boldsymbol{\ell}-\boldsymbol{\ell}_1)^2}C^\psi_{\ell_1}\\ &= C_\ell\left[1+\frac{\ell^2}{4\pi}\int_\ell^\infty d\ell_1\,\ell_1 C^\psi_{\ell_1}\left(\ell_1^2-\ell^2\right)\right].\end{aligned} \tag{7.39}$$

For the last equals sign we have performed the angular integral that is derived in Ex. 7.3.

The integral in Eq. (7.39) is, in general, small, of $\mathcal{O}(10^{-3})$, which is significantly smaller than $\ell^2 R^\psi$. Hence the two terms of Eq. (7.37) cancel to a large extent for (nearly) scale-invariant spectra. Note also that the spectrum at a scale ℓ is affected by the lensing power from scales smaller than ℓ only. If the lensing power vanished above a certain value ℓ_0, a scale-invariant spectrum would not be modified by lensing for ℓ's larger than ℓ_0. A large-scale lensing mode magnifies and demagnifies small-scale fluctuations, which has no effect if the fluctuations are scale invarant. The effect of CMB lensing is important because of the acoustic oscillations and Silk damping on small scales that break scale invariance.

7.3.2 Arbitrary Deflection Angles

As we argued at the beginning of this chapter, for $\ell > 3000$, the deflection angle is comparable to the angular separations that contribute mainly to C_ℓ. A gradient expansion in the deflection angle is therefore no longer justified.

Let us first consider very small scales, $\ell \gg 3000$. On these scales the primordial anisotropies are nearly completely wiped out by Silk damping and are very small.

Even though the deflection angle is larger than the scale in consideration, we may approximate $\mathcal{M}(\mathbf{x}+\boldsymbol{\alpha}) \sim \mathcal{M}(\mathbf{x}) + \nabla\psi \cdot \nabla\mathcal{M}$. Setting the intrinsic $C_\ell = 0$, we obtain

$$\tilde{C}_\ell \simeq \int \frac{d^2\ell'}{(2\pi)^2} C_{\ell'} \left[\boldsymbol{\ell}' \cdot (\boldsymbol{\ell} - \boldsymbol{\ell}')\right]^2 C^\psi_{|\boldsymbol{\ell}-\boldsymbol{\ell}'|} \simeq C^\psi_\ell \int \frac{d^2\ell'}{(2\pi)^2} C_{\ell'} \left[\boldsymbol{\ell}' \cdot \boldsymbol{\ell}\right]^2$$
$$= \ell^2 C^\psi_\ell \int_0^\infty \frac{d\ell'}{4\pi} \ell'^3 C_{\ell'}. \tag{7.40}$$

On very small scales, where intrinsic anisotropies are negligible, the lensed anisotropy power spectrum is given by the power of the deflection angle on this scale multiplied with the integrated anisotropy power on all scales.

To determine the general formula for the lensed CMB anisotropy spectrum, we consider the correlation function. As before, we set the lensed temperature anisotropy equal to

$$\widetilde{\mathcal{M}}(\mathbf{x}) = \mathcal{M}(\mathbf{x} + \boldsymbol{\alpha}(\mathbf{x})),$$

where $\boldsymbol{\alpha} = \nabla\psi$ is the deflection angle. For $\mathbf{r} = \mathbf{x} - \mathbf{x}'$, $r = |\mathbf{r}|$ the lensed correlation function $\tilde{\xi}(r)$ is given by

$$\tilde{\xi}(r) = \langle \widetilde{\mathcal{M}}(\mathbf{x}) \widetilde{\mathcal{M}}(\mathbf{x}') \rangle = \langle \mathcal{M}(\mathbf{x}+\boldsymbol{\alpha}) \mathcal{M}(\mathbf{x}'+\boldsymbol{\alpha}') \rangle$$
$$= \int \frac{d^2\boldsymbol{\ell}}{2\pi} \int \frac{d^2\boldsymbol{\ell}'}{2\pi} \left\langle e^{-i\boldsymbol{\ell}\cdot(\mathbf{x}+\boldsymbol{\alpha})} e^{i\boldsymbol{\ell}'\cdot(\mathbf{x}'+\boldsymbol{\alpha}')} \right\rangle \left\langle \mathcal{M}(\boldsymbol{\ell}) \mathcal{M}(\boldsymbol{\ell}') \right\rangle$$
$$= \int \frac{d^2\boldsymbol{\ell}}{(2\pi)^2} C_\ell e^{-i\boldsymbol{\ell}\mathbf{r}} \left\langle e^{i\boldsymbol{\ell}\cdot(\boldsymbol{\alpha}'-\boldsymbol{\alpha})} \right\rangle. \tag{7.41}$$

Here we have used the fact that the CMB anisotropies and the deflection angle are virtually uncorrelated and we can therefore write the expectation value of the product $\left\langle e^{-i\boldsymbol{\ell}\cdot(\mathbf{x}+\boldsymbol{\alpha})} e^{i\boldsymbol{\ell}'\cdot(\mathbf{x}'+\boldsymbol{\alpha}')} \mathcal{M}(\boldsymbol{\ell}) \mathcal{M}(\boldsymbol{\ell}') \right\rangle$ as the product of the expectation values.

We assume that linear perturbations are Gaussian so that $\boldsymbol{\alpha}$ is a Gaussian field. Hence $\boldsymbol{\ell} \cdot (\boldsymbol{\alpha} - \boldsymbol{\alpha}')$ is a Gaussian random variable with mean $\langle \boldsymbol{\ell} \cdot (\boldsymbol{\alpha} - \boldsymbol{\alpha}') \rangle = 0$ and variance $\langle [\boldsymbol{\ell} \cdot (\boldsymbol{\alpha} - \boldsymbol{\alpha}')]^2 \rangle$. The expectation value of its exponential is given by (see Exercise 7.4)

$$\langle e^{i\boldsymbol{\ell}\cdot(\boldsymbol{\alpha}'-\boldsymbol{\alpha})} \rangle = \exp\left(-\frac{1}{2} \left\langle \left[\boldsymbol{\ell} \cdot (\boldsymbol{\alpha}' - \boldsymbol{\alpha})\right]^2 \right\rangle \right).$$

To calculate the variance of $\boldsymbol{\ell} \cdot (\boldsymbol{\alpha} - \boldsymbol{\alpha}') = \boldsymbol{\ell} \cdot (\boldsymbol{\alpha}(\mathbf{x}) - \boldsymbol{\alpha}(\mathbf{x}+\mathbf{r}))$ we define

$$A_{ij}(\mathbf{r}) = \langle \alpha_i(\mathbf{x}) \alpha_j(\mathbf{x}+\mathbf{r}) \rangle = \langle \nabla_i \psi(\mathbf{x}) \nabla_j \psi(\mathbf{x}+\mathbf{r}) \rangle = \int \frac{d^2\boldsymbol{\ell}}{(2\pi)^2} \ell_i \ell_j C^\psi_\ell e^{i\mathbf{r}\cdot\boldsymbol{\ell}}. \tag{7.42}$$

By statistical homogeneity and isotropy, for fixed $r = |\mathbf{r}|$, this symmetric matrix depends on directions only via $\mathbf{r}$. Therefore it is of the form

$$A_{ij}(\mathbf{r}) = \frac{1}{2}A_0(r)\delta_{ij} - A_2(r)\left[\hat{\mathbf{r}}_i\hat{\mathbf{r}}_j - \frac{1}{2}\delta_{ij}\right]. \tag{7.43}$$

To determine the functions A_0 and A_2 we first take the trace of A_{ij}. This yields

$$A_0(r) = \int_0^\infty \frac{d\ell\,\ell^3}{(2\pi)^2}C_\ell^\psi \int_0^{2\pi} d\phi e^{i\ell r\cos\phi} = \int_0^\infty \frac{d\ell\,\ell^3}{2\pi}C_\ell^\psi J_0(r\ell). \tag{7.44}$$

For the last equals sign we made use of Eq. (5.106). We then contract A_{ij} with $\hat{\mathbf{r}} = \mathbf{r}/r$,

$$\begin{aligned}
A_{ij}(\mathbf{r})\hat{\mathbf{r}}_i\hat{\mathbf{r}}_j = \frac{1}{2}(A_0(r) - A_2(r)) &= \int_0^\infty \frac{d\ell\,\ell^3}{(2\pi)^2}C_\ell^\psi \int_0^{2\pi} d\phi \cos^2\phi e^{i\ell r\cos\phi} \\
&= \int_0^\infty \frac{d\ell\,\ell^3}{(2\pi)^2}C_\ell^\psi \int_0^{2\pi} d\phi \frac{1}{2}[1+\cos(2\phi)]e^{i\ell r\cos\phi} \\
&= \frac{1}{2}\int_0^\infty \frac{d\ell\,\ell^3}{2\pi}C_\ell^\psi (J_0(r\ell) - J_2(r\ell)).
\end{aligned} \tag{7.45}$$

We have again used Eq. (5.106) for the last equality. Together with Eq. (7.44) this determines $A_2(r)$,

$$A_2(r) = \int_0^\infty \frac{d\ell\,\ell^3}{2\pi}C_\ell^\psi J_2(r\ell). \tag{7.46}$$

Inserting these results in the variance, we find (using that $A_2(0) = 0$)

$$\begin{aligned}
\left\langle \left[\boldsymbol{\ell}\cdot(\boldsymbol{\alpha}' - \boldsymbol{\alpha})\right]^2 \right\rangle &= 2\ell_i\ell_j\left(\langle\alpha_i\alpha_j\rangle - \langle\alpha_i'\alpha_j\rangle\right) \\
&= \ell^2\left[A_0(0) - A_0(r) + A_2(r)\cos(2\phi)\right].
\end{aligned}$$

Inserting this in the correlation function for the lensed anisotropies yields

$$\tilde{\xi}(r) = \int \frac{d^2\boldsymbol{\ell}}{(2\pi)^2}C_\ell \exp[-i\ell r\cos\phi]\exp\left[-\frac{\ell^2}{2}\left[A_0(0) - A_0(r) + A_2(r)\cos(2\phi)\right]\right]. \tag{7.47}$$

This expression is exact. We now use it to determine the lensed anisotropy power spectrum to first order in the lensing power spectrum. With the flat sky relation (5.107),

$$\tilde{C}_{\ell'} = \frac{1}{4\pi}\int_0^\infty r\,dr\,J_0(r\ell')\tilde{\xi}(r),$$

we can obtain the lensed power spectrum from the correlation function.

To perform the ϕ–integration in Eq. (7.47) we use that Bessel functions of imaginary arguments are related to the modified Bessel function (see Appendix 4, Section A4.3),

$$\exp(-y\cos\phi) = J_0(iy) + 2\sum_{n=1}^{\infty} i^n J_n(iy)\cos(n\phi)$$
$$= I_0(y) + 2\sum_{n=1}^{\infty}(-1)^n I_n(y)\cos(n\phi), \tag{7.48}$$

so that

$$\frac{1}{2\pi}\int_0^{2\pi} d\phi\, \exp(-y\cos\phi)\cos(n\phi) = (-1)^n I_n(y). \tag{7.49}$$

With this and Eq. (5.105), we can perform the angular integration of (7.47) with the result

$$\tilde{\xi}(r) = \int \frac{\ell\, d\ell}{2\pi} C_\ell \exp\left[-\frac{\ell^2}{2}\left[A_0(0) - A_0(r)\right]\right]$$
$$\times \left(I_0(r\ell) + 2\sum_{n=1}^{\infty} I_n(\ell^2 A_2(r)/2) J_{2n}(r\ell)\right). \tag{7.50}$$

Note that even though the modified Bessel functions grow exponentially $I_n(r) \to e^r/\sqrt{2\pi r}$ for large arguments, the combination $\exp[-\frac{\ell^2}{2}[A_0(0) - A_0(r)]]$ $I_n(\ell^2 A_2(r)/2) \to 0$ for large ℓ, since $A_0(0) - A_0(r) > A_2(r)$ for all values of r (see Fig. 7.2).

Since $A_2(0) = 0$ and $I_n(0) = \delta_{n0}$, the variance of the lensed CMB anisotropies remains unchanged,

$$\tilde{\xi}(0) = \int \frac{\ell\, d\ell}{2\pi} C_\ell = \xi(0). \tag{7.51}$$

Weak lensing only alters photon directions and hence the spatial structure of the correlation function. The power is redistributed by weak lensing but no power is lost.

A simpler but also accurate expression for the correlation function can be obtained if we approximate the exponential in Eq. (7.47):

$$\exp\left[-\frac{\ell^2}{2}\left[A_0(0) - A_0(r) + A_2(r)\cos(2\phi)\right]\right]$$
$$\simeq \exp\left[-\frac{\ell^2}{2}\left[A_0(0) + A_0(r)\right]\right]\left(1 - \frac{\ell^2}{2}A_2(r)\cos(2\phi)\right).$$

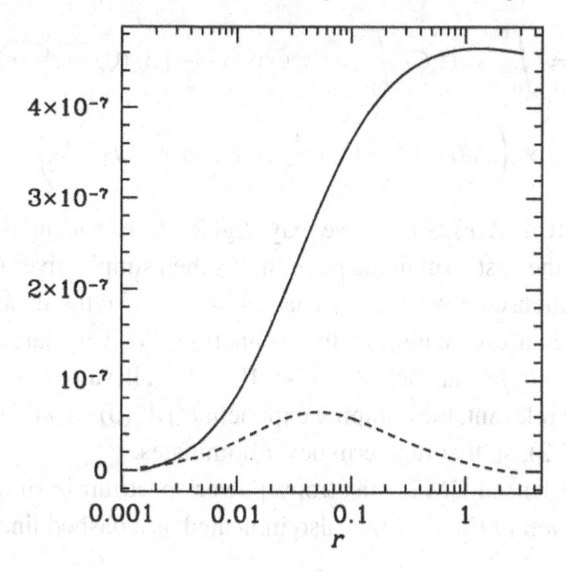

Fig. 7.2 The functions $A_0(0) - A_0(r)$ (solid) and $A_2(r)$ (dashed) are shown as functions of the separation angle r (in radians). The underlying cosmological model is a typical concordance model.

Note that this is not an expansion in the deflection angle $\boldsymbol{\alpha}$. The longitudinal part of the correlation function, $\langle \boldsymbol{\alpha} \cdot \boldsymbol{\alpha}' \rangle$, is fully taken into account and we have expanded only the traceless part A_2. A change in the direction of $\boldsymbol{\alpha}'$ with respect to the direction of $\boldsymbol{\alpha}$ contributes to this part. But since the deflection angle has most power on large scales $\ell \sim 50$, this change is small for scales corresponding to $\ell \gtrsim 1000$. As one sees in Fig. 7.2, the function $A_2(r)$ is much smaller than $A_0(0) - A_0(r)$ on all scales and it peaks at $r \sim 0.05$, which corresponds to $\ell \sim \pi/r \sim 60$, after which it decays like a power law.

Inserting this expansion in Eq. (7.47) we find

$$\begin{aligned}\tilde{\xi}(r) &\simeq \int \frac{d^2\boldsymbol{\ell}}{(2\pi)^2} C_\ell \exp[-i\ell r \cos\phi] \exp\left[-\frac{\ell^2}{2}\left[A_0(0) - A_0(r)\right]\right] \\ &\quad \times \left(1 - \frac{\ell^2}{2} A_2(r)\cos(2\phi)\right) \\ &= \int_0^\infty \frac{\ell\, d\ell}{2\pi} C_\ell \exp\left[-\frac{\ell^2}{2}[A_0(0) - A_0(r)]\right]\left(J_0(r\ell) + \frac{\ell^2}{2} A_2(r) J_2(r\ell)\right).\end{aligned} \tag{7.52}$$

Equation (7.52) is a very good approximation that can be used for all ℓ's for which CMB lensing is relevant. The $\tilde{C}_\ell$'s can be obtained from Eq. (7.52) with the help of Eq. (5.107),

$$\tilde{C}_{\ell'} = \int_0^\infty \ell\, d\ell\, C_\ell \int_0^\infty r dr\, \exp\left[-\frac{\ell^2}{2}\left[A_0(0) - A_0(r)\right]\right]$$
$$\times \left(J_0(r\ell') J_0(r\ell) + \frac{\ell^2}{2} A_2(r) J_0(r\ell') J_2(r\ell) \right). \tag{7.53}$$

Observing that $A(0) - A(r) \lesssim 10^{-6}$, we may neglect the exponential for small ℓ. The integral over r of the first term in the parentheses then simply gives $\ell^{-1}\,\delta(\ell - \ell')$ and reproduces the unlensed spectrum. For larger values of ℓ the exponential reduces power and induces a broadening of the δ-function. For very large ℓ's the second term also becomes relevant, but $A_2(r) < 10^{-7}$ for all values of r. Therefore, if $\ell^2 A_2(r)$ becomes relevant, the damping exponent $\ell^2[A_0(0) - A_0(r)]$ is several times bigger (see Fig. 7.2), so that this term never dominates.

In Fig. 7.3 the lensed CMB anisotropy power spectrum is shown. The large ℓ approximation given in Eq. (7.40) is also indicated as a dashed line.

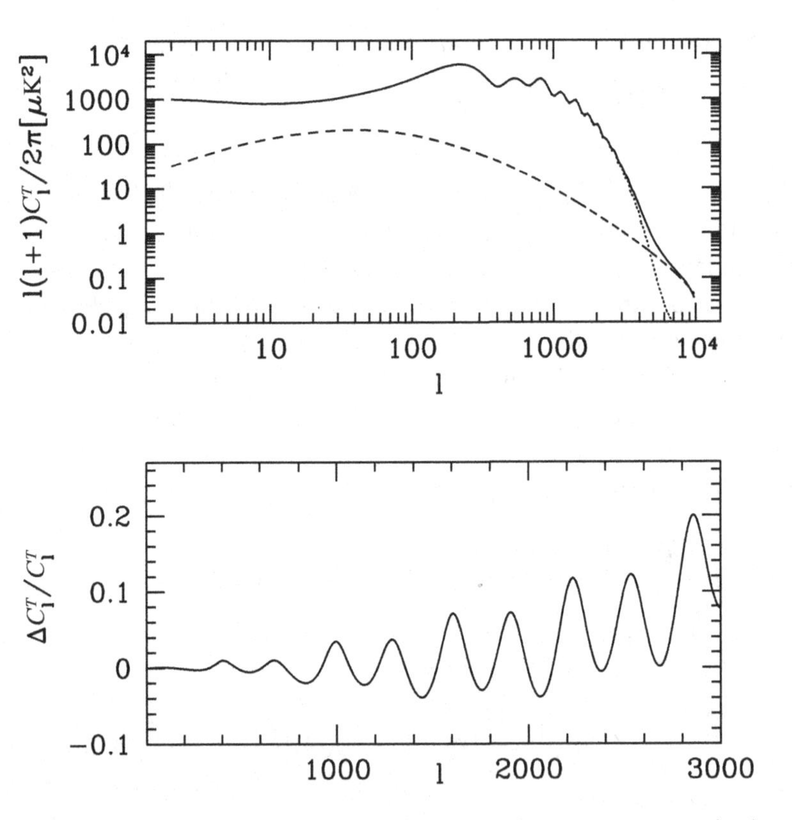

Fig. 7.3 Top panel: the lensed CMB temperature anisotropy spectrum is shown (solid). Underlaid is the unlensed spectrum (dotted). The large ℓ approximation for the lensed CMB spectrum is also indicated (dashed).
Bottom panel: the fractional difference between the lensed and unlensed CMB spectrum.

As anticipated, lensing is not relevant for large scales, $\ell \lesssim 100$, where the CMB anisotropy spectrum has nearly no structure. But it becomes very important for $\ell > 1000$ and actually dominates the signal for $\ell > 3000$. In the analyses or recent experiments like Planck (see Chapter 9), taking into account the full exponential $\exp\left[-\ell^2 \left[A_0(0) - A_0(r) + A_2(r)\cos(2\phi)\right]/2\right]$ and not just its first-order approximation is relevant to model the data sufficiently precisely. For this reason, workers in the field have started to study lensing beyond leading order in more generality (Pratten and Lews, 2016; Marozzi *et al.*, 2017).

7.4 Lensing of the CMB Polarization

In this section we study how polarization is affected by lensing. We work again in the flat sky approximation, which is sufficient for $\ell \gtrsim 20$. There are, in principle, two contributions. First, like for temperature anisotropies, the direction $\mathbf{n}$ in which a given photon is received has been deflected by the deflection angle $\boldsymbol{\alpha}$ from the direction in which it has been emitted, $\mathbf{n}' = \mathbf{n} + \boldsymbol{\alpha}$. Second, the polarization tensor is parallel-transported along the perturbed photon geodesics. To lowest order this means that the orientation of the polarization in the observed direction $\mathbf{n}$ and in the lensed direction $\mathbf{n}'$ is the same if it is determined w.r.t. a basis that is parallel-transported from $\mathbf{n}$ to $\mathbf{n}'$. Since the distance between $\mathbf{n}$ and $\mathbf{n}'$ is already first order, we may neglect the perturbation of the gravitational field along the geodesic from $\mathbf{n}$ to $\mathbf{n}'$. In the flat sky approximation, this simply means that we have to measure polarization with respect to the same basis $\boldsymbol{\epsilon}^{(1)}$ and $\boldsymbol{\epsilon}^{(2)}$ in both points. With this, the second effect is automatically taken care of, to first order.

We introduce, as in Chapter 5; see Eq. (5.6),

$$\mathbf{e}_\pm = \frac{1}{\sqrt{2}}(\boldsymbol{\epsilon}^{(1)} \pm i\boldsymbol{\epsilon}^{(2)}) \quad \text{and}$$

$$\mathcal{P} \equiv 2\mathrm{e}^i_+\mathrm{e}^j_+\mathcal{P}_{ij} = \mathcal{Q} + i\mathcal{U}, \text{ so that } \mathcal{P}^* \equiv 2\mathrm{e}^{*i}_+\mathrm{e}^{*j}_+\mathcal{P}_{ij} = 2\mathrm{e}^i_-\mathrm{e}^j_-\mathcal{P}_{ij} = \mathcal{Q} - i\mathcal{U}.$$

Expanding $\mathcal{Q} \pm i\mathcal{U}$ in Fourier space and direction, we have; see Eq. (5.58)

$$\mathcal{Q} \pm i\mathcal{U} = \int \frac{d^3k}{(2\pi)^3} \sum_{\ell=2}^{\infty} \sum_{m=-2}^{2} \times \left(\mathcal{E}_\ell^{(m)}(t,\mathbf{k}) \pm i\mathcal{B}_\ell^{(m)}(t,\mathbf{k})\right) {}_{\pm 2}G_{\ell m}(\mathbf{x},\mathbf{n}), \tag{7.54}$$

with

$${}_sG_{\ell m}(\mathbf{x},\mathbf{n}) = (-i)^\ell \sqrt{\frac{4\pi}{2\ell+1}}\, e^{i\mathbf{k}\cdot\mathbf{x}}\, {}_sY_{\ell m}(\mathbf{n}). \tag{7.55}$$

The spin-2 spherical harmonics are given by (see Appendix 4, Section A4.2.6) ${}_{\pm 2}Y_{\ell m}(\mathbf{x}) = 2\sqrt{(\ell-2)!/(\ell+2)!}\nabla_{\mathbf{e}_\pm}\nabla_{\mathbf{e}_\pm}Y_{\ell m}$. As we have already seen in Chapter 5, in the flat sky approximation they become

$$ {}_{\pm 2}Y_{\ell m} = {}_{\pm 2}Y_{\boldsymbol{\ell}}(\mathbf{x}) = \frac{2}{\ell^2}\mathbf{e}^i_\pm \mathbf{e}^j_\pm \nabla_i \nabla_j e^{i\boldsymbol{\ell}\cdot\mathbf{x}} = -e^{\pm 2i\phi} e^{i\boldsymbol{\ell}\cdot\mathbf{x}}. \tag{7.56} $$

Here ϕ denotes the angle that $\boldsymbol{\ell}$ encloses with the x-axis. The relation of the polarization field with its Fourier transforms $\mathcal{E}(\boldsymbol{\ell})$ and $\mathcal{B}(\boldsymbol{\ell})$ is given in Chapter 5 in Eqs. (5.108)–(5.111).

7.4.1 The Lensed Polarization Power Spectrum

We again start by expanding the polarization tensor in the deflection angle up to second order. This is a good approximation only when considering angular scales that are much larger than the deflection angle, that is, up to about $\ell \sim 1000$. We will have to do better in a second approach, but this approximation helps us to develop an intuition for the modifications of CMB polarization by lensing.

We shall see that even if, initially, perturbations are purely scalar and therefore do not have B-modes, the lensed polarization will develop B-modes. This is the most important effect from lensing: it generates B-modes from scalar perturbations so that B-modes are no longer an unambiguous sign of gravitational waves.

7.4.2 Approximation for Small Deflection Angles

As for the temperature anisotropies, we Taylor expand the polarization tensor to second order,

$$ \begin{aligned} \tilde{P}_{ij}(\mathbf{x}) &= P_{ij}(\mathbf{x} + \nabla\psi) \\ &\simeq P_{ij}(\mathbf{x}) + \nabla^m\psi\nabla_m P_{ij}(\mathbf{x}) + \frac{1}{2}\nabla^m\psi\nabla^n\psi\nabla_n\nabla_m P_{ij}(\mathbf{x}). \end{aligned} $$

Since parallel-transporting in the flat sky just means keeping the polarization basis $\mathbf{e}_\pm$ constant, the same expansion is also valid for $\mathcal{P} = \mathcal{Q} + i\mathcal{U}$ and $\mathcal{P}^* = \mathcal{Q} - i\mathcal{U}$. Fourier transforming the above expression leads to the same convolution integrals as we obtained for the lensed temperature anisotropies in Eq. (7.35). With the help of Eqs. (5.108)–(5.111) we find

$$ \begin{aligned} \left(\tilde{\mathcal{E}}(\boldsymbol{\ell}) \pm i\tilde{\mathcal{B}}(\boldsymbol{\ell})\right) e^{2i\phi_\ell} &\simeq (\mathcal{E}(\boldsymbol{\ell}) \pm i\mathcal{B}(\boldsymbol{\ell}))\, e^{2i\phi_\ell} \\ &\quad - \int \frac{d^2\ell'}{2\pi} \boldsymbol{\ell}' \cdot (\boldsymbol{\ell} - \boldsymbol{\ell}')\psi(\boldsymbol{\ell} - \boldsymbol{\ell}')[\mathcal{E}(\boldsymbol{\ell}') \pm i\mathcal{B}(\boldsymbol{\ell}')]e^{2i\phi'_\ell} \end{aligned} $$

$$-\frac{1}{2}\int\frac{d^2\boldsymbol{\ell}_1}{2\pi}\int\frac{d^2\boldsymbol{\ell}_2}{2\pi}\boldsymbol{\ell}_1\cdot(\boldsymbol{\ell}_1+\boldsymbol{\ell}_2-\boldsymbol{\ell})\boldsymbol{\ell}_1\cdot\boldsymbol{\ell}_2$$
$$\times\,[\mathcal{E}(\boldsymbol{\ell}_1)\pm i\mathcal{B}(\boldsymbol{\ell}_1)]e^{2i\phi_{\ell_1}}\times\psi(\boldsymbol{\ell}_2)\psi^*(\boldsymbol{\ell}-\boldsymbol{\ell}_1-\boldsymbol{\ell}_2). \tag{7.57}$$

In the flat sky approximation, the E- and B-polarization spectra and the T–E cross polarization spectrum are of the form

$$\langle\mathcal{E}(\boldsymbol{\ell})\mathcal{E}^*(\boldsymbol{\ell}')\rangle=\delta^2(\boldsymbol{\ell}-\boldsymbol{\ell}')C_\ell^{(\mathcal{E})},\qquad\langle\mathcal{B}(\boldsymbol{\ell})\mathcal{B}^*(\boldsymbol{\ell}')\rangle=\delta^2(\boldsymbol{\ell}-\boldsymbol{\ell}')C_\ell^{(\mathcal{B})},$$
$$\langle\mathcal{E}(\boldsymbol{\ell})\mathcal{M}^*(\boldsymbol{\ell}')\rangle=\delta^2(\boldsymbol{\ell}-\boldsymbol{\ell}')C_\ell^{(\mathcal{EM})}.$$

Multiplying Eq. (7.57) with its complex conjugate, with itself, or with the expression for lensed temperature anisotropies in $\boldsymbol{\ell}$-space, Eq. (7.35), and keeping only lowest-order expressions in C_ℓ^ψ, we obtain

$$\tilde{C}_\ell^{(\mathcal{E})}+\tilde{C}_\ell^{(\mathcal{B})}=C_\ell^{(\mathcal{E})}+C_\ell^{(\mathcal{B})}+\int\frac{d^2\boldsymbol{\ell}'}{(2\pi)^2}[\boldsymbol{\ell}'\cdot(\boldsymbol{\ell}-\boldsymbol{\ell}')]^2C^\psi_{|\boldsymbol{\ell}-\boldsymbol{\ell}'|}\left[C_{\ell'}^{(\mathcal{E})}+C_{\ell'}^{(\mathcal{B})}\right]$$
$$-\left[C_\ell^{(\mathcal{E})}+C_\ell^{(\mathcal{B})}\right]\int\frac{d^2\boldsymbol{\ell}'}{(2\pi)^2}(\boldsymbol{\ell}'\cdot\boldsymbol{\ell})^2C^\psi_{\ell'}, \tag{7.58}$$

$$\tilde{C}_\ell^{(\mathcal{E})}-\tilde{C}_\ell^{(\mathcal{B})}=C_\ell^{(\mathcal{E})}-C_\ell^{(\mathcal{B})}$$
$$+\int\frac{d^2\boldsymbol{\ell}'}{(2\pi)^2}[\boldsymbol{\ell}'\cdot(\boldsymbol{\ell}-\boldsymbol{\ell}')]^2e^{4i(\phi_{\ell'}-\phi_\ell)}C^\psi_{|\boldsymbol{\ell}-\boldsymbol{\ell}'|}\left[C_{\ell'}^{(\mathcal{E})}-C_{\ell'}^{(\mathcal{B})}\right]$$
$$-\left[C_\ell^{(\mathcal{E})}-C_\ell^{(\mathcal{B})}\right]\int\frac{d^2\boldsymbol{\ell}'}{(2\pi)^2}(\boldsymbol{\ell}'\cdot\boldsymbol{\ell})^2C^\psi_{\ell'}, \tag{7.59}$$

$$\tilde{C}_\ell^{(\mathcal{EM})}=C_\ell^{(\mathcal{EM})}+\int\frac{d^2\boldsymbol{\ell}'}{(2\pi)^2}[\boldsymbol{\ell}'\cdot(\boldsymbol{\ell}-\boldsymbol{\ell}')]^2e^{2i(\phi_{\ell'}-\phi_\ell)}C^\psi_{|\boldsymbol{\ell}-\boldsymbol{\ell}'|}C_{\ell'}^{(\mathcal{EM})}$$
$$-C_\ell^{(\mathcal{EM})}\int\frac{d^2\boldsymbol{\ell}'}{(2\pi)^2}(\boldsymbol{\ell}'\cdot\boldsymbol{\ell})^2C^\psi_{\ell'}. \tag{7.60}$$

For these results we have made use of the fact that $\mathcal{B}$ is uncorrelated with both $\mathcal{E}$ and $\mathcal{M}$. In the angular integration of Eqs. (7.59) and (7.60), only the real part contributes. Also noting that the scalar product $\boldsymbol{\ell}'\cdot\boldsymbol{\ell}=\ell'\ell\cos(\phi_{\ell'}-\phi_\ell)$ only depends on the angle difference $\phi\equiv\phi_{\ell'}-\phi_\ell$ the angular integral in (7.50) is of the form

$$\int_0^{2\pi}d\phi\,f(\cos\phi)e^{4i\phi}$$
$$=\int_0^{2\pi}d\phi\,f(\cos\phi)\big[\cos(4\phi)+4i\sin\phi\cos\phi(\cos^2\phi-\sin^2\phi)\big]$$
$$=\int_0^{2\pi}d\phi\,f(\cos\phi)\cos(4\phi).$$

The imaginary part of such an integral vanishes since $\int_0^{2\pi} f(\cos\phi)\sin\phi\, d\phi = \int_{-\pi}^{\pi} f(\cos\phi)\sin\phi\, d\phi = 0$ for arbitrary functions of $\cos\phi$. Correspondingly we have

$$\int_0^{2\pi} d\phi\, f(\cos\phi)e^{2i\phi} = \int_0^{2\pi} d\phi\, f(\cos\phi)[\cos(2\phi) + 2i\sin\phi\cos\phi]$$
$$= \int_0^{2\pi} d\phi\, f(\cos\phi)\cos(2\phi).$$

We may therefore replace $e^{4i\phi} \to \cos(4\phi) = \cos^2(2\phi) - \sin^2(2\phi)$ and $e^{2i\phi} \to \cos(2\phi)$. With the definition (7.38) of the mean square deflection angle, $R^{\psi} = (4\pi)^{-1}\int_0^{\infty} d\ell\, \ell^3 C_\ell^{\psi}$, we then find

$$\tilde{C}_\ell^{(\mathcal{E})} = (1 - \ell^2 R^{\psi})C_\ell^{(\mathcal{E})} + \int \frac{d^2\ell'}{(2\pi)^2}[\boldsymbol{\ell}' \cdot (\boldsymbol{\ell} - \boldsymbol{\ell}')]^2 C^{\psi}_{|\boldsymbol{\ell}-\boldsymbol{\ell}'|}$$
$$\times \left[C_{\ell'}^{(\mathcal{E})}\cos^2 2(\phi_{\ell'} - \phi_\ell) + C_{\ell'}^{(\mathcal{B})}\sin^2 2(\phi_{\ell'} - \phi_\ell)\right], \qquad (7.61)$$

$$\tilde{C}_\ell^{(\mathcal{B})} = (1 - \ell^2 R^{\psi})C_\ell^{(\mathcal{B})} + \int \frac{d^2\ell'}{(2\pi)^2}[\boldsymbol{\ell}' \cdot (\boldsymbol{\ell} - \boldsymbol{\ell}')]^2 C^{\psi}_{|\boldsymbol{\ell}-\boldsymbol{\ell}'|}$$
$$\times \left[C_{\ell'}^{(\mathcal{B})}\cos^2 2(\phi_{\ell'} - \phi_\ell) + C_{\ell'}^{(\mathcal{E})}\sin^2 2(\phi_{\ell'} - \phi_\ell)\right], \qquad (7.62)$$

$$\tilde{C}_\ell^{(\mathcal{E}\mathcal{M})} = (1 - \ell^2 R^{\psi})C_\ell^{(\mathcal{E}\mathcal{M})}$$
$$+ \int \frac{d^2\ell'}{(2\pi)^2}[\boldsymbol{\ell}' \cdot (\boldsymbol{\ell} - \boldsymbol{\ell}')]^2 C^{\psi}_{|\boldsymbol{\ell}-\boldsymbol{\ell}'|} C_{\ell'}^{(\mathcal{E}\mathcal{M})}\cos 2(\phi_{\ell'} - \phi_\ell). \qquad (7.63)$$

As we see from Eq. (7.62), even if there is no unlensed B-mode, $C^{(\mathcal{B})} = 0$, like for purely scalar perturbations, the lensing deflection induces a nonzero B-spectrum, $\tilde{C}^{(\mathcal{B})} \neq 0$. On relatively large scales, $\ell \ll \ell'$, the lensed B-mode induced by a pure primordial E-mode is roughly[1]

$$\tilde{C}_\ell^{(\mathcal{B})} \sim \int \frac{d^2\ell'}{(2\pi)^2}\ell'^4 C_{\ell'}^{\psi} C_{\ell'}^{(\mathcal{E})}\sin^2 2(\phi_{\ell'} - \phi_\ell) = \int \frac{d\ell'}{4\pi}\ell'^5 C_{\ell'}^{\psi} C_{\ell'}^{(\mathcal{E})}. \qquad (7.64)$$

This is an ℓ-independent constant. On large scales, the B-mode power spectrum induced by lensing is white noise. The contribution to the power per logarithmic interval, $d\log(\ell) = d\ell/\ell$, is

$$d\tilde{C}_\ell^{(\mathcal{B})} = \frac{1}{2}\ell^4 C_\ell^{\psi}\frac{\ell^2 C_\ell^{(\mathcal{E})}}{2\pi},$$

[1] We integrate over ℓ' and so, in principle, the inequality $\ell \ll \ell'$ does not strictly make sense. What we mean, of course, is that ℓ is much smaller than those values of ℓ', which mainly contribute to the above integral. In the same sense we shall use $\ell \gg \ell'$ below.

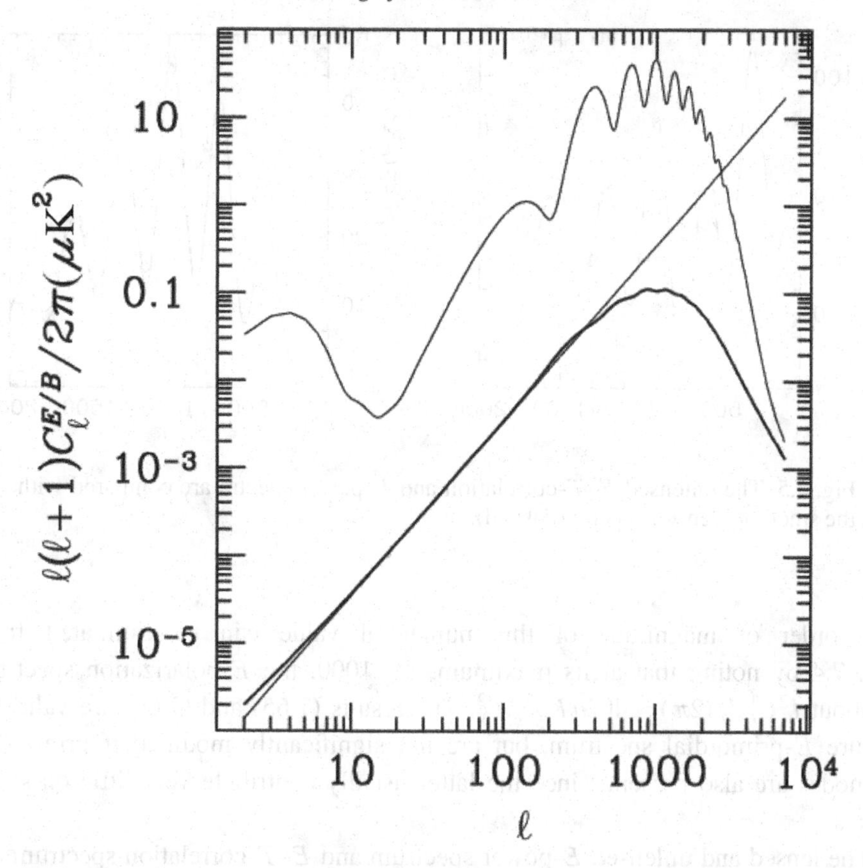

Fig. 7.4 The B-mode power spectrum induced from a pure E-mode by lensing is shown (thick solid curve). The lensed E-power spectrum (thin solid curve) is also indicated. The thin straight line traces the white noise approximation (7.64), which is excellent for $\ell < 200$.

half the product of the power of the deflection angle and the E-polarization; see Fig. 7.4.

At very small scales, $\ell \gg \ell'$ we can approximate the lensed E- and B-spectra by

$$\tilde{C}_\ell^{(\mathcal{E})} \simeq C_\ell^\psi \int \frac{d^2\ell'}{(2\pi)^2}[\boldsymbol{\ell}' \cdot \boldsymbol{\ell}]^2 C_{\ell'}^{(\mathcal{E})} \cos^2 2(\phi_{\boldsymbol{\ell}'} - \phi_{\boldsymbol{\ell}}) = \frac{1}{2}\ell^2 C_\ell^\psi R_E, \tag{7.65}$$

$$\begin{aligned}\tilde{C}_\ell^{(\mathcal{B})} &\simeq C_\ell^\psi \int \frac{d^2\ell'}{(2\pi)^2}[\boldsymbol{\ell}' \cdot \boldsymbol{\ell}]^2 C_{\ell'}^{(\mathcal{E})} \sin^2 2(\phi_{\boldsymbol{\ell}'} - \phi_{\boldsymbol{\ell}}) = \frac{1}{2}\ell^2 C_\ell^\psi R_E \\ &= \tilde{C}_\ell^{(\mathcal{E})},\end{aligned} \tag{7.66}$$

where we have introduced the variance of the gradient of polarization,

$$R_E = \frac{1}{4\pi}\int d\ell\, \ell^3 C_\ell^{(\mathcal{E})} = \langle|\nabla\mathcal{Q}|^2\rangle = \langle|\nabla\mathcal{U}|^2\rangle \sim \frac{2\times 10^7(\mu K)^2}{T_0^2}. \tag{7.67}$$

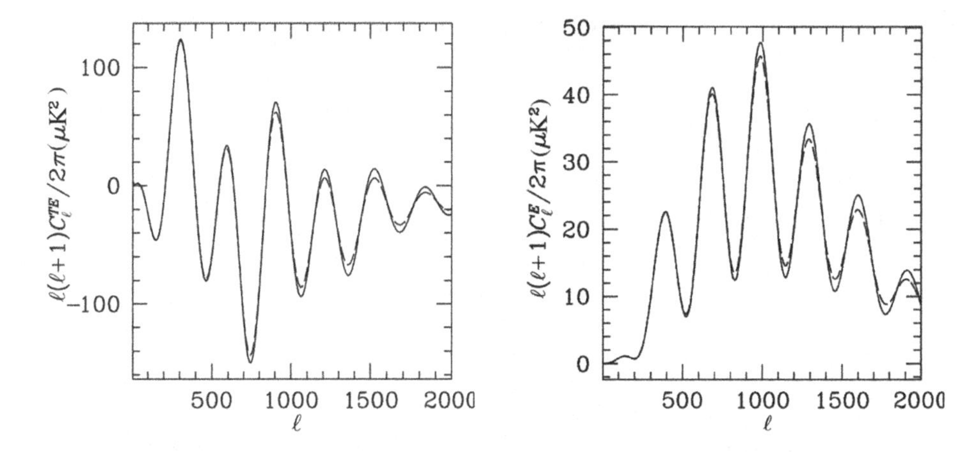

Fig. 7.5 The unlensed E–T-correlation and E-power spectra are compared with the smoother lensed spectra (dashed).

The order of magnitude of this numerical value can be estimated from Fig. 7.4 by noting that at its maximum, $\ell \sim 1000$, the E-polarization spectrum is about $\ell^2 C_\ell^{(\mathcal{E})}/(2\pi) \sim 40(\mu K)^2/T_0^2$. The results (7.65) and (7.66) are valid for a pure E-primordial spectrum, but are not significantly modified if primordial B-modes are also present, since the latter usually contribute very little on small scales.

The lensed and unlensed E-power spectrum and E–T-correlation spectrum are compared in Fig. 7.5. The lensed E- and B-power spectra from purely scalar primordial perturbations are shown in Fig. 7.6.

7.4.3 Arbitrary Deflection Angles

We now derive an expression that is also valid for large ℓ's where the deflection angle is no longer smaller than the scale of interest. As for the temperature anisotropies, we study the modification of the correlation function by lensing.

As in Chapter 5, we consider two points $\mathbf{x}$ and $\mathbf{x}'$ and define the polarization basis along $\mathbf{r} = \mathbf{x} - \mathbf{x}'$. The rotated polarization is given by

$$\mathcal{P}_r(\mathbf{x}) = e^{-2i\phi_r}\mathcal{P}(\mathbf{x}).$$

The correlation functions of this rotated polarization are denoted ξ_+ and ξ_-, and its correlation with the temperature anisotropy is $\xi_\times$. These functions are defined in Eqs. (5.112)–(5.114). The calculation of the lensed correlation function $\tilde{\xi}_+$ is exactly analogous to that for the temperature anisotropy; one just has to replace C_ℓ by $C_\ell^{(\mathcal{E})} + C_\ell^{(\mathcal{B})}$. Doing so we have

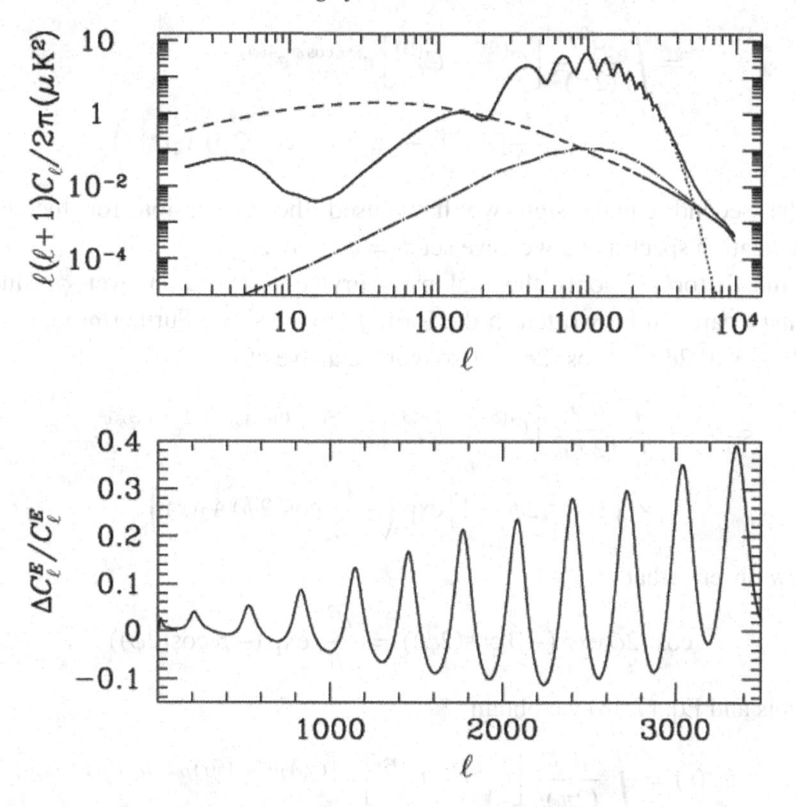

Fig. 7.6 The lensed E- (solid) and the induced B- (long dashed) power spectra are shown. The deflection angle spectrum (short dashed) and the unlensed E-power spectrum (dotted) are also indicated. The bottom panel shows the relative difference of the lensed and unlensed E-spectra.

$$\begin{aligned}\tilde{\xi}_+(r) &= \left\langle \mathcal{P}^*(\mathbf{x}+\boldsymbol{\alpha})\mathcal{P}(\mathbf{x}'+\boldsymbol{\alpha}')\right\rangle \\ &= \frac{1}{2\pi}\int \ell\, d\ell \left[C_\ell^{(\mathcal{E})} + C_\ell^{(\mathcal{B})}\right] e^{-(\ell^2/2)(A_0(0)-A_0(r))} \\ &\quad \times \left(I_0(\ell^2 A_2(r)/2) J_0(r\ell) + 2\sum_{n=1}^{\infty} I_n(\ell^2 A_2(r)/2) J_{2n}(r\ell)\right). \end{aligned} \tag{7.68}$$

The lensing of ξ_- and $\xi_\times$ is somewhat different because of the exponentials $e^{im\phi_r}$. Inserting the Fourier transform of $\mathcal{P}(\mathbf{x})$ given in Eq. (5.108) in the expression for ξ_-, we find

$$\begin{aligned}\tilde{\xi}_-(r) &= \left\langle \mathcal{P}(\mathbf{x}+\boldsymbol{\alpha})\mathcal{P}(\mathbf{x}'+\boldsymbol{\alpha}')e^{-i4\phi_r}\right\rangle \\ &= \int \frac{d^2\ell}{2\pi}\frac{d^2\ell'}{2\pi} \langle [\mathcal{E}(\boldsymbol{\ell}) + i\mathcal{B}(\boldsymbol{\ell})][\mathcal{E}^*(\boldsymbol{\ell}') + i\mathcal{B}^*(\boldsymbol{\ell}')]\rangle \\ &\quad \times e^{i(\boldsymbol{\ell}\cdot\mathbf{x}-\boldsymbol{\ell}'\cdot\mathbf{x}')} e^{i(2\phi_\ell+2\phi_{\ell'}-4\phi_r)} \langle e^{i(\boldsymbol{\ell}\cdot\boldsymbol{\alpha}-\boldsymbol{\ell}'\cdot\boldsymbol{\alpha}')}\rangle\end{aligned}$$

$$= \int \frac{d^2\ell}{(2\pi)^2}\left[C_\ell^{(\mathcal{E})} - C_\ell^{(\mathcal{B})}\right] e^{ir\ell\cos\phi} e^{4i\phi} \times \exp\left(-\frac{\ell^2}{2}[A_0(0) - A_0(r) + \cos(2\phi)A_2(r)]\right).$$

For the second equals sign we have used the expression for the E- and B-polarization spectra and we have set $\phi = \phi_\ell - \phi_r$.

Of the factor $e^{4i\phi}$ only the real part survives integration over ϕ, since the imaginary part can be written in the form $f(\cos\phi)\sin\phi$. Furthermore, $\cos 4\phi = \cos^2 2\phi - \sin^2 2\phi = 2\cos^2 2\phi - 1$ so that we arrive at

$$\tilde{\xi}_-(r) = \int \frac{d^2\ell}{(2\pi)^2}\left[C_\ell^{(\mathcal{E})} - C_\ell^{(\mathcal{B})}\right] e^{-\ell^2(A_0(0)-A_0(r))/2} e^{ir\ell\cos\phi} \times \left[2\cos^2 2\phi - 1\right]\exp\left(-\frac{\ell^2}{2}\cos(2\phi)A_2(r)\right).$$

We now observe that

$$\cos^2 2\phi \exp\left(-\beta\cos(2\phi)\right) = \frac{d^2}{d\beta^2}\exp\left(-\beta\cos(2\phi)\right).$$

With this and Eq. (7.48) we obtain

$$\tilde{\xi}_-(r) = \int \frac{d^2\ell}{(2\pi)^2}\left[C_\ell^{(\mathcal{E})} - C_\ell^{(\mathcal{B})}\right] e^{-\ell^2(A_0(0)-A_0(r))/2} e^{ir\ell\cos\phi} \times \left(2I_0''(\ell^2A_2(r)/2) - I_0(\ell^2A_2(r)/2) + 2\sum_{n=1}^{\infty}(-1)^n \times [2I_n''(\ell^2A_2(r)/2) - I_n(\ell^2A_2(r)/2)]\cos(2n\phi)\right).$$

Here primes indicate the derivative with respect to the argument. Integration over ϕ finally yields

$$\tilde{\xi}_-(r) = \int \frac{\ell\, d\ell}{2\pi}\left[C_\ell^{(\mathcal{E})} - C_\ell^{(\mathcal{B})}\right] e^{-\ell^2(A_0(0)-A_0(r))/2} \times \left([2I_0''(\ell^2A_2(r)/2) - I_0(\ell^2A_2(r)/2)]J_0(\ell r) + 2\sum_{n=1}^{\infty}[2I_n''(\ell^2A_2(r)/2) - I_n(\ell^2A_2(r)/2)]J_{2n}(\ell r)\right), \tag{7.69}$$

$$\simeq \int \frac{\ell\, d\ell}{2\pi}\left[C_\ell^{(\mathcal{E})} - C_\ell^{(\mathcal{B})}\right] e^{-\ell^2(A_0(0)-A_0(r))/2} \times \left(J_4(\ell r) + \frac{\ell^2}{4}A_2(r)[J_2(\ell r) + J_6(\ell r)]\right). \tag{7.70}$$

For the last line we have expanded the general expression to first order in $A_2(r)$. To obtain an accuracy of better than cosmic variance one has to also include the term $\propto A_2(r)^2$; see Challinor and Lewis (2005). This is indeed relevant for recent experiments. A similar calculation gives the cross correlation function,

$$\tilde{\xi}_\times(r) = \int \frac{\ell\, d\ell}{2\pi} C_\ell^{(\mathcal{EM})} e^{-\ell^2(A_0(0)-A_0(r))/2} \times \left(I_0'(\ell^2 A_2(r)/2) J_0(\ell r) + 2\sum_{n=1}^{\infty} (-1)^n I_n'(\ell^2 A_2(r)/2) J_{2n}(\ell r) \right), \quad (7.71)$$

$$\simeq \int \frac{\ell\, d\ell}{2\pi} C_\ell^{(\mathcal{EM})} e^{-\ell^2(A_0(0)-A_0(r))/2} \left(J_2(\ell r) + \frac{\ell^2}{4} A_2(r)[J_0(\ell r) + J_4(\ell r)] \right). \quad (7.72)$$

Like for the temperature anisotropy, the polarization power spectra and correlation functions are related via two-dimensional Fourier transforms. The relations between polarization spectra and correlation functions are the same as those for the unlensed quantities given in Chapter 5.

In Fig. 7.7 we show the lensed E- and B-mode spectra for a fixed spectral index and a ΛCDM model. The B-mode spectra from tensors for $r = 0.1$ and $r = 10^{-3}$ are also indicated.

Considering the B-polarization induced by lensing of E-polarization as the only (Gaussian) noise in an all-sky polarization experiment, one finds that the primordial tensor B-mode is detectable for $r \geq 10^{-3}$. If $r < 10^{-3}$, a method must be found to subtract the lensing contribution to the B-polarization spectrum. This is not impossible. At small scales, $\ell > 1000$, the B-mode is nearly entirely due to lensing and can therefore help to determine the spectrum of the lensing potential. Once we know the latter, we can, in principle, invert our expressions for the lensed spectra to obtain the "delensed" primordial spectra. The procedure can be applied iteratively. In a first step, one may assume that the B-spectrum is purely due to lensing, but neglect the effect of lensing in the E-spectrum. Determining the spectrum of the lensing potential in this approximation, one can now calculate the first-order delensed E-spectrum and from it the new lensed B-spectrum. The difference of this and the measured B-spectrum is a first estimate for the primordial B-spectrum. I suppose that this procedure converges rapidly, but this has not been shown in detail. Present methods have measured the lensing signal with about 10% accuracy on scales up to $\ell \sim 1000$ directly from the reconstructed lens map; see Fig. 9.8.

The lensing potential comes dominantly from low redshift and may also be determined or at least constrained by other observations, such as, for example, weak lensing of galaxy surveys, especially on small scales. On large scales, where perturbations are linear, we can obtain a first approximation to the lensing potential from

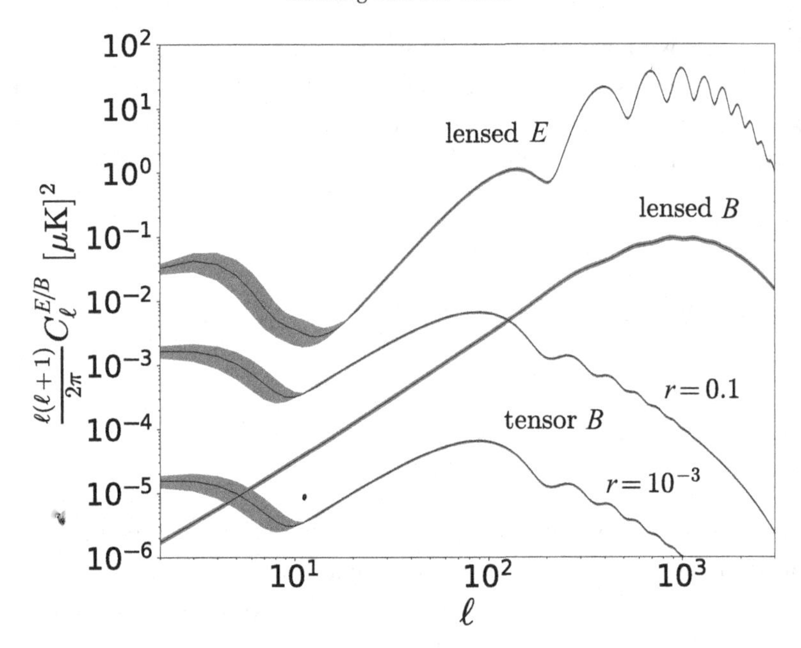

Fig. 7.7 The 95% confidence regions for the polarization spectra from a compilation of the CMB and large-scale structure data available in 2020. The optical depth τ, the amplitude A_s, the matter density parameter Ω_m, and the spectral index n_s are varied over their 95% confidence range. Varying also the other parameters of the flat ΛCDM model does not enlarge the line thickness of the curves. The tensor B-mode spectra from tensors for $r = 0.1$ and 10^{-3} are also indicated.

parameter estimation, neglecting lensing, which determines the Bardeen potentials Φ, Ψ and then via Eq. (7.21) the lensing potential.

All calculations done in this chapter approximate the sky as flat. This approximation is very good within patches of the size of the deflection angle and up to a few degrees. But if we want to determine the modifications of the low C_ℓ's by lensing, sky curvature has to be taken into account. The result from such a curved sky treatment for the correlation between two directions $\mathbf{n}_1$ and $\mathbf{n}_2$ is very similar to our formulas (7.50), (7.68), (7.70), and (7.72). Only the sums over the modified Bessel functions become double sums over m and m' and the ordinary Bessel functions of r are replaced by elements $D^\ell_{mm'}$ of the representation matrix of the rotation that turns $\mathbf{n}_1$ into $\mathbf{n}_2$. The detailed expressions can be found in Lewis and Challinor (2006).

7.5 Non-Gaussianity

There is also another method that may help to single out the lensing contribution to the CMB power spectra. This is statistical in nature: we usually assume that primordial fluctuations are Gaussian. However, lensing, being a second-order effect, is not Gaussian. Furthermore, since correlations between the lensing potential ψ and the CMB anisotropies and polarization can be neglected, to lowest order lensing does not induce a bispectrum but only a trispectrum, a nonvanishing connected 4-point correlation function. This also offers a possibility to identify the lensing part of the CMB anisotropies and polarization, which has been extensively used in the Planck data analysis [Planck Coll. XV (2016)]. The connected 4-point correlation function from lensing can be calculated and can be used to construct a "quadratic estimator" for the lensing potential. However, care has to be applied, since all forms of noise, foregrounds, instrumental noise, and so forth, are, in general, non-Gaussian; see Okamoto and Hu (2003) for details.

7.6 Other Second-Order Effects

Lensing affects CMB anisotropies and polarization at second order. At $\ell \gtrsim 3000$, lensing induces changes of order unity and more. Only for $\ell \lesssim 400$, neglecting lensing is accurate to better than half a percent. This naturally brings up the question: are there other second-order effects that are similarly important? The answer to this question is "probably not, at least not up to $\ell \sim 2000$."

Several second-order effects have been considered in the literature, but a systematic study of second-order CMB anisotropies and polarization is still lacking. Here I briefly present the physical effects that have been studied so far, but we do not enter into their calculation. We shall, however, discuss the Sunyaev–Zel'dovich (SZ) effect in Chapter 10 in some detail.

- **Lensing by clusters.** So far we have discussed lensing mainly using the linear lensing potential. However, CMB photons are also lensed by nonlinear structures such as individual clusters. Statistically this effect can be taken into account by using the nonlinear lensing potential on small scales.
- **Ostriker–Vishniac effect.** This is a second-order Doppler term that comes from the fact that the optical depth is proportional to the electron density and the Doppler term therefore has a second-order contribution of the form $\mathbf{n} \cdot \mathbf{V}_b D_b$, where D_b is the baryon density fluctuation, $\mathbf{V}_b$ the baryon velocity, and $\mathbf{n}$ the photon direction. It has been argued in Hu and White (1996) that this term is less affected by Silk damping than the first-order Doppler term and may thus become important on small scales, $\ell \gg 1000$. Calculations show, however, that the effect is smaller than the lensing contribution for all $\ell \lesssim 3000$; see Lewis and Challinor (2006).
- **SZ effect.** Clusters contain a hot plasma with a temperature of several keVs. As we shall see in Chapter 10, whenever CMB photons pass through a plasma with hot electrons, their spectrum is modified in a well-defined way. This is

the *thermal SZ effect*. It can be distinguished quite easily from primordial CMB anisotropies by its spectral signature. In the Planck satellite experiment, this effect has been routinely used to discover several hundred clusters; see Planck Coll. XXIV (2016). However, clusters usually have a coherent peculiar velocity and CMB photons that scatter off hot electrons of a cluster also acquire a Doppler shift, the so-called *kinetic SZ effect*, which has exactly the same spectrum as ordinary CMB anisotropies. The kinetic SZ effect is much smaller than the thermal SZ effect, typically of the same order as the Vishniac effect.

- **Rees–Sciama effect.** The gravitational potential from linear perturbation theory is constant in a CDM background and decaying in a late ΛCDM background. However, once density perturbations become nonlinear, the gravitational potential also starts growing. This leads to a late integrated Sachs–Wolfe effect on very small scales. Estimations show that this effect is probably subdominant on all scales (Seljak, 1996a).
- **Patchy reionization.** As we shall discuss in Chapter 9, reionization is supposed to be caused by the radiation of the first, probably very massive stars. It is reasonable to expect that these stars have not formed everywhere at the same time and with the same number density. Therefore, reionization was probably earlier in some patches of the sky than in others, leading to more rescattering and therefore damping of CMB anisotropies and regeneration of polarization in some places than in others. It is not clear what and how strong the signature of this "patchy reionization" is, but it probably is relevant only on very small scales where it may be comparable with the kinetic SZ and Ostriker–Vishniac effects. A recent study can be found in Feng and Holder (2019). While the signal is still out of reach in present experiments, it should be measurable in the near future.

Exercises

(The exercises marked with an asterisk are solved in Appendix 11 which is not in this printed book but can be found online.)

7.1 **The deflection angle from a point mass***
Consider a point mass M with gravitational potential $\Psi = GM/r$. Approximate the Schwarzschild metric for this mass by

$$ds^2 = -(1 + 2\Psi)\, dt^2 + (1 - 2\Psi)\, d\mathbf{x}^2.$$

Show that the light deflection in this metric to first order in Ψ is given by $\boldsymbol{\alpha} = \varphi \mathbf{e}$, where $\mathbf{e}$ is the normal to the original photon direction $\mathbf{n}$ in the plane defined by $\mathbf{n}$ and the position of the mass, and

$$\varphi = \frac{4GM}{d}, \tag{7.73}$$

where d is the impact parameter of the photon trajectory (i.e., its closest distance to the mass M).

7.2 **The lensing shear**
Show that $\gamma_\pm = \gamma_1 \pm i\gamma_2$ [see Eq. (7.24) and Eq. (7.25)] is obtained from the lensing potential by acting with the spin raising and spin lowering operator on the lensing potetial,

$$\gamma_+ = \frac{1}{2}(\eth^*)^2\psi \tag{7.74}$$

$$\gamma_- = \frac{1}{2}\eth^2\psi. \tag{7.75}$$

7.3 **Lensing of a scale-invariant power spectrum**
In Section 7.3.1 we show that lensing of a scale-invariant spectrum by a small deflection angle, $|\boldsymbol{\alpha}| \ll \pi/\ell$, can be approximated by

$$\tilde{C}_\ell \simeq (1 - \ell^2 R^\psi)C_\ell + \ell^2 C_\ell \int \frac{d^2\ell_1}{(2\pi)^2} \frac{[(\boldsymbol{\ell}_1 - \boldsymbol{\ell}) \cdot \boldsymbol{\ell}_1]^2}{(\boldsymbol{\ell} - \boldsymbol{\ell}_1)^2} C^\psi_{\ell_1}.$$

Bring the above integral into the form

$$\int \frac{d^2\ell_1}{(2\pi)^2} \frac{[(\boldsymbol{\ell}_1 - \boldsymbol{\ell}) \cdot \boldsymbol{\ell}_1]^2}{(\boldsymbol{\ell} - \boldsymbol{\ell}_1)^2} C^\psi_{\ell_1} = \int_0^\infty \frac{\ell_1\, d\ell_1}{(2\pi)^2} C^\psi_{\ell_1} \int_0^{2\pi} d\phi\, \frac{(\ell_1^2 - \ell_1\ell\cos\phi)^2}{\ell_1^2 + \ell^2 - 2\ell_1\ell\cos\phi}.$$

Using complex integration, show that the angular integral gives

$$\int_0^{2\pi} d\phi \frac{(\ell_1^2 - \ell_1\ell\cos\phi)^2}{\ell_1^2 + \ell^2 - 2\ell_1\ell\cos\phi} = \pi\left(\ell_1^2 + \theta(\ell_1 - \ell)(\ell_1^2 - \ell^2)\right).$$

Here θ is the Heaviside function,

$$\theta(x) = \begin{cases} 1 & \text{if } x \geq 0 \\ 0 & \text{if } x < 0. \end{cases}$$

This implies the result (7.39).

7.4 **Expectation values of Gaussian variables**
Show that for a Gaussian variable X with mean zero and variance σ we have

$$\langle e^{iX} \rangle = e^{-\sigma^2/2}.$$

8

Observations of Large-Scale Structure

8.1 Introduction

In addition to the CMB, the large-scale distribution of matter in the Universe [large-scale structure (LSS)] is an interesting observable that is widely used to determine not only the properties of our Universe but also to test the theory of gravitation, General Relativity itself. In this chapter we discuss observations of LSS from a fully relativistic point of view. We first make contact with the standard nonrelativistic treatment and briefly discuss its merits and its shortcomings. Then we develop a relativistic analysis that has much in common with our study of the CMB, the main topic of this book. In the last section of this chapter we briefly also discuss intensity mapping, a new, promising technique to observe LSS or, more generally, the distribution of neutral hydrogen in the Universe.

LSS is more complicated than the CMB because we usually observe the distribution of galaxies – discrete, highly over-dense, small regions in the sky – which we approximate as points in this context. On the other hand, we calculate the matter over density, and the relation between these two quantities is what we call "bias." Galaxies are a discrete biased tracer of the matter distribution. We have good reasons to believe that on large scales bias is linear and scale independent, but we expect it to depend on redshift. In this book we concentrate on these large scales, since we treat the problem within linear perturbation theory, which is valid only on sufficiently large scales and sufficiently high redshift. Furthermore, the relativistic treatment, which is the novelty of this text, is relevant mainly on large scales.

In the past, observers usually surveyed a rather small region in the sky and considered the observed galaxy number density in some volume element $r^2 dr d\Omega$ as proportional to the matter density $\rho(t)\delta(\mathbf{x},t)$. They then performed a discrete Fourier transform on this dataset to infer the power spectrum. The details of this procedure can be found, for example, in Peebles (1980). For small regions this is actually sufficient (apart from redshift space distortions, which we discuss in the text that follows and that are today also included in the analysis). However, when we go out to large redshifts and observe large patches in the sky, we have to take into account that observations are made on the background lightcone and not in a spatial volume. We also have to take into account that with the perturbed metric, this background lightcone is also perturbed.

What we truly observe of a galaxy is its direction in the sky, $-\mathbf{n}$ (like in Section 2.5, $\mathbf{n}$ is the propagation direction of the incoming photon), and its redshift. The measured over-density is therefore a quantity of the form $\Delta(\mathbf{n}, z)$ and in this chapter we want to compute it and to relate it to the matter density fluctuation $\delta(\mathbf{x}, t)$ and other perturbation variables.

Let $N(\mathbf{n}, z)$ be the number of galaxies in a small solid angle $d\Omega$ around $\mathbf{n}$ and in a redshift bin $[z, z + dz]$. We define the number count fluctuation as

$$\Delta(\mathbf{n}, z) = \frac{N(\mathbf{n}, z) - \bar{N}(z)}{\bar{N}(z)}. \tag{8.1}$$

Here $4\pi f_{\rm sky} \bar{N}(z)$ is the total number of galaxies observed in the redshift bin $[z, z + dz]$ and $f_{\rm sky}$ denotes the observed sky fraction. $\Delta(\mathbf{n}, z)$ is a truly observable quantity that vanishes in the Friedmann background universe and therefore its expression within linear perturbation theory is gauge invariant (Stewart lemma).

We expand the angular dependence of Δ in terms of spherical harmonics,

$$\Delta(\mathbf{n}, z) = \sum_{\ell,m} a_{\ell m}(z) Y_{\ell m}(\mathbf{n}). \tag{8.2}$$

The corresponding power spectra are

$$\langle a_{\ell m}(z) a^*_{\ell' m'}(z') \rangle = C_\ell(z, z') \delta_{\ell\ell'} \delta_{mm'}. \tag{8.3}$$

Like for the CMB, the Kronecker deltas are a consequence of statistical isotropy, but contrary to the CMB we have a density field at arbitrary redshift and different redshifts are not uncorrelated. As we shall see in the text that follows, the correlation of different redshifts is an excellent means to determine the lensing convergence κ introduced in Eq. (7.19).

8.2 Redshift Space Distortion and Lensing

Before we derive the full relativistic expression for the number count fluctuation we consider the quasi-Newtonian situation. In this section we also do not pay attention to gauge issues and the result we derive here is actually not gauge invariant. The only relativistic term we take into account in this section is the deflection of light coming from our galaxies, that is, lensing. We consider objects (e.g., galaxies or a certain class of galaxies) with a density that is proportional to the matter density. Neglecting first this proportionality factor (the bias), its fluctuation is given by

$$N(\mathbf{n}, z) = \rho(\mathbf{n}, z)V(\mathbf{n}, z) = \bar{\rho}\bar{V}\left(1 + \delta_z + \frac{\delta V}{V}\right), \tag{8.4}$$

where δ_z denotes the density fluctuation at fixed redshift. In addition to the naively expected fluctuation of the observed number in a small volume V, we also have to take into account the fluctuation of the volume element itself. For a given direction at the observer, $-\mathbf{n}$, and observed redshift z, the volume element is

$$V = r^2(z)d\Omega_{\mathbf{n}}dr = r^2(z)\frac{dr}{dz}d\Omega_{\mathbf{n}}dz. \tag{8.5}$$

8.2.1 Redshift Space Distortion

In an unperturbed Friedmann universe, r is simply the comoving distance of the emitter at redshift z, $r(z) = \int_0^z H^{-1}(z')dz'$ and $dr/dz = H^{-1}(z) = a/\mathcal{H}$; see Eq. (1.47). In a perturbed Universe, both r and z acquire perturbations. In a Newtonian setting only z is perturbed by the Doppler effect and we have $z = \bar{z} + \delta z$ with [see Eq. (2.236)]

$$\frac{\delta z}{1 + \bar{z}} = -\mathbf{V}(z) \cdot \mathbf{n} = V_r. \tag{8.6}$$

Here we neglect Sachs–Wolfe and integrated Sachs–Wolfe effects that are taken into account in Eq. (2.236) and that we shall also consider in the next section.

In our derivative in Eq. (8.5) we have to insert $dz = (1 + d\delta z/d\bar{z})d\bar{z}$ or

$$\frac{dr}{dz} = \left(1 - \frac{d\delta z}{d\bar{z}}\right)\frac{dr}{d\bar{z}} = \frac{1}{(1+z)\mathcal{H}}\left(1 + \mathbf{V}\cdot\mathbf{n} + \frac{d(\mathbf{V}\cdot\mathbf{n})/dr}{\mathcal{H}}\right). \tag{8.7}$$

In the last term we have converted d/dz into d/dr. At small to intermediate scales, the term $\mathbf{V}\cdot\mathbf{n}$ is usually neglected, as it is a factor $\mathcal{H}/k$ smaller than the last term, which contains an additional derivative. Neglecting the subdominant middle term in Eq. (8.7), this leads to a radial volume distortion of

$$\frac{\delta V}{V} = \frac{\delta\left(\frac{dr}{dz}\right)}{\frac{dr}{dz}} = \frac{d(\mathbf{V}\cdot\mathbf{n})/dr}{\mathcal{H}}. \tag{8.8}$$

Let us now consider a relatively small survey of galaxies positioned in a global mean direction $-\mathbf{n}$ from the observer and at observed redshift z. To take into account redshift space distortions (RSDs) we have to take into account the radial volume distortion and replace the observed $\delta_g = b\delta$ by

$$\Delta = \frac{\delta N}{N} = \delta_g(\mathbf{x}, z) + \frac{\delta V}{V} = b\delta(\mathbf{x}, z) - \mathcal{H}^{-1}\mathbf{n}\cdot\nabla\left(\mathbf{n}\cdot\mathbf{V}(\mathbf{x}, z)\right), \tag{8.9}$$

where we have neglected the subdominant term of Eq. (8.7) and δ_g denotes the galaxy density fluctuation that we assume to be linearly related to the matter density fluctuation. We have inserted $\mathbf{x} = -r\mathbf{n}$ and we have used that $\partial_r = -\mathbf{n}\cdot\nabla$. The quantity b is a bias factor that depends on the chosen "tracer" and is in general redshift dependent, but we assume it to be scale independent. We consider scalar perturbations such that $\mathbf{V} = -\nabla V_s$ for a velocity potential V_s. Fourier transforming Eq. (8.9) we find

$$\Delta(\mathbf{k}, z) = b\delta(\mathbf{k}, z) - \mu^2 k^2 V_s(\mathbf{k}, z)/\mathcal{H}, \tag{8.10}$$

where $\mu = \hat{\mathbf{k}}\cdot\mathbf{n}$ is the direction cosine between the incoming photon and the wave number $\mathbf{k}$. We now use the Newtonian continuity equation in Fourier space,

$$\dot{\delta} + k^2 V_s = 0. \tag{8.11}$$

This corresponds to the first of Eqs. (2.116) or (2.118) for $w = c_s^2 = 0$, neglecting Γ, Π, and gauge issues and setting $V_s = V/k$. We set $\delta(k,t) = D_1(t)\delta(k,t_0)$, where $D_1(t)$ is the deterministic linear growth factor that we normalize to 1 today. For pure matter perturbations D_1 does not depend on the wave number; see Eq. (2.186). We also introduce the logarithmic growth rate, called the growth function,

$$f = \frac{\dot{D}_1}{D_1\mathcal{H}} \equiv \frac{d\ln D_1}{d\ln a}. \tag{8.12}$$

We then obtain for the observed power spectrum of Δ

$$P_{\rm obs}(\mathbf{k}, z) = P_\delta(k, z)\left[b + \mu^2 f\right]^2. \tag{8.13}$$

This is the very interesting result first derived by Kaiser (1987). It shows that even in a statistically isotropic universe the observed power spectrum is not isotropic due to observational effects. Furthermore, isolating the term proportional to μ^2 or μ^4 allows us to measure the growth function f which depends sensitively on the expansion history of the Universe.

The term $b^2 P_\delta(k,z)$ exhibits the so-called Baryon Acoustic Oscillations "BAOs" which are the acoustic oscillations of the baryon-photon fluid prior to decoupling left over in the baryons. Since baryons make only a small contribution to the total matter, the amplitude of BAOs is much smaller than the one of the acoustic oscillations in the CMB, but they have unambiguously been detected in the data and are routinely being used to measure a distance out to redshift z.

Let us denote the comoving wavelength of the nth BAO peak by λ_n. It is roughly given by the peaks of $\cos(k_n r_s)$, where r_s is the comoving "drag scale." The computation of the drag scale is somewhat subtle. First of all, at $z_{\rm dec}$, when photons decouple from baryons, the baryons still remain coupled until nearly $z \sim 100$, which is why their temperature still equals roughly the photon temperature; see

Section 1.3.2. Nevertheless, the baryon sound speed drops significantly during this epoch so that the baryon sound horizon, which roughly corresponds to the drag scale, grows only very little. An additional effect is the so-called velocity overshoot that is due to the change in the evolution of perturbations at the equality scale [see Montanari and Durrer (2012) for details]. Numerically one finds a comoving drag scale of $r_s \simeq 147$ Mpc. In transversal directions the nth BAO peak is seen under an angle $\theta_n(z) = \lambda_n/[(1+z)d_A(z)]$. In radial directions they are seen at a redshift difference $\Delta z_n = \lambda_n H(z)$. In an angular average the physical distance $d_V(z)$ that is estimated from these oscillations is given by

$$d_V(z) = \left(\frac{d_A^2(z)}{(1+z)H(z)}\right)^{1/3}. \tag{8.14}$$

If we can measure the radial and transversal BAOs independently, by determining Δz_n and θ_n in the angular power spectrum we can measure

$$F(z) \equiv \frac{\Delta z_n(z)}{(1+z)\theta_n(z)} = H(z)d_A(z) = H(z)\int_0^z \frac{dz'}{H(z')}. \tag{8.15}$$

This is the so-called Alcock–Paczynski test (Alcock and Paczynski, 1979), which we can perform whenever we can see the same physical scale, here λ_n, radially and transversally. Forecasts of how well this can be measured in the angular correlation function with future surveys have been studied in the literature (Montanari and Durrer, 2012; Lepori *et al.*, 2016). At present, no experimental results on the Alcock–Paczynski test with BAOs are yet available. Some attempts have been made to use the test for stacked voids; see, for example, Mao *et al.* (2017).

One usually expands Eq. (8.13) in Legendre polynomials [see Appendix 4, Section A4.1], which we denote by L_ℓ in this chapter (in order to avoid too many P's),

$$P_{\rm obs}(\mathbf{k}, z) = P_\delta(k)D_1^2(z)\left[\beta_0(z)L_0(\mu) + \beta_2(z)L_2(\mu) + \beta_4(z)L_4(\mu)\right], \tag{8.16}$$

with

$$\beta_0 = b^2 + \frac{2bf}{3} + \frac{f^2}{5} \tag{8.17}$$

$$\beta_2 = \frac{4bf}{3} + \frac{4f^2}{7} \tag{8.18}$$

$$\beta_4 = \frac{8f^2}{35}. \tag{8.19}$$

$P_\delta(k)$ is the linear density fluctuation spectrum today.

The products $\beta_0 P_\delta(k)$ and $\beta_2 P_\delta(k)$ have been determined by observations at good accuracy, but $\beta_4 P_\delta(k)$ has not yet been positively detected. Only when we are able to measure all three quantities will we be able to break the degeneracy and isolate both b and f. In a ΛCDM or open universe $f(z) \simeq [\Omega_m(z)]^{0.55}$. For $\Omega_m(z=0) \sim 0.3$ we find that at low redshift $\beta_4 \sim 0.06$ while β_0 is of order unity or larger. At higher redshift, $z \gtrsim 2$, we have $f \simeq 1$ so that $\beta_4(z \geq 2) \simeq 8/35 \simeq 0.23$. Unfortunately, presently there are no data available at these redshifts and as we shall see in the text that follows contributions from lensing cannot be neglected at $z = 2$ and larger; see, for example, Montanari and Durrer (2015).

By Fourier transforming the power spectrum we obtain the correlation function,

$$\xi_{\rm obs}(\mathbf{d}, z) = D_1^2(z)\left[\beta_0\xi_0(d)L_0(\mu) - \beta_2\xi_2(d)L_2(\mu) + \beta_4\xi_4(d)L_4(\mu)\right], \tag{8.20}$$

where here μ denotes the direction cosine between the outward normal $-\mathbf{n}$ and the vector $\mathbf{d}$ connecting the correlated galaxies and

$$\xi_n(d) = \int \frac{k^2 dk}{2\pi^2} P_\delta(k) j_n(kd). \tag{8.21}$$

The details of this are derived in Exercise 8.1. As we denote the comoving distance out to redshift z by $r(z)$, we here use $\mathbf{d}$ for the distance vector connecting the two "pixels" that we correlate in ξ.

Before we go on we introduce an important quantity often used to characterize the amplitude of observed fluctuations. It is the mean square of number count fluctuations inside a ball of radius R, defined by

$$\langle(\Delta N/\bar{N})^2\rangle_R = \left\langle\left(\int W_R(\mathbf{y}) b\delta(\mathbf{y})\, d^3y\right)^2\right\rangle \tag{8.22}$$

$$= \int W_R(\mathbf{y}) W_R(\mathbf{x}) \xi_g(|\mathbf{y} - \mathbf{x}|)\, d^3y d^3x\,. \tag{8.23}$$

Here W_R is a window function of size R. The most common shapes for window functions are a Gaussian or a "top hat." The galaxy correlation function ξ_g is proportional to the monopole of $\xi_{\rm obs}$. Using the fact that the Fourier transform of a convolution is the product of the Fourier transforms, we have

$$\int W_R(\mathbf{x} - \mathbf{y})\delta(\mathbf{y})\, d^3y = \int \frac{d^3k}{(2\pi)^3} e^{i\mathbf{k}\cdot\mathbf{x}} W_R(k)\delta(k), \tag{8.24}$$

so that

$$\langle(\Delta N/\bar{N})^2\rangle = \int \frac{d^3k}{(2\pi)^3} |W_R(k)|^2 b^2 P_\delta(k)\,. \tag{8.25}$$

If the bias is relatively close to unity, the average galaxy number fluctuations inside a ball of radius R can therefore give the normalization of the matter power spectrum. One usually introduces the matter density fluctuation inside a ball of radius $R = 8$ Mpc to characterize the amplitude of density fluctuations:

$$\sigma_8^2 = \int \frac{d^3k}{(2\pi)^3} W^2_{8\text{Mpc}}(k) P_\delta(k) \,. \tag{8.26}$$

The mean galaxy number fluctuations inside a ball of radius $R = 8$ Mpc are related to σ_8 by the bias. While the monopole of the observed correlation function is mainly sensitive to $b^2\sigma_8^2$, the quadrupole is also sensitive to $b\sigma_8^2 f$ and the hexadecapole measures $\sigma_8^2 f^2$. More precisely, the multipoles are measuring $\beta_\ell \sigma_8^2$.

8.2.2 Lensing

In the previous section we considered radial volume distortions; in this section we consider transversal distortions due to lensing.

The observed transverse surface element is $r^2 \sin\vartheta_o d\vartheta_o d\varphi_o$. The transverse surface element at emission, that is, at the source, is $r^2 \sin\vartheta_s d\vartheta_s d\varphi_s$. Inserting $\vartheta_s = \vartheta_o + \delta\vartheta$, $\varphi_s = \varphi_o + \delta\varphi$, the ratio is to first order in the perturbations

$$\frac{\sin\vartheta_s}{\sin\vartheta_0}\left|\frac{\partial(\vartheta_s,\varphi_s)}{\partial(\vartheta_o,\varphi_o)}\right| = 1 + (\cot\vartheta_o + \partial_\vartheta)\delta\vartheta + \partial_\varphi\delta\varphi. \tag{8.27}$$

Here we used that $\det(\mathbb{I} + \epsilon M) = 1 + \epsilon \mathrm{Tr} M$ to first order in ϵ, where Tr denotes the trace. Inserting $\delta\vartheta$ and $\delta\varphi$ from Eqs. (7.13) and (7.14), and using the definition of the lensing potential (7.21) we find

$$\frac{\sin\vartheta_s}{\sin\vartheta_o}\left|\frac{\partial(\vartheta_s,\varphi_s)}{\partial(\vartheta_o,\varphi_o)}\right| = 1 - (\cot\vartheta_o\,\partial_\vartheta + \partial_\vartheta^2 + \frac{1}{\sin^2\vartheta_o}\partial_\varphi^2)\psi(\mathbf{n},z) \tag{8.28}$$

$$= 1 - \Delta_\Omega\psi(\mathbf{n},z) \;=\; 1 - 2\kappa(\mathbf{n},z). \tag{8.29}$$

Here Δ_Ω is the Laplacian on the 2-sphere w.r.t. the observed direction $-\mathbf{n} \equiv (\vartheta_o, \varphi_o)$. The only difference w.r.t. the CMB calculation is that here the convergence κ is to be taken at the source redshift z and not at the fixed CMB redshift z_*. Adding also this transversal volume fluctuation (and again neglecting fluctuations in r) we finally obtain

$$\Delta_{\text{obs}}(\mathbf{x},z) = b\delta_z + \frac{\delta V}{V} = b\delta(\mathbf{x},z) - \mathcal{H}^{-1}\mathbf{n}\cdot\nabla\left(\mathbf{n}\cdot\mathbf{V}(\mathbf{x},z)\right) - 2\kappa(\mathbf{n},z), \tag{8.30}$$

This is the correct result if we see all galaxies (of the type considered in a given survey). But a telescope has a finite sensitivity and cannot see objects that emit light below a given flux limit F_* depending on the telescope. This flux limit is usually given in terms of a so-called apparent magnitude limit,

$$m_* = -\frac{5}{2}\log_{10} F_* + \text{const.}, \tag{8.31}$$

where the constant is traditionally defined such that the star Vega has apparent magnitude zero. Note, the lower the magnitude of a galaxy the brighter it is. If galaxies are too faint, they are not observed in the given survey. However, due to the lensing magnification κ some galaxies that would be intrinsically too faint are amplified above the flux limit and make it into our survey. Denoting the mean number of galaxies with observed magnitude below m_*, flux higher than F_*, or intrinsic luminosity above $L_*(z)$ by $\bar{n}_g(z, L_*)$, the observed number of galaxies below this magnitude in a given direction $-\mathbf{n}$ at redshift z is corrected by

$$n_g(z, L_*, \mathbf{n}) = \bar{n}_g(z, L_*) + \left.\frac{\partial \bar{n}(z, L)}{\partial \ln L}\right|_{L=L_*} \frac{\delta L}{L}. \tag{8.32}$$

Neglecting other relativistic effects apart from the focusing of light by lensing, we have $\delta L/L = \mu - 1 = 2\kappa$; see Eq. (7.20). Introducing the magnification bias

$$s(z, m_*) = \frac{2}{5}\left.\frac{\partial \bar{n}(z, L)}{\partial \ln L}\right|_{L=L_*(z)} \tag{8.33}$$

we obtain for the number counts

$$\begin{aligned}\Delta_{\text{obs}}(\mathbf{n}, z) &= b\delta_z + \frac{\delta V}{V} + \left.\frac{\partial \bar{n}, L}{\partial \ln L}\right|_{L=L_*} \frac{\delta L}{L} \\ &= b\delta(r(z)\mathbf{n}, z) - \mathcal{H}^{-1}\mathbf{n}\cdot\nabla\left(\mathbf{n}\cdot\mathbf{V}(r(z)\mathbf{n}, z)\right) - (2 - 5s)\kappa(\mathbf{n}, z).\end{aligned} \tag{8.34}$$

The strange prefactor in the definition of s comes from the fact that it was originally defined as a derivative w.r.t the apparent magnitude m_*.

This is the formula that includes RSD and the two effects of lensing. It was considered for the first time by Matsubara (2000). The increase of the transversal volume reduces the number count per volume element while focusing enhances the number of galaxies that make it into a given survey. The two lensing terms therefore have opposite signs (as $\bar{n}_g(m)$ is monotonically growing with increasing m, s is always positive) and, depending on the value of s one or the other term may dominate. At low redshift we usually see most galaxies and s is small. At high redshift, however, we see only the brightest objects and s can become quite large. Depending on the survey, the prefactor $(2 - 5s)$ can change sign at a given redshift. To take the lensing effect correctly into account we therefore have to measure $s(z)$ for the given sample of galaxies. This can be done by choosing m_* slightly higher than the true limiting magnitude and counting $n_g(m_* - dm)$ and $n_g(m_* + dm)$.

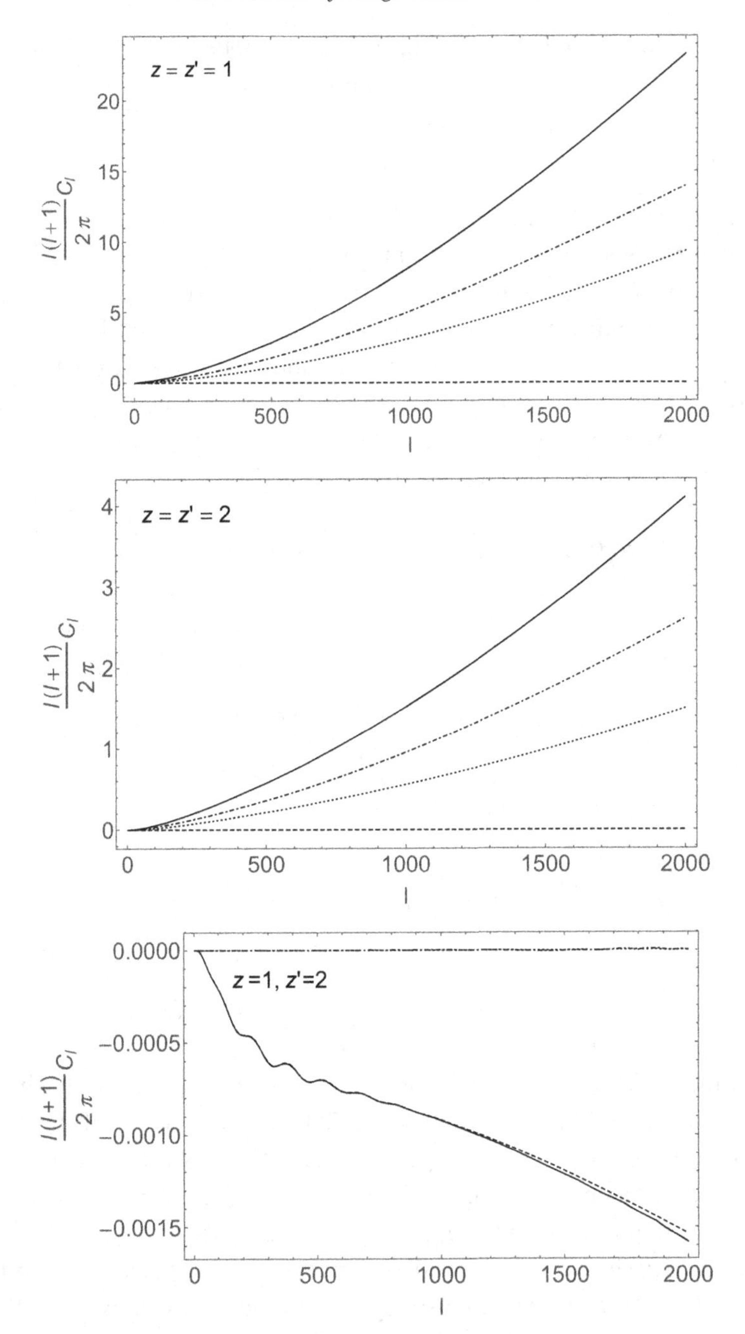

Fig. 8.1 We show the observed matter angular power spectrum for $z = z' = 1$ (top panel), $z = z' = 2$ (middle panel), and $z = 1, z' = 2$ (bottom panel). The solid line is the full result, the dotted line is the density term only, the dash-dotted line shows the RSD and RSD-density correlation, and the dashed line shows all terms containing κ. We have set $b = 1$ and $s = 0$ in these plots.

In Eq. (8.34) the lensing term $\kappa(\mathbf{n}, z) = \Delta\psi(\mathbf{n}, z)$ does not simply depend on $\mathbf{x} = -r\mathbf{n}$. Like the lensing potential ψ it is given as an integral along the line of sight. Therefore, there is no simple, straightforward way to convert the expression (8.34) into a power spectrum in Fourier space. It is much more natural to consider it as a redshift-dependent correlation function or angular power spectrum:

$$\Delta_{\text{obs}}(\mathbf{n}, z) = \sum_{\ell m} a_{\ell m}(z) Y_{\ell m}(n) \tag{8.35}$$

$$\langle a_{\ell m}(z) a^*_{\ell' m'}(z') \rangle = C_\ell(z, z') \delta_{\ell\ell'} \delta_{m,m'} \tag{8.36}$$

$$\langle \Delta_{\text{obs}}(\mathbf{n}, z) \Delta_{\text{obs}}(\mathbf{n}', z') \rangle = \frac{1}{4\pi} \sum_\ell (2\ell + 1) C_\ell(z, z') L_\ell(\mathbf{n} \cdot \mathbf{n}'). \tag{8.37}$$

In Fig. 8.1 we show the power spectra including density, RSD, and lensing for $z = z' = 1$ (top panel), $z = z' = 2$ (middle panel), and $z = 1, z' = 2$ (lower panel). Interestingly, for both equal redshift correlations the RSD contributions are larger than the density term. This is due to our choice of $b = 1$ in the plot. (Note also that the "RSD contribution" contains both the RSD$\times$density and the RSD$\times$RSD terms.) Furthermore, the lensing term is negligible in equal redshift correlations for $z \leq 2$. However, for $z = 1$ and $z' = 2$ the lensing term largely dominates the result and the density and RSD contributions are negligible. Of course, this extreme case of an off-diagonal spectrum has a very low amplitude and it will be very difficult to observe it.

From the angular power spectrum, which is well defined also for wide angle surveys, the growth function f cannot be readily extracted. Therefore, it is also very useful to measure the correlation function given in Eq. (8.20). For small angular separation we can define a common direction $\bar{\mathbf{n}}$ and split the distance $\mathbf{d} = r(z)\mathbf{n} - r(z')\mathbf{n}'$ into a radial and a transversal part, $\mathbf{d} = (r(z) - r(z'))\bar{\mathbf{n}} + \mathbf{d}_\perp$. For small redshift differences, $z = \bar{z} + \Delta z/2$, $z' = \bar{z} - \Delta z/2$; the correlation function can then be understood as a function of $d = |\mathbf{d}|$, $\mu = (r(z) - r(z'))/d$ and $\bar{z} = (z + z')/2$. As long as lensing can be neglected, the μ-dependence expressed in Legendre polynomials is proportional[1] to the terms β_0, $-\beta_2$, and β_4.

8.3 The Fully Relativistic Angular Matter Power Spectrum

In this section we determine the fully relativistic matter power spectrum. The derivation is now more involved as we take into account the full perturbed metric and also perturbations of the radial distance. Even though the final formula is

[1] When going from the power spectrum to the correlation function the terms that are not of the form $4n$ acquire a minus sign due to the expansion of the exponential in Legendre polynomials and spherical Bessel functions; see Eq. (2.259) and Exercise 8.1.

significantly longer than Eq. (8.34), the new terms are suppressed by at least one power of $\mathcal{H}/k$ w.r.t. the density, RSD, and lensing. For typical surveys with redshifts up to $z \sim 3$ they can be safely neglected for $\ell > 10$. For single-tracer surveys in a parity symmetric universe, terms with odd powers of $k/\mathcal{H}$ cannot be correlated with terms containing even powers so that the power spectrum of contribution with large-scale terms is suppressed at least by $(\mathcal{H}(z)/k)^2$. For $\ell = 10$ and redshift $z \gtrsim 1$, this gives a suppression factor of about $\ell^{-2} = 0.01$. Nevertheless, we want to compute them here. First of all, as we shall see, at very low redshifts some relativistic terms become relevant. Also, if ever we can go to significantly higher redshifts, for example, $z \sim 20$ or so, for example, with intensity mapping, some terms become relevant. Furthermore, they show how galaxy number counts are in principle sensitive not only to the density and velocity fields but also to metric perturbations. Therefore, they can be used to test the consistency of LSS with General Relativity. For this, we do not even need the large-scale relativistic effects that we now determine; we can determine the lensing potential at arbitrary redshift already with the dominant lensing term. Finally, the relativistic terms that we compute below, when converted to a power spectrum, lead to an upturn on very large scales, exactly like the effect of a primordial non-Gaussianity as described in Dalal *et al.* (2008). Therefore, neglecting it might lead to a false "discovery" of primordial non-Gaussianity.

8.3.1 Derivation of the Relativistic Number Counts

Let us present the fully relativistic derivation. We shall do the derivation in longitudinal (Newtonian) gauge. Since all multipoles with $\ell \geq 2$ vanish in the background they are gauge invariant (Stewart's lemma). The monopole and dipole, however, are gauge dependent and we shall not consider them. For this reason we also disregard terms at the observer that contribute only to the monopole or the dipole. We present the derivation for vanishing spatial curvature so that $\chi(t) = r = t_0 - t$. The expression for number counts in a spatially curved universe can be found in Di Dio *et al.* (2016).

We first note that $\delta_z = \delta\rho/\bar{\rho}|_z$, that is, the density fluctuation at fixed redshift is related to the density fluctuation at fixed time t by

$$\delta_z = \frac{\delta\rho}{\bar{\rho}} + \frac{d\bar{\rho}}{dz}\delta z = D_s - 3\frac{\delta z}{1+z}. \tag{8.38}$$

Here D_s is the matter density fluctuation in longitudinal gauge and δz is the redshift perturbation in longitudinal gauge given in Eq. (2.236). Let us also determine the relativistic volume perturbation. The three-dimensional (spatial) volume element has to be defined w.r.t. an observer moving with 4-velocity u^μ as

$$\begin{aligned} dV &= \sqrt{-g}\epsilon_{\mu\nu\alpha\beta}u^\mu dx^\nu dx^\alpha dx^\beta \\ &= \sqrt{-g}\ \epsilon_{\mu\nu\alpha\beta}u^\mu \frac{\partial x^\nu}{\partial z}\frac{\partial x^\alpha}{\partial \vartheta_s}\frac{\partial x^\beta}{\partial \varphi_s}\left|\frac{\partial(\vartheta_s,\varphi_s)}{\partial(\vartheta_o,\varphi_o)}\right| dz d\vartheta_o d\varphi_o \\ &\equiv v(z,\vartheta_o,\varphi_o)dz d\Omega_o\,, \end{aligned} \tag{8.39}$$

where $d\Omega_o = \sin\vartheta_o d\vartheta_o d\varphi_o$, z is the source redshift, and we have introduced the density v that defines the volume perturbation,

$$\frac{\delta V}{V} = \frac{v-\bar{v}}{\bar{v}} = \frac{\delta v}{\bar{v}}.$$

As previously, a suffix o denotes the observer position while a suffix s denotes the source (galaxy) position. In addition to the Jacobian of the transformation from the angles at the source to the angles at the observer, which we already had in the previous section, $\left|\frac{\partial(\vartheta_s,\varphi_s)}{\partial(\vartheta_o,\varphi_o)}\right|$, there are now terms coming from $\sqrt{-g}$ and the perturbations of the radial distance. Equation (8.39) is still exact. To first order the perturbed angles at the source, $\vartheta_s = \vartheta_o + \delta\vartheta$ and $\varphi_s = \varphi_o + \delta\varphi$, have been determined in Eqs. (7.13) and (7.14). As in the previous section, at first order in the perturbations, the Jacobian determinant is

$$\left|\frac{\partial(\vartheta_s,\varphi_s)}{\partial(\vartheta_o,\varphi_o)}\right| = 1 + \frac{\partial\delta\vartheta}{\partial\vartheta} + \frac{\partial\delta\varphi}{\partial\varphi}. \tag{8.40}$$

Using the first-order expression for the metric determinant, $\sqrt{-g} = a^4(1+\Psi-3\Phi)$, and the 4-velocity of the source, $(u^\mu) = \frac{1}{a}(1-\Psi, V^i)$, we find to first order

$$v = a^3(1+\Psi-3\Phi)\left[\frac{dr}{dz}r^2\frac{\sin\vartheta_s}{\sin\vartheta_o}\left(1+\frac{\partial\delta\vartheta}{\partial\vartheta}+\frac{\partial\delta\varphi}{\partial\varphi}\right)(1-\Psi+V_r)\right]. \tag{8.41}$$

Here dr/dz is to be understood as the derivative of the comoving distance r with respect to the redshift along the photon geodesic. At linear order we can write (the distinction between the true z and the background $\bar{z}$ is relevant only for background quantities)

$$\frac{dr}{dz} = \frac{d\bar{r}}{d\bar{z}} + \frac{d\delta r}{d\bar{z}} - \frac{d\delta z}{d\bar{z}}\frac{d\bar{r}}{d\bar{z}} = \left(\frac{d\bar{r}}{dt} + \frac{d\delta r}{d\lambda} - \frac{d\delta z}{d\lambda}\frac{d\bar{r}}{d\bar{z}}\right)\frac{dt}{d\bar{z}}, \tag{8.42}$$

where we have used that for first-order quantities we can set $dt = d\lambda$ when we have to take the derivative along the photon geodesic. The last term of Eq. (8.42) contains the redshift space distortion discussed in the previous section. To lowest order along a photon geodesic $-d\bar{r}/d\bar{z} = dt/d\bar{z} = -H^{-1} = -a/\mathcal{H}$. With this the volume element becomes

$$v = \frac{a^4\bar{r}^2}{\mathcal{H}}\left[1 - 3\Phi + \left(\cot\vartheta_o + \frac{1}{\partial\vartheta}\right)\delta\vartheta + \frac{\partial\delta\varphi}{\partial\varphi} - \mathbf{V}\cdot\mathbf{n} + \frac{2\delta r}{r} - \frac{d\delta r}{d\lambda} + \frac{a}{\mathcal{H}}\frac{d\delta z}{d\lambda}\right]. \tag{8.43}$$

From this we subtract the unperturbed part $\bar{v}(z)$ evaluated at the observed redshift, $z = \bar{z} + \delta z$,

$$\bar{v}(z) = \bar{v}(\bar{z}) + \frac{d\bar{v}}{d\bar{z}}\delta z.$$

With the unperturbed expression, $a = 1/(\bar{z}+1)$,

$$\bar{v}(\bar{z}) = \frac{\bar{r}^2}{(1+\bar{z})^4\mathcal{H}} \tag{8.44}$$

and we obtain

$$\begin{aligned}\frac{\delta v}{\bar{v}}(\mathbf{n}, z) &= \frac{v(z) - \bar{v}(z)}{\bar{v}(z)}\\ &= -3\Phi + \left(\cot\vartheta_o + \frac{\partial}{\partial\vartheta}\right)\delta\vartheta + \frac{\partial\delta\varphi}{\partial\varphi} - \mathbf{V}\cdot\mathbf{n} + \frac{2\delta r}{r} - \frac{d\delta r}{d\lambda}\\ &\quad + \frac{1}{\mathcal{H}(1+\bar{z})}\frac{d\delta z}{d\lambda} - \left(-4 + \frac{2}{\bar{r}\mathcal{H}} + \frac{\dot{\mathcal{H}}}{\mathcal{H}^2}\right)\frac{\delta z}{1+\bar{z}}.\end{aligned} \tag{8.45}$$

To compute the radial perturbation δr we have to integrate the photon geodesic from the source to the observer. We use

$$\frac{dx^\mu}{d\lambda} = n^\mu, \tag{8.46}$$

where n^μ denotes the photon 4-velocity. Neglecting perturbations at the observer position that give raise to unobservable monopole and dipole terms, using Eqs. (7.5) and (7.6) for vanishing curvature we find

$$\delta x^i(t_s) = -2\int_0^{r_s} dr(\Psi + \Phi)\mathbf{n}^i - \int_0^{r_s} dr(r_s - r)\left((\Psi + \Phi)_{,i} + (\dot{\Psi} + \dot{\Phi})\mathbf{n}^i\right), \tag{8.47}$$

where we have used $r_s - r(\lambda) = \lambda$ and $dr = -d\lambda$ to lowest order. From this we obtain

$$\delta r \equiv \delta x^i n_{ri} = \int_0^{r_s} dr(\Phi + \Psi). \tag{8.48}$$

We have also used that $\mathbf{n} = -\mathbf{n}_r$, $\mathbf{n}^i\partial_i + \partial_t = \frac{d}{d\lambda} = \frac{d}{dt}$ and $r_s = t_0 - t_s$ to lowest order. For the derivative of δr we obtain

$$\frac{d\delta r}{d\lambda} = -(\Phi + \Psi). \tag{8.49}$$

Inserting Eqs. (7.13) and (7.14) for the angular contribution to the volume we find, as in the previous section,

$$\begin{aligned}(\cot\vartheta + \partial_\vartheta)\delta\vartheta + \partial_\varphi\delta\varphi &= -\int_0^{r_s} dr \frac{(r_s - r)}{r r_s}\Delta_\Omega(\Phi + \Psi) \\ &= -\int_0^{r_s} dr \frac{(r_s - r)r}{r_s}\Delta_\perp(\Phi + \Psi) = -2\kappa,\end{aligned} \tag{8.50}$$

where Δ_Ω denotes the angular part of the Laplacian and $\Delta_\perp \equiv r^{-2}\Delta_\Omega$,

$$\Delta_\Omega \equiv \left(\cot\vartheta\,\partial_\vartheta + \partial_\vartheta^2 + \frac{1}{\sin^2\vartheta}\partial_\varphi^2\right). \tag{8.51}$$

Adding all the contributions of Eq. (8.45) together we obtain

$$\begin{aligned}\frac{\delta v}{v} &= -2(\Psi + \Phi) - 4\mathbf{V}\cdot\mathbf{n} + \frac{1}{\mathcal{H}}\left[\dot{\Phi} + \partial_r\Psi - \frac{d(\mathbf{V}\cdot\mathbf{n})}{d\lambda}\right] \\ &+ \left(\frac{\dot{\mathcal{H}}}{\mathcal{H}^2} + \frac{2}{r_s\mathcal{H}}\right)\left(\Psi + \mathbf{V}\cdot\mathbf{n} + \int_0^{r_s} dr(\dot{\Phi} + \dot{\Psi})\right) \\ &- 3\int_0^{r_s} dr(\dot{\Phi} + \dot{\Psi}) + \frac{2}{r_s}\int_0^{r_s} dr(\Phi + \Psi) - \frac{1}{r_s}\int_0^{r_s} dr\frac{r_s - r}{r}\Delta_\Omega(\Phi + \Psi).\end{aligned} \tag{8.52}$$

Adding this to the density perturbation in redshift space given in Eq. (8.38), we obtain the number count fluctuations to first order first derived by Bonvin and Durrer (2011),

$$\begin{aligned}\Delta(\mathbf{n}, z) &= D_s - 2\Phi + \Psi + \frac{1}{\mathcal{H}}\left[\dot{\Phi} + \partial_r(\mathbf{V}\cdot\mathbf{n})\right] \\ &+ \left(\frac{\dot{\mathcal{H}}}{\mathcal{H}^2} + \frac{2}{r_s\mathcal{H}}\right)\left(\Psi + \mathbf{V}\cdot\mathbf{n} + \int_0^{r_s} dr(\dot{\Phi} + \dot{\Psi})\right) \\ &+ \frac{1}{r_s}\int_0^{r_s} dr\left[2 - \frac{r_s - r}{r}\Delta_\Omega\right](\Phi + \Psi).\end{aligned} \tag{8.53}$$

Here we have also used the momentum conservation equation (2.119) for pressureless matter,

$$\mathbf{n} \cdot \dot{\mathbf{V}} + \mathcal{H}\mathbf{n} \cdot \mathbf{V} - \partial_r \Psi = 0,$$

in order to remove the term $\dot{\mathbf{V}}$ in $d\mathbf{V}/d\lambda = \dot{\mathbf{V}} + n^i \partial_i \mathbf{V}$. The terms in the integrals have always to be evaluated at the positions $\mathbf{x} = -r\mathbf{n}, t = t_0 - r$, while the source position is $\mathbf{x}_s = -r_s\mathbf{n}, t_s = t_0 - r_s$ and we set $\mathbf{x}_o \equiv 0$.

Equation (8.53) is the observable linear matter density fluctuation in angular and redshift space. In Bonvin and Durrer (2011) it is also shown that this expression is gauge invariant. Note that we did not use Einstein's equation in this derivation, which is therefore valid for all metric theories of gravity, that is, theories in which photons and dark matter particles move along geodesics.

The term $\mathcal{H}^{-1}\partial_r(\mathbf{V} \cdot \mathbf{n})$ is the well-known RSD, while the last term of Eq. (8.53) is simply the convergence κ, that is, the trace of the Jacobian of the lens map, which we have obtained in the previous section,

$$-2\kappa = -\Delta_\Omega \psi = -\Delta_\Omega \int_0^{r_s} dr \frac{r_s - r}{r_s r} (\Phi + \Psi) . \tag{8.54}$$

As already discussed, galaxies are biased tracers of the matter density fluctuations. In relativistic perturbation theory there are different gauge-invaiant definitions of the matter density fluctuation, and we have to decide which one might be linearly related to the galaxy density. It is physically most sensible to assume that a linear relation exists between the matter density and the galaxy density in *comoving gauge*, that is, in the gauge where matter is at rest. The matter density in this gauge is D, which is related to the density D_s in longitudinal gauge by [see Eq. (2.87)]

$$D = D_s - \frac{\dot{\rho}}{\rho} V_s = D_s + 3\mathcal{H} V_s, \tag{8.55}$$

where V_s is the velocity potential introduced in the previous section, $\mathbf{V} = -\nabla V_s$. (Note that V_s in real space defined in this way has the dimension of a length. It is related to the dimensionless V in Fourier space defined in Eq. (2.85) via $V(k) = kV_s(k)$). We assume that in comoving gauge the galaxy number density fluctuation is proportional to the matter density fluctuation,

$$\delta_g = bD, \tag{8.56}$$

where b is a bias factor that generically depends on redshift. Bias can also be more complicated, scale-dependent, nonlinear, stochastic etc., but we do not consider these possibilities in our discussion.

Furthermore, the comoving galaxy number density may increase due to the formation of new galaxies (or decrease due to mergers), so that the physical number density of galaxies decays slower (or faster) than the mean matter density. We model this as

$$\frac{\dot{N}}{N} = (1 - b_e/3)\frac{\dot{\rho}}{\rho}, \tag{8.57}$$

where b_e is called "evolution bias." Therefore we have to replace D_s not simply by $bD - 3\mathcal{H}V_s$ but by

$$D_s^{\rm obs} = bD - (3 - b_e)\mathcal{H}V_s. \tag{8.58}$$

In addition, in Eq. (8.38) we have used that $d\bar{\rho}_m/dz = -3\bar{\rho}_m/(1+z)$. If galaxies are generated as modeled with the evolution bias b_e in Eq. (8.57), we have to replace the term $-3\delta z/(1+z)$ by $(-3+b_e)\delta z/(1+z)$. This adds a term $-b_e$ in the parentheses $(\dot{\mathcal{H}}/\mathcal{H}^2 + 2/(r_s\mathcal{H}))$ of Eq. (8.53).

Finally, we also want to take into account magnification bias. As in the previous section, the number count fluctuations up to a limiting flux F_* are given by

$$\Delta_g(\mathbf{n}, z, m_*) = \Delta_g(\mathbf{n}, z) + \left.\frac{\partial \ln \bar{n}_g(z, L)}{\partial \ln L}\right|_{L=L_*} \frac{\delta L}{\bar{L}} = \Delta_g(\mathbf{n}, z) + \frac{5s}{2}\frac{\delta L}{\bar{L}}, \tag{8.59}$$

where s is as defined in Eq. (8.33). In a relativistic treatment we have to be more careful in the determination of the luminosity perturbation. In terms of the fluctuation of the luminosity distance it is given by $\delta L/\bar{L} = -2\delta D_L/\bar{D}_L$. In Appendix 10 we calculate the perturbation of the luminosity distance. Inserting the result given in Eq. (A10.27) and putting all the biasing effects together, we find the following result first derived by Challinor and Lewis (2011):

$$\begin{aligned}\Delta_g(\mathbf{n}, z, m_*) = {} & bD - (3 - b_e)\mathcal{H}V + \frac{1}{\mathcal{H}}\left[\dot{\Phi} + \partial_r(\mathbf{V}\cdot\mathbf{n})\right] \\ & + \left(\frac{\dot{\mathcal{H}}}{\mathcal{H}^2} + \frac{2-5s}{r_s\mathcal{H}} + 5s - b_e\right)\left(\Psi + \mathbf{V}\cdot\mathbf{n} + \int_0^{r_s} dr(\dot{\Phi} + \dot{\Psi})\right) \\ & - (2-5s)\Phi + \Psi + \frac{2-5s}{2r_s}\int_0^{r_s} dr\left[2 - \frac{r_s - r}{r}\Delta_\Omega\right](\Phi + \Psi).\end{aligned} \tag{8.60}$$

Like in Eq. (8.34), also in Eq. (8.60) s enters mainly in the combination $2 - 5s$. The first term is a transversal volume distortion. Focusing increases the angular separation of two points at a given transverse distance and hence lets the volume appear larger and the density smaller. On the other hand, focusing also enhances the luminosity of sources and galaxies that otherwise would be too faint to make it into our surveys, leading to an enhanced density. Depending on the sign of $2 - 5s$ one or the other effect wins. As we shall see in Section 8.5, for intensity maps the two effects exactly cancel. If we do not count individual sources but the intensity coming from a certain area, the area "appears" larger due to focusing exactly by the increase in the luminosity coming from it so that the surface brightness is conserved. Therefore for intensity mapping we can simply set $s = 2/5$ and there is no lensing effect at first order in perturbation theory. This is also the case for CMB observations where lensing is a second-order effect, as we have seen in Chapter 7.

8.3.2 The Power Spectrum

Let us now compute the relativistic angular matter power spectrum for the case of purely scalar adiabatic fluctuations. We assume that the initial fluctuations are given in Fourier space by the initial curvature fluctuation $\zeta(\mathbf{k})$ defined in Eq. (2.145) that has been generated during inflation with some power spectrum

$$k^3\langle\zeta(\mathbf{k})\zeta^*(\mathbf{k}')\rangle = (2\pi)^3\delta^3(\mathbf{k}-\mathbf{k}')\mathcal{P}_\zeta(k). \tag{8.61}$$

The star indicates complex conjugation.

In the simplest models of adiabatic perturbations all scalar perturbations at later times are determined by the random variable $\zeta(\mathbf{k})$ via a deterministic transfer function,

$$X(\mathbf{k},z) = T_X(k,z)\zeta(\mathbf{k}), \qquad \mathbf{V}(\mathbf{k},z) = i\hat{\mathbf{k}}T_V(k,z)\zeta(\mathbf{k}). \tag{8.62}$$

The first equation applies for scalar quantities while the second one applies for (spatial) vectors; $\hat{\mathbf{k}}$ is the unit vector in direction $\mathbf{k}$. Note that within first-order perturbation theory, z in these perturbation variables can be related to t via the background Friedmann model. In Fourier space therefore the vector $\mathbf{V} = i\hat{\mathbf{k}}V$ and the potential V have the same dimension. [As already mentioned, the Fourier transform of the velocity potential $V_s(\mathbf{x})$ is not $V(\mathbf{k})$ but $k^{-1}V(\mathbf{k})$.] The transfer functions depend on the content of the Universe and on the theory of gravity. We shall see in the next chapter that measuring CMB anisotropies and polarization, but also the galaxy power spectrum under the assumption of simple initial conditions, allows us to estimate cosmological parameters.

To determine the number count power spectrum $C_\ell(z,z')$ and the angular correlation function

$$\xi(\theta,z,z') = \frac{1}{4\pi}\sum_\ell(2\ell+1)C_\ell(z,z')P_\ell(\cos\theta), \tag{8.63}$$

we make use of Eq. (2.259). A short calculation using Eq. (8.60) gives (see Bonvin and Durrer, 2011)

$$C_\ell(z,z') = \frac{2}{\pi}\int\frac{dk}{k}\mathcal{P}_\zeta(k)F_\ell(k,z)F_\ell(k,z'), \tag{8.64}$$

where $r_s \equiv r(z) = t_0 - t(z)$ is the comoving distance of the source and

$$F_\ell(k,z) = j_\ell(kr_s)\left[bT_D - (3-b_e)\frac{\mathcal{H}}{k}T_V + (2-5s)T_\Phi + \frac{1}{\mathcal{H}}\dot{T}_\Phi \right.$$
$$\left. + \left(1 - \frac{\dot{\mathcal{H}}}{\mathcal{H}^2} + \frac{2-5s}{r_s\mathcal{H}} + 5s - b_e\right)T_\Psi\right]$$
$$+ j_\ell'(kr_s)\left(\frac{\dot{\mathcal{H}}}{\mathcal{H}^2} + \frac{2-5s}{r_s\mathcal{H}} + 5s - b_e\right)T_V + \frac{k}{\mathcal{H}}T_V j_\ell''(kr_s)$$
$$+ \frac{2-5s}{2r_s}\int_0^{r_s} j_\ell(kr)\left(2 + \frac{r_s - r}{r}\ell(\ell+1)\right)(T_\Psi + T_\Phi)dr$$
$$+ \left(\frac{\dot{\mathcal{H}}}{\mathcal{H}^2} + \frac{2-5s}{r_s\mathcal{H}} + 5s - b_e\right)\int_0^{r_s} j_\ell(kr)(\dot{T}_\Psi + \dot{T}_\Phi)dr\,. \tag{8.65}$$

The prime in the spherical Bessel functions denotes the derivative w.r.t. the argument. Let us briefly estimate the order of magnitude of the different terms in a standard ΛCDM cosmology. Neglecting anisotropic stresses and using the perturbed Einstein equations, Eqs. (2.105) and (2.107) in ΛCDM, it is easy to express all the transfer functions in terms of T_Ψ,

$$T_\Phi = T_\Psi \tag{8.66}$$

$$T_D = -\frac{2a}{3\Omega_m}\left(\frac{k}{\mathcal{H}_0}\right)^2 T_\Psi \tag{8.67}$$

$$T_V = \frac{2a}{3\Omega_m}\frac{k}{\mathcal{H}_0^2}\left(\mathcal{H}T_\Psi + \dot{T}_\Psi\right). \tag{8.68}$$

Here $\mathcal{H}_0 = H_0$ is the present Hubble parameter and Ω_m is the present matter density parameter. On scales $k \gg \mathcal{H}_0$ the density term $\propto (k/\mathcal{H}_0)^2 T_\Psi$ and the RSD term $\propto (k/\mathcal{H})T_V \propto (k/\mathcal{H}_0)^2 T_\Psi + \cdots$ clearly dominate (the $\cdots$ denote additional subdominant contributions). Furthermore, considering that for source redshift of order unity and more, $r_s \sim \mathcal{H}_0^{-1}$, so that a given angular scale $\ell \simeq \pi/\theta \simeq kr_s \simeq k/\mathcal{H}_0$, we find that for sufficiently high redshifts, $z \gtrsim 1$; also the κ-contribution to the lensing term is of the same order. However, the line of sight integral of the lensing term smoothes overdensities and underdensities so that the first two terms are dominant for correlations at equal redshifts, $z = z'$. The RSD is the dominant radial volume distortion while the κ or magnification term is the dominant transversal volume distortion.

When the redshift difference is substantial or when a wide redshift window is used, the lensing term can be as large or even larger than the standard density and RSD terms (see Montanari and Durrer, 2015). The remaining gravitational potential and Doppler terms are relevant only on very large scales and special techniques such as multitracer methods are required to render them observable (see, e.g., Alonso and Ferreira, 2015; Irsic *et al.*, 2015).

All the terms multiplied by $2-5s$ appear also in the relativistic weak lensing calculation, in the perturbation of the determinant of the magnification matrix given in Eq. (7.19). The term $(2/r_s)\int_0^{r_s}(\Phi+\Psi)dr$ is the Shapiro time delay coming from the prolongation of the photon path when it passes through a potential well. In addition to this there are the integrated Sachs–Wolfe term (ISW), the Doppler term, and the value of the gravitational potential at the source; see also Appendix 10.

Contrary to the density fluctuation $D(\mathbf{x},t)$, the observable number count fluctuations $\Delta(\mathbf{n},z)$ also have contributions from vector and tensor fluctuations. We do not derive these here. They can be found in the literature; for example, the power spectrum $C_\ell(z,z')$ from tensor fluctuations is derived in Bonvin and Durrer (2011) while the one for vector perturbation is given in Durrer and Tansella (2016).

In Fig. 8.2 we show the large-scale relativistic corrections to the $C_\ell(z,z')$ power spectra. For $z=z'$ where $z=2$ or 1, these never exceed a fraction of 10^{-3} of

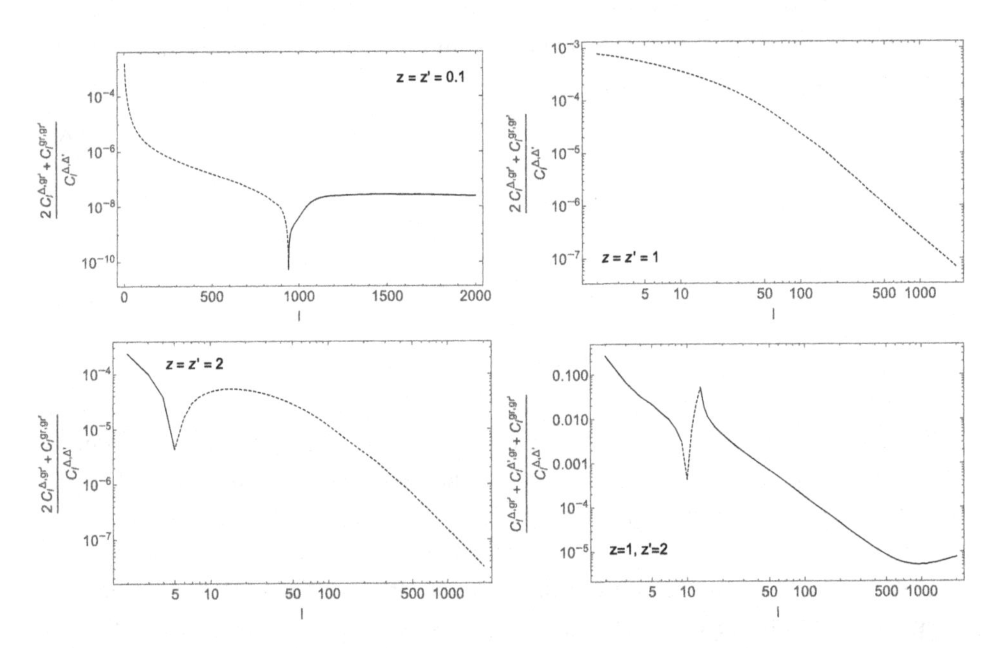

Fig. 8.2 The large-scale relativistic corrections to the power spectrum from the gravitational potential terms are shown for $z=z'=0.1$ (top, left panel), $z=z'=1$ (top, right panel), $z=z'=2$ (bottom, left panel), and $z=1, z'=2$ (bottom, right panel). The dashed lines show negative contributions (in log scale). The contributions are always significantly less than 1% except at very low ℓ in the case $1=z\neq z'=2$.

the total result. For low redshifts, $z \simeq 0.1$, the relativistic contributions are much larger, as we shall see when studying the correlation function. Here we have chosen a delta function window, or perfect resolution, in redshift. Smoothing over a wider redshift window can enhance the fractional contribution of the relativistic terms as it reduces the density and RSD contributions. Also for $z \neq z'$ (see bottom right panel of Fig. 8.2) the relativistic terms can make up to 20% on large scales.

At present, observers do not yet systematically measure the full angular power spectrum of number counts. So far they mostly used the flat sky approximation and determined the power spectrum from a small-angle patch of the sky or the correlation function, which we discuss in the next section. Exceptions are recent photometric surveys such as, for example, DES in Abbott *et al.* (2018). But also in this analysis the large-scale relativistic contributions and, more relevant, also the lensing term are neglected.

8.4 The Correlation Function

The observable angular-redshift power spectrum $C_\ell(z, z')$ is routinely calculated with fast codes, as we have them for the CMB angular power spectrum. The presently most popular CMB codes CAMB and CLASS have been extended to compute also these spectra see Challinor and Lewis (2011) and Di Dio *et al.* (2013). This is very useful as these spectra contain all the observable information. However, to compute only the C_ℓ's is not really optimal for spectroscopic redshift surveys. These surveys can observe tens of millions of galaxies with a redshift resolution of about $\Delta z = 10^{-3}$ over a redshift interval $z \in [0.5, 2.5]$ that amounts to about 2000 redshift bins. The full computation of all possible $C_\ell(z, z')$ therefore comprises more than a million spectra. As we shall see in the next chapter, when using angular power spectra to estimate cosmological parameters we proceed via the so-called Markov chain Monte Carlo technique, which requires the computation of about 10^5 spectra per chain. For the number counts this would be equivalent to 10^{11} CMB spectra, which is simply forbidding even if highly parallelized.

Furthermore, with 1000 bins there are only about $N \sim 10^3$ to 10^4 galaxies per bin, which implies significant "shot noise," that is, the fact that we have a finite number N of galaxies to probe this function, which induces fluctuations of order $\sqrt{N}/N$ on all scales. $C_\ell(z, z')$ is a function of three variables, which is much harder to determine by observations than a simple power spectrum $P_\delta(k)$ or, including RSD, $\beta_0 P_\delta(k)$ together with β_2/β_0 and β_4/β_0. Therefore, shot noise is usually the limiting factor, especially for high ℓ's.

As we have seen in the previous section, beyond $\ell \sim 20$ only three terms are really important: the density, redshift space distortions, and the lensing term. Furthermore, density and RSD generate only a monopole, quadrupole, $(n = 2)$

and hexadecapole ($n = 4$) in the correlation function. The lensing, however, also generates higher (even) multipoles. As we have seen in Section 8.2, if RSD is the dominating contribution to the quadrupole ($n = 2$) and hexadecapole ($n = 4$), their measurements can be used to isolate the bias and the growth function.

We shall see in the text that follows, that for close redshifts, $|z - z'| \ll 1$, we can define a fully relativistic correlation function that in the limit of small d and small redshifts reduces to the one determined in Section 8.2. We first introduce the angular correlation function as for the CMB in Eq. (2.253),

$$\xi(\theta, z, z') = \frac{1}{4\pi}\sum_{\ell}(2\ell+1)C_\ell(z,z')L_\ell(\cos\theta). \tag{8.69}$$

Using the cosine law, the distance d between the two pixels that we correlate is given by ($r = r(z)$, $r' = r(z')$):

$$d = \sqrt{r^2 + r'^2 - 2rr'\cos\theta}. \tag{8.70}$$

Note that the value $r(z)$ depends on the cosmological parameters. Only for very small z do we have $r(z) = H_0^{-1}z$ [see Eq. (1.47)], and we can absorb the dependence of r on H_0 by measuring distances in units of h^{-1}Mpc. For redshifts of order unity and more, $r(z)$ depends also on Ω_m, Ω_Λ (or whatever parameterizes dark energy) and on the curvature Ω_K (which is set to 0 in expression (8.70).

We introduce also

$$\mu = \frac{r - r'}{d} = \frac{d_\parallel}{d} \qquad \text{and} \qquad d_\perp = \sqrt{d^2 - d_\parallel^2}. \tag{8.71}$$

Elementary geometry shows that μ is the cosine of the angle α in Fig. 8.3. The definition of μ requires the measurement of θ, z, z' and the choice of a cosmology that determines $r(z)$, $r(z')$ and via Eq. (8.70) also d. So far, observers have used somewhat different definitions of μ, for example, the angle between the vector $\mathbf{d}$ and the radial line dividing the angle θ or the one dividing d, but the results are very similar for most choices [see Tansella *et al.* (2018) for a study of this].

Setting $\bar{z} = (z + z')/2$ and $\Delta z = (z - z')/2$ we have $\Delta r = (r - r')/2 = \mathcal{H}(\bar{z})^{-1}\Delta z + \mathcal{O}(\Delta z^2)$, hence $\Delta z = \mu d\mathcal{H}(\bar{z})/2$ and

$$z = \bar{z} + \mu d\mathcal{H}(\bar{z})/2 \tag{8.72}$$

$$z' = \bar{z} - \mu d\mathcal{H}(\bar{z})/2 \tag{8.73}$$

$$\cos\theta = \sqrt{\frac{r^2 + r'^2 - d^2}{2rr'}} = \left(\frac{2\bar{r}^2 - d^2 + \frac{1}{2}\mu^2 d^2}{2\bar{r}^2 - \frac{1}{2}\mu^2 d^2}\right)^{1/2} \tag{8.74}$$

$$= \left(\frac{2\bar{r}^2 - d_\perp^2 - \frac{1}{2}d_\parallel^2}{2\bar{r}^2 - \frac{1}{2}d_\parallel^2}\right)^{1/2} \equiv c(\bar{z}, d, \mu). \tag{8.75}$$

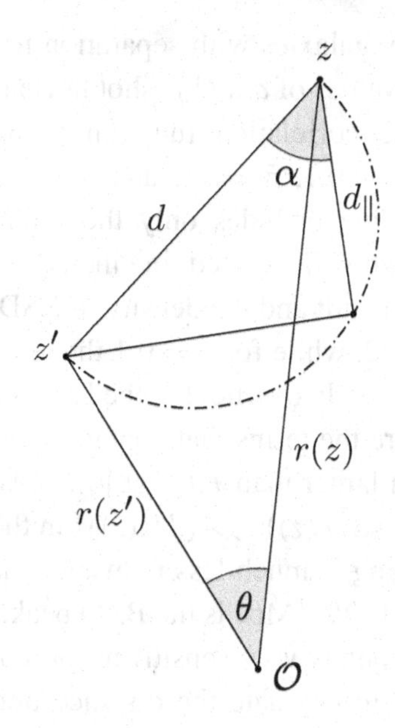

Fig. 8.3 The variable μ is the cosine of the angle α between the line of length $d_\parallel$ that intersects the Thales circle over d and d itself.

With this we can now write the correlation function as a function of the separation d, the direction cosine μ, and the mean redshift $\bar{z}$,

$$\xi(d,\mu,\bar{z}) = \frac{1}{4\pi}\sum_\ell (2\ell+1)C_\ell(z,z')L_\ell\left(c(\bar{z},d,\mu)\right), \tag{8.76}$$

where c is given by Eq. (8.75) and z, z' by Eqs. (8.72) and (8.73). This form is valid only for $|z-z'| \ll 1$ at first order in $|z-z'|$. Alternatively, we may express $\cos\theta$ as a function of z, z', and d using Eq. (8.74) to obtain $\xi(d,z,z')$, which is valid also for large redshift differences $|z-z'|$.

It is important to always keep in mind that the step from the angular to the distance correlation function requires the assumption of a cosmological model. Contrary to $\xi(\theta,z,z')$, the correlation function $\xi(d,\mu,\bar{z})$ or $\xi(d,z,z')$ is model dependent. When using it to constrain cosmological parameters this has to be taken into account. We shall discuss this in Section 9.8.

The advantage of the correlation function $\xi(d,\mu,\bar{z})$ w.r.t. the angular power spectrum is that in the small scale, small redshift, small angle limit it reduces to the nonrelativistic expression (8.20). We can therefore use it in this limit to determine the growth function f and the bias b directly from its quadrupole and hexadecapole. Of course, this information is also contained in the angular power spectrum, but there it is mixed together with other parameters. Another advantage of the correlation function is that within a sizable redshift bin $[\bar{z}-\Delta z, \bar{z}+\Delta z]$

we can expect to find many galaxies with separation in a small bin around d and around μ so that for many values of d and μ shot noise is not a serious problem.

In Fig. 8.4 we plot the correlation function as a function of d for fixed $\mu = 0.95$ at redshifts $\bar{z} = 0.1$, $\bar{z} = 1$, and $\bar{z} = 2$. The solid line includes all the terms, the dashed line includes only the "standard terms," density and redshift space distortions, while the dotted line includes also the lensing term. The difference between the full result and the density + RSD + lensing terms is nearly invisible for $z = 1$ and $z = 2$, while for $z = 0.1$ the lensing term is negligible (the dashed and the dotted lines nearly overlie) but the large scale relativistic corrections are clearly visible. These are the terms that contain a factor $1/r(z)$ (especially the Doppler term) that is much larger than $\mathcal{H}(z)$ at low redshift. These terms are then suppressed only by a factor $(d/r(z))^2 \gg (d\mathcal{H}(z))^2$ in the correlation function. For small values of μ, the lensing is much less relevant as is clear from Fig. 8.5. The pronounced feature at $d \simeq 100h^{-1}$Mpc is the BAO peak. Its position is quite stable under nonlinearities but depends very sensitively on cosmological parameters. It is therefore routinely used to estimate the distance out to a given redshift z, as we shall discuss in Section 9.8. While the relativistic terms make the correlation function at $z = 0.1$ more negative at large distance, the lensing terms contribute positively, so that for $z = 2$ the correlation function even becomes positive again at $d \simeq 380h^{-1}$Mpc.

In Fig. 8.5 the correlation function is shown at fixed $d = 350h^{-1}$Mpc as a function of μ. In the forward direction, $\mu \sim 1$, for redshifts $z = 1$ and 2 the lensing term is very important, even dominant, while for $\mu < 0.6$ it is nearly irrelevant. For $\mu \to 1$, the lensing contribution is negative at $z = 1$ while it is positive at $z = 2$. This is due to the fact that at low redshift, $z \lesssim 1.5$, the negative cross term $D \cdot \kappa$ dominates while at higher redshift the positive $\kappa \cdot \kappa$ term dominates. The sign on the $D \cdot \kappa$ is given by $-(2 - 5s)$; hence this term can become positive also if s is very large. In our figures we have chosen $s = 0$. The significant difference of the standard terms from the relativistic expression at very low redshift, $z = 0.1$, comes from the Doppler term, $(r_s\mathcal{H})^{-1}(2 - 5s)\mathbf{V} \cdot \mathbf{n}$, which at $d = 350h^{-1}$Mpc dominates the signal.

In Fig. 8.6 we plot the multipoles of the correlation function with (left panels) and without (right panels) the lensing and large-scale relativistic terms. The standard terms only generate $n = 0$, 2, and 4 multipoles, while lensing and other relativistic terms also lead to higher multipoles. Especially at $z = 2$, the $n = 6$ multipole that comes from lensing is of the same order as the hexadecapole ($n = 4$). At large scales, $d > 200h^{-1}$Mpc, it is even dominant. Clearly, ignoring the lensing term produces very significant errors in the multipoles that cannot be tolerated in the analysis of future galaxy surveys.

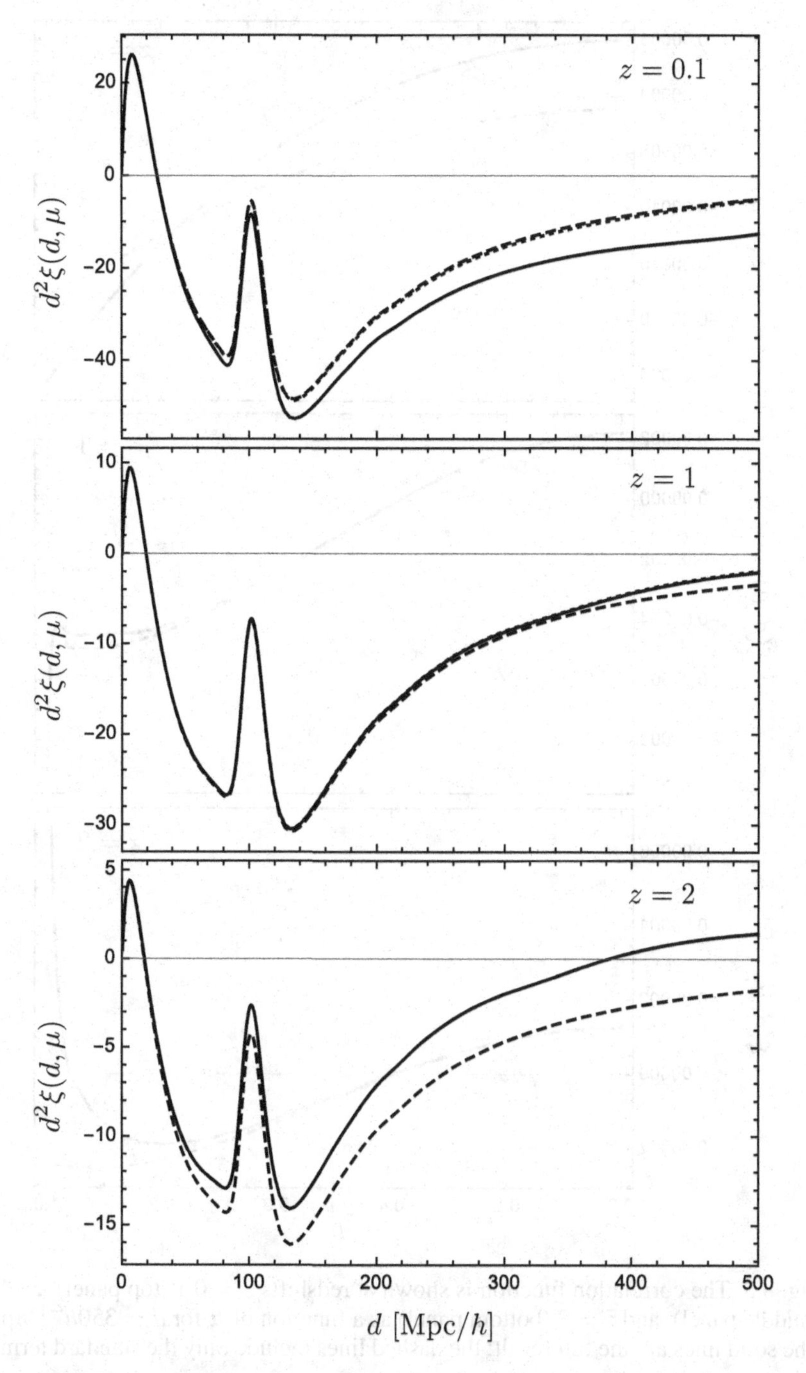

Fig. 8.4 The correlation function is shown at redshifts $\bar{z} = 0.1$ (top panel), $\bar{z} = 1$ (middle panel), and $\bar{z} = 2$ (bottom panel) as a function of d for $\mu = 0.95$. The solid lines are the full result, the dashed lines include only the standard terms, and the dotted lines include also the lensing term.

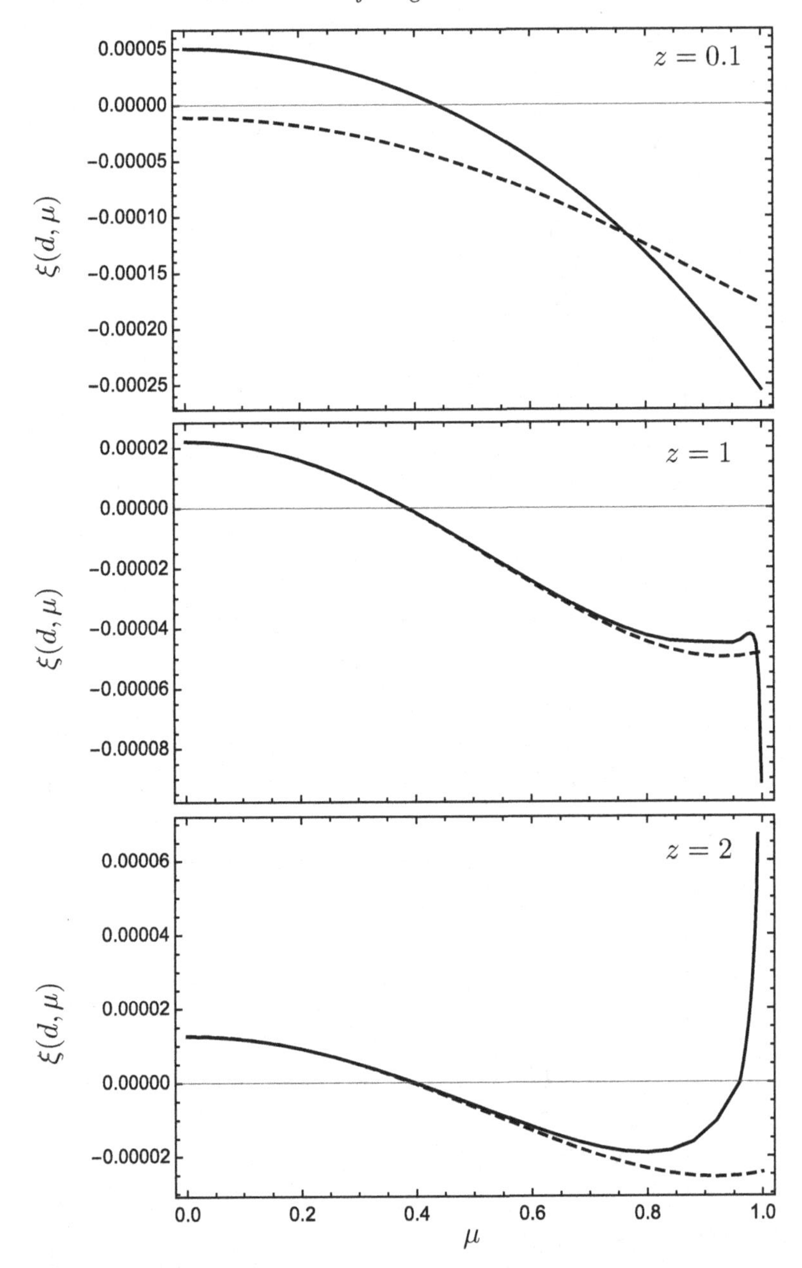

Fig. 8.5 The correlation function is shown at redshifts $\bar{z} = 0.1$ (top panel), $\bar{z} = 1$ (middle panel), and $\bar{z} = 2$ (bottom panel) as a function of μ for $d = 350h^{-1}$Mpc. The solid lines are the full result; the dashed lines include only the standard terms.

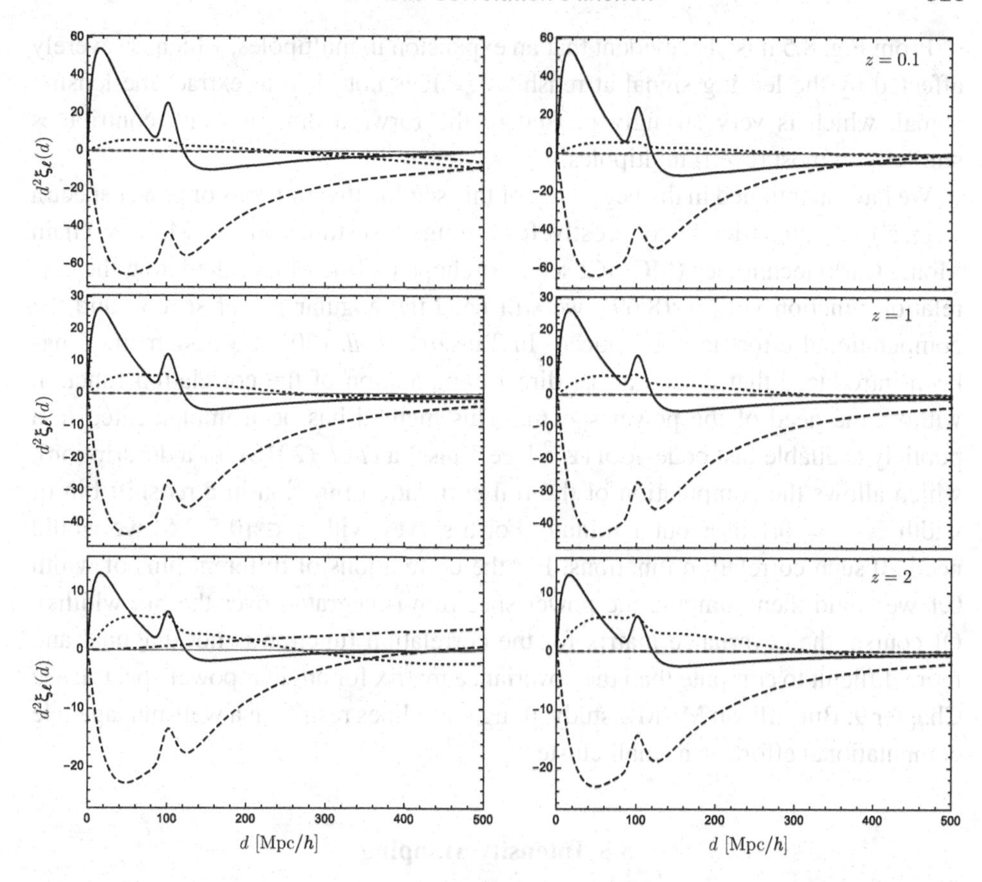

Fig. 8.6 The correlation function multipoles $n = 0$ (solid), $n = 2$ (dashed), $n = 4$ (dotted), and $n = 6$ (dot-dashed) are shown at redshifts $\bar{z} = 0.1$ (top, left panel), $\bar{z} = 1$ (middle left panel), and $\bar{z} = 2$ (bottom, left panel) as a function of d. For comparison we plot the standard multipoles (density and RSD) in the right panels. Note that on very large scales, for $z = 0.1$ and for $z = 2$ the multipole $n = 6$ is comparable in amplitude to $n = 4$.

In the monopole and the quadrupole of the correlation function the BAO peak is again very pronounced. Note that, when including relativistic effects, the monopole becomes positive again at very large scales whereas it remains negative when only the standard terms are considered. The same is true for the quadrupole at $z = 2$. It is also interesting that at higher redshifts the hexadecapole that comes purely from velocities (and lensing) is less suppressed w.r.t. the monopole than at low redshift. This is due to the fact that velocities decay less rapidly with increasing redshift than density perturbations, which dominate the monopole. The amplitude of the negative quadrupole is even larger than the monopole at $z = 1$ and 2. This of course strongly depends on the choice $b = 1$ for this plot, which is not very realistic.

From Fig. 8.5 it is also evident that an expansion in multipoles, which is severely affected by the lensing signal at redshifts $z \gtrsim 1$, is not ideal to extract the lensing signal, which is very strongly peaked in the forward direction and contributes similarly to most (even) multipoles.

We have mentioned in the beginning of this section that millions of power spectra $C_\ell(z, z')$ are numerically too costly for parameter estimation via Markov chain Monte Carlo techniques (MCMC; see next chapter). But when calculating the correlation function via Eq. (8.69), we still need the angular power spectra and the computational effort is not reduced. In Tansella *et al.* (2018) a new method has been introduced that allows a fast direct computation of the correlation function without the need of the power spectra. This method has been implemented in a publicly available fast code "COFFE"[2] [see Tansella *et al.* (2018) for a description], which allows the computation of the full correlation function in a redshift bin of width $2\Delta z = 0.1$ in about 1 minute. For a survey with $z \in [0.5, 2.5]$ we would need 20 such correlation functions. For the correlations of different bins of width 0.1 we could then compute the power spectrum (integrated over the bin widths). Of course, the covariance matrix for the correlation function is not diagonal and more difficult to compute than the covariance matrix for angular power spectra; see Chapter 9. But still, an MCMC study along these lines results in a well-manageable computational effort on a small cluster.

8.5 Intensity Mapping

Hydrogen is the most abundant element in the Universe, making up about 75% of all baryons; see Section 1.4. After the recombination of hydrogen, the Universe is neutral until about $z \sim 7$, when it gets reionized by UV radiation from the first stars, a process that we shall discuss in Section 9.3. Neutral hydrogen is denoted by HI in astrophysics while ionized hydrogen or proton gas is denoted HII. After reionization, at $z \lesssim 6$, neutral hydrogen is found mainly in proto-galaxies, that is, regions where baryons have clustered significantly to allow for cooling and for the recombination of protons and electrons into neutral hydrogen.

A very distinctive line of neutral hydrogen is the 21 cm line from the hyperfine transition of aligned proton and electron spins to proton and electron spins with opposite orientation. The proton spin generates a magnetic dipole field to which the electron is subjected. This leads to a contribution to the Hamiltonian of the form

$$\Delta H_{hf} = \frac{\gamma_p e^2}{m_e m_p} \left\{ \frac{1}{r^3} \left[3(\mathbf{S}_p \cdot \hat{\mathbf{r}})(\mathbf{S}_e \cdot \hat{\mathbf{r}}) - (\mathbf{S}_p \cdot \mathbf{S}_e) \right] + \frac{8\pi}{3} (\mathbf{S}_p \cdot \mathbf{S}_e) \delta(\mathbf{r}) \right\}. \quad (8.77)$$

[2] The code can be found on https://github.com/JCGoran/coffe

Here $\mathbf{S}_{p,e}$ denotes the (normalized) proton and electron spin respectively, $m_{p,e}$ denotes their masses, and γ_p is a numerical factor relating the proton spin to its magnetic moment. It is known experimentally to be $\gamma_p = g_p/2 = 2.7928$; we have set $c = 1$ as usual. In the ground state of the hydrogen atom, due to spherical symmetry, only the second term contributes. It leads to a splitting between the state with aligned proton and electron spins, $F = 1$, and anti-aligned proton and electron spins, $F = 0$ (F is the total angular momentum of the hydrogen atom) given by

$$E_{10} = 2\frac{8\pi\gamma_p e^2}{3m_e m_p}|\psi(0)|^2 \simeq 5.88 \times 10^{-6}\text{eV} = h\nu_{10} = hc/\lambda_{10}, \tag{8.78}$$

$$\nu_{10} = 1.420 \times 10^9 \text{ Hz}, \qquad \lambda_{10} = 21.10 \text{ cm}. \tag{8.79}$$

Here ψ is the ground state wave function of the electron [see Park (1974) or any other quantum mechanics book for more details on the hyperfine splitting].

Observing this 21 cm line during reionization, $10 > z > 6$, will allow to separate HI and HII regions and to study the evolution and clumpiness of the reionization process. Observing the 21 cm line before or after reionization will allow us to study baryon or matter density fluctuations in a way similar to galaxy number counts. However, instead of counting sources of 21 cm emission (or absorption), we can also just measure the intensity of the line coming from different directions at some fixed redshift, without resolving individual sources. This is the 21 cm intensity mapping technique that we discuss in this section. Experiments that plan to apply 21 cm intensity mapping in the near future are discussed in Furlanetto *et al.* (2006) and, more recently in Kovetz *et al.* (2017). Observations at high redshift, $z \sim 20$ to 100, would be especially exciting, as they open an entirely new window to the early phases of cosmic structure formation before luminous structures form, the so-called dark ages.

The HI density is proportional to the neutral hydrogen density, $n_b x_{\rm HI}$, where n_b is the baryon density and $x_{\rm HI}$ is the HI fraction of baryons. If $x_{\rm HI}$ is assumed to be independent of the fluctuation amplitude (which is probably a good approximation before and after the reionization epoch), the HI density fluctuation is given by the baryon density fluctuation,

$$\delta_{\rm HI} = \delta_b. \tag{8.80}$$

To emit a 21 cm photon, the hydrogen atom has to be in the excited state. It is usually excited by low-energy CMB photons, by collisions, or by Lyman-α photons over the "Wouthuysen–Field effect," which we discuss in the text that follows.

To derive the first-order perturbation equation for HI intensity fluctuations, we consider the brightness I_ν of HI emission of a given cloud of hydrogen. I_ν is the power emitted per Hz, per steradian, and per cm^2; it has units of erg/sec/Hz/sr/cm^2.

There exists exactly one blackbody temperature, called the brightness temperature T_b, that has a given intensity I_ν at frequency ν. The brightness temperature, T_b, is given in terms of the brightness by

$$I_\nu = \frac{d\rho}{d\nu d\Omega} = \frac{2}{(2\pi)^2}\frac{T_b^3 x^3}{e^x - 1}, \tag{8.81}$$

where we have used (1.54) with $N_b = 2$ and $x = p/T_b = 2\pi\nu/T_b$. Note that in our units Planck's constant $h = 2\pi$. In the Rayleigh–Jeans limit, $x \ll 1$, which is most relevant for 21 cm photons, we can approximate $e^x - 1 \simeq x$ and obtain

$$I_\nu = 2T_b\nu^2 \quad \text{or} \quad T_b = \frac{I_\nu}{2\nu^2}. \tag{8.82}$$

As the brightness scales like $(1+z)^3$ with redshift, this temperature scales as $T_b \propto (1+z)$, like the CMB temperature. The brightness temperature of a hydrogen cloud is

$$T_b = T_S(1 - e^{-\tau_\nu}) + T_R(\nu)e^{-\tau_\nu}. \tag{8.83}$$

Here T_S is the spin temperature defined by the ratio of excited hydrogen atoms in the spin-1 state, n_1, and the hydrogen atoms in the ground state with spin 0, n_0, by

$$\frac{n_1}{n_0} = \frac{g_1}{g_0}e^{-E_{10}/T_S}, \tag{8.84}$$

where $g_1 = 3$ and $g_0 = 1$ denote the multiplicities of the corresponding spin states. $T_R(\nu)$ is the brightness temperature of an external radiation field incident on the cloud (e.g., the CMB) at frequency ν and τ_ν is the optical depth through the cloud, $\tau_\nu = \int_{\text{cloud}} \alpha_\nu d\ell$, where α_ν is the absorption coefficient. In astrophysical applications usually $T_S \gg E_{10}$ so that $n_1 \simeq 3n_0$ and the absorption coefficient must include a correction for stimulated emission. Hence

$$\tau_\nu = \int d\ell\sigma_{01}\left(1 - e^{-E_{10}/T_S}\right)\phi(\nu)n_0 \simeq \sigma_{01}\frac{h\nu}{T_S}(N_{\rm HI}/4)\phi(\nu). \tag{8.85}$$

Here $N_{\rm HI}$ is the column density of neutral hydrogen of the cloud and we used that $n_0 \simeq n_{\rm HI}/4$, and $\phi(\nu)$ is the line profile, $\phi(\nu) = h(1+z)|d\lambda/dz|/L$, where λ is the affine parameter along the photon geodesic; z is the redshift of the line and L is the thickness of the cloud. The cross section σ_{01} for 21 cm absorption is given by

$$\sigma_{01} = \frac{2A_{10}}{8\pi\nu^2}, \tag{8.86}$$

where A_{10} is the spontaneous emission coefficient of the 21 cm line. Setting $N_{\rm HI} = Ln_{\rm HI} = Ln_b x_{\rm HI}$ and putting it all together we obtain the following expression for the brightness temperature [see Hall *et al.* (2012)]:

$$T_b = \frac{3}{32\pi}\frac{h^3 A_{10}}{E_{10}} n_b x_{\rm HI}(1+z)\left|\frac{d\lambda}{dz}\right| + T_R e^{-\tau_\nu}, \tag{8.87}$$

where again $h = 2\pi$ is Planck's constant. The study of linear perturbations of T_b, assuming that $x_{\rm HI}$ is constant and $n_b \propto \rho$, is presented in detail in Hall *et al.* (2012). It leads exactly to Eq. (8.60) with $s = 2/5$. Here we do not repeat this derivation, as the result is actually not very surprising. For a brightness temperature or an intensity, the reduction of the density by the increased transverse area is exactly compensated by the increase of the number of photons due to focusing. This is a consequence of the photon number conservation, which also is the reason that there are no lensing terms in the CMB at first order.

To measure the brightness temperature we can contrast lines of sight through a hydrogen cloud that is irradiated only by the CMB, $T_R = T_\gamma$, with lines of sight to "clear CMB." This yields

$$\delta T_b = T_b - T_0 = \frac{T_S - T_\gamma(z)}{1+z}\left(1 - e^{-\tau_\nu}\right) \simeq \frac{T_S - T_\gamma(z)}{1+z}\tau_\nu$$
$$= \left[1 - \frac{T_\gamma(z)}{T_S}\right]\frac{3}{32\pi}\frac{h^3 A_{10}}{E_{21}} n_b x_{\rm HI}\left|\frac{d\lambda}{dz}\right|. \tag{8.88}$$

This expression is saturated for $T_S \gg T_\gamma$ but it can become arbitrarily negative for small spin temperature, $T_S < T_\gamma$. A negative δT_b just means that we see the line in absorption while for a $T_S > T_\gamma$ we see it in emission. To decide whether we see the 21 cm line in emission or absorption, we have to determine the spin temperature T_S.

At early times, $z \geq 200$, the free electrons remaining after recombination keep the baryon fluid in thermal equilibrium with the CMB and we have $T_B = T_\gamma$, where T_B denotes the kinetic baryon temperature. Collisions also keep the spin temperature at this value so that $T_S - T_\gamma = 0$ and we cannot detect the 21 cm line. At $z \sim 150$, the heating by Thomson scattering of the remaining electrons drops out of equilibrium; see Exercise 8.2. After that time, the baryon temperature decays like $(1+z)^2$; see Eq. (1.93) for $\sigma = 0$. Initially the spin temperature is in equilibrium with the kinetic baryon temperature due to collisions, $T_S = T_B$, and we can (in principle) see the 21 cm line in absorption at redshifts $30 < z < 150$. However, if there is no additional cooling of the spin temperature, at $z \sim 30$ also collisions drop out of equilibrium and the spin temperature rises back to the CMB temperature so that $\delta T_b \to 0$. However, when the first structures form, the so-called

Wouthuysen–Field effect (Wouthuysen, 1952; Field, 1959) drives again $T_S \to T_B$. This effect simply takes into account that an electric dipole transitions can change the total angular momentum by 1 or 0. Therefore, Lyman alpha photons can induce a transition from $1_0S_{1/2}$ to $2_1P_{1/2}$ that can then decay into $1_1S_{1/2}$, and similarly with the $1_0S_{1/2}$ to $2_1P_{3/2}$ transition of hydrogen. Here the first number is n, the principle quantum number; the first index is F, the total angular momentum of the atom; the letter indicates the orbital angular momentum of the electron; S means 0 while P corresponds to 1 and the second index is J, the total angular momentum of the electron. The Wouthuysen–Field effect couples the hydrogen kinetic temperature and its spin temperature. The latter is expected to always remain somewhat higher than the former, but by how much depends on the model; Furlanetto *et al.* (2006).

Recently, the detection of T_b at a redshift centered around $z \simeq 17$ in absorption was announced by Bowman *et al.* (2018). However, the effect seems to be at least a factor of 2 larger than the most optimistic estimate with $T_S = T_K$. This very difficult experiment, which "fishes" a ~ -0.5 K signal out of a several thousand K background, certainly needs confirmation. The theoretically expected value would have been around -0.1 to -0.2 K.

After reionization, at $z \lesssim 6$ inside structures (proto-galaxies and galaxies), the density becomes again large enough so that the kinetic baryon temperature is much higher than the CMB temperature and roughly equal to the spin temperature. In these structures it will be possible to see T_b in emission and its value is independent of the spin temperature [see Eq. (8.88)] in the limit $T_S \gg T_\gamma$. To study the angular fluctuations of T_b, we fix a "clear CMB" direction and simply study

$$\Delta T_b = \delta T_b(\mathbf{n}) - \delta T_b(\mathbf{n}') = \delta \bar{T}_b[\Delta_T(\mathbf{n}) - \Delta_T(\mathbf{n}')], \tag{8.89}$$

like for the CMB; see Eq. (2.239). In this expression, the somewhat ill-defined "clear CMB" direction drops out and we may use (8.88) in the limit $T_S \gg T_\gamma$. At low redshift, $z \lesssim 6$, we therefore expect fluctuations of the brightness temperature given by Eq. (8.60), with $s = 2/5$ and its own bias $b(z)$ and evolution bias $b_e(z)$.

Summarizing this section, we have found that the 21 cm emission line is interesting for at least three different reasons:

(1) It can pave a way to observe the baryon density and its fluctuation at $150 > z > 50$ when there are not yet any structures emitting photons in the Universe, the so-called dark ages.
(2) It is sensitive to the neutral hydrogen fraction $x_{\rm HI}$ which has fluctuations of order unity during reionization. These in principle allow us to study in detail the process of reionization at redshift $10 > z > 6$.

(3) At lower redshifts, $z \lesssim 6$, neutral hydrogen is predominantly in structures where also the spin temperature is much higher than the CMB temperature. At these redshifts we can detect 21 cm radiation in emission and we expect it to be a very useful additional trace of large-scale structure.

Since the frequency of the 21 cm is so well defined, all these observations have exquisite redshift resolution.

Of course, the 21 cm line is not the only line that can be observed. It may also be interesting to study other hydrogen lines, for example, the Lyman-α lines or lines of heavier elements such as carbon that are generated in cosmic structures; see Kovetz *et al.* (2017) for a recent review in which preliminary detections of a rotational carbon-monoxide (CO) line, a CII fine-structure line, Lyman-α and H_α lines, as well as low redshift 21 cm measurements are described. So far these detections have been made by correlating intensity mapping measurements with galaxy surveys, but clearly this is just the beginning...

Exercises

(The exercises marked with an asterisk are solved in Appendix 11 which is not in this printed book but can be found online.)

8.1 **From the power spectrum to the correlation function***

Consider a power spectrum that depends not only on k but also on its direction cosine w.r.t. a fixed direction $\mathbf{n}$. Expanding this dependence in Legendre polynomials,

$$P(k,\mu) = \sum_n P_n(k) L_n(\mu), \qquad \mu = \mathbf{n}\cdot\mathbf{k}/k, \tag{8.90}$$

show that the correlation function is given by

$$\xi(r,\mu) = \sum_n (i)^n \xi_n(r) L_n(\mu), \qquad \mu = \mathbf{n}\cdot\mathbf{r}/r, \tag{8.91}$$

where

$$\xi_n(r) = \int \frac{dkk^2}{2\pi^2} P_n(k) j_n(kr). \tag{8.92}$$

Hint: Use the addition theorem of spherical harmonics.

Therefore, if both the correlation function and the power spectrum are real, only even powers of μ are possible.

8.2 Decoupling of the baryon temperature from the CMB

Using the expression for the fraction of ionized electrons after recombination,

$$x_R = 1.2 \times 10^{-5} \Omega_m^{1/2} / (\Omega_B h)$$

discussed in Section 1.3.2, show that heating by Thomson scattering of these electrons with CMB photons drops out of thermal equilibrium at $z_\gamma \simeq 150$. For this, use that the Thomson cooling (heating) rate is

$$\Gamma = \frac{x_R}{t_\gamma} \qquad \text{with } t_\gamma = \frac{m_e}{8 \sigma_T \rho_\gamma}. \tag{8.93}$$

9

Cosmological Parameter Estimation

9.1 Introduction

In the previous chapters we calculated the CMB anisotropies and polarization. Generically the resulting spectra show a series of acoustic oscillations that present a snapshot of the CMB sky at the moment when photons decouple from electrons. The details of these spectra depend on the one hand on the initial fluctuations and on the other hand on the background cosmological parameters that determine the evolution of fluctuations.

If we make no hypothesis on the initial fluctuations, a given observed spectrum can be obtained by a nearly arbitrary choice of cosmological parameters. For a given initial power spectrum $P_m(k)$ of scalar ($m = 0$), vector ($m = \pm 1$), and tensor ($m = \pm 2$) perturbations, under the assumption of statistical homogeneity and isotropy, the resulting CMB power spectrum is generically of the form

$$C_\ell = \sum_{m=-2}^{2} \int dk\, T_m^2(\ell, k) P_m(k), \tag{9.1}$$

where T_m is the CMB "transfer function" that depends on the cosmological parameters. Therefore, for a nearly arbitrary transfer function $T_m(\ell, k)$ and arbitrary C_ℓ's one can find initial power spectra P_m such that Eq. (9.1) holds. Since vector perturbations decay, the vector transfer function $T_{\pm 1}$ is very small. If the initial perturbations are sufficiently small for linear perturbation theory to hold, vector perturbations will not show up in the CMB spectrum. We shall therefore neglect them in our discussion.

"Sources" form an exception to this rule. A source is an inhomogeneous and anisotropic component of the energy–momentum tensor that is too small to contribute to the background, but that sources perturbations in all fluids. We have discussed this case in Section 2.7. If sources are relevant, some parts of the perturbations are generated at late time by a stochastic source term. In this case, vector perturbations can also be important.

We have to keep in mind that when determining cosmological parameters with the CMB, we are not really "measuring" them directly, but we are "estimating" them under certain, usually well-motivated but very restrictive assumptions on the initial power spectrum. On the other hand, whenever we make a physical measurement we are using prior knowledge, that is, that our apparatus obeys Maxwell's equation. However, the apparatus has usually been tested by some other measurements, while only the CMB and very few other data sets contain experimental information about the initial conditions of the fluctuations in the Universe. Therefore, in the "ideal" world we would want to measure the cosmological parameters by other means and then with the well-known transfer functions at hand, use the CMB to determine the initial fluctuations, which help us to understand the physics of inflation, the physics at very high energies, probably close to the Planck scale, energies that are not available in any laboratory on Earth.

However, the real world is not ideal, and since CMB fluctuations are the most accurate and theoretically the best understood cosmological data set, we use them to determine the parameters both of the background cosmology and the initial fluctuations. For this to work, we have to assume that the initial power spectrum depends only on a few parameters, for example, scalar perturbations given by $k^3 P_\zeta = 2\pi^2 \Delta_\zeta = 2\pi^2 A_s (k/H_0)^{n_s - 1}$ and tensor perturbations $P_{\pm 2} = 0$. These initial fluctuations are described by only two parameters, A_s and n_s. The better the data, the more parameters can be fitted. However, up to now, several attempts of fitting models with more complicated initial power spectra to the data have shown that adding more parameters does not significantly improve the fit. The "Occam's razor" argument (see Section 9.5.2) then tells us that we should stick to the simpler model with only two parameters. Nevertheless, there are good physical arguments that at some level we may expect, for example, tensor perturbations or running of the spectral index n_s. But present data are not sensitive to these small effects.

Naively one might think that the knowledge of 2000 C_ℓ's to percent accuracy allows us to determine as many parameters with similar accuracy. But this is not so, since the CMB power spectrum can usually be well fitted with a function of only a few parameters. Also, it is sensitive only to a certain combination of cosmological parameters. This leads to *degeneracies* that we shall discuss in Section 9.6.

But before we enter the technical details of parameter estimation, we briefly want to discuss the physics of their influence on the CMB spectrum. This helps us to develop a good intuition for the parameters which can be estimated with the CMB to high accuracy and those which cannot.

9.2 The Physics of Parameter Dependence

9.2.1 The Acoustic Peaks

The first acoustic peak corresponds to the comoving wavelength $\lambda_1 = \pi/k_1$, which has undergone exactly one compression since entering the horizon and whose fluctuations are at a maximum at decoupling. As we have discussed in Chapter 2, this scale is determined by $k_1 t_s(z_{\rm dec}) = \pi$ for adiabatic perturbations [$k_1 t_s(z_{\rm dec}) = 3\pi/2$ for isocurvature perturbations]. Here t_s is the sound horizon,

$$t_s(t) = \int_0^t c_s(t')dt'.$$

Subsequent peaks are at $k_n t_s(z_{\rm dec}) = n\pi$ for adiabatic perturbations [and $k_n t_s(z_{\rm dec}) = (2n+1)\pi/2$ for isocurvature perturbations]. The angle onto which these peaks are projected in the sky is

$$\theta_n = n\theta_1, \qquad \theta_1 \simeq \frac{\lambda_1 a(z_{\rm dec})}{d_A(z_{\rm dec})}, \qquad \ell_n \simeq \pi/\theta_n, \tag{9.2}$$

where $d_A(z)/a(z)$ denotes the comoving angular diameter distance to an event at time $t(z)$ and $\ell(\theta) = \pi/\theta$ is the harmonics, which corresponds roughly to the angle θ. The redshift of decoupling, $z_{\rm dec}$, and also $t_{\rm dec}$ depend mainly on the baryon density of the Universe, while the angular diameter distance depends strongly on curvature, but also on the cosmological constant and the dark matter density. Apart from the CMB temperature T_0, the angle θ_1 is the best measured variable in cosmology. The most accurate measurement from the Planck satellite reports (Planck Coll. VI, 2018)

$$\theta_1 = (1.04089 \pm 0.00031) \times 10^{-2}. \tag{9.3}$$

This angle is directly measured and not very model dependent. However, once we translate it together with other measurements into a value, for example, for Ω_Λ, the result does depend on whether we assume a model with or without curvature. In a universe containing radiation, matter, and a cosmological constant we have (see Chapter 1)

$$t_s(z_{\rm dec}) = \frac{1}{a_0 H_0}\int_{z_{\rm dec}}^{\infty} \frac{c_s(z)dz}{\left[\Omega_r(z+1)^4 + \Omega_m(z+1)^3 + \Omega_\Lambda + \Omega_K(z+1)^2\right]^{1/2}}, \tag{9.4}$$

$$\frac{d_A(z_{\rm dec})}{a(z_{\rm dec})} = \chi_K\left(\frac{1}{H_0 a_0}\int_0^{z_{\rm dec}} \frac{dz}{\left[\Omega_r(z+1)^4 + \Omega_m(z+1)^3 + \Omega_\Lambda + \Omega_K(z+1)^2\right]^{1/2}}\right), \tag{9.5}$$

$$\ell_n \simeq \frac{\pi d_A(z_{\rm dec})}{a(z_{\rm dec})\lambda_n}$$
$$= \frac{n\pi\,\chi_K\left(\frac{1}{H_0 a_0}\int_0^{z_{\rm dec}}\frac{dz}{\left[\Omega_r(z+1)^4+\Omega_m(z+1)^3+\Omega_\Lambda+\Omega_K(z+1)^2\right]^{1/2}}\right)}{\frac{1}{a_0 H_0}\int_{z_{\rm dec}}^{\infty}\frac{c_s(z)dz}{\left[\Omega_r(z+1)^4+\Omega_m(z+1)^3+\Omega_\Lambda+\Omega_K(z+1)^2\right]^{1/2}}}. \tag{9.6}$$

Here

$$\chi_K(r) = \begin{cases} r & \text{if} \quad K=0 \\ \frac{1}{\sqrt{K}}\sin\left(\sqrt{K}r\right) & \text{if} \quad K>0 \\ \frac{1}{\sqrt{|K|}}\sinh\left(\sqrt{|K|}r\right) & \text{if} \quad K<0, \end{cases} \tag{9.7}$$

and we have used

$$\lambda_n = n t_s(z_{\rm dec}).$$

It is evident, that, via χ_K, the position ℓ_n strongly depends on curvature. Since curvature and the cosmological constant are irrelevant at high redshift, the denominator of Eq. (9.6) is nearly independent of them. It depends mainly on Ω_r, Ω_m, and on the baryon density via the sound speed c_s,

$$c_s^2 = \frac{1}{3}\frac{4\Omega_\gamma(1+z)}{4\Omega_\gamma(1+z)+3\Omega_b}, \tag{9.8}$$

where $h = H_0/100\,\text{km s}^{-1}\text{Mpc}^{-1}$. Since $\Omega_\gamma h^2$, which is proportional to the present photon energy density, hence to T_0^4, is very well known, the sound speed provides a measure of $\Omega_b h^2$. Note that $\Omega_\gamma h^2$ is much better known than $\Omega_\gamma = 8\pi G a_{\rm SB} T_0^4/3H_0^2$. The latter contains the considerable uncertainty of the Hubble constant.

In Fig. 9.1 we show the dependence of ℓ_1 on Ω_Λ. Since we have to satisfy the Friedmann constraint,

$$\Omega_\Lambda + \Omega_K + \Omega_m + \Omega_r = 1, \tag{9.9}$$

we cannot simply vary Ω_Λ. In Fig. 9.1 one also sees that when varying Ω_Λ, the resulting peak position strongly depends on what has been kept fixed during the variation. We can fix Ω_K and h but not Ω_m: since increasing Ω_Λ just increases the dimensionless angular diameter distance, $H_0 d_A$ due to the decrease in Ω_m, while $H_0\lambda_1$ is not affected. The opposite is true if we let h and Ω_m vary but keep fixed $\Omega_m h^2$ and Ω_K. Then the peak position is reduced with growing Ω_Λ. This comes from the fact that $H_0\lambda_1$ now increases with increasing Hubble parameter due to the decrease in Ω_r. This effect more than compensates the increase in $H_0 d_A$

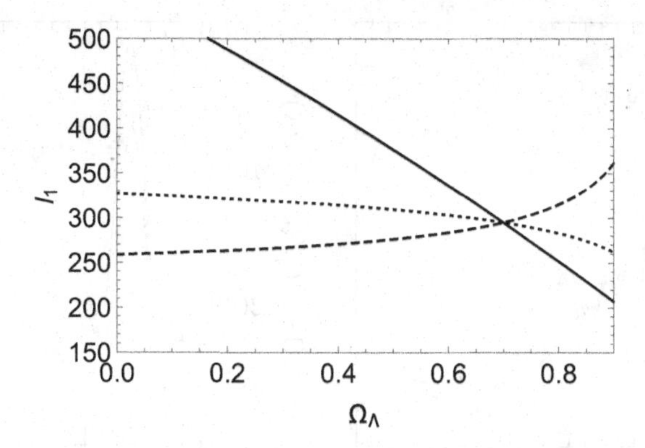

Fig. 9.1 We show the position of the first peak as a function of Ω_Λ. In the solid line we vary Ω_K, leaving all other parameters fixed. The two other lines are with $\Omega_K = 0$. For the dashed line we vary Ω_Λ and Ω_m at fixed h and Ω_K, and for the dotted line we vary Ω_Λ and Ω_m at fixed $\Omega_m h^2$ and Ω_K. The fixed parameters take the values $\Omega_K = 0$, $h = 0.7$, $\Omega_b h^2 = 0.022$, $\Omega_m = 0.3$. Therefore, all the curves cross at $\Omega_\Lambda = 0.70$. Notice the strong dependence of $\ell_{\rm peak}$ on curvature (solid line).

due to the decrease in Ω_m. Note that $\Omega_r h^2$, like $\Omega_\gamma h^2$, is determined by the CMB temperature T_0 and therefore always remains fixed. (We neglect neutrino masses.)

As we have seen in Chapter 2, dark matter fluctuations grow only logarithmically during the radiation-dominated era. Once the Universe becomes matter dominated, they start growing like the scale factor. Therefore, the amplitude of the gravitational potential, which is determined mainly by the dark matter density, depends on $\Omega_m h^2$. Especially, the ratio of the height of the first acoustic peak and the Sachs–Wolfe plateau is sensitive to this parameter.

The baryon density also enters CMB physics via the asymmetry of even and odd acoustic peaks. As we have said before, the first peak at scale ℓ_1 is a contraction peak, an over-density. Correspondingly, the second peak is an expansion peak, an under-density. If the oscillating fluid consists solely of massless photons it would undergo perfectly harmonic oscillations and the amplitudes of contraction and expansion peaks would be equal. However, the massive baryons reinforce contraction via their self-gravity and their reaction to the gravitational potential of dark matter. Correspondingly they reduce expansion (see Fig. 9.2).

On small scales, the fluctuation amplitudes decay due to Silk damping. Again the strength of the damping depends on the baryon density, hence on $\Omega_b h^2$. The larger the baryon density, the smaller the collision time and hence the larger the damping time scale $t_d \propto t_c^{-1}$. Therefore for larger $\Omega_b h^2$ Silk damping already starts at smaller multipoles.

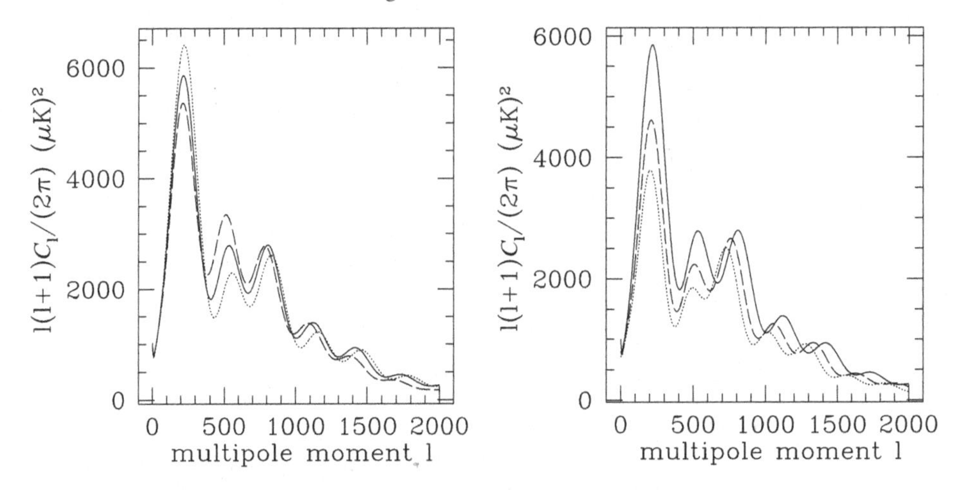

Fig. 9.2 In the left-hand panel we show the asymmetry of even and odd peaks and its dependence on $\Omega_b h^2$. The temperature anisotropy spectrum is plotted for $\Omega_b h^2 = 0.02$ (solid line), $\Omega_b h^2 = 0.03$ (dotted), and $\Omega_b h^2 = 0.01$ (dashed). On the right-hand side $\Omega_b h^2 = 0.02$ is fixed and three different values for the matter density are chosen: $\Omega_m h^2 = 0.12$ (solid), $\Omega_m h^2 = 0.2$ (dashed), and $\Omega_m h^2 = 0.3$ (dotted). Note that higher values of $\Omega_m h^2$ also lead to a stronger peak asymmetry. In addition, a smaller value of $\Omega_m h^2$ boosts the height, especially of the first peak, due to the stronger contribution from the early integrated Sachs–Wolfe effect. The peaks are also somewhat shifted since d_A depends on Ω_m.

9.2.2 Neutrinos

As we saw in Chapter 1, neutrinos decouple when the Universe has a temperature of about $T_\nu \sim 1.4\,\text{MeV}$, which corresponds to a redshift of $z_\nu \sim 0.6 \times 10^{10}$. After that, weak interactions no longer can keep them in thermal equilibrium with the other constituents, electrons, baryons, photons, and also cold dark matter. (If dark matter interactions with neutrinos were stronger than weak interactions, we would have detected dark matter in the laboratory long ago.)

After that, neutrinos propagate freely, described by the Liouville equation (see Section 4.7). As long as their masses can be neglected, they build up anisotropic stresses by free streaming and their energy density dilutes like that of radiation. As soon as their masses become relevant, their pressure and anisotropic stresses decay and they behave like dark matter. This changes their effects on CMB anisotropies and thereby leads to a way of measuring their mass with the CMB. The significance is not very high, since neutrino anisotropic stresses contribute only about 5% to the CMB fluctuations and since massive neutrinos have a signature similar to that of cold dark matter in the CMB. In Fig. 9.3 we compare the CMB spectrum for three sorts of degenerate neutrinos, all with mass $m_\nu = 2$ eV, with the spectrum of

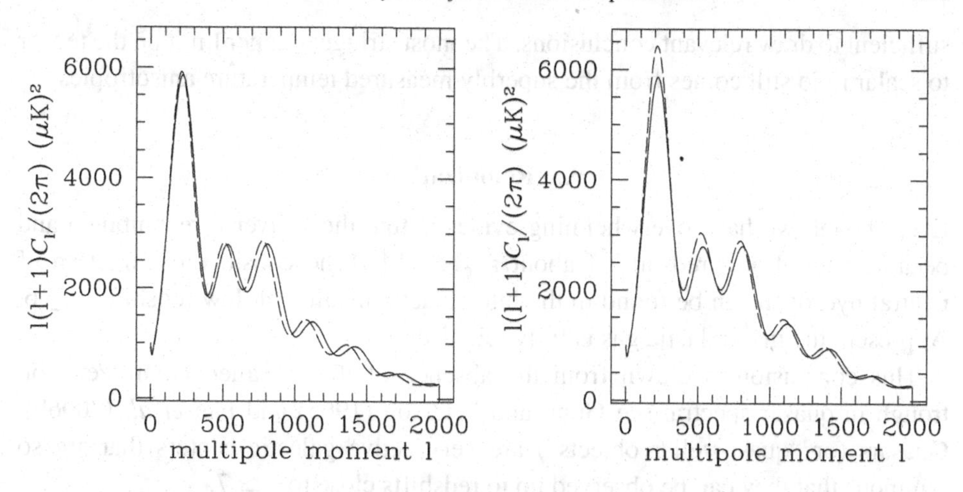

Fig. 9.3 We show the CMB power spectrum for massless neutrinos and neutrinos with mass $m_\nu = 2$ eV (dashed lines). In the left-hand panel $\Omega_{cdm}h^2 = 0.12$ is fixed while on the right-hand side $\Omega_m h^2 = 0.144$ is fixed. In all curves $\Omega_{\rm tot} = 1$, $\Omega_b h^2 = 0.022$, and $h = 0.7$. Keeping $\Omega_m h^2$ fixed, adding neutrinos acts a bit like a lower-matter density, since the neutrinos are not yet fully nonrelativistic at decoupling.

massless neutrinos, once by fixing the total matter density and once by fixing the cold dark matter density.

The neutrinos in the CMB also cannot be replaced simply by some relativistic fluid. It is clear that they are free-streaming particles developing anisotropic stress and beyond, and the higher multipoles in their distributions function are clearly detected in present CMB data see Sellentin and Durrer (2015).

9.2.3 Gravitational Waves

The CMB anisotropies from gravitational waves are significant on scales that are super-Hubble before decoupling, which corresponds to $\ell \lesssim 100$. On smaller scales they decay. If it is small, such a signal is difficult to disentangle from a slightly red ($n_s < 1$) spectrum of scalar fluctuations that simply has somewhat more power on the Sachs–Wolfe plateau than a Harrison–Zel'dovich spectrum with $n_s = 1$.

However, as we saw in Chapter 5, scalar perturbations do not generate B-mode polarization. Therefore, the detection of B-mode polarization would be a fingerprint of gravitational waves from inflation. In Chapter 7 we have, however, seen that second-order effects, especially lensing, also generate a B-mode signal from scalar perturbations, and for a tensor to scalar ratio $r < 0.01$ these lensing B-modes are larger than the ones from the inflationary gravitational wave background for $\ell \geq 10$. So far, even though the lensing B-modes have been detected, the data are not sufficient to draw relevant conclusions. The most stringent upper limit on the tensor to scalar ratio still comes from the superbly measured temperature anisotropies.

9.3 Reionization

Even though we have overwhelming evidence that the Universe recombined and became neutral at a redshift of about $z_{\rm rec} \simeq 1400$, no considerable fraction of neutral hydrogen can be found in the intergalactic medium at low redshift, $z \leq 6$. At present the intergalactic gas is fully reionized.

This conclusion is drawn from the absence of the so-called Gunn–Peterson trough in quasar spectra; see Gunn and Peterson (1965) and Fan *et al.* (2006b). Quasars (or "quasi-stellar objects") are very active galactic centers that are so luminous that they can be observed up to redshifts close to $z \simeq 7$.

Gunn and Peterson (1965) calculated that even a modest density of neutral hydrogen would lead to a significant absorption trough in the part of the quasar spectra that is bluer than Lyman-α at emission and redder at absorption. These photons have, at some moment during their propagation from the quasar to us, exactly Lyman-α frequency and are then resonantly absorbed by neutral hydrogen. Inserting numbers one finds (Peacock, 1999) that the neutral hydrogen density in the intergalactic medium amounts to less than $\Omega_H h \lesssim 10^{-8}$.

There are, however, so-called Lyman-α clouds, that is, clouds of neutral hydrogen that intervene the lines of sight of quasars and lead to a "forest" of absorption lines in quasars, the Lyman-α forest; see Section 9.8.2. But even integrating the total optical depth of the Lyman-α forest one infers a neutral hydrogen density of only $\Omega_H h \simeq 10^{-5}$. It is very unlikely that galaxy formation has been so efficient as to sweep up 99.9% of all the hydrogen in the Universe. We are therefore led to the conclusion that the present intergalactic hydrogen is ionized.

During the last decade or so, the Lyman-α and Lyman-β troughs have been found in very high redshift quasars with $z > 6$; see Fan *et al.* (2006a, 2006b). This confirms that at these high redshifts, some neutral hydrogen has been present that has been reionized later on, probably by UV light from the first burst or star formation.

Reionization was terminated at redshift $z_{\rm ri} \simeq 6$, but it is not clear when the process started and whether it was very fast or slow. The unknown reionization history of the Universe affects CMB anisotropies and polarization. Once the Universe is reionized, CMB photons can in principle scatter again with the free electrons. Since the electron density is significantly lower than at decoupling, the scattering probability or optical depth is rather low and the effect is probably on the level of 5–10%.

Rescattering of electrons leads to additional polarization on a scale that corresponds to the sound horizon at reionization, $\lambda_{\rm ri} \simeq c_s(z_{\rm ri})t(z_{\rm ri})$. In addition, due to the much larger free-streaming scale of electrons after reionization, there is some additional Silk damping on scales up to λ_{ri}.

Usually, reionization is parameterized simply by a reionization redshift z_{ri} or by the optical depth τ_{ri} to reionization,

$$\tau_{ri}(z_{ri}) = \sigma_T \int_{t(z_{ri})}^{t_0} a(t) n_e(t)\, dt. \tag{9.10}$$

Of course, the reionization history can be more complicated than this, for example, it does make a difference whether reionization is instantaneous or slow. However, present data are not sufficient to determine more than the optical depth.

9.4 CMB Data

So far, we have mainly discussed the theoretical aspects of the CMB. Of course these are very interesting, mainly since we have high-quality data to compare with our theoretical models. On the other hand, good quality CMB data are so valuable, since, for a given cosmological model and specified initial fluctuations, we can calculate the CMB anisotropies and polarization with high accuracy.

This interplay of theory and data, which makes physics so fascinating, acts in its most beautiful way in CMB physics: observing the largest structures in the Universe, the anisotropy patterns in the CMB, we learn not only a lot about the parameters of the Universe but also about physics at the highest energies corresponding to the smallest scales. The largest patterns in the cosmos turn out to be an imprint of quantum physics!

In the previous chapters we learned how to calculate the CMB anisotropy and polarization spectrum for a given model. A theoretical model does not predict the CMB anisotropy or polarization amplitude in a given position (θ_0, φ_0) in the sky. However, this is what an experiment measures.[1]

Let us assume that we are given a temperature fluctuation map $\Delta T_s(\mathbf{n}) = T_s(\mathbf{n}) - \bar{T}_s$ from an experiment. Here $\bar{T}_s$ is the mean temperature and the suffix s stands for "signal" and by construction, $\langle \Delta T_s(\mathbf{n}) \rangle = 0$. The correlation function $\langle \Delta T_s(\mathbf{n}_1) \Delta T_s(\mathbf{n}_2) \rangle$ is a measure for the mean temperature difference,

$$\langle (T_s(\mathbf{n}_1) - T_s(\mathbf{n}_2))^2 \rangle = 2 \left(\langle \Delta T_s^2 \rangle - \langle \Delta T_s(\mathbf{n}_1) \Delta T_s(\mathbf{n}_2) \rangle \right). \tag{9.11}$$

When we put brackets around observed quantities such as $\langle \Delta T_s(\mathbf{n}_1) \Delta T_s(\mathbf{n}_2) \rangle$, we understand an averaging over directions $\mathbf{n}_1$ and $\mathbf{n}_2$ with fixed opening angle

[1] The experiment actually measures voltage differences as a function of time. We shall not enter into the rather involved process of how an optimal map $T(\theta, \varphi)$ is obtained from these time-ordered data streams [for an introduction, see Dodelson (2003) and for more details see Planck Coll. VI (2015)].

$\cos\theta = \mathbf{n}_1 \cdot \mathbf{n}_2$. In this section we simply equate such averages to theoretical ensemble averages. This is an implicit assumption of statistical isotropy. In the next section we shall discuss additional limitations of this procedure that go under the name "cosmic variance."

The measured temperature $T_s(\mathbf{n})$ is obtained from the true sky temperature by convolution with a beam profile $B(\mathbf{n},\mathbf{n}')$ centered at $\mathbf{n}$,

$$T_s(\mathbf{n}) = \int B(\mathbf{n},\mathbf{n}')T(\mathbf{n}')d\Omega'. \tag{9.12}$$

We can relate the correlation function of the measured temperature fluctuations between two directions, $\mathbf{n}_1$ and $\mathbf{n}_2$, to the power spectrum by

$$\begin{aligned}\left\langle \frac{\Delta T_s(\mathbf{n}_1)\Delta T_s(\mathbf{n}_2)}{\bar{T}^2} \right\rangle &= \frac{1}{\bar{T}^2}\int B(\mathbf{n}_1,\mathbf{n}_1')B(\mathbf{n}_2,\mathbf{n}_2')\langle \Delta T(\mathbf{n}_1')\Delta T(\mathbf{n}_2')\rangle \, d\Omega_1' \, d\Omega_2' \\ &= \sum_{\ell,m,\ell',m'} \langle a_{\ell m} a^*_{\ell' m'} \rangle \\ &\quad \times \int B(\mathbf{n}_1,\mathbf{n}_1')B(\mathbf{n}_2,\mathbf{n}_2')Y_{\ell m}(\mathbf{n}_1')Y^*_{\ell' m'}(\mathbf{n}_2') \, d\Omega_1' \, d\Omega_2' \\ &= \sum_\ell \frac{2\ell+1}{4\pi} C_\ell W_\ell(\mathbf{n}_1,\mathbf{n}_2),\end{aligned} \tag{9.13}$$

where we have inserted $\langle a_{\ell m} a^*_{\ell' m'} \rangle = C_\ell \, \delta_{\ell\ell'} \, \delta_{mm'}$. We then made use of the addition theorem for spherical harmonics, and we have defined the window function $W_\ell(\mathbf{n}_1,\mathbf{n}_2)$,

$$W_\ell(\mathbf{n}_1,\mathbf{n}_2) = \int B(\mathbf{n}_1,\mathbf{n}_1')B(\mathbf{n}_2,\mathbf{n}_2')P_\ell(\mathbf{n}_1' \cdot \mathbf{n}_2') \, d\Omega_1' \, d\Omega_2'. \tag{9.14}$$

Beam patterns are usually translation invariant so that $B(\mathbf{n},\mathbf{n}')$ depends only on the angle between $\mathbf{n}$ and $\mathbf{n}'$. Expanding the beam pattern in Legendre polynomials,

$$B(\mathbf{n}\cdot\mathbf{n}') = \sum_\ell \frac{2\ell+1}{4\pi} B_\ell P_\ell(\mathbf{n}\cdot\mathbf{n}') \tag{9.15}$$

Eq. (9.14) implies

$$W_\ell(\mathbf{n}_1,\mathbf{n}_2) = B_\ell^2 P_\ell(\mathbf{n}_1\cdot\mathbf{n}_2), \tag{9.16}$$

and, not surprisingly, the window function depends only on $\cos\theta = \mathbf{n}_1 \cdot \mathbf{n}_2$. Also the mean temperature difference and $\langle T_s(\mathbf{n}_1)T_s(\mathbf{n}_2)\rangle$ depend only on θ. This simply reflects statistical isotropy. Actually, since we determine the expectation value $\langle \bullet \rangle$ by averaging over directions that include the same angle, we obtain a result that depends only on this angle by construction.

9.4.1 Example Window Functions

As an illustration we calculate the window function for two examples of beam patterns. For this we observe that the beam is usually very narrow, a few degrees or less, so that, for the beam pattern $B(\mathbf{n},\mathbf{n}')$ we can approximate the sphere by a plane orthogonal to $\mathbf{n}$ in the regime where the beam is nonvanishing. The planar vectors $\mathbf{x}_i$ then correspond to the angle between $\mathbf{n}$ and $\mathbf{n}'_i$. Setting $\mathbf{n}'_i = \frac{\mathbf{n}_i+\mathbf{x}_i}{\sqrt{1+x_i^2}}$,

$$\begin{aligned}\mathbf{n}'_1\cdot\mathbf{n}'_2 &= (\mathbf{n}_1\cdot\mathbf{n}_2+\mathbf{n}_1\cdot\mathbf{x}_2+\mathbf{n}_2\cdot\mathbf{x}_1+\mathbf{x}_1\cdot\mathbf{x}_2)/\sqrt{(1+x_1^2)(1+x_2^2)}\\ &\simeq \mathbf{n}_1\cdot\mathbf{n}_2\left(1-\frac{1}{2}(x_1^2+x_2^2)\right)+\mathbf{n}_1\cdot\mathbf{x}_2+\mathbf{n}_2\cdot\mathbf{x}_1+\mathbf{x}_1\cdot\mathbf{x}_2.\end{aligned}$$

For the approximation we have used $x_i \equiv |\mathbf{x}_i| \ll 1$ and we have included terms up to order x_i^2. The beam function B is negligibly small if this is not satisfied. Note that even though the directions $\mathbf{n}'_i$ are close to the directions $\mathbf{n}_i$, we cannot apply the flat sky approximation since the $\mathbf{n}_i$ themselves subtend arbitrary, maybe large, angles. The generic expression for the window function then becomes

$$W_\ell(\mathbf{n}_1\cdot\mathbf{n}_2) = \int B(\mathbf{n}_1,\mathbf{x}_1)B(\mathbf{n}_2,\mathbf{x}_2)P_\ell(\mathbf{n}'_1\cdot\mathbf{n}'_2)\,dx_1^2\,dx_2^2. \tag{9.17}$$

We first simplify this formula for $\mathbf{n}_1 = \mathbf{n}_2 = \mathbf{n}$. (In this case we could apply the flat sky approximation.) Then, since $\mathbf{x}_1$ and $\mathbf{x}_2$ are normal to $\mathbf{n}$, the scalar product $\mathbf{n}'_1\cdot\mathbf{n}'_2$ becomes $\mathbf{n}'_1\cdot\mathbf{n}'_2 \simeq 1-\frac{1}{2}(x_1^2+x_2^2)+\mathbf{x}_1\cdot\mathbf{x}_2 \simeq \cos(|\mathbf{x}_1-\mathbf{x}_2|)$. Furthermore, for small values of $|\mathbf{x}_1-\mathbf{x}_2|/\ell$ and sufficiently large ℓ, we can approximate (see Appendix 4, Section A4.1)

$$\begin{aligned}P_\ell\left(\cos(|\mathbf{x}_1-\mathbf{x}_2|)\right) &\to J_0(\ell|\mathbf{x}_1-\mathbf{x}_2|)\\ &= \frac{1}{\pi}\int_0^\pi d\phi\,\exp[-i\ell|\mathbf{x}_1-\mathbf{x}_2|\cos\phi]\\ &= \frac{1}{2\pi}\int_0^{2\pi} d\phi\,\exp[-i\ell|\mathbf{x}_1-\mathbf{x}_2|\cos\phi],\end{aligned} \tag{9.18}$$

where we have used Eq. (A4.135) for the first equality. Let us define the planar vector $\boldsymbol{\ell}$ as the vector with length ℓ that points at an angle ϕ from $\mathbf{x}_1-\mathbf{x}_2$ so that $\exp[-i\ell|\mathbf{x}_1-\mathbf{x}_2|\cos\phi] = \exp[-i\boldsymbol{\ell}\cdot\mathbf{x}]$ and

$$\tilde{B}(\boldsymbol{\ell}) = \int B(\mathbf{x})e^{-i\boldsymbol{\ell}\mathbf{x}}\,d^2x, \tag{9.19}$$

the two-dimensional Fourier transform of the beam pattern. Equation (9.17) together with Eq. (9.18) then yields

$$W_\ell(1) = \frac{1}{2\pi}\int_0^{2\pi} d\phi\,|\tilde{B}(\boldsymbol{\ell})|^2. \tag{9.20}$$

The window function for $\mathbf{n}_1 = \mathbf{n}_2$ is the angular average of the square of the Fourier transformed beam pattern.

To find an expression for the window function for $\mathbf{n}_2 \neq \mathbf{n}_1$, let us expand $P_\ell(\mathbf{n}'_1 \cdot \mathbf{n}'_2)$ to second order in $\mathbf{x}_1$ and $\mathbf{x}_2$ also if $\mathbf{n}_1 \neq \pm\mathbf{n}_2$. Setting

$$\mathbf{n}'_1 \cdot \mathbf{n}'_2 = \mathbf{n}_1 \cdot \mathbf{n}_2 + \epsilon$$

up to order x_i^2 we have

$$\epsilon = -\frac{1}{2}\mathbf{n}_1 \cdot \mathbf{n}_2 \left(x_1^2 + x_2^2\right) + \mathbf{n}_1 \cdot \mathbf{x}_2 + \mathbf{n}_2 \cdot \mathbf{x}_1 + \mathbf{x}_1 \cdot \mathbf{x}_2,$$

so that

$$P_\ell(\mathbf{n}'_1 \cdot \mathbf{n}'_2) \simeq P_\ell(\mathbf{n}_1 \cdot \mathbf{n}_2) + \epsilon P'_\ell(\mathbf{n}_1 \cdot \mathbf{n}_2) + \frac{1}{2}\epsilon^2 P''_\ell(\mathbf{n}_1 \cdot \mathbf{n}_2). \tag{9.21}$$

This approximation is sufficient if P_ℓ does not vary too much in an interval ϵ, hence if $\ell\epsilon < 1$. Inserting it in Eq. (9.17) and keeping only terms up to second order in x_i, we find

$$W_\ell(z) = P_\ell(\mathbf{n}_1 \cdot \mathbf{n}_2) + \frac{1}{2}\int d^2x_1\, d^2x_2\, B(\mathbf{n}_1, \mathbf{x}_1) B(\mathbf{n}_2, \mathbf{x}_2)$$
$$\times \left[-2z\left(x_1^2 + x_2^2\right) P'_\ell(z) + ((\mathbf{n}_1 \cdot \mathbf{x}_2)^2 + (\mathbf{n}_2 \cdot \mathbf{x}_1)^2) P''_\ell(z)\right], \tag{9.22}$$

with $z = \mathbf{n}_1 \cdot \mathbf{n}_2 = \cos\theta$. We have assumed that the beam is spherically symmetric around its center and we have dropped all terms that were linear in the vectors $\mathbf{x}_i$ and therefore integrate to zero. Decomposing the vectors $\mathbf{x}_i$ into a component that lies in the $(\mathbf{n}_1, \mathbf{n}_2)$-plane $(\mathbf{e}_i)$ and a component orthogonal to it $(\mathbf{m})$, $\mathbf{x}_i = \mathbf{e}_i \cos\varphi_i + \mathbf{m} \sin\varphi$, we find

$$\mathbf{n}_1\mathbf{x}_2 = \sin\theta \cos\varphi_2, \quad \mathbf{n}_2\mathbf{x}_1 = \sin\theta\cos\varphi_1.$$

Inserting this above, integration over angles gives

$$W_\ell(z) = P_\ell(z) + \frac{(2\pi)^2}{4}\int_0^\infty dx_1\, dx_2\, x_1 x_2 B(\mathbf{n}_1, \mathbf{x}_1) B(\mathbf{n}_2, \mathbf{x}_2)$$
$$\times \left[-2z\left(x_1^2 + x_2^2\right) P'_\ell(z) + \left(x_2^2 + x_1^2\right)(1 - z^2) P''_\ell(z)\right]. \tag{9.23}$$

We now use the fact that $(1-z^2)P''_\ell(z) - 2zP'_\ell(z) = -\ell(\ell+1)P_\ell(z)$ (see Appendix 4, Section A4.1) and the normalization of the beam, $2\pi\int_0^\infty dx\, x B(\mathbf{x}) = 1$. Furthermore, we define the width of the beam

$$\sigma^2 \equiv \pi \int_0^\infty dx\, x^3 B(\mathbf{x}). \tag{9.24}$$

With this, the window function simply becomes

$$W_\ell(z) = P_\ell(z)\left(1 - \sigma^2\ell(\ell+1)\right), \quad \text{for} \quad \sigma\ell \ll 1. \tag{9.25}$$

This last condition is necessary to ensure that $P_\ell(z+\epsilon)$ is well approximated by Eq. (9.21) for all $|\epsilon| \lesssim \sigma$. A reasonable approximation for all values of ℓ might therefore be $W_\ell(z) = P_\ell(z)\exp(-\sigma^2\ell(\ell+1))$, which is (nearly) the result of the Gaussian beam, as we see in the text that follows.

Gaussian Beam

For a Gaussian beam,

$$B(\mathbf{x}) = \frac{1}{2\pi\sigma^2}\exp\left(-\frac{x^2}{2\sigma^2}\right), \tag{9.26}$$

with Fourier transform

$$B(\boldsymbol{\ell}) = \exp\left(-\ell^2\sigma^2/2\right); \tag{9.27}$$

hence

$$W_\ell(1) = e^{-\ell^2\sigma^2}. \tag{9.28}$$

For $z \neq 0$ and $\sigma\ell \ll 1$ we reproduce Eq. (9.25) and

$$W_\ell(z) = P_\ell(z)\exp\left(-\sigma^2\ell(\ell+1)\right) \tag{9.29}$$

is an excellent approximation to the numerical result for all values of z and ℓ as long as $\sigma \ll 1$. The only difference in $W_\ell(1)$ is that $\ell(\ell+1)$ becomes ℓ^2, which comes from the fact that we have approximated the sphere by a plane normal to $\mathbf{n}$. This is an irrelevant difference for sufficiently large values of ℓ.

Differencing Beam

The disadvantage of the Gaussian beam is the fact that we have to subtract the mean temperature to relate it to the theoretical power spectrum of the CMB anisotropies. We subtract two large numbers to obtain a small result. A notoriously dangerous procedure to perform on noisy data. Therefore, instead of the Gaussian beam one usually utilizes a beam pattern with mean zero,

$$\int d^2\mathbf{x}\, B(\mathbf{x}) = 0.$$

Often one simply adds the signals coming from different directions with weights that add up to 0. Let us analyze the simplest case of a single subtraction. We choose the line connecting the two beam centers to be the x-axis in the $\mathbf{x} = (x, y)$ plane and define

$$B(x, y) = \frac{1}{2\pi\sigma^2}\left[\exp\left[-\frac{(x-x_0)^2+y^2}{2\sigma^2}\right] - \exp\left[-\frac{(x+x_0)^2+y^2}{2\sigma^2}\right]\right]. \tag{9.30}$$

This is the difference of two Gaussian beams separated by a "throw" of $2x_0$. A calculation similar to the one for the Gaussian beam leads to

$$W_\ell(1) = e^{-\sigma^2\ell^2}\,(1 - P_\ell(\cos(2x_0)))\,. \tag{9.31}$$

In Fig. 9.4 we plot the window function for $\sigma = 1° = \pi/180$ for two different values of the throw. As for the Gaussian beam, we cannot measure fluctuations with $\ell\sigma \gg 1$. But, what is new for the differencing beam, we are also not sensitive to fluctuations on scales much larger than the throw of the beam pattern, since on these scales the beam averages to zero.

With the known window function, we can now relate the measured temperature fluctuations to the theoretical C_ℓ's. As always when doing an experiment, we want to know the best estimate for C_ℓ and its error, or better, its probability distribution.

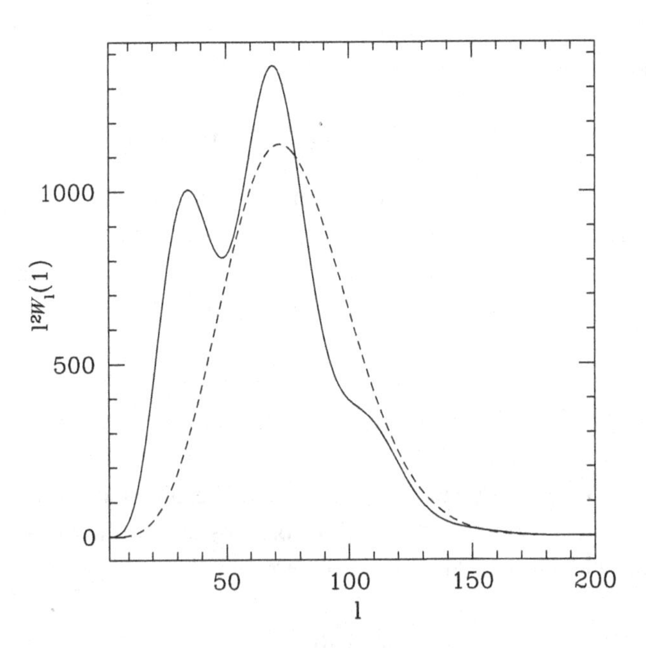

Fig. 9.4 We show the window function $\ell^2 W_\ell(1)$ for a differencing beam with width $\sigma = 1°$ for two different values of the throw, $x_0 = 4°$ (solid) and $x_0 = 1°$ (dashed). The peak at $\ell \simeq 70$ is due to the beam size. If throw and beam size differ, a second peak appears on the left for the larger value of the throw.

Before exploring statistical methods to analyse CMB data, we discuss an error that is always present in cosmological experiments.

9.4.2 Cosmic Variance

Only one CMB sky is at our disposition for observation. Therefore, when we measure the mean fluctuation in large angular patches, not many statistically independent patches are available in the sky and we expect relatively large statistical fluctuations.

Let us calculate this fluctuation under the assumption that the initial fluctuations are Gaussian. Then, the coefficients $a_{\ell m}$ are Gaussian variables and, in the optimal case when our data cover all sky, we can determine $2\ell+1$ statistically independent[2] $a_{\ell m}$'s for a given value of ℓ. We want to determine the variance

$$\sigma_\ell = \sqrt{\frac{\langle (C_\ell^o - C_\ell)^2 \rangle}{C_\ell^2}}.$$

Here $C_\ell^o = (2\ell+1)^{-1}\sum_m |a_{\ell m}|^2$ is the "random" variable that we obtain when averaging over the $2\ell+1$ measured $a_{\ell m}$'s, and $C_\ell = \langle |a_{\ell m}|^2 \rangle$ is the statistical expectation value. The square variance, σ^2 of $2\ell+1$ independent Gaussian variables is simply $1/(2\ell+1)$. For the squares of these variables, we expect σ^2 to double, by simple error propagation. This is exactly what we will find now with a more thorough calculation,

$$\langle (C_\ell^o - C_\ell)^2 \rangle = \frac{1}{(2\ell+1)^2} \left\langle \left(\sum_m \left[|a_{\ell m}|^2 - \langle |a_{\ell m}|^2 \rangle \right] \right)^2 \right\rangle$$
$$= \frac{1}{(2\ell+1)^2} \sum_{m,m'} \left(\langle |a_{\ell m}|^2 |a_{\ell m'}|^2 \rangle - \langle |a_{\ell m}|^2 \rangle \langle |a_{\ell m'}|^2 \rangle \right).$$

The second term in the above sum is simply C_ℓ^2. For the first term we apply Wick's theorem, which states that for a set of Gaussian variables, the $2n$-point correlation function is given by the sum of all the possible 2-point correlation functions that can be formed from it (see Appendix 7). Hence

$$\langle |a_{\ell m}|^2 |a_{\ell m'}|^2 \rangle = \langle a_{\ell m} a_{\ell m}^* a_{\ell m'} a_{\ell m'}^* \rangle$$
$$= \langle a_{\ell m} a_{\ell m}^* \rangle \langle a_{\ell m'} a_{\ell m'}^* \rangle + \langle a_{\ell m} a_{\ell m'}^* \rangle \langle a_{\ell m'} a_{\ell m}^* \rangle + \langle a_{\ell m} a_{\ell m'} \rangle \langle a_{\ell m'}^* a_{\ell m}^* \rangle$$
$$= C_\ell^2 + \delta_{m,m'} C_\ell^2 + \delta_{m,-m'} C_\ell^2.$$

[2] Even though the $a_{\ell m}$ are complex, the reality of $\Delta T/T$ requires that $a_{\ell -m} = a_{\ell m}^*$ so that we have $2\ell+1$ real coefficients.

For the last equals sign we have used the fact that $a_{\ell m}$ and $a^*_{\ell m'} = a_{\ell,-m'}$ are independent random variables if $m \neq m'$. Summation over m and m' gives now

$$\langle (C^o_\ell - C_\ell)^2 \rangle = \frac{2}{2\ell+1} C^2_\ell \quad \text{so that}$$

$$\sigma_\ell = \sqrt{\frac{\langle (C^o_\ell - C_\ell)^2 \rangle}{C^2_\ell}} = \sqrt{\frac{2}{2\ell+1}}.$$

This is the absolutely minimal error on the CMB temperature or polarization power spectrum that can be achieved from one sky. It is a principle causality limit that cannot be escaped. To it we have to add instrumental noise, foregrounds, atmospheric noise, and so forth.

Even if there were a faraway civilization at a cosmological distance that would undertake similar measurements and then send us their results, this would not really help. By the time it takes for their information to arrive on Earth, the CMB sky has grown by so much, that the region they have observed is now also inside our Hubble horizon and we can observe it in our CMB experiments. On the other hand, if they sent us the data long ago, the sky they could observe at this time is now also inside our Hubble horizon. The problem is, of course, causality. If the Universe is not inflating, we can by no means obtain any information about a region outside our Hubble horizon that corresponds roughly to the CMB sky.

If we observe a fraction $f_{\rm sky} < 1$ of the sky, the error increases, since we now cannot determine all the $a_{\ell m}$'s, and $2\ell + 1$ is replaced by $(2\ell + 1) f_{\rm sky}$. This is roughly the number of independent $a_{\ell m}$'s which can be measured in a fraction $f_{\rm sky}$ of the sky. The variance then increases to

$$\sigma_\ell = \sqrt{\frac{2}{(2\ell+1) f_{\rm sky}}}.$$

Finally, we note that in real full sky CMB experiments, the data close to the galactic plane are strongly contaminated by foregrounds and it is safest not to use them at all. One therefore usually cuts out a region of about 20° around the galactic plane. In this cut sky, the spherical harmonics no longer form an orthonormal basis of function and one has to conceive a new method to define such a basis. This can, in principle, be done by applying a Cauchy–Schwartz orthogonalization procedure on the old basis; see Gorski (1994).

In Figs. 9.5–9.8 we show the presently (April 2020) available CMB data in terms of the temperature anisotropy, polarization, and cross correlation spectra. There are some more data on temperature anisotropies available in small scales, $\ell > 2000$, especially from the South Pole Telescope (SPT) but also from the Atacama Cosmology Telescope (ACT) but on these scales CMB anisotropy are

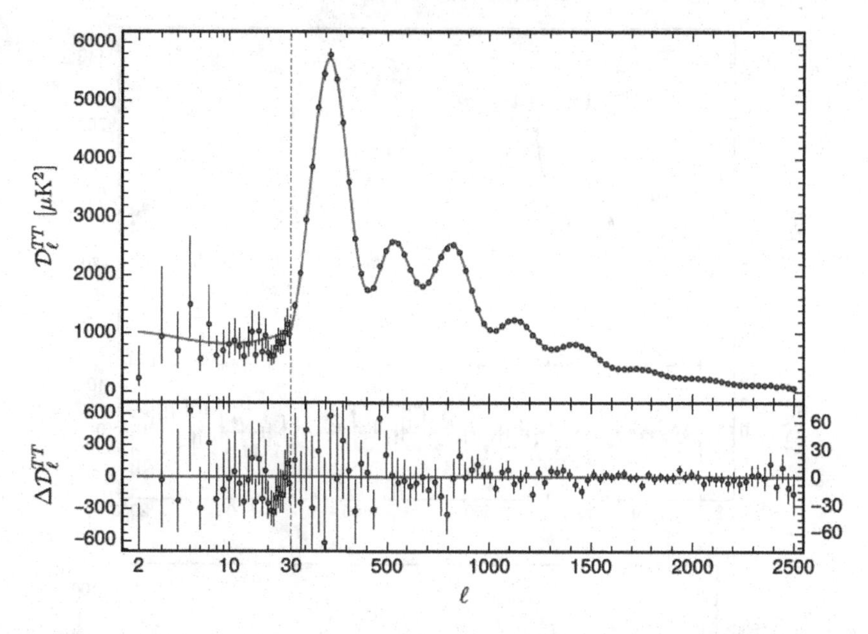

Fig. 9.5 The observed CMB anisotropy spectrum from Planck [figure from Planck Coll. VI (2018)]. The vertical axis shows $\mathcal{D}_\ell^{TT} \equiv \ell(\ell+1)C_\ell/(2\pi)$. The line draws the best-fitting ΛCDM model. For $\ell < 30$ (dashed vertical line) the horizontal axis is logarithmic while for $\ell > 30$ it is linear. The bottom inset shows the difference between the data and the model.

dominated by "secondary effects" that we briefly mention at the end of this chapter but that we do not discuss in detail.

9.5 Statistical Methods

To extract the optimal information from data one needs to apply the best statistical methods. However, as a rule of thumb, results that strongly depend on the statistical method applied are not to be trusted.

Some elementary statistical tools that are used in this chapter are presented in Appendix 7.

9.5.1 Bayes' Theorem and the Likelihood Function

To estimate cosmological parameters from CMB data one uses a simple result from probability theory that goes under the name of Bayes' theorem. In a probability space, consider two sets A and B and their intersection $A \cap B$. The probability that a given event of which we know that it is in A is also in B is

$$\frac{P[A \cap B]}{P[A]} =: P[B|A]. \tag{9.32}$$

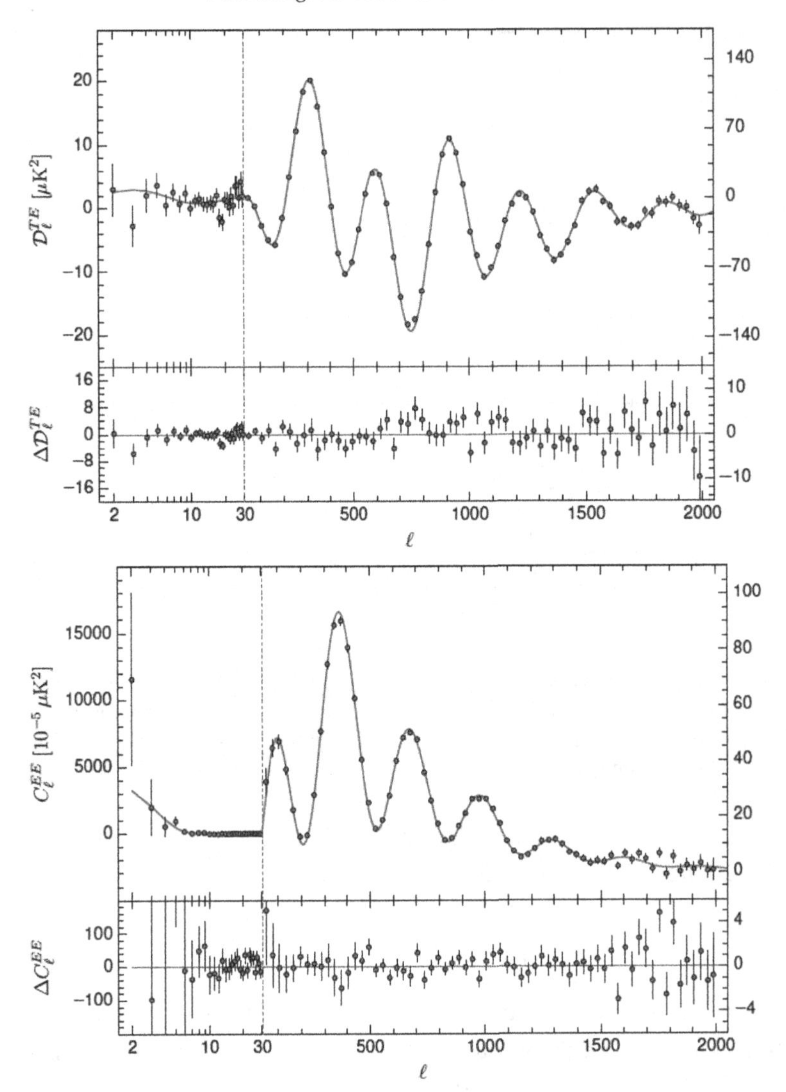

Fig. 9.6 The CMB temperature–polarization cross-correlation and the E-polarization spectra obtained by Planck (figure from Planck Coll. VI (2018)). The vertical axes show $\mathcal{D}_\ell^{TE} \equiv \ell(\ell+1)C_\ell^{TE}/(2\pi)$ and C_ℓ^{EE}. The line draws the best-fitting ΛCDM model. For $\ell < 30$ (dashed vertical line) the horizontal axis is logarithmic while for $\ell > 30$ it is linear. The bottom inset shows the difference between the data and the model.

$P[B|A]$ denotes the conditional probability of B given A. Exchanging A and B we obtain the conditional probability of A given B. Hence

$$P[A \cap B] = P[B|A]P[A] = P[A|B]P[B]. \tag{9.33}$$

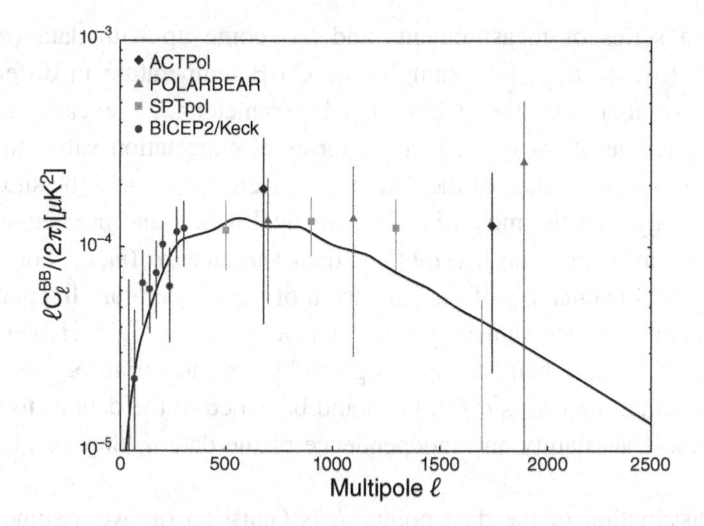

Fig. 9.7 The measured BB polarization spectrum from different experiments. The curve is the B-polarization spectrum from lensing expected from the best-fit Planck ΛCDM model. For more details see Louis *et al.* (2017), from which this figure is adapted

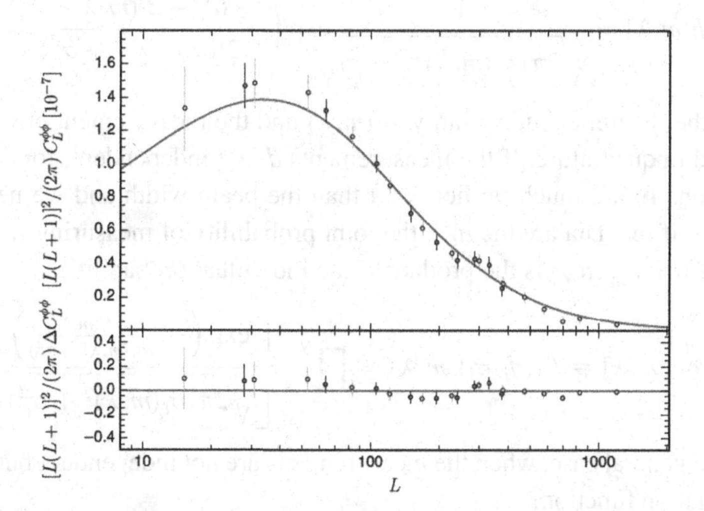

Fig. 9.8 The lensing power spectrum from the Planck experiment. The solid line is the theoretical lensing spectrum for the best-fit Planck ΛCDM model. [Figure from Planck Coll. VI (2018)]

The last equation that is here written for the probabilities of sets is, of course, also true for the corresponding probability densities. What is the relevance of this simple statement for parameter estimation? Let us fix a cosmological model m that is described by a set of parameters $(\lambda_1, \ldots, \lambda_M) = \lambda$. Our experiment has made a series of measurements and has come up with data $(d_1, \ldots, d_N)$ with errors $(\sigma_1, \ldots, \sigma_N)$, for example, the CMB temperature in different directions $(\mathbf{n}_1, \cdots, \mathbf{n}_N)$. For the given model parameters λ we can calculate the predicted outcome of the experiment in terms of expectation value and variance $(d_i(m,\lambda), \sigma_i(m,\lambda))$ for each of the data points. Here we assume the situation, as it is in cosmology, that the model is of a statistical nature and predicts expectation values for the measurements $d_i(m,\lambda)$ and their variances, $\sigma_i(m,\lambda)$. For example, if the d_i's are coefficients $a_{\ell m}$ of the expansion of the temperature fluctuations, then their expectation values vanish and the variances are the C_ℓ's. However, if your data are the C_ℓ's then their variance is given by cosmic variance. It is sometimes not clear whether the errors $\sigma_i(m,\lambda)$ should be added to the data or to the model. If we assume Gaussianity and independence of the data d_i they can be added in quadrature.

If the distribution of the data points d_i is Gaussian (as we assume the CMB temperature fluctuations to be), the probability of measuring d_i in model m with parameter values λ taking into account the measurement uncertainty σ_i is given by (see Appendix 7)

$$P[d_i|m,\lambda] = \frac{1}{\sqrt{2\pi(\sigma_i(m,\lambda)^2 + \sigma_i^2)}} \exp\left(-\frac{(d_i - d_i(m,\lambda))^2}{2(\sigma_i(m,\lambda)^2 + \sigma_i^2)}\right). \quad (9.34)$$

Note that the theoretical uncertainty, $\sigma_i(m,\lambda)$ and the measurement error, σ_i, have been added in quadrature. If the measurements d_i are independent, for example, if the directions $\mathbf{n}_i$ are much farther apart than the beam width and we neglect correlations, or if the data are the $a_{\ell m}$, the joint probability of measuring $(d_1, \ldots, d_N)$ with errors $(\sigma_1, \ldots, \sigma_N)$ is the product of the individual probabilities,

$$P[\{d_i,\sigma_i\}|m,\lambda] \equiv \mathcal{L}(\{d_i,\sigma_i\},m,\lambda) = \prod_{i=1}^{N} \left[\frac{\exp\left(-\frac{(d_i - d_i(m,\lambda))^2}{2(\sigma_i(m,\lambda)^2+\sigma_i^2)}\right)}{\sqrt{2\pi(\sigma_i(m,\lambda)^2 + \sigma_i^2)}}\right]. \quad (9.35)$$

In the more general case, when the measurements are not independent but Gaussian with correlation function,

$$\langle d_i d_j \rangle = C_{ij}, \quad (9.36)$$

Eq. (9.35) becomes

$$\mathcal{L}(\{d_i,\sigma_i\},m,\lambda) = \frac{1}{\sqrt{\det C(2\pi)^N}} \exp\left(-\frac{d_i C_{ij}^{-1} d_j}{2}\right), \quad (9.37)$$

(see Appendix 7). This expression is called the likelihood function. It gives us the likelihood of the data $\{d_i,\sigma_i\}$ given the model m with parameters λ. But since it is the data that we know and the model that we would like to know, we would be more interested in the probability of a model (m,λ) given the data. And here, Bayes' theorem comes to our rescue. According to Eq. (9.33),

$$P[m,\lambda|\{d_i,\sigma_i\}] = P[\{d_i,\sigma_i\}|m,\lambda]\frac{P[m,\lambda]}{P[\{d_i,\sigma_i\}]}. \tag{9.38}$$

Here $P[m,\lambda]$ is called "the prior" and the denominator is called "the evidence." The left-hand side is the "posterior distribution" or simply the "posterior." Ideally it can be used as "prior" for the next experiment. The evidence $P[\{d_i,\sigma_i\}]$ is unimportant for parameter estimation since it does not depend on the model parameters; hence the ratio of the posteriors for two different sets of parameters, $P[m,\lambda^{(1)}|\{d_i,\sigma_i\}]/P[m,\lambda^{(2)}|\{d_i,\sigma_i\}]$ is independent of the evidence. Furthermore, it can be eliminated, noting that when integrating over the entire space of model parameters, the left-hand side must be normalized to 1,

$$P[\{d_i,\sigma_i\}] = \int d^M\lambda \, P[m,\lambda]\mathcal{L}(\{d_i,\sigma_i\},m,\lambda).$$

The prior $P[m,\lambda]$ depends on our prior knowledge of the model space. For example, if previous experiments have shown us that curvature is not large, it makes sense to choose $P[m,\lambda] = 0$ for parameter values $|\Omega_K| > \frac{1}{2}$. But how shall we choose $P[m,\lambda]$ for $|\Omega_K| < \frac{1}{2}$? We may opt for a "flat" prior, that is, the same probability for all values of Ω_K. But this is as much a special case as choosing a flat prior for Ω_K^3 or $\exp(\Omega_K)$. This is the weakest point of Bayesian analysis: the probability of model parameters (m,λ) for given data $\{d_i,\sigma_i\}$ depends on our prior.

Things are not as hopeless as it might seem. First of all, physics often tells us to some extent what the distribution of the prior should be. As a rule of thumb, parameters that add to the data should have a flat prior while parameters that multiply the data (scaling parameters) more naturally have a logarithmic prior, that is, a flat prior in $\log(\lambda)$.

If the data have sufficiently small error bars, most priors are relatively flat in the relevant interval and the resulting maximum likelihood is nearly independent of the prior. However, if the data are weak and errors are large, the Bayesian posterior $P[m,\lambda|\{d_i,\sigma_i\}]$ (i.e., the likelihood of the model parameters given the data) may well be strongly prior dependent. In this case, the data are simply not good enough to determine the parameters of the model.

The model parameters with the highest probability to be correct are given by the maximum of $P[m,\lambda|\{d_i,\sigma_i\}]$. If the data are sufficiently good to choose model parameters, that is, if the prior $P[m,\lambda]$ is sufficiently flat around the maximum of $P[m,\lambda|\{d_i,\sigma_i\}]$, then the latter is close to the maximum of the likelihood function, $\mathcal{L}(\{d_i,\sigma_i\},m,\lambda)$. When estimating parameters, we therefore can simply search for a maximum of the likelihood function.

9.5.2 Model Comparison

Sometimes, we would like to compare two models with different parameter sets and decide which one of them is more probable given the data. We might want to answer questions like: since we have not seen any tensor modes in the CMB data yet, is it improbable that they are there at all or do we have to continue searching for r? In this case the answer is well known: our present limit is about $r \lesssim 0.05$ and there are well-motivated models with $r \sim 10^{-3}$ to 10^{-2}; hence we have to improve our measurements. Clearly, this answer is given by physical modeling and not by statistics. Another question is as follows: I have a dark energy model with a free function f (e.g., the potential of a scalar field or the kinetic term in a k-essence model) that I can choose such that my model fits the data better than ΛCDM. Does that mean that I have ruled out ΛCDM and people should rather study my model? Of course, having an entire free function that can be modeled with many parameters, my model will contain many more parameters than simple ΛCDM.

When comparing models, at first we certainly want to look at the likelihood functions and compare those. Models with a larger maximum likelihood fit the data better. But on the other hand, if model m_2 has many more parameters than model m_1, we are not surprised if m_2 fits the data better than m_1 and may still decide in favor of m_1 with the argument that m_1 is "more physical" and more "predictive" or, certainly, more economical. The latter criterion is often called "Occam's razor": we should explain the data by the simplest possible model. Can these seemingly subjective criteria be made objective? We shall now see that under certain assumptions the answer is yes.

To be more specific let us consider two models that we want to compare; m_1 with parameters $\lambda^{(1)} = (\lambda_1^{(1)}, \ldots, \lambda_{M_1}^{(1)})$ and m_2 with parameters $\lambda^{(2)} = (\lambda_1^{(2)}, \ldots, \lambda_{M_2}^{(2)})$. We assume that the priors are fixed in both cases. (In Bayesian statistics the priors are just part of the game and cannot be ignored!) Bayes' theorem then gives us the probability of some set of parameters for given data D by

$$P[\lambda^{(1)}|D, m_1] = \frac{P[D|m_1, \lambda^{(1)}]P[\lambda^{(1)}|m_1]}{P[D|m_1]}, \tag{9.39}$$

$$P[\lambda^{(2)}|D, m_2] = \frac{P[D|m_2, \lambda^{(2)}]P[\lambda^{(2)}|m_2]}{P[D|m_2]}. \tag{9.40}$$

Our present notation indicates that our parameter choice $\lambda^{(1)}$ or $\lambda^{(2)}$ assumes the model m_1 or m_2 and, more importantly, that the evidence depends on the model under consideration. The prior is now of the form of a probability for the parameters $\lambda^{(1)}$ and $\lambda^{(2)}$ respectively, given the model m_1 and m_2 respectively. From Bayes' theorem we also obtain

$$P[m_i|D] \propto P[D|m_i]P[m_i]. \tag{9.41}$$

Here, $P[m_i]$ is the total prior we assign to model m_i while $P[D|m_i]$ is simply the evidence from above. If we have no idea which model should be preferred, we can simply set $P[m_i] = \frac{1}{2}$. The probability of a model is then proportional to its evidence, and more importantly, their *ratio* is equal to the ratio of the evidences. But the evidence is given by the integral

$$P[D|m_i] = \int P[D|m_i,\lambda^{(i)}]P[\lambda^{(i)}|m_i]\,d^{M_i}\lambda^{(i)}. \tag{9.42}$$

For many problems the posterior $P[\lambda^{(i)}|D,m_i] = P[D|m_i,\lambda^{(i)}]P[\lambda^{(i)}|m_i]$ is strongly peaked around some best-fitting value $\bar{\lambda}^{(i)}$ with widths $\sigma^{(i)}(D) = (\sigma_1^{(i)},\ldots,\sigma_{M_i}^{(i)})$ (let us assume, for simplicity, that the parameters are uncorrelated). We may then evaluate the integral (9.42) by multiplying the peak height with the width,

$$\Sigma^{(i)}(D) = \Pi_{j=1}^{M_i}\sigma_j^{(i)}.$$

The evidence can then be approximated by

$$P[D|m_i] \simeq P[D|m_i,\bar{\lambda}^{(i)}]P[\bar{\lambda}^{(i)}|m_i]\Sigma^{(i)}(D). \tag{9.43}$$

Furthermore, let us assume that the prior of model i has some (large) total width $\Sigma^{(i)}$ and that $\bar{\lambda}^{(i)}$ is nicely inside the prior distribution. Then, due to normalization, $P[\bar{\lambda}^{(i)}|m_i] \simeq 1/\Sigma^{(i)}$ and the evidence for model m_i becomes

$$P[D|m_i] \simeq P[D|m_i,\bar{\lambda}^{(i)}]\frac{\Sigma^{(i)}(D)}{\Sigma^{(i)}}. \tag{9.44}$$

Hence the first guess, the maximum of the likelihood, is multiplied by the so-called Occam factor $\Sigma^{(i)}(D)/\Sigma^{(i)}$, that is, the ratio of the parameter space "occupied by the data" divided by the volume of parameter space allowed by the prior. Models that are not predictive at all are penalized by a small Occam factor, and can in this way lose against a model even if they allow for a parameter choice $\bar{\lambda}^{(i)}$ with higher likelihood. Of course if the prior is constraining, but does not have a significant overlap with the posterior from the data, this will also disfavor a model; it will render $\Sigma^{(i)}(D)$ very small. These situations are illustrated in the panels (a)–(c) of Fig. 9.9. As becomes clear from this example, for models that do allow a good fit to the data, the evidence strongly depends on the prior.

It is, however, important to note that the Occam factor enters the model probability just as a power law, while the offset of the measured parameter from the model prediction reduces the model probability exponentially.

A situation that is often encountered in cosmology are so-called nested models. Two models are called "nested" if one of them is obtained by fixing one or several parameters of the other model. For example, a model with vanishing tensor

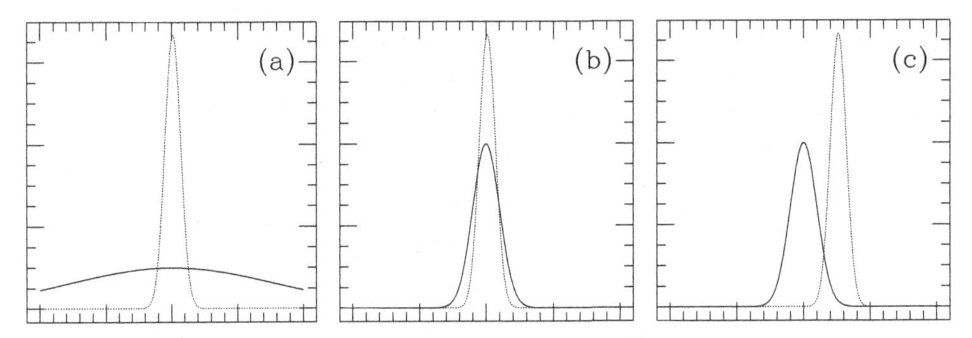

Fig. 9.9 Three 1-parameter models with their prior (solid) and posterior (dotted) distributions shown. They all have a comparable maximum likelihood. In (a) the prior distribution is much wider than the posterior. In models (b) and (c) the widths are comparable, but model (c) does not provide a good fit. The evidence for model (b) is by far the largest.

contribution is nested inside the more general models that allow for tensors, by setting the tensor to scalar ratio $r = 0$. Other examples are models that do not allow for curvature that are nested inside models with curvature. Let us concentrate on these to illustrate the significance of the prior for model selection.

As we shall see, present data yield the constraints $\Omega_K \simeq -0.0106 \pm 0.0065$ from Planck data alone, or even $\Omega_K \simeq -0.0007 \pm 0.0019$ when including BAO data. Does this mean that a universe with vanishing curvature is preferred? To discuss this let us approximate the marginalized posterior distribution of the parameter Ω_K by a Gaussian with width $\sigma_K = 0.002$. Cosmologist A says that the Universe might well have a curvature in the full range $\Omega_K \in [-1, 1]$, and therefore models allowing for curvature are penalized by an Occam factor of 0.002; hence models without curvature are strongly preferred.

Cosmologist B argues differently. She says that we know from inflation, which is in good agreement with all other observations, that curvature must be small, say $\Omega_K \in [-0.005, 0.005]$, and therefore there is no significant Occam factor, and models with nonvanishing but small curvature are as plausible as vanishing curvature.

This example makes it clear: for model selection the prior is crucial. It is of course a very different statement to say that since curvature is small it has little effect on all the other parameters and we therefore set it to zero in our analysis (which somewhat speeds up the CMB codes). This is simply a practical statement and does not address the problem of model selection.

If the posterior probability distribution is non-Gaussian, the foregoing approximations no longer hold, and we have to resort to more complicated methods to evaluate the posterior distribution and the model probabilities like MCMC;

see Section 9.5.4. But the principal arguments of the foregoing discussion remain valid. For more details and more advanced methods of model selection see Kunz *et al.* (2006) and Trotta (2017).

9.5.3 Fisher Matrix and Parameter Estimation

9.5.3.1 Best-Fitting Parameters

We now fix some model m with parameters $\lambda = (\lambda_1, \ldots, \lambda_M)$. The best-fitting parameters $\bar{\lambda}$ correspond to the maximum of the likelihood function,

$$\left.\frac{d\mathcal{L}(\lambda)}{d\lambda}\right|_{\bar{\lambda}} = 0. \tag{9.45}$$

They are most easily determined with a root-finder method applied to $d\mathcal{L}(\lambda)/d\lambda$. However, for a Gaussian distribution the likelihood function is an exponential. Therefore, its logarithm is often better suited to a numerical root finder. As $\mathcal{L}$ is nonnegative and the logarithm is a monotonic function, we can in full generality maximize $\ln \mathcal{L}$. To do this one starts at some first guess value $\lambda^{(0)}$ and then approximates the derivative $d \ln \mathcal{L}/d\lambda$ to first order,

$$\frac{\partial \ln \mathcal{L}}{\partial \lambda_i}(\bar{\lambda}) \simeq \frac{\partial \ln \mathcal{L}}{\partial \lambda_i}(\lambda^{(0)}) + (\bar{\lambda}_j - \lambda_j^{(0)})\frac{\partial^2 \ln \mathcal{L}}{\partial \lambda_i \partial \lambda_j}(\lambda^{(0)}).$$

Setting $\frac{\partial \ln \mathcal{L}}{\partial \lambda}(\bar{\lambda}) = 0$, we obtain to first order

$$(\bar{\lambda}_j - \lambda_j^{(0)}) \simeq -\left(\frac{\partial^2 \ln \mathcal{L}}{\partial \lambda_j \partial \lambda_i}\right)^{-1} \frac{\partial \ln \mathcal{L}}{\partial \lambda_i}(\lambda_0) \equiv \delta\lambda_j. \tag{9.46}$$

For the next step we can replace $\lambda^{(0)}$ by $\lambda^{(1)} = \lambda^{(0)} + \delta\lambda$ and iterate the procedure until it converges. In directions in which the likelihood function is very flat, we make large steps while in directions along which it is steep, the steps are small.

However, this is not how it is usually done. There is a simplification that can be made without a great loss of accuracy. For the CMB anisotropies and polarization, we expect the likelihood function to be (nearly) Gaussian and hence of the form (9.37) so that

$$\frac{\partial \ln \mathcal{L}}{\partial \lambda_i} = -\frac{1}{2}\frac{\partial}{\partial \lambda_i}\left[\ln(\det \mathcal{C}) + d^T \mathcal{C}^{-1} d\right]. \tag{9.47}$$

For notational simplicity we now denote $\partial/\partial\lambda_i \equiv \partial_i$. With $\ln(\det \mathcal{C}) = \mathrm{Tr}(\ln(\mathcal{C}))$ and $\partial_i \mathcal{C}^{-1} = -\mathcal{C}^{-1}(\partial_i \mathcal{C})\mathcal{C}^{-1}$ we obtain

$$\frac{\partial \ln \mathcal{L}}{\partial \lambda_i} = \frac{1}{2}\left[d^T \mathcal{C}^{-1}(\partial_i \mathcal{C})\mathcal{C}^{-1} d - \mathrm{Tr}\left(\mathcal{C}^{-1}(\partial_i \mathcal{C})\right)\right]. \tag{9.48}$$

For the second derivative we find after similar manipulations

$$\frac{\partial^2 \ln \mathcal{L}}{\partial \lambda_i \partial \lambda_j} = -\frac{1}{2}\Big[d^T C^{-1}(\partial_i C)C^{-1}(\partial_j C)C^{-1}d$$

$$+ d^T C^{-1}(\partial_j C)C^{-1}(\partial_i C)C^{-1}d - \mathrm{Tr}\left(C^{-1}(\partial_j C)C^{-1}(\partial_i C)\right)$$

$$- d^T C^{-1}(\partial^2_{ij} C)C^{-1}d + \mathrm{Tr}\left(C^{-1}(\partial^2_{ij} C)\right)\Big]. \tag{9.49}$$

The Fisher matrix is defined as the expectation value

$$-F_{ij} \equiv \left\langle \frac{\partial^2 \ln \mathcal{L}}{\partial \lambda_i \partial \lambda_j} \right\rangle. \tag{9.50}$$

To determine it we use $\langle d_i d_j \rangle = C_{ij}$ so that

$$\begin{aligned}\langle d^T C^{-1}(\partial_i C)C^{-1}(\partial_j C)C^{-1}d\rangle &= \langle d_m C^{-1}_{mn}(\partial_i C)_{np}C^{-1}_{pq}(\partial_j C)_{qr}C^{-1}_{rs}d_s\rangle \\ &= C^{-1}_{mn}(\partial_i C)_{np}C^{-1}_{pq}(\partial_j C)_{qr}C^{-1}_{rs}C_{sm} \\ &= \mathrm{Tr}\left((\partial_i C)C^{-1}(\partial_j C)C^{-1}\right) \\ &= \mathrm{Tr}\left((\partial_j C)C^{-1}(\partial_i C)C^{-1}\right) \\ &= \mathrm{Tr}\left(C^{-1}(\partial_j C)C^{-1}(\partial_i C)\right).\end{aligned}$$

For the last two equals signs we use the fact that the trace is invariant under cyclic permutations. Equivalently

$$\langle d^T C^{-1}(\partial^2_{ij} C)C^{-1}d\rangle = \mathrm{Tr}\left(C^{-1}(\partial^2_{ij} C)\right).$$

Inserting these results to calculate the expectation value of (9.49) we obtain

$$F_{ij} = \frac{1}{2}\mathrm{Tr}\left(C^{-1}(\partial_j C)C^{-1}(\partial_i C)\right). \tag{9.51}$$

We assume the Fisher matrix to be nonsingular. Otherwise not all the parameters λ_i can be determined by the data, since then the Fisher matrix has a vanishing eigenvalue for some eigenvector $\mu \neq 0$. The covariance matrix and hence the likelihood function are then independent of the linear combination $\mu_i \lambda_i$, which therefore cannot be estimated by the experiment. By definition, F is symmetric and $F_{ij}\lambda^i\lambda^j = \frac{1}{2}\mathrm{Tr}\left((C^{-1}(\lambda_j \partial_j C))^2\right) > 0$ for all $\lambda \neq 0$. The Fisher matrix is a positive-definite symmetric matrix.

Instead of dividing by the true curvature of the likelihood function in Eq. (9.46) to determine the next estimator for the parameters, one divides by the corresponding element of the Fisher matrix,

$$\lambda_i^{(1)} = \lambda_i^{(0)} + \frac{1}{2}F^{-1}_{ij}\left[d^T C^{-1}(\partial_j C)C^{-1}d - \mathrm{Tr}\left(C^{-1}(\partial_j C)\right)\right](\lambda^{(0)}). \tag{9.52}$$

The advantage of this method is that for a given starting parameter the Fisher matrix $F_{ij}(\lambda^{(0)})$ is readily calculated from the covariance matrix alone, without having to know the data. The Fisher matrix can also be used to forecast the precision with which a given experimental setup (hence given covariance matrix) is able to estimate parameters, if they are situated not too far from a first guess $\lambda^{(0)}$.

The above estimated parameters $\lambda^{(1)}$ depend quadratically on the data vector d. For this reason they are called a "quadratic estimator." The Fisher matrix actually is a local Gaussian approximation to the distribution of the parameters in the vicinity of the parameter values $\lambda^{(0)}$. Dividing by the true curvature of the likelihood function and not its expectation value would not have provided a quadratic estimator. It can be shown that the estimator described here is actually an optimal quadratic estimator in the sense of the Cramér-Rao bound [see, e.g., Kendall and Stuart (1969)].

We can now take the values $\lambda^{(1)}$ as our first (or rather second) guess and repeat the above procedure. In practice, relatively few iterations are needed to achieve convergence. This is very important because we often search in a parameter space of 10 or more dimensions. A modest grid of 10 points per side would already lead to 10^{10} evaluations of the likelihood function. Each of these requires one run of a fast CMB code that in an optimized code takes about a second. One evaluation of the likelihood function then requires a computational time of roughly 1 s. Hence 10^{10} evaluations take in the optimal case $10^{10}\,\mathrm{s} \simeq 300\,\mathrm{years}$ – not a time span in which we can comfortably wait for the output of a computation. Therefore, it is imperative that we use an iterative procedure and not a grid to estimate parameters.

The method described so far still has several problems, some of which we briefly want to address.

- *What if we end up in a shallow local maximum of the likelihood function and are stuck there?*
  To avoid this problem, one usually adds a small random fluctuation to the obtained $\delta\lambda$, a "temperature," so that one can leave a shallow local maximum.
- *What if there are several local maxima, some of them quite steep and separated by deep ridges?*
  To check this, one performs not only one but many iteration chains with different starting points. One can then compare the height of the different maxima. A procedure along these lines is the Markov chain Monte Carlo method (MCMC) discussed in the text that follows. It is presently the method of choice for CMB analysis. A publicly available MCMC code and more details of the method can be found in the paper by Lewis and Bridle (2002).
- *What if the maximum is somewhere at the border of the parameter space?*
  The border of the parameter space is given by the prior. If the data are best fitted by parameter values lying at the border of what is allowed by the prior, this

hints that either the prior is wrong or the model is incorrect. This is one of the most important drawbacks of the Fisher matrix technique. It provides relatively rapidly the best-fitting model under consideration, but it works independently of whether this model is actually a good fit to the data or not. For this an evaluation of the likelihood function at the best-fitting parameter values has to be performed. If the likelihood function is very small, this is either a sign that the model is wrong or that the real errors are much smaller than those assumed.

9.5.3.2 Estimating Errors

So far we have only studied the problem of how to find the best fit. But we also want error bars on the estimated parameters. A good first estimate for 1σ error bars are the diagonal elements of the Fisher matrix at the maximum $\bar{\lambda}$. To see this, we use the expression (9.52) for the deviation $\delta\lambda$:

$$
\begin{aligned}
\langle \delta\lambda_i \delta\lambda_j \rangle &= \frac{1}{4} F^{-1}_{im} F^{-1}_{jn} \left\langle \left[d^T C^{-1} (\partial_m C) C^{-1} d - \mathrm{Tr}\left(C^{-1} (\partial_m C) \right) \right] \right. \\
&\quad \left. \times \left[d^T C^{-1} (\partial_n C) C^{-1} d - \mathrm{Tr}\left(C^{-1} (\partial_n C) \right) \right] \right\rangle \\
&= \frac{1}{4} F^{-1}_{im} F^{-1}_{jn} \left[\langle d_k d_l d_p d_q \rangle C^{-1}_{kr} (\partial_m C)_{rs} C^{-1}_{sl} C^{-1}_{pv} (\partial_n C)_{vw} C^{-1}_{wq} \right. \\
&\quad \left. - \mathrm{Tr}\left(C^{-1} (\partial_m C) \right) \mathrm{Tr}\left(C^{-1} (\partial_n C) \right) \right].
\end{aligned}
$$

For the second term we made use of $\langle d_p d_q \rangle = C_{pq}$. With this, the mixed terms become equal to the pure trace term,

$$
\langle d^T C^{-1} (\partial_m C) C^{-1} d \rangle \mathrm{Tr}\left(C^{-1} (\partial_n C) \right) = \mathrm{Tr}\left(C^{-1} (\partial_m C) \right) \mathrm{Tr}\left(C^{-1} (\partial_n C) \right).
$$

For the first term we apply Wick's theorem (see Appendix 7),

$$
\langle d_k d_l d_p d_q \rangle = C_{kl} C_{pq} + C_{kp} C_{lq} + C_{kq} C_{lp}.
$$

Inserting this above, the first term results again in $\mathrm{Tr}\left(C^{-1} (\partial_m C) \right) \times \mathrm{Tr}\left(C^{-1} (\partial_n C) \right)$ while the second and third terms give rise to twice the Fisher matrix F_{mn}. We therefore finally obtain

$$
\langle \delta\lambda_i \delta\lambda_j \rangle = F^{-1}_{im} F^{-1}_{jn} F_{mn} = F^{-1}_{ij}, \tag{9.53}
$$

where we have used the symmetry of the Fisher matrix for the last equals sign. Therefore, the diagonal elements of the Fisher matrix usually give reasonable errors for the parameters. Of course, these are the true 1σ errors only if the distribution is Gaussian *in the parameters* λ, which usually it is not. But since the log of the likelihood function peaks at $\bar{\lambda}$ it will *locally* be of the form of a Gaussian,

$$\mathcal{L}(\bar{\lambda} + \delta\lambda) \simeq \mathcal{L}(\bar{\lambda}) \exp\left(-\frac{1}{2} \delta\lambda^T F(\bar{\lambda})\, \delta\,\lambda \right),$$

for small enough $\delta\lambda$. The Fisher matrix defines an ellipse around $\bar{\lambda}$ in the parameter space via the equation,

$$\delta\lambda^T F(\bar{\lambda}) \delta\lambda = 1. \tag{9.54}$$

The principal directions of this error ellipse are parallel to the eigenvectors of F and their half length is given by the square root of the eigenvalues of F^{-1}.

According to Eq. (9.54), the total width of the error ellipse in a given direction λ_i at the center $\bar{\lambda}$ is $2/\sqrt{F_{ii}}$. Therefore, $1/\sqrt{F_{ii}}$ is the error of the parameter λ_i if all other parameters are known and are equal to $\bar{\lambda}$. However, realistically, we do not know the other parameters any better than λ_i. Therefore, the true error in λ_i is given by the size of the projection of the error ellipse onto the i-axis; see Fig. 9.10. These are the so-called marginalized errors, which we obtain when integrating over all the other parameters.

We now show that the marginalized error of λ_i is given by $\sqrt{F^{-1}_{ii}}$, that is, the diagonal element of the inverse of the Fisher matrix. To find the value of $\delta\lambda_i$ at the boundary of the ellipse we solve the quadratic equation Eq. (9.54) for $\delta\lambda_i$,

$$\delta\lambda_i = \frac{1}{F_{ii}} \left(-\sum_{j\neq i} F_{ji}\delta\lambda_j \pm \sqrt{\left(\sum_{j\neq i} F_{ji}\delta\lambda_j \right)^2 - F_{ii} \left(\sum_{jk\neq i} F_{jk}\delta\lambda_j\delta\lambda_k - 1 \right)} \right).$$

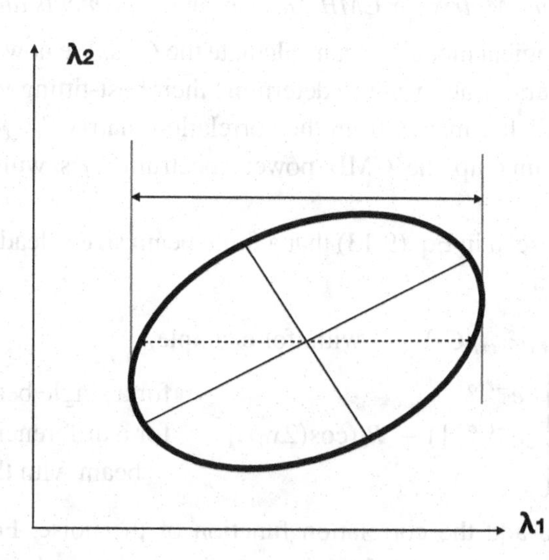

Fig. 9.10 The error ellipse is shown in a two-dimensional example. The widths $2/\sqrt{F_{11}}$ (dashed double arrow) and $2\sqrt{F^{-1}_{11}}$ (solid double arrow) are indicated.

This equation expresses $\delta\lambda_i$ as a function of all the other parameters $\delta\lambda_j$. To determine *the maximum* of $\delta\lambda_i$, we set the gradient of this function to zero. From the derivative w.r.t. λ_j we obtain

$$0 = \frac{1}{F_{ii}}\left(-F_{ji} \pm \frac{F_{ji}\sum_{k\neq i} F_{ki}\delta\lambda_k - F_{ii}\sum_{k\neq i} F_{jk}\delta\lambda_k}{\sqrt{\cdots}}\right),$$

where the square root $\sqrt{\cdots}$ is the same as in the previous equation. Multiplying by $\mp\sqrt{\cdots} = F_{ii}\delta\lambda_i + \sum_{k\neq i} F_{ki}\delta\lambda_k$ we find

$$0 = F_{ji}\delta\lambda_i + \sum_{k\neq i} F_{jk}\delta\lambda_k. \tag{9.55}$$

This equation must hold for all $j \neq i$. Inserting $\delta\lambda_k = aF_{ki}^{-1}$ and $\delta\lambda_i = aF_{ii}^{-1}$ we obtain $0 = a\delta_{ij}$ which is certainly true since $j \neq i$. Since the Fisher matrix is nonsingular the above solution is the only possibility. Inserting it in Eq. (9.54) determines $a = 1/\sqrt{F_{ii}^{-1}}$, so that we arrive at the important result

$$\delta\lambda_i^{(\mathrm{marg})} = \sqrt{F_{ii}^{-1}}. \tag{9.56}$$

These error ellipses are useful for forecasting if the errors are roughly Gaussian. However, if the true error contours have a very different shape, for example, "bananas" as in Fig. 9.19, lower left panel, the Fisher matrix approximation can severely underestimate the parameter errors.

9.5.3.3 The Fisher Matrix for CMB Anisotropy Experiments and Forecasting

For a given cosmological model we can calculate the C_ℓ's. We now consider *them* as our model parameters λ_ℓ and want to determine their best-fitting values and errors. We determine the Fisher matrix from the correlation matrix $\mathcal{C}_{\ell m,\ell' m'} = \langle a_{\ell m} a^*_{\ell' m'}\rangle$. (Take care not to mix up the CMB power spectrum C_ℓ's with the correlation function $\mathcal{C}$'s.)

We have already seen in Eq. (9.13) that a finite beam size σ leads to a correlation function of the form

$$\mathcal{C}_{\ell m,\ell' m'} = \delta_{\ell\ell'}\delta_{m m'} C_\ell W_\ell \quad \text{with, for example,}$$

$$W_\ell = \begin{cases} e^{-\ell^2\sigma^2} & \text{for a single beam,} \\ e^{-\ell^2\sigma^2}\left[1 - P_\ell(\cos(2x_0))\right] & \text{for a differencing beam with throw } x_0. \end{cases}$$

To this we have to add the correlation function of the noise. For simplicity we assume isotropic noise of amplitude σ_n in each pixel and a pixel size $\Delta\Omega$ in radians, in other words, a noise correlation function of the form

$$\frac{1}{T^2}\langle \Delta T^{(n)}(\mathbf{n})\Delta T^{(n)}(\mathbf{n}')\rangle = \begin{cases} \sigma_n^2 & \text{if } \mathbf{n} \text{ and } \mathbf{n}' \text{ are in the same pixel} \\ 0 & \text{else,} \end{cases}$$
$$= \frac{1}{4\pi}\sum_\ell (2\ell+1)C_\ell^{(n)} P_\ell(\mathbf{n}\cdot\mathbf{n}'). \tag{9.57}$$

We have already taken into account that the noise is isotropic and therefore $\left\langle a_{\ell m}^{(n)} a_{\ell' m'}^{*\,(n)} \right\rangle = \delta_{\ell\,\ell'}\delta_{m\,m'}C_\ell^{(n)}$. To isolate $C_{\ell_1}^{(n)}$ we set $\mu = \mathbf{n}\cdot\mathbf{n}'$, multiply the above equation with $P_{\ell_1}(\mu)$, and integrate over μ. Defining

$$f(\mu) = \begin{cases} \sigma_n^2 & \text{if } 1-\mu < \Delta\mu \\ 0 & \text{else,} \end{cases}$$

we have

$$\int_{-1}^{1} d\mu\, P_{\ell_1}(\mu) f(\mu) \simeq \Delta\mu\sigma_n^2$$
$$\simeq \Delta\vartheta\,\sin(\Delta\vartheta/2)\sigma_n^2, \quad \text{for } \ell_1 < \frac{1}{\Delta\mu}.$$

Here we have used the fact that $P_\ell(\mu) \simeq 1$ for $\mu \simeq 1$ and $\Delta\mu = -\Delta\cos\vartheta \simeq \Delta\vartheta\,\sin(\Delta\vartheta/2)$. If $\ell \gtrsim 1/\Delta\mu$ this approximation breaks down. But it is clear that with an experiment of pixel size corresponding to $\mathbf{n}\cdot\mathbf{n}' \leq 1-\Delta\mu$ we cannot measure C_ℓ's with $\ell > 1/\Delta\mu$. We are therefore not considering these values. In other words, we are only determining the C_ℓ's for values of ℓ with $\ell < \ell_{\max} \simeq 1/(2\Delta\mu)$. Also multiplying the left-hand side of Eq. (9.57) with P_{ℓ_1} and using $\int P_\ell P_{\ell_1} = \delta_{\ell\ell_1} 2/(2\ell+1)$ we obtain

$$C_{\ell_1}^{(n)} = 2\pi\,\Delta\vartheta\,\sin(\Delta\vartheta/2)\sigma_n^2 = \Delta\Omega\sigma_n^2. \tag{9.58}$$

It is reasonable to assume that the signal and the noise are uncorrelated so that we can simply add their correlation functions and arrive at

$$\left\langle a_{\ell m}^{(n)} a_{\ell' m'}^{*\,(n)} \right\rangle = \delta_{\ell\,\ell'}\,\delta_{m\,m'}\left[C_\ell W_\ell + w^{-1}\right], \tag{9.59}$$

where we have introduced the width $w = (\Delta\Omega\sigma_n^2)^{-1}$. With this correlation function at hand, we can now calculate the Fisher matrix. Denoting the derivative with respect to C_ℓ by ∂_ℓ, Eq. (9.51) yields

$$F_{\ell\ell'} = \frac{1}{2}\mathrm{Tr}\left(C^{-1}(\partial_\ell C)C^{-1}(\partial_{\ell'}C)\right)$$
$$= \frac{1}{2}C^{-1}_{\ell_1 m_1,\ell_2 m_2}(\partial_\ell C)_{\ell_2 m_2,\ell_3 m_3}C^{-1}_{\ell_3 m_3,\ell_4 m_4}(\partial_{\ell'}C)_{\ell_4 m_4,\ell_1 m_1}$$

$$= \frac{\frac{1}{2}\delta_{\ell_1\ell_2}\delta_{m_1m_2}}{\left[C_{\ell_1}W_{\ell_1}+w^{-1}\right]}\delta_{\ell_2\ell_3}\,\delta_{m_2m_3}W_{\ell_2}\,\delta_{\ell_2\ell}\,\frac{\delta_{\ell_3\ell_4}\delta_{m_3m_4}}{\left[C_{\ell_3}W_{\ell_3}+w^{-1}\right]}\delta_{\ell_4\ell_1}\delta_{m_4m_1}W_{\ell'}\,\delta_{\ell_4\ell'}$$

$$= \delta_{\ell\ell'}\,\frac{(2\ell+1)W_\ell^2}{2[C_\ell W_\ell+w^{-1}]^2} = \delta_{\ell\ell'}\,\frac{(2\ell+1)}{2[C_\ell+(wW_\ell)^{-1}]^2}. \tag{9.60}$$

The factor $2\ell+1$ comes from the summation over the m's,

$$\sum_{m_1m_2m_3m_4}\delta_{m_1m_2}\,\delta_{m_2m_3}\,\delta_{m_3m_4}\,\delta_{m_4m_1} = \sum_{m_1}\delta_{m_1m_1} = 2\ell_1+1,$$

while the summation over the ℓ's requires $\ell_1 = \ell_2 = \ell = \ell_3 = \ell_4 = \ell'$ and therefore leads to $\delta_{\ell\ell'}$.

Using the C_ℓ's as our "parameters" has the big advantage that the Fisher matrix is diagonal and the marginalized errors are simply given by $1/\sqrt{F_{\ell\ell}} = \sqrt{F_{\ell\ell}^{-1}}$,

$$\delta C_\ell = \sqrt{\frac{2}{2\ell+1}}[C_\ell+(wW_\ell)^{-1}]. \tag{9.61}$$

We recognize the first term as cosmic variance while the second term is the experimental error. For a given window function, pixel size, and pixel noise, this gives a good error estimate for the accuracy with which the C_ℓ's can be determined in a full sky experiment. If the experiment covers only a fraction $f_{\rm sky}$ of the sky, a good approximation for the error is again

$$\delta C_\ell = \sqrt{\frac{2}{(2\ell+1)f_{\rm sky}}}[C_\ell+(wW_\ell)^{-1}]. \tag{9.62}$$

Of course, the model parameters that we really want to estimate are not the C_ℓ's, but rather cosmological parameters like $\Omega_m h^2$, Ω_Λ, and the curvature. For this we can now start the process over again, considering the C_ℓ's as our "data" with correlation matrix $F_{\ell\ell'}^{-1}$ and calculate the Fisher matrix of the cosmological parameters, $\tilde F(\lambda)$. The only disadvantage here is that contrary to the $a_{\ell m}$'s, the C_ℓ's do not obey a Gaussian distribution. But for high ℓ's the distribution is nearly Gaussian due to the central limit theorem (see Appendix 7), and for low ℓ's the error is relatively large due to cosmic variance, so that treating the distribution as Gaussian does not usually induce large errors. Of course $\tilde F(\lambda)$ is by no means diagonal and the errors in the cosmological parameters alone are not Gaussian at all. If we want to estimate not only errors of the parameters, which can be obtained from the inverse of the Fisher matrix $\tilde F^{-1}$, but also the full marginalized probability distribution, we have to use a more sophisticated method, like MCMC discussed in the text that follows. The probability distributions evaluated with an MCMC method for the Planck satellite experiment are shown in Fig. 9.11. This

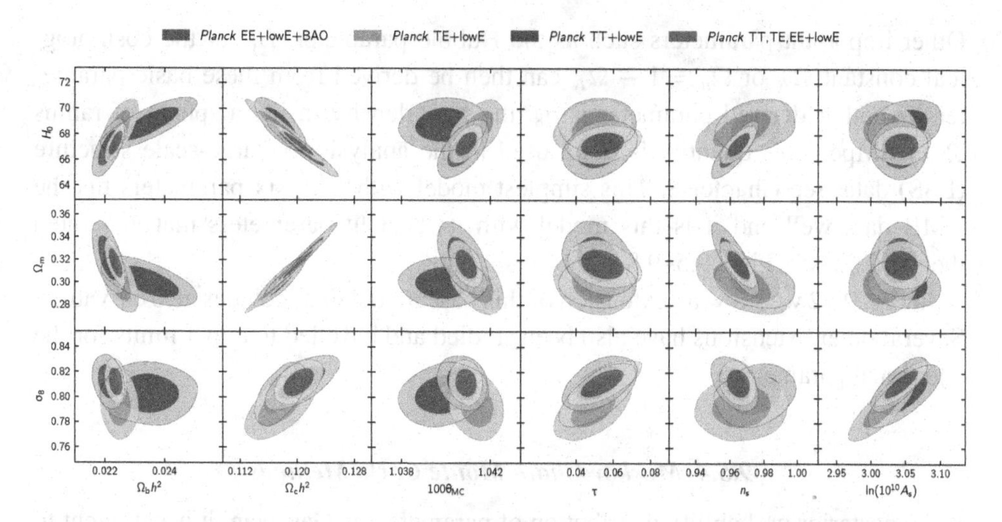

Fig. 9.11 The marginalized two-parameter distributions from the Planck experiment for the minimal six-parameter ΛCDM model. A_s is the amplitude of scalar metric perturbations at $k_0 = 0.05\,\mathrm{Mpc}^{-1}$, $A_s = \Delta_\zeta(k_0)$, and τ is the optical depth. The acoustic scale is cast as θ_{MC}. Differently shaded contours include more and more of the data. Starting from polarization and BAO data only for the largest contours, then TE-correlations and temperature anisotropies. The smallest contours include the full Planck data. These parameter distributions have been obtained with an MCMC routine using the Metropolis–Hastings algorithm discussed in Section 9.5.4.1. Figure from Planck Coll. VI (2018)

figure shows the parameter distributions for the minimal ΛCDM model, which is a flat cosmology with vanishing tensor perturbations and scalar perturbations given by a primordial power spectrum of the form $\Delta_\zeta = A_s(k/k_0)^{n_s-1}$. The sum of the neutrino masses is assumed to be the minimal value required from neutrino oscillation experiments (Tanabashi *et al.*, 2019) which is about

$$\sum m_\nu = 0.06\mathrm{eV}. \tag{9.63}$$

The parameters of this so called base model are A_s and n_s together with the baryon density, $\Omega_b h^2$; the cold dark matter density, $\Omega_c h^2$; the optical depth due to reionization, $\tau_{\rm ri}$; and θ_{MC}, which is a parametric approximation to the angular scale of the sound horizon at decoupling, θ_* defined by

$$\theta_* = \frac{t_s(z_*)}{\chi_K(z_*)}, \tag{9.64}$$

where the redshift z_* is defined by the optical depth to it reaching unity (neglecting reionization),

$$1 = \tau(z_*) = \int_0^{z_*} \frac{\sigma_T n_e(z)}{\mathcal{H}(z)} \frac{dz}{(1+z)^2}.$$

Other important parameters such as the Hubble parameter, H_0, or the cosmological constant Ω_Λ or $\Omega_m = 1 - \Omega_\Lambda$ can then be derived from these basic parameters. Another derived parameter is σ_8, the over-density inside a sphere of radius $R = 8$Mpc. This quantity is often used in the analysis for large-scale structure (LSS) data; see Chapter 8. This simplest model with only six parameters fits the CMB data well and it is this model with its best fit parameters that generated the solid lines in Figs. 9.5–9.8.

In Fig. 9.12 we show an extension of this base model that includes also curvature. Several other extensions have also been studied and have led to upper limits for the additional parameters.

9.5.4 Markov Chain Monte Carlo Methods

If the posterior probability distribution of parameters is Gaussian, it is sufficient to give its mean and its width in the direction of the principal axis, that is, the Fisher matrix; this contains all the statistical information. However, even if the distribution is Gaussian in the data, it is very often far from Gaussian in the cosmological parameters of interest to us.

Given our model with parameters λ and the data D, we can evaluate the probability density $P(\lambda) \equiv P[\lambda|D]$ up to a constant simply by Eq. (9.33),

$$P[\lambda|D] = P[D|\lambda]\frac{P[\lambda]}{P[D]}, \tag{9.65}$$

where usually, the distribution of the data, $D = (d_1, \ldots, d_N)$ is a Gaussian with some covariance matrix $\mathcal{C}$,

$$P[D|\lambda] = \mathcal{L}(D,\lambda)d^N d = \frac{1}{\sqrt{(2\pi)^N \det \mathcal{C}}} \exp\left(-\frac{d_i \mathcal{C}_{ij}^{-1} d_j}{2}\right) d^N d. \tag{9.66}$$

Therefore, once we have fixed a prior $P[\lambda]$, we can evaluate the probability $P[\lambda|D] \equiv p(\lambda)d^M\lambda$ of a given choice of parameters, $\lambda = (\lambda_1, \ldots, \lambda_M)$, up to a constant, the evidence $P[D]$. Let us define this not normalized distribution by

$$P^*(\lambda) \equiv P[\lambda|D]P[D] = P[D|\lambda]P[\lambda]. \tag{9.67}$$

We would like to answer the following questions.

- What is the shape of the probability density $p(\lambda)$ in the full parameter space, and what are the densities of some arbitrary subset of parameters, $\mu = (\lambda_{i_1}, \ldots, \lambda_{i_K})$, $K < M$ marginalized over all the other parameters. We are especially interested in the cases of $K = 1$ and 2, which are easy to visualize and that indicate how strongly the parameters λ_{i_1} and λ_{i_2} are correlated; see Figs. 9.11 and 9.12.

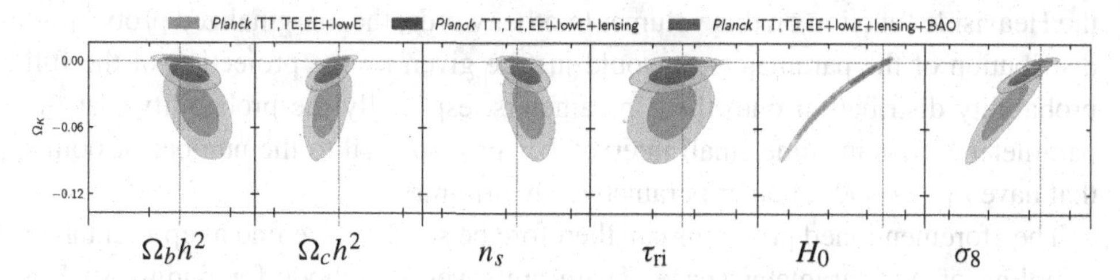

Fig. 9.12 The marginalized 2-parameter distribution from the Planck experiment for a ΛCDM model including curvature. The parameter on the vertical axis is Ω_K. The parameters on the horizontal axis are from left to right the baryon density, $\Omega_b h^2$; the cold dark matter density, $\Omega_c h^2$; the scalar spectral index, n_s; the optical depth τ; the Hubble parameter, H_0; and the matter density fluctuations inside a ball of radius $R = 8h^{-1}$Mpc, σ_8. The largest contours are the Planck data not including lensing, note the strong degeneracy of Ω_K with H_0. The smaller contours include also CMB lensing and the smallest contours include also BAOs from large-scale structure. Figure from Planck Coll. VI (2018)

- We would also like to compute the expectation value and variance of derived parameters,

$$\langle h \rangle = \int h(\lambda) p(\lambda)\, d^M \lambda,$$

and

$$\left\langle (h(\lambda) - \langle h \rangle)^2 \right\rangle = \int (h(\lambda) - \langle h \rangle)^2 p(\lambda)\, d^M \lambda;$$

in brief, integrals of the form $\int f(\lambda) p(\lambda)\, d^M \lambda$.

If we had a *representative* (or *fair*) *sampling* of parameter space, $S = (\lambda^{(1)}, \lambda^{(2)}, \ldots, \lambda^{(R)})$, that is, a sampling in which the number of points $\lambda \in S$ in a small volume V of parameter space is proportional to $\int_V p(\lambda)\, d^M \lambda$, we could approximate

$$\langle f \rangle = \int f(\lambda) p(\lambda)\, d^M \lambda \simeq \frac{1}{R} \sum_{i=1}^{R} f\left(\lambda^{(i)}\right).$$

Furthermore, marginalized probability densities at $\mu = (\lambda_{i_1}, \ldots, \lambda_{i_K})$ would be proportional to the sum of all the points for which the parameters of interest are in a given infinitesimal volume around μ, that is,

$$p_K(\mu) = \int \left(\Pi_{j \neq i_r} d\lambda_j\right) p(\lambda_1, \ldots, \lambda_M) \simeq \frac{1}{R} \sum_{i=1}^{R} \Theta_{V_\mu}\left(\lambda^{(i)}\right),$$

where V_μ is the volume in parameter space where the parameter values λ_{i_1} to λ_{i_K} are very close to the value μ while all other parameters are arbitrary, and Θ_{V_μ} is the Heaviside function on this volume. In other words, the marginalized probability distribution of the parameters μ would just be given by the projection of the full probability distribution onto these parameters, especially the probability that the parameter λ_1 lies in some small interval I is proportional to the number of points that have $\lambda_1 \in I$ and all other parameters are arbitrary.

The aforementioned problems can therefore be solved if we find a representative sampling of our parameter space. There are several methods for finding such a sampling that all have their advantages and disadvantages [see, e.g., Gamerman (1997) and MacKay (2003)]. For high-dimensional problems, the Markov chain Monte Carlo (MCMC) methods are especially useful. Of these, we concentrate here on the Metropolis–Hastings algorithm that is dominantly in use for CMB analysis. We shall mention also the Hamiltonian Monte Carlo method. Full proofs that these algorithms really converge are found in the aforementioned monographs.

9.5.4.1 Metropolis–Hastings Algorithm

Let us start with some arbitrary point $\lambda^{(1)}$, and some "proposal density" $Q\big(\lambda^{(2)}, \lambda^{(1)}\big)$ for a new value $\lambda^{(2)}$ that depends on $\lambda^{(1)}$. We call $\lambda^{(1)}$ the "current point" and $\lambda^{(2)}$ the "proposal." For the moment, let Q be arbitrary but simple enough so that we can easily (with little numerical investment) sample it. We shall soon be more precise. To generate our sampling S we start at some arbitrary point $\lambda^{(1)}$. With probability $Q(\lambda, \lambda^{(1)})$ we now determine a proposal λ. To decide whether to accept this point as the next element of our sampling (which now becomes a chain, since it is ordered), we compute

$$r = \frac{P^*(\lambda) Q\left(\lambda^{(1)}, \lambda\right)}{P^*\big(\lambda^{(1)}\big) Q\left(\lambda, \lambda^{(1)}\right)}. \tag{9.68}$$

If $r \geq 1$, we accept λ as the next member of our chain, $\lambda^{(2)} = \lambda$, if $r < 1$ we assign $\lambda^{(2)} = \lambda$ with probability r and $\lambda^{(2)} = \lambda^{(1)}$ with probability $1 - r$. And so on, we generate our Markov chain, $S = (\lambda^{(1)}, \ldots, \lambda^{(R)})$.

If the proposal density Q is symmetric in its arguments, as, for example, a Gaussian centered on the current point, the factor $Q\left(\lambda^{(1)}, \lambda\right) / Q\left(\lambda, \lambda^{(1)}\right)$ drops out and we only have to calculate $P^*(\lambda)/P^*\left(\lambda^{(1)}\right)$. We now concentrate on this case, which is often simply called the "Metropolis algorithm." The proposal density Q is then only needed to suggest the next point, but is not involved in the decision whether it is accepted or not.

It can be shown that for strictly positive Q, the distribution of the points in S always tends to the posterior distribution $P(\lambda) = P^*(\lambda)/P[D]$ for $R \to \infty$. How long will this take? Or in other words, how large do we have to choose R so that S becomes a fair sample of P? This is a difficult question, but it is relatively easy to find a lowest estimate for R. In order to have a reasonable acceptance rate r we do not want to choose a too large step size $\epsilon = |\lambda - \lambda^{(i)}|$. Hence the proposal density Q has to be sufficiently narrow. A reasonable step size is probably of the order of the smallest widths σ_s of the 1-parameter distributions (of course, strictly speaking we do not yet know these widths, but in practice we can make an educated guess and revise it if necessary). Now, the distance the chain has to travel is at least equal to the width of the largest 1-parameter distribution, σ_l. Since our chain performs a random walk in parameter space, the number of steps it needs to travel a given distance is $R_{\min} \propto (\sigma_l/\sigma_s)^2$. The proportionality factor will be roughly the inverse of the mean acceptance probability, since, if the next point is not accepted, the chain does not move forward at all.

To obtain a reasonably fair sample, one certainly has to cover the high-probability part of the parameter volume several times and choose chain lengths of a few times $R_{\min}$. This all works if our probability distribution has only one high-probability region. If it has several, even though, usually a given chain will rapidly find one of them, it is very hard to cross from one of these regions into another. Therefore, instead of generating just one chain, one usually generates several (tens of) chains. On the other hand, the first few (of order 20) points of a chain do not sample the posterior distribution but depend mainly on the random initial point $\lambda^{(1)}$. This "burn in phase" is usually discarded. Furthermore, the points in the chain are not independent. As usual for Markov chains, each point depends on its predecessor. This is not really a hindrance, as the mathematical theorem shows, but nevertheless analysts often "thin out" their chains, that is, they use only every tenth or so point for the posterior distribution.

Since the "burn in phase" of the chain is useless, it is not economical to use many very short chains. On the other hand, since the probability distribution can have several high-probability islands in parameter space that are separated by deep canyons, it is not advisable to use only one very long chain. As so often, the "golden middle" of many reasonably long chains is usually the best. However, if the likelihood function turns out to have more than one maximum to which chains may converge, care is required. The probability of a maximum is not simply proportional to the number of chains that converge to it. This number can be large, not because the maximum is large, but simply because its "basin of attraction" is large. In this case, one has to analyze the shape of the likelihood function in more detail.

It is important to have a relatively good guess for the proposal density Q to start with. A simple possibility is a Gaussian with the correlation matrix given by the inverse of the Fisher matrix.

In practice, to test whether the chains have converged one just adds 10% more steps and investigates if the results change. To test whether the series of chains represents a fair sampling one adds a couple more independent chains and checks the effect on the results. If the results are not affected, or only well within error bars, one is usually confident that the procedure has converged. Often, one simply compares the variance of the parameters obtained from the chain with the error bars from the data.

In Figs. 9.11 and 9.12 we show the results obtained by this analysis method with the best currently available CMB data (Planck data) for the simplest flat ΛCDM model and including curvature. In addition to the cosmological parameters the Planck analysis varies more than 30 other parameters that describe uncertainties in the foreground and the experimental apparatus. To speed up the analysis, these parameters are combined into directions that lead to a (nearly) diagonal Fisher matrix, so-called orthogonal directions. Several other methods to speed up the MCMC calculations are also employed; see Lewis (2013) for details.

There are several other Monte Carlo methods that some people begin use for cosmological data sets, like the slice sampling (MacKay, 2003), but the principle behind them is always the same: to obtain a Markov chain that produces a representative sampling of an underlying probability distribution proportional to $P^*(\lambda)$, we have to find a transition probability $T(\lambda, \lambda')$ that leaves P^* invariant, that is,

$$P^*(\lambda) = \int T(\lambda, \lambda') P^*(\lambda')\, d^M\lambda'. \tag{9.69}$$

Choosing, as in the Metropolis algorithm, $T(\lambda, \lambda') = V_{\rm tot}^{-1} P^*(\lambda)/P^*(\lambda')$, is clearly a transition probability that obeys Eq. (9.69). Here $V_{\rm tot}$ is the total volume of parameter space that has to be introduced in order for T to be correctly normalized.

9.5.5 Hamiltonian Monte Carlo

Finally, let us briefly mention a method that is at present not widely used in cosmology (for a first attempt to study its usefulness for CMB data analysis, see Hajian, 2007). This method uses not only $P^*(\lambda)$, but also its gradient. Evaluating P^* with reasonable accuracy requires one run of a fast CMB code. Evaluating the gradient with respect to M parameters requires $2M$, or for a stable numerical derivative, $3M$ to $5M$ evaluations. As this is the costly part of CMB analysis, its usefulness has to be checked in detail. However, as we explain now, the Hamiltonian Monte Carlo might reduce the number of points needed for a representative sampling from $R_{\rm min}$ to order $\sqrt{R_{\rm min}}$. Therefore, if the chains have difficulties in converging, it might be useful.

The basic idea is very simple. Let us write

$$P^*(\lambda) = e^{-V(\lambda)}.$$

Maximizing the probability is then equivalent to minimizing the potential V. Therefore, it is useful to take the next step in the direction $-\nabla V$. But this is exactly what Hamiltonian dynamics does. Therefore, we introduce momentum variables $\pi = (\pi_1, \ldots, \pi_M)$ and define the Hamiltonian

$$H(\lambda, \pi) = V(\lambda) + K(\pi),$$

where K is the kinetic energy, for example $K = \frac{1}{2}\pi^T \pi$. We then define the non-normalized probability

$$P_H^*(\lambda, \pi) = e^{-H(\lambda,\pi)} = e^{-V(\lambda)} e^{-K(\pi)}.$$

We now sample P_H^* in the following way. We first choose some initial value $\lambda^{(1)}$ and draw a momentum $\pi^{(1)}$ from the Gaussian distribution $e^{-K(\pi)}/Z_K$. For the next, dynamical proposal, the present momentum $\pi^{(1)}$ decides the displacement of λ and the gradient of the present potential $V(\lambda^{(1)})$ decides the change in the momentum, via the canonical equations

$$\dot{\lambda} = \pi, \quad \text{and} \quad \dot{\pi} = -\nabla V(\lambda).$$

We then advance this system for some (fixed) number of steps to arrive at the proposal (λ, π), which is then accepted or rejected according to the Metropolis rule (9.68) for P_H^* (and some symmetrical Q that is no longer needed).[3]

The big advantage of this method is that the distance covered by the parameters $\lambda^{(i)}$ is now proportional to the computer time per step and not only to its square root.

From the representative sampling of P_H^* obtained in this way, we obtain a fair sampling of $P^*(\lambda)$ by simply ignoring the momentum variables in the chains.

9.6 Degeneracies

To explain the problem of degeneracies, let us first consider parameters that obey a Gaussian distribution with some Fisher matrix $F(\bar{\lambda})$ at the maximum likelihood parameters $\bar{\lambda}$. Its inverse is the correlation matrix of the parameters, $C_{ij} = \langle(\lambda_i - \bar{\lambda}_i)(\lambda_j - \bar{\lambda}_j)\rangle = F_{ij}^{-1}(\bar{\lambda})$. If one (or several) of the eigenvalues of the Fisher matrix is (are) very small, the variance of the parameters in the corresponding direction is very large. Hence this linear combination of parameters cannot be determined accurately. In the limiting case, when the eigenvalue vanishes, the linear combination that defines the eigenvector with vanishing eigenvalue is not at all constrained by the data.

[3] If the simulation is perfect, the proposal is accepted every time since $H = V + K$ is a constant of motion and so r is always equal to 1. In practice, however, the inaccuracy in the numerical evaluation of the gradient will lead to some rejections.

Such directions are called degenerate or nearly degenerate directions. It is very useful to identify them and to express the results of the experiment in terms of quantities that are well determined and have small errors, that is, that are orthogonal to the degenerate directions. Usually, degenerate directions have a simple physical interpretation. They can be lifted either by improving the data (if they are only nearly degenerate) or by considering complementary data. Here we discuss the main examples.

9.6.1 Curvature

The most prominent example of a degeneracy of CMB anisotropies comes from the fact that they are strongly dependent on $\Omega_b h^2$, $\Omega_m h^2$ and the angular size of the sound horizon, $\theta_{\rm dec}$, or equivalently the angular diameter distance $d_A(z_{\rm dec})$. Considering a ΛCDM model, only three combinations of the parameters of interest, Ω_m, Ω_b, Ω_Λ, and h, are well determined by CMB anisotropies. This is especially important for the determination of the curvature $\Omega_K = 1 - \Omega_\Lambda - \Omega_m$. Changes in the curvature that keep d_A, $\Omega_m h^2$ and $\Omega_b h^2$ fixed have nearly no effect on CMB anisotropies (see Fig. 9.13).

The main effect in the CMB that breaks this degeneracy is CMB lensing. The reason for this is that the lensing kernel dominates at redshift $z_{\rm lens} \sim 3 \ll z_{\rm dec}$ and therefore gives us access to the angular diameter distance at roughly $z_{\rm lens}$, which breaks this degeneracy; see Fig. 9.14. Observing BAOs in a large-scale structure (see Chapter 8) at a much better defined redshift breaks this degeneracy even better and leads to a very precise determination of curvature when marginalized over all the other parameters. Including BAOs, as described in Section 9.8, the error on the curvature decreases by more than a factor 3,

$$\Omega_K = \begin{cases} -0.0106 \pm 0.0065 & \text{CMB only} \\ 0.0007 \pm 0.0019 & \text{CMB and BAO} \end{cases} \tag{9.70}$$

Of course also fixing h breaks the curvature degeneracy. But as we shall see in Section 9.9.1, direct measurements of the Hubble constant are not in good agreement with the CMB results and it may therefore not be consistent to combine these data.

9.6.2 Scalar Spectral Index, Tensor Component, and Related Degeneracies

The tensor contribution to the CMB anisotropies significantly adds to the C_ℓ's only for $\ell \lesssim 80$. For $\ell \gtrsim 80$ it rapidly decays and can be neglected (see Fig. 2.5). But a slight enhancement of power on large scales can also be produced by reducing somewhat the spectral index n_s. A slightly redder spectrum also has somewhat more power on large scales; see Fig. 9.15.

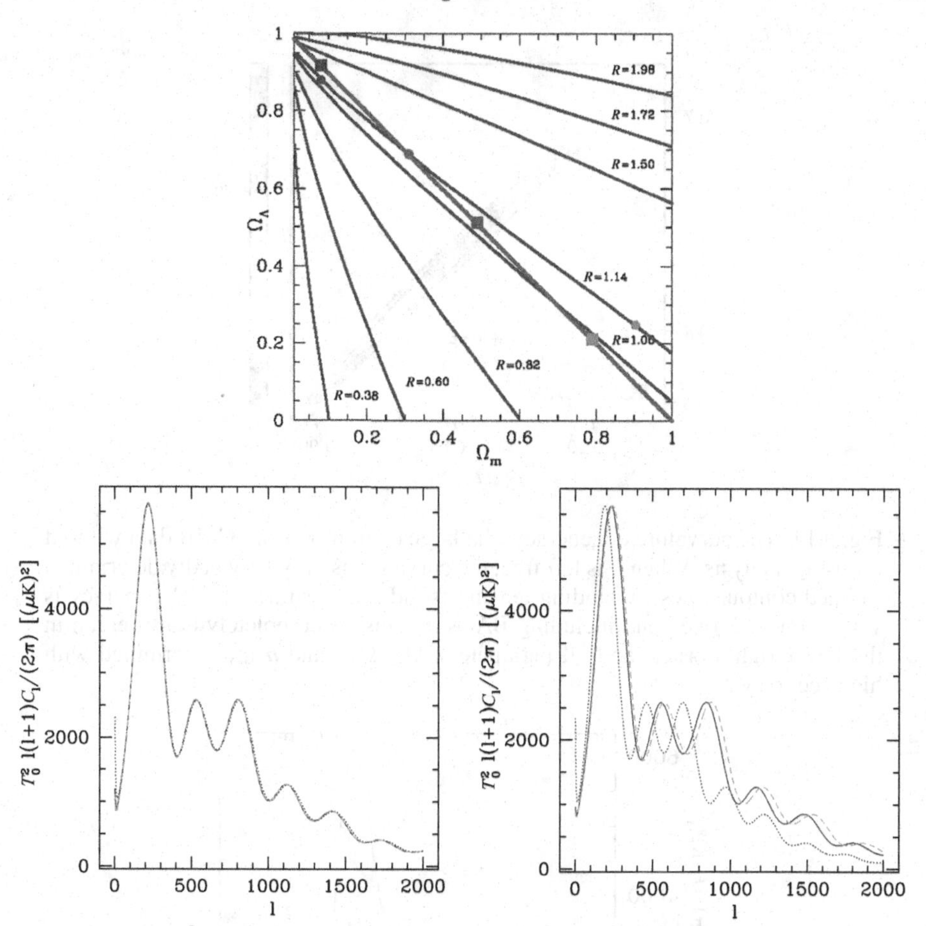

Fig. 9.13 In the upper panel lines of equal angular diameter distance are indicated. The number R is the ratio of the angular diameter distance of the model to the one of a concordance model with $\Omega_\Lambda = 0.7$, $\Omega_m = 0.3$, $h = 0.7$. The lines of constant curvature are parallel to the diagonal, which is also drawn. In the lower left-hand panel we show CMB anisotropy spectra with $\Omega_K > 0$ (dashed), $\Omega_K < 0$ (dotted), and $\Omega_K = 0$ (solid), which have identical angular diameter distance, matter density, and baryon density. They correspond to the dots indicated in the upper panel. The spectra overlay so precisely that we can hardly distinguish them by eye. On the lower right-hand panel we show three spectra with curvature zero, identical matter density and baryon density, but with different angular diameter distances (the squares indicated in the upper panel on the $K = 0$ line). The spectra are significantly different. Note that in these plots CMB lensing is neglected.

Another parameter that affects the power on large scales is the optical depth to the last scattering surface, τ. Enhancing it leads to more damping on larger scales, which in turn can be compensated by a tensor component.

This degeneracy is lifted by including the high-quality polarization data from Planck. Reionization leads to the rescattering of photons at late times and induces a

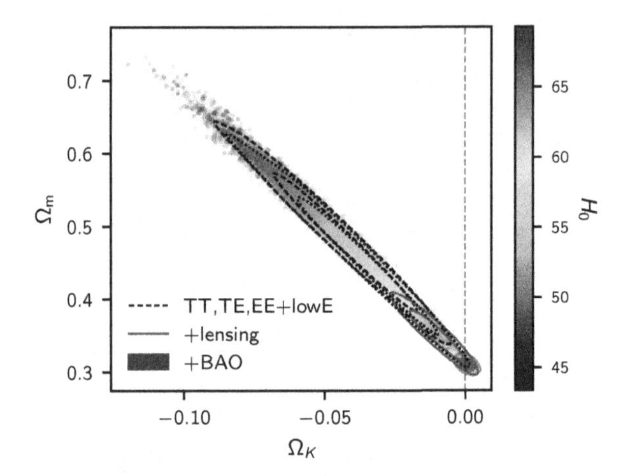

Fig. 9.14 The curvature degeneracy: the large contour is from CMB data without a lensing analysis. When h is left free, the curvature is only very badly determined (dashed contour lines). Including lensing (solid contour lines) this degeneracy is very strongly broken and including BAOs it vanishes completely (dark region in the lower right corner) and all parameters, Ω_m, Ω_Λ, and h are determined with high accuracy.

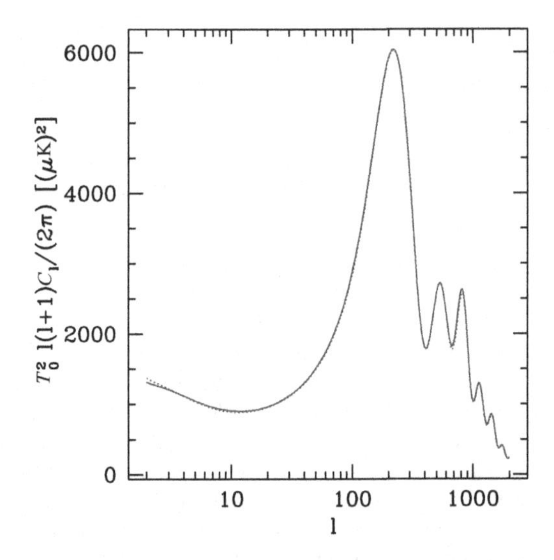

Fig. 9.15 A spectrum of purely scalar perturbations with $n_s = 0.96$ (solid) is compared to one with a tensor contribution of $r = 0.3$ (dotted). The cosmological parameters of the two models are given by ($h = 0.73, \Omega_b h^2 = 0.0225, \Omega_m h^2 = 0.135, \tau_{\rm ri} = 0.1, n_s = 0.96$) for the purely scalar model and ($h = 0.8, \Omega_b h^2 = 0.023, \Omega_m h^2 = 0.118, \tau = 0.1, n_s = 1.0, r = 0.3, n_t = 0$) for the model with tensor contribution. Clearly, these two models cannot be distinguished from their temperature anisotropies alone.

small amount of polarization on large scales that would not be seen without reionisation. The polarization data from Planck provides the best estimate of the optical depth: $\tau_{\rm ri} = 0.0544 \pm 0.0073$, obtained assuming a vanishing tensor component.

On the other hand, a significant tensor component can be (nearly) unambiguously determined by a measurement of the B-mode of polarization. This is the next "quantum leap" to be expected from CMB data: good B-polarization data that even might allow the testing of the slow roll consistency condition (3.105),

$$\frac{A_t}{A_s} = -8n_t .$$

9.6.3 Initial Conditions

As we discussed in Chapter 3, simple inflationary models generate perturbations with adiabatic initial conditions. In the simplest case, neglecting a possible tensor component, the initial conditions are characterized by the two parameters (A_s, n_s). However, in principle, other initial conditions are also possible. We have discussed the mixture of adiabatic and CDM isocurvature perturbations. Also allowing for the two neutrino modes (the neutrino density and the neutrino velocity mode), one obtains four essentially different modes. Together with arbitrary correlations this gives a 4×4 symmetric, positive semidefinite matrix of initial conditions, hence 10 parameters. The spectral indices of each component allow for additional parameters (see, however, Exercise 9.2).

Introducing these many additional parameters leads to serious additional degeneracies. In particular, the first peak of the acoustic oscillations for isocurvature perturbations no longer determines the angular diameter distance to the last scattering surface, since it also depends strongly on the initial conditions.

However, isocurvature perturbations, while contributing to the Sachs–Wolfe plateau, do not induce significant density fluctuations. But the normalization of CMB fluctuations on large scales leads to about the right amplitude for density fluctuations in the purely adiabatic case. Therefore, combining CMB fluctuations with the galaxy power spectrum leads to stringent constraints for the isocurvature contribution (Trotta *et al.*, 2001/3). An analysis of the Planck data limits a possible isocurvature fraction on large scale to 2% for the CDM isocurvature mode and to 7.4% and 6.8% for the neutrino density and neutrino velocity isocurvature modes respectively; see Planck Coll. X (2018).

Very roughly, isocurvature modes have an effect similar to a tensor component: they contribute to the CMB anisotropies but not to the galaxy power spectrum. However, since they contribute to the CMB not only on very large scales, but also on smaller scales where the data have smaller error bars, they are better constrained than a tensor contribution.

Finally, polarization information also helps to break the degeneracies from isocurvature modes (see Bucher *et al.*, 2001 and Planck Coll. X, 2018). This comes first of all from the fact that the spacing between the acoustic peaks in ℓ-space depends only on the cosmological parameters and not on the initial conditions.

Similar limits as for iso-curvature perturbations can also be derived for topological defects (see Section 2.7). For example, in Lizarraga *et al.* (2016) it is found that cosmic strings from a $U(1)$ Abelian Higgs model can contribute at most

$$f_{10} = \frac{C_{10}^{\rm string}}{C_{10}^{\rm adi}} \leq 0.014, \tag{9.71}$$

where the C_ℓ's are marginalized over the six parameters of the base ΛCDM model. This leads to a limit for the symmetry breaking scale M of $4\pi GM^2 = \epsilon < 4\times10^{-7}$.

Even when considering simple adiabatic perturbations, one can allow for additional features in the initial power spectrum that will influence the estimated cosmological parameters. As long as these are parameterized by a few numbers, we can test good enough data against them, but as mentioned in the beginning of this chapter, if we do not make any simplifying assumptions on the initial power spectrum, we must know the cosmological parameters that then fix the transfer function in order to determine the initial power spectrum. Without assumptions we cannot determine them both.

One often allows for a so-called running of the spectral index. For this one fits for a "running parameter" $\alpha = dn_s/d\ln(k)$, which is assumed to be constant. An initial spectrum, with running, is determined by three parameters. First one has to fix some "pivot scale" k_* that is arbitrary, but has to be a scale where the power spectrum is well constrained by the data. One then sets

$$k\Delta_\zeta(k) = A_*(k/k_*)^{n_s-1+\alpha\ln(k/k_*)}. \tag{9.72}$$

The resulting amplitude A_* and spectral index n_s, in general, depend on the pivot scale. Only if $\alpha = 0$ does n_s not depend on k_*. If $\alpha = 0$ and $n_s = 1$ as well A_* is independent of k_*. For the pivot scale $k_* = 0.05h/$Mpc the Planck experiment has published the following results:

$$n_s = 0.9641 \pm 0.0044, \qquad \frac{dn_s}{d\ln k} = -0.0045 \pm 0.0067, \tag{9.73}$$

hence perfectly compatible with no running but clearly deviating from an HZ spectrum with $n_s = 1$. If one allows also for a tensor contribution, the limits on running become weaker by about a factor of 2.

One can of course allow for more complicated features like one or several kinks in the power spectrum, that is, sudden changes of the spectral index. There are

Table 9.1 *Parameters for the basic, flat ΛCDM model and their marginalized 1σ (or rather 68% probability) errors from Planck alone (middle column) and including BAO data (right column).*

Parameter	Value ±68%	
	Planck only	Planck and BAO
$\Omega_b h^2$	0.02237 ± 0.00015	0.02242 ± 0.00014
$\Omega_c h^2$	0.1200 ± 0.0012	0.11933 ± 0.00091
$100\theta_{MC}$	1.04092 ± 0.00031	1.04101 ± 0.00029
τ_{ri}	0.0544 ± 0.0073	0.0561 ± 0.0071
$\ln(10^{10} A_s)$	3.043 ± 0.014	3.047 ± 0.014
n_s	0.9649 ± 0.0042	0.9665 ± 0.0038
$\Omega_m h^2$	0.1430 ± 0.0011	0.14240 ± 0.00087
H_0 [km/sec/Mpc]	67.36 ± 0.54	67.66 ± 0.42
Ω_Λ	0.6847 ± 0.0073	0.6889 ± 0.0056
Age [Gyr]	13.797 ± 0.023	13.787 ± 0.020
σ_8	0.8111 ± 0.0060	0.8102 ± 0.0060
$S_8 \equiv \sigma_8(\Omega_m/0.3)^{0.5}$	0.832 ± 0.013	0.825 ± 0.011
z_{re}	7.67 ± 0.73	7.82 ± 0.71
$100\theta_{dec}$	1.04110 ± 0.00031	1.04119 ± 0.00029
z_{dec}	1089.92 ± 0.25	1089.80 ± 0.21
r_s [Mpc]	147.09 ± 0.26	147.21 ± 0.23

Above the horizontal line are the six base parameters while the parameters below the horizontal line are derived. Even though the BAO data are not good enough to significantly reduce the error bars, they are in good agreement with the CMB. Note the superb precision of the acoustic scale θ_{dec} which is $\delta\theta_{dec}/\theta_{dec} \simeq 3 \times 10^{-4}$. Data from Planck Coll. VI (2018)

inflationary models that predict such features, for example, models where inflation is driven by several scalar fields (Adams *et al.*, 1997; Achucarro *et al.*, 2013).

A comparison of the cosmological parameters obtained by CMB data alone and CMB in conjunction with BAO data is given in Table 9.1. At the present level of accuracy, complementary data sets do not reduce the error bars considerably, but it is important that they are consistent. However, when we extend the ΛCDM base model, complementary observations often become important in order to break degeneracies. Several examples are listed in Table 9.2.

In the w_0-line of Table 9.2 the dark energy sector has been extended in the simplest way by allowing an equation of state $P = w_0 \rho$. Even though the ΛCDM value $w_0 = -1$ is allowed, the errors are still substantial. To constrain dark energy, SNIa data can help. Allowing for a simple phenomenological ansatz for the equation of state, of the form

$$w = w_0 + (1 - a) w_a, \tag{9.74}$$

Table 9.2 *Several one-parameter extensions of the basic ΛCDM model and their marginalized 2σ (or rather 95% probability) errors from Planck alone (middle column) and including BAO data.*

	Value ±95%	
Parameter	Planck only	Planck and BAO
Ω_K	-0.01 ± 0.013	0.0007 ± 0.0037
$r_{0.002}$	<0.101	0.106
$\sum m_\nu$[eV]	<0.241	<0.12
$N_{\rm eff}$	2.89 ± 0.38	2.99 ± 0.34
w_0	-1.57 ± 0.5	-1.04 ± 0.1
$dn_s/d\log k$	-0.005 ± 0.013	-0.004 ± 0.013

The tensor to scalar ratio, $r_{0.002}$, is evaluated at the pivot scale $k = 0.002/\text{Mpc}$. Data from Planck Coll. VI (2018).

one obtains the constraints (Planck Coll. VI, 2018)

$$w_0 = \begin{cases} -0.961 \pm 0.077 & \text{Planck, BAO, SNIa} \\ -0.76 \pm 0.20 & \text{Planck, BAO/RSD, weak lensing} \end{cases} \tag{9.75}$$

$$w_a = \begin{cases} -0.28^{+0.31}_{-0.27} & \text{Planck, BAO, SNIa} \\ -0.72^{+0.62}_{-0.54} & \text{Planck, BAO/RSD, weak lensing} \end{cases} \tag{9.76}$$

Here RSD refers to redshift space distortions observations shown in Fig. 9.18. A discussion of RSD and weak lensing data is presented in Section 9.8. Even though error-bars are large, SNIa data clearly help in constraining dark energy.

9.7 Non-Gaussianity

To check the CMB anisotropies and polarization for non-Gaussianities, the Planck analysis team studied both the CMB bispectrum and trispectrum. Rather than deriving estimators for these N-spectra, quadratic estimators for the different $f_{\rm nl}$'s themselves and for $g_{\rm nl}$ have been derived. Also this is quite challenging, and different methods have been used that go beyond the scope of this book. We just present an example. One can define an inner product between bispectra of the form

$$\langle b^A, b^B \rangle = \sum_{\ell_i} b^A_{\ell_1\ell_2\ell_3} b^B_{\ell_1\ell_2\ell_3} h^2_{\ell_1\ell_2\ell_3} C^{-1}_{\ell_1} C^{-1}_{\ell_2} C^{-1}_{\ell_3}, \tag{9.77}$$

where $h^2_{\ell_1\ell_2\ell_3}$ is given by Eq. (6.83),

$$h_{\ell_1\ell_2\ell_3} = \sqrt{\frac{(2\ell_1+1)(2\ell_2+1)(2\ell_3+1)}{4\pi}} \begin{pmatrix} \ell_1 & \ell_2 & \ell_3 \\ 0 & 0 & 0 \end{pmatrix}. \tag{9.78}$$

Table 9.3 *The bispectrum and trispectrum measurements from Planck have been summarized by constraining three shapes for the bispectrum that are parameterized by* $f_{\rm nl}^{\rm (local)}$, $f_{\rm nl}^{\rm (equil)}$, $f_{\rm nl}^{\rm (ortho)}$, *and a purely local trispectrum determined by the parameter* $g_{\rm nl}$; *see Chapter 6.*

Variable	T only	T and P data
$f_{\rm nl}^{\rm (local)}$	2.5 ± 5.7	0.8 ± 5.0
$f_{\rm nl}^{\rm (equil)}$	-16 ± 70	-4 ± 43
$f_{\rm nl}^{\rm (ortho)}$	-34 ± 33	-26 ± 21
$g_{\rm nl}$	$(-9 \pm 7.7) \times 10^4$	—

Data from Planck Coll. XVII (2016). All values are compatible with zero and hence non-Gaussianities have not been detected by the Planck satellite

An estimator for $f_{\rm nl}$ is then

$$\hat{f}_{\rm nl} = \frac{\langle b^{\rm (obs)}, b^{\rm (th)} \rangle}{\langle b^{\rm (th)}, b^{\rm (th)} \rangle}, \tag{9.79}$$

where "obs" indicates the observed bispectrum while "th" is the theoretical one. Apart from this relatively straightforward estimator several others can be used. But already Eq. (9.79) is numerically quite intensive, as it requires a sum over three ℓ's and already the estimation of $b^{\rm (obs)}_{\ell_1\ell_2\ell_3}$ requires a sum of $(2\ell_1+1)(2\ell_2+1)(2\ell_3+1)$ values $a_{\ell m}$ (We ignore the partial sky coverage here, which of course also has to be taken into account.). The scalar product (9.77) can easily be generalized to include also polarization and the temperature–polarization cross correlation. In the Planck analysis several different estimators were used, also one derived from Minkowski functionals; see Section 6.4.2. The finally published values for the $f_{\rm nl}$'s and $g_{\rm nl}$ are given in Table 9.3.

9.8 Large-Scale Structure Observations

As discussed in Section 8.2.1, observations of the baryon acoustic oscillation (BAO) peak in the monopole of the correlation function or of the 3D power spectrum of galaxy surveys determine the angle subtended by the comoving drag scale r_s, given by

$$\theta_s(z) = \frac{(1+z)r_s}{d_V(z)}, \qquad d_V(z) = \left(\frac{d_A^2(z)}{(1+z)H(z)}\right)^{1/3}. \tag{9.80}$$

Here d_V is the 3D directional mean of the radial and the angular distance for a correlation function that is averaged over directions. The factor $(1+z)$ is needed

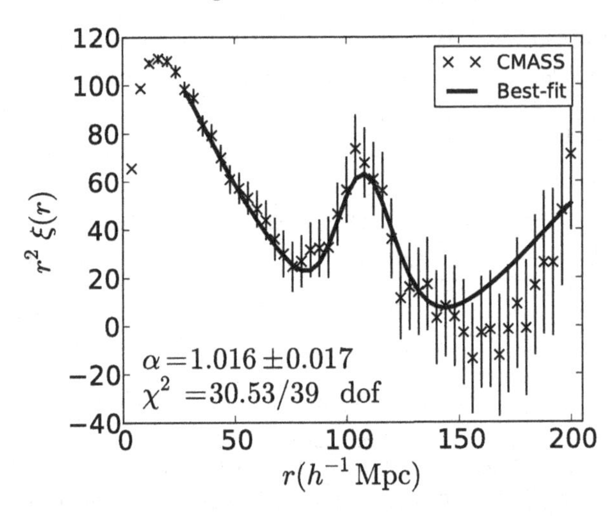

Fig. 9.16 The galaxy correlation function from the CMASS sample, which contains 264,283 massive galaxies covering 3275 square degrees at a mean redshift $z = 0.57$ and covering the redshift range $0.43 < z < 0.7$. Figure from Anderson *et al.* (2012), where more details can be found

since $d_V(z)$ is a physical distance while r_s as given in Table 9.1 is comoving. In Fig. 9.16 we show the reconstructed correlation function from the BOSS data Release 9 (Anderson *et al.*, 2012). The BAO feature is clearly visible in the data. The "BAO data" used in the Planck analysis are summarized in Fig. 9.17. Clearly LSS data are in excellent agreement with the Planck experiment, but its error bars are still considerable. This is expected to improve drastically within the next decade.

In the future, when the radial distance $\Delta z/[(1+z)H(z)]$ and the angular distance $\theta d_A(z)$ can be measured independently with good accuracy, the BAOs in the LSS can be used for an Alcock–Paczynski test as outline in Chapter 8.

Also the quadrupole of the LSS correlation function due to RSD has been measured and it has been used to determine $\beta = f/b$ as discussed in Section 8.2.1. More precisely, the quadrupole of the correlation function determines $f\sigma_8$, where f is the growth function and σ_8 denotes the fluctuation amplitude inside a ball of radius $8h^{-1}$Mpc . The present measurements of $f(z)\sigma_8(z)$ at different redshifts from different experiments are summarized in Fig. 9.18.

9.8.1 Shear Measurements, Weak Lensing

So far we have considered galaxies as "points" in a survey and concentrated on the fluctuations in their number. But galaxies are extended objects of a typical size of arc minutes. They are easily resolved and we can measure their shape, which, for elliptical galaxies, is elliptical.

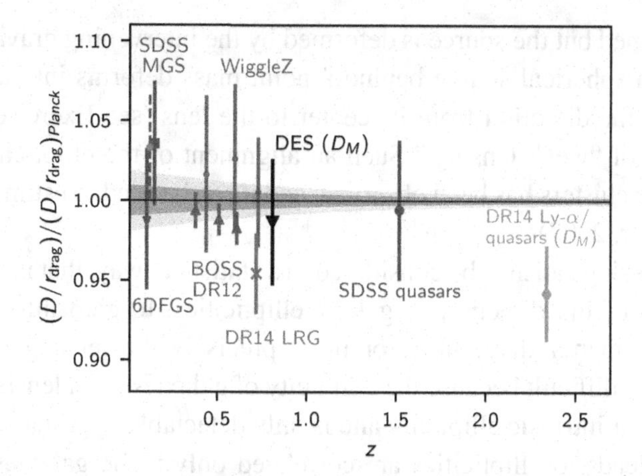

Fig. 9.17 The ratio of θ_s^{-1} obtained from various LSS experiments to the value inferred from the best fit base model from Planck. For the points annotated with (D_M), θ_s is determined by using the transverse angular distance d_A. The dark and light regions around 1 are the 1 and 2σ error regions from Planck. Figure from Planck Coll. VI (2018), where more details and the corresponding references can be found

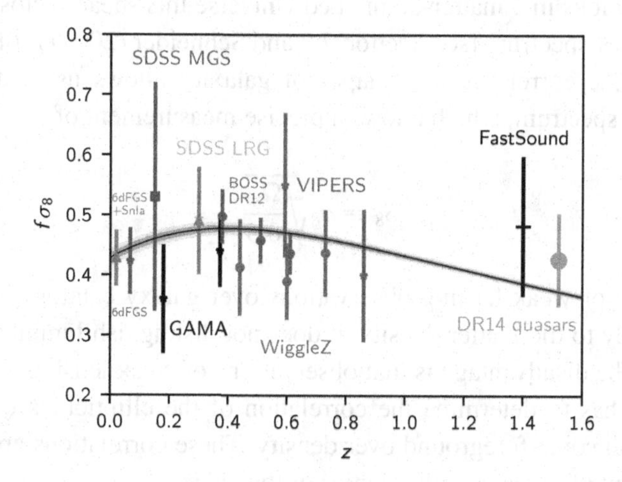

Fig. 9.18 The value of $f\sigma_8$ inferred from the redshift space distortions (RSDs) measured in different galaxy surveys at different redshifts. The dark and light gray regions around 1 are the 1 and 2σ error regions from Planck. Figure from Planck Coll. VI (2018), where more details and the corresponding references can be found.

The deflection of light from a source behind a mass concentration can lead to the formation of multiple images. This effect is called lensing, or more precisely, "strong lensing." If the impact parameter of the light ray connecting the source to the observer is too large, or if the intervening mass is too small, no separate images are formed but the source is deformed by the intervening gravitational field. For example, a spherical source behind a point mass deforms into an ellipse that is squashed in the direction from its center to the lens (see Exercise 9.4). In this case we speak of "weak lensing." Such an alignment of the ellipticity of galaxies behind massive clusters has been observed, and it can be used to estimate the cluster mass (Schneider, 2007).

But weak lensing can also be considered in a statistical way, that is, by measuring the correlation of the direction of galaxy ellipticities, to gain information about the foreground matter distribution or more precisely the lensing potential. The observations are difficult because the ellipticity of galaxies from lensing is typically about 1% of their intrinsic ellipticity and is only detectable by a statistical analysis: weak lensing leads to ellipticities are correlated only if the galaxies are close in angular position even if they are very far apart in physical space, that is, at different redshifts. In this way, the correlation of ellipticities from lensing can in principle be separated from physical alignment (intrinsic alignment) of galaxies that may come from the process of galaxy formation. These galaxy ellipticities measure the shear $\gamma = \gamma_1 + i\gamma_2$ of the deformation matrix A_{ab}; see Sections 7.1 and 7.2. For purely scalar perturbations in a matter-dominated Universe this shear is closely related to the matter power spectrum [see Section 7.2 and Schneider (2007) for more details].

Measuring the correlations of shapes of galaxies allows us to determine the lensing power spectrum, which allows a precise measurement of

$$S_8 \equiv \sigma_8 \sqrt{\frac{\Omega_m}{0.3}}. \tag{9.81}$$

The advantage of weak lensing observations over galaxy catalogs is that lensing responds purely to the matter density, it does not distinguish luminous matter and dark matter. The disadvantage is that observations of weak lensing are much more difficult. One has to determine the correlation of the ellipticities of background galaxies behind some foreground over density. These correlations are on the level of a few percent of the actual ellipticities of the galaxies.

In Chapter 7, we have developed some of the beautiful theory of weak lensing. Here we simply recall that, for linear scalar perturbations, lensing can be expressed in terms of the gravitational potential $\Psi \propto \Omega_m D \propto \Omega_m \sigma_8$. What really enters in weak lensing is the line-of-sight integral of the gravitational potential, which reduces the sensitivity to Ω_m to roughly $\sqrt{\Omega_m}\sigma_8$. Constraints from weak lensing lead to the typical "banana-shaped" contours in the Ω_m–σ_8 plane; see Fig. 9.19, lower left panel. The best limit from Abbott *et al.* (2018) is

$$S_8 = 0.783^{+0.021}_{-0.025}. \tag{9.82}$$

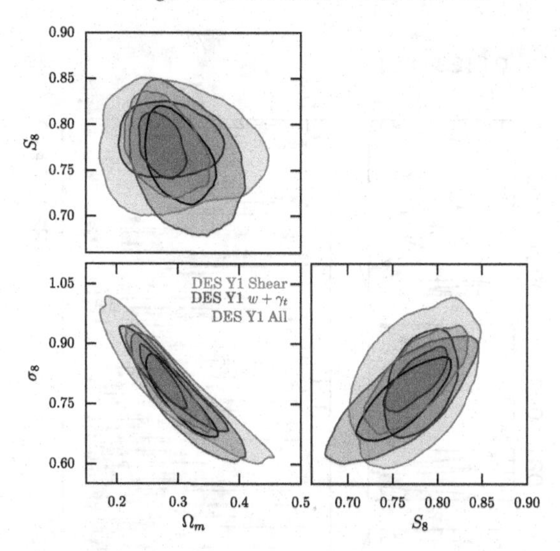

Fig. 9.19 The Ω_m–S_8 constraints from the first year DES (Dark Energy Survey) analysis of lensing and galaxy number counts. The lighter, larger contours are for shear alone while the inner contours also include galaxy clustering (middle contours) and the number count-shear cross correlations (innermost contours). The results are is some tension with the Planck results, but these are still early days and improvements in the shear analysis, especially a more thorough treatment of intrinsic alignment, are expected in the near future. Figure from Abbott *et al.* (2018)

The Planck value, given in Table 9.1, inferred from the best fit base-model is $S_8^{(\mathrm{Planck})} = 0.832 \pm 0.013$, which is in some tension with the weak lensing result. Even though this tension is interesting, the analysis is still not solid enough to claim a breakdown of ΛCDM.

For the future, one plans to measure the lensing power spectrum with much higher precision, for example, with the Euclid satellite to be launched in 2022 (see Amendola *et al.*, 2018) or at the Rubin observatory (see Abate *et al.*, 2012). This very powerful tool, which can, in principle, measure the lensing power spectrum as a function of redshift, is theoretically nearly as simple as the CMB. If observational and systematic difficulties, especially intrinsic alignment, can be overcome, this will provide a most valuable complementary tool for parameter estimation; for a review of weak lensing see Schneider (2007). Comparing the galaxy number count power spectrum with the lensing one can even perform tests of General Relativity (Zhang *et al.*, 2007; Dizgah and Durrer, 2016).

9.8.2 The Lyman-α Forest

The light from a high-redshift quasar, on its way from emission into our detector, not only propagates through the reionized intergalactic medium, but also crosses

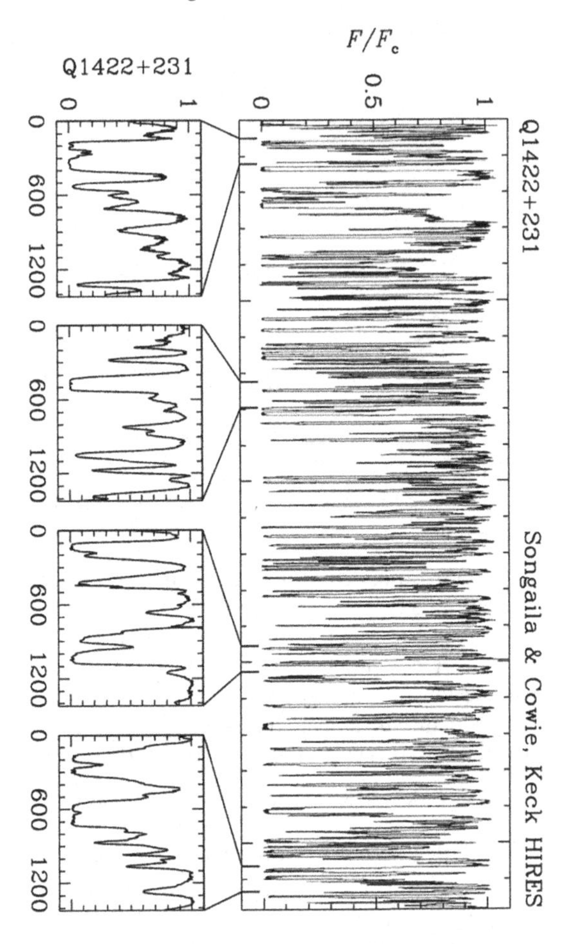

Fig. 9.20 The Ly-α forest region in the Keck HIRES spectrum of the Quasar QSO 1422 + 231 at $z = 3.61$. From Songaila and Cowie (1996)

through clouds of hydrogen. These are regions where the matter density is relatively high and baryons are relatively cool so that a fraction of them have recombined into hydrogen. These hydrogen clouds are not considered as isolated "proto-galaxies," but simply as regions of relatively high density, where collisions that allow cooling processes can occur. When quasar light passes through them, the Lyman-α (Ly-α) photons are absorbed by the neutral hydrogen, leading to a 'forest' of absorption lines in the quasar spectrum; see Fig. 9.20.

This Ly-α forest is related to the one-dimensional distribution of neutral hydrogen, which is in turn related to the matter distribution along the line of sight. The depth of the lines is a measure of the hydrogen density. The observations are usually presented in terms of the power spectrum of the transmitted flux fraction, $P_F(k,z)$, as $F(\lambda) = \exp(-\tau(\lambda))$, where $\tau(\lambda)$ is the optical depth to Ly-α averaged over the scale λ.

To relate this to the matter density power spectrum $P_D(k, z)$, we have to make assumptions about cooling and recombination; see, for example, McDonald *et al.* (2005). With such a relation at hand (which is usually nonlinear and given by hydrodynamical simulations), the correlations of the lines provide, in principle, a measure of the matter power spectrum. If the underlying density fluctuations were still linear, we could relate them to the initial fluctuation by the deterministic matter transfer function $T_D(z)$ via $D(\mathbf{x}, z) = T_D(z)D_{\rm in}(\mathbf{x})$. The transfer function depends only on the background cosmology, that is, on the cosmological parameters. The correlation between linear density fluctuations in some fixed direction $\mathbf{n}$ at redshifts z_1 and z_2 is simply

$$\begin{aligned}\langle D(z_1)D(z_2)\rangle &= \langle D\left(\mathbf{x}_0 - \mathbf{n}(t_0 - t(z_1)), t(z_1)\right) D\left(\mathbf{x}_0 - \mathbf{n}(t_0 - t(z_2)), t(z_2)\right)\rangle \\ &= T(z_1)T(z_2)C_{\rm in}(|t(z_1) - t(z_2)|), \end{aligned} \tag{9.83}$$

where $C_{\rm in}$ is the initial density correlation function, that is, the Fourier transform of the initial power spectrum. Typically $z_1, z_2 \sim$ 2–3 and $\Delta z = |z_1 - z_2| < 0.01$, so that $|t(z_1) - t(z_2)| \simeq \mathcal{H}^{-1}(z)(\Delta z/(z+1))$. Here z is the mean redshift and we have approximated $|t(z_1)-t(z_2)|$ to first order in Δz. These length scales are small, hence the Ly-α forest explores the power spectrum at relatively small scales. For example, for $z = 2.5$ we have $\mathcal{H}^{-1}(z)/(z+1)\big|_{z=2.5} \simeq 2850\, h^{-1}\,$Mpc so that we find $|t(z) - t(z + \Delta z)|_{z=2.5} \simeq 2.8\, h^{-1}\,$Mpc for $\Delta z = 0.001$.

On the other hand, at $z \sim$ 2–3 perturbations on these scales are already nonlinear, so that time evolution and correlation do not simply factorize as in Eq. (9.83). To use the Ly-α forest for parameter estimation, we have to rely on N-body simulations. The main sensitivity of the Ly-α forest is therefore to the amplitude of fluctuations on a small scale, which is sensitive to both σ_8 and the spectral index n_s, as well as to a possible contribution of massive neutrinos.

If the theoretical difficulties can be overcome, the Ly-α forests of quasars provide a very interesting data set. They are our most promising tool for estimating the linear power spectrum on small scales at high redshift. In that sense they help enormously to extend the "lever arm" of our knowledge on the primordial power spectrum. The length of this lever arm is crucial for a precise estimation of the scalar spectral index n_s or of a possible "running" of the spectral index, $dn/d\log k \neq 0$. An analysis of the Ly-α forest from quasars in the SDSS is presented in McDonald *et al.* (2005) and Palanque-Delabrouille *et al.* (2015); there a neutrino mass limit of $\sum m_\nu < 0.14\,$eV when combined with CMB data. From the Ly-α forest alone one obtains a much poorer mass limit of about $\sum m_\nu < 1.1$ eV. This analysis, however, illustrates the potential of the method. With the BOSS survey, Ly-α forest data have also been obtained on larger scales and led to a determination of the BAO scale, θ_s at redshift $z \simeq 2.3$, which is in reasonable agreement with the Planck value at this redshift; see Bautista *et al.* (2017).

9.8.3 Galaxy Clusters

Finally, we want to address briefly the relevance of galaxy clusters for cosmological parameters. Rich clusters with masses $M \sim 10^{14.5}\, h^{-1}\, M_\odot$ are the largest bound structures in the Universe ($M_\odot$ is the mass of the Sun). Clusters can vary from small groups of tens of galaxies to more than 1000 galaxies. They represent over-densities of more than unity and are thus nonlinear. For Gaussian fluctuations the probability to measure an over- (or under-) density $\delta\rho = \rho_m D$ on a scale λ is

$$P(D,z) = \frac{1}{\sqrt{2\pi}\sigma(\lambda,z)} \exp\left(-\frac{D^2}{2\sigma^2(\lambda,z)}\right). \tag{9.84}$$

Here $\sigma^2(\lambda,z)$ is the variance of the density fluctuations on the scale λ at redshift z. Assuming that a relative over-density D_c is needed for an object to collapse, the probability that an over-density on scale λ has collapsed into a cluster becomes

$$P(D > D_c|\lambda,z) = 1 - \mathrm{erf}\left(\frac{D_c}{\sqrt{2}\sigma(\lambda,z)}\right), \tag{9.85}$$

where $\mathrm{erf}(x)$ denotes the error function. This is the basic ingredient of the Press–Schechter formalism (Press and Schechter, 1974).

The spherical collapse model (see, e.g., Peebles, 1993) requires $D_c \simeq 1.69$. Assigning the total mass M inside a sphere of radius λ to the collapsed object, allows one to determine the number density of clusters of mass larger than M for different redshifts. This quantity is very sensitive to Ω_m, which determines the growth function of linear density fluctuations and thereby $\sigma(\lambda,z)$, which enters exponentially in the cluster abundance. The study of the abundance of large clusters was one of the first observations indicating $\Omega_m \sim 0.3$ (see Bahcall and Cen, 1992).

Furthermore, clusters usually form at a fixed velocity dispersion. The kinetic energy has to be smaller than the gravitational potential energy for a bound structure to form. Therefore, the cluster density also constrains the velocity power spectrum, $P_V \propto \Omega_m^{1.2}\sigma_8^2$ [see Eq. (2.251)]. Comparing observations with numerical simulations gives (Böhringer *et al.*, 2014)

$$\sigma_8\left(\frac{\Omega_m}{0.3}\right)^{0.57} = 0.753 \pm 0.03.$$

Comparing lensing observations from clusters that are sensitive to the total mass to X-ray emission, which depends on the baryon density in clusters, one can estimate the ratio $\Omega_b/\Omega_m \sim 0.1$. Several assumptions go into this value. First of all, X-ray emission is proportional to the line-of-sight integral of ρ_b^2 and assuming $\langle\rho_b^2\rangle \simeq \langle\rho_b\rangle^2$ is not trivial at all, since baryons are strongly clustered on small scales. Furthermore, relating the baryon to mass ratio in clusters to Ω_b/Ω_m assumes that this ratio in clusters is similar to its mean in the total Universe. We know, for example, that this is not so in the central parts of galaxies. However, clusters seem to have a sufficiently low mean density, so that hydrodynamical processes that affect baryons, but not dark matter, do not significantly modify the ratio Ω_b/Ω_m in clusters.

Concluding, we state that clusters, being the largest bound structures in the Universe, are an interesting tracer of the mass distribution that should not be ignored. It is very reassuring that the cosmological parameters inferred from clusters fit well with the results from CMB and other observations. However, since clusters are nonlinear structures, with highly nonlinear physics determining their X-ray emission and probably also their mass, we cannot expect to achieve the degree of accuracy that we obtain from probes of linear and quasilinear perturbations.

9.9 Complementary Observations

CMB and LSS observations are not the only cosmological observations at our disposal. The main reason why they are so useful is that they are relatively easy to calculate to good accuracy. On a wide range of scales, we do not expect any complicated physics to obscure the relation between data and theory. Nevertheless, it would be a waste not to also consider other available data and, especially in view of the degeneracies, we need other data to confidently interpret the CMB. Here we only briefly introduce the most important complementary observations.

9.9.1 The Hubble Parameter $H(z)$

A notoriously difficult quantity to measure is not only the function $H(z)$, but more basically the value of the present Hubble parameter, $H_0 = H(0)$. The main difficulty lies in the measurement of cosmological distances. The history of astrophysics and cosmology is marked by repeated underestimations of distances. E. Hubble originally overestimated his parameter by a factor of about 7 (Hubble, 1929). Even though it is relatively straightforward to measure the redshift of a cosmological source, how can we find its distance? The main tools are standard candles or standard rulers. If we know the intrinsic size or luminosity of a distant source, we can use this to determine its angular diameter or luminosity distance. To lowest order in the redshift z, these are simply $H_0^{-1}z$. For redshift higher than $z \simeq 0.1$, one has to take into account the full expression for the distance as derived in Chapter 1. The full function $H(z)$ determines not only the present Hubble parameter but, via the Friedmann constraint equation, also the matter content and the curvature of the Universe.

A very promising method is to determine $H(z)$ with the observation of supernovae of type Ia. These are supernovae without hydrogen lines. They are extremely luminous and can be seen out to redshift of 2 and maybe more. The idea is that they come from white dwarfs that accrete material, for example, from a companion star, until they pass over the Chandrasekhar mass limit of about 1.4 $M_\odot$ (Chandrasekhar, 1939). At this moment they become unstable and explode. Most probably this leads to the formation of a neutron star. The intrinsic luminosity of this explosion is quite constant and the mild variation is strongly correlated with the width of the light curve. Correcting for this variation with a phenomenological formula, one can obtain very small variations in the corrected intrinsic luminosity (about 0.1 magnitude). This allows an accurate measurement of the luminosity distance to these explosions.

At present, luminosity distances to supernovae with redshifts up to 1.7 have been determined. These are used not only to measure H_0 but especially to determine $H(z)/H_0$, which can be obtained with much better accuracy. These measurements have provided the first clear indication that the expansion of the Universe is currently dominated by a cosmological constant or some form of dark energy with strong negative pressure leading to acceleration (see Chapter 1). For this discovery S. Perlmutter, A. Riess, and B. Schmidt were awarded the Nobel Prize in 2011 (Riess *et al.*, 1998; Schmidt *et al.*, 1998; Perlmutter *et al.*, 1999). Up to this day more than 1000 SNIas have been observed. However, at present the CMB alone gives tighter constraints on Ω_Λ, Ω_m and when combined with BAOs, adding SNIa data does not lead to a significant improvement. Nevertheless, it is an important cross-check, and if we could master the difficult systematics that plague SNIa data at higher redshifts, they might help us in our aim to reveal the nature of dark energy. Is it simply a cosmological constant with $w = -1$ or is it dynamical, for example, a scalar field with a time-dependent equation of state, $w(z)$? Or is it even some "phantom matter" with $w < -1$?

Unfortunately, the luminosity distance directly measures only the integral [see Chapter 1, Eqs. (1.39) and (1.51)]

$$d_L(z) = \frac{(1+z)}{\sqrt{|\Omega_k|}H_0}\chi_K\left(\sqrt{|\Omega_k|}H_0\int_0^z \frac{dz}{(1+z)H(z)}\right).$$

The equation of state parameter w, on the other hand, enters at the level of the derivative $H'(z)$. To determine it, two derivatives from relatively noisy data have to be taken, a very difficult task. It has also been shown that the dipole of the luminosity distance which is, like the CMB dipole, due to our motion with respect to the Friedmann background, allows a direct measure of $H(z)$ so that only one additional derivative is needed to arrive at the equation of state (Bonvin *et al.*, 2006b). On the other hand, to accurately determine the amplitude of the dipole, many SNIas in a given redshift bin are needed. It remains to be seen whether this approach will bear fruit.

To measure the Hubble constant, one usually employs SNIa at low redshift, $z < 0.1$ to avoid degeneracies with other cosmological parameters. The most recent value measured with SNIa reports

$$H_0 = (73.5 \pm 1.4)\text{km/s/Mpc} \tag{9.86}$$

A similar value has obtained from time delays in strongly lensed systems (Wong *et al.*, 2019) $H_0 = (74.2 \pm 1.8)$km/s/Mpc. Another analysis using "the tip of the red giant branch" as the standard candle has obtained $H_0 = (69.8 \pm 1.9)$km/s/Mpc. See Riess (2019) for a review of the present situation. In particular the SNIa result (9.86) exhibits a slightly more than 4σ tension with the Planck value of $H_0 = (67.36 \pm 0.54)$km/s/Mpc. This is presently the strongest discrepancy in the base ΛCDM model. It is not easy to accommodate it with, for example, massive neutrinos or curvature, and so forth without spoiling anything else. The present parameter values are already very tight.

There are already several hundreds of papers written about this discrepancy that either solve it with some exotic physics or with unaccounted for systematics in one or the other experiment. Here we do not contribute to this debate but advise the reader to also study it from the point of view of an interesting sociological event in the scientific community.

9.9.2 Nucleosynthesis

As we have seen in Chapter 1, by an analysis of what has happened during nucleosynthesis, that is, at $T \simeq 0.1$ MeV and $z \simeq 2.3 \times 10^9$, we can calculate the light element abundance as a function of the baryon density, $\Omega_b h^2$; see Fig. 1.10. Comparing with the observations of these abundances yields the nucleosynthesis value of the baryon density. How to estimate the primordial abundance from the present abundance of light elements is entirely nontrivial, an art that we do not discuss here any further. An estimate from the most sensitive deuterium abundance gives (Tanabashi *et al.*, 2019)

$$0.021 \le \Omega_b h^2 \le 0.024 \quad \text{(at 95\% confidence)}\,. \tag{9.87}$$

The agreement of this result with the CMB estimate is most remarkable. Both values are based on completely different physics. Such agreements give us confidence in the standard cosmological model.

The abundance of helium generated during nucleosynthesis is sensitive to the number of relativistic degrees of freedom at the time of nucleosynthesis, which determines the expansion rate during nucleosynthesis. The photon and three types of neutrinos (at their somewhat lower temperature) lead to a good fit to the observed helium abundance. The nucleosynthesis constraint on the content of relativistic particles at the time of nucleosynthesis is usually formulated as a constraint on the number N_ν of light neutrino species. The data require (Tanabashi *et al.*, 2019)

$$2.3 < N_\nu < 3.4 \quad \text{(at 95\% confidence)}. \tag{9.88}$$

In very good agreement with the value from the width of Z-decay obtained at accelerators (Tanabashi *et al.*, 2019),

$$N_\nu = 2.991 \pm 0.007.$$

Even though the cosmological result is older, the accelerator result has become much more accurate. It is also much more accurate than the CMB constraint given in Table 9.2.

The only data that are in disagreement with the standard cosmological model of nucleosynthesis is the Li^7 abundance, which is more than a factor of 3 lower than the predicted value. However, during early star formation lithium is mainly destroyed and it is difficult to infer the primordial value from the observed one. Nevertheless, the value of $\Omega_b h^2$ inferred from the deuterium abundance is in an about 5σ discrepancy with the value inferred from the lithium abundance. It is still under debate whether this discrepancy is due to modeling of stellar evolution or whether new physics is needed to resolve it. See Tanabashi *et al.* (2019) for a brief overview of the situation.

Exercises

(Exercises marked with an asterisk are solved in Appendix 11 which is not in this printed book but can be found online.)

9.1 **Optical depth from reionization**

Calculate the optical depth $\tau_{\rm ri}(z_{\rm ri})$ as a function of the reionization redshift, $z_{\rm ri}$, for a pure matter universe, $\Omega_{\rm tot} = \Omega_m = 1$ and for a Λ-dominated universe with $\Omega_\Lambda = 0.7$ and $\Omega_m = 0.3$. Express the result as a function of $\Omega_b h^2$.

Hint: For the Λ-dominated case you may assume $z_{\rm ri} \gtrsim 6$ and neglect the influence of the cosmological constant for $z > 2$ and the contribution to $\tau_{\rm ri}$ for $z < 2$. Estimate your error. Consider two cases.

(1) The universe ionized suddenly at redshift $z_{\rm ri}$.
(2) Ionization started at redshift $z_{\rm ri} > 6$ and was completed at $z = 6$. In the reionization interval, $6 \le z \le z_{\rm ri}$, the free electron fraction, x, rises linearly with the scale factor, $1 - x = (z - 6)/(z_{\rm ri} - 6)$.

9.2 **Isocurvature initial conditions***
Let us denote $X_1 = D_\gamma$, $X_2 = D_m$, $X_3 = D_\nu$, and $X_4 = V_\nu$. We parameterize the initial conditions by

$$C_{ij} = \langle X_i(\mathbf{k})X_j^*(\mathbf{k}')\rangle = A_{ij}(k/H_0)^{n_{ij}}\delta(\mathbf{k}-\mathbf{k}').$$

Show that C_{ij} is positive semidefinite for all values of k if and only if the matrix A_{ij} is positive semidefinite and $n_{ii} \le n_{ij} \le n_{jj}$ or $n_{jj} \le n_{ij} \le n_{ii}$ for all i, j with $A_{ij} \neq 0$.

9.3 **The shape parameter**
Show that the comoving Hubble scale at equality $H_0 t_{eq} \propto (\Omega_m h)^{-1}$. How does the matter power spectrum depend on this scale?

9.4 **Weak lensing**
Consider a point mass M at distance d_L in front of a circular source with radius r_s at distance D_S, the center of which passes the lens with impact parameter b. Using the small angle and small deflection approximation calculate the shape of the image. Show that the ellipticity is parallel to the radial direction. Calculate the ellipticity for a source distance, $d_S = 30$ Mpc; lens distance, $d_L = 25$ Mpc; impact parameter $b = 0.1$ Mpc; source radius $r_s = 0.03$ Mpc; and mass $M = 10^{15}\, M_\odot$.
Hint: Approximate the gravitational potential by $\Phi = GM/r$. Calculate the impact parameter of a point on the circular border of the source as a function of $d_{LS} = d_S - d_L$ (neglect the expansion of the Universe). Determine now the image position of this point.

10

The Frequency Spectrum of the CMB

The observed frequency power spectrum of the CMB is perfectly approximated by a Planck spectrum; see Fig. 1.7. In our units, $\hbar = c = 1$, it is given by

$$I(\omega) = 4\pi f(\omega) = \frac{1}{\pi^2} \frac{\omega^3}{e^{\omega/T} - 1}. \tag{10.1}$$

No deviation from this spectrum has been observed so far.

In this chapter we discuss physical processes that might lead to spectral distortions. We first introduce the collisional processes relevant at temperatures $T < m_e$. These are Compton scattering, double Compton scattering, and Bremsstrahlung. We derive the Boltzmann equation for these processes and calculate the relevant timescales. In Section 10.2 we analyze how the injection of high-energy photons, for example, by the decay of a long-lived unstable particle or Silk damping modifies the CMB spectrum. In the final section, we discuss what happens when the CMB photons pass through a hot electron gas affected only by Compton scattering. We shall see that this leads, in general, to a so-called Compton-y distortion of the spectrum. We estimate the effect from the passage of CMB photons through a cluster of galaxies and discuss observations.

10.1 Collisional Processes in the CMB

10.1.1 Generalities

At very high temperature many collisional processes keep the cosmic background radiation in thermal equilibrium with itself and all other particles. As we saw in Section 1.4, at $T \simeq 1.4$ MeV weak interactions drop out of equilibrium and neutrinos cease to interact. They are not heated by the decay of electron–positron pairs, which takes place at $T \sim m_e \simeq 500$ keV. Below that temperature, but before recombination, Compton scattering, $e + \gamma \to e + \gamma$; double Compton scattering,

$e + \gamma \rightarrow e + 2\gamma$; and Bremsstrahlung, $e + X \rightarrow e + X + \gamma$ keep the CMB thermalized. Here X denotes an atomic nucleus, usually a proton or a helium-4 nucleus.

As we shall see in the text that follows, at a redshift of about $z_\mu \simeq 10^7$ also Bremsstrahlung and double Compton drop out of equilibrium and only Compton scattering is still active. This can still redistribute the CMB photons in energy, but it does not change their number density. Therefore, energy injection after z_μ leads to a Bose–Einstein distribution with a nonvanishing chemical potential.

In this section we derive the equations that govern the evolution of the photon distribution function in the temperature range $m_e > T > T_{\rm rec}$. We also calculate the timescales for the aforementioned processes and the redshifts above which they are faster than the expansion timescale, that is, above which they efficiently keep the photon distribution in thermal and "chemical" equilibrium (by the latter we mean that no chemical potential is developed).

10.1.2 Compton Scattering and the Kompaneets Equation

We want to derive a differential equation that describes the thermalization of photons when they interact with electrons that also have a thermal distribution but may be at a different temperature $T_e \neq T$. We consider a photon with initial energy ω and final energy ω' and a nonrelativistic electron with initial velocity $v \ll 1$ and final velocity $v' \ll 1$. Hence we must also require $\omega, \omega' \ll m_e$. In the center of the mass system, energy and momentum of the two particles remain unchanged in a two-body interaction. This is a simple consequence of energy and momentum conservation. However, in the laboratory frame, if initially the electron momentum is much larger than the photon momentum, after the collision the photon will have gained energy while the electron has lost and vice versa. Let us first study the process in the frame in which the initial electron momentum vanishes, so that by energy and momentum conservation

$$\omega \mathbf{n} = m_e \mathbf{v}' + \omega' \mathbf{n}' \quad \text{and} \quad \omega = \frac{1}{2} m_e v'^2 + \omega' = \frac{1}{2m_e}(\omega \mathbf{n} - \omega' \mathbf{n}')^2 + \omega', \quad (10.2)$$

where $\mathbf{n}$ and $\mathbf{n}'$ denote the photon direction before and after the collision. If we neglect all terms of order ω/m_e, we find $\omega = \omega'$. This would imply no change in the photon frequency. In this approximation we obtain nonrelativistic Thomson scattering where only the direction of the photon but not its energy is affected. The energy difference is of the order ω^2/m_e, where ω is either ω or ω'. Taking this difference into account to lowest order we may set $\omega^2 = \omega'^2 = \omega\omega'$ in the term proportional to $1/(2m_e)$. With this Eq. (10.2) yields

$$\frac{\omega}{\omega'} = 1 + \left(\frac{\omega}{m_e}\right)(1 - \cos\vartheta), \tag{10.3}$$

where ϑ is the scattering angle, $\cos\vartheta = \mathbf{n}\cdot\mathbf{n}'$. To first order in ω/m_e this can be written as

$$\frac{\omega' - \omega}{\omega} \equiv \frac{\Delta\omega}{\omega} = -\left(\frac{\omega}{m_e}\right)(1 - \cos\vartheta). \tag{10.4}$$

In a generic frame, denoting the photon and electron 4-momentum before and after the scattering process by $p_\gamma = \omega(1,\mathbf{n})$ and $p_e = (E,\mathbf{p})$ respectively $p'_\gamma = \omega'(1,\mathbf{n}')$ and $p'_e = (E',\mathbf{p}')$, energy–momentum conservation implies

$$(p'_e)^2 = m_e^2 = (p_e + p_\gamma - p'_\gamma)^2 = m_e^2 + 2p_e(p_\gamma - p'_\gamma) - 2p_\gamma p'_\gamma,$$

which yields

$$\begin{aligned} 0 &= E(\omega - \omega') - \mathbf{p}\cdot(\omega\mathbf{n} - \omega'\mathbf{n}') - \omega\omega' + \omega\omega'\mathbf{n}\cdot\mathbf{n}' \\ &= E(\omega - \omega') - \mathbf{p}\cdot\mathbf{n}'(\omega - \omega') + \omega(\omega - \omega')(1 - \mathbf{n}\cdot\mathbf{n}') \\ &\quad + \omega\mathbf{p}\mathbf{n}' - \omega^2(1 - \mathbf{n}\cdot\mathbf{n}') - \omega\mathbf{p}\cdot\mathbf{n}. \end{aligned}$$

Defining $x_e \equiv \omega/T_e$, we obtain for the energy difference ($\frac{T_e}{E} \cong \frac{p^2}{2m_e^2} = \frac{v^2}{2}$)

$$\Delta \equiv \frac{\omega' - \omega}{T_e} = \frac{x_e\mathbf{p}(\mathbf{n}' - \mathbf{n}) - x_e^2 T_e(1 - \mathbf{n}\cdot\mathbf{n}')}{E - \mathbf{p}\cdot\mathbf{n}' + x_e T_e(1 - \mathbf{n}\cdot\mathbf{n}')} \simeq \frac{x_e\mathbf{p}(\mathbf{n}' - \mathbf{n})}{m_e}. \tag{10.5}$$

For the $\simeq$ sign we have neglected terms of higher order in $p/E \simeq v \ll 1$. The energy transfer, $\omega' - \omega$ is of order ωv, that is, it is suppressed by a factor v.

To calculate how the photon distribution $f(\omega)$ changes by Compton scattering we write the Boltzmann equation. Neglecting the expansion of the Universe and perturbations we have (see Section 4.5)

$$\frac{\partial f}{\partial t}(\omega) = \frac{df_+}{dt}(\omega) - \frac{df_-}{dt}(\omega) \equiv C[f](\omega), \tag{10.6}$$

where df_+/dt denotes the phase space density of photons that are scattered into the energy range $[\omega, \omega + d\omega]$ per unit time and df_-/dt denotes the density of photons scattered out of this energy range. We assume that f is independent of direction and position. Contrary to the situation in Section 4.5 we now consider a distribution of electrons in momentum space, which we denote by $f_e(E)$. The collision integral now becomes

$$\begin{aligned} &C[f](\omega) \\ &= \int d^3p \int d\Omega_{\mathbf{n}'} \frac{d\sigma}{d\Omega}\left\{f(\omega')[1 + f(\omega)]f_e(E) - f(\omega)[1 + f(\omega')]f_e(E')\right\}. \end{aligned} \tag{10.7}$$

Here d^3p denotes integration over electron momenta with $E = \sqrt{m_e^2 + p^2} \simeq m_e + p^2/2m_e$ and $d\Omega_{\mathbf{n}'}$ denotes integration over photon directions. The variables $\omega' = \omega + T_e\Delta$ and $E' = E - T_e\Delta$ are eliminated via Eq. (10.5). The factors $1 + f$ take into account the quantum effect of stimulated emission for photons that are bosons. For the electrons we should, in principle, include a factor $[1 - f_e]$ due to their fermionic nature, but we assume that the electron gas is sufficiently diluted, $f_e \ll 1$, so that we may neglect this quantum correction. The Compton scattering cross section is given by Eq. (4.109)

$$\frac{d\sigma}{d\Omega} = \frac{3}{16\pi}\sigma_T\left(1 + (\mathbf{n}\cdot\mathbf{n}')^2\right).$$

Strictly speaking, this is the cross section in the electron rest frame and when transforming it to the laboratory frame the photon directions and $(\mathbf{n}\cdot\mathbf{n}')^2$ change due to aberration. But as we shall argue, this effect can be neglected for nonrelativistic electrons (as it is of order v^2).

We now expand the integrand of Eq. (10.7) to second order in the small energy transfer $\omega' - \omega = E - E' \sim \mathcal{O}(\omega v)$:

$$f(\omega') = f(\omega) + \Delta\frac{\partial f}{\partial x_e} + \frac{\Delta^2}{2}\frac{\partial^2 f}{\partial x_e^2} + \cdots$$

$$f_e(E') = f_e(E) - T_e\Delta\frac{\partial f_e}{\partial E} + \frac{T_e^2\Delta^2}{2}\frac{\partial^2 f_e}{\partial E^2} + \cdots$$

$$= f_e(E)\left[1 + \Delta + \frac{\Delta^2}{2} + \cdots\right].$$

For the last equality sign we have assumed that the electrons obey a Maxwell distribution, $f_e(E) \propto \exp(-E/T)$. Inserting this expansion in Eq. (10.7), we find

$$\frac{\partial f}{\partial t}(\omega) = \left[\frac{\partial f}{\partial x_e} + f(1+f)\right]I_1 + \frac{1}{2}\left[\frac{\partial^2 f}{\partial x_e^2} + 2(1+f)\frac{\partial f}{\partial x_e} + f(1+f)\right]I_2, \tag{10.8}$$

with

$$I_1 = \int d^3p \int d\Omega_{\mathbf{n}'}\frac{d\sigma}{d\Omega}f_e(E)\Delta, \tag{10.9}$$

$$I_2 = \int d^3p \int d\Omega_{\mathbf{n}'}\frac{d\sigma}{d\Omega}f_e(E)\Delta^2. \tag{10.10}$$

We want to calculate these integrals up to order v^2. The second integral is readily performed. Since the lowest-order approximation to Δ is already of order v we can simply set $\Delta^2 = (x_e/m_e)^2(\mathbf{p}\cdot(\mathbf{n}'-\mathbf{n}))^2$. Inserting this in Eq. (10.10) and choosing the p_3-direction along $\mathbf{n}'-\mathbf{n}$ we obtain

$$\begin{aligned} I_2 &= \frac{x_e^2}{m_e^2}\int \frac{d\sigma}{d\Omega}d\Omega \int d^3p f_e(E)p^2(\mathbf{n}'-\mathbf{n})^2\cos^2\theta \\ &= \frac{4\pi x_e^2}{3m_e^2}\int \frac{d\sigma}{d\Omega}(\mathbf{n}'-\mathbf{n})^2 d\Omega \int_0^\infty dp\, p^4 f_e(E) \\ &= \frac{T_e n_e x_e^2}{m_e}\int \frac{d\sigma}{d\Omega}(\mathbf{n}'-\mathbf{n})^2. \end{aligned}$$

We have used the fact that to lowest order in v, $f_e' = -(p/m_eT_e)f_e$ so that $p^4 f_e = -m_eT_ep^3f_e' = m_eT_e[-(p^3f_e)' + 3p^2f_e]$. The p-integral over the first term in the square bracket does not contribute while the integral over the second term gives $3n_e/(4\pi)$, where n_e denotes the electron density. We still have to integrate over scattering angles,

$$\begin{aligned} \int \frac{d\sigma}{d\Omega}(\mathbf{n}'-\mathbf{n})^2 d\Omega &= \frac{3\sigma_T}{16\pi}\int d\Omega(1+\cos^2\vartheta)(2-2\cos\vartheta) \\ &= \frac{3\sigma_T}{4}\int_{-1}^{1} d(\cos\vartheta)[1+\cos^2\vartheta] = 2\sigma_T. \end{aligned}$$

We finally obtain

$$I_2 = 2n_e\sigma_T\frac{T_e}{m_e}x_e^2. \tag{10.11}$$

Note that $T_e/m_e \sim p^2/(2m_e^2) \sim v^2$; hence the term is of the required order of magnitude.

The calculation of I_1 is trickier. To lowest order $\Delta \propto \mathbf{p}(\mathbf{n}'-\mathbf{n})$, so that the integral over d^3p vanishes. (Were this not the case, this term that is of order v would by far dominate all other contributions and also I_2.) We therefore have to include the next order. Expanding Δ to the next order gives

$$\Delta \simeq \frac{x_e}{m_e}\mathbf{p}\cdot(\mathbf{n}'-\mathbf{n})\left[1+\frac{\mathbf{p}\cdot\mathbf{n}'}{m_e}\right] - \frac{x_e^2}{m_e}T_e(1-\mathbf{n}\cdot\mathbf{n}').$$

In addition, we have to take into account the fact that the photon density seen by the electron (in its rest frame) is not $4\pi f\omega^2\,d\omega$ but $4\pi f\omega^2\,d\omega\,(1-\mathbf{p}\cdot\mathbf{n}/m_e)^3 \simeq 4\pi f\omega^2\,d\omega\,(1-3\mathbf{p}\cdot\mathbf{n}/m_e)$, to lowest order in $v \simeq p/m_e$. Therefore, the integral that we really have to compute is not the one given in Eq. (10.9) but

$$I_1 = \int d^3p \int d\Omega_{\mathbf{n}'}\frac{d\sigma}{d\Omega}f_e(E)(1-3\mathbf{p}\cdot\mathbf{n}/m_e)\Delta, \tag{10.12}$$

with

$$(1 - 3\mathbf{p}\cdot\mathbf{n}/m_e)\Delta = \frac{x_e}{m_e}\mathbf{p}\cdot(\mathbf{n}' - \mathbf{n})\left[1 + \frac{\mathbf{p}\cdot\mathbf{n}'}{m_e} - \frac{3\mathbf{p}\cdot\mathbf{n}}{m_e}\right]$$
$$- \frac{x_e^2}{m_e}T_e(1 - \mathbf{n}\cdot\mathbf{n}') + \mathcal{O}(v^3).$$

Of course this correction also applies to I_2, but there it is subdominant and does not enter up to order v^2. To order v^2, in principle, aberration also has to be taken into account. We should replace $\mathbf{n}\cdot\mathbf{n}'$ by the corresponding scalar product in the electron rest frame, $\mathbf{n}_R(\mathbf{n},\mathbf{v})\cdot\mathbf{n}'_R(\mathbf{n}',\mathbf{v})$ in $d\sigma/d\Omega$. Here a subscript R denotes the electron rest frame (see Exercise 10.2). However, the resulting expression will always be symmetrical in $\mathbf{n}$ and $\mathbf{n}'$. Since to lowest order it is multiplied with the antisymmetrical factor $\mathbf{p}\cdot(\mathbf{n}' - \mathbf{n})$, its angular integral vanishes.

Using as before that

$$\int d^3p\,(\mathbf{p}\cdot\mathbf{n})^2 f_e(E) = m_e T_e n_e, \qquad \int d^3p\, f_e = n_e,$$

and

$$\int d^2p\,(\mathbf{p}\cdot\mathbf{n})(\mathbf{p}\cdot\mathbf{n}')f_e(E) = (\mathbf{n}\cdot\mathbf{n}')\frac{4\pi}{3}\int_0^\infty dp\,p^4 f_e = (\mathbf{n}\cdot\mathbf{n}')m_e T_e n_e,$$

we find

$$\int d^3p\, f_e(1 - 3\mathbf{p}\cdot\mathbf{n}/m_e)\Delta = \left[4\frac{T_e n_e x_e}{m_e} - \frac{x_e^2 T_e n_e}{m_e}\right](1 - \mathbf{n}\cdot\mathbf{n}').$$

The integral over photon directions gives

$$\int d\Omega\,\frac{d\sigma}{d\Omega}(1 - \mathbf{n}\cdot\mathbf{n}') = \int d\Omega\,\frac{d\sigma}{d\Omega} = \sigma_T.$$

Putting this together we obtain

$$I_1 = \frac{\sigma_T n_e T_e}{m_e}x_e(4 - x_e).$$

Inserting the results for I_1 and I_2 in Eq. (10.8), we obtain the Kompaneets equation

$$\frac{m_e}{T_e}\frac{1}{n_e\sigma_T}\frac{\partial f}{\partial t} = \frac{1}{x_e^2}\frac{\partial}{\partial x_e}\left[x_e^4\left(\frac{\partial f}{\partial x_e} + f + f^2\right)\right]. \tag{10.13}$$

Solving the time-dependent Kompaneets equation in order to study "Comptonization" of a photon distribution on thermal electrons in full generality can only be achieved numerically. However, there are important situations in which meaningful analytical results can be obtained.

First of all, as it should, the photon number density remains unchanged by evolution under the Kompaneets equation,

$$\frac{dn_\gamma}{dt} \propto \int dx_e\, x_e^2 \frac{\partial f}{\partial t} \propto \int dx_e \frac{\partial}{\partial x_e}\left[x_e^4\left(\frac{\partial f}{\partial x_e} + f + f^2\right)\right] = 0. \tag{10.14}$$

Furthermore, a Bose–Einstein distribution with temperature T_e, hence $f_{BE} = \left(e^{(\omega/T_e+\mu)} - 1\right)^{-1}$, is the (unique) equilibrium solution of this equation, $\frac{d}{dt} f_{BE} = 0$. Let us also write the equation in terms of the photon energy ω instead of $x_e = \omega/T_e$,

$$\frac{m_e}{n_e\sigma_T}\frac{\partial f}{\partial t} = \frac{1}{\omega^2}\frac{\partial}{\partial\omega}\left[\omega^4\left(T_e\frac{\partial f}{\partial\omega} + f + f^2\right)\right]. \tag{10.15}$$

We finally write the Kompaneets equation in the form

$$\frac{\partial f}{\partial t} = \frac{1}{\tau_K}\frac{1}{x_e^2}\frac{\partial}{\partial x_e}\left[x_e^4\left(\frac{\partial f}{\partial x_e} + f + f^2\right)\right], \tag{10.16}$$

with

$$\tau_K = \frac{m_e}{T_e}\frac{1}{n_e\sigma_T} \simeq 10^{28}\,\mathrm{s}\,(1 - Y_{He}/2)^{-1}(\Omega_b h^2)^{-1}\frac{T}{T_e}(1+z)^{-4}, \tag{10.17}$$

where we have used

$$n_e = n_p = n_B(1 - Y_{He}/2) = \frac{3\Omega_b H_0^2(1 - Y_{He}/2)}{8\pi G m_p}.$$

Furthermore, we have set $T = T_0(1+z)$ with present photon temperature $T_0 = 2.7\,\mathrm{K}$ and m_p is the proton mass.

In equilibrium, $f = f_{BE} = 1/(e^{x_e+\mu} - 1)$, it is easy to see that the electron temperature is

$$T_e = \frac{1}{4}\frac{\int d\omega\,\omega^4 f(f+1)}{\int d\omega\,\omega^3 f}. \tag{10.18}$$

This is the electron temperature whenever the electrons are in thermal equilibrium with the photons, even if the photons are not in thermal equilibrium with the electrons. The timescale for Compton scattering of the electrons is of the order

$$t_e \simeq \frac{n_e}{n_\gamma} t_K \simeq 10^{-10} t_K,$$

which is much shorter than all other timescales involved. Therefore, instead of including a kinetic equation for the electrons, we shall always assume that they are in thermal equilibrium with the photons and therefore follow a Boltzmann distribution at temperature T_e given by Eq. (10.18). Note that it needs only a fraction of about 10^{-10} of all photons for one scattering event on each electron. This huge mismatch, $n_\gamma \gg n_e$, leads to the somewhat unusual behavior, that electrons are much more rapidly thermalized than photons.

10.1.3 Thermal Bremsstrahlung

According to the Larmor formula (see Jackson, 1975) an accelerated electron emits electromagnetic radiation. For nonrelativistic electrons, the energy emitted per unit time is

$$\frac{d\mathcal{E}}{dt} = \frac{2\alpha}{3}|\mathbf{a}(t)|^2, \tag{10.19}$$

where $\mathbf{a}(t)$ denotes the acceleration and α is the fine structure constant; see Appendix 1, Section A1.2. The radiation spectrum is obtained by Fourier transforming $\mathbf{a}$. If the period of acceleration is a short interval Δt, over which the velocity changes by an amount Δv, one obtains (see Padmanabhan, 2000)

$$\frac{d\mathcal{E}}{d\omega} = \begin{cases} \frac{2}{3\pi}\alpha(\Delta v)^2, & \omega \ll (\Delta t)^{-1} \\ 0, & \omega \gg (\Delta t)^{-1}. \end{cases}$$

We now consider an electron that passes by an ion X of charge Ze, with impact parameter b and initial velocity v. We assume the change in the velocity to be small so that we may take into account the component normal to the initial velocity only and integrate the equation of motion, $\mathbf{a} \simeq \mathbf{a}_\perp = \mathbf{e}_\perp \alpha/(m_e r^2)$, along the unperturbed path. With $|\mathbf{e}_\perp| = b/\sqrt{b^2 + (vt)^2}$, this gives the velocity change

$$\Delta v = \frac{Z\alpha}{m_e}\int_{-\infty}^{\infty} \frac{b}{[b^2 + (vt)^2]^{3/2}}\, dt = \frac{2Z\alpha}{m_e b v}.$$

The acceleration is important during a time interval of about $\Delta t = b/v$ around the closest encounter at $t = 0$. Inserting this in the above expression for the radiated energy spectrum we obtain

$$\frac{d\mathcal{E}}{d\omega} = \begin{cases} \frac{8Z^2\alpha^3}{3\pi m_e^2 v^2 b^2}, & \text{if} \quad \omega \ll v/b \\ 0, & \text{if} \quad \omega \gg v/b. \end{cases} \tag{10.20}$$

We now want to determine the energy emitted by an ion density n_i and an electron density n_e. The electron flux incident on one ion is simply $n_e v$ and the surface area with a given impact parameter b is $2\pi b\, db$. Multiplying by the ion density and integrating over the impact parameter, we obtain the energy emitted per volume per unit time and per frequency,

$$\frac{d\mathcal{E}(\omega, v)}{dV\, dt\, d\omega} = \frac{16\alpha^3 Z^2}{3m_e^2 v} n_i n_e \int_{b_{\min}}^{b_{\max}} \frac{db}{b} = \frac{16\alpha^3 Z^2}{3m_e^2 v} n_i n_e \log\left(\frac{b_{\max}}{b_{\min}}\right). \tag{10.21}$$

Here $b_{\max}$ is determined by the maximal impact parameter, which can still produce photons with frequency ω, $b_{\max} = v/\omega$. The minimum impact parameter can be estimated in two ways. First we can take it as the smallest value for which the straight line approximation that we have used to determine Δv is still reasonable; this is roughly when $\Delta v \sim v$. Inserting this in the above expression for Δv yields

$$b_{\min,1} = \frac{2Z\alpha}{m_e v^2}.$$

On the other hand, the uncertainty principle requires $p = m_e v > \hbar/b$ so that

$$b_{\min,2} = \frac{\hbar}{m_e v}.$$

The correct value for $b_{\min}$ is whichever of the two values is larger.

In general, one casts the uncertainty of this logarithmic term in a so-called Gaunt factor defined by

$$g_{\rm ff} = \frac{\sqrt{3}}{\pi} \log\left(\frac{b_{\max}}{b_{\min}}\right). \tag{10.22}$$

The correct expression for the Gaunt factor has to be obtained by a quantum mechanical treatment (see, e.g., Padmanabhan, 2000).

We now want to average Eq. (10.21) over electron velocities that follow a Maxwell distribution, $f_e \propto \exp(-m_e v^2/(2T_e))$

$$\frac{d\mathcal{E}(\omega, T)}{dV\,dt\,d\omega} = \frac{\int_{v_{\min}}^{\infty} dv \frac{d\mathcal{E}(\omega,v)}{dV\,dt\,d\omega} v^2 \exp(-m_e v^2/(2T_e))}{\int_0^{\infty} dv\, v^2 \exp(-m_e v^2/(2T_e))}. \tag{10.23}$$

Here $v_{\min}$ is the minimal velocity that can generate a photon of energy ω, $\omega = m_e v_{\min}^2/2$. Neglecting the weak velocity dependence of the Gaunt factor, the integral in the numerator is elementary and the one in the denominator simply gives the mean square velocity, $2T_e/m$, so that we arrive at

$$\frac{d\mathcal{E}(\omega, T)}{dV\,dt\,d\omega} = \frac{16\alpha^3 Z^2}{3m_e}\left(\frac{2\pi}{3m_e}\right)^{1/2} T_e^{-1/2} n_i n_e e^{-\omega/T_e} \bar{g}_{\rm ff}(T, \omega). \tag{10.24}$$

The modification of this formula obtained from a correct quantum treatment of Bremsstrahlung can be absorbed in the dimensionless Gaunt factor $\bar{g}_{\rm ff}$.

This emitted Bremsstrahlung changes the photon energy spectrum. Since, according to Eq. (1.54),

$$\frac{d\mathcal{E}_\gamma}{dV d\omega} = \frac{d\rho_\gamma}{d\omega} = \frac{1}{\pi^2}\omega^3 f,$$

we can translate Eq. (10.24) into an equation for the change of the photon distribution function. With $\sigma_T = 8\pi\alpha^2/(3m_e^2)$ we find

$$\left[\frac{df}{dt}\right]_{\rm ff\,em} = \frac{\sigma_T\, n_e}{x_e^3\, e^{x_e}}\, Q\, \bar{g}_{\rm ff}, \quad \text{with} \tag{10.25}$$

$$Q = 2\pi\sqrt{\frac{2\pi}{3}}\,\frac{\alpha}{T_e^3}\sqrt{\frac{m_e}{T_e}}\sum_i Z_i^2 n_i, \tag{10.26}$$

where we now sum over the contributions from different ion species. The "em" in the subscript in Eq. (10.25) stands for "emission." We discuss free–free absorption below. In the early Universe, we can approximate the ions by hydrogen and helium-4 only so that $\sum_i Z_i^2 n_i = n_B$. In the energy range of interest, we can approximate the Gaunt factor by (Rybicki and Lightman, 1979)

$$\bar{g}_{\rm ff} \simeq \begin{cases} \frac{\sqrt{3}}{\pi}\ln\left(0.37\, e^{\pi/\sqrt{3}}/x_e\right), & x_e \le 0.37 \\ 1, & x_e \ge 0.37. \end{cases} \tag{10.27}$$

Of course, there is not only Bremsstrahlung emission but also absorption. We can either calculate the latter directly or obtain it by the argument of detailed balance: emission and absorption have to cancel exactly in equilibrium; that is, if the photon distribution is a Planck distribution at temperature $T = T_e$, $f = 1/(e^{x_e} - 1)$. With this and using the fact that absorption is proportional to f, we obtain

$$\left[\frac{df}{dt}\right]_{\rm ff\,ab} = -\left[\frac{df}{dt}\right]_{\rm ff\,em} f(e^{x_e} - 1).$$

The kinetic equation for Bremsstrahlung then becomes

$$\left[\frac{df}{dt}\right]_{\rm ff} = \frac{\sigma_T\, n_e}{x_e^3\, e^{x_e}}\, Q\, \bar{g}_{\rm ff}\left[1 - f(e^{x_e} - 1)\right]. \tag{10.28}$$

It is convenient to write this equation as

$$\left[\frac{df}{dt}\right]_{\rm ff} = \frac{1}{\tau_{\rm ff}}\frac{\bar{g}_{\rm ff}}{x_e^3\, e^{x_e}}\left[1 - f(e^{x_e} - 1)\right], \tag{10.29}$$

with

$$\tau_{\rm ff}^{-1} = 2\pi\sqrt{\frac{2\pi}{3}}\, n_e\, \sigma_T\, n_B\, \frac{\alpha}{T_e^3}\sqrt{\frac{m_e}{T_e}} \tag{10.30}$$

$$\tau_{\rm ff} \simeq 2.3 \times 10^{23}\,{\rm s}\,(1 - Y_{He}/2)^{-1}(\Omega_b h^2)^{-2}\left[\frac{T_e}{T}\right]^{7/2}(z+1)^{-5/2}.$$

Here $T = (1+z)T_0$ is the CMB temperature.

10.1.4 Double Compton Scattering

It actually turns out that in most cosmological circumstances double Compton scattering is more efficient than Bremsstrahlung, even though the double Compton process $e + \gamma' \leftrightarrow e + \gamma_1 + \gamma$ is second order. We shall see that it is most efficient for very small energies of the second photon γ and therefore we neglect the small energy transfer, assuming that the photons γ' and γ_1 have the same frequency. The angle integrated double Compton cross section then gives (Jauch and Rorlich, 1976)

$$\frac{d\sigma_{2\gamma}}{d\omega_1\, d\omega} = \frac{4\alpha}{3\pi}\sigma_T \left(\frac{\omega'}{m_e}\right)^2 \frac{1}{\omega}\,\delta(\omega_1 - \omega'). \tag{10.31}$$

Here ω' is the energy of the incoming photon, ω_1 is the energy of the incoming photon after the collision, and ω is the energy of the photon generated by the collision.

To derive an equation for the distribution function f, we use the fact that the number density of photons in the energy interval ω and $\omega + d\omega$ is given by

$$dn(\omega) = \frac{1}{\pi^2} f(\omega)\omega^2\, d\omega.$$

With Eq. (10.31) we then obtain for double Compton emission

$$\frac{\omega^2}{\pi^2}\frac{\partial f(\omega)}{\partial t} = n_e \int d\omega_1\, d\omega' \frac{d\sigma_{2\gamma}}{d\omega_1 d\omega}\frac{\omega'^2}{\pi^2} f(\omega') \tag{10.32}$$

$$\frac{\partial f(\omega)}{\partial t} = \frac{4\alpha}{3\pi}\sigma_T\, n_e \int d\omega' \left(\frac{\omega'}{m_e}\right)^2 \frac{\omega'^2}{\omega^3} f(\omega')[f(\omega') + 1][f(\omega) + 1].$$

In the second equation we have multiplied by the factor $[f(\omega_1) + 1][f(\omega) + 1] = [f(\omega') + 1][f(\omega) + 1]$ to take into account stimulated emission.

To obtain the contribution from double Compton absorption, $\gamma + \gamma_1 + e \to \gamma' + e$, we can again use detailed balance and the fact that absorption is proportional to $f(\omega) f(\omega_1)[f(\omega') + 1]$. The full equation for double Compton scattering then takes the form

$$\left[\frac{\partial f(\omega)}{\partial t}\right]_{2\gamma} = \frac{4\alpha}{3\pi}\sigma_T\, n_e \int d\omega' \left(\frac{\omega'}{m_e}\right)^2 \frac{\omega'^2}{\omega^3}\Bigg\{ f(\omega')[f(\omega') + 1][f(\omega) + 1] - \exp\left(\frac{\omega}{T_e}\right) f(\omega') f(\omega)[f(\omega') + 1]\Bigg\}.$$

With $x_e = \omega/T_e$ and $x'_e = \omega'/T_e$ this yields

$$\left[\frac{\partial f}{\partial t}\right]_{2\gamma} = \frac{4\alpha}{3\pi}\sigma_T n_e \left(\frac{T_e}{m_e}\right)^2 \frac{1}{x_e^3}[1 - f(e^{x_e} - 1)]I$$

$$= \frac{1}{\tau_{2\gamma}} \frac{1}{x_e^3}[1 - f(e^{x_e} - 1)]I, \qquad (10.33)$$

with

$$I = \int dx'_e\, (x'_e)^4 f(x'_e)[1 + f(x'_e)],$$

and

$$\tau_{2\gamma} = \frac{3\pi}{4\alpha\, \sigma_T n_e}\left(\frac{m_e}{T_e}\right)^2 \simeq 7 \times 10^{39}\,\mathrm{s}\, \frac{1}{(1 - Y_{He}/2)\Omega_b h^2}\left(\frac{T}{T_e}\right)^2 (1+z)^{-5}. \qquad (10.34)$$

Note that for a Planck distribution I is easily obtained by integration by parts,

$$I_P = \int dx \frac{x^4 e^x}{(e^x - 1)^2} = -\int dx x^4 \partial_x \frac{1}{e^x - 1} = 4\int \frac{dx\, x^3}{e^x - 1} = 4I_b(3) = \frac{4\pi^4}{15}, \qquad (10.35)$$

where we have used Eq. (1.56) for the last two equals signs.

Strictly speaking, Eq. (10.33) is valid only for $x_e < 1$, since the double Compton cross section assumes that the energy of the photon generated by the collision is much smaller than the kinetic energy of the electron. But double Compton scattering becomes very inefficient at high energy, so that we can simply set $\left[\partial f(x_e)/\partial t\right]_{2\gamma} = 0$ for $x_e \geq 1$.

10.1.5 Timescales and Redshifts

In this section we follow Hu and Silk (1993).

Taking into account Compton scattering, double Compton scattering, and Bremsstrahlung, the kinetic equation for the photon distribution becomes

$$\frac{\partial f}{\partial t} = \left[\frac{\partial f}{\partial t}\right]_K + \left[\frac{\partial f}{\partial t}\right]_{\mathrm{ff}} + \left[\frac{\partial f}{\partial t}\right]_{2\gamma}.$$

In this equation, cosmic expansion is not taken into account. Accounting also for the dilution of photons due to expansion, we have to add a term $-3Hf$ on the right-hand side and we must take care to distinguish between cosmic time τ and conformal time t. Of course, the foregoing are derivatives with respect to cosmic time so that the kinetic equation in the expanding Universe becomes

$$\frac{\partial f}{\partial \tau} = -\frac{3}{\tau_{\rm exp}} f + \left[\frac{\partial f}{\partial t}\right]_K + \left[\frac{\partial f}{\partial t}\right]_{\rm ff} + \left[\frac{\partial f}{\partial t}\right]_{2\gamma}. \tag{10.36}$$

Here $\tau_{\rm exp} = 1/H$ is the expansion timescale, which can be approximated by

$$\tau_{\rm exp}(z) = 1/H(z) \simeq 5 \times 10^{19}\,{\rm s}\,(z + z_{\rm eq})^{-1/2}(1+z)^{-3/2}, \tag{10.37}$$

where $z_{\rm eq} = 2.4 \times 10^4 \Omega_m h^2$ is the redshift where the matter and radiation densities are equal; see Exercise 10.1. Equation (10.37) holds in a matter/radiation universe, that is, as long as the cosmological constant (or any other dark energy component) and curvature are negligible.

Let us first compare the expansion time and the timescale for Compton scattering, τ_K. Since $\left[\partial f/\partial t\right]_K$ is not simply proportional to $\tau_K^{-1} f$ one has to compare the term on the right-hand side of the Kompaneets equation (10.16) with $3f/\tau_{\rm exp}$. For Compton scattering to be efficient, one typically requires (Hu and Silk, 1993) $\tau_K/4 \le \tau_{\rm exp}$, which implies

$$z \gtrsim z_K \sim 10^4(\Omega_b h^2)^{-1/2}. \tag{10.38}$$

Here we have assumed $z_K \gg z_{\rm eq}$. At redshifts below z_K, Compton scattering drops out of equilibrium and only Thomson scattering, which does not change the photon energies, is relevant.

Also of interest is the redshift at which double Compton scattering becomes more important than Bremsstrahlung (i.e., free–free). Considering the redshift dependence of $\tau_{\rm ff}$ and $\tau_{2\gamma}$ it is clear that at high redshift double Compton is more efficient. For our estimates, we may replace the integral I in Eq. (10.33) by its value for a Planck spectrum, $I \simeq I_P \simeq 26$. Furthermore, both processes are most efficient at small $x_e \ll 1$, where Eqs. (10.29) and (10.33) with $I \simeq 26$ yield $\left[df/dt\right]_{2\gamma} \simeq 26/(\tau_{2\gamma} x_e^3)$ and $\left[df/dt\right]_{\rm ff} \simeq \bar{g}_{\rm ff}(x_e)/(\tau_{\rm ff} x_e^3)$. Equating these two terms, neglecting the small temperature difference, $T_e \simeq T$, yields

$$z_{{\rm ff},2\gamma} \sim 10^6 \left(\Omega_b h^2 \bar{g}_{\rm ff}(x_e)\right)^{2/5}, \quad x_e \ll 1. \tag{10.39}$$

Above this redshift, double Compton scattering is more efficient than free–free.

We also determine the energy at which double Compton scattering or Bremsstrahlung become as efficient as Compton scattering. For this we use the fact that

$$\left[I_P^{-1} \tau_{2\gamma} \frac{x_e^3}{e^{x_e} - 1}\right]\left[\frac{\partial f}{\partial t}\right]_{2\gamma} \sim \frac{1}{e^{x_e} - 1} - f \quad \text{and}$$

$$\left[\frac{e^{x_e}}{\bar{g}_{\rm ff}} \tau_{\rm ff} \frac{x_e^3}{e^{x_e} - 1}\right]\left[\frac{\partial f}{\partial t}\right]_{\rm ff} \sim \frac{1}{e^{x_e} - 1} - f.$$

Hence, within the timescale in square brackets, double Compton and Bremsstrahlung respectively are able to establish a Planck spectrum. Equating this timescale to the Compton timescale, $\tau_K/4$, we obtain a redshift dependent energy $x_c = \omega_c/T_e$, below which double Compton and Bremsstrahlung respectively are efficient. Assuming $x_c \ll 1$ we obtain

$$I_P^{-1}\tau_{2\gamma}x_{c,2\gamma}^2 = \tau_K/4 \tag{10.40}$$

so that with Eq. (10.35)

$$x_{c,2\gamma}(z) = \frac{\pi^2}{\sqrt{15}}\left(\frac{\tau_K}{\tau_{2\gamma}}\right)^{1/2} \simeq 3\times 10^{-6}\sqrt{z+1}, \tag{10.41}$$

$$x_{c,\text{ff}}(z) \simeq 77\,(z+1)^{-3/4}(\Omega_b h^2)^{1/2}. \tag{10.42}$$

At energies below $x_c \equiv \sqrt{x_{c,2\gamma}^2 + x_{c,\text{ff}}^2}$, photon number changing processes are efficient and a Planck spectrum can be established, if $z > z_K$ so that Compton scattering is still efficient. At energies above x_c, photons settle into a Bose–Einstein distribution, if $z > z_K$.

Correspondingly, comparing double Compton scattering and Bremsstrahlung with expansion, we find that at a given redshift z double Compton scattering or Bremsstrahlung is still efficient only for photon energies with $x_e < x_{\text{exp},2\gamma}(z)$ or $x_e < x_{\text{exp},\text{ff}}(z)$ respectively with

$$x_{\text{exp},2\gamma}(z) = 4.3\times 10^{-10}\frac{z^{7/4}}{(z+z_{\text{eq}})^{1/4}}(\Omega_b h^2)^{1/2}(1-Y_{He}/2)^{1/2}, \tag{10.43}$$

$$x_{\text{exp},\text{ff}}(z) = 1.1\times 10^{-2}\frac{z^{1/2}}{(z+z_{\text{eq}})^{1/4}}\Omega_b h^2(1-Y_{He}/2)^{1/2}. \tag{10.44}$$

The energies below which photon number changing processes are still faster than expansion and can lead to establishing a thermal equilibrium are given by $x_e < x_{\text{exp}} \equiv \sqrt{x_{\text{exp},2\gamma}^2 + x_{\text{exp},\text{ff}}^2}$.

In Fig. 10.1 we plot z_K, $z_{\text{ff},2\gamma}(x_e)$ as well as $x_c(z)$ and $x_{\text{exp}}(z)$. When $x_c < 0.1$, say if $z < 10^8$, a Planck spectrum is established rapidly only for very small energies, $\omega < 0.1\times T_e$, while at larger energies we first obtain a Bose–Einstein distribution. Nevertheless, if there is a short period of injection of photons at a redshift where $x_{\text{exp}}(z) \gtrsim 1$, a Planck spectrum will be established eventually for the relevant regime of energies with $x_e \lesssim 1$. We may therefore say that such processes still are fully thermalized if they happen sufficiently earlier than $z \sim 10^7$. However, if energy injection happens at $z \leq 10^7$, the Planck spectrum cannot be established anymore for energies with $x_e > x_{\text{exp}}(z) \leq 1$.

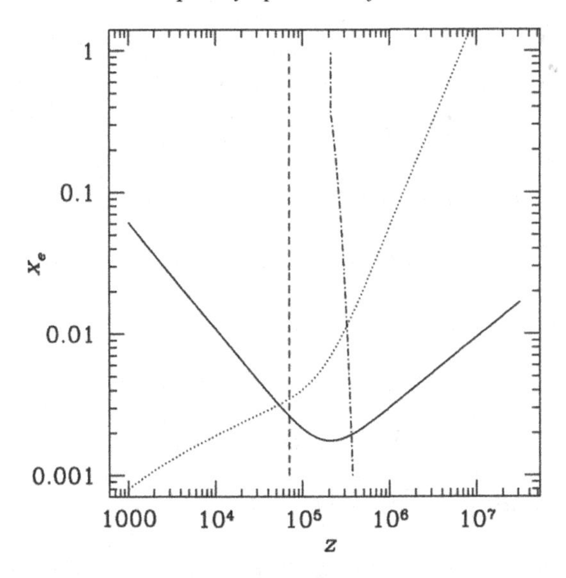

Fig. 10.1 The redshifts z_K (dashed) and $z_{\rm ff,2\gamma}(x_e)$ (dash-dotted) are plotted together with the energies $x_c = \omega_c/T_e$ (solid) and $x_{\rm exp} = \omega_{\rm exp}/T_e$ (dotted).

It is also interesting to note that in the regime where Bremsstrahlung is more efficient than double Compton scattering, $z < z_{\rm ff,2\gamma} \simeq 2 \times 10^5$, these processes are very inefficient anyway and can thermalize the spectrum only for $x_e < x_{\rm exp}(2 \times 10^5) \simeq 0.01$; the bulk part of the spectrum remains Bose–Einstein. Therefore, it is a good approximation to disregard Bremsstrahlung entirely in these considerations.

10.2 A Chemical Potential

Observational studies of the CMB spectrum have shown that it is very close to a blackbody, that is, Planck spectrum; see Fig. 1.7. Up to this day, no deviations from a blackbody have been detected. The experimental bound for the reduced chemical potential comes mainly from the FIRAS experiment aboard the COBE satellite. It limits μ to (Fixsen *et al.*, 1996)

$$|\mu| \leq 9 \times 10^{-5} \text{ at 95\% confidence.} \tag{10.45}$$

This seems very small, but as we shall see, a quite violent event is needed to generate such a chemical potential. This comes from the fact that there are many more photons in the Universe than baryons or (most probably) dark matter particles. Therefore, producing about one photon per dark matter particle will not induce a large chemical potential.

As we have seen in the previous section, if energy is injected into the CMB at a redshift $z_1 > z_K$ with $x_{\rm exp}(z_1) < 1$, we expect a spectral distortion that leads to a chemical potential on energies with $\omega/T_e \equiv x_e > x_{c,2\gamma}(z_1)$. In this section we want to estimate not only the chemical potential produced by a given energy input $\delta\rho$, but also the timescale at which this chemical potential is established. First we assume that the energy input happens rapidly at some redshift z_1. Furthermore, we neglect the part of the integrals over photon energies in which the spectrum has been able to relax to a Planck spectrum, that is, $x < x_{\rm exp}(z_1)$. We must therefore assume $z_1 \lesssim z_\mu \simeq 6 \times 10^6$ (see Fig. 10.1).

This can happen, for example via a "long" lived, unstable particle that decays at redshift z_1. Some models of supersymmetry predict the existence of a "next-to-lightest" supersymmetric particle that is rather long lived and decays quite late into the lightest supersymmetric particle that then plays the role of dark matter. But also the annihilation of a very light particle with mass $m \simeq T_0(1+z_1) \simeq 2.3 \times 10^{-4}(1+z_1)$ eV, when the temperature drops below its mass threshold, can induce a chemical potential.

According to what we have learned in the previous section, since $z_1 > z_K$, at $x_e > x_c$ a Bose–Einstein distribution is established rapidly at some temperature T_e and with chemical potential μ. The temperature T_e and the chemical potential μ are determined by

$$\rho = \frac{1}{\pi^2}T_e^4 \int \frac{x_e^3\,dx_e}{e^{x_e+\mu}-1} = \rho_\gamma + \delta\rho = \frac{\pi^2}{15}T^4(1+\epsilon), \quad \text{and} \tag{10.46}$$

$$n = \frac{1}{\pi^2}T_e^3 \int \frac{x_e^2\,dx_e}{e^{x_e+\mu}-1} = n_\gamma + \delta n = \frac{2\zeta(3)}{\pi^2}T^3(1+\alpha). \tag{10.47}$$

For the last equals signs we have introduced $\epsilon \equiv \delta\rho/\rho_\gamma$ and $\alpha \equiv \delta n/n_\gamma$. For ρ_γ and n_γ we use the expressions for a Planck spectrum given in Eqs. (1.54) and (1.62). For $z_1 \ll 10^7$, photon number changing processes are no longer active and $\alpha \equiv 0$. But here we keep the expressions general. Experimentally we know that $\mu \ll 1$. Expanding the integrals in Eqs. (10.46) and (10.47) to first order in μ yields

$$\rho = \frac{T_e^4}{\pi^2}\left[I_b(3) - 3\mu I_b(2)\right] = \frac{\pi^2}{15}T^4(1+\epsilon) \tag{10.48}$$

$$n = \frac{T_e^4}{\pi^2}\left[I_b(2) - 2\mu I_b(1)\right] = \frac{2\zeta(3)}{\pi^2}T^3(1+\alpha). \tag{10.49}$$

Here $I_b(n) = \int_0^\infty dx\,\frac{x^n}{e^x-1} = \Gamma(n+1)\zeta(n+1)$ as defined in Eq. (1.55). Inserting $I_b(3) = \pi^4/15$, $I_b(2) = 2\zeta(3)$ and $I_b(1) = \pi^2/6$, we can write these equations as

$$\mu = \frac{6\zeta(3)}{\pi^2}\left[1 - \left(\frac{T}{T_e}\right)^3(1+\alpha)\right] \quad \text{and} \quad 1 - \left(\frac{T}{T_e}\right)^4(1+\epsilon) = \frac{90\zeta(3)}{\pi^4}\mu.$$

Since $|\mu| \ll 1$ and $\zeta(3) \simeq 1.2$, we must have $|1 - T/T_e| \ll 1$. We therefore expand $T_e/T = 1 + \delta$ with $|\delta| \ll 1$ so that $(T_e/T)^n \simeq 1 + n\delta$. Inserting this approximation above, we can determine δ and μ in terms of ϵ and α. The above relations give (we neglect the second-order terms $\propto \delta\epsilon$ and $\delta\alpha$)

$$\mu = \frac{18\zeta(3)}{\pi^2}(\delta - \alpha/3)$$

$$4\delta - \epsilon = \frac{90\zeta(3)}{\pi^4}\mu,$$

so that

$$\delta = \frac{\epsilon - (540\zeta(3)^2/\pi^6)\alpha}{4[1 - 405\zeta(3)^2/\pi^6]} \simeq 0.64\frac{\delta\rho}{\rho} - 0.52\frac{\delta n}{n}, \tag{10.50}$$

$$\mu = \frac{3\zeta(3)}{2\pi^2[1 - 405\zeta(3)^2/\pi^6]}(3\epsilon - 4\alpha) \simeq 0.46\left(3\frac{\delta\rho}{\rho_\gamma} - 4\frac{\delta n}{n_\gamma}\right). \tag{10.51}$$

First of all, we note that when photon number changing processes are still very rapid, the photon number will change so that $\delta n/n_\gamma = (3/4)(\delta\rho/\rho_\gamma)$ and no chemical potential is generated. The temperature is then modified by the injection of energy to $T \to T(1 + \delta) = T_e = T(1 + \epsilon/4)$, which is evident since in this case $\rho/\rho_\gamma = (T_e/T)^4$.

The situation is very different if $z_1 < z_\mu$ and photon number changing processes are no longer active. Then $\delta n = 0$ and

$$\mu \simeq 1.4\frac{\delta\rho}{\rho_\gamma}. \tag{10.52}$$

First of all, the chemical potential generated by such an energy injection is always positive. This is good, since a distribution with a negative chemical potential is not well defined for frequencies $\omega \le \omega_c = -T_e\mu$. But since we know that double Compton scattering is still active at very low frequencies, this is not a real problem, as at these low frequencies a Planck spectrum would be established anyway.

Let us now estimate the chemical potential generated by the decay of a species of nonrelativistic particles that contributes an energy density $\rho_X/\rho_{c0} = \Omega_X (1 + z)^3$ before they decay. Here ρ_{c0} is the critical energy density today. Since photons contribute the energy density $\rho_\gamma/\rho_{c0} = \Omega_\gamma(1 + z)^4$, assuming that a fraction f of the energy of these particles is heating up the CMB we have

$$\mu = 1.4\frac{\delta\rho}{\rho_\gamma} = 1.4\frac{f\Omega_X}{\Omega_\gamma(1 + z_1)}, \tag{10.53}$$

where z_1 denotes the redshift of the decay. This formula is of course valid only if $z_1 < z_\mu \sim 10^7$, since an energy injection at higher redshift is still fully thermalized.

Using $\Omega_\gamma = 5 \times 10^{-5}$ (see Appendix 1, Section A1.3), the limit on the chemical potential can be translated into a limit for $f\Omega_X$,

$$f\Omega_X \leq 3 \times 10^{-3} \left(\frac{1+z_1}{10^6}\right), \quad z_1 < 10^7. \tag{10.54}$$

This might appear as a small number; nevertheless it is more than the entire mass density in stars. If the decay product of the particle species X is supposed to be the dark matter, we need $\Omega_X \simeq 0.3$. The above bound then implies $f < 0.01(z_1/10^6)$; hence only a small fraction of the energy may be injected into standard model particles (other than neutrinos). If a particle decays at a redshift $z_1 \simeq 10^7$ partial thermalization, especially at small frequencies, $x_e \lesssim 1$ can still be achieved. In this case, the Boltzmann equation (10.36) with a source term describing the injection has to be solved numerically and the resulting "chemical potential" depends on the frequency.

For a particle species that decays into photons when the temperature goes below its mass threshold we would expect $\delta\rho/\rho \sim 0.1$–1, so that it would produce a chemical potential of order unity. This shows that no particle with mass $m < T_0(1+z_\mu) \simeq 230$ eV that interacts significantly with photons can exist. Of course such a particle would also be produced in accelerators, so that this does not come as a surprise.

Let us now study the situation in somewhat more detail also for low frequencies. We know that at $z > z_K$, at low frequency double Compton scattering is still active and a Planck spectrum is established while at high frequencies a chemical potential μ develops. In complete generality we may express the distribution function with a frequency-dependent chemical potential $\tilde{\mu}(x_e)$,

$$f(x_e) = \frac{1}{\exp[x_e + \tilde{\mu}(x_e)] - 1}. \tag{10.55}$$

Once the equilibrium is established we have (neglecting Bremsstrahlung which never dominates)

$$0 = \left[\frac{\partial f}{\partial t}\right]_K + \left[\frac{\partial f}{\partial t}\right]_{2\gamma} \tag{10.56}$$

$$= \frac{1}{\tau_K}\frac{1}{x_e^2}\frac{\partial}{\partial x_e}\left[x_e^4\left(\frac{\partial f}{\partial x_e} + f + f^2\right)\right] + \frac{1}{\tau_{2\gamma}}\frac{1}{x_e^3}[1 - f(e^{x_e} - 1)]I, \tag{10.57}$$

or, inserting our ansatz for f,

$$\frac{d}{dx_e}\left[x_e^4 \frac{\exp[x_e + \tilde{\mu}(x_e)]}{(\exp[x_e + \tilde{\mu}(x_e)] - 1)^2}\frac{d\tilde{\mu}}{dx_e}\right] = \frac{\tau_K}{\tau_{2\gamma}} I \frac{e^{x_e}}{x_e} \frac{\exp[\tilde{\mu}(x_e)] - 1}{\exp[x_e + \tilde{\mu}(x_e)] - 1}. \tag{10.58}$$

If we neglect the chemical potential in I we may replace $I\tau_K/\tau_{2\gamma}$ by $4x_c^2$ according to Eq. (10.40). We want to solve this equation for small $x_e \ll 1$ and to lowest order in $\tilde{\mu} \ll x_e$. In this approximation Eq. (10.58) becomes

$$2x_e \frac{d\tilde{\mu}}{dx_e} + x_e^2 \frac{d^2\tilde{\mu}}{dx_e^2} = 4\frac{x_c^2}{x_e^2}\tilde{\mu}, \tag{10.59}$$

which is solved by

$$\tilde{\mu}(x_e) = \mu \exp(-2x_c/x_e). \tag{10.60}$$

The integration constant μ is determined by the asymptotic regime, $x_e \gg x_c$, where we know that a Bose–Einstein distribution with chemical potential μ establishes and we have chosen the boundary condition $\tilde{\mu}(0) = 0$.

This shows how the chemical potential develops from very small energies, $x_e \ll x_c$, where a Planck spectrum is established to high energies, $x_e \gg 2x_c$ where a Bose–Einstein spectrum with chemical potential μ is formed.

Let us also estimate the time it takes for the Bose–Einstein spectrum to decay again due to double Compton scattering; that is, let us derive an evolution equation for μ after energy injection. For this we take the time derivatives of Eqs. (10.48) and (10.49),

$$\dot{\rho} = \left(4\frac{\dot{T}_e}{T_e} - \frac{3I_b(2)\dot{\mu}}{I_b(3) - 3\mu I_b(2)}\right)\rho \tag{10.61}$$

$$\dot{n} = \left(3\frac{\dot{T}_e}{T_e} - \frac{2I_b(1)\dot{\mu}}{I_b(2) - 2\mu I_b(1)}\right)n. \tag{10.62}$$

We neglect expansion here. We could introduce expansion-corrected temperatures, $T_e \to aT_e$; particle density, $n \to a^3 n$; and energy density, $\rho \to a^4\rho$, but we shall simply not consider expansion in what follows. Therefore, when there is no energy injection ρ no longer changes, $\dot{\rho} = 0$, and we can solve the above equations for $\dot{\mu}$;

$$\dot{\mu} = 4\frac{\dot{n}}{nB}, \qquad B = \frac{9I_b(2)}{I_b(3) - 3\mu I_b(2)} - \frac{8I_b(1)}{I_b(2) - 2\mu I_b(1)} \simeq 2.14, \tag{10.63}$$

where we have neglected μ in B. Inserting

$$f = \frac{1}{\exp[x_e + \mu e^{2x_c/x_e}] - 1} \quad \text{in} \quad \partial_\tau n = \frac{T_e^3}{\pi^2}\int dx_e x_e^2 \left[\partial_\tau f\right]_{2\gamma} \tag{10.64}$$

we find to lowest order in μ

$$\partial_\tau n/n = -\mu \frac{I}{I_b(2)\tau_{2\gamma}} J_{BE} \tag{10.65}$$

$$J_{BE}(x_c) = \int_0^1 \frac{dx}{x}\frac{e^x \exp[-2x_c/x]}{e^x - 1}. \tag{10.66}$$

We perform the integral only until $x = 1$ because it is dominated at small x. Note that for $x_c = 0$ the integral actually diverges. In Exercise 10.3 we show that for small $x_c \ll 1$, $J_{BE}(x_c) \simeq (2x_c)^{-1}$ up to logarithmic corrections. Using also $I = 4I_b(3)$ we find

$$4\frac{\partial_\tau n}{nB} = -\frac{8I_b(3)}{BI_b(2)}\frac{1}{x_c\tau_{2\gamma}}\mu = -\frac{\mu}{\tau_\mu} \tag{10.67}$$

$$\tau_\mu = \frac{BI_b(2)}{8I_b(3)}x_c\tau_{2\gamma} \simeq 2.1 \times 10^{33} s\frac{(1+z)^{-9/2}}{(1-Y_{He}/2)\Omega_b h^2}. \tag{10.68}$$

Hence

$$\partial_\tau \mu = -\frac{\mu}{\tau_\mu}. \tag{10.69}$$

In the timescale τ_μ the chemical potential is significantly damped. For low redshift, this timescale is much larger than the age of the Universe, $\tau_0 \sim 3 \times 10^{17}$s, but at high redshift, $z > 10^7$, this happens relatively fast. Let us consider a redshift z_*, in the radiation-dominated era at which a chemical potential μ_* is induced. We want to determine the remaining chemical potential today, at $z = 0$. In the radiation-dominated era, $\tau \propto a^2 \propto 1/(1+z)^2$ and we can integrate Eq. (10.69) to

$$\mu(0) \simeq \mu_* \exp\left[-\left(\frac{z_*}{z_\mu}\right)^{5/2}\right], \qquad z_\mu \simeq 4 \times 10^5[(1-Y_{He}/2)\Omega_b h^2]^{-2/5}. \tag{10.70}$$

Hence if $z_* \lesssim z_\mu$, the chemical potential survives while if $z_* \gg z_\mu$ it is washed out by subsequent double Compton scattering at low energy and upscattering of the low energy photons by Compton scattering.

10.2.1 The Chemical Potential from Silk Damping

If there is continuous energy injections leading to a chemical potential and wash out by double Compton scattering, the chemical potential evolves as

$$\partial_\tau \mu = -\frac{\mu}{\tau_\mu} + 1.4\frac{Q}{\rho}. \tag{10.71}$$

Here Q is the energy injection rate and we have used Eq. (10.52). Using the method of "variation of the constant" we obtain the solution to this equation from the homogeneous solution (10.70),

$$\mu(z) = 1.4\int_0^{\tau(z)} d\tau\frac{Q}{\rho}\exp\left[-\left(\frac{z_*}{z_\mu}\right)^{5/2}\right]. \tag{10.72}$$

There is a process that injects energy into the photon–baryon plasma during cosmic evolution: Silk damping; see Section 4.6. The energy in the small-scale fluctuations that are removed by Silk damping is redistributed to the baryon-photon plasma during this process. We now estimate the chemical potential induced by Silk damping. According to Eq. (4.171), a k-mode of the photon–baryon plasma evolves according to

$$D(k,t) = \Delta_0(k)\cos(\omega_R t)\exp\left(-\int_0^t dt'\omega_I(t')\right), \tag{10.73}$$

with

$$\omega_R = c_s k = \frac{k}{\sqrt{3(R+1)}} \qquad \omega_I(t) = \frac{k^2 t_c}{6}\frac{R^2 + \frac{4}{5}(R+1)}{(R+1)^2}, \tag{10.74}$$

where $t_c^{-1} = n_e\sigma_T a$ is the Thomson scattering rate and $R \equiv 3\rho_b/4\rho_r$. The energy density in a plane acoustic wave is given by $\epsilon \simeq c_s^2\rho\langle D^2\rangle$, where $\rho = \rho_b + \rho_r$. Since μ distortions are generated predominantly in the radiation-dominated era, we can set $R = 0$ and $c_s^2 = 1/3$ so that

$$\langle D^2\rangle \simeq \frac{\Delta_0^2(k)}{2}\exp[-(k/k_D(t))^2], \tag{10.75}$$

where

$$k_D^{-2}(t) = \frac{4}{15\sigma_T}\int_0^t \frac{dt}{n_e a}. \tag{10.76}$$

The full energy injection rate from all the modes is

$$\frac{Q}{\rho} \simeq -c_s^2\int\frac{d^3k}{(2\pi)^3}\partial_\tau\langle D^2\rangle. \tag{10.77}$$

In the radiation-dominated era $1/(n_e a) \propto t^2$; hence $k_D^{-2}(t) \propto t^3$, which implies

$$\partial_\tau\langle D^2\rangle = a^{-1}\partial_t\langle D^2\rangle = -\frac{3k^2}{atk_D^2}\langle D^2\rangle, \tag{10.78}$$

so that

$$\frac{Q}{\rho} \simeq \frac{3c_s^2}{2atk_D^2}\int\frac{d^3k}{(2\pi)^3}k^2\Delta_0^2\exp[-(k/k_D(t))^2]. \tag{10.79}$$

Let us determine $\Delta_0^2(k)$. For this we consider very early times, when Silk damping is not yet relevant and $D = \Delta_0\cos(c_s kt)$. In the radiation-dominated era [Eq. (2.150)] with $q = 1$ and neglecting the decaying mode gives

$$\Psi = \frac{A}{a}j_1(c_s kt) \simeq \frac{A}{ac_s kt}\cos(c_s kt), \tag{10.80}$$

where the $\simeq$ sign holds well inside the Hubble scale, $c_s kt \gg 1$. The Einstein 00 constraint equation (2.105) for a flat universe then gives

$$\Delta_0 = -\frac{2k^2}{3\mathcal{H}^2}\frac{A}{ac_s kt}. \tag{10.81}$$

In a radiation-dominated universe where $a \propto t$ and $\mathcal{H} = 1/t$, this is a (in general k-dependent) constant. To relate it to the amplitude of the primordial power spectrum we consider also the limit $t \to 0$ of Eq. (10.80),

$$\lim_{t\to 0} \Psi = \frac{Ac_s kt}{3a},$$

which is again time independent. Introducing the primordial power spectrum of Ψ we obtain the relation

$$k^3 P_\Psi = A_\Psi (k/k_*)^{n_s-1} = \left(\frac{Ac_s kt}{3a}\right)^2 k^3. \tag{10.82}$$

Introducing this back in Δ_0 we find

$$k^3 \Delta_0^2 = 36 A_\Psi (k/k_*)^{n_s-1}. \tag{10.83}$$

Now we can perform the k-integration in Eq. (10.79) with the result

$$\frac{Q}{\rho} \simeq \frac{36}{4\pi^2 at} A_\Psi (k_D(t)/k_*)^{(n_s-1)} \Gamma\left(\frac{n_s+1}{2}\right). \tag{10.84}$$

The integration over $d\tau = adt$ can be approximated by an evaluation of the function at the decay redshift z_μ and a multiplication by $at(z_\mu)$ that leads to the result

$$\mu_{\text{Silkdamping}} \simeq 10^{-8}\left(\frac{k_D(z_\mu)}{k_*}\right)^{n_s-1}. \tag{10.85}$$

Here we have also introduced $A_\Psi = 2\pi^2(9/25)A_s$ and $A_s \simeq 2.1 \times 10^{-9}$.

This result is nearly independent of the Silk damping scale, but let us nevertheless determine $k_D(t)$. The integral (10.76) is elementary and yields

$$k_D^{-2}(t) = \frac{4}{45\sigma_T}\frac{t}{n_e a} = \frac{4}{45\sigma_T}\frac{1}{\mathcal{H} n_e a}. \tag{10.86}$$

Inserting $\mathcal{H} = H_0\sqrt{\Omega_r}(1+z)$ and $n_e a = (1 - Y_{He}\Omega_b h^2)(1+z)^2$ we obtain

$$k_D(z) \simeq 0.36 \times 10^{-4}(1 - Y_{He}\Omega_b h^2)^{1/2}(1+z)^{3/2}\text{Mpc}^{-1}. \tag{10.87}$$

The detection of this chemical potential would not only be expected, but it would also confirm that inflation has generated a nearly scale-invariant spectrum up to

$$k_D(z_\mu) \simeq 10^4\text{Mpc}^{-1}. \tag{10.88}$$

This is presently the only known method to test the primordial power spectrum on these very small scales. A measurement of the chemical potential of the CMB to the precision of 10^{-8} would require a satellite experiment. Several such experiments are presently under study; see Chluba *et al.* (2019) for a brief review.

10.3 The Sunyaev–Zel'dovich effect

Clusters of galaxies are permeated by a hot plasma of electrons and nuclei at a temperature of several keV. This is much hotter than CMB photons at redshifts $z \lesssim 1$. Therefore, in the Kompaneets equation (10.15) the first term on the right-hand side dominates.

Furthermore, the plasma is optically thin to Compton scattering so that we can neglect multiple scattering. The change of the photon distribution when passing through a cluster is then simply

$$\delta f = y\omega^{-2}\frac{\partial}{\partial\omega}\left(\omega^4\frac{\partial f}{\partial\omega}\right), \tag{10.89}$$

where we have introduced the Compton-y parameter

$$y \equiv \sigma_T \int n_e \frac{T_e}{m}\, dr. \tag{10.90}$$

The integral extends through the cluster, and the approximation holds, if the optical depth $\tau = \sigma_T \int n_e\, dr \ll 1$. Inserting a blackbody spectrum, $f \propto (\exp(\omega/T)-1)^{-1}$, we obtain with $x \equiv \omega/T \neq \omega/T_e$

$$\frac{\delta f}{f} = -y\frac{xe^x}{e^x-1}\left[4 - x\coth\left(\frac{x}{2}\right)\right] \simeq \begin{cases} -2y, & \text{if } x \ll 1 \\ yx^2, & \text{if } x \gg 1. \end{cases} \tag{10.91}$$

This is the frequency dependent Sunyaev–Zel'dovich (SZ) effect. With the help of

$$\frac{\delta T}{T} = \frac{f}{T}\frac{\delta f}{f}\left(\frac{df}{dT}\right)^{-1} = \frac{e^x-1}{xe^x}\frac{\delta f}{f},$$

we can translate it in a frequency-dependent temperature shift

$$\frac{\delta T}{T} = -y\left[4 - x\coth\left(\frac{x}{2}\right)\right] \simeq \begin{cases} -2y, & \text{if } x \ll 1 \\ yx, & \text{if } x \gg 1. \end{cases} \tag{10.92}$$

When passing through a hot thermal plasma, the low-energy Rayleigh–Jeans regime of the photons' spectrum is depleted and the spectrum is enhanced at high energies, in the Wien tail. Photons are on average up-scattered in energy. The spectral change vanishes at $x_0 \simeq 3.8$ given by $0 = 4 - x_0\coth(x_0/2)$; see Fig. 10.2. For a CMB temperature of $T = 2.726$ this corresponds to a frequency of $\nu = 217$ GHz.

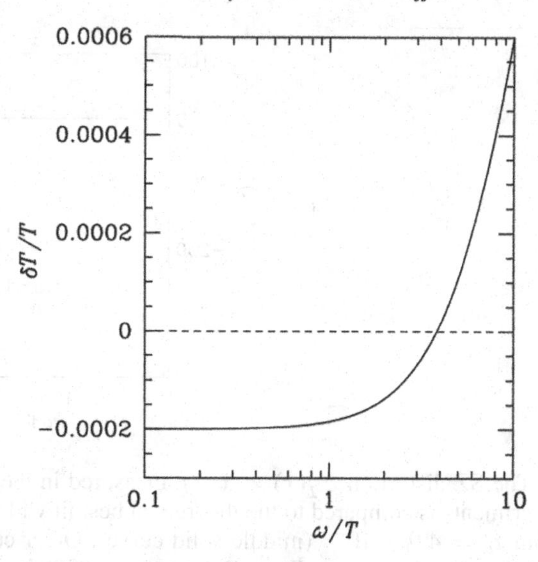

Fig. 10.2 The function $\delta T/T$ given in Eq. (10.92) is shown for $y = 10^{-4}$. Note that it passes through zero roughly at $x = \omega/T \simeq 3.8$.

Numbers for a typical cluster are $T_e \simeq 7$ keV, $n_e \sim 10^{-2}\,\text{cm}^{-3}$, and the diameter is of the order of $R \simeq 300$ kpc. Estimating the Compton-y parameter by $y \sim y_c = (\sigma_T T_e/m_e)n_e R$ we obtain $y_c \simeq 10^{-4}$. This is the order of magnitude of the SZ effect in clusters. Of course the true value can deviate substantially and depends on the details of the cluster (De Petris *et al.*, 2002; LaRoque *et al.*, 2006). At present, SZ observers use the effect to generate cluster maps of the quantity $T_e n_e$ integrated along the line of sight. Not surprisingly, one of the best studied clusters so far is the Coma cluster, which is closest to us (De Petris *et al.*, 2002; Planck Coll. X, 2013). The SZ results for the Coma cluster are shown in Fig. 10.3. In general, the situation is complicated because clusters are complicated objects that may have different "electron populations," one of them thermal and others not. Also the electron density might be a complicated function of position, and so forth. Here we are not entering into all of these interesting difficulties of cluster physics, but we just mention some important points.

- The SZ effect in clusters can add to the CMB anisotropies on very small scales and may, depending on the cluster number density which is not well known, even dominate it above $\ell \sim 3000$. Fortunately it can be distinguished from primordial anisotropies due to its spectral signature; see Fig. 10.2.
- The motion of clusters induces in addition to the thermal SZ effect discussed above a temperature shift due to the bulk motion of the cluster,

$$\delta T/T = v_c \sigma_T \int n_e\, dr = v_c \tau_e,$$

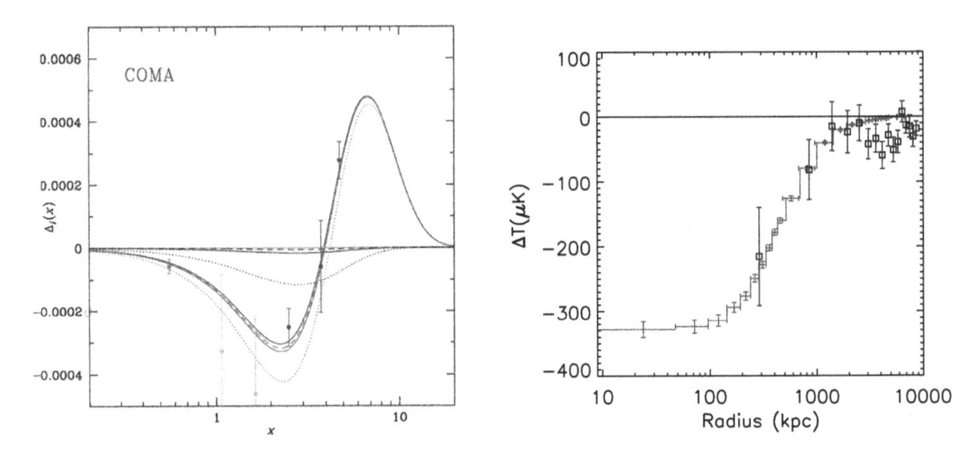

Fig. 10.3 *Left:* The SZ distortion $\Delta_j(x) = x^3\,\delta f$ measured in the Coma cluster by different experiments is compared to the theoretical best fit with $T_e = 8.2$ keV and optical depth $\tau_e = 4.9 \times 10^{-3}$ (middle solid curve). Other curves that add possible other effects are also shown. From Colafrancesco (2007).
Right: Data points from or the y-profile of the Coma cluster. $\Delta T = -yT_e$ are plotted as function of the distance from the center. From Planck Coll. X (2013)

where v_c is the bulk velocity in the direction of the line of sight and τ_e is the optical depth of the cluster. This effect, which is spectrally identical to primordial CMB anisotropies, is typically several times smaller than the thermal SZ effect. It goes under the name "kinetic SZ effect."

- The average effect from all clusters should contribute a mean Compton-y parameter in the Universe. Its amplitude strongly depends on the cluster distribution, but is estimated to be of the order of $\bar{y} \sim 10^{-7}$. This number is within reach of planned CMB spectrum experiments (Singal *et al.*, 2002; Kogut *et al.*, 2007).
- Clearly the SZ effect in clusters is very interesting for cluster physics and the distribution of clusters in redshift. Together with X-ray observations that probe the square of the electron density, it allows us, in principle, to gain detailed information about the electron density and temperature distribution inside the clusters. Furthermore, the SZ effect, which represents the "shadow" of the cluster in the CMB, is independent of redshift and has led to the detection of many new clusters with high redshift that are too faint to be seen in optical or X-ray telescopes Planck Coll. XXII (2016).
- Using the specific frequency signature of the thermal SZ, the Planck team had even produced sky maps and the full sky angular power spectrum of the SZ signal in the CMB Planck Coll. XXII (2016).

The fact that the observed average Compton-y parameter is so small, $y \leq 10^{-5}$, leads to a limit on early reionization. Let us derive this limit for a simple toy model.

We assume that the Universe is reionized at some redshift $z_{\rm ri}$. Then during the ionization process, the electrons also gain some kinetic energy that we estimate to be typically in the 10 eV range. This seems reasonable, if we do not want to assume that the reionizing photons have exactly the reionization energy, 13.6 eV, but some energy in this ballpark. The remaining energy is then simply absorbed by the electron as kinetic energy. As the Universe evolves, since the electron momenta are redshifted $p \propto 1/a$, the temperature, which is a measure of the kinetic energy of the electrons, is also redshifted, $T_e \simeq p^2/2m \propto a^{-2}$. Already in Chapter 1 in our discussion below Eq. (1.93) we have seen that the temperature of nonrelativistic particles decays like $1/a^2$. Denoting the electron temperature at reionization by $T_{\rm ri}$ we find that reionization should induce a Compton-y parameter given by

$$y = \frac{\sigma_T T_{\rm ri} n_e(t_0)}{m_e(z_{\rm ri}+1)^2} \int_{t_{\rm ri}}^{t_0} dt\,(z+1)^5. \tag{10.93}$$

We expect $z_{\rm ri}$ to lie in the matter-dominated phase of expansion and before the cosmological constant becomes relevant. During matter domination the Friedmann equation yields

$$dt = \frac{-dz}{(1+z)^{5/2}} \frac{1}{H_0\sqrt{\Omega_m}},$$

so that we obtain

$$\begin{aligned} y &= \frac{\sigma_T T_{\rm ri} n_e(t_0)}{m_e H_0 \sqrt{\Omega_m}(z_{\rm ri}+1)^2} \int_0^{z_{\rm ri}} dz\,(z+1)^{5/2} = \frac{2(1+z_{\rm ri})^{3/2}\sigma_T T_{\rm ri} n_e(t_0)}{7 m_e H_0\sqrt{\Omega_m}} \\ &\simeq 5\times 10^{-7} \frac{\Omega_b h^2}{\sqrt{\Omega_m h^2}}(1+z_{\rm ri})^{3/2}\left(\frac{T_{\rm ri}}{10\,{\rm eV}}\right). \end{aligned} \tag{10.94}$$

For $\Omega_m h^2 = 0.13$ and $\Omega_b h^2 \cong 0.022$ the limit $y < 10^{-5}$ translates into the reionization redshift

$$z_{\rm ri} < 50\left(\frac{T_{\rm ri}}{10\,{\rm eV}}\right)^{2/3}. \tag{10.95}$$

This is a truly interesting number and it reduces by a factor $(10^5 y)^{2/3}$ if we lower the limit on the y parameter. Already a y parameter $y < 10^{-6}$ would require a reionization redshift of $z_{\rm ri} < 10 \times (T_{\rm ri}/10\text{ eV})^{2/3}$. As discussed in Chapter 9 the CMB polarization spectrum favors $z_{\rm ri} \sim 8$. Reducing the assumed electron temperature by a factor of 10 can help, but when $y < 10^{-7}$ we have no simple way out. Therefore, according to our understanding of the reionization process that took place probably at $6.5 < z_{\rm ri} \simeq 8$, this should have led to a global Compton-y parameter of the order of $y \simeq 10^{-7}$–10^{-6}.

Exercises

(The exercise marked with an asterisk is solved in Appendix 11 which is not in this printed book but can be found online.)

10.1 **The Hubble parameter in a matter/radiation universe and the collision times**

Using the Friedmann equation show that in a spatially flat universe containing only matter and radiation, the Hubble parameter is given by

$$H^2(z) = H_0^2\Omega_r(1+z)^3(z+z_{\text{eq}}+2), \tag{10.96}$$

$$H \simeq 2\times 10^{-20}\,\text{s}^{-1}(2+z+z_{\text{eq}})^{1/2}(1+z)^{3/2}, \tag{10.97}$$

where $1+z_{\text{eq}} = \Omega_m/\Omega_r \simeq 2.4\times 10^4\Omega_m h^2$ is the redshift where the matter and radiation densities are equal. We use the relativistic density parameter for photons and three species of massive neutrinos, $\Omega_r = 4.19\times 10^{-5}\,h^{-2}$ and $H_0 = 3.24\times 10^{-18}\,h\,\text{s}^{-1}$, given in Appendix 1, Sections A1.2 and A1.3.

Using $n_e = (1-Y_{He}/2)n_B$ also verify Eqs. (10.17), (10.30), and (10.34).

10.2 **Aberration**

For a given nonrelativistic electron velocity $\mathbf{v}$ and incoming and outgoing photon directions $\mathbf{n}$ and $\mathbf{n}'$ in the laboratory frame determine the scalar product $\mathbf{n}_R\cdot\mathbf{n}'_R$ of the photon directions in the electron rest frame.

10.3 **An integral***

Show that

$$J_{BE}(x_c) = \int_0^1 \frac{dx}{x}\frac{e^x\exp[-2x_c/x]}{e^x-1} = \frac{1}{2x_c} - \frac{1}{2}\log(x_c) + \text{higher order}, \tag{10.98}$$

where “higher order” denotes terms which remain finite when $x_c \to 0$.

Final Remarks

The goal of this book has been to give you an overview of one of the most successful and fascinating topics of cosmology, the physics of the cosmic microwave background. Its success is best illustrated by the two Nobel Prizes the subject led to: Penzias and Wilson (1978) for the discovery of the CMB and Mather and Smoot (2006) for the detailed measurement of its spectrum and for the discovery of the fluctuations.

I have concentrated on the theoretical side of the topic not only because this is my expertise, but also because I believe that this subject is mature enough for a textbook. On the experimental side, certainly there is much to tell and tremendous progress has been made in the past 15 years, but I think, hope, that this is not the end of it. There will be much more to come and therefore a book on CMB experiments could be only a snapshot of the present situation. The theory of the CMB, on the other hand, is in many of its aspects basically complete, so that I can hope that this book may have some lasting value for students who want to learn about the topic and also for researchers in the field who want to obtain a rather detailed overview.

I am afraid that despite a big effort there are still some misprints or errors in the book. If you, dear reader, have spotted one, please let me know (ruth.durrer@unige.ch) so that I can correct it in forthcoming editions.

Appendix 1

Fundamental Constants, Units and Relations

Here we summarize some useful relations between units and the values of physical constants that are used throughout this book.

A1.1 Conversion Factors, Units

In a system of units where $\hbar = c = k_{\text{Boltzmann}} = 1$, as is often used in this book, all units can be expressed in terms of a unit of energy such as, for example, the GeV or a length scale such as, for example, cm. We then have

$$
\begin{aligned}
1\,\text{GeV} &= 1.6022 \times 10^{-3}\,\text{erg} \\
&= 1.1605 \times 10^{13}\,\text{K} \\
&= 1.7827 \times 10^{-24}\,\text{g} \\
&= 5.0677 \times 10^{13}\,\text{cm}^{-1} \\
&= 1.5192 \times 10^{24}\,\text{s}^{-1}
\end{aligned}
$$

The relation I always remember by heart for order of magnitude estimates is $1 = 200$ MeV fm. Here "fm" is 1 femtometer (or fermi); 1 fm $= 10^{-15}$ m.

Other useful relations are

$$
\begin{aligned}
1\ \text{parsec (pc)} &= 3.2612\ \text{light years} \ = 3.0856 \times 10^{18}\,\text{cm} \\
1\ \text{Mpc} &= 10^{6}\,\text{pc} \simeq 3 \times 10^{24}\text{cm} \simeq 10^{14}\,\text{s} \\
1\,\text{g}\,\text{cm}^{-3} &= 4.3102 \times 10^{-18}\,\text{GeV}^{4} \\
\text{(Astronomical unit)}\ \ 1\ \text{AU} &= 1.4960 \times 10^{13}\,\text{cm} \\
1\ (\text{Gauss})^{2}/8\pi &= 1.9084 \times 10^{-40}\,\text{GeV}^{4} \\
\text{(Jansky)}\ \ 1\ \text{Jy} &= 10^{-23}\,\text{erg cm}^{-2}\,\text{s}^{-1}\,\text{Hz}^{-1} \\
&= 2.4730 \times 10^{-48}\,\text{GeV}^{3} \\
1\ \text{yr} &\simeq \pi \times 10^{7}\,\text{s} \\
\text{(Radian)}\ \ 1\ \text{rad} &= (180/\pi)\ \text{degrees} = 57.296\ \text{degrees} \\
\text{(Steradian)}\ \ 1\ \text{sr} &= 1\,\text{rad}^{2} \ = \ 3.283 \times 10^{3}\ \text{degrees}^{2}
\end{aligned}
$$

A1.2 Constants

A1.2.1 Fundamental Constants

Planck's constant $\hbar = 1 \; = h/(2\pi)$

$= 1.0546 \times 10^{-27}\,\mathrm{cm^2\,g\,s^{-1}}$

Speed of light $c = 1 \; = \; 2.9979 \times 10^{10}\,\mathrm{cm\,s^{-1}}$

Fine structure constant $\alpha \equiv \dfrac{e^2}{4\pi} = 1/137.036$

Gravitational constant $G = \; 6.673 \times 10^{-8}\,\mathrm{cm^3 g^{-1} s^{-2}}$

Planck mass $m_P = \; 1.2211 \times 10^{19}\,\mathrm{GeV}$

$= \; 2.1768 \times 10^{-5}\,\mathrm{g}$

Planck length $\ell_P = \; 8.189 \times 10^{-20}\,\mathrm{GeV^{-1}}$

$= \; 1.616 \times 10^{-33}\,\mathrm{cm}$

Planck time $\tau_P = \; 8.189 \times 10^{-20}\,\mathrm{GeV^{-1}}$

$= \; 5.3904 \times 10^{-44}\,\mathrm{s}$

Electron mass $m_e = \; 0.5110\,\mathrm{MeV}$

Proton mass $m_p = \; 938.27\,\mathrm{MeV}$

Neutron mass $m_n = \; 939.57\,\mathrm{MeV}$

Rydberg $1\,\mathrm{Ry} = \alpha^2 m_e/4\pi = 13.606\,\mathrm{eV}$

Thomson cross section $\sigma_T \equiv 8\pi\alpha^2/3m_e^2 = 6.65246 \times 10^{-25}\,\mathrm{cm^2}$

Bohr radius $a_0 \equiv \dfrac{1}{\alpha m_e} = 5.2918 \times 10^{-9}\,\mathrm{cm}$

Bohr magneton $\mu_0 \equiv \dfrac{e}{2m_e} = 5.7884 \times 10^{-18}\,\dfrac{\mathrm{GeV}}{\mathrm{Gauss}}$

Avogadro's number $N_A = \; 6.022 \times 10^{23}$

Stefan–Boltzmann constant $a_{\mathrm{SB}} \equiv \pi^2/15 = 0.658$

$= 7.566 \times 10^{-15}\,\mathrm{erg\,cm^{-3}\,K^{-4}}$

A1.2.2 Important Constants

Solar mass $M_\odot = \; 1.989 \times 10^{33}\,\mathrm{g} = 1.116 \times 10^{57}\,\mathrm{GeV}$

Solar radius $R_\odot = \; 6.9598 \times 10^{10}\,\mathrm{cm} = 3.527 \times 10^{24}\,\mathrm{GeV^{-1}}$

Luminosity of the Sun $L_\odot = \; 3.90 \times 10^{33}\,\mathrm{erg\,s^{-1}} = 1.6 \times 10^{12}\,\mathrm{GeV^2}$

Mass of Earth $M_\oplus = \; 5.977 \times 10^{27}\,\mathrm{g}$

$= \; 3.357 \times 10^{51}\,\mathrm{GeV}$

Solar magnitude $m_\odot = \; -26.85$ (apparent)

$\mathcal{M}_\odot = \; 4.72$ (absolute)

Distance modulus $m - \mathcal{M} = 5\log(D/10\,\mathrm{pc})$

Hubble constant $H_0 = \; 100\,h\,\mathrm{km\,s^{-1}\,Mpc^{-1}}$

$= \; 2.1332\,h \times 10^{-42}\,\mathrm{GeV}$

where $0.65 < h < 0.75$

Hubble time, distance	$H_0^{-1} =$	$3.0856 \times 10^{17}\, h^{-1}$ s
	$=$	$9.7776 \times 10^{9}\, h^{-1}$ yr
	$=$	$2997.9\, h^{-1}$ Mpc
	$=$	$9.2503 \times 10^{27} h^{-1}$ cm
Critical density	$\rho_c = 3H_0^2/8\pi G =$	$1.8791\, h^2 \times 10^{-29}$ g cm^{-3}
	$=$	$8.0992\, h^2 \times 10^{-47}$ GeV^4
	$=$	$1.0540\, h^2 \times 10^{4}$ eV cm^{-3}
	$=$	$11.2\, h^2$ (proton masses)/m^3
	$=$	$0.277 \times 10^{12} M_\odot/\mathrm{Mpc}^3$
CMB temperature	$T_0 =$	2.72548 ± 0.00057 K
	$=$	2.35×10^{-13} GeV
Neutrino temperature	$T_\nu =$	$1.945\,\mathrm{K} = (4/11)^{1/3} T_0$

A1.3 Useful Relations

Photons

Number density	$n_\gamma =$	411 cm^{-3}
Entropy density	$s_\gamma =$	1480 cm^{-3} $= 3.602 n_\gamma$
En ergy density	$\rho_\gamma =$	2.01×10^{-51} GeV^4
Density parameter	$\Omega_\gamma h^2 =$	2.48×10^{-5}

Neutrino (per species)

Number density	$n_\nu =$	112 cm^{-3}
Entropy density	$s_\nu =$	470 cm^{-3} $= 4.202 n_\nu$
Energy density (massless)	$\rho_\nu =$	4.565×10^{-52} GeV^4
Density parameter (massless)	$\Omega_\nu h^2 =$	5.63×10^{-6}
Density parameter (massive)	$\Omega_\nu h^2 =$	$m_\nu/(94\mathrm{eV})$

Relativistic entropy	$s_0 =$	2900 cm^{-3} $= s_\gamma + 3s_\nu$
Relativistic density parameter	$\Omega_{\gamma 3\nu} h^2 =$	4.17×10^{-5}
Baryon density	$\Omega_B h^2 =$	$3.639 \times 10^{7} \eta_B$
	$=$	0.02230 ± 0.00014
Baryons per photon $n_B/n_\gamma =$	$\eta_B =$	$(6.0 \pm 0.5) \times 10^{-10}$
Age of the Universe		(conformal time is for $a_0 = 1$)
for $T > T_{\mathrm{eq}}$:	$\tau =$	$2.42\,\mathrm{s} \times (1\,\mathrm{MeV}/T)^2/\sqrt{g_{\mathrm{eff}}}$
	$=$	$0.30118 (m_P/T^2)/\sqrt{g_{\mathrm{eff}}}$
	$t =$	$1.7 \times 10^{10}\mathrm{s} \times (1\,\mathrm{MeV}/T) g_{\mathrm{eff}}^{-1/6}$
	$=$	$0.489 \dfrac{m_P}{T_0 T} g_{\mathrm{eff}}^{-1/6}$
for $T < T_{\mathrm{eq}}$		

$$\text{(neglecting } \Lambda \text{ and } K\text{):} \quad \tau = \frac{2.057 \times 10^{17}}{\sqrt{\Omega_m h^2}}(1+z)^{-3/2}\,\text{s}$$

$$= \frac{7.504 \times 10^{11}}{\sqrt{\Omega_m h^2}}(T/1\,\text{eV})^{-3/2}\,\text{s}$$

$$t = \frac{6.17 \times 10^{17}}{\sqrt{\Omega_m h^2}}(1+z)^{-1/2}\,\text{s}$$

$$= \frac{9.46 \times 10^{15}}{\sqrt{\Omega_m h^2}}(T/1\,\text{eV})^{-1/2}\,\text{s}$$

Matter density $\Omega_m h^2 = 0.142 \pm 0.0015$

Equivalence redshift $z_{\text{eq}} = 2.4 \times 10^4(\Omega_m h^2)$

Equivalence temperature $T_{\text{eq}} = 5.6\,\text{eV}(\Omega_m h^2)$

Decoupling redshift $z_{\text{dec}} \simeq 1090 \pm 0.3$

Decoupling temperature $T_{\text{dec}} \simeq 2974 \pm 1\,\text{K} = 0.26\,\text{eV}$

Decoupling time $\tau_{\text{dec}} \simeq 2.65 \times 10^{13}(0.14/\Omega_m h^2)^{1/2}\,\text{s}$

$t_{\text{dec}} \simeq 2.9 \times 10^{16}(0.14/\Omega_m h^2)^{1/2}\,\text{s}$

Recombination redshift $z_{\text{rec}} \simeq 1360$

Nucleosynthesis temperature $T_{\text{nuc}} \simeq 0.08\,\text{MeV} = 9 \times 10^8\,\text{K}$

Time of nucleosynthesis $\tau_{\text{nuc}} \simeq 206\,\text{s}$

$t_{\text{nuc}} \simeq 1.35 \times 10^9\,\text{s}$

Age of the Universe $\tau_0 = (1.3807 \pm 0.03) \times 10^{10}\,\text{yr}$

$t_0 = (4.659 \pm 0.08) \times 10^{10}\,\text{yr}$

Appendix 2

General Relativity

Throughout this book it is assumed that the reader is familiar with the basics of general relativity as presented, for example, in Wald (1984). This appendix does not present an introduction to general relativity but just fixes the notation used throughout this book. Furthermore, we calculate the curvature tensor for an FL universe.

A2.1 Notation

We consider a four-dimensional pseudo-Riemannian spacetime given by a manifold $\mathcal{M}$ and a metric g with signature $(-,+,+,+)$. For a given choice of coordinates $(x^\mu)_{\mu=0}^3$ the metric is given by the 10 components of a 4×4 symmetric tensor,

$$g = ds^2 = g_{\mu\nu}\,dx^\mu\,dx^\nu. \qquad \text{(A2.1)}$$

Contra- and covariant tensor fields on a pseudo-Riemannian manifold are equivalent. Their indices can be lowered and raised with the metric, for example,

$$g_{\beta\nu}T^{\alpha\nu} = T^\alpha_{\ \beta} = g^{\alpha\mu}T_{\mu\beta}. \qquad \text{(A2.2)}$$

Here $g^{\alpha\mu}$ is the inverse of the metric such that $g^{\alpha\mu}g_{\mu\beta} = \delta^\alpha_\beta$, and we adopt Einstein's summation convention: indices that appear as subscripts and superscripts are summed over.

The Christoffel symbols are defined by

$$\Gamma^\mu_{\alpha\beta} = \frac{1}{2}g^{\mu\nu}\left[\partial_\alpha g_{\nu\beta} + \partial_\beta g_{\nu\alpha} - \partial_\nu g_{\alpha\beta}\right]. \qquad \text{(A2.3)}$$

Here ∂_μ indicates a partial derivative w.r.t. the coordinate x^μ, this is sometimes also simply denoted by a comma, $\partial_\mu f \equiv f_{,\mu}$. Covariant derivatives are indicated by a semicolon, or by the symbol ∇.

A geodesic $\gamma(t)$ with $X = \dot\gamma$ is a solution to the differential equation

$$\nabla_X X = 0, \quad X^\mu\partial_\mu X^\nu + \Gamma^\nu_{\alpha\beta}X^\alpha X^\beta = \frac{d^2\gamma^\mu}{ds^2} + \Gamma^\mu_{\alpha\beta}\frac{d\gamma^\alpha}{ds}\frac{d\gamma^\beta}{ds} = 0, \qquad \text{(A2.4)}$$

where the second equation expresses the first equation in components. The vector field $X = \dot\gamma$ is given by $X = X^\mu\partial_\mu = \frac{d}{ds}$. We often conveniently identify a vector field with the partial derivative in its direction. A tensor field T of rank (p,q) is parallel transported along the vector field X if

$$\nabla_X T = 0, \quad X^\mu T^{\alpha_{i_1}\cdots\alpha_{i_p}}_{\beta_{j_1}\cdots\beta_{j_q};\mu} = 0. \tag{A2.5}$$

Covariant derivatives of a tensor field are given by

$$T^{\alpha_{i_1}\cdots\alpha_{i_p}}_{\beta_{j_1}\cdots\beta_{j_q};\mu} = T^{\alpha_{i_1}\cdots\alpha_{i_p}}_{\beta_{j_1}\cdots\beta_{j_q},\mu} + \Gamma^{\alpha_{i_1}}_{\mu\sigma} T^{\sigma\cdots\alpha_{i_p}}_{\beta_{j_1}\cdots\beta_{j_q}} + \cdots - \Gamma^{\sigma}_{\mu\beta_{j_1}} T^{\alpha_{i_1}\cdots\alpha_{i_p}}_{\sigma\cdots\beta_{j_q}} - \cdots. \tag{A2.6}$$

The Riemann curvature tensor is defined by

$$R^\alpha{}_{\beta\mu\nu} = \Gamma^\alpha_{\nu\beta,\mu} - \Gamma^\alpha_{\mu\beta,\nu} + \Gamma^\rho_{\beta\nu}\Gamma^\alpha_{\mu\rho} - \Gamma^\rho_{\beta\mu}\Gamma^\alpha_{\nu\rho}. \tag{A2.7}$$

The tensor $R_{\alpha\beta\mu\nu} = g_{\alpha\sigma} R^\sigma_{\beta\mu\nu}$ is antisymmetric in the first $(\alpha\beta)$ and second $(\mu\nu)$ pair of indices and symmetric in the exchange of the pairs, $(\alpha\beta) \leftrightarrow (\mu\nu)$. The Bianchi identities read

$$\Sigma_{(\beta\mu\nu)} R^\alpha{}_{\beta\mu\nu} = 0 \qquad \text{1st Bianchi identity.} \tag{A2.8}$$

$$\Sigma_{(\mu\nu\sigma)} R^\alpha{}_{\beta\mu\nu;\sigma} = 0 \qquad \text{2nd Bianchi identity.} \tag{A2.9}$$

Here $\Sigma_{(\beta\mu\nu)}$ denotes the sum over all cyclic permutations of these three indices.

The Ricci tensor and the Riemann scalar are given by

$$R_{\mu\nu} = R^\alpha{}_{\mu\alpha\nu}, \qquad R = R^\mu_\mu = R_{\mu\nu} g^{\mu\nu}. \tag{A2.10}$$

With these sign conventions, the curvature of the sphere is positive, and changing the order of covariant derivatives of a vector field X yields

$$\nabla_\mu \nabla_\nu X^\alpha - \nabla_\nu \nabla_\mu X^\alpha = R^\alpha{}_{\sigma\mu\nu} X^\sigma. \tag{A2.11}$$

The Einstein tensor is defined as

$$G_{\mu\nu} = R_{\mu\nu} - \frac{1}{2} g_{\mu\nu} R. \tag{A2.12}$$

The second Bianchi identity and the symmetries of the Riemann tensor imply $G^\nu_{\mu;\nu} = 0$.

The field equations of general relativity relate the curvature to the energy–momentum tensor $T_{\mu\nu}$ via Einstein's equation,

$$G_{\mu\nu} = 8\pi G T_{\mu\nu}, \tag{A2.13}$$

where G denotes Newton's constant, $G = m_P^{-2}$. The second Bianchi identity ensures that $T_{\mu\nu}$ is covariantly conserved, $T^\nu_{\mu;\nu} = 0$. Equation (A2.13) can also be derived from an action principle with

$$S = S_{\text{grav}} + S_{\text{mat}}.$$

Here S_{mat} is the usual matter action and

$$S_{\text{grav}} = \frac{m_P^2}{16\pi} \int d^4x \sqrt{-g} R \tag{A2.14}$$

is the Hilbert action. A somewhat tedious but standard calculation gives (see, e.g., Wald, 1984)

$$\delta S_{\rm grav} = -\frac{m_P^2}{16\pi}\int d^4x\,\sqrt{-g}G^{\mu\nu}\delta g_{\mu\nu}. \tag{A2.15}$$

The Einstein equation implies then that the energy–momentum tensor can be obtained by varying the matter action w.r.t. the metric,

$$\sqrt{-g}T^{\mu\nu} = 2\frac{\delta S_{\rm mat}}{\delta g_{\mu\nu}}.$$

By construction, this energy–momentum tensor is always symmetric, but it does, in general, not agree with the canonical energy–momentum tensor. Of course the conserved quantities (if any!) are the same for both definitions.

The Weyl tensor specifies the degrees of freedom of the Riemann tensor that are not determined by the Ricci tensor (or Einstein tensor). It is the traceless part of $R^\alpha_{\beta\mu\nu}$. In n dimensions, $n \geq 3$, it is given by

$$C_{\alpha\beta\mu\nu} = R_{\alpha\beta\mu\nu} - \frac{2}{n-2}\left(g_{\alpha[\mu}R_{\nu]\beta} + g_{\beta[\mu}R_{\nu]\alpha}\right) - \frac{2}{(n-1)(n-2)}Rg_{\alpha[\mu}g_{\nu]\beta}. \tag{A2.16}$$

Here $[\mu\nu]$ denotes antisymmetrization in the indices μ and ν. The Weyl tensor has the same symmetries as the Riemann tensor but all its traces vanish. It describes the degrees of freedom of the curvature (gravitational field) in source-free spacetime; hence it describes gravitational waves.

An introduction to general relativity can be found, for example, in the books by Straumann (2004) or Wald (1984).

A2.2 The Lie Derivative

For a vector field X with flow ϕ_t^X the Lie derivative of a tensor field T of arbitrary rank is defined by

$$L_XT = \lim_{\epsilon\to 0}\frac{1}{\epsilon}\left(\left(\phi_\epsilon^X\right)^* T - T\right). \tag{A2.17}$$

Here $\left(\phi_\epsilon^X\right)^*$ denotes the pullback of the map $\phi_t^X : \mathcal{M} \to \mathcal{M} : p \mapsto \gamma_p(t)$, where γ_p is the integral curve to X with starting point p. The existence and uniqueness of solutions to ordinary differential equations tells us that for sufficiently small t, ϕ_t^X is a local diffeomorphism. If $T(t)$ denotes the value of the tensor field T at the position $\gamma_p(t)$ we also have

$$L_XT(p) = \left.\frac{d}{dt}\right|_{t=0} T(t). \tag{A2.18}$$

Hence the Lie derivative in direction X vanishes if the tensor field T is conserved along integral curves of X. Furthermore, for small t we have

$$\left(\phi_t^X\right)^* T = T + tL_XT + \mathcal{O}(t^2). \tag{A2.19}$$

In coordinates the Lie derivative becomes (see, e.g., Wald, 1984)

$$L_X T^{\alpha_{i_1}\cdots\alpha_{i_p}}_{\beta_{j_1}\cdots\beta_{j_q}} = X^\mu T^{\alpha_{i_1}\cdots\alpha_{i_p}}_{\beta_{j_1}\cdots\beta_{j_q}},_\mu - X^{\alpha_{i_1}},_\sigma T^{\sigma\cdots\alpha_{i_p}}_{\beta_{j_1}\cdots\beta_{j_q}} - \cdots$$
$$+ X^\sigma,_{\beta_{j_1}} T^{\alpha_{i_1}\cdots\alpha_{i_p}}_{\sigma\cdots\beta_{j_q}} + \cdots . \qquad \text{(A2.20)}$$

For a vector field Y this reads

$$(L_X Y)^\alpha = X^\mu \partial_\mu Y^\alpha - Y^\mu \partial_\mu X^\alpha \equiv [X, Y]. \qquad \text{(A2.21)}$$

A2.3 Friedmann Metric and Curvature

The Friedmann metric is given by

$$ds^2 = g_{\mu\nu}\, dx^\mu\, dx^\nu = -d\tau^2 + a^2(\tau)\gamma_{ij}\, dx^i\, dx^j = a^2(t)[-dt^2 + \gamma_{ij}\, dx^i\, dx^j]. \qquad \text{(A2.22)}$$

The Christoffel symbols with respect to cosmic or conformal time are

	Cosmic time τ	Conformal time t	
$\Gamma^0_{00} =$	0	$\frac{\dot a}{a}$,	(A2.23)
$\Gamma^i_{00} =$	0	0,	(A2.24)
$\Gamma^0_{i0} =$	0	0,	(A2.25)
$\Gamma^i_{j0} =$	$\frac{a'}{a}\delta^i_j = H\delta^i_j$	$\frac{\dot a}{a}\delta^i_j = \mathcal{H}\delta^i_j$,	(A2.26)
$\Gamma^0_{ij} =$	$a' a \gamma_{ij}$	$\frac{\dot a}{a}\gamma_{ij}$,	(A2.27)
$\Gamma^k_{ij} = {}^{(3)}\Gamma^k_{ij} =$	$\frac{1}{2}\gamma^{km}\left(\gamma_{im,j} + \gamma_{jm,i} - \gamma_{ij,m}\right)$	${}^{(3)}\Gamma^k_{ij}$,	(A2.28)

where ${}^{(3)}\Gamma^k_{ij}$ denotes the three-dimensional Christoffel symbols of the metric γ that depend on the coordinate system chosen on the spatial slices. The overdot indicates a derivative w.r.t. conformal time t while the prime indicates a derivative w.r.t. cosmic time τ.

The nonvanishing components of the Riemann and Ricci curvature tensors in **cosmic time** τ are then given by

$$R^0_{i0j} = a'' a \gamma_{ij}, \qquad \text{(A2.29)}$$

$$R^i_{00j} = \frac{a''}{a}\delta^i_j, \qquad \text{(A2.30)}$$

$$R^i_{jkm} = {}^{(3)}R^i_{jkm} + (a')^2\left(\delta^i_k \gamma_{jm} - \delta^i_m \gamma_{jk}\right), \qquad \text{(A2.31)}$$

$$R_{00} = -3\frac{a''}{a}, \qquad \text{(A2.32)}$$

$$R_{ij} = \left[a'' a + 2\left(a'^2 + K\right)\right]\gamma_{ij}, \qquad \text{(A2.33)}$$

$$R = 6\left[\frac{a''}{a} + H^2 + \frac{K}{a^2}\right], \qquad \text{(A2.34)}$$

while in **conformal time** t we have

$$R^0_{i0j} = \left(\frac{\dot{a}}{a}\right)^{\cdot} \gamma_{ij} = \dot{\mathcal{H}}\gamma_{ij}\,, \tag{A2.35}$$

$$R^i_{00j} = \left(\frac{\dot{a}}{a}\right)^{\cdot} \delta^i_j = \dot{\mathcal{H}}\delta^i_j, \tag{A2.36}$$

$$R^i_{\ jkm} = {}^{(3)}R^i_{\ jkm} + \mathcal{H}^2\left(\delta^i_k\gamma_{jm} - \delta^i_m\gamma_{jk}\right), \tag{A2.37}$$

$$R_{00} = -3\left(\frac{\dot{a}}{a}\right)^{\cdot} = -3\dot{\mathcal{H}}, \tag{A2.38}$$

$$R_{ij} = \left[\dot{\mathcal{H}} + 2\left(\mathcal{H}^2 + K\right)\right]\gamma_{ij}, \tag{A2.39}$$

$$R = \frac{6}{a^2}\left[\dot{\mathcal{H}} + \mathcal{H}^2 + K\right]. \tag{A2.40}$$

The curvature of the metric γ_{ij} on the three-dimensional slices of constant time is given by

$${}^{(3)}R^i_{\ jkm} = K\left(\delta^i_k\gamma_{jm} - \delta^i_m\gamma_{jk}\right), \tag{A2.41}$$

$${}^{(3)}R_{ij} = 2K\gamma_{ij} \qquad \text{and} \tag{A2.42}$$

$${}^{(3)}R = 6K. \tag{A2.43}$$

Appendix 3

Perturbations

In this appendix we present the intermediate results in the calculation of the perturbed Einstein equations for a given "Fourier mode" k. We also determine the Weyl tensor. All the results are for conformal time t.

A3.1 Scalar Perturbations

We work in the longitudinal gauge,

$$ds^2 = a^2\left(-(1+2\Psi Q^{(S)})dt^2 + (1-2\Phi Q^{(S)})\gamma_{ij}\,dx^i\,dx^j\right). \qquad \text{(A3.1)}$$

Here $Q^{(S)}$ is an eigenfunction of the spatial Laplacian with eigenvalue $-k^2$ (see Section 2.2.2).

A3.1.1 The Christoffel Symbols

$$\delta\Gamma^0_{00} = \dot\Psi Q^{(S)}, \quad \delta\Gamma^0_{0j} = -k\Psi Q^{(S)}_j, \qquad \text{(A3.2)}$$

$$\delta\Gamma^j_{00} = -k\Psi Q^{(S)j}, \quad \delta\Gamma^j_{i0} = -\dot\Phi\delta^j_i Q^{(S)}, \qquad \text{(A3.3)}$$

$$\delta\Gamma^0_{ij} = \left[-2\mathcal{H}(\Psi+\Phi) - \dot\Phi\right]Q^{(S)}\gamma_{ij}, \qquad \text{(A3.4)}$$

$$\delta\Gamma^j_{im} = k\Phi\left[\delta^j_i Q^{(S)}_m + \delta^j_m Q^{(S)}_i - \gamma_{im}Q^{(S)j}\right]. \qquad \text{(A3.5)}$$

A3.1.2 The Riemann Tensor

$$\delta R^0_{00j} = \delta R^0_{0ij} = 0, \qquad \text{(A3.6)}$$

$$\delta R^0_{i0j} = -\left[2\dot{\mathcal{H}}(\Psi+\Phi) + \mathcal{H}(\dot\Psi+\dot\Phi) + \ddot\Phi - \frac{k^2}{3}\Psi\right]\gamma_{ij}Q^{(S)} - k^2\Psi Q^{(S)}_{ij}, \qquad \text{(A3.7)}$$

$$\delta R^0_{ijm} = -k\left[\mathcal{H}\Psi + \dot\Phi\right]\left(\gamma_{ij}Q^{(S)}_m - \gamma_{im}Q^{(S)}_j\right), \qquad \text{(A3.8)}$$

$$\delta R^i_{00j} = \left[\frac{k^2}{3}\Psi - \mathcal{H}(\dot\Psi+\dot\Phi) - \ddot\Phi\right]\delta^i_j Q^{(S)} - k^2\Psi Q^{(S)i}_j, \qquad \text{(A3.9)}$$

$$\delta R^i_{0jm} = -k\left[\dot{\Phi} + \mathcal{H}\Psi\right]\left(\delta^i_j Q^{(S)}_m - \delta^i_m Q^{(S)}_j\right), \tag{A3.10}$$

$$\delta R^i_{j0m} = k\left[\mathcal{H}\Psi + \dot{\Phi}\right]\left(\delta^i_m Q^{(S)}_j - \gamma_{jm} Q^{(S)i}\right), \tag{A3.11}$$

$$\begin{aligned}\delta R^i_{jmn} = &-2\left[\mathcal{H}^2(\Psi + \Phi) + \mathcal{H}\dot{\Phi} + \frac{1}{3}k^2\Phi\right]\left(\delta^i_m \gamma_{jn} - \delta^i_n \gamma_{jm}\right) Q^{(S)} \\ &- k^2\Phi\left(\delta^i_n Q^{(S)}_{jm} - \delta^i_m Q^{(S)}_{jn} + Q^{(S)i}_n \gamma_{jm} - Q^{(S)i}_m \gamma_{jn}\right).\end{aligned} \tag{A3.12}$$

A3.1.3 The Ricci and Einstein Tensors

The perturbation if the Ricci tensor is

$$\delta R_{00} = \left[3\mathcal{H}(\dot{\Psi} + \dot{\Phi}) - k^2\Psi + 3\ddot{\Phi}\right] Q^{(S)}, \tag{A3.13}$$

$$\delta R_{0j} = -2k\left[\mathcal{H}\Psi + \dot{\Phi}\right] Q^{(S)}_j, \tag{A3.14}$$

$$\begin{aligned}\delta R_{ij} = &\left[-2(\dot{\mathcal{H}} + 2\mathcal{H}^2)(\Psi + \Phi) - \mathcal{H}\dot{\Psi} + \frac{k^2}{3}\Psi - \ddot{\Phi} - 5\mathcal{H}\dot{\Phi}\right. \\ &\left. - \frac{4}{3}k^2\Phi\right]\gamma_{ij} Q^{(S)} + k^2(\Phi - \Psi) Q^{(S)}_{ij}.\end{aligned} \tag{A3.15}$$

The perturbation of the Riemann scalar then becomes

$$\delta R = -\frac{2}{a^2}\left[6(\dot{\mathcal{H}} + \mathcal{H}^2)\Psi + 3\mathcal{H}\dot{\Psi} - k^2\Psi + 9\mathcal{H}\dot{\Phi} + 3\ddot{\Phi} + 2(k^2 - 3K)\Phi\right] Q^{(S)}. \tag{A3.16}$$

For the Einstein tensor we find

$$\delta G^0_0 = \frac{2}{a^2}\left[3\mathcal{H}^2\Psi + 3\mathcal{H}\dot{\Phi} + (k^2 - 3K)\Phi\right] Q^{(S)}, \tag{A3.17}$$

$$\delta G^0_j = \frac{2}{a^2}k\left[\mathcal{H}\Psi + \dot{\Phi}\right] Q^{(S)}_j, \tag{A3.18}$$

$$\delta G^j_0 = -\frac{2}{a^2}k\left[\mathcal{H}\Psi + \dot{\Phi}\right] Q^{(S)j}, \tag{A3.19}$$

$$\begin{aligned}\delta G^i_j = &\frac{2}{a^2}\left[(2\dot{\mathcal{H}} + \mathcal{H}^2)\Psi + \mathcal{H}\dot{\Psi} - \frac{k^2}{3}\Psi + \ddot{\Phi} + 2\mathcal{H}\dot{\Phi} + \left(\frac{k^2}{3} - K\right)\Phi\right]\delta^i_j Q^{(S)} \\ &+ \frac{k^2}{a^2}(\Phi - \Psi) Q^{(S)i}_j.\end{aligned} \tag{A3.20}$$

A3.1.4 The Weyl Tensor

The Weyl tensor from scalar perturbations only has an "electric" component, that is, all the components are determined by

$$C^0_{i0j} \equiv -E_{ij} = \frac{k^2}{2}(\Phi + \Psi) Q^{(S)}_{ij}. \tag{A3.21}$$

More precisely we have

$$C_{0i0j} = a^2 E_{ij}, \tag{A3.22}$$
$$C_{0ijk} = 0, \tag{A3.23}$$
$$C_{ijk\ell} = g_{ik} E_{j\ell} + g_{j\ell} E_{ik} - g_{jk} E_{i\ell} - g_{i\ell} E_{jk}. \tag{A3.24}$$

A3.2 Vector Perturbations

We work in the vector gauge defined in Eq. (2.60),

$$ds^2 = a^2 \left(-dt^2 + 2\sigma Q_i^{(V)}\, dt dx^i + \gamma_{ij}\, dx^i\, dx^i\right), \tag{A3.25}$$

where $\sigma^i = \sigma Q^{(V)i}$ is divergence-free and $Q_{ij}^{(V)} = -\frac{1}{2k}\left(Q_{i|j}^{(V)} + Q_{j|i}^{(V)}\right)$.

A3.2.1 *The Christoffel Symbols*

$$\delta\Gamma^0_{00} = 0, \qquad \delta\Gamma^0_{0j} = \mathcal{H}\sigma Q_j^{(V)}, \tag{A3.26}$$
$$\delta\Gamma^j_{00} = [\dot\sigma + \mathcal{H}\sigma] Q^{(V)j}, \quad \delta\Gamma^j_{i0} = \frac{1}{2}\sigma\left(Q^{(V)j}_{\ |i} - Q_i^{(V)|j}\right), \tag{A3.27}$$
$$\delta\Gamma^0_{ij} = k\sigma Q_{ij}^{(V)}, \qquad \delta\Gamma^j_{im} = -\mathcal{H}\sigma\gamma_{im} Q^{(V)j}. \tag{A3.28}$$

A3.2.2 *The Riemann Tensor*

$$\delta R^0_{\ 00j} = \dot{\mathcal{H}}\sigma Q^{(V)j}, \qquad \delta R^0_{\ 0ij} = 0, \tag{A3.29}$$
$$\delta R^0_{\ i0j} = k\,[\dot\sigma + \mathcal{H}\sigma]\, Q_{ij}^{(V)}, \tag{A3.30}$$
$$\delta R^0_{\ ijm} = -k\sigma\left(Q^{(V)}_{ij|m} - Q^{(V)}_{im|j}\right), \tag{A3.31}$$
$$\delta R^i_{\ 00j} = k\,[\dot\sigma + \mathcal{H}\sigma]\, Q_j^{(V)i}, \tag{A3.32}$$
$$\begin{aligned}\delta R^i_{\ 0jm} &= \left[K + \mathcal{H}^2\right]\sigma\left(\delta^i_j Q_m^{(V)} - \delta^i_m Q_j^{(V)}\right)\\ &\quad + k\sigma\left[\left(Q_m^{(V)i}\right)_{|j} - \left(Q_j^{(V)i}\right)_{|m}\right],\end{aligned} \tag{A3.33}$$
$$\begin{aligned}\delta R^i_{\ j0m} &= -\sigma\left[\mathcal{H}^2\left(\delta^i_m Q_j^{(V)} - \gamma_{jm} Q^{(V)i}\right) + \dot{\mathcal{H}}\gamma_{jm} Q^{(V)i}\right.\\ &\quad \left. -\frac{1}{2}\left(Q_j^{(V)|i} - Q^{(V)i}_{\ |j}\right)_{|m}\right],\end{aligned} \tag{A3.34}$$
$$\delta R^i_{\ jmn} = k\mathcal{H}\sigma\left(\delta^i_m Q_{jn}^{(V)} - \delta^i_n Q_{jm}^{(V)} + Q_m^{(V)i}\gamma_{jn} - Q_n^{(V)i}\gamma_{jm}\right). \tag{A3.35}$$

A3.2.3 The Ricci and Einstein Tensors

The perturbation if the Ricci tensor is

$$\delta R_{00} = 0, \tag{A3.36}$$

$$\delta R_{0j} = \left[K + \frac{1}{2}k^2 + 2\mathcal{H}^2 + \dot{\mathcal{H}}\right]\sigma Q_j^{(V)}, \tag{A3.37}$$

$$\delta R_{ij} = k\left[\dot{\sigma} + 2\mathcal{H}\sigma\right] Q_{ij}^{(V)}. \tag{A3.38}$$

The vector perturbation of the Riemann scalar of course vanishes. For the Einstein tensor we find

$$\delta G_0^0 = 0, \tag{A3.39}$$

$$\delta G_j^0 = \frac{2K - k^2}{2a^2}\sigma Q_j^{(V)}, \tag{A3.40}$$

$$\delta G_0^j = \frac{1}{a^2}\left[2(\mathcal{H}^2 - \dot{\mathcal{H}}) + K + \frac{k^2}{2}\right]\sigma Q^{(V)j}, \tag{A3.41}$$

$$\delta G_j^i = \frac{k}{a^2}\left[\dot{\sigma} + 2\mathcal{H}\sigma\right] Q_j^{(V)i}. \tag{A3.42}$$

A3.2.4 The Weyl Tensor

$$C^0{}_{i0j} = -\frac{k}{2}\dot{\sigma}^{(V)} Q_{ij}^{(V)} \equiv -E_{ij}^{(V)}, \tag{A3.43}$$

$$C_{ijk\ell} = g_{ik}E_{j\ell}^{(V)} + g_{j\ell}E_{ik}^{(V)} - g_{jk}E_{i\ell}^{(V)} - g_{i\ell}E_{jk}^{(V)}, \tag{A3.44}$$

$$C^0{}_{jlm} \equiv \epsilon_{lmi} B^{(V)i}{}_j$$

$$= \frac{1}{2}\sigma\left[Q_{l|jm}^{(V)} - Q_{m|jl}^{(V)} - \frac{k^2}{2}\gamma_{jl}Q_m^{(V)} + \frac{k^2}{2}\gamma_{jm}Q_l^{(V)}\right]. \tag{A3.45}$$

All other components are determined by symmetry.

A3.3 Tensor Perturbations

The metric is given by

$$ds^2 = a^2\left(-dt^2 + \left(\gamma_{ij} + 2HQ_{ij}^{(T)}\right)dx^i\,dx^i\right), \tag{A3.46}$$

where $H_{ij} = HQ_{ij}^{(T)}$ is symmetric, traceless, and divergence free. For tensor perturbations all scalar- and vector-type quantities vanish and we shall not write them down here. The non-vanishing tensor perturbations are given in Sections A3.3.1 to A3.3.3.

A3.3.1 The Christoffel Symbols

$$\delta\Gamma^i_{0j} = \dot{H}Q_j^{(T)i}, \qquad \delta\Gamma^0_{ij} = (2\mathcal{H}H + \dot{H})Q_{ij}^{(T)}, \tag{A3.47}$$

$$\delta\Gamma^i_{jm} = H\left(Q_{j|m}^{(T)i} + Q_{m|j}^{(T)i} - Q_{mj}^{(T)|i}\right). \tag{A3.48}$$

A3.3.2 The Riemann Tensor

$$\delta R^0_{i0j} = \left[\ddot{H} + \mathcal{H}\dot{H} + 2\dot{\mathcal{H}}H\right] Q^{(T)}_{ij}, \tag{A3.49}$$

$$\delta R^0_{ijm} = -\dot{H}\left(Q^{(T)}_{ij|m} - Q^{(T)}_{im|j}\right), \tag{A3.50}$$

$$\delta R^i_{00j} = \left[\ddot{H} + \mathcal{H}\dot{H}\right] Q^{(T)i}_{j}, \tag{A3.51}$$

$$\delta R^i_{0jm} = \dot{H}\left(Q^{(T)i}_{m|j} - Q^{(T)i}_{j|m}\right), \tag{A3.52}$$

$$\delta R^i_{j0m} = -\dot{H}\left(Q^{(T)|i}_{jm} - Q^{(T)i}_{m|j}\right), \tag{A3.53}$$

$$\begin{aligned}\delta R^i_{jmn} &= 2\mathcal{H}^2 H\left(\delta^i_m Q^{(T)}_{jn} - \delta^i_n Q^{(T)}_{jm}\right)\\ &\quad + H\left(Q^{(T)i}_{j|nm} - Q^{(T)i}_{j|mn} + Q^{(T)i}_{n|jm} - Q^{(T)i}_{m|jn} + Q^{(T)|i}_{jm}{}_{|n} - Q^{(T)|i}_{jn}{}_{|m}\right)\\ &\quad + \mathcal{H}\dot{H}\left(\delta^i_m Q^{(T)}_{jn} - \delta^i_n Q^{(T)}_{jm} - Q^{(T)i}_{n}\gamma_{jm} + Q^{(T)i}_{m}\gamma_{jn}\right).\end{aligned} \tag{A3.54}$$

A3.3.3 The Ricci and Einstein Tensors

$$\delta R_{ij} = \left[\ddot{H} + 2\mathcal{H}\dot{H} + (2\dot{\mathcal{H}} + 4\mathcal{H}^2 + k^2 + 6K)H\right] Q^{(T)}_{ij}, \tag{A3.55}$$

$$\delta G^i_j = a^{-2}\left[\ddot{H} + 2\mathcal{H}\dot{H} + (k^2 + 2K)H\right] Q^{(T)i}_{j}. \tag{A3.56}$$

A3.3.4 The Weyl Tensor

$$C^0{}_{i0j} \equiv -E^{(T)}_{ij} = -\frac{1}{2}(\partial_t^2 - k^2)HQ^{(T)}_{ij}, \tag{A3.57}$$

$$C_{ijk\ell} = g_{ik}E^{(T)}_{j\ell} + g_{j\ell}E^{(T)}_{ik} - g_{jk}E^{(T)}_{i\ell} - g_{i\ell}E^{(T)}_{jk}, \tag{A3.58}$$

$$C^0{}_{jlm} \equiv \epsilon_{lmk}B^{(T)k}{}_j = -\dot{H}\left[Q^{(T)}_{jl|m} - Q^{(T)}_{jm|l}\right]. \tag{A3.59}$$

All other components are determined by symmetry.

Appendix 4

Special Functions

A4.1 Legendre Polynomials and Legendre Functions

The **Legendre polynomials** form an orthonormal set of polynomials on the interval $[-1, 1]$. The lowest-order polynomials are $P_0 = 1$ and $P_1 = x$. The higher-order polynomials can then be obtained via the Gram–Schmidt orthogonalization procedure starting from the monomial x^n. They obey the normalization condition

$$\int_{-1}^{1} dx\, P_\ell(x) P_{\ell'}(x) = \frac{2}{2\ell+1}\,\delta_{\ell\ell'}. \tag{A4.1}$$

The Legendre polynomials can also be obtained via the recursion relation

$$(\ell+1)P_{\ell+1}(x) = (2\ell+1)x P_\ell(x) - \ell P_{\ell-1}(x). \tag{A4.2}$$

They obey the differential equation

$$(1-x^2)P_\ell'' - 2x P_\ell' + \ell(\ell+1)P_\ell = 0. \tag{A4.3}$$

The Legendre polynomials can also be defined via Rodrigues' formula:

$$P_\ell(x) = \frac{1}{2^\ell \ell!}\frac{d^\ell}{dx^\ell}\left(x^2-1\right)^\ell. \tag{A4.4}$$

The lowest-order Legendre polynomials are given by

$$P_0 = 1, \tag{A4.5}$$

$$P_1 = x, \tag{A4.6}$$

$$P_2 = \frac{1}{2}(3x^2 - 1), \tag{A4.7}$$

$$P_3 = \frac{1}{2}(5x^3 - 3x), \tag{A4.8}$$

$$P_4 = \frac{1}{8}(35x^4 - 30x^2 + 3). \tag{A4.9}$$

Clearly $P_\ell(-x) = (-1)^\ell P_\ell(x)$. Via induction, using Eq. (A4.2), one finds that

$$P_\ell(1) = 1. \tag{A4.10}$$

The Legendre polynomials obey the limiting relation

$$\lim_{\ell\to\infty} P_\ell(\cos(\theta/\ell)) = J_0(\theta). \tag{A4.11}$$

Here J_0 is the Bessel function of order zero (see Section A4.3).

The associated **Legendre functions** are defined by

$$P_{\ell\,m}(x) = (1-x^2)^{m/2}\frac{d^m P_\ell(x)}{dx^m} = (1-x^2)^{m/2}\frac{1}{2^\ell \ell!}\frac{d^{\ell+m}}{dx^{\ell+m}}\left(x^2-1\right)^\ell, \tag{A4.12}$$

for $0 \le m \le \ell$. [We use the notation of Abramowitz and Stegun (1970) with $P_{\ell\,m} = (-1)^m P_\ell^m$.] The associated Legendre functions are nonvanishing for with $-\ell \le m \le \ell$. Functions with the opposite sign of m are related via

$$P_{\ell\,-m} = (-1)^m\frac{(\ell-m)!}{(\ell+m)!}P_{\ell\,m}. \tag{A4.13}$$

From the above definition and Eq. (A4.10) one obtains

$$P_{\ell\,m}(1) = \delta_{m0}. \tag{A4.14}$$

The Legendre functions solve the differential equation

$$(1-x^2)P''_{\ell\,m} - 2xP'_{\ell\,m} + \left[\ell(\ell+1) - \frac{m^2}{1-x^2}\right]P_{\ell\,m} = 0. \tag{A4.15}$$

They are in principle defined for arbitrary complex degree ℓ and order m as (meromorphic) functions of complex variables x. We shall need them only for integer m and nonnegative integer ℓ's with $|m| \le \ell$ and $x \in [-1,1]$. In this interval and with these values of order and degree they are singularity free and analytic.

The Legendre functions satisfy the following (and several more) recurrence relations:

$$P_{\ell\,m+1} = \frac{2mx}{\sqrt{1-x^2}}P_{\ell\,m} - [\ell(\ell+1) - m(m+1)]\,P_{\ell\,m-1}, \tag{A4.16}$$

$$xP_{\ell\,m} = \frac{\ell+m}{2\ell+1}P_{\ell-1\,m} + \frac{\ell-m+1}{2\ell+1}P_{\ell+1\,m}, \tag{A4.17}$$

$$\frac{dP_{\ell\,m}}{dx} = \frac{1}{2\sqrt{1-x^2}}\left[P_{\ell\,m+1} - (\ell+m)(\ell-m+1)P_{\ell\,m-1}\right], \tag{A4.18}$$

$$(x^2-1)\frac{dP_{\ell\,m}}{dx} = \ell xP_{\ell\,m} + (m+\ell)P_{\ell-1\,m}, \tag{A4.19}$$

$$P_{\ell+1\,m} = P_{\ell-1\,m} + (2\ell+1)\sqrt{1-x^2}P_{\ell\,m-1}. \tag{A4.20}$$

The parity relation of the associated Legendre function is a simple consequence of their definition,

$$P_{\ell\,m}(-x) = (-1)^{\ell+m}P_{\ell\,m}(x). \tag{A4.21}$$

Also of importance for us is the orthogonality relation

$$\int_{-1}^{1} P_{\ell\,m}(x)P_{\ell'\,m}(x)\,dx$$

$$= \int_0^\pi P_{\ell\,m}(\cos\vartheta)P_{\ell'\,m}(\cos\vartheta)\sin\vartheta\,d\vartheta = \frac{2}{2\ell+1}\frac{(\ell+m)!}{(\ell-m)!}\delta_{\ell\,\ell'}. \tag{A4.22}$$

The derivation of most of these results and more can be found in Arfken and Weber (2001).

A4.2 Spherical Harmonics

A4.2.1 The Irreducible Representations of the Rotation Group

Here we briefly repeat some basics about the rotation group and its irreducible representations. Much more can be found in most quantum mechanics books, for example, Sakurai (1993). Here we are interested only in ordinary (i.e., not projective) representations and therefore integer values of the angular momentum. For a function Ψ on the sphere we define its transformation under rotations $R \in SO(3)$ by

$$[\mathcal{U}(R)\Psi](\mathbf{n}) \equiv \Psi\left(R^{-1}\mathbf{n}\right) \quad \forall\, \mathbf{n} \in \mathbb{S}^2. \tag{A4.23}$$

This is clearly a unitary representation of the rotation group on $\mathcal{L}^2(\mathbb{S}^2)$, that is, the Hilbert space of square integrable functions on the sphere.

The one-parameter subgroup of rotations around a given axis $\mathbf{e}$ is

$$R(\mathbf{e},\alpha)\mathbf{n} = \cos\alpha\mathbf{n} + [1-\cos\alpha](\mathbf{e}\cdot\mathbf{n})\mathbf{e} + \sin\alpha\mathbf{e}\wedge\mathbf{n}. \tag{A4.24}$$

Its generator is defined by

$$\Omega(\mathbf{e})\mathbf{n} = \left.\frac{d}{d\alpha}R(\mathbf{e},\alpha)\mathbf{n}\right|_{\alpha=0}.$$

With Eq. (A4.24) we obtain

$$\Omega(\mathbf{e})_{ij} = \mathbf{e}_k I^k_{ij}, \quad \text{where} \quad I^k_{ij} = -\epsilon_{ijk}. \tag{A4.25}$$

The generator of $\mathcal{U}(R(\mathbf{e},\alpha))$ is the angular momentum in direction $\mathbf{e}$:

$$\left.\frac{d}{d\alpha}\mathcal{U}(R(\mathbf{e},\alpha))\right|_{\alpha=0} \equiv \mathcal{U}_*(I^j)\mathbf{e}_j = \frac{i}{\hbar}L^j\mathbf{e}_j,$$

with

$$\mathbf{L} = -i\hbar\mathbf{x}\wedge\nabla. \tag{A4.26}$$

In spherical coordinates (r,ϑ,φ) one finds

$$\mathbf{L} = i\hbar\begin{pmatrix} \sin\varphi\cot\vartheta\,\partial_\varphi + \cos\varphi\partial_\vartheta \\ \cos\varphi\cot\vartheta\,\partial_\varphi - \sin\varphi\partial_\vartheta \\ -\partial_\varphi \end{pmatrix}. \tag{A4.27}$$

One easily verifies that the matrices I_k and the operators L_k satisfy the commutation relations

$$\left[I_j, I_k\right] = \epsilon_{jkl}I_l, \tag{A4.28}$$

$$\left[L_j, L_k\right] = i\hbar\epsilon_{jkl}L_l. \tag{A4.29}$$

Introducing also $L_\pm = L_1 \pm iL_2$ and $\mathbf{L}^2 = L_1^2 + L_2^2 + L_3^2$ one finds the commutation relations

$$\left[\mathbf{L}^2, L_j\right] = 0 = \left[\mathbf{L}^2, L_\pm\right], \tag{A4.30}$$

$$\left[L_3, L_\pm\right] = \pm\hbar L_\pm, \quad \text{and} \tag{A4.31}$$

$$L_\pm L_\mp = \mathbf{L}^2 - L_3^2 \pm \hbar L_3. \tag{A4.32}$$

Let us now consider a representation of the rotation group on some finite-dimensional vector space $\mathcal{V}$. Since $\mathbf{L}^2$ and L_3 are commuting hermitian operators, we can find an orthonormal basis of simultaneous eigenvectors of $\mathbf{L}^2$ and L_3. We order them according to their eigenvalue of L_3, so that the eigenvalue of ψ_1 is maximal. Let us call it $\hbar a$. Furthermore, $\hbar^2 b$ denotes the corresponding eigenvalue of $\mathbf{L}^2$. Hence $L_3\psi_1 = \hbar a\psi_1$ and $\mathbf{L}^2\psi_1 = \hbar^2 b\psi_1$. Equation (A4.31) gives $L_3 L_+\psi_1 = \hbar(a+1)L_+\psi_1$. Hence $L_+\psi_1$ is an eigenvector of L_3 with eigenvalue $\hbar(a+1)$ or zero. Since $\hbar a$ is maximal, this implies $L_+\psi_1 = 0$. With Eq. (A4.32) therefore $0 = (\mathbf{L}^2 - L_3^2 - \hbar L_3)\psi_1 = \hbar^2(b - a(a+1))\psi_1$, so that $b = a(a+1)$. Applying L_- on ψ_1 and using again Eq. (A4.31), we find that $L_-\psi_1$ is also an eigenvector of L_3 with eigenvalue $\hbar(a-1)$. Repeated application of L_- shows that $(L_-)^m\psi_1$ is an eigenvector of L_3 with eigenvalue $\hbar(a-m)$. We finally arrive at the eigenvector with the lowest eigenvalue $\hbar(a-n)$ of L_3; let us call it

$$\psi_{n+1} = (L_-)^n\psi_1 / \parallel (L_-)^n\psi_1 \parallel .$$

Necessarily, $L_-\psi_{n+1} = 0$ since it would otherwise have an even lower eigenvalue of L_3. From this we conclude

$$0 = L_+(L_-)^{n+1}\psi_1 = (\mathbf{L}^2 - L_3^2 + \hbar L_3)(L_-)^n\psi_1 = \hbar^2(b - (a-n)^2 + a - n)(L_-)^n\psi_1,$$

so that $b = (a-n)^2 + n - a$. Together with the previous identity, $b = a(a+1)$, this implies $a = n/2$. Therefore, a must be an integer or half-integer number, the representation with $a = \ell$ is denoted by $D^{(\ell)}$ and has dimension $n + 1 = 2\ell + 1$. The induced representation of the generators (the Lie algebra) defines the angular momentum, $L_j = i\hbar D_*^{(\ell)}(I_j)$. The vector space that carries $D^{(\ell)}$ is denoted by $\mathcal{V}^{(\ell)}$. The vectors $\left((L_-^n)\psi_1 / \|(L_-^n)\psi_1\|\right)_{n=0}^{2\ell}$ form an orthonormal basis of eigenvectors of L_3 and $\mathbf{L}^2$, the so-called canonical basis. The eigenvalues of L_3 are $\hbar\ell$, $\hbar(\ell-1)$, $\ldots$, $-\hbar\ell$. The operator $\mathbf{L}^2$ is constant on $\mathcal{V}^{(\ell)}$ with eigenvalue $\hbar^2\ell(\ell+1)$. $D^{(0)}$ is the trivial representation, $D^{(0)}(R) = \mathbb{I}$, which is irreducible only on a one-dimensional space; and $D^{(1)}$ is the identical representation, $D^{(1)}(R) = R$.

In the next section, when we realize these representations on $\mathcal{L}^2(\mathbb{S}^2)$ we shall see that only the representations $D_*^{(\ell)}$ with integer ℓ can be lifted to representations of the rotation group. Half-integer ℓ's give projective representation with $D^{(\ell)}(R_1)D^{(\ell)}(R_2) = \pm D^{(\ell)}(R_1R_2)$ that are relevant in quantum mechanics where a state is defined only up to a constant phase. The existence of particles with half-integer spin, fermions, is a purely quantum mechanical phenomenon.

Acting with $L_\pm$, we can pass from one basis vector to every other in $\mathcal{V}^{(\ell)}$. Hence the representation $D^{(\ell)}$ is irreducible, that is, $\mathcal{V}^{(\ell)}$ contains no invariant subspaces. We have obtained all irreducible representations of the rotation group in this way.

A4.2.2 The Clebsch–Gordan Decomposition

The tensor product, $\mathcal{V}^{(\ell)} \otimes \mathcal{V}^{(\ell')}$ carries the tensor representation $D^{(\ell)} \otimes D^{(\ell')}$. In general this representation is not irreducible but can be decomposed into a sum of irreducible representations. We show that

$$D^{(\ell)} \otimes D^{(\ell')} = \sum_{j=|\ell-\ell'|}^{\ell+\ell'} D^{(j)}. \qquad \text{(A4.33)}$$

This sum is called the Clebsch–Gordan series.

Without loss of generality, we assume $\ell \geq \ell'$. The Leibnitz rule implies that the induced representation on the generators is $(D^{(\ell)} \otimes D^{(\ell')})_* = (D_*^{(\ell)} \otimes \mathbb{I}) \oplus (\mathbb{I} \otimes D_*^{(\ell')})$. We denote the canonical basis on $\mathcal{V}^{(\ell)}$ by $(\psi_{\ell\, m})_{m=-\ell}^{\ell}$. L_3 takes once the maximal value on the state $\psi_{\ell\,\ell} \otimes \psi_{\ell'\,\ell'}$, where we have $L_3(\psi_{\ell\,\ell} \otimes \psi_{\ell'\,\ell'}) = (L_3\psi_{\ell\,\ell}) \otimes \psi_{\ell'\,\ell'} + \psi_{\ell\,\ell} \otimes (L_3\psi_{\ell'\,\ell'}) = \hbar(\ell + \ell')\psi_{\ell\,\ell} \otimes \psi_{\ell'\,\ell'}$. Hence $D^{(\ell)} \otimes D^{(\ell')}$ contains $D^{(j)}$ with $j = \ell + \ell'$ exactly once and it does not contain any higher angular momentum. However, there are two states with $L_3\phi = \hbar(\ell + \ell' - 1)\phi$, namely the states $\psi_{\ell\,\ell-1} \otimes \psi_{\ell'\,\ell'}$ and $\psi_{\ell\,\ell} \otimes \psi_{\ell'\,\ell'-1}$; hence $D^{(\ell)} \otimes D^{(\ell')}$ must also contain $D^{(\ell+\ell'-1)}$ (except if $\ell' = 0$). Furthermore, there are three states with $L_3\phi = \hbar(\ell + \ell' - 2)\phi$, namely the states $\psi_{\ell\,\ell-2} \otimes \psi_{\ell'\,\ell'}$, $\psi_{\ell\,\ell-1} \otimes \psi_{\ell'\,\ell'-1}$, and $\psi_{\ell\,\ell} \otimes \psi_{\ell'\,\ell'-2}$; hence $D^{(\ell)} \otimes D^{(\ell')}$ must in addition contain $D^{(\ell+\ell'-2)}$. This goes on until the eigenvalue $\ell + \ell' - m$, with $m = 2\ell'$ is reached, which has an eigenspace of dimension $m + 1$ and that also implies that the representation $D^{(\ell-\ell')}$ is contained. For higher values of m the dimension of the eigenspace is reduced by 1 at each step and is therefore just sufficient to contain the eigenvectors of each of the representations $D^{(j)}$ already inferred. This can also be concluded from the fact that $(2\ell + 1)(2\ell' + 1) = \sum_{j=\ell-\ell'}^{\ell+\ell'}(2j + 1)$ and therefore the dimension of the total space agrees with the sum of the dimensions of all the irreducible representations already defined. This proves the Clebsch–Gordan series.

The matrix that induces the change of basis from the tensor product basis to the canonical basis on each of the irreducible pieces of $\mathcal{V}^{(\ell)} \otimes \mathcal{V}^{(\ell')}$ defines the Clebsch–Gordan coefficients in the following way. We have seen that

$$\mathcal{V}^{(\ell)} \otimes \mathcal{V}^{(\ell')} = \mathcal{W}^{(\ell+\ell')} \oplus \mathcal{W}^{(\ell+\ell'-1)} \oplus \cdots \oplus \mathcal{W}^{(|\ell-\ell'|)},$$

where $\mathcal{W}^{(j)}$ carries the representation $D^{(j)}$ of the rotation group. Let us denote the canonical basis in $\mathcal{W}^{(j)}$ by $(\phi_{j\,m})_{m=-j}^{j}$. The transformation from the basis $(\psi_{\ell\,m} \otimes \psi_{\ell'\,m'})_{m,m'=-\ell,-\ell'}^{\ell,\ell'}$ to the basis $\left((\phi_{j\,m})_{m=-j}^{j}\right)_{j=|\ell-\ell'|}^{\ell+\ell'}$ is of the form

$$\phi_{j\,m_j} = \sum_{m,m'} \langle \ell, \ell'; m, m' | j, m_j \rangle \psi_{\ell\,m} \otimes \psi_{\ell'\,m'}. \tag{A4.34}$$

The complex coefficients $\langle \ell, \ell'; m, m' | j, m_j \rangle$ are called Clebsch–Gordan coefficients. In the literature they are often denoted by $\langle \ell, \ell'; m, m' | j, m_j \rangle \equiv \langle \ell, \ell'; m, m' | \ell, \ell'; j, m_j \rangle$. We shall not repeat the redundant numbers ℓ, ℓ' in the second argument. The Clebsch–Gordan coefficients present an orthogonal basis transformation. Using the general formula that under such a basis transformation S, the matrix $D(R)$ transforms into $S^T D(R) S$ we obtain the corresponding transformation for the representations matrices,

$$D^{(\ell)}_{mm'} D^{(\ell_1)}_{m_1 m_1'} = \sum_{j\,m_j\,m_j'} \langle \ell, \ell_1; m, m_1 | j, m_j \rangle D^{(j)}_{m_j m_j'} \langle \ell, \ell_1; m', m_1' | j, m_j' \rangle. \tag{A4.35}$$

From the foregoing discussion it is clear that $\langle \ell, \ell'; m, m' | j, m_j \rangle \neq 0$ only if $m_j = m + m'$ and $j \in \{\ell + \ell', \ell + \ell' - 1, \ldots, |\ell - \ell'|\}$. The general formula for these coefficients is given below (Abramowitz and Stegun, 1970). They can also be computed with Mathematica.

Table A4.1 *The nonvanishing Clebsch–Gordan coefficients for* $\ell_2 = 1$.

j	$\langle \ell, 1; m, m_2 \vert j, m + m_2 \rangle$ $m_2 = 1$	$m_2 = 0$	$m_2 = -1$
$\ell + 1$	$\sqrt{\frac{(\ell+m+1)(\ell+m+2)}{(2\ell+1)(2\ell+2)}}$	$\sqrt{\frac{(\ell+m+1)(\ell-m+1)}{(2\ell+1)(\ell+1)}}$	$\sqrt{\frac{(\ell-m+1)(\ell-m+2)}{(2\ell+1)(2\ell+2)}}$
ℓ	$-\sqrt{\frac{(\ell+m+1)(\ell-m)}{2\ell(2\ell+1)}}$	$\frac{m}{\sqrt{\ell(\ell+1)}}$	$\sqrt{\frac{(\ell-m+1)(\ell+m)}{2\ell(\ell+1)}}$
$\ell - 1$	$\sqrt{\frac{(\ell-m-1)(\ell-m)}{2\ell(2\ell+1)}}$	$-\sqrt{\frac{(\ell+m)(\ell-m)}{\ell(2\ell+1)}}$	$\sqrt{\frac{(\ell+m)(\ell+m-1)}{2\ell(2\ell+1)}}$

$$
\begin{aligned}
&\langle \ell_1, \ell_2; m_1, m_2 | j, m_1 + m_2 \rangle \\
&= \sqrt{\frac{(\ell_1 + \ell_2 - j)!\,(j + \ell_1 - \ell_2)!\,(j + \ell_2 - \ell_1)!\,(2j+1)}{(\ell_1 + \ell_2 + j + 1)!}} \\
&\times \sum_k \left[\frac{(-1)^k \sqrt{(\ell_1 + m_1)!\,(\ell_1 - m_1)!\,(\ell_2 + m_2)!\,(\ell_2 - m_2)!}}{k!\,(\ell_1 + \ell_2 - j - k)!\,(\ell_1 - m_1 - k)!} \right. \\
&\times \left. \frac{\sqrt{(j + m_1 + m_2)!\,(j - m_1 - m_2)!}}{(\ell_2 + m_2 - k)!\,(j - \ell_2 + m_1 + k)!\,(j - \ell_1 - m_2 + k)!} \right].
\end{aligned} \tag{A4.36}
$$

In the sum over k only the terms with a finite denominator contribute; hence $k \geq \max\{0, \ell_2 - j - m_1, \ell_1 - j + m_2\}$ and $k \leq \min\{\ell_1 + \ell_2 - j, \ell_1 - m_1, \ell_2 + m_2\}$.

In Chapter 5 we need the Clebsch–Gordan coefficients $\langle \ell_1, \ell_2; m_1, m_2 | j, m_1 + m_2 \rangle$ for $\ell_2 \leq 2$. We therefore give the nonvanishing ones of these coefficients in Tables A4.1 and A4.2. Of course $\langle \ell_1, 0; m_1, 0 | \ell, m \rangle = \delta_{\ell_1 \ell}\, \delta_{m_1 m}$.

A4.2.3 Spherical Harmonics of Spin-0

The spherical harmonics are functions on the sphere. For a unit vector $\mathbf{n}$ defined by its polar angles (ϑ, φ) the spherical harmonics are given by

$$
Y_{\ell m}(\mathbf{n}) = (-1)^m \sqrt{\frac{2\ell + 1}{4\pi} \frac{(\ell - m)!}{(\ell + m)!}}\, e^{im\varphi} P_{\ell m}(\mu), \qquad \mu = \cos\vartheta. \tag{A4.37}
$$

From the properties and the orthogonality of the associated Legendre functions, Eqs. (A4.13) and (A4.22), we conclude $Y_{\ell -m} = (-1)^m Y^*{}_{\ell m}$ and

$$
\int Y_{\ell m}(\mathbf{n}) Y^*{}_{\ell' m'}(\mathbf{n}) d\Omega_{\mathbf{n}} = \delta_{\ell \ell'} \delta_{m m'}. \tag{A4.38}
$$

Note also that under parity, $\mathbf{n} \to -\mathbf{n}$, the spherical harmonics transform as

$$
Y_{\ell m}(-\mathbf{n}) = (-1)^\ell Y_{\ell m}(\mathbf{n}). \tag{A4.39}
$$

We now show that the spherical harmonics $(Y_{\ell m})_{m=-\ell}^{\ell}$ carry the representation $D^{(\ell)}$.

From Eq. (A4.27) we know $L_3 = -i\hbar \partial_\varphi$. Therefore, the set of functions $f_{\ell m}$ that forms a canonical basis for the representation $D^{(\ell)}$ must be of the form $f_{\ell m} = \exp(im\varphi) g_{\ell m}(\mu)$. Furthermore, Eq. (A4.27) implies

Table A4.2 *The nonvanishing Clebsch–Gordan coefficients for $\ell_2 = 2$.*

j	$\langle \ell, 2; m, m_2 \vert j, m + m_2 \rangle$	
	$m_2 = 2$	$m_2 = 1$
$\ell + 2$	$\sqrt{\frac{(\ell+m+1)(\ell+m+2)(\ell+m+3)(\ell+m+4)}{(2\ell+1)(2\ell+2)(2\ell+3)(2\ell+4)}}$	$\sqrt{\frac{(\ell-m+1)(\ell+m+3)(\ell+m+2)(\ell+m+1)}{(2\ell+1)(\ell+1)(2\ell+3)(\ell+2)}}$
$\ell + 1$	$-\sqrt{\frac{(\ell+m+1)(\ell+m+2)(\ell+m+3)(\ell-m)}{2\ell(\ell+1)(\ell+2)(2\ell+1)}}$	$-(\ell - 2m)\sqrt{\frac{(\ell+m+2)(\ell+m+1)}{2\ell(2\ell+1)(\ell+1)(\ell+2)}}$
ℓ	$\sqrt{\frac{3(\ell+m+1)(\ell+m+2)(\ell-m-1)(\ell-m)}{(2\ell-1)2\ell(\ell+1)(2\ell+3)}}$	$(3 - 2m)\sqrt{\frac{3(\ell-m)(\ell+m+1)}{\ell(2\ell-1)(2\ell+2)(2\ell+3)}}$
$\ell - 1$	$-\sqrt{\frac{(\ell+m+1)(\ell-m-2)(\ell-m-1)(\ell-m)}{2(\ell-1)\ell(\ell+1)(2\ell+1)}}$	$(\ell + 2m + 1)\sqrt{\frac{(\ell-m)(\ell-m-1)}{\ell(\ell-1)(2\ell+1)(2\ell+2)}}$
$\ell - 2$	$\sqrt{\frac{(\ell-m-3)(\ell-m-2)(\ell-m-1)(\ell-m)}{(2\ell-2)(2\ell-1)2\ell(2\ell+1)}}$	$-\sqrt{\frac{(\ell-m)(\ell-m-1)(\ell-m-2)(\ell+m)}{(\ell-1)(2\ell-1)\ell(2\ell+1)}}$

j	$m_2 = 0$	$m_2 = -1$
$\ell + 2$	$\sqrt{\frac{3(\ell-m+2)(\ell-m+1)(\ell+m+2)(\ell+m+1)}{(2\ell+1)(2\ell+2)(2\ell+3)(\ell+2)}}$	$\sqrt{\frac{(\ell-m+3)(\ell-m+2)(\ell-m+1)(\ell+m+1)}{(2\ell+1)(\ell+1)(2\ell+3)(\ell+2)}}$
$\ell + 1$	$m\sqrt{\frac{3(\ell-m+1)(\ell+m+1)}{\ell(2\ell+1)(\ell+1)(\ell+2)}}$	$(\ell + 2m)\sqrt{\frac{(\ell-m+2)(\ell-m+1)}{2\ell(2\ell+1)(\ell+1)(\ell+2)}}$
ℓ	$\frac{3m^2-\ell(\ell+1)}{\sqrt{(2\ell-1)\ell(2\ell+1)(\ell+1)}}$	$(2m - 1)\sqrt{\frac{3(\ell-m+1)(\ell+m)}{\ell(2\ell-1)(2\ell+2)(2\ell+3)}}$
$\ell - 1$	$-m\sqrt{\frac{3(\ell-m)(\ell+m)}{(\ell-1)\ell(2\ell+1)(\ell+1)}}$	$-(\ell - 2m + 1)\sqrt{\frac{(\ell+m)(\ell+m-1)}{\ell(\ell-1)(2\ell+1)(2\ell+2)}}$
$\ell - 2$	$\sqrt{\frac{3(\ell-m)(\ell-m-1)(\ell+m)(\ell+m-1)}{(2\ell-2)(2\ell-1)(2\ell+1)\ell}}$	$-\sqrt{\frac{(\ell-m)(\ell+m)(\ell+m-1)(\ell+m-2)}{(\ell-1)(2\ell-1)\ell(2\ell+1)}}$

j	$m_2 = -2$
$\ell + 2$	$\sqrt{\frac{(\ell-m+1)(\ell-m+2)(\ell-m+3)(\ell-m+4)}{(2\ell+1)(2\ell+2)(2\ell+3)(2\ell+4)}}$
$\ell + 1$	$\sqrt{\frac{(\ell-m+1)(\ell-m+2)(\ell-m+3)(\ell+m)}{\ell(2\ell+1)(\ell+1)(2\ell+4)}}$
ℓ	$\sqrt{\frac{3(\ell-m+1)(\ell-m+2)(\ell+m-1)(\ell+m)}{\ell(2\ell-1)(2\ell+2)(2\ell+3)}}$
$\ell - 1$	$\sqrt{\frac{(\ell-m+1)(\ell+m-2)(\ell+m-1)(\ell+m)}{(\ell-1)\ell(2\ell+1)(2\ell+2)}}$
$\ell - 2$	$\sqrt{\frac{(\ell+m-3)(\ell+m-2)(\ell+m-1)(\ell+m)}{(2\ell-2)(2\ell-1)2\ell(2\ell+4)}}$

$$\mathbf{L}^2 = L_1^2 + L_2^2 = L_3^2 = -\hbar^2 \left[\frac{1}{\sin\vartheta} \partial_\vartheta \, \sin\vartheta \, \partial_\vartheta + \frac{1}{\sin^2\vartheta} \partial_\varphi^2 \right] = -\hbar^2 \Delta,$$

where Δ denotes the Laplacian on the 2-sphere. For $f_{\ell m} = \exp(im\varphi) g_{\ell m}(\mu)$ we obtain

$$\Delta f_{\ell m} = \left[(1-\mu^2)\frac{d^2}{d\mu^2} - 2\mu \frac{d}{d\mu} - \frac{m^2}{1-\mu^2} \right] g_{\ell m}(\mu) \exp(im\varphi).$$

With $\mathbf{L}^2 = \hbar^2 \ell(\ell+1)$ it follows that $g_{\ell m}(\mu)$ satisfies the differential equation of the associated Legendre function, Eq. (A4.15); hence $g_{\ell m} = c_{\ell m} P_{\ell m}(\mu)$. The constants $c_{\ell m}$ are chosen to normalize the functions $f_{\ell m}$. Furthermore, since $f_{\ell m}$ and $f_{\ell' m'}$ are eigenfunctions with different eigenvalues for some hermitian operator (L_3 if $m \neq m'$ and $\mathbf{L}^2$ if $\ell \neq \ell'$), they are certainly orthogonal. Hence the functions $f_{\ell m}$ are proportional to the spherical harmonics and obey the same normalization condition, that is, they are the spherical harmonics $Y_{\ell m}$.

We can relate the spherical harmonic $Y_{\ell m}(\mathbf{n})$ to the matrix element $D^{(\ell)}_{m0}(R)$, where R is a rotation that turns $\mathbf{e}_z$ into $\mathbf{n}$. To do this we observe that the spherical harmonics of order ℓ form an orthonormal basis for the $(2\ell+1)$-dimensional space of functions on the sphere that transform with the representation $D^{(\ell)}$ under rotation. Let f be such a function and (f_m) be its coefficients in the basis $Y_{\ell m}$. In other words,

$$f(\mathbf{n}) = \sum_m f_m Y_{\ell m}(\mathbf{n}).$$

Under a rotation, the vector (f_m) transforms with $D^{(\ell)}_{m_1 m_2}$ so that

$$f(R^{-1}\mathbf{n}) = \sum_{m_1} \left(\sum_{m_2} D^{(\ell)}_{m_1 m_2}(R) f_{m_2} \right) Y_{\ell m_1}(\mathbf{n}).$$

Considering the function with $f_{m_2} = \delta_{m m_2}$ this yields

$$Y_{\ell m}(R^{-1}\mathbf{n}) = \sum_{m_1} D^{(\ell)}_{m_1 m}(R) Y_{\ell m_1}(\mathbf{n}). \tag{A4.40}$$

Let us now consider $\mathbf{n} = \mathbf{e}_z$. Using $Y_{\ell m}(\mathbf{e}_z) = \delta_{0m}\sqrt{(2\ell+1)/4\pi}$ we arrive at

$$Y_{\ell m}(R^{-1}\mathbf{e}_z) = \sqrt{\frac{2\ell+1}{4\pi}} D^{(\ell)}_{0m}(R). \tag{A4.41}$$

If R is an (otherwise arbitrary) rotation that turns $\mathbf{n}$ into $\mathbf{e}_z$, so that $R^{-1}\mathbf{e}_z = \mathbf{n}$, we therefore have

$$Y_{\ell m}(\mathbf{n}) = \sqrt{\frac{2\ell+1}{4\pi}} D^{(\ell)}_{0m}(R). \tag{A4.42}$$

Usually one chooses for R the rotation with Euler angles $(0, -\vartheta, -\varphi)$ for the unit vector $\mathbf{n}$ with polar angles (ϑ, φ). We denote the representation matrix of the rotation by Euler angles (α, β, γ) by $D^{(\ell)}_{mn}(\alpha, \beta, \gamma)$: first a rotation by angle γ around the z-axis, then a rotation by angle β around the y-axis, and finally a rotation by angle α around the (new) z-axis. The inverse of the rotation (α, β, γ) is the rotation with Euler angles $(-\gamma, -\beta, -\alpha)$. Observing that the representation $D^{(\ell)}$ is unitary we find $D^{(\ell)}_{0m}(0, -\vartheta, -\varphi) = D^{(\ell)^{-1}}_{0m}(\varphi, \vartheta, 0) = D^{*(\ell)}_{m0}(\varphi, \vartheta, 0)$ so that we can also write

$$Y_{\ell m}(\mathbf{n}) = \sqrt{\frac{2\ell+1}{4\pi}} D^{*(\ell)}_{m0}(\varphi, \vartheta, 0). \tag{A4.43}$$

With this it is now easy to show the addition theorem of spherical harmonics. Consider two directions $\mathbf{n}_1$ and $\mathbf{n}_2$ separated by an angle γ, $\cos\gamma = \mathbf{n}_1 \cdot \mathbf{n}_2$. We denote the rotation with Euler angles $(\varphi_1, \vartheta_1, 0)$, which rotates $\mathbf{e}_z$ into $\mathbf{n}_1$ by R_1. With Eqs. (A4.40) and (A4.43) we have

$$Y_{\ell 0}(R_1^{-1}\mathbf{n}_2) = \sum_m D^{(\ell)}_{m0}(R_1) Y_{\ell\, m}(\mathbf{n}_2) = \sqrt{\frac{4\pi}{2\ell+1}} \sum_m Y^*_{\ell m}(\mathbf{n}_1) Y_{\ell m}(\mathbf{n}_2). \tag{A4.44}$$

But since R_1^{-1} rotates $\mathbf{n}_1$ into $\mathbf{e}_z$, the polar angle ϑ of $R_1^{-1}\mathbf{n}_2$ is simply the angle between $\mathbf{n}_1$ and $\mathbf{n}_2$, so that $Y_{\ell 0}(R_1^{-1}\mathbf{n}_2) = \sqrt{(2\ell+1)/4\pi}\, P_\ell(\mathbf{n}_1 \cdot \mathbf{n}_2)$. Inserting this above yields the addition theorem for spherical harmonics,

$$\frac{2\ell+1}{4\pi} P_\ell(\mathbf{n}_1 \cdot \mathbf{n}_2) = \sum_{m=-\ell}^{\ell} Y^*_{\ell m}(\mathbf{n}_1) Y_{\ell m}(\mathbf{n}_2). \tag{A4.45}$$

The lowest ℓ spherical harmonics are given by

$$\ell = 0 \quad Y_{00} = \frac{1}{\sqrt{4\pi}}, \tag{A4.46}$$

$$\ell = 1 \begin{cases} Y_{11} = & -\sqrt{\frac{3}{8\pi}} \sin\vartheta\, e^{i\varphi}, \\ Y_{10} = & \sqrt{\frac{3}{4\pi}} \cos\vartheta, \end{cases} \tag{A4.47}$$

$$\ell = 2 \begin{cases} Y_{22} = & \sqrt{\frac{15}{32\pi}} \sin^2\vartheta\, e^{2i\varphi}, \\ Y_{21} = & -\sqrt{\frac{15}{8\pi}} \sin\vartheta \cos\vartheta\, e^{i\varphi}, \\ Y_{20} = & \sqrt{\frac{5}{4\pi}} \left(\frac{3}{2}\cos^2\vartheta - \frac{1}{2}\right), \end{cases} \tag{A4.48}$$

$$\ell = 3 \begin{cases} Y_{33} = & -\sqrt{\frac{35}{64\pi}} \sin^3\vartheta\, e^{3i\varphi}, \\ Y_{32} = & \sqrt{\frac{105}{32\pi}} \sin^2\vartheta \cos\vartheta\, e^{2i\varphi}, \\ Y_{31} = & -\sqrt{\frac{21}{16\pi}} \sin\vartheta \left(\frac{5}{2}\cos^2\vartheta - \frac{1}{2}\right) e^{i\varphi}, \\ Y_{30} = & \sqrt{\frac{7}{4\pi}} \cos\vartheta \left(\frac{5}{2}\cos^2\vartheta - \frac{3}{2}\right), \end{cases} \tag{A4.49}$$

$$Y_{\ell\, -m} = (-1)^m Y^*_{\ell m}. \tag{A4.50}$$

A4.2.4 Integrals of Spherical Harmonics (of Spin-0)

The Clebsch–Gordan coefficients also allow us to derive formulas for integrals of products of spherical harmonics. Using

$$Y_{\ell_3 m_3} = \sum_{m_1 m_2} \langle \ell_1, \ell_2; m_1, m_2 | \ell_3, m_3 \rangle Y_{\ell_2 m_2} Y_{\ell_1 m_1} \tag{A4.51}$$

and, especially its inverse,

$$Y_{\ell_2 m_2} Y_{\ell_1 m_1} = \sum_{\ell_3 m_3} \sqrt{\frac{(2\ell_2+1)(2\ell_1+1)}{4\pi(2\ell_3+1)}} \langle \ell_2, \ell_1; 0, 0 | \ell_3 0 \rangle \times \langle \ell_2, \ell_1; m_2, m_1 | \ell_3 m_3 \rangle Y_{\ell_3 m_3}(\mathbf{n}), \tag{A4.52}$$

we can reduce integrals of three and more spherical harmonics to two spherical harmonics that we know are orthonormal. One usually writes these not in terms of the Clebsch–Gordan symbols but in the somewhat more symmetric Wigner 3J symbols, which are defined by

$$\langle \ell_1, \ell_2; m_1, m_2 | \ell_3 m_3 \rangle = (-1)^{-\ell_1+\ell_2-m_3} \sqrt{2\ell_3+1} \begin{pmatrix} \ell_1 & \ell_2 & \ell_3 \\ m_1 & m_2 & -m_3 \end{pmatrix}. \tag{A4.53}$$

Equation (A4.53) implies from the corresponding properties of the Clebsch–Gordan coefficients that the Wigner 3J symbols vanish unless the ℓ_i satisfy the triangle inequality and the m_i add up to zero, that is,

$$|\ell_1 - \ell_2| \le \ell_3 \le \ell_1 + \ell_2, \qquad m_1 + m_2 - m_3 = 0. \tag{A4.54}$$

Even though this is not evident at first sight, it is easy to check that the first of the above relations is symmetric in the ℓ_i.

The transformation of the representation matrices, Eq. (A4.35), in terms of the Wigner 3J symbols is

$$D^{(\ell)}_{mm'} D^{(\ell_1)}_{m_1 m_1'} = \sum_{LMM'} (-1)^{M+M'} (2L+1) \times \begin{pmatrix} \ell & \ell_1 & L \\ m & m_1 & -M \end{pmatrix} D^{(L)}_{MM'} \begin{pmatrix} \ell & \ell_1 & L \\ m' & m_1' & -M' \end{pmatrix}. \tag{A4.55}$$

Some properties of the Wigner 3J symbols are

$$\begin{pmatrix} \ell_1 & \ell_2 & \ell_3 \\ m_1 & m_2 & m_3 \end{pmatrix} = \begin{pmatrix} \ell_2 & \ell_3 & \ell_1 \\ m_2 & m_3 & m_1 \end{pmatrix} = \begin{pmatrix} \ell_3 & \ell_1 & \ell_2 \\ m_3 & m_1 & m_2 \end{pmatrix} \tag{A4.56}$$

$$\begin{pmatrix} \ell_1 & \ell_2 & \ell_3 \\ m_1 & m_2 & m_3 \end{pmatrix} = (-1)^{\ell_1+\ell_2+\ell_3} \begin{pmatrix} \ell_2 & \ell_1 & \ell_3 \\ m_2 & m_1 & m_3 \end{pmatrix} \tag{A4.57}$$

$$\begin{pmatrix} \ell_1 & \ell_2 & \ell_3 \\ m_1 & m_2 & m_3 \end{pmatrix} = (-1)^{\ell_1+\ell_2+\ell_3} \begin{pmatrix} \ell_1 & \ell_2 & \ell_3 \\ -m_1 & -m_2 & -m_3 \end{pmatrix} \tag{A4.58}$$

$$(2\ell_3+1) \sum_{m_1, m_2} \begin{pmatrix} \ell_1 & \ell_2 & \ell_3 \\ m_1 & m_2 & m_3 \end{pmatrix} \begin{pmatrix} \ell_1 & \ell_2 & \ell_3' \\ m_1 & m_2 & m_3' \end{pmatrix} = \begin{cases} \delta_{\ell_2, \ell_3'} \delta_{m_3, m_3'} & \text{if } \ell_1 + \ell_2 \ge \ell_3 \ge |\ell_1 - \ell_2| \\ 0 & \text{else.} \end{cases} \tag{A4.59}$$

$$\sum_{\ell_3 m_3} (2\ell_3+1) \begin{pmatrix} \ell_1 & \ell_2 & \ell_3 \\ m_1 & m_2 & m_3 \end{pmatrix} \begin{pmatrix} \ell_1 & \ell_2 & \ell_3 \\ m_1' & m_2' & m_3 \end{pmatrix} = \delta_{m_1, m_1'} \delta_{m_2 m_2'}. \tag{A4.60}$$

$$\sum_{m_1 m_2 m_3} \left[\begin{pmatrix} \ell_1 & \ell_2 & \ell_3 \\ m_1 & m_2 & m_3 \end{pmatrix} \right]^2 = 1, \tag{A4.61}$$

if $\ell_1 + \ell_2 \ge \ell_3 \ge |\ell_1 - \ell_2|$.

Corresponding identities can be established for the Clebsch–Gordan coefficients. Note also that Eq. (A4.58) implies that

$$\begin{pmatrix} \ell_1 & \ell_2 & \ell_3 \\ 0 & 0 & 0 \end{pmatrix} = 0 \text{ unless } \ell_1 + \ell_2 + \ell_3 \text{ is even.} \tag{A4.62}$$

A useful special value is

$$\begin{pmatrix} \ell_1 & \ell_2 & 0 \\ m_1 & m_2 & 0 \end{pmatrix} = \delta_{\ell_1 \ell_2} \delta_{m_1 \, -m_2} (-1)^{\ell_1 - m_1} \sqrt{\frac{1}{2\ell_1 + 1}}. \tag{A4.63}$$

For the integral of up to three spherical harmonics we obtain[1]

$$\int d\Omega \, Y_{\ell m}(\mathbf{n}) = \sqrt{4\pi} \, \delta_{\ell 0} \delta_{m0} \tag{A4.64}$$

$$\int d\Omega \, Y_{\ell_1 m_1}(\mathbf{n}) Y_{\ell_2 m_2}(\mathbf{n}) = (-1)^{m_2} \delta_{\ell_1 \ell_2} \delta_{m_1, -m_2} \tag{A4.65}$$

$$\int d\Omega \, Y_{\ell_1 m_1}(\mathbf{n}) Y_{\ell_2 m_2}(\mathbf{n}) Y_{\ell_3 m_3}(\mathbf{n}) = \sqrt{\frac{(2\ell_1 + 1)(2\ell_2 + 1)(2\ell_3 + 1)}{4\pi}}$$

$$\times \begin{pmatrix} \ell_1 & \ell_2 & \ell_3 \\ 0 & 0 & 0 \end{pmatrix} \begin{pmatrix} \ell_1 & \ell_2 & \ell_3 \\ m_1 & m_2 & m_3 \end{pmatrix} \tag{A4.66}$$

$$\equiv \mathcal{G}_{\ell_1 \ell_2 \ell_3}^{m_1 m_2 m_3} \tag{A4.67}$$

$$= \int d\Omega \, Y^*_{\ell_1 m_1}(\mathbf{n}) Y^*_{\ell_2 m_2}(\mathbf{n}) Y^*_{\ell_3 m_3}(\mathbf{n}). \tag{A4.68}$$

The integral (A4.66) is sometimes also called the "Gaunt integral" $\mathcal{G}_{\ell_1 \ell_2 \ell_3}^{m_1 m_2 m_3}$ (not to be confused with the Gaunt factor for Bremsstrahlung discussed in Chapter 10). For the last equals sign we made use of the fact that $Y^*_{\ell m} = (-1)^m Y^*_{\ell -m}$ and we used the property (A4.58) of the Wigner 3J symbols. Another way to see this is to observe that the Gaunt integrals are real, that is, they vanish when $m_1 + m_2 + m_3 \neq 0$.

Note also that $\mathcal{G}_{\ell_1 \ell_2 \ell_3}^{m_1 m_2 m_3}$ is an invariant tensor carrying the representation $D^{(\ell_1)} \otimes D^{(\ell_2)} \otimes D^{(\ell_3)}$ under rotations. In other words,

$$\mathcal{G}_{\ell_1 \ell_2 \ell_3}^{m_1 m_2 m_3} = \sum_{m_1' m_2' m_3'} D^{(\ell_1)}_{m_1 m_1'}(R) D^{(\ell_2)}_{m_2 m_2'}(R) D^{(\ell_3)}_{m_3 m_3'}(R) \mathcal{G}_{\ell_1 \ell_2 \ell_3}^{m_1' m_2' m_3'} \tag{A4.69}$$

for an arbitrary rotation R. But the representations $D^{(\ell)}$ are irreducible. Therefore, an arbitrary invariant tensor carrying this representation is in its m_i dependence proportional to $\mathcal{G}_{\ell_1 \ell_2 \ell_3}^{m_1 m_2 m_3}$ or equivalently to $\begin{pmatrix} \ell_1 & \ell_2 & \ell_3 \\ m_1 & m_2 & m_3 \end{pmatrix}$ with a proportionality factor that depends only on the ℓ_i.

For the Legendre polynomials Eq. (A4.66) implies

$$\frac{1}{2} \int_{-1}^{1} dx \, P_{\ell_1}(x) P_{\ell_2}(x) P_{\ell_3}(x) = \begin{pmatrix} \ell_1 & \ell_2 & \ell_3 \\ 0 & 0 & 0 \end{pmatrix}^2. \tag{A4.70}$$

[1] In this book we adopt the phase convention $Y_{\ell -m} = (-1)^m Y^*_{\ell m}$ like, for example, Jackson (1975), Arfken and Weber (2001) and also Mathematica. This is the origin of the factors $(-1)^m$ in the formulas.

Integrals of four spherical harmonics can also be obtained from Eq. (A4.52) with the result

$$\int d\Omega\, Y_{\ell_1 m_1}(\mathbf{n}) Y_{\ell_4 m_4}(\mathbf{n}) Y_{\ell_3 m_3}(\mathbf{n}) Y_{\ell_2 m_2}(\mathbf{n})$$

$$= \sum_{\ell'',m''} (-1)^{m''} \frac{\sqrt{(2\ell_1+1)(2\ell_4+1)(2\ell+1)(2\ell_2+1)(2\ell''+1)^2}}{4\pi} \begin{pmatrix} \ell_1 & \ell_4 & \ell'' \\ 0 & 0 & 0 \end{pmatrix}$$

$$\times \begin{pmatrix} \ell_3 & \ell_2 & \ell'' \\ 0 & 0 & 0 \end{pmatrix} \begin{pmatrix} \ell_1 & \ell_4 & \ell'' \\ m_1 & m_4 & -m'' \end{pmatrix} \begin{pmatrix} \ell_3 & \ell_2 & \ell'' \\ m_3 & m_2 & m'' \end{pmatrix}. \tag{A4.71}$$

In full generality, from the integral of n spherical harmonics one can derive the integral of $n+1$ spherical harmonics by using Eq. (A4.52) to reduce it to a sum of products with n factors.

A4.2.5 Invariant Functions on the Sphere

This section follows largely Mitsou *et al.* (2019). Here we study the properties of functions on the sphere, depending on several variables, which are invariant under arbitrary common rotations $R \in SO(3)$ of all variables,

$$f(\mathbf{n}_1, \cdots \mathbf{n}_N) = f(R\mathbf{n}_1, \cdots R\mathbf{n}_N). \tag{A4.72}$$

The interest of this comes of course from our study of N-point functions of the CMB, which have exactly this property. Let us expand f in spherical harmonics,

$$f(\mathbf{n}_1, \cdots \mathbf{n}_N) = \sum_{\ell_i m_i} f^{m_1 \cdots m_N}_{\ell_1 \cdots \ell_N} Y_{\ell_1 m_1}(\mathbf{n}_1) \cdots Y_{\ell_N m_N}(\mathbf{n}_N). \tag{A4.73}$$

Here the $\sum_{\ell_i m_i}$ symbolically indicates the sum over all $0 \le \ell_i < \infty$ and $-\ell_i \le m_i \le \ell_i$. We shall show that the m_i dependence of the coefficients $f^{m_1 \cdots m_N}_{\ell_1 \cdots \ell_N}$ is fully fixed by the invariance property (A4.72), and we can re-write Eq. (A4.73) in terms of coefficients that depend only on the ℓ_i and certain auxilary L_i. First we use that

$$f^{m_1 \cdots m_N}_{\ell_1 \cdots \ell_N} = \int \prod_{i=1}^{N} \left[d\Omega_i Y^*_{\ell_i m_i}(\mathbf{n}_i) \right] f(\mathbf{n}_1, \cdots \mathbf{n}_N). \tag{A4.74}$$

Now we act with an arbitrary rotation on $f(\mathbf{n}_1, \cdots \mathbf{n}_N)$ and use the fact that it is invariant,

$$f^{m_1 \cdots m_N}_{\ell_1 \cdots \ell_N} = \int \prod_{i=1}^{N} \left[d\Omega_i Y^*_{\ell_i m_i}(\mathbf{n}_i) \right] f(R\mathbf{n}_1, \cdots R\mathbf{n}_N) \tag{A4.75}$$

$$= \int \prod_{i=1}^{N} \left[d\Omega_i Y^*_{\ell_i m_i}(R^{-1}\mathbf{n}_i) \right] f(\mathbf{n}_1, \cdots \mathbf{n}_N) \tag{A4.76}$$

$$= \int \prod_{i=1}^{N} \left[d\Omega_i D^{(\ell_i)}_{m'_i m_i}(R) Y^*_{\ell_i m'_i}(\mathbf{n}_i) \right] f(\mathbf{n}_1, \cdots \mathbf{n}_N) \tag{A4.77}$$

$$= \prod_{i=1}^{N} \left[D^{(\ell_i)}_{m'_i m_i}(R) \right] f^{m'_1 \cdots m'_N}_{\ell_1 \cdots \ell_N}. \tag{A4.78}$$

Here we have used that $D^{(\ell_i)}_{m_i m'_i}(R^{-1}) = D^{(\ell_i)*}_{m'_i m_i}(R)$ and the sum over the m'_i is understood. We introduce the coefficients

$$I^{\ell_1\cdots\ell_N}_{m_1\cdots m_N;m'_1\cdots m'_N} = \int_{SO(3)} dR \prod_{i=1}^{N}\left[D^{(\ell_i)}_{m'_i m_i}(R)\right], \tag{A4.79}$$

where dR is the normalized Haar measure (Wigner, 1959) of the rotation group. In terms of the Euler angles,

$$\int dRF(R(\alpha,\beta,\gamma)) = \frac{1}{8\pi^2}\int_0^{2\pi} d\alpha \int_0^{\pi} d\beta \sin\beta \int_0^{2\pi} d\gamma F(\alpha,\beta,\gamma). \tag{A4.80}$$

Since the Haar mesure is invariant under rotations, so are the coefficients $I^{\ell_1\cdots\ell_N}_{m_1\cdots m_N;m'_1\cdots m'_N}$. Integrating Eq. (A4.78) over the rotation group we then have the relation

$$f^{m_1\cdots m_N}_{\ell_1\cdots\ell_N} = I^{\ell_1\cdots\ell_N}_{m_1\cdots m_N;m'_1\cdots m'_N} f^{m'_1\cdots m'_N}_{\ell_1\cdots\ell_N}. \tag{A4.81}$$

Let us first evaluate this relation for the case $N = 2$. For this we use also that the matrix elements of the representation matrices, $D^{(\ell)}_{mm'}$, form a complete orthonormal system of functions on $SO(3)$. Together with the relation

$$D^{(\ell)*}_{m'm} = (-1)^{m+m'} D^{(\ell)}_{-m'-m} \tag{A4.82}$$

we obtain for $N = 2$

$$I^{\ell_1\ell_2}_{m_1 m_2;m'_1 m'_2} = \frac{(-1)^{m_1+m'_1}}{2\ell_1+1}\delta_{\ell_1\ell_2}\delta_{-m_1 m_2}\delta_{-m'_1 m'_2}. \tag{A4.83}$$

Hence for $N = 2$ the only nonvanishing coefficients are of the form $f^{-mm}_{\ell\ell}$ and Eq. (A4.81) implies that they are independent of the value of m. Setting

$$f_\ell = (-1)^m f^{-m\,m}_{\ell} \tag{A4.84}$$

and using the addition theorem we find

$$f(\mathbf{n}_1,\mathbf{n}_2) = \sum_\ell f_\ell P_\ell(\mathbf{n}_1\cdot\mathbf{n}_2). \tag{A4.85}$$

This is the well-known relation that we used intuitively for the CMB temperature 2-point function. To go to arbitrary N, we use Eq. (A4.55) to reduce N on the right-hand side of Eq. (A4.79) by 1 and proceed by induction. A lengthy but straightforward analysis yields the following result:

$$I^{\ell_1\cdots\ell_N}_{m_1\cdots m_N;m'_1\cdots m'_N} = \sum_{L_i} W^{\ell_1\ldots\ell_N|L_1\ldots L_{N-3}}_{m_1\ldots m_N} W^{\ell_1\ldots\ell_N|L_1\ldots L_{N-3}}_{m'_1\ldots m'_N}, \tag{A4.86}$$

where we have introduced the generalized Wigner symbols

$$W_{m_1\ldots m_N}^{\ell_1\ldots\ell_N|L_1\ldots L_{N-3}} \equiv \sum_{M_i}\begin{pmatrix}\ell_1 & \ell_2 & L_1\\ m_1 & m_2 & -M_1\end{pmatrix}$$
$$\times\left[\prod_{k=1}^{N-4}(-1)^{L_k+M_k}\sqrt{2L_k+1}\begin{pmatrix}L_k & \ell_{k+2} & L_{k+1}\\ M_k & m_{k+2} & -M_{k+1}\end{pmatrix}\right]$$
$$\times(-1)^{L_{N-3}+M_{N-3}}\sqrt{2L_{N-3}+1}\begin{pmatrix}L_{N-3} & \ell_{N-1} & \ell_N\\ M_{N-3} & m_{N-1} & m_N\end{pmatrix} \tag{A4.87}$$

for $N \geq 4$,

$$W_{m_1\ldots m_3}^{\ell_1\ldots\ell_3} \equiv \begin{pmatrix}\ell_1 & \ell_2 & \ell_3\\ m_1 & m_2 & m_3\end{pmatrix} \quad \text{for } N=3. \tag{A4.88}$$

The sums over the M_i's run, as usual, from $-L_i$ to L_i.

We now introduce the following invariant functions on $(\mathbb{S}^2)^N$,

$$Y_{\ell_1\ldots\ell_N|L_1\ldots L_{N-3}}(\mathbf{n}_1,\cdots,\mathbf{n}_N) \equiv W_{m_1\ldots m_N}^{\ell_1\ldots\ell_N|L_1\ldots L_{N-3}}\prod_{k=1}^{N}Y_{\ell_k m_k}(\mathbf{n}_k), \tag{A4.89}$$

where also here the sum over the m_i is understood. It is easy to check that the generalized Wigner symbols transform inversely to the $Y_{\ell m}$'s so that their product is invariant under a global rotation. With the same sum convention as in Eq. (A4.89) we introduce also the rotation invariant coefficients

$$f^{\ell_1\ldots\ell_N|L_1\ldots L_{N-3}} \equiv W_{m_1\ldots m_N}^{\ell_1\ldots\ell_N|L_1\ldots L_{N-3}} f_{\ell_1\cdots\ell_N}^{m_1\cdots m_N}. \tag{A4.90}$$

So the invariant function can be written as

$$f(\mathbf{n}_1,\cdots,\mathbf{n}_N) = \sum f^{\ell_1\ldots\ell_N|L_1\ldots L_{N-3}}Y_{\ell_1\ldots\ell_N|L_1\ldots L_{N-3}}(\mathbf{n}_1,\cdots,\mathbf{n}_N), \tag{A4.91}$$

where the sum now is taken over all $\ell_i \geq 0$ and $L_i \geq 0$. Each coefficient and each function is now manifestly invariant under rotation. It is an easy exercise to show that for $N=1$ only $\ell=0$ survives while for $N=2$, we obtain $Y_{\ell_1,\ell_2}(\mathbf{n}_1,\mathbf{n}_2) \propto \delta_{\ell_1\ell_2}P_\ell(\mathbf{n}_1\cdot\mathbf{n}_2)$. Due to the Wigner symbol factors, the triples (ℓ_1,ℓ_2,L_1), (L_1,ℓ_3,L_2), (L_2,ℓ_4,L_3), ..., $(L_{N-3},\ell_{N-1},\ell_N)$ have to satisfy the triangle relations. Graphically this is represented in Fig. A4.1. For all the triangles appearing in this figure, the triangle relation must be satisfied; otherwise the generalized Wigner symbol $W_{m_1\ldots m_N}^{\ell_1\ldots\ell_N|L_1\ldots L_{N-3}}$ vanishes. Note also that in (A4.87) all the M_i's appear with a positive and a negative sign so that the generalized Wigner symbols also vanish if $m_1+\cdots m_N \neq 0$.

A4.2.6 Spherical Harmonics of Spin s

We now consider tensor fields on the sphere[2]. We can express their components in terms of the standard "real" orthonormal basis, $\mathbf{e}_1=\mathbf{e}_\vartheta=\partial_\vartheta$, $\mathbf{e}_2=\mathbf{e}_\varphi=\frac{1}{\sin\vartheta}\partial_\varphi$, or in terms of the helicity basis

[2] A more detailed treatment of spin weighted spherical harmonics can be found in Goldberg (1967) and Newman and Penrose (1966).

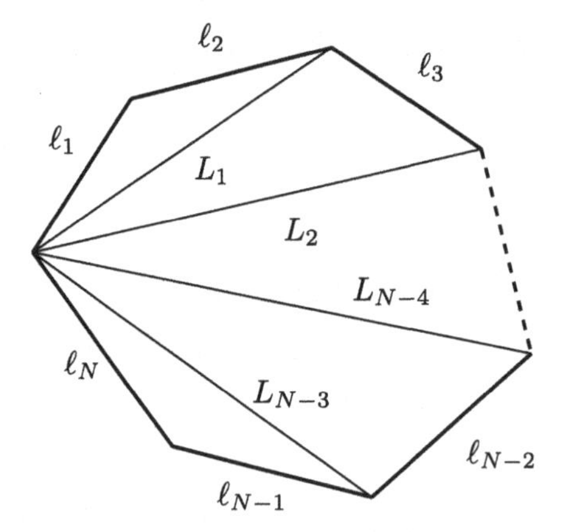

Fig. A4.1 The triangles that satisfy the triangle inequality for the generalized Wigner symbols.

$$\mathbf{e}_+ = \frac{1}{\sqrt{2}}(\mathbf{e}_1 - i\mathbf{e}_2), \quad \mathbf{e}_- = \frac{1}{\sqrt{2}}(\mathbf{e}_1 + i\mathbf{e}_2). \tag{A4.92}$$

Here we identify, as is often done, a vector with the derivative in a given direction. A vector is then defined by its action on functions: $\mathbf{e}_1 f = \partial_\vartheta f$ and $\mathbf{e}_2 f = \frac{1}{\sin\vartheta}\partial_\varphi f$. Note also that the metric $ds^2 = d\vartheta^2 + \sin^2\vartheta\, d\varphi^2$ on the sphere has the components $g_{-+} = g_{+-} = 1$, and $g_{++} = g_{--} = 0$ in the helicity basis.

Under a rotation of the "real" basis, $\mathbf{e}_1 \to \cos\gamma\mathbf{e}_1 - \sin\gamma\mathbf{e}_2$, $\mathbf{e}_2 \to \cos\gamma\mathbf{e}_2 + \sin\gamma\mathbf{e}_1$, the helicity basis transforms as $\mathbf{e}_+ \to e^{-i\gamma}\mathbf{e}_+$, $\mathbf{e}_- \to e^{i\gamma}\mathbf{e}_-$.

The components of a tensor field of rank r in the helicity basis transform under a rotation by

$$T^{\overbrace{+\cdots+}^{s}\overbrace{-\cdots-}^{r-s}} \to e^{i(2s-r)\gamma}\, T^{\overbrace{+\cdots+}^{s}\overbrace{-\cdots-}^{r-s}}. \tag{A4.93}$$

For example, the components of a vector transform as $V^+ \to e^{i\gamma}V^+$ and $V^- \to e^{-i\gamma}V^-$. (The vector itself $V = V^+\mathbf{e}_+ + V^-\mathbf{e}_-$ is invariant; hence the components transform "contragradient" to the basis.) Components that transform with $e^{is\gamma}$ are called components of spin $|s|$ and of helicity s.

We are mainly interested in symmetric (real) rank-2 tensors like the polarization. These have one spin-0 component $T^{+-} = T^{-+} = \frac{1}{2}\mathrm{tr}\, T \equiv I$; one spin-2 component with positive helicity, $T^{++} = \frac{1}{2}(T_{11} - T_{22}) - iT_{12}$; and one spin-2 component with negative helicity, $T^{--} = \frac{1}{2}(T_{11} - T_{22}) + iT_{12}$ (see Exercise A4.1).

We can write a symmetric rank-2 tensor in these components as

$$T = \frac{1}{2}\left[I\mathbb{1} + T^{++}\sigma_+ + T^{--}\sigma_-\right], \tag{A4.94}$$

where $\sigma_\pm = \sigma^3 \pm i\sigma^1$ are given by the Pauli matrices,

$$\sigma^3 = \begin{pmatrix} 1 & 0 \\ 0 & -1 \end{pmatrix}, \quad \sigma^1 = \begin{pmatrix} 0 & 1 \\ 1 & 0 \end{pmatrix}.$$

To expand a spin-s component of a tensor field on the sphere, one employs the spin weighted spherical harmonics. These are spin-s components of tensor fields on the 2-sphere. In the basis $(\mathbf{e}_\vartheta, \mathbf{e}_\varphi)$ they are given in terms of the irreducible representations of the rotation group by

$$ {}_sY_{\ell m}(\vartheta,\varphi) \equiv \sqrt{\frac{2\ell+1}{4\pi}} D^{*(\ell)}_{m-s}(\varphi,\vartheta,0), \tag{A4.95}$$

$$= (-1)^m \sqrt{\frac{(2\ell+1)}{4\pi}\frac{(\ell+m)!\,(\ell-m)!}{(\ell+s)!\,(\ell-s)!}}(\sin\vartheta/2)^{2\ell}e^{im\varphi} \times \sum_r \binom{\ell-s}{r}\binom{\ell+s}{r+s-m}(-1)^{\ell-r-s}(\cot\vartheta/2)^{2r+s-m}. \tag{A4.96}$$

Here the sum over r goes over those values for which the binomial coefficients are nonvanishing; this means $\max\{0, m-s\} \le r \le \min\{\ell-s, \ell+s\}$. (Remember $\binom{0}{0} = 1$.) The spin weighted spherical harmonics are defined for $|s| \le \ell$ and $|m| \le \ell$. For each fixed spin s they form a complete set of orthonormal functions on the sphere, so that

$$\int d\Omega_{\mathbf{n}}\, {}_sY^*_{\ell m}(\mathbf{n})\, {}_sY_{\ell' m'}(\mathbf{n}) = \delta_{\ell\ell'}\delta_{mm'}, \tag{A4.97}$$

and

$$\sum_{\ell m} {}_sY^*_{\ell m}(\mathbf{n})\, {}_sY_{\ell m}(\mathbf{n}) = \delta(\varphi-\varphi')\delta(\cos\vartheta - \cos\vartheta'). \tag{A4.98}$$

An initial rotation with angle ψ around the z-axis simply multiplies the matrix element $D^{(\ell)}_{m\,-s}$ by a factor $e^{-is\psi}$, so that

$$D^{*(\ell)}_{m\,-s}(\varphi,\vartheta,\psi) = \sqrt{\frac{4\pi}{2\ell+1}}\, {}_sY_{\ell m}(\vartheta,\varphi)e^{is\psi}. \tag{A4.99}$$

Now let R_1 be the rotation with Euler angles $(\varphi_1, \vartheta_1, 0)$ that rotates $\mathbf{e}_z$ into $\mathbf{n}_1$ and R_2 the rotation with Euler angles $(\varphi_2, \vartheta_2, 0)$ that rotates $\mathbf{e}_z$ into $\mathbf{n}_2$. Let (α, β, γ) be the Euler angles of the rotation $R_1^{-1}R_2$, which first rotates $\mathbf{n}_2$ into $\mathbf{e}_z$ and then $\mathbf{e}_z$ into $\mathbf{n}_1$. We then have

$$\begin{aligned} D^{(\ell)}_{m\,-s}(\alpha,\beta,\gamma) &= \sum_{m'} D^{(\ell)-1}_{mm'}(\varphi_1,\vartheta_1,0)D^{(\ell)}_{m'\,-s}(\varphi_2,\vartheta_2,0) \\ &= \sum_{m'} D^{*(\ell)}_{m'm}(\varphi_1,\vartheta_1,0)D^{(\ell)}_{m'\,-s}(\varphi_2,\vartheta_2,0) \\ &= \frac{4\pi}{2\ell+1}\sum_{m'} {}_sY^*_{\ell m'}(\mathbf{n}_2)\, {}_{-m}Y_{\ell m'}(\mathbf{n}_1). \end{aligned} \tag{A4.100}$$

Using Eq. (A4.99) we find

$$\sqrt{\frac{4\pi}{2\ell+1}}\sum_{m'} {}_sY_{\ell m'}(\vartheta_2,\varphi_2)\, {}_{-m}Y^*_{\ell m'}(\vartheta_1,\varphi_1) = {}_sY_{\ell m}(\beta,\alpha)e^{-is\gamma}. \tag{A4.101}$$

This is the generalized addition theorem for spin weighted spherical harmonics.

In analogy to $L_\pm$ as raising and lowering operators for the "magnetic quantum number" m, we introduce the spin raising and lowering operators $\eth$ and $\eth^*$. They are defined by $\mu = \cos\vartheta$,

$$\eth\, {}_sY_{\ell m} = \left(s\,\mathrm{ctg}\vartheta - \partial_\vartheta - \frac{i}{\sin\vartheta}\partial_\varphi\right) {}_sY_{\ell m} \tag{A4.102}$$

$$= \left(\frac{s\mu}{\sqrt{1-\mu^2}} + \sqrt{1-\mu^2}\partial_\mu - \frac{i}{\sqrt{1-\mu^2}}\partial_\varphi\right) {}_sY_{\ell m} \tag{A4.103}$$

$$= -(1-\mu^2)^{\frac{s}{2}}\left(\partial_\vartheta + \frac{i\partial_\varphi}{\sqrt{1-\mu^2}}\right)[(1-\mu^2)^{-s/2}\, {}_sY_{\ell m}],$$

$$\eth^*\, {}_sY_{\ell m} = \left(-s\,\mathrm{ctg}\vartheta - \partial_\vartheta + \frac{i}{\sin\vartheta}\partial_\varphi\right) {}_sY_{\ell m} \tag{A4.104}$$

$$= \left(\frac{-s\mu}{\sqrt{1-\mu^2}} + \sqrt{1-\mu^2}\partial_\mu + \frac{i}{\sqrt{1-\mu^2}}\partial_\varphi\right) {}_sY_{\ell m} \tag{A4.105}$$

$$= -(1-\mu^2)^{\frac{-s}{2}}\left(\partial_\vartheta - \frac{i\partial_\varphi}{\sqrt{1-\mu^2}}\right)[(1-\mu^2)^{s/2}\, {}_sY_{\ell m}].$$

The interest of these operators is that they allow us to construct the spin weighted spherical harmonics directly from the spin-0 harmonics and, inversely, we can use them to build spin-0 quantities from spin weighted harmonics. One can actually show [e.g., by using Eq. (A4.96)] that

$$\eth\,({}_sY_{\ell m}) = \sqrt{(\ell-s)(\ell+s+1)}\, {}_{s+1}Y_{\ell m}, \tag{A4.106}$$

$$\eth^*\,({}_sY_{\ell m}) = -\sqrt{(\ell+s)(\ell-s+1)}\, {}_{s-1}Y_{\ell m}, \tag{A4.107}$$

and therefore

$$\eth^2\,({}_{-2}Y_{\ell m}) = \sqrt{\frac{(\ell+2)!}{(\ell-2)!}}\,Y_{\ell m}, \tag{A4.108}$$

$$(\eth^*)^2\,({}_2Y_{\ell m}) = \sqrt{\frac{(\ell+2)!}{(\ell-2)!}}\,Y_{\ell m}. \tag{A4.109}$$

On the other hand, the spin-2 spherical harmonics can be obtained from the ordinary spherical harmonics by acting twice with the differential operators $\eth$ or $\eth^*$,

$$(\eth)^2 Y_{\ell m} = \sqrt{\frac{(\ell+2)!}{(\ell-2)!}}\, {}_2Y_{\ell m}, \tag{A4.110}$$

$$(\eth^*)^2 Y_{\ell m} = \sqrt{\frac{(\ell+2)!}{(\ell-2)!}}\, {}_{-2}Y_{\ell m}. \tag{A4.111}$$

To see that $\eth\, {}_sf$ has spin $s+1$ and $\eth^*\, {}_sf$ has spin $s-1$, for an arbitrary component ${}_sf$ with spin weight s, we show that $\eth$ is proportional to a covariant derivative in direction $g^{+-}\mathbf{e}_-$ and correspondingly $\eth^* \propto g^{-+}\nabla_{\mathbf{e}_+}$. Since $\eth^*$ is the adjoint of $-\eth$, it is sufficient if we show the first identity. For $s=0$, using $\mathbf{e}_+ = \frac{1}{\sqrt{2}}(\mathbf{e}_\vartheta - i\mathbf{e}_\varphi) = \frac{1}{\sqrt{2}}(\partial_\vartheta - i\frac{1}{\sin\vartheta}\partial_\varphi)$, we obtain

$\eth^* f = -\sqrt{2}\mathbf{e}_+ f$. For $s \neq 0$ tensor fields we have to compute the Christoffel symbols in order to determine the covariant derivatives. In terms of the helicity basis, the canonical metric on the 2-sphere takes the form $ds^2 = 2\theta^+\theta^-$, where $\theta^\pm$ denote the 1-forms dual to the vector fields $\mathbf{e}_\pm$ defined by $\theta^+(\mathbf{e}_+) = \theta^-(\mathbf{e}_-) = 1$ and $\theta^-(\mathbf{e}_+) = \theta^+(\mathbf{e}_-) = 0$. Hence

$$\theta^\pm = \frac{1}{\sqrt{2}}\,(d\vartheta \pm i \sin\vartheta\, d\varphi)\,.$$

Therefore, the metric components are simply $g_{++} = g_{--} = 0$ and $g_{-+} = g_{+-} = 1$. A careful evaluation of the Christoffel symbols defined by $\nabla_{\mathbf{e}_k}\mathbf{e}_j = \Gamma^i_{kj}\mathbf{e}_i$ in the helicity basis gives[3]

$$\Gamma^+_{-+} = \Gamma^-_{+-} = -\Gamma^+_{++} = -\Gamma^-_{--} = -\frac{1}{\sqrt{2}}\frac{\cos\vartheta}{\sin\vartheta} = -\frac{1}{\sqrt{2}}\mathrm{ctg}\vartheta. \tag{A4.112}$$

All other Christoffel symbols vanish. With this we find for the covariant derivatives of the spin-s components of a tensor

$$\begin{aligned} T^{+\cdots+;+} = T^{+\cdots+}_{;-} &= \mathbf{e}_-(T^{+\cdots+}) - \frac{s}{\sqrt{2}}\mathrm{ctg}\vartheta\, T^{+\cdots+}, \\ &= \frac{1}{\sqrt{2}}\left(\partial_\vartheta + \frac{i}{\sin\vartheta}\partial_\varphi\right)(T^{+\cdots+}) - \frac{s}{\sqrt{2}}\mathrm{ctg}\vartheta\, T^{+\cdots+}, \\ &= \frac{1}{\sqrt{2}}\left[\partial_\vartheta - s\,\mathrm{ctg}\vartheta + \frac{i}{\sin\vartheta}\partial_\varphi\right]T^{+\cdots+} \\ &= \frac{-1}{\sqrt{2}}\eth T^{+\cdots+}. \end{aligned}$$

In the same way one obtains

$$T^{+\cdots+;-} = \frac{-1}{\sqrt{2}}\eth^* T^{+\cdots+}.$$

In other words,

$$\eth = -\sqrt{2}\,\nabla_{\mathbf{e}_-} \text{ and } \eth^* = -\sqrt{2}\,\nabla_{\mathbf{e}_+}. \tag{A4.113}$$

Since for an arbitrary rank-s tensor field the component $T^{+\cdots+}$ has helicity s and $T^{+\cdots+;+}$ has helicity $s+1$ while $T^{+\cdots+;-}$ has helicity $s-1$, this shows that $\eth$ and $\eth^*$ are spin raising and lowering operators. Correspondingly one obtains $\frac{-1}{\sqrt{2}}\eth^* T^{-\cdots-} = T^{-\cdots-;-}$ and $\frac{-1}{\sqrt{2}}\eth T^{-\cdots-} = T^{-\cdots-;+}$, which have spin weight $-s-1$ and $-s+1$ respectively.

The spin-2 spherical harmonics are therefore just the doubly covariant derivatives of the usual spherical harmonics,

$${}_{-2}Y_{\ell m} = 2\sqrt{\frac{(\ell-2)!}{(\ell+2)!}}\nabla_{\mathbf{e}_+}\nabla_{\mathbf{e}_+}Y_{\ell m}, \tag{A4.114}$$

$${}_{+2}Y_{\ell m} = 2\sqrt{\frac{(\ell-2)!}{(\ell+2)!}}\nabla_{\mathbf{e}_-}\nabla_{\mathbf{e}_-}Y_{\ell m}. \tag{A4.115}$$

[3] These Christoffel symbols are most easily calculated using the Cartan formalism, which can be found in most modern books on general relativity, for example, Straumann (2004).

Finally, we want to interpret the spin-0 quantities $\eth\eth T^{--}$ and $\eth^*\eth^* T^{++}$ of a symmetric traceless spin-2 tensor $T = T^{++}\mathbf{e}_+ \otimes \mathbf{e}_+ + T^{--}\mathbf{e}_- \otimes \mathbf{e}_-$. In Exercise A4.2.2 we find that for a vector with components V^+ and V^-, the divergence and curl are given by

$$V^+_{;-} + V^-_{;+} = V^i_{;i} = \text{div}\, V \quad\text{and}\quad V^+_{;-} - V^-_{;+} = -i\epsilon_{ij}V^{i;j} = -i\,\text{rot}\, V.$$

Here ϵ_{ij} is the totally antisymmetric tensor in two dimensions,

$$\epsilon_{\vartheta\vartheta} = \epsilon_{\varphi\varphi} = 0, \qquad \epsilon_{\vartheta\varphi} = -\epsilon_{\varphi\vartheta} = \sqrt{\det g} = \sin\vartheta. \tag{A4.116}$$

In our helicity basis this tensor is

$$\epsilon_{++} = \epsilon_{--} = 0, \qquad \epsilon_{+-} = -\epsilon_{-+} = i. \tag{A4.117}$$

One also finds

$$\frac{1}{2}\left(\eth^{*2}T^{++} + \eth^2 T^{--}\right) = T^{++;--} + T^{--;++}$$
$$= T^{ij}{}_{;ij} = \text{div}(\text{div}(T)), \tag{A4.118}$$
$$\frac{1}{2}\left(\eth^{*2}T^{++} - \eth^2 T^{--}\right) = T^{++;--} - T^{--;++}$$
$$= -\frac{i}{2}\epsilon_{ik}\left[(T^{ij}{}_{;j})^{;k} + (T^{ij;k})_{;j}\right]$$
$$= -\frac{i}{2}\left[\text{rot}(\text{div}(T)) + \text{div}(\text{rot}(T))\right]. \tag{A4.119}$$

These results are verified most easily by just writing out both sides in components. Hence the sum of the two scalars $T^{++;--}$ and $T^{--;++}$ gives the double divergence while their difference gives the curl of the divergence of T. Note that in two dimensions the curl of a vector is a (pseudo-)scalar. It is a three-dimensional (pseudo-)vector with a purely radial component that as a field on the tangent space of the sphere has scalar character. The curl of a 2-tensor is a (pseudo-)vector.

Exercises

A4.2.1 Tensor components in the helicity basis
Show that the components T^{ij} of a 2-tensor on the sphere are related to the components in the helicity basis $\mathbf{e}_+$ and $\mathbf{e}_-$ via

$$T^{\pm,\pm} = \frac{1}{2}\left(T^{11} - T^{22} \mp i(T^{12} + T^{21})\right),$$
$$T^{\pm,\mp} = \frac{1}{2}\left(T^{11} + T^{22} \pm i(T^{12} - T^{21})\right).$$

In particular, if T is symmetric $T^{+-} = T^{-+} = \frac{1}{2}\text{tr}\, T$.

A4.2.2 Divergence and curl in the helicity basis Show that for a vector field on the sphere given by

$$V = V^+ e_+ + V^- e_- = V^\vartheta \partial_\vartheta + V^\varphi \partial_\varphi,$$

we have

$$V^{+;-} + V^{-;+} = V^{\vartheta}_{,\vartheta} + V^{\varphi}_{,\varphi} + \mathrm{ctg}\vartheta\, V^{\vartheta} = \mathrm{div}V \tag{A4.120}$$

$$V^{+;-} - V^{-;+} = -i\left(\frac{1}{\sin\vartheta}V^{\vartheta}_{,\varphi} - \sin\vartheta\, V^{\varphi}_{,\vartheta} - 2\cos\vartheta\, V^{\varphi}\right),$$

$$= \frac{-i}{\sin\vartheta}\left(V^{\vartheta}_{,\varphi} - (\sin^2\vartheta\, V^{\varphi})_{,\vartheta}\right) = -i\,\mathrm{rot}V. \tag{A4.121}$$

Solution

We first calculate the Christoffel symbols with respect to the coordinate basis (ϑ,φ). Using $ds^2 = d\vartheta^2 + \sin^2\vartheta\, d\varphi^2$ quickly gives

$$\Gamma^{\vartheta}_{\varphi\varphi} = -\cos\vartheta\sin\vartheta, \quad \Gamma^{\varphi}_{\varphi\vartheta} = \Gamma^{\varphi}_{\vartheta\varphi} = \mathrm{ctg}\,\vartheta,$$

and zero for all other Christoffel symbols, so that

$$V^{\vartheta}_{;\vartheta} = V^{\vartheta}_{,\vartheta}, \tag{A4.122}$$

$$V^{\vartheta}_{;\varphi} = V^{\vartheta}_{,\varphi} - \cos\vartheta\sin\vartheta\, V^{\varphi}, \tag{A4.123}$$

$$V^{\varphi}_{;\varphi} = V^{\varphi}_{,\varphi} + \mathrm{ctg}\vartheta\, V^{\vartheta}, \tag{A4.124}$$

$$V^{\varphi}_{;\vartheta} = V^{\varphi}_{,\vartheta} + \mathrm{ctg}\vartheta\, V^{\varphi}. \tag{A4.125}$$

so that

$$\mathrm{div}V = V^{\vartheta}_{;\vartheta} + V^{\varphi}_{;\varphi} = V^{\vartheta}_{,\vartheta} + V^{\varphi}_{,\varphi} + \mathrm{ctg}\vartheta\, V^{\vartheta}, \tag{A4.126}$$

$$\mathrm{rot}V = \sqrt{\det g}(V^{\vartheta;\varphi} - V^{\varphi;\vartheta}) = \sin\vartheta\left(\frac{1}{\sin^2\vartheta}V^{\vartheta}_{;\varphi} - V^{\varphi}_{;\vartheta}\right)$$

$$= \frac{1}{\sin\vartheta}V^{\vartheta}_{,\varphi} - \sin\vartheta\, V^{\varphi}_{,\vartheta} - 2\cos\vartheta\, V^{\varphi}. \tag{A4.127}$$

On the other hand, we have

$$V^{+;-} \pm V^{-;+} = V^{+}_{;+} \pm V^{-}_{;-} = V^{+}_{,+} \pm V^{-}_{,-} + \Gamma^{+}_{++}V^{+} \pm \Gamma^{-}_{--}V^{-}. \tag{A4.128}$$

Inserting $V^{\pm} = \frac{1}{\sqrt{2}}(V^{\vartheta} \pm i\sin\vartheta\, V^{\varphi})$ and $V^{\bullet}_{,\pm} = \frac{1}{\sqrt{2}}(\partial_{\vartheta} \mp \frac{i}{\sin\vartheta}\partial_{\varphi})V^{\bullet}$ together with $\Gamma^{+}_{++} = \Gamma^{-}_{--} = \frac{1}{\sqrt{2}}\mathrm{ctg}\,\vartheta$ we find the above result.

A4.2.3 The Laplacian in the helicity basis

In the helicity basis the Laplacian on the sphere is given by

$$\Delta = g^{ab}\nabla_b\nabla_a = g^{-+}\nabla_-\nabla_+ + g^{+-}\nabla_+\nabla_- = \nabla_-\nabla_+ + \nabla_+\nabla_- = \frac{1}{2}\left(\eth\eth^* + \eth^*\eth\right).$$

Apply this formula to the spherical harmonics $Y_{\ell m}$ to show that $\Delta Y_{\ell m} = -\ell(\ell+1)Y_{\ell m}$.

A4.3 Bessel Functions

A4.3.1 Bessel Functions of Integer Order

The Bessel functions $J_\nu(x)$ and $Y_\nu(x)$ are real solutions to the differential equation

$$x^2\frac{d^2f}{dx^2}+x\frac{df}{dx}+(x^2-\nu^2)f=0. \tag{A4.129}$$

We only consider $\nu \in \mathbb{R}$; hence we may consider $\nu \geq 0$. Actually J_ν and $J_{-\nu}$ satisfy the same differential equation but one defines another independent solution Y_ν that is related to $J_{-\nu}$ via

$$J_{-\nu}=\cos(\nu\pi)J_\nu-\sin(\nu\pi)Y_\nu.$$

The Bessel functions J_ν and Y_ν are also analytic in the order ν. For $\nu \geq 0$, J_ν is regular at $x=0$, $J_\nu(x) \propto x^\nu$ for $|x| \ll \nu$. On the other hand, Y_ν diverges, $Y_\nu(x) \propto x^{-\nu}$ for $|x| \ll \nu$ and $\nu > 0$, while $Y_0(x) \propto \ln(x)$. For large values of $|x|$ both functions oscillate with a period of approximately 2π and decay like $1/\sqrt{x}$. We sometimes also use the Hankel functions defined by

$$H_\nu^{(1)}=J_\nu+iY_\nu, \quad H_\nu^{(2)}=J_\nu-iY_\nu. \tag{A4.130}$$

All of these functions satisfy the recurrence relations

$$F_{\nu-1}+F_{\nu+1}=\frac{2\nu}{x}F_\nu, \tag{A4.131}$$

$$F_{\nu-1}-F_{\nu+1}=2F'_\nu, \tag{A4.132}$$

$$F_{\nu-1}-\frac{\nu}{x}F_\nu=F'_\nu, \tag{A4.133}$$

$$-F_{\nu+1}+\frac{\nu}{x}F_\nu=F'_\nu. \tag{A4.134}$$

The Bessel functions are well defined in the complex plane (with suitably chosen cuts) even for complex values ν. The Bessel functions J_n, $n \in \mathbb{N}$ can be represented as the integral

$$J_n(x)=\frac{(-i)^n}{\pi}\int_0^\pi e^{ix\cos\theta}\cos(n\theta)\,d\theta. \tag{A4.135}$$

Table A4.3 *The spin-2 spherical harmonics with $\ell = 2$.*

m	${}_0Y_{2m}$	${}_{\pm 2}Y_{2m}$
2	$\frac{1}{4}\sqrt{\frac{15}{2\pi}}\sin^2\vartheta e^{2i\varphi}$	$\frac{1}{8}\sqrt{\frac{5}{\pi}}(1-\cos\vartheta)^2 e^{2i\varphi}$
1	$\sqrt{\frac{15}{8\pi}}\sin\vartheta\cos\vartheta e^{i\varphi}$	$\frac{1}{4}\sqrt{\frac{5}{\pi}}\sin\vartheta(1-\cos\vartheta)e^{i\varphi}$
0	$\frac{1}{2}\sqrt{\frac{5}{4\pi}}(3\cos^2\vartheta-1)$	$\frac{3}{4}\sqrt{\frac{5}{6\pi}}\sin^2\vartheta$
-1	$-\sqrt{\frac{15}{8\pi}}\cos\vartheta\sin\vartheta e^{-i\varphi}$	$\frac{1}{4}\sqrt{\frac{5}{\pi}}\sin\vartheta(1+\cos\vartheta)e^{-i\varphi}$
-2	$\frac{1}{4}\sqrt{\frac{15}{2\pi}}\sin^2\vartheta e^{-2i\varphi}$	$\frac{1}{8}\sqrt{\frac{5}{\pi}}(1+\cos\vartheta)^2 e^{-2i\varphi}$

With this one finds the useful expansion

$$e^{iy\cos\phi} = J_0(y) + 2\sum_{n=1}^{\infty} i^n J_n(y)\cos(n\phi) = \sum_{n=-\infty}^{\infty} i^n J_n(y) e^{in\phi}. \tag{A4.136}$$

We shall also employ the modified Bessel functions, which are defined by

$$I_\nu(x) = (-i)^\nu J_\nu(ix), \tag{A4.137}$$

$$K_\nu(x) = \frac{i\pi}{2}(i)^\nu H_\nu^{(1)}(ix) = -\frac{i\pi}{2}(-i)^\nu H_\nu^{(2)}(-ix). \tag{A4.138}$$

A4.3.2 Spherical Bessel Functions

The spherical Bessel (Hankel) functions are Bessel (Hankel) functions of half-integer order,

$$j_n(x) = \sqrt{\frac{\pi}{2x}} J_{n+1/2}(x), \tag{A4.139}$$

$$y_n(x) = \sqrt{\frac{\pi}{2x}} Y_{n+1/2}(x), \tag{A4.140}$$

$$h_n^{(1)} = j_n + i y_n, \tag{A4.141}$$

$$h_n^{(2)} = j_n - i y_n. \tag{A4.142}$$

They are solutions of the differential equation

$$x^2 \frac{d^2 f}{dx^2} + 2x \frac{df}{dx} + (x^2 - n(n+1)) f = 0. \tag{A4.143}$$

The spherical Bessel/Hankel functions satisfy the recurrence relations

$$\frac{f_n}{x} = \frac{1}{2n+1}\left(f_{n-1} + f_{n+1}\right), \tag{A4.144}$$

$$f_n' = \frac{1}{2n+1}\left(n f_{n-1} - (n+1) f_{n+1}\right). \tag{A4.145}$$

Expressing the three-dimensional Laplace operator in polar coordinates, and observing that the spherical part of the Laplacian applied to a spherical harmonic function gives $\Delta_{\vartheta\varphi} Y_{\ell m} = -\ell(\ell+1) Y_{\ell m}$, we find that $j_\ell(rk) Y_{\ell m}(\hat{\mathbf{x}})$ as well as $y_\ell(rk) Y_{\ell m}(\hat{\mathbf{x}})$ is a solution of

$$\left(\Delta + k^2\right) f = 0$$

for arbitrary values of ℓ and $-\ell \le m \le \ell$. Only the j_ℓ's are regular at $r = 0$. On the other hand, the exponential function, $e^{i\mathbf{x}\cdot\mathbf{k}}$, for arbitrary $\mathbf{k}$ with modulus $|\mathbf{k}| = k$ also solves this equation. Since the spherical harmonics form a complete system of functions on the sphere, there must exist an expansion

$$e^{i\mathbf{x}\cdot\mathbf{k}} = e^{ikr\hat{\mathbf{x}}\cdot\hat{\mathbf{k}}} = \sum_{\ell m} c_{\ell m} j_{\ell m}(rk) Y_{\ell m}(\hat{\mathbf{x}}).$$

This represents the decomposition of the exponential into its contributions of orbital angular momentum ℓ. To determine the coefficients $c_{\ell m}$, we choose the z-axis in the direction of $\mathbf{k}$, so that the function is independent of φ and only the terms with $m = 0$ contribute. Setting $\mu = \cos\vartheta$ this yields

$$e^{irk\mu} = \sum_{\ell} c_{\ell 0} j_\ell(rk) Y_{\ell 0}(\mu) = \sum_{\ell} c_\ell j_\ell(rk) P_\ell(\mu),$$

where we have set $c_\ell = \frac{2\ell+1}{4\pi} c_{\ell 0}$ and made use of Eq. (A4.37). The coefficients c_ℓ are now obtained by taking the nth derivative with respect to rk, multiplying with P_n, and integrating over μ (for more details see Arfken and Weber, 2001). One finds $c_\ell = i^\ell(2\ell+1)$ so that

$$e^{i\mathbf{k}\cdot\mathbf{n}r} = \sum_{\ell=0}^{\infty}(2\ell+1) i^\ell j_\ell(kr) P_\ell(\mu). \tag{A4.146}$$

Let us employ this representation to derive the so-called closure relation for spherical Bessel function. For this we use

$$\int d^3x \exp\left(i(\mathbf{k}_1 - \mathbf{k}_2)\mathbf{x}\right) = (2\pi)^3 \delta^{(3)}(\mathbf{k}_1 - \mathbf{k}_2). \tag{A4.147}$$

Representing the exponentials as Bessel series as in Eq. (A4.146) and integrating over the directions of $\mathbf{x}$ we find

$$(4\pi)^2 \sum_{\ell m} \int_0^\infty dx\, x^2 j_\ell(k_1 x) j_\ell(k_2 x) Y_{\ell m}(\widehat{\mathbf{k}}_1) Y^*_{\ell m}(\widehat{\mathbf{k}}_2) = (2\pi)^3 \delta^{(3)}(\mathbf{k}_1 - \mathbf{k}_2). \tag{A4.148}$$

Now we multiply both sides with $Y^*_{\ell_1 m_1}(\widehat{\mathbf{k}}_2)$ and integrate over directions of $\mathbf{k}_2$. The right-hand side of (A4.148) then becomes $(2\pi)^3\delta(k_1 - k_2)/k_2^2 Y_{\ell m}(\widehat{\mathbf{k}}_1)$ and the left-hand side simply gets a $\delta_{\ell\ell_1}\delta_{m m_1}$ so that we obtain

$$\frac{2}{\pi}\int_0^\infty dx\, x^2 j_\ell(k_1 x) j_\ell(k_2 x) = \frac{\delta(k_1 - k_2)}{k_1^2}. \tag{A4.149}$$

This result is relevant for the calculation of the local bispectrum in Chapter 6.

Another important Bessel function integral that we use especially in Chapter 4 is

$$\begin{aligned} I(p,\ell) \equiv \frac{2}{\pi}\int_0^\infty dx\, x^p j_\ell^2(x) &= \int_0^\infty dx\, x^{p-1} J^2_{\ell+1/2}(x) \\ &= \frac{\Gamma(1-p)\Gamma\left(\ell + \frac{p+1}{2}\right)}{2^{1-p}\Gamma^2(1-p/2)\Gamma\left(\ell + \frac{3-p}{2}\right)}. \end{aligned} \tag{A4.150}$$

Here Γ denotes the Gamma function, i.e., $\Gamma(n) = (n-1)!$ for positive integers. The integral converges for $1 > p > -2\ell - 1$. Of special interest is the case $p = -1$ that occurs in the calculation of the CMB spectrum for scale-invariant fluctuations,

$$I(-1,\ell) = \frac{1}{\pi\,\ell(\ell+1)}. \tag{A4.151}$$

Finally we shall use also the following approximate relation, the so-called Limber approximation (Limber, 1959; Lo Verde and Afshordi, 2008), which states that for an arbitrary slowly varying function we have

$$\frac{2}{\pi}\int dkk^2 f(k) j_\ell(kr) j_\ell(kr') \simeq \frac{\delta(r-r')}{r^2} f\left(\frac{\ell+1/2}{r}\right). \tag{A4.152}$$

Note that for f = constant this approximation is exact and reproduces Eq. (A4.149). Approximation (A4.152) is especially useful when discussing lensing.

Appendix 5

Entropy Production and Heat Flux

Here we show that the perturbation variable Γ defined in Eq. (2.84) is related to the divergence of the entropy flux. We consider a system that deviates slightly from thermal equilibrium.

A5.1 Thermal Equilibrium

We first recollect some important relations in thermal equilibrium. We consider an arbitrary mix of different (relativistic and nonrelativistic) particles that may or may not be conserved. The only *total* thermodynamical quantities then are temperature T, entropy S, energy E, pressure P, and volume V. We shall also use the densities $s = dS/dV$ and $\rho = dE/dV$. Certain conserved species may have a chemical potential, but we are not interested in this "fine structure" here. The corresponding treatment for one conserved particle species can be found in Straumann (1984), Appendix B.

We start with the Gibbs relation

$$T\,dS = dE + P\,dV, \quad \text{or} \quad T\frac{dS}{dV} = Ts = \rho + P. \tag{A5.1}$$

S and E are extensive quantities. Locally they are simply given by $S = sV$ and $E = \rho V$. Inserting this in the Gibbs relation we obtain

$$TV\,ds + Ts\,dV = V\,d\rho + \rho\,dV + P\,dV \quad \text{hence} \quad T\,ds = d\rho. \tag{A5.2}$$

Defining the entropy 4-velocity field by U^μ, the entropy flux is then given by $S^\mu = sU^\mu = T^{-1}(\rho + P)U^\mu$. In thermal equilibrium the entropy velocity coincides with the energy flux $u^\mu = U^\mu$, so that $T^{\mu\nu}U_\mu = -\rho U^\nu$. In thermal equilibrium we therefore have

$$S^\mu = -\frac{1}{T}U_\nu T^{\mu\nu} + \frac{P}{T}U^\mu. \tag{A5.3}$$

In an FL background $(U^\mu) = (u^\mu) = a^{-1}(1, \mathbf{0})$ with $U^\mu{}_{;\mu} = 3\dot{a}/a^2$, so that entropy conservation becomes $0 = S^\mu{}_{;\mu} = a^{-1}\dot{s} + 3(\dot{a}/a^2)s$, which results in the well-known law of adiabatic expansion, $\dot{s} = -3(\dot{a}/a)s$. Furthermore, with Eq. (A5.2) small variations of the entropy flux at fixed velocity field U^μ are given by

$$dS^\mu = U^\mu\,ds = \frac{1}{T}U^\mu\,d\rho = -\frac{1}{T}U_\nu\,dT^{\mu\nu}. \tag{A5.4}$$

A5.2 Small Departures from Thermal Equilibrium

We now proceed to the study of small deviations from equilibrium. There is some arbitrariness in fitting the actual state with an equilibrium state plus small deviations. Following Israel and Stewart (1980), we approximate the actual state with the thermal equilibrium at the same energy density ρ and entropy velocity field U^μ. We neglect all second-order quantities, taking into account only first-order deviations from thermal equilibrium and/or from the FL background. We specify the deviation of the energy–momentum tensor from thermal equilibrium, $\delta T^{\mu\nu}$, by the following ansatz:

$$T^{\mu\nu} = (\rho + P_{\rm eq})U^\mu U^\nu + P_{\rm eq} g^{\mu\nu} + \delta T^{\mu\nu} \,. \tag{A5.5}$$

Here $P_{\rm eq}$ is the pressure of the thermal equilibrium state with energy density ρ. Setting $\rho = \bar\rho + \delta\rho$ we therefore have $P_{\rm eq} = \bar P + \delta P$ with $\delta P = c_s^2 \delta\rho$.

On the other hand, the energy flux 4-velocity u^μ is defined by (2.64) as the timelike eigenvector of the energy–momentum tensor and $T^{\mu\nu}$ can also be written in the form

$$T^{\mu\nu} = (\rho + P)u^\mu u^\nu + P g^{\mu\nu} + \Pi^{\mu\nu} = \rho u^\mu u^\nu + \tau^{\mu\nu}, \tag{A5.6}$$

where τ is the stress tensor normal to u^μ given in Eq. (2.68),

$$\tau^{\mu\nu} = P(u^\mu u^\nu + g^{\mu\nu}) + \Pi^{\mu\nu}, \quad \Pi^\lambda_\lambda = 0. \tag{A5.7}$$

The tensor $\Pi^{\mu\nu}$ is orthogonal to u^μ, $\Pi^{\mu\nu}u_\nu = 0$. Defining Q^μ by

$$u^\mu = U^\mu + Q^\mu,$$

we can rewrite (A5.6) in the following manner:

$$\begin{aligned} T^{\mu\nu} &= (\rho + P)U^\mu U^\nu + P g^{\mu\nu} + U^\mu q^\nu + U^\nu q^\mu + \Pi^{\mu\nu} \\ &= (\rho + P_{\rm eq})U^\mu U^\nu + P_{\rm eq} g^{\mu\nu} + U^\mu q^\nu + U^\nu q^\mu \\ &\quad + (P - P_{\rm eq})(U^\mu U^\nu + g^{\mu\nu}) + \Pi^{\mu\nu} \,, \end{aligned} \tag{A5.8}$$

where we have introduced

$$q^\mu = (\rho + p)Q^\mu \,. \tag{A5.9}$$

$\Pi_{\mu\nu}$, $\delta T_{\mu\nu}$, Q^μ, and therefore also q^μ vanish in the background; they are of first order. Since $u^2 = U^2 = -1$, we have to first order $q \cdot U = 0$, $q \cdot u = 0$.

Identifying $\delta T^{\mu\nu}$ by comparing Eq. (A5.8) with the definition given in Eq. (A5.5), we obtain to first order

$$\delta T^{\mu\nu} = U^\mu q^\nu + U^\nu q^\mu + (P - P_{\rm eq})(U^\mu U^\nu + g^{\mu\nu}) + \Pi^{\mu\nu}, \tag{A5.10}$$

and

$$\delta T^{\mu\nu} U_\mu = -q^\nu - \Pi^{\mu\nu} Q_\mu = -q^\nu,$$

since $\Pi^{\mu\nu}$ and Q_μ are both first order and normal to U^μ. With Eq. (A5.4) the perturbed entropy flux $S^\mu = S^\mu_{\rm eq} + \delta S^\mu$ becomes

$$S^\mu = sU^\mu - \frac{1}{T}\delta T^{\mu\nu} U_\nu = sU^\mu + \frac{1}{T} q^\mu. \tag{A5.11}$$

This equation shows that q^μ represents the heat flux.

From $P = \bar{P}(1+\pi_L)$ and $P_{\rm eq} = \bar{P}(1+\frac{c_s^2}{w}\delta)$, $\delta = \delta\rho/\bar{\rho}$ we find with Eq. (2.84)

$$P - P_{\rm eq} = \bar{P}\left(\pi_L - \frac{c_s^2}{w}\delta\right) = \bar{P}\Gamma\,. \qquad \text{(A5.12)}$$

Taking the divergence of Eq. (A5.11) we find

$$S^{\mu}{}_{;\mu} = s_{,\mu}U^{\mu} + sU^{\mu}{}_{;\mu} + \frac{T_{,\mu}}{T^2}\delta T^{\mu\nu}U_{\nu} - \frac{1}{T}\delta T^{\mu\nu}{}_{;\mu}U^{\nu} - \frac{1}{T}\delta T^{\mu\nu}U_{(\nu;\mu)}, \qquad \text{(A5.13)}$$

where $(\nu;\mu)$ denotes symmetrization, $U_{(\nu;\mu)} = \frac{1}{2}(U_{\mu;\nu} + U_{\nu;\mu})$. In the last term we have used the fact that $\delta T^{\mu\nu}$ is symmetric. To evaluate the fourth term on the right-hand side we make use of energy–momentum conservation in the form

$$0 = U_{\nu}T^{\mu\nu}{}_{;\mu} = U_{\nu}[(\rho + P_{\rm eq})U^{\mu}U^{\nu} + P_{\rm eq}g^{\mu\nu}]_{;\mu} + U_{\nu}\delta T^{\mu\nu}{}_{;\mu}.$$

Expanding the derivative of the square bracket leads to

$$(\rho + P_{\rm eq})U^{\mu}{}_{;\mu} + \rho_{,\mu}U^{\mu} = \delta T^{\mu\nu}{}_{;\mu}U_{\nu}. \qquad \text{(A5.14)}$$

With Eq. (A5.1), the first term on the left-hand side of Eq. (A5.14) cancels the second term on the right-hand side of Eq. (A5.13), and Eq. (A5.2) implies $s_{,\mu} = T^{-1}\rho_{,\mu}$, so that the second term of Eq. (A5.14) cancels the first term in Eq. (A5.13). The fourth term on the right-hand side of Eq. (A5.13) therefore cancels the first two and we are left with

$$S^{\mu}{}_{;\mu} = -\frac{T_{,\mu}}{T^2}q^{\mu} - \frac{1}{T}\delta T^{\mu\nu}U_{(\nu;\mu)}. \qquad \text{(A5.15)}$$

To evaluate $\delta T^{\mu\nu}U_{(\nu;\mu)}$ we define the projector $h_{\mu}{}^{\nu}$ onto the three-dimensional subspace of tangent space normal to U^{μ}:

$$h_{\mu}{}^{\nu} = U^{\mu}U_{\nu} + \delta_{\mu}{}^{\nu},$$

and the acceleration

$$a_{\mu} = U^{\nu}U_{\mu;\nu}.$$

With this it is easy to show that

$$U_{\mu;\nu} = h_{\mu}{}^{\alpha}h_{\nu}{}^{\beta}U_{\alpha;\beta} - a_{\mu}U_{\nu}.$$

Defining the expansion tensor $\theta_{\mu\nu} = \theta_{\nu\mu} = h_{\mu}{}^{\alpha}h_{\nu}{}^{\beta}U_{(\alpha;\beta)}$, Eq. (A5.15) now becomes

$$S^{\mu}{}_{;\mu} = -\frac{1}{T}\left(\frac{T_{,\mu}}{T} + a_{\mu}\right)q^{\mu} - \frac{1}{T}\,\delta T^{\mu\nu}\theta_{\nu\mu}. \qquad \text{(A5.16)}$$

The acceleration a_{μ} is of first order, and to lowest order $T_{,\mu}/T \propto U_{\mu}$ so that $(\frac{T_{,\mu}}{T} + a_{\mu})q^{\mu}$ vanishes to first order. Furthermore, $\delta T^{\mu\nu}$ is of first order. To determine the divergence of the entropy flux to first order, it is therefore sufficient to determine $\theta_{\nu\mu}$ to zeroth order. But to zeroth order

$$h_0^0 = h_i^0 = h_0^i = 0 \quad \text{and} \quad h_i^j = \delta_i^j.$$

Furthermore, $(U_{\mu}) = -a(1,0)$ so that

$$U_{i;j} = -\dot{a}\delta_{ij} \quad \text{and} \quad U_{0;0} = U_{0;j} = U_{i;0} = 0.$$

Inserting this in the definition of $\theta_{\mu\nu}$, we obtain to lowest order

$$\theta_{ij} = \frac{\dot{a}}{a^2} g_{ij} \quad \text{and} \quad \theta_{00} = \theta_{i0} = \theta_{0j} = 0. \tag{A5.17}$$

With Eq. (A5.16) the divergence of the entropy flux then becomes to first order

$$S^{\mu}{}_{;\mu} = \frac{1}{T}\,\delta T^{\mu\nu}\theta_{\nu\mu} = \frac{1}{T}\,\delta T^{ij}\theta_{ij} = 3\frac{\dot{a}}{a^2}\frac{P - P_{\rm eq}}{T} = 3\frac{\dot{a}}{a^2}\frac{\bar{P}}{T}\Gamma. \tag{A5.18}$$

For the last equals sign we have used Eq. (A5.12). This demonstrates that the entropy production rate is proportional to Γ.

A5.3 Phenomenological Coefficients

Finally, for completeness, we want to introduce the heat conductivity coefficient as well as bulk and shear viscosities. We now no longer assume the energy–momentum tensor to be nearly homogeneous and isotropic, but we allow only for small departures from thermal equilibrium that we take into account to first order only. It is then easy to check that Eq. (A5.16) is still valid. We now define $\eta^{\mu\nu}$ by

$$\delta T^{\mu\nu} = \eta^{\mu\nu} + U^{\mu}q^{\nu} + q^{\mu}U^{\nu}, \quad \eta \equiv \eta^{\mu}_{\mu}. \tag{A5.19}$$

From Eq. (A5.10) we have $U_{\mu}\eta^{\mu\nu} = 0$; hence $h^{\mu}_{\alpha}h^{\nu}_{\beta}\eta^{\alpha\beta} = h^{\mu}_{\alpha}\eta^{\alpha\nu} = \eta^{\mu\nu}$. Therefore $\eta^{\mu\nu}$ "lies" in the hypersurface of tangent space normal to U^{μ},

$$S^{\mu}{}_{;\mu} = -\frac{1}{T^2}q^{\mu}h^{\nu}_{\mu}\left(T_{,\nu} + Ta_{\nu}\right) - \frac{1}{T}\eta^{\mu\nu}\theta_{\mu\nu}. \tag{A5.20}$$

Let us also introduce the traceless part of $\eta^{\mu\nu}$,

$$\hat{\eta}^{\mu\nu} = \eta^{\mu\nu} - \frac{1}{3}\eta\, h^{\mu\nu}.$$

With the shear tensor defined as the traceless part of $\theta_{\mu\nu}$,

$$\sigma_{\mu\nu} = \theta_{\nu\mu} - \frac{1}{3}\theta h_{\nu\mu}, \quad \theta = \theta^{\mu}_{\mu}, \tag{A5.21}$$

Equation (A5.16) can now be written as

$$S^{\mu}{}_{;\mu} = -\frac{1}{T^2}q^{\mu}h^{\nu}_{\mu}\left(T_{,\mu} + Ta_{\mu}\right) - \frac{1}{T}\hat{\eta}^{\mu\nu}\sigma_{\mu\nu} - \frac{1}{3T}\theta\eta. \tag{A5.22}$$

The three terms in Eq. (A5.22) are independent. The second law of thermodynamics, $S^{\mu}{}_{;\mu} \geq 0$, requires that each of them be nonnegative. This is possible only if

$$q^{\mu} = -\chi h^{\mu\nu}\left(T_{,\nu} + Ta_{\nu}\right), \quad \chi \geq 0 \tag{A5.23}$$

$$\hat{\eta}^{\mu\nu} = -2\zeta\sigma^{\mu\nu}, \quad \zeta \geq 0 \tag{A5.24}$$

$$\eta = -3\xi\theta, \quad \xi \geq 0. \tag{A5.25}$$

The quantities ζ and ξ are called the shear and bulk viscosity respectively and χ is the heat conductivity coefficient. For small deviations from thermal equilibrium, the second law of thermodynamics requires that $\delta T^{\mu\nu}$ be fully determined by these three coefficients and be of the form

$$\delta T^{\mu\nu} = -\chi\left[U^{\mu}h^{\nu\lambda}\left(T_{,\lambda} + Ta_{\lambda}\right) + U^{\nu}h^{\mu\lambda}\left(T_{,\lambda} + Ta_{\lambda}\right)\right] - 2\zeta\sigma^{\mu\nu} - \xi\theta h^{\mu\nu}. \quad \text{(A5.26)}$$

Eq. (A5.23) is the relativistic generalization of Fourier's law, $\mathbf{q} = -\chi\nabla T$. In terms of the phenomenological coefficients χ, ζ, and ξ the entropy production is given by

$$S^{\mu}{}_{;\mu} = \frac{1}{\chi T^2}q^{\mu}q_{\mu} + \frac{2\zeta}{T}\sigma^{\mu\nu}\sigma_{\mu\nu} + \frac{\xi}{T}\theta^2 \geq 0. \quad \text{(A5.27)}$$

Each of these contributions is nonnegative because q^{μ} is a spatial vector and $\sigma^{\mu\nu}$ is a spatial tensor. They are understood as entropy production due to heat flux, shear viscosity, and bulk viscosity respectively. Only the third term, the bulk viscosity, contributes to first order in deviations from homogeneity and isotropy, since in this case q^{μ} and $\sigma^{\mu\nu}$ are small.

Appendix 6

Mixtures

In this appendix we derive Eqs. (2.137) and (2.139). Let us first recall the definitions of the difference variables,

$$S_{\alpha\beta} = \left[\frac{D_{g\alpha}}{1+w_\alpha} - \frac{D_{g\beta}}{1+w_\beta}\right], \tag{A6.1}$$

$$V_{\alpha\beta} = V_\alpha - V_\beta, \tag{A6.2}$$

$$\Gamma_{\alpha\beta} = \frac{w_\alpha}{1+w_\alpha}\Gamma_\alpha - \frac{w_\beta}{1+w_\beta}\Gamma_\beta, \tag{A6.3}$$

$$\Pi_{\alpha\beta} = \frac{w_\alpha}{1+w_\alpha}\Pi_\alpha - \frac{w_\beta}{1+w_\beta}\Pi_\beta. \tag{A6.4}$$

We now calculate the derivative of one of the terms in $S_{\alpha\beta}$.

$$\begin{aligned}\left(\frac{\dot{D}_{g\alpha}}{1+w_\alpha}\right) &= \frac{\dot{D}_{g\alpha}}{1+w_\alpha} - \frac{\dot{w}_\alpha}{1+w_\alpha}\frac{D_{g\alpha}}{1+w_\alpha} \\ &= (1+w_\alpha)^{-1}\left[\dot{D}_{g\alpha} + 3\mathcal{H}(c_\alpha^2 - w_\alpha)D_{g\alpha}\right] \\ &= -kV_\alpha - 3\mathcal{H}\frac{w_\alpha}{1+w_\alpha}\Gamma_\alpha.\end{aligned} \tag{A6.5}$$

For the second equals sign we have used $\dot{w}_\alpha = -3\mathcal{H}(c_\alpha^2 - w_\alpha)(1+w_\alpha)$ and for the last equals sign we have inserted Eq. (2.131). With the definition (A6.1) we now simply obtain Eq. (2.137):

$$\dot{S}_{\alpha\beta} = -kV_{\alpha\beta} - 3\mathcal{H}\Gamma_{\alpha\beta}. \tag{A6.6}$$

To find a differential equation for $V_{\alpha\beta}$ we first take the difference of Eq. (2.131) for the component α with the same equation for component β. This leads to

$$\begin{aligned}\dot{V}_{\alpha\beta} + \mathcal{H}V_{\alpha\beta} - 3\mathcal{H}(c_\alpha^2 V_\alpha - c_\beta^2 V_\beta) = 3k(c_\alpha^2 - c_\beta^2)\Phi + kc_\alpha^2\frac{D_{g\alpha}}{1+w_\alpha} \\ - kc_\beta^2\frac{D_{g\beta}}{1+w_\beta} + k\Gamma_{\alpha\beta} - \frac{2k}{3}\left(1 - \frac{3K}{k^2}\right)\Pi_{\alpha\beta}.\end{aligned} \tag{A6.7}$$

We now write

$$\begin{aligned}\frac{D_{g\alpha}}{1+w_\alpha} &= \frac{D_g}{1+w} + \frac{D_{g\alpha}}{1+w_\alpha} - \frac{D_g}{1+w} \\ &= \frac{D_g}{1+w} + \sum_\gamma \frac{\rho_\gamma + P_\gamma}{\rho + P}\left(\frac{D_{g\alpha}}{1+w_\alpha} - \frac{D_{g\gamma}}{1+w_\gamma}\right), \\ &= \frac{D_g}{1+w} + \sum_\gamma \frac{\rho_\gamma + P_\gamma}{\rho + P} S_{\alpha\gamma}. \end{aligned} \tag{A6.8}$$

Inserting this and $\frac{kD_g}{1+w} = \frac{kD}{1+w} - 3k\Phi - 3\mathcal{H}V$ in Eq. (A6.7) yields

$$\begin{aligned}&\dot{V}_{\alpha\beta} + \mathcal{H}V_{\alpha\beta} - 3\mathcal{H}\left[c_\alpha^2(V_\alpha - V) - c_\beta^2(V_\beta - V)\right] \\ &\quad = \frac{k\left(c_\alpha^2 - c_\beta^2\right)}{1+w}D + k\sum_\gamma \frac{\rho_\gamma + P_\gamma}{\rho + P}\left(c_\alpha^2 S_{\alpha\gamma} - c_\beta^2 S_{\beta\gamma}\right) + k\Gamma_{\alpha\beta} - \frac{2k}{3}\left(1 - \frac{3K}{k^2}\right)\Pi_{\alpha\beta}. \end{aligned} \tag{A6.9}$$

Note also that $V_\alpha - V = \sum_\gamma \frac{\rho_\gamma + P_\gamma}{\rho+P}(V_\alpha - V_\gamma) = \sum_\gamma \frac{\rho_\gamma+P_\gamma}{\rho+P}V_{\alpha\gamma}$. Furthermore, $(\rho_\gamma + P_\gamma)S_{\alpha\gamma} = \frac{1}{2}(\rho_\gamma + P_\gamma)[S_{\alpha\gamma} + S_{\beta\gamma}] + \frac{1}{2}(\rho_\gamma + P_\gamma)[S_{\alpha\gamma} - S_{\beta\gamma}]$. Using $S_{\alpha\gamma} - S_{\beta\gamma} = S_{\alpha\beta}$ this gives

$$\sum_\gamma (\rho_\gamma + P_\gamma)S_{\alpha\gamma} = \frac{1}{2}\sum_\gamma(\rho_\gamma + P_\gamma)[S_{\alpha\gamma} + S_{\beta\gamma}] + \frac{1}{2}(\rho + P)S_{\alpha\beta}.$$

The same identity holds with $S_{..}$ replaced by $V_{..}$. Inserting this in Eq. (A6.9) we finally obtain Eq. (2.138):

$$\begin{aligned}&\dot{V}_{\alpha\beta} + \mathcal{H}V_{\alpha\beta} - \frac{3}{2}\mathcal{H}\left(c_\alpha^2 + c_\beta^2\right)V_{\alpha\beta} - \frac{3}{2}\mathcal{H}\left(c_\alpha^2 - c_\beta^2\right)\sum_\gamma \frac{\rho_\gamma + P_\gamma}{\rho + P}\left(V_{\alpha\gamma} + V_{\beta\gamma}\right) \\ &\quad = k\left[\frac{c_\alpha^2 - c_\beta^2}{1+w}D + \frac{c_\alpha^2 + c_\beta^2}{2}S_{\alpha\beta} + \frac{c_\alpha^2 - c_\beta^2}{2}\sum_\gamma \frac{\rho_\gamma + P_\gamma}{\rho + P}\left(S_{\alpha\gamma} + S_{\beta\gamma}\right)\right. \\ &\qquad \left. + \Gamma_{\alpha\beta} - \frac{2}{3}\left(1 - \frac{3K}{k^2}\right)\Pi_{\alpha\beta}\right]. \end{aligned} \tag{A6.10}$$

Appendix 7

Statistical Utensils

A7.1 Gaussian Random Variables

A7.1.1 Introduction

A random variable is a real function X on a probability space $(\Omega, d\mu)$. The set Ω is a measurable space with normalized measure μ, that is, $\int_\Omega d\mu = 1$. The integral

$$\int_\Omega X\, d\mu = \langle X \rangle$$

is called the expectation value or simply the mean of X, while

$$\int_\Omega (X - \langle X \rangle)^2\, d\mu = \langle X^2 \rangle - \langle X \rangle^2$$

is called the variance of X and its (positive) square root is the standard deviation. If $\langle X \rangle = 0$, we call X a fluctuation. We are mainly interested in fluctuations. We sometimes call Ω the "space of realizations" or the "ensemble." A random variable is strongly continuous if the derivative of its probability distribution $d\mu/dX \equiv p$ is an integrable function[1] on $\mathbb{R}$. Then we can write

$$\langle X \rangle = \int_\Omega X\, d\mu = \int_\mathbb{R} x p(x)\, dx. \tag{A7.1}$$

The probability distribution satisfies the normalization condition

$$\int_\mathbb{R} p(x)\, dx = 1.$$

The distribution function $p(x)$, also called the probability density, fully determines the random variable.

Definition 7.1 *A random variable with probability distribution*

$$p(x) = \frac{1}{\sqrt{2\pi}\,\sigma} e^{-(x-x_0)^2/2\sigma^2}$$

[1] In the more general case, p is a distribution in the sense of Schwartz, that is, a functional on some space of functions on $\mathbb{R}$.

is called a Gaussian random variable (normal distribution) with mean x_0 and variance σ^2.

The Gaussian distribution with mean 0 and variance 1 is called the standard normal distribution.

The moments of a random variable with mean x_0 are defined by

$$V_n \equiv \int_{-\infty}^{\infty} (x - x_0)^n p(x)\, dx. \tag{A7.2}$$

The moments of a Gaussian random variable are given by

$$V_0 = 1, V_n = \begin{cases} 0 & \text{if } n \quad \text{is odd} \\ \sigma^n (n-1)!! & \text{if } n > 0, \quad \text{is even,} \end{cases} \tag{A7.3}$$

where the double factorial of a number is defined by $m!! = m(m-2)(m-4)\cdots$. This statement is evident for n odd. The proof of the even case is easiest done by induction and is left as an exercise.

A7.1.2 The Central Limit Theorem

Let X_i be independent random variables with means x_i and variances σ_i. Independence means that $\langle (X_i - x_i)(X_j - x_j)\rangle = \delta_{ij}\sigma_i^2$. Then the sums

$$S_n = \frac{\sum_{i=1}^n X_i - x_i}{\sqrt{n}\sum_{i=1}^n \sigma_i}$$

converge (weakly) to the standard normal distribution. A proof of this important theorem can be found in most texts on probability theory.[2]

In physics it means that an experimental error that comes from many independent sources is often close to Gaussian.

For the CMB its main relevance is that the observed C_ℓ's, which are given by the average $C_\ell^o = \frac{1}{2\ell+1}\sum_{-\ell}^{\ell} |a_{\ell m}|^2$ even though by themselves not Gaussian, tend to Gaussian variables with mean C_ℓ and variance C_ℓ^2/ℓ for large ℓ. As we have seen in Section 9.4.2, the variance of the variable $|a_{\ell m}|^2$ is $2C_\ell^2$; therefore the central limit theorem implies that

$$\frac{\sqrt{2}}{\sqrt{(2\ell+1)C_\ell}} \sum_{-\ell}^{\ell} (|a_{\ell m}|^2 - C_\ell),$$

converges to the standard normal distribution. Hence C_ℓ^o converges to a Gaussian distribution with mean C_ℓ and variance C_ℓ^2/ℓ which becomes small with increasing ℓ.

A7.1.3 A Collection of Gaussian Random Variables

A collection $X_1 \cdots X_N$ of random variables is called Gaussian if their joint probability density is given by

[2] There is a technical condition that has to be fulfilled for the theorem to hold. It ensures that the series is not dominated by one (or a small number) of variables. This is certainly fulfilled if the variables X_i are equally distributed. But the central limit theorem applies in much more general cases.

$$p(\mathbf{x}) = \frac{1}{\sqrt{(2\pi)^N \det(C)}} \exp\left(-\frac{1}{2}\mathbf{x}^T C^{-1}\mathbf{x}\right), \qquad \mathbf{x} \in \mathbb{R}^N, \tag{A7.4}$$

where C is a real, positive-definite, symmetric $N \times N$ matrix.

First we show that

$$\langle X_i X_j \rangle \equiv \frac{1}{\sqrt{(2\pi)^N \det(C)}} \int x_i x_j \exp\left(-\frac{1}{2}\mathbf{x}^T C^{-1}\mathbf{x}\right) dx^N = C_{ij}. \tag{A7.5}$$

To prove this, we use the fact that symmetric matrices can be diagonalized. In other words, there exists an orthogonal matrix S, $SS^T = 1$ so that

$$S^T C^{-1} S = D = \begin{pmatrix} \lambda_1\, 0 \cdots & \\ 0\, \lambda_2\, 0 \cdots & \\ & \vdots \\ 0 \cdots 0\, \lambda_N & \end{pmatrix}. \tag{A7.6}$$

Since $|\det S| = 1$, $1/\det C = \det D = \lambda_1 \lambda_2 \cdots \lambda_N$ and for $y = Sx$ we have $d^N x = d^N y$. This yields

$$\begin{aligned} \langle X_i X_j \rangle &= S^T_{ik} S^T_{jl} \frac{\sqrt{\Pi_{n=1}^N \lambda_n}}{(2\pi)^{N/2}} \int y_k y_l \exp\left(-\frac{1}{2}\sum_{n=1}^{N} y_n^2 \lambda_n\right) dx^N \\ &= S^T_{ik} (\lambda_k)^{-1} \delta_{kl} S_{lj} = (S^T D^{-1} S)_{ij} = C_{ij}. \end{aligned} \tag{A7.7}$$

Here the sum over repeated indices is understood and we have made use of Eq. (A7.3) for the case $n = 1$ if $k \neq l$ and $n = 2$ if $k = l$. For obvious reasons, the matrix C is called the correlation matrix.

It is easy to see that arbitrary linear combinations of a collection of Gaussian random variables result again in a collection of Gaussian random variables, whereas powers of Gaussian random variables are not Gaussian. Actually, the sum of the squares of n independent Gaussian random variables with the same distribution results in a χ^2-distributed variable with n degrees of freedom.

A7.1.4 Wick's Theorem

Wick's theorem provides a general formula for the m-point correlator of a collection of Gaussian variables. In this sense it is a generalization of Eq. (A7.3). It is clear that for m odd the result vanishes. For even $m = 2n$ we obtain the m-point correlator by summing all the possible products of 2-point correlators made from the variables $X_{i_1}, \ldots, X_{i_m}$,

$$\langle X_{i_1} \ldots X_{i_{2n}} \rangle = \sum_{\{j_1,\ldots j_{2n}\}=\{i_1,\ldots i_{2n}\}} C_{j_1 j_2} \cdots C_{j_{2n-1} j_{2n}}. \tag{A7.8}$$

Here the sum is not over all permutations, but only over those that give rise to different pairs. Since $C_{ij} = C_{ji}$ we could also simply sum over all permutations of $(i_1, \ldots, i_{2n})$ and divide by $2^n n!$, since for each collection into pairs, there are $2^n n!$ permutations that give rise to the same pairs. The factor 2^n stems from the fact that in each of the pairs we can interchange the factors and $n!$ permutations simply interchange some of the pairs.

For example, $2n = 4$ admits $4!/2^2 2 = 3$ different pairings, namely $\langle X_1X_2\rangle\langle X_3X_4\rangle$, $\langle X_1X_3\rangle\langle X_2X_4\rangle$, and $\langle X_1X_4\rangle\langle X_3X_2\rangle$.

This theorem is extremely important not only in probability theory but also in quantum field theory and statistical mechanics. It means that for a collection of Gaussian random variables, all n-point correlators are determined by the 2-point correlator alone. Since its proof is not easily found in texts on probability but more often in the infinite-dimensional context of field theory where it is more complicated, we present it here for those who are interested.

Proof We consider only the case of a diagonal covariance matrix C. The general case can then be obtained by diagonalization of C in the same way as in Eq. (A7.7). We show Eq. (A7.8) by induction. The case $2n = 2$ is nothing other than Eq. (A7.7). For the step from $2n$ to $2n+2$ we use the fact that for an exponentially decaying function $\int_{-\infty}^{\infty} dx\, \frac{df}{dx} = 0$; hence

$$\begin{aligned} 0 &= \int dx^N \frac{d^2}{x^j x^k}\left(x_{i_1}x_{i_2}\cdots x_{i_{2n}}e^{-\sum_{m=1}^N x_m^2\lambda_m/2}\right) \\ &= \int dx^N\Bigg(\sum_{r\neq s=1}^{2n}\delta_{i_r j}\delta_{i_s k}x_{i_1}\cdots\check{x}_{i_r}\cdots\check{x}_{i_s}\cdots x_{i_{2n}} - \lambda_j\delta_{jk}x_{i_1}\cdots x_{i_{2n}} \\ &\qquad + \lambda_j\lambda_k x_j x_k x_{i_1}\cdots x_{i_{2n}} - \sum_r \lambda_j x_j \delta_{k i_r}x_{i_1}\cdots\check{x}_{i_r}\cdots x_{i_{2n}} \\ &\qquad - \sum_r \lambda_k x_k\delta_{j i_r}x_{i_1}\cdots\check{x}_{i_r}\cdots x_{i_{2n}}\Bigg)e^{-\sum_{m=1}^N x_m^2\lambda_m/2}. \end{aligned}$$

Here a check over a variable x_m means that this variable is omitted in the product. In the above integral, all except the term proportional to $\lambda_j\lambda_k$ are $2n$- or $(2n-2)$-point correlators. We can therefore express the $(2n+2)$-point correlator proportional to $\lambda_j\lambda_k$ in terms of $2n$- and $(2n-2)$-point correlators. Furthermore, we shall use the fact that $\lambda_k^{-1}\delta_{jk} = \langle X_jX_k\rangle$. Dividing the above equation by $\lambda_j\lambda_k$ therefore gives

$$\begin{aligned} \langle X_jX_kX_{i_1}\cdots X_{i_{2n}}\rangle = &-\sum_{r\neq s=1}^{2n}\langle X_{i_r}X_j\rangle\langle X_{i_s}X_k\rangle\langle X_{i_1}\cdots\check{X}_{i_r}\cdots\check{X}_{i_s}\cdots X_{i_{2n}}\rangle \\ &+\langle X_jX_k\rangle\langle X_{i_1}\cdots X_{i_{2n}}\rangle \\ &+\sum_r\langle X_kX_{i_r}\rangle\langle X_jX_{i_1}\cdots\check{X}_{i_r}\cdots X_{i_{2n}}\rangle \\ &+\sum_r\langle X_jX_{i_r}\rangle\langle X_kX_{i_1}\cdots\check{X}_{i_r}\cdots X_{i_{2n}}\rangle. \end{aligned}$$

Wick's theorem for $2n$ implies that

$$\langle X_kX_{i_1}\cdots\check{X}_{i_r}\cdots X_{i_{2n}}\rangle = \sum_s\langle X_kX_{i_s}\rangle\langle X_{i_1}\cdots\check{X}_{i_r}\cdots\check{X}_{i_s}\cdots X_{i_{2n}}\rangle.$$

Therefore, the last sum cancels with the double sum and we end up with

$$\langle X_j X_k X_{i_1} \cdots X_{i_{2n}} \rangle = \langle X_j X_k \rangle \langle X_{i_1} \cdots X_{i_{2n}} \rangle + \sum_r \langle X_k X_{i_r} \rangle \langle X_j X_{i_1} \cdots \check{X}_{i_r} \cdots X_{i_{2n}} \rangle . \qquad \text{(A7.9)}$$

But since the $2n$-point correlators $\langle X_{i_1} \cdots X_{i_{2n}} \rangle$ and $\langle X_j X_{i_1} \cdots \check{X}_{i_r} \cdots X_{i_{2n}} \rangle$ are the sum of all possible products of 2-point correlators, this represents simply the sum of all possible products of 2-point correlators of $X_j, X_k, X_{i_1}, \ldots, X_{i_{2n}}$; hence Wick's theorem is proven.

A7.2 Random Fields

A random field is an application $X : S \to \{\text{random variables}\} : \mathbf{n} \mapsto X(\mathbf{n})$ that assigns to each point $\mathbf{n}$ in the space S a random variable $X(\mathbf{n})$. Here the space S can be $\mathbb{R}^n$, the sphere, or some other space. We think mainly of S being the CMB sky, hence the sphere on which our random fields are, for example, the temperature fluctuations $\Delta T(\mathbf{n})$ or the polarization. Another example is three-dimensional-space (either $\mathbb{R}^3$, the 3-sphere, or the 3-pseudo-sphere) where, for example, the density and velocity fluctuations are random fields of interest to us.

> **Definition 7.2** *A random field is called Gaussian if arbitrary (finite) collections $X(\mathbf{n}_1), \ldots, X(\mathbf{n}_N)$ are Gaussian random variables according to* (A7.4). *The correlator*
>
> $$\langle X(\mathbf{n}_1) X(\mathbf{n}_2) \rangle = C(\mathbf{n}_1, \mathbf{n}_2)$$
>
> *is called the correlation function or the 2-point function.*

For Gaussian random fields, the n-point function is given by the sum of all possible different products of 2-point functions.[3]

Formally one can write

$$\langle X(\mathbf{n}_1) X(\mathbf{n}_2) \rangle = \frac{1}{\sqrt{\det(2\pi C)}} \int \delta X X(\mathbf{n}_1) X(\mathbf{n}_2) \times \exp\left(-\frac{1}{2} \int d\mathbf{n}_1 d\mathbf{n}_2 X(\mathbf{n}_1) C^{-1}(\mathbf{n1}, \mathbf{n2}) X(\mathbf{n2}) \right), \qquad \text{(A7.10)}$$

where here C is to be understood as an operator on (a certain space of) functions on S given by

$$X \to X' = CX, X'(\mathbf{n}_1) = \int d\mathbf{n}_2 C(\mathbf{n}_1, \mathbf{n}_2) X(\mathbf{n}_2). \qquad \text{(A7.11)}$$

The integral δX is a functional integral over the infinite dimensional space of functions on S. It is, however, not clear that the determinant of such an operator exists even if we

[3] Physicists are often more familiar with quantum field theory. In (Euclidean) quantum field theory the 2-point function is called the propagator and the theory is Gaussian if and only if it is trivial. Only in the absence of interactions are all n-point functions determined by the propagators and the so-called connected part, which is the n-point function subtracted by the Gaussian result, vanishes.

require it to be positive definite. The situation becomes much simpler if we find a linear transformation of our variables X that renders the operator C diagonal. Let us assume that there is such a linear invertible transformation, of the form

$$Y(m) = \int d\mathbf{n} L(m,\mathbf{n}) Y(\mathbf{n}) \tag{A7.12}$$

such that $C(m,m') = C_m \delta(m - m')$. In terms of the variable Y the inverse of C is simply $C_m^{-1}\delta(m - m')$ and

$$\langle Y_m Y_{m'} \rangle = \delta(m - m') \frac{1}{\sqrt{2\pi C_m}} \int dY_m Y_m^2 \exp(-Y_m^2/(2C_m), \tag{A7.13}$$

like in the one-dimensional case. Since C is a symmetric positive operator, such a diagonal basis always exists, but in general it might not be easy to find it.

However, symmetries can help in this case. If we decompose our random variables into components that transform irreducibly under a symmetry group of the problem, different components cannot be correlated with each other, in order not to break the symmetry. We now show that this symmetry property allows us in cosmology to find the diagonal basis that corresponds to a decomposition into irreducible components under the group of rotations or translations.

A7.2.1 Statistical Homogeneity and Isotropy

A random field respects a symmetry group G of the space S if the correlation function is invariant under transformations $\mathbf{n} \mapsto R\mathbf{n}$ for all $R \in G$. In other words,

$$C(R\mathbf{n}_1, R\mathbf{n}_2) = C(\mathbf{n}_1, \mathbf{n}_2).$$

For this it is not necessary that $X(\mathbf{n}) = X(R\mathbf{n})$, but the transformed variable must have identical statistical properties. In cosmology, we expect the CMB sky to be statistically isotropic, that is, invariant under rotations. This means that the correlation function is a function only of the scalar product $\mu = \mathbf{n}_1 \cdot \mathbf{n}_2$. Therefore, we can expand it in terms of Legendre polynomials,

$$C(\mathbf{n}_1, \mathbf{n}_2) = C(\mu) = \frac{1}{4\pi} \sum_\ell (2\ell + 1) C_\ell P_\ell(\mu). \tag{A7.14}$$

We shall show now, that when expanding a statistically isotropic random variable on the CMB sky in spherical harmonics,

$$X(\mathbf{n}) = \sum_{\ell m} a_{\ell m} Y_{\ell m}(\mathbf{n}), \tag{A7.15}$$

the coefficients $a_{\ell m}$ satisfy

$$\langle a_{\ell_1 m_1} a^*_{\ell_2 m_2} \rangle = \delta_{m_1 m_2} \delta_{\ell_1 \ell_2} C_{\ell_1}. \tag{A7.16}$$

To see this we write the correlation function as

$$\sum_{\ell m} C_\ell Y_{\ell m}(\mathbf{n}_1) Y^*_{\ell m}(\mathbf{n}_2) = \sum_{\ell m \ell' m'} \langle a_{\ell m} a^*_{\ell' m'} \rangle Y_{\ell m}(\mathbf{n}_1) Y^*_{\ell' m'}(\mathbf{n}_2). \tag{A7.17}$$

For the left-hand side of this equation we have used the addition theorem for spherical harmonics (see Appendix 4, Section A4.2.3) to replace the P_ℓ's in Eq. (A7.14) and for the right-hand side we simply used the expansion (A7.15). Multiplying this equation by $Y^*_{\ell_1 m_1}(\mathbf{n}_1)Y_{\ell_2 m_2}(\mathbf{n}_2)$ and integrating over $\mathbf{n}_1$ and $\mathbf{n}_2$ we obtain Eq. (A7.16).

Therefore, for statistically isotropic random fields, the expansion in terms of spherical harmonics diagonalizes the correlation function. This is why it is so useful to determine the $a_{\ell m}$'s measured by an experiment. If the underlying fluctuations are Gaussian, these are independent Gaussian variables.

Let us now turn to random fields on three-dimensional space. For simplicity we consider Euclidean space. For a random field $X(\mathbf{x})$ we expect its statistical properties to be independent of translations and rotations. Therefore, the correlation function $C(\mathbf{x}_1, \mathbf{x}_2)$ depends only on the distance $|\mathbf{x}_1 - \mathbf{x}_2| \equiv r$. We now show that for statistically homogeneous random fields, the power spectrum is simply the Fourier transform of the correlation function. Statistical isotropy implies that the latter depends only on the modulus of $\mathbf{k}$. In Chapter 2 we have defined the power spectrum of a statistically homogeneous and isotropic random field X by

$$\langle X(\mathbf{k})X^*(\mathbf{k}')\rangle = (2\pi)^3\delta(\mathbf{k}-\mathbf{k}')P_X(k). \tag{A7.18}$$

We shall now see that $\langle X(\mathbf{k})X^*(\mathbf{k}')\rangle$ is indeed of this form and that

$$P_X(k) = \hat{C}(k). \tag{A7.19}$$

We simply insert the definition

$$\begin{aligned}
\langle X(\mathbf{k})X^*(\mathbf{k}')\rangle &= \int d^3x\, d^3x'\, \langle X(\mathbf{x})X(\mathbf{x}')\rangle e^{i(\mathbf{k}\cdot\mathbf{x}-\mathbf{k}'\cdot\mathbf{x}')} \\
&= \int d^3x\, d^3x'\, C(\mathbf{x}-\mathbf{x}')e^{i(\mathbf{k}\cdot(\mathbf{x}-\mathbf{x}')-(\mathbf{k}'-\mathbf{k})\cdot\mathbf{x}')} \\
&= \int d^3z\, d^3x'\, C(\mathbf{z})e^{i(\mathbf{k}\cdot\mathbf{z}-(\mathbf{k}'-\mathbf{k})\cdot\mathbf{x}')} \\
&= (2\pi)^3\,\delta(\mathbf{k}-\mathbf{k}')\hat{C}(k).
\end{aligned}$$

For the third equals sign we made the variable transform $\mathbf{z} = \mathbf{x} - \mathbf{x}'$ and for the last equals sign we used the fact that the integral of $e^{i\mathbf{k}\cdot\mathbf{x}}$ is a delta function. More precisely,

$$\int e^{i(\mathbf{k}'-\mathbf{k})\cdot\mathbf{x}}d^3x = \int e^{-i(\mathbf{k}'-\mathbf{k})\cdot\mathbf{x}}d^3x = (2\pi)^3\delta(\mathbf{k}'-\mathbf{k}). \tag{A7.20}$$

This proves the ansatz (A7.18) and Eq. (A7.19). This shows that for statistically homogeneous fields, the Fourier coefficients $X(\mathbf{k})$ are independent random variables and are therefore especially useful for statistical analysis and parameter estimation.

Appendix 8

Approximation for the Tensor C_ℓ Spectrum

In this appendix we derive in a self-consistent way the approximation (2.278) for the C_ℓ power spectrum of tensor perturbations in an FL background with vanishing curvature, $K = 0$. We start with the power spectrum of tensor perturbations of the metric,

$$\left\langle H_{ij}^{(T)}(\mathbf{k},t) H_{mn}^{(T)*}(\mathbf{k}',t') \right\rangle = (2\pi)^3 P_{ijmn}(\mathbf{k},t,t') \delta^3(\mathbf{k}-\mathbf{k}').$$

$H_{ij}^{(T)}(\mathbf{k},t)$ is symmetric, traceless, and transverse, and since its Fourier transform is real we have $H_{ij}^{(T)*}(\mathbf{k},t) = H_{ij}^{(T)}(-\mathbf{k},t)$; hence $P_{ijmn}(\mathbf{k},t,t') = P_{mnij}(-\mathbf{k},t',t)$.

We define the projection tensor onto the plane normal to $\mathbf{k}$,

$$P_{ij} = \delta_{ij} - k^{-2} k_i k_j. \tag{A8.1}$$

The most generic tensor P_{ijmn}, which has the above properties, is isotropic,[1] and is also invariant under parity, $P_{ijmn}(\mathbf{k},t,t') = P_{ijmn}(-\mathbf{k},t,t')$, is to be of the form[2]

$$P_{ijmn}(\mathbf{k},t,t') = \mathcal{H}(k,t,t') \mathcal{M}_{ijmn}(\mathbf{k}) \quad \text{with} \tag{A8.2}$$

$$\begin{aligned} \mathcal{M}_{ijmn}(\mathbf{k}) &\equiv P_{im} P_{jn} + P_{in} P_{jm} - P_{ij} P_{mn} \\ &= [\delta_{im}\delta_{jn} + \delta_{in}\delta_{jm} - \delta_{ij}\delta_{mn} + k^{-2}(\delta_{ij} k_m k_n \\ &\quad + \delta_{mn} k_i k_j - \delta_{im} k_j k_n - \delta_{in} k_m k_j - \delta_{jm} k_i k_n - \delta_{jn} k_m k_i) \\ &\quad + k^{-4} k_i k_j k_m k_n]. \end{aligned} \tag{A8.3}$$

The Fourier transform on Eq. (2.243) gives

$$\left(\frac{\Delta T(\mathbf{n},\mathbf{k})}{T}\right)^{(T)} = -\int_i^f dt \exp(i\mathbf{k}\cdot\mathbf{n}(t_0 - t))\, \partial_t H_{ij}(t,\mathbf{k}) n^i n^j. \tag{A8.4}$$

[1] This means that only $\mathbf{k}$ and invariant tensors like δ_{ij} or ϵ_{ijm} enter its construction. No external given vector or tensor that is not invariant under rotations is allowed.

[2] If we do not require parity invariance, an additional term proportional to $\mathcal{A}_{ijlm} = k^{-1} k_q (P_{jm}\epsilon_{ilq} + P_{il}\epsilon_{jmq} + P_{im}\epsilon_{jlq} + P_{jl}\epsilon_{imq})$ can be added. But this term changes sign under parity and can be shown to be proportional to the difference of the amplitudes of the two polarization states (Caprini *et al.*, 2004). When it is present, parity violating terms (like corrolators between the temperature anisotropy and B-polarization (see Caprini *et al.*, 2004 and Chapter 5) appear. We neglect this possibility in this appendix.

With this and Eq. (A8.2) we obtain

$$\left\langle \frac{\Delta T}{T}(\mathbf{n},\mathbf{x})\frac{\Delta T}{T}(\mathbf{n}',\mathbf{x})\right\rangle$$
$$\equiv \left(\frac{1}{2\pi}\right)^6 \int d^3k\, d^3k' \left\langle \frac{\Delta T}{T}(\mathbf{n},\mathbf{k})\frac{\Delta T}{T}(\mathbf{n}',\mathbf{k}')\right\rangle \exp(-i\mathbf{x}(\mathbf{k}-\mathbf{k}'))$$
$$= \left(\frac{1}{2\pi}\right)^3 \int k^2\, dk\, d\Omega_{\hat{\mathbf{k}}} \int_{t_{\rm dec}}^{t_0} dt \int_{t_{\rm dec}}^{t_0} dt' \exp(i\mathbf{k}\cdot\mathbf{n}(t_0-t))\exp(-i\mathbf{k}\cdot\mathbf{n}'(t_0-t'))$$
$$\times \frac{\partial^2}{\partial t\partial t'} P_{ijlm}(\mathbf{k},t,t')n_i n_j n'_l n'_m. \tag{A8.5}$$

Here $d\Omega_{\hat{\mathbf{k}}}$ denotes the integral over directions in $\mathbf{k}$ space and we made use of the δ-function to get rid of the integral over d^3k'.

We now introduce the form (A8.2) of P_{ijlm}. We further assume that the perturbations have been created at some early epoch, for example, during an inflationary phase, after which they evolved deterministically. The function $\mathcal{H}(k,t,t')$ is thus a product of the form

$$\mathcal{H}(k,t,t') = H(k,t)\cdot H^*(k,t'), \tag{A8.6}$$

where $H(k,t)$ is the growing mode solution of Eq. (2.109) with the correct initial spectrum, $\langle H_{ij}(\mathbf{k},t_{\rm in})H^*_{ij}(\mathbf{k}',t_{\rm in})\rangle = 4|H(k,t_{\rm in})|^2\delta(\mathbf{k}-\mathbf{k}')$ and $\langle \dot{H}_{ij}(\mathbf{k},t_{\rm in})\dot{H}^*_{ij}(\mathbf{k}',t_{\rm in})\rangle = 4|\dot{H}(k,t_{\rm in})|^2\delta(\mathbf{k}-\mathbf{k}')$. Introducing this form of P_{ijlm} in Eq. (A8.5) yields

$$\left\langle \frac{\Delta T}{T}(\mathbf{n})\frac{\Delta T}{T}(\mathbf{n}')\right\rangle$$
$$= \left(\frac{1}{2\pi}\right)^3 \int k^2\, dk\, d\Omega_{\hat{\mathbf{k}}} \left[2(\mathbf{n}\cdot\mathbf{n}')^2 - 1 + \mu'^2 + \mu^2 - 4\mu\mu'(\mathbf{n}\cdot\mathbf{n}') + \mu^2\mu'^2\right]$$
$$\times \int_{t_{\rm dec}}^{t_0} dt \int_{t_{\rm dec}}^{t_0} dt' \left[\dot{H}(k,t)\dot{H}^*(k,t')\exp(ik\mu(t_0-t))\exp(-ik\mu'(t_0-t'))\right], \tag{A8.7}$$

where $\mu = (\mathbf{n}\cdot\hat{\mathbf{k}})$, $\mu' = (\mathbf{n}'\cdot\hat{\mathbf{k}})$ and $\dot{H} = \partial_t H$. To proceed, we use the identity (Abramowitz and Stegun, 1970)

$$\exp(ik\mu(t_0-t)) = \sum_{r=0}^{\infty}(2r+1)i^r j_r(k(t_0-t))P_r(\mu). \tag{A8.8}$$

Here j_r denotes the spherical Bessel function of order r and P_r is the Legendre polynomial of degree r.

Furthermore, we replace each factor of μ in Eq. (A8.7) by a derivative of the exponential $\exp(ik\mu(t_0-t))$ with respect to $k(t_0-t)$ and correspondingly with μ'. We then obtain

$$\left\langle \frac{\Delta T}{T}(\mathbf{n})\frac{\Delta T}{T}(\mathbf{n}')\right\rangle$$
$$= \left(\frac{1}{2\pi}\right)^3 \sum_{r,r'}(2r+1)(2r'+1)i^{(r-r')} \int k^2\, dk\, d\Omega_{\hat{\mathbf{k}}}\, P_r(\mu)P_{r'}(\mu')\times$$

$$\times \Big[2(\mathbf{n}\cdot\mathbf{n}')^2 \int dt\,dt'\; j_r(k(t_0-t))\,j_{r'}(k(t_0-t'))\dot{H}(k,t)\dot{H}^*(k,t')$$
$$-\int dtdt'[j_r(k(t_0-t))\,j_{r'}(k(t_0-t')) + j_r''(k(t_0-t))\,j_{r'}(k(t_0-t'))$$
$$+ j_r(k(t_0-t))\,j_{r'}''(k(t_0-t')) - j_r''(k(t_0-t))\,j_{r'}''(k(t_0-t'))]\dot{H}(k,t)\dot{H}^*(k,t')$$
$$-4(\mathbf{n}\cdot\mathbf{n}')\int dt\,dt'\; j_r'(k(t_0-t))\,j_{r'}'(k(t_0-t'))\dot{H}(k,t)\dot{H}^*(k,t')\Big]. \tag{A8.9}$$

The primes on the Bessel functions denote derivatives with respect to the argument of the function. In the expression (A8.9) only the Legendre polynomials, $P_r(\mu)$ and $P_{r'}(\mu')$, depend on the direction $\hat{\mathbf{k}}$. To perform the integration $d\Omega_{\hat{\mathbf{k}}}$, we use the addition theorem for spherical harmonics (see Appendix 4, Section A4.2.3),

$$P_r(\mu) = \frac{4\pi}{(2r+1)}\sum_{s=-r}^{r} Y_{rs}(\mathbf{n})Y_{rs}^*(\hat{\mathbf{k}}). \tag{A8.10}$$

The orthogonality of spherical harmonics (see Appendix 4, Section A4.2.3) then yields

$$\begin{aligned}(2r+1)(2r'+1)&\int d\Omega_{\hat{\mathbf{k}}}\, P_r(\mu)P_{r'}(\mu')\\ &= 16\pi^2\delta_{rr'}\sum_{s=-r}^{r} Y_{rs}(\mathbf{n})Y_{rs}^*(\mathbf{n}')\\ &= (2r+1)4\pi\,\delta_{rr'}P_r(\mathbf{n}\cdot\mathbf{n}').\end{aligned} \tag{A8.11}$$

In Eq. (A8.9) the integration over $d\Omega_{\hat{\mathbf{k}}}$ leads to terms of the form $(\mathbf{n}\cdot\mathbf{n}')P_r(\mathbf{n}\cdot\mathbf{n}')$ and $(\mathbf{n}\cdot\mathbf{n}')^2P_r(\mathbf{n}\cdot\mathbf{n}')$. To reduce them, we use recursion relations for Legendre polynomials like

$$xP_r(x) = \frac{r+1}{2r+1}P_{r+1} + \frac{r}{2r+1}P_{r-1}. \tag{A8.12}$$

Applying this and its iteration for $x^2P_r(x)$, we obtain

$$\begin{aligned}&\left\langle \frac{\Delta T}{T}(\mathbf{n})\frac{\Delta T^*}{T}(\mathbf{n}')\right\rangle\\
&= \frac{1}{2\pi^2}\sum_r (2r+1)\int k^2dk\int dtdt'\dot{H}(k,t)\dot{H}^*(k,t')\\
&\times\left\{\left[\frac{2(r+1)(r+2)}{(2r+1)(2r+3)}P_{r+2} + \frac{1}{(2r-1)(2r+3)}P_r + \frac{2r(r-1)}{(2r-1)(2r+1)}P_{r-2}\right]\right.\\
&\times j_r(k(t_0-t))\,j_r(k(t_0-t')) - P_r[\,j_r(k(t_0-t))\,j_r''(k(t_0-t'))\\
&+ j_r(k(t_0-t'))\,j_r''(k(t_0-t)) - j_r''(k(t_0-t))\,j_{r'}''(k(t_0-t'))]\\
&\left.-4\left[\frac{r+1}{2r+1}P_{r+1} + \frac{r}{2r+1}P_{r-1}\right]j_r'(k(t_0-t))\,j_r'(k(t_0-t'))\right\},\end{aligned} \tag{A8.13}$$

where the argument, $\mathbf{n}\cdot\mathbf{n}'$, of the Legendre polynomials has been suppressed. Also using the relation

$$j_r' = -\frac{r+1}{2r+1}j_{r+1} + \frac{r}{2r+1}j_{r-1} \tag{A8.14}$$

for Bessel functions, and its iteration for j'', we can rewrite Eq. (A8.13) in terms of the Bessel functions j_{r-2} to j_{r+2}.

To proceed we use the definition of C_ℓ:

$$\left\langle \frac{\Delta T}{T}(\mathbf{n}) \cdot \frac{\Delta T}{T}(\mathbf{n}') \right\rangle_{(\mathbf{n}\cdot\mathbf{n}')=\cos\theta} = \frac{1}{4\pi}\Sigma(2\ell+1)C_\ell P_\ell(\cos\theta). \tag{A8.15}$$

If we expand

$$\frac{\Delta T}{T}(\mathbf{n}) = \sum_{\ell,m} a_{\ell m} Y_{\ell m}(\mathbf{n}) \tag{A8.16}$$

and use the orthogonality of the spherical harmonics as well as the addition theorem, Eq. (A8.10), we find

$$C_\ell = \langle a_{\ell m} a^*_{\ell m} \rangle. \tag{A8.17}$$

We thus have to determine the correlators

$$\langle a_{\ell m} a^*_{\ell' m'} \rangle = \int d\Omega_{\mathbf{n}} \int d\Omega_{\mathbf{n}'} \left\langle \frac{\Delta T^*}{T}(\mathbf{n}) \frac{\Delta T}{T}(\mathbf{n}') \right\rangle Y^*_{\ell m}(\mathbf{n}) Y_{\ell' m'}(\mathbf{n}'). \tag{A8.18}$$

Inserting our result (A8.13), we obtain the somewhat lengthy expression

$$\begin{aligned}
\langle a_{\ell m} a^*_{\ell' m'} \rangle = {} & \frac{2}{\pi}\delta_{\ell\ell'}\,\delta mm' \int dk\, k^2 \int dt\, dt'\, \dot{H}(k,t)\dot{H}^*(k,t') \\
& \times \Bigg\{ j_\ell(k(t_0-t))\, j_\ell(k(t_0-t')) \\
& \times \left(\frac{1}{(2\ell-1)(2\ell+3)} + \frac{2(2\ell^2+2\ell-1)}{(2\ell-1)(2\ell+3)} + \frac{(2\ell^2+2\ell-1)^2}{(2\ell-1)^2(2\ell+3)^2} \right. \\
& \left. - \frac{4\ell^3}{(2\ell-1)^2(2\ell+1)} - \frac{4(\ell+1)^3}{(2\ell+1)(2\ell+3)^2} \right) \\
& - \left[j_\ell(k(t_0-t))\, j_{\ell+2}(k(t_0-t')) + j_{\ell+2}(k(t_0-t))\, j_\ell(k(t_0-t')) \right] \frac{1}{2\ell+1} \\
& \times \left(\frac{2(\ell+2)(\ell+1)(2\ell^2+2\ell-1)}{(2\ell-1)(2\ell+3)^2} + \frac{2(\ell+1)(\ell+2)}{(2\ell+3)} - \frac{8(\ell+1)^2(\ell+2)}{(2\ell+3)^2} \right) \\
& - \left[j_\ell(k(t_0-t))\, j_{\ell-2}(k(t_0-t')) + j_{\ell-2}(k(t_0-t))\, j_\ell(k(t_0-t')) \right] \\
& \times \frac{1}{2\ell+1} \left(\frac{2\ell(\ell-1)(2\ell^2+2\ell-1)}{(2\ell-1)^2(2\ell+3)} + \frac{2\ell(\ell-1)}{(2\ell-1)^2} - \frac{8\ell^2(\ell-1)}{(2\ell-1)^2} \right) \\
& + j_{\ell+2}(k(t_0-t))\, j_{\ell+2}(k(t_0-t')) \\
& \times \left(\frac{2(\ell+2)(\ell+1)}{(2\ell+1)(2\ell+3)} - \frac{4(\ell+1)(\ell+2)^2}{(2\ell+1)(2\ell+3)^2} + \frac{(\ell+1)^2(\ell+2)^2}{(2\ell+1)^2(2\ell+3)^2} \right) \\
& + j_{\ell-2}(k(t_0-t))\, j_{\ell-2}(k(t_0-t')) \\
& \times \left(\frac{2\ell(\ell-1)}{(2\ell-1)(2\ell+1)} - \frac{4\ell(\ell-1)^2}{(2\ell-1)^2(2\ell+1)} + \frac{\ell^2(\ell-1)^2}{(2\ell-1)^2(2\ell+1)^2} \right) \Bigg\}.
\end{aligned} \tag{A8.19}$$

An analysis of the coefficient of each term reveals that this expression is equivalent to

$$C_\ell^{(T)} = \frac{2}{\pi}\int dk\, k^2 \left| \int_{t_{\rm dec}}^{t_0} dt\, \dot{H}(t,k) \frac{j_\ell(k(t_0-t))}{(k(t_0-t))^2} \right|^2 \frac{(\ell+2)!}{(\ell-2)!}\,. \qquad \text{(A8.20)}$$

To obtain this result we have used the identity

$$\frac{j_{\ell+2}(k(t_0-t))}{(2\ell+3)(2\ell+1)} + \frac{2j_\ell(k(t_0-t))}{(2\ell+3)(2\ell-1)} + \frac{j_{\ell-2}(k(t_0-t))}{(2\ell+1)(2\ell-1)} = \frac{j_\ell(k(t_0-t))}{(k(t_0-t))^2}. \qquad \text{(A8.21)}$$

Appendix 9
Boltzmann Equation in a Universe with Curvature

In this appendix we discuss the changes of the Boltzmann equation in the case of nonvanishing curvature $K = H_0^2(\Omega_0 - 1) \neq 0$. We closely follow the treatment in Abbott and Schaefer (1982) and Hu *et al.* (1998). We write the metric of the unperturbed FL universe as

$$ds^2 = a^2\gamma_{\mu\nu}\, dx^\mu\, dx^\nu, \tag{A9.1}$$

with

$$\gamma_{00} = -1, \tag{A9.2}$$

$$\gamma_{0i} = 0, \tag{A9.3}$$

$$\gamma_{ij}\, dx^i\, dx^j = \frac{1}{|K|}\left(d\chi^2 + \sin_K^2(\chi)(d\theta^2 + \sin^2\theta\, d\phi^2)\right). \tag{A9.4}$$

Here we have set

$$\sin_K \chi = \begin{cases} \sin\chi & \text{for} \quad K > 0 \\ \chi & \text{for} \quad K = 0 \\ \text{sh}\chi & \text{for} \quad K < 0, \end{cases} \tag{A9.5}$$

where sh denotes the hyperbolic sine (correspondingly we shall denote the hyperbolic cosine by ch).

A9.1 The Boltzmann Equation

We now derive the Boltzmann equation. The collision term is not affected by curvature since it is purely local. If we use a "quasi-orthonormal" spatial basis, also the gravitational source term coming from $\delta\Gamma^i_{\alpha\beta}n^\alpha n^\beta \gamma_{ij}n^j v^2(d\bar{f}/dv)$ is not modified. Here "quasi-orthonormal" means that $n^i n^j \gamma_{ij} = 1$ and $(p^\mu) = (p, pn^i)$. Again $v = pa$ denotes the redshift corrected photon energy, the only variable on which the background distribution function $\bar{f}$ depends. The only modification comes from the fact that on the left-hand side we have to add a term due to the unperturbed three-dimensional Christoffel symbols,

$$(\partial_t + n^i\partial_i)\mathcal{M} \to \left(\partial_t + n^i\partial_i - \bar{\Gamma}^i_{jl}n^j n^l \frac{\partial}{\partial n^i}\right)\mathcal{M}. \tag{A9.6}$$

We now show that this simply corresponds to replacing the partial derivatives ∂_i in the second term by covariant derivatives w.r.t. the spatial background metric γ_{ij}. To see this we expand $\mathcal{M}$ in moments of n^i,

$$\mathcal{M}(t,\mathbf{x},\mathbf{n}) = \sum Q(t,\mathbf{x})^{(m)}_{i_1\cdots i_m} n^{i_1}\cdots n^{i_m}. \tag{A9.7}$$

Here the sum goes from 1 to 3 for each of the indices i_l and the number m of indices goes from zero to infinity. For uniqueness, we require that $Q_{i_1\cdots i_m}$ be a traceless totally symmetric tensor. Since $\gamma_{jl}n^j n^l = 1$, a trace in $Q^{(m)}_{i_1\cdots i_{m-2}jl}$ contributes a term $Q^{(m-2)}_{i_1\cdots i_{m-2}}\gamma_{jl}n^{i_1}\cdots n^{i_{m-2}}n^j n^l = Q^{(m-2)}_{i_1\cdots i_{m-2}}$ and can be absorbed in $Q^{(m-2)}$. With this ansatz we find

$$\begin{aligned}
&\left(n^s\partial_s - \bar\Gamma^j_{rs}n^r n^s\frac{\partial}{\partial n^j}\right)\mathcal{M} \\
&= \sum n^s\partial_s Q(t,\mathbf{x})^{(m)}_{i_1\cdots i_m}n^{i_1}\cdots n^{i_m} - \bar\Gamma^j_{rs}Q^{(m)}_{i_1\cdots i_m}n^r n^s(\delta_{ji_1}n^{i_2}\cdots n^{i_m} \\
&\qquad\qquad + \cdots + n^{i_1}\cdots n^{i_{m-1}}\delta_{ji_m}) \\
&= \sum n^s\left(\partial_s Q(t,\mathbf{x})^{(m)}_{i_1\cdots i_m}n^{i_1}\cdots n^{i_m} - \bar\Gamma^j_{rs}Q^{(m)}_{j\cdots i_m}n^r n^{i_2}\cdots n^{i_m}\right. \\
&\qquad \left. - \cdots - \bar\Gamma^j_{rs}Q^{(m)}_{i_1\cdots j}n^r n^{i_1}\cdots n^{i_{m-1}}\right) \\
&= \sum n^s\left(\partial_s Q(t,\mathbf{x})^{(m)}_{i_1\cdots i_m}n^{i_1}\cdots n^{i_m} - \bar\Gamma^j_{i_1 s}Q^{(m)}_{j\cdots i_m}n^{i_1}n^{i_2}\cdots n^{i_m}\right. \\
&\qquad \left. - \cdots - \bar\Gamma^j_{i_m s}Q^{(m)}_{i_1\cdots j}n^{i_1}\cdots n^{i_m}\right) \\
&= \sum n^s Q(t,\mathbf{x})^{(m)}_{i_1\cdots i_m|s}n^{i_1}\cdots n^{i_m} \equiv n^s\mathcal{M}_{|s}.
\end{aligned} \tag{A9.8, A9.9}$$

Note that the last expression really is a definition. It tells us how we have to interpret a covariant derivative for a function in momentum space. With this, the Boltzmann equation in spaces with nonvanishing curvature becomes simply

$$\partial_t\mathcal{V} + n^i\mathcal{V}_{|i} = C[\mathcal{V}] + \begin{pmatrix} G[h_{\mu\nu}] \\ 0 \\ 0 \end{pmatrix}, \tag{A9.10}$$

where $C[\mathcal{V}]$ denotes the collision term given in Eq. (5.52) and

$$\mathcal{V} = \begin{pmatrix} \mathcal{M} \\ \mathcal{E}+i\mathcal{B} \\ \mathcal{E}-i\mathcal{B} \end{pmatrix}. \tag{A9.11}$$

The gravitational term is

$$G[h_{\mu\nu}]\begin{cases} -n^i(\Psi+\Phi)_{|i} & \text{for scalar perturbations,} \\ -\sigma^{(V)}_{ij}n^i n^j & \text{for vector perturbations,} \\ -\dot H_{ij}n^i n^j & \text{for tensor perturbations.} \end{cases} \tag{A9.12}$$

A9.2 Line-of-Sight Integration

The homogeneous part of the Boltzmann equation (A9.10),

$$\partial_t\mathcal{V} + n^i\mathcal{V}_{|i} = 0, \tag{A9.13}$$

simply represents free streaming. If the source term can be neglected, the temperature fluctuations and the polarization are modified by photon free streaming, which in this case means that the photons move along spatial geodesics of the unperturbed metric γ_{ij}. Unlike in flat space, $\mathbf{n}$ is not constant, but varies along a geodesic.

Let us denote by $\mathbf{y}(\mathbf{x}, \mathbf{n}(\lambda), \lambda)$ the (spatial) geodesic of the unperturbed metric γ_{ij} that arrives at $\mathbf{x}$ at time λ and then moves on in direction $-\mathbf{n}(\lambda)$, so that $\mathbf{x} = \mathbf{y}(\mathbf{x}, \mathbf{n}, \lambda)$. In flat space, $\mathbf{y}(\mathbf{x}, \mathbf{n}, \lambda)(t) = \mathbf{x} - (t-\lambda)\mathbf{n}$. The general solution to Eq. (A9.13) with initial condition $\mathcal{V}(t_{\text{in}}, \mathbf{x}, \mathbf{n}) = \mathcal{V}_1(\mathbf{x}, \mathbf{n})$ is then simply,

$$\mathcal{V}(t, \mathbf{x}, \mathbf{n}) = \mathcal{V}_1(\mathbf{y}(\mathbf{x}, \mathbf{n}(t - t_{\text{in}}), t - t_{\text{in}}), \mathbf{n}(t - t_{\text{in}})). \tag{A9.14}$$

To verify this we use that both $\mathbf{y}$ and $\mathbf{n}$ depend on time so that $\partial_t \mathcal{V}(t, \mathbf{x}, \mathbf{n}) = -n^i \partial_i \mathcal{V}(t, \mathbf{x}, \mathbf{n}) + \dot{n}^i \frac{\partial}{\partial n^i} \mathcal{V}(t, \mathbf{x}, \mathbf{n})$. But since n moves along a geodesic, $\dot{n}^i = \bar{\Gamma}^i_{rs} n^r n^s$, so that we end up with

$$\partial_t \mathcal{V}(t, \mathbf{x}, \mathbf{n}) + n^i \partial_i \mathcal{V}(t, \mathbf{x}, \mathbf{n}) - \bar{\Gamma}^i_{rs} n^r n^s \frac{\partial}{\partial n^i} \mathcal{V}(t, \mathbf{x}, \mathbf{n}) = 0.$$

This corresponds to the expression after the arrow in Eq. (A9.6), which is equivalent to Eq. (A9.13) according to our definition (A9.9). This observation allows us also to formally solve (A9.10) with a line-of-sight integration as in the flat case: for an arbitrary source term, $S(t, \mathbf{x}, \mathbf{n})$, on the right-hand side of Eq. (A9.13) the solution is given by

$$\mathcal{V}(t, \mathbf{x}, \mathbf{n}) = \mathcal{V}_1(\mathbf{y}(\mathbf{x}, \mathbf{n}, t), \mathbf{n}) + \int_{t_{\text{in}}}^{t} dt' \; S(t', \mathbf{y}(\mathbf{x}(t'), \mathbf{n}(t'), t'), \mathbf{n}(t')). \tag{A9.15}$$

Here $\mathbf{x}(t')$ is the geodesic that ends at $\mathbf{x} = \mathbf{x}(t)$ at time t with velocity $-\mathbf{n}(t)$.

So far this is only a formal solution. In the case of the Boltzmann equation the source on the right-hand side depends on the left-hand side. Furthermore, the geodesics $\mathbf{y}(\mathbf{x}, \mathbf{n}(t'), t')$ are not given explicitly. However, this is not a serious problem, since they are the solutions to well-known ordinary differential equations.

A9.3 Mode Functions, Radial Functions

In flat space, we have expanded Eq. (A9.10) in terms of mode functions

$$_sG_{\ell m}(\mathbf{x}, \mathbf{n}) = (-i)^{\ell} \sqrt{\frac{4\pi}{2\ell+1}} \, _sY_{\ell m}(\mathbf{n}) \exp(i\mathbf{k} \cdot \mathbf{x}), \quad K = 0. \tag{A9.16}$$

Here we have to replace the exponentials by eigenfunctions of the Laplacian in curved space. The functions $Q^{(m)}_{i_1 \cdots i_{|m|}}$ with

$$(\Delta_K + k^2) Q^{(m)}_{i_1 \cdots i_{|m|}} = 0, \tag{A9.17}$$

where $k^2 > (|m|+1)|K|$ and $k^2 = (p(p+2)+|m|)K, p \in \mathbb{N}$, form a complete set of basis functions for $K < 0$ and $K > 0$ respectively (see Vilenkin and Smorodinskii, 1964). Here, like in Chapter 2, $Q^{(m)}$ is a totally symmetric traceless tensor with helicity m and rank $|m|$.

For convenience we set

$$q = \sqrt{p(p+2)+1+|m|}, \quad K > 0 \tag{A9.18}$$

$$q = \sqrt{\frac{k^2}{|K|} - 1 - |m|}, \quad K < 0, \tag{A9.19}$$

and we shall use this dimensionless number to denominate the mode functions. The functions $p = 0$ and $p = 1$ are not of interest for us. They represent a simple constant ($p = 0$) and ($p = 1$) a pure dipole contribution that is gauge dependent. We therefore consider $q = 2, 3, \ldots$ for $m = 0$, $K > 0$.

As we have done for the exponential, we want to expand the function $Q(\mathbf{x})$ in its orbital angular momentum. For a given mode function Q we orient the coordinate system such that the angular dependence is given by Y_{L0} alone. (In flat space this corresponds to choosing the z direction parallel to $\mathbf{k}$.) We can then write

$${}_sG_{\ell m} = \left(4\pi \sum_L \sqrt{\frac{2L+1}{2\ell+1}} i^{\ell-L} \phi_{q\,L}(\chi) Y_{L0}(\hat{\mathbf{x}})\right) {}_sY_{\ell m}. \tag{A9.20}$$

Each angular momentum component $Q = \phi_{q\,L}(\chi) Y_{L0}(\hat{\mathbf{x}})$ satisfies

$$\Delta Q = \gamma^{rs} Q_{|rs} - \gamma^{rs}\Gamma^1_{rs} Q_{,1} + \gamma^{11} Q_{,11} = -|K|(q^2 \mp 1) Q,$$

where 1 denotes the χ direction and rs stand for the ϑ and φ directions. In $q^2 \mp 1$, the minus sign is for $K > 0$ and the plus sign for $K < 0$. Denoting $\sin_K = \sin$ for $K > 0$ and $\sin_K = \mathrm{sh}$ for $K < 0$, we have

$$\Gamma^1_{rs} = -\frac{\sin'_K(\chi)}{\sin_K(\chi)} \gamma_{rs},$$

and $\gamma^{rs} Q_{|rs} = |K| \sin_K^{-2}(\chi) \Delta_{\vartheta,\varphi} Q = -|K| L(L+1) \sin_K^{-2} Q$. With this we obtain the following differential equation for $\phi_{q\,L}$:

$$\frac{d^2\phi_{q\,L}}{d\chi^2} + 2\frac{\cos\chi}{\sin\chi}\frac{d\phi_{q\,L}}{d\chi} + \left(q^2 - 1 - \frac{L(L+1)}{\sin^2\chi}\right)\phi_{q\,L} = 0.$$

In this form, the equation is valid for $K > 0$. For $K < 0$ one has to replace $\sin\chi$ by $\mathrm{sh}\chi$ and $\cos\chi$ by $\mathrm{ch}\chi$ as well as $q^2 - 1$ by $q^2 + 1$. The solutions to this equation, which are regular at $\chi = 0$, are

$$\phi_{q\,L}(\chi) \propto \begin{cases} -q^{-2}(\sin\chi)^L \frac{d^{L+1}}{d(\cos\chi)^{L+1}} \cos(q\chi), & \text{for } K > 0 \\ -q^{-2}(\mathrm{sh}\chi)^L \frac{d^{L+1}}{d(\mathrm{ch}\chi)^{L+1}} \cos(q\chi), & \text{for } K < 0. \end{cases} \tag{A9.21}$$

These functions can also be expressed in terms of associated Legendre functions; see Abramowitz and Stegun (1970),

$$\phi_{q\,L}(\chi) \propto \begin{cases} (\sin\chi)^{-1/2} P_{\ell\,m}(\cos\chi), \text{ with} \\ \quad \ell = -1/2 - q, \ m = -\frac{1}{2} - L, \quad \text{for } K > 0 \\ (\mathrm{sh}\chi)^{-1/2} P_{\ell\,m}(\mathrm{ch}\chi), \text{with} \\ \quad \ell = -1/2 + iq, \ m = -\frac{1}{2} - L, \quad \text{for } K < 0. \end{cases} \tag{A9.22}$$

This is easily verified by deriving the differential equation for $(\sin\chi)^{1/2}\phi_{q\,L}(\chi)$ and comparing it with the one for associated Legendre functions given in Eq. (A4.15). The hyperspherical Bessel functions $\phi_{q\,L}$ satisfy the recurrence relations

$$\frac{d}{d\chi}\phi_{q\,L} = \frac{1}{2L+1}\left[L\sqrt{q^2 - L^2}\phi_{q\,L-1} - (L+1)\sqrt{q^2-(L+1)^2}\phi_{q\,L+1}\right],$$

$$\cot\chi\,\phi_{q\,L} = \frac{1}{2L+1}\left[\sqrt{q^2 - L^2}\phi_{q\,L-1} + \sqrt{q^2-(L+1)^2}\phi_{q\,L+1}\right], \tag{A9.23}$$

for $K > 0$. For negative curvature, the terms $q^2 - n^2$ have to be replaced by $q^2 + n^2$ and $\cot\chi$ by $\coth(\chi)$. These relations define the hyperspherical Bessel functions in terms of the first member

$$\phi_{q\,0}(\chi) = \begin{cases} \frac{\sin(q\chi)}{q\sin\chi} & \text{for } K > 0 \\ \frac{\sin(q\chi)}{q\mathrm{sh}\chi} & \text{for } K < 0. \end{cases} \tag{A9.24}$$

The normalization is chosen such that $\lim_{K\to 0}\phi_{q\,L}(\chi) = j_L(kr)$.

We do not need all the details of the mode functions ${}_sG_{\ell m}$, but we have to calculate $n^i[{}_sG_{\ell m}]_{|i}(\mathbf{x},\mathbf{n})$ that enters the Boltzmann equation. To obtain the Boltzmann hierarchy we have to express this derivative in terms of ${}_sG_{\ell+1\,m}$, ${}_sG_{\ell m}$, and ${}_sG_{\ell-1\,m}$. This is obtained most easily if we consider $\mathbf{n} = -\hat{\mathbf{x}}$ so that $\mathbf{x} = -\sqrt{|K|}\chi\mathbf{n}$. We then have

$$n^i[{}_sG_{\ell m}]_{|i}(-\sqrt{|K|}\chi\mathbf{n},\mathbf{n}) = -\sqrt{|K|}\frac{d}{d\chi}[{}_sG_{\ell m}](-\sqrt{|K|}\chi\mathbf{n},\mathbf{n}). \tag{A9.25}$$

To calculate this derivative we expand ${}_sG_{\ell m}(-\sqrt{|K|}\chi\mathbf{n},\mathbf{n})$ in its total angular momentum.

$$\begin{aligned} {}_sG_{\ell m}(-\sqrt{|K|}\chi\mathbf{n},\mathbf{n}) &= \left(4\pi\sum_L\sqrt{\frac{2L+1}{2\ell+1}}i^{\ell-L}\phi_{q\,L}(\chi)Y_{L0}(\mathbf{n})\right){}_sY_{\ell m}(\mathbf{n}) \\ &= \sum_j(-i)^j\sqrt{4\pi(2j+1)}\,{}_sf_j^{(\ell m)}(\chi)Y_{jm}(\mathbf{n}). \end{aligned} \tag{A9.26}$$

For $\ell = 0$, we immediately obtain

$${}_0f_j^{(00)}(\chi,q) \equiv \alpha_j^{(00)}(\chi) = \phi_{q\,j}(\chi). \tag{A9.27}$$

The other coefficients can in principle be obtained with the help of the Clebsch–Gordan series for the products $Y_{L0}(\mathbf{n})\,{}_sY_{\ell m}(\mathbf{n})$ and the recurrence relations for the hyperspherical Bessel functions $\phi_{q\,L}$. This straightforward but cumbersome calculation has never appeared in print, and we do not want to break with this tradition here. It is much easier to use the fact that ${}_0G_{m\,m}$ are given by [see, e.g., Thorne (1980) and Maggiore (2007)].

$${}_0G_{m\,m} = n^{i_1}\cdots n^{i_m}\,Q^{(m)}_{i_1\cdots i_m},$$

and

$${}_{\pm 2}G_{2\,m} \propto \mathbf{e}^{\pm}_{i_1}\mathbf{e}^{\pm}_{i_2}Q^{(m)}_{i_1 i_2},$$

for $0 \le |m| \le 2$. With this it is relatively easy to derive relations between the hyperspherical Bessel functions and the coefficients ${}_s f_j^{(\ell m)}(\chi)$. Most importantly,

$$\alpha_j^{(11)}(\chi,q) = \sqrt{\frac{j(j+1)}{2(q^2-1)}}\,\mathrm{sc}(\chi)\,\phi_{q\,j}(\chi),$$

$$\alpha_j^{(22)}(\chi,q) = \sqrt{\frac{3}{8}\frac{(j+2)(j^2-1)j}{(q^2-4)(q^2-1)}}\,\mathrm{sc}^2(\chi)\,\phi_{q\,j}(\chi). \tag{A9.28}$$

Similarly for

$${}_{\pm 2}f_j^{(2m)} = \epsilon_j^{(m)} \pm i\beta_j^{(m)},$$

one finds

$$\epsilon_j^{(0)}(\chi,q) = \sqrt{\frac{3}{8}\frac{(j+2)(j^2-1)j}{(q^2-4)(q^2-1)}}\,\mathrm{sc}^2(\chi)\phi_{qj}(\chi),$$

$$\epsilon_j^{(1)}(\chi,q) = \frac{1}{2}\sqrt{\frac{(j-1)(j+2)}{(q^2-4)(q^2-1)}}\,\mathrm{sc}\chi\left[\cot(\chi)\phi_{q\,j}(\chi) + \phi'_{qj}(\chi)\right], \tag{A9.29}$$

$$\epsilon_j^{(2)}(\chi,q) = \frac{1}{4}\sqrt{\frac{1}{(q^2-4)(q^2-1)}}\left[\phi''_{q\,j}(\chi) + 4\cot(\chi)\phi'_{q\,j}(\chi) - \left(q^2+1-2\cot^2\chi\right)\phi_{q\,j}(\chi)\right],$$

and

$$\beta_j^{(0)}(\chi,q) = 0,$$

$$\beta_j^{(1)}(\chi,q) = \frac{1}{2}\sqrt{\frac{(j-1)(j+2)q^2}{(q^2-4)(q^2-1)}}\,\mathrm{sc}(\chi)\phi_{q\,j}(\chi), \tag{A9.30}$$

$$\beta_j^{(2)}(\chi,q) = \frac{1}{2}\sqrt{\frac{q^2}{(q^2-4)(q^2-1)}}\left[\phi'_{q\,j}(\chi) + 2\cot(\chi)\phi_{q\,j}(\chi)\right],$$

for $m > 0$. For $m < 0$, $\beta_j^{(-m)} = -\beta_j^{(m)}$ while $\epsilon_j^{(m)} = \epsilon_j^{(-m)}$ and $\alpha_j^{(m)} = \alpha_j^{(-m)}$. The formulas presented are for positive curvature. For negative curvature all terms of the form $q^2 - n^2$ have to be replaced by $q^2 + n^2$, and the trigonometric functions have to be replaced by hyperbolic functions, for example, sc $\chi \equiv 1/\sin\chi$ becomes sch$\chi \equiv 1/\mathrm{sh}\,\chi$. The overall normalization of the modes is chosen such that

$$ {}_sf_j^{(\ell m)}(0,q) = \frac{1}{2j+1}\delta_{j\ell}. \tag{A9.31}$$

From the recurrence relation for the hyperspherical Bessel functions we obtain the following relation for the coefficients ${}_sf_j^{(\ell m)}$ defined so far:

$$\frac{d}{d\chi}[{}_sf_j^{(\ell m)}] = \frac{q}{2j+1}\left[{}_s\theta_j^m\,{}_sf_{j-1}^{(\ell m)} - {}_s\theta_{j+1}^m\,{}_sf_{j+1}^{(\ell m)}\right] - i\frac{qms}{j(j+1)}\,{}_sf_j^{(\ell m)}, \tag{A9.32}$$

where

$$ {}_s\theta_j^m = \sqrt{\left[\frac{(j^2-m^2)(j^2-s^2)}{j^2}\right]\left(\frac{j^2}{q^2}\pm 1\right)}; \tag{A9.33}$$

here the + sign is for negative curvature and the − sign is for $K > 0$.

Since the relation (A9.32) is independent of ℓ it is valid for all ℓ's and can also be used to define ${}_sf_{j+1}^{(\ell m)}$ from ${}_sf_j^{(\ell m)}$ and ${}_sf_{j-1}^{(\ell m)}$.

With the expansion Eqs. (A9.26) and (A9.32) we can now write the derivative (A9.25) as

$$n^i[{}_sG_{\ell m}]_{|i} = -\frac{\sqrt{|K|}q}{2\ell+1}\left[{}_s\theta_\ell^m\,{}_sG_{\ell-1\,m} - {}_s\theta_{\ell+1}^m\,{}_sG_{\ell+1\,m}\right] + i\frac{\sqrt{|K|}qms}{\ell(\ell+1)}\,{}_sG_{\ell m} \tag{A9.34}$$

Like in flat space, we expand the CMB anisotropy and polarization as

$$\mathcal{M}(t,\mathbf{x},\mathbf{n}) = \sum\!\!\!\!\!\!\int \frac{d^3q}{(2\pi)^3}\sum_\ell\sum_{m=-2}^{2}\mathcal{M}_\ell^{(m)}\,{}_0G_{\ell m}, \tag{A9.35}$$

$$(Q\pm iU)(t,\mathbf{x},\mathbf{n}) = \sum\!\!\!\!\!\!\int \frac{d^3q}{(2\pi)^3}\sum_\ell\sum_{m=-2}^{2}(\mathcal{E}_\ell^{(m)}\pm i\mathcal{B}_\ell^{(m)})\,{}_{\pm 2}G_{\ell m}. \tag{A9.36}$$

The symbol $\sum\!\!\!\!\!\int$ indicates that for positive curvature the integral over q has to be replaced by a sum.

For the coefficients $\mathcal{M}_\ell^{(m)}(t,q)$, $\mathcal{E}_\ell^{(m)}(t,q)$, and $\mathcal{B}_\ell^{(m)}(t,q)$ we now obtain the desired Boltzmann hierarchy:

$$\dot{\mathcal{M}}_\ell^{(m)} - q\sqrt{|K|}\left[{}_0\theta_\ell^m\mathcal{M}_{\ell-1}^{(m)} - {}_0\theta_{\ell+1}^m\mathcal{M}_{\ell+1}^{(m)}\right] = S_\ell^{(m)} + \dot\kappa\left[P_\ell^{(m)} - \mathcal{M}_\ell^{(m)}\right], \tag{A9.37}$$

with

$$S_\ell^{(0)} = -k(\Psi + \Phi)\delta_{\ell 1}, \tag{A9.38}$$

$$S_\ell^{(\pm 1)} = -\frac{\sqrt{3}}{3k}\sqrt{k^2 - 2K}\,\sigma_\pm\,\delta_{\ell 2}, \tag{A9.39}$$

$$S_\ell^{(\pm 2)} = \frac{1}{\sqrt{3}}\dot{H}_{\pm 2}\,\delta_{\ell 2}, \tag{A9.40}$$

$$P_\ell^{(0)} = \mathcal{M}_0^{(0)}\delta_{\ell 0} + V^{(b)}\delta_{\ell 1} + \frac{1}{10}[\mathcal{M}_2^{(0)} - \sqrt{6}E_2^{(0)}]\delta_{\ell 2}, \tag{A9.41}$$

$$P_\ell^{(\pm 1)} = V_b^{(\pm 1)}\delta_{\ell 1} + \frac{1}{10}[\mathcal{M}_2^{(\pm 1)} - \sqrt{6}E_2^{(\pm 1)}]\delta_{\ell 2}, \tag{A9.42}$$

$$P_\ell^{(\pm 2)} = \frac{1}{10}[\mathcal{M}_2^{(\pm 2)} - \sqrt{6}E_2^{(\pm 2)}]\delta_{\ell 2}. \tag{A9.43}$$

As in Chapter 5, the superscript (m) indicates scalar perturbations for $m = 0$, vector perturbations for $m = \pm 1$, and tensor perturbations for $m = \pm 2$. For Eqs. (A9.38) and (A9.39) we made use of

$$-n^i[{}_0G_{00}]_{|i} = k\,{}_0G_{10} \quad \text{and}$$

$$n^i n^j Q_{i|j}^{(m)} = n^i[{}_0G_{1\,m}]_{|i} = \frac{\sqrt{3}}{3}\sqrt{k^2 - 2K}\;{}_0G_{2m},$$

for $m = \pm 1$.

The Boltzmann hierarchy for E- and B-polarization becomes

$$\dot{\mathcal{E}}_\ell^{(m)} = q\sqrt{|K|}\left[\frac{{}_2\theta_\ell^m}{(2\ell-1)}\mathcal{E}_{\ell-1}^{(m)} - \frac{2m}{\ell(\ell+1)}B_\ell^{(m)} - \frac{{}_2\theta_{\ell+1}^m}{(2\ell+3)}\mathcal{E}_{\ell+1}^{(m)}\right]$$
$$- \dot{\kappa}\left(\mathcal{E}_\ell^{(m)} + \frac{\sqrt{6}}{10}\left[\mathcal{M}_2^{(m)} - \sqrt{6}\mathcal{E}_2^{(m)}\right]\delta_{\ell,2}\right), \tag{A9.44}$$

$$\dot{\mathcal{B}}_\ell^{(m)} = q\sqrt{|K|}\left[\frac{{}_2\theta_\ell^m}{(2\ell-1)}\mathcal{B}_{\ell-1}^{(m)} + \frac{2m}{\ell(\ell+1)}\mathcal{E}_\ell^{(m)} - \frac{{}_2\theta_{\ell+1}^m}{(2\ell+3)}\mathcal{B}_{\ell+1}^{(m)}\right] - \dot{\kappa}\mathcal{B}_\ell^{(m)}. \tag{A9.45}$$

A fast Boltzmann code (such as CMBfast) calculates only the lowest few (about 10) modes with the Boltzmann hierarchy and then uses the results $\ell = 0$, 1, and 2 as input for the integral solutions. These are obtained exactly like in the flat case (see Chapter 5) by replacing the flat radial functions by the ones obtained for curved spaces.

$$\frac{\mathcal{M}_\ell^{(m)}(t_0, q)}{\ell+1} = \int_0^{t_0} dt\, e^{-\kappa}\sum_{j=0}^{2}[S_j^{(m)} + \dot{\kappa}P_j^{(m)}]\alpha_\ell^{(jm)}(\sqrt{|K|}t, q), \tag{A9.46}$$

$$\frac{\mathcal{E}_\ell^{(m)}(t_0, q)}{\ell+1} = \int_0^{t_0} dt\, e^{-\kappa}\dot{k}\frac{\sqrt{6}}{10}\left[\mathcal{M}_2^{(m)} - \sqrt{6}\mathcal{E}_2^{(m)}\right]\epsilon_\ell^{(m)}(\sqrt{|K|}t, q), \tag{A9.47}$$

$$\frac{\mathcal{B}_\ell^{(m)}(t_0, q)}{\ell+1} = \int_0^{t_0} dt\, e^{-\kappa}\dot{k}\frac{\sqrt{6}}{10}\left[\mathcal{M}_2^{(m)} - \sqrt{6}\mathcal{E}_2^{(m)}\right]\beta_\ell^{(m)}(\sqrt{|K|}t, q). \tag{A9.48}$$

A9.4 The Energy–Momentum Tensor

The perturbations of the photon energy–momentum tensor that enter the Einstein equations are obtained from their definitions by integration over directions **n**,

$$D_g^{(\gamma)} = 4\mathcal{M}_0^{(0)}, \tag{A9.49}$$

$$V_\gamma^{(m)} = \mathcal{M}_1^{(m)}, \tag{A9.50}$$

$$\sqrt{1-\frac{3K}{k^2}}\,\Pi_\gamma^{(0)} = \frac{12}{5}\mathcal{M}_2^{(0)}, \tag{A9.51}$$

$$\sqrt{1-\frac{2K}{k^2}}\,\Pi_\gamma^{(1)} = \frac{8\sqrt{3}}{5}\mathcal{M}_2^{(1)}, \tag{A9.52}$$

$$\Pi_\gamma^{(2)} = \frac{8}{5}\mathcal{M}_2^{(2)}. \tag{A9.53}$$

A9.5 Power Spectra

In the derivation of the power spectra the only change w.r.t. flat space is that for positive curvature the integral over q has to be replaced by a sum. Note also that our variable q is dimensionless and therefore so are our amplitudes $X_\ell^{(m)}$,

$$(2\ell+1)^2 C_\ell^{XY} = \frac{2}{\pi}\sum\!\!\!\!\!\!\int \frac{dq}{q}\sum_{m=-2}^{2} q^3 P_{\ell m}^{(XY)}, \tag{A9.54}$$

where X, Y are $\mathcal{M}$, $\mathcal{E}$, or $\mathcal{B}$ and the power spectra are defined like in the flat case,

$$\langle \mathcal{M}_\ell^{(m)}(q)\mathcal{M}_\ell^{(m)*}(q')\rangle \equiv (2\pi)^3\delta_{q,q'}M_\ell^{(m)}(q), \tag{A9.55}$$

$$\langle \mathcal{E}_\ell^{(m)}(q)\mathcal{E}_\ell^{(m)*}(q')\rangle \equiv (2\pi)^3\delta_{q,q'}E_\ell^{(m)}(q), \tag{A9.56}$$

$$\langle \mathcal{B}_\ell^{(m)}(q)\mathcal{B}_\ell^{(m)*}(q')\rangle \equiv (2\pi)^3\delta_{q,q'}B_\ell^{(m)}(q), \tag{A9.57}$$

$$\langle \mathcal{E}_\ell^{(m)}(q)\mathcal{M}_\ell^{(m)*}(q')\rangle \equiv (2\pi)^3\delta_{q,q'}F_\ell^{(m)}(q). \tag{A9.58}$$

The formulas here are written for positive curvature. For negative curvature the Kronecker delta becomes a Dirac delta-function like in flat space. Hence $P_{\ell m}^{(\mathcal{MM})}(q) = M_\ell^{(m)}(q)$, $P_{\ell m}^{(\mathcal{EE})}(q) = E_\ell^{(m)}(q)$, $P_{\ell m}^{(\mathcal{BB})}(q) = B_\ell^{(m)}(q)$, and $P_{\ell m}^{(\mathcal{EM})}(q) = F_\ell^{(m)}(q)$. Due to their different parity, $\mathcal{M}$ and $\mathcal{B}$ as well as $\mathcal{E}$ and $\mathcal{B}$ are uncorrelated.

Appendix 10

Perturbations of the Luminosity Distance

In this appendix we derive the linear perturbation of the luminosity distance in a perturbed Friedmann Universe with vanishing curvature, $K = 0$. For this we first derive a general equation for the so-called area distance, which in an FL spacetime is the same as the angular diameter distance. We then express this distance in terms of the observed redshift z. We can finally use the Etherington relation (see, e.g., Schneider *et al.*, 1993)

$$d_L(z) = (1+z)^2 d_A(z) \tag{A10.1}$$

to obtain the luminosity distance.

Let us consider a faraway source from which many neighboring photons reach us. We want to relate the angular size of the source to a distance measure. We parametrize the photon 4-velocity vectors as $n(\lambda, s)$, where λ is the affine parameter of the photon geodesic and s parameterizes different geodesics at fixed λ in a direction on the source. We can choose s such that the vector $\eta = \partial_s$ is normal to n. Since $n = \partial_\lambda$ [we choose coordinates (λ, s, x, y) in a small patch] we have $L_n\eta = [n, \eta] = [\partial_\lambda, \partial_s] = 0$, and the normality condition reads $(n, \eta) = n^\mu \cdot \eta_\mu = 0$. Using

$$L_n\eta = [n, \eta] = \nabla_n\eta - \nabla_\eta n = 0$$

we find

$$\nabla_n^2\eta = \nabla_n\nabla_\eta n - \nabla_\eta\nabla_n n = R(n, \eta)n \tag{A10.2}$$

in coordinates

$$\left(\nabla_n^2\eta\right)^\mu = R^\mu_{\alpha\beta\gamma} n^\beta \eta^\gamma n^\alpha . \tag{A10.3}$$

This is the Sachs equation (see, e.g., Straumann, 2004; Schneider *et al.*, 1993), which determines the evolution of a small deviation vector η along a bundle of photon geodesics; that is, it determines the deformation of an image along its path.

We now choose a basis at the observer that is given by the observer 4-velocity u_o, the photon 4-velocity n, and two spacelike normal unit vectors e_a that satisfy

$$(n, e_a) = (u_o, e_a) = 0 \quad \text{and } (e_a, e_b) = \delta_{ab}. \tag{A10.4}$$

We parallel transport these vectors backwards along the photon geodesic. This defines $u_o(\lambda)$, $e_a(\lambda)$ with

$$\nabla_n u_o = \nabla_n e_a = 0. \tag{A10.5}$$

Note that $(n, u_o) \neq 0$; hence our basis (n, e_1, e_2, u_o), which we now have defined along the photon geodesics, is not orthonormal. In our basis we have

$$\eta^\mu = \eta^a e_a^\mu + \eta^0 n^\mu \tag{A10.6}$$

[since $(n, \eta) = 0$ η cannot have a component along u_o]. Rewriting the Sachs equation (A10.3) with this ansatz and using Eq. (A10.5) we obtain

$$\frac{d^2}{d\lambda^2}\eta^b = R^\mu_{\alpha\beta\gamma} n^\alpha n^\beta e_a^\gamma e^b_\mu \eta^a = \mathcal{R}^b{}_a \eta^a, \tag{A10.7}$$

where

$$\mathcal{R}^b{}_a = R^\mu_{\alpha\beta\gamma} n^\alpha n^\beta e_a^\gamma e^b_\mu. \tag{A10.8}$$

This is a linear differential equation. Since we want all the geodesics to converge at the observer position we set $\eta(\lambda_o) = \eta(0) = 0$. For definiteness, we have chosen $\lambda_o = 0$, so that the affine parameter of the photon before impact at the observer is negative. The first derivative

$$\left.\frac{d\eta^a}{d\lambda}\right|_{\lambda=0} = \theta^a \tag{A10.9}$$

defines the direction in which the photon enters at the observer. Hence η^b/θ^b is the angular diameter distance in direction θ^b. In an isotropic universe like FL, this distance does not depend on the chosen direction. The general solution of the Sachs equation is now of the form

$$\eta^b = \mathcal{D}^b{}_a \theta^a \tag{A10.10}$$

and the generic area distance is defined by

$$d_A^2 = \det \mathcal{D}^b{}_a. \tag{A10.11}$$

A source covering a solid angle Ω then has a surface $d_A^2 \Omega$, like in flat spacetime. The so called Jacobi matrix $\mathcal{D}^b{}_a$ satisfies the differential equation

$$\frac{d^2}{d\lambda^2}\mathcal{D}^b{}_a = \mathcal{R}^b{}_c \mathcal{D}^c{}_a. \tag{A10.12}$$

Note that so far our treatment has been completely generic. It can be applied to arbitrary spacetimes; however, special care is needed at caustics, when photon geodesics cross.

We want to compute $\det \mathcal{D}^b{}_a$ to first order in a spatially perturbed FL universe. We use, like in Section 2.5, that photon geodesics are conformally invariant and we can disregard expansion in a first step. We consider an observer at $r = 0$ and a radially incoming geodesic. We set $\mathcal{D}_{ab} = \mathcal{D}^{(0)}_{ab} + \mathcal{D}^{(1)}_{ab}$. Note that the directional indices are raised and lowered with δ_{ab}, since the e_a are normalized spacelike vectors normal to ∂_r. To zeroth order we are in Minkowski spacetime and

$$\mathcal{D}^{(0)}_{ab} = -\lambda \delta_{ab} = r\delta_{ab}, \tag{A10.13}$$

where r denotes the (conformal) distance to the source. We want to compute

$$\det \mathcal{D}_{ab} = r^2 \det(\delta_{ab} + r^{-1}\mathcal{D}^{(1)}_{ab}) = r^2(1 + r^{-1}\mathcal{D}^{(1)}_{aa}) \tag{A10.14}$$

to first order for scalar perturbations. We shall transform the parameter λ first into conformal time and then into the radial coordinate via the relation

$$\frac{dt}{d\lambda} = n^0(\lambda) = 1 + (\Phi - \Psi)|_s^o - \int_{\lambda_s}^{\lambda_o} d\lambda\, \mathbf{n}\cdot\nabla(\Psi + \Phi). \tag{A10.15}$$

For the second equals sign we used Eq. (2.233) and perfomed an integration, $\dot{\Phi} + \dot{\Psi} = d/d\lambda(\Phi + \Psi) - \mathbf{n}\cdot\nabla(\Psi + \Phi)$. With this, the first-order part of Eq. (A10.12) becomes

$$\frac{d^2}{dr^2}\mathcal{D}_{ab}^{(1)} = \frac{d\delta n^0}{dr}\delta_{ab} + r\mathcal{R}_{ab}^{(1)} \tag{A10.16}$$

$$\frac{d}{dr}\mathcal{D}_{ab}^{(1)} = \delta n^0 \delta_{ab} + \int_0^r dr' r' \mathcal{R}_{ab}^{(1)}(r') \tag{A10.17}$$

$$\mathcal{D}_{ab}^{(1)} = \int_0^r dr' \delta n^0(r')\delta_{ab} + \int_0^r dr'(r - r')r'\mathcal{R}_{ab}^{(1)}(r') \tag{A10.18}$$

$$= \left[-\int_0^r dr'(\Phi - \Psi) + \int_0^r dr'(r - r')\partial_{r'}(\Phi + \Psi)\right]\delta_{ab} + \int_0^r dr'(r - r')r'\mathcal{R}_{ab}^{(1)}(r'). \tag{A10.19}$$

Here we have systematically neglected terms at the observer that contribute only a monopole or a dipole term to the distance fluctuations. The monopole term is not gauge invariant, as we can add a constant to the Bardeen potentials such that $\Psi(t_o, 0) = \Phi(t_o, 0) = 0$. The dipole term that we shall obtain is determined by the local velocity, which cannot be computed with linear perturbation theory. We have also used that $\mathbf{n}\cdot\nabla = -\partial_r$ at lowest order. For the two terms that would contain double integrals we have performed the following integration by parts to reduce them to single integrals:

$$\int_0^r dr' \int_0^{r'} dr'' f(r'') \equiv \int_0^r dr'(r - r')f(r').$$

Inserting the perturbed Riemann tensor in Minkowski spacetime obtained from Eqs. (A3.9) and (A3.12) by setting $\mathcal{H} = 0$, we obtain

$$\mathcal{R}_{ab} = -R_{i00j}e_a^i e_b^j - R_{ilmj}n^l n^m e_a^i e_b^j = -e_a^i e_b^j \partial_i\partial_j(\Phi + \Psi) - \frac{d^2}{d\lambda^2}\Phi. \tag{A10.20}$$

Inserting also δn^0 from Eq. (A10.15) yields

$$D_{aa}^{(1)} = 2\int_0^r dr'(\Psi - \Phi) + 2\int_0^r dr'(r - r')\partial_r(\Psi + \Phi) - 2\int_0^r dr'(r - r')r'\frac{d^2\Phi}{dr'^2} - \int_0^r dr'(r - r')r' e_a^i e_b^j \partial_i\partial_j(\Phi + \Psi). \tag{A10.21}$$

The third term of this equation can be converted via integration by parts into

$$-2\int_0^r dr'(r - r')r'\frac{d^2\Phi}{dr'^2} = 4\int_0^r dr'\Phi - 2r\Phi(r). \tag{A10.22}$$

To simplify the last term of Eq. (A10.21) we note that the transverse (angular) Laplacian is given on the one hand by

$$\nabla_\perp^2 = \left(\partial_i - n_i(n^j\partial_j)\right)^2 = \partial_i\partial^i - n^i n^j \partial_i\partial_j + \frac{2}{r}n^i\partial_i = r^{-2}\Delta_\Omega, \tag{A10.23}$$

where Δ_Ω denotes the Laplacian on the 2-sphere and $n^i\partial_i = -\partial_r$. On the other hand, the e_a and ∂_r form an orthonormal basis such that $\partial_i\partial^i = \partial_r^2 + e_a^i e_a^j \partial_i\partial_j$. Hence

$$\nabla_\perp^2 = e_a^i e_a^j \partial_i\partial_j - \frac{2}{r}\partial_r. \tag{A10.24}$$

With this, the second and the last terms combine into an angular Laplacian. Inserting all this in Eq. (A10.21) we obtain

$$\begin{aligned}\det \mathcal{D}_{ab}(r) &= r^2\left[1 - 2\Phi(r) + \frac{2}{r}\int_0^r dr'(\Phi+\Psi) - \int_0^r dr'\frac{r-r'}{rr'}\Delta_\Omega(\Phi+\Psi)\right]\\ &= r^2\left[1 + 2\frac{d_A^{(1)}(r)}{r}\right].\end{aligned} \tag{A10.25}$$

The last of these terms corresponds to κ which we have found in the lens equation (7.19). The first term comes from a change of size at the source due to the perturbed metric, while the second term is the Shapiro time delay. On small angular scales (high multipoles) only the last term, $\propto \ell(\ell+1)\Psi$, is relevant. The first two terms are large-scale relativistic corrections to this expression.

Equation A10.25 is the expression for the determinant of the Jacobi map in perturbed Minkowski spacetime. Now we want to go back to a perturbed FL spacetime. The difference there is that $d_A \to a(r)d_A = d_A/(1+\bar z) = \tilde d_A$, where $\bar z$ is the background redshift. However, we can only measure the true redshift of the source, $z = \bar z + \delta z$. Also, measuring z the corresponding background comoving distance r is

$$r(\bar z) = r(z) - \delta z \frac{dr}{dz}.$$

Using that $dr/dz = [\mathcal{H}(1+z)]^{-1}$ we obtain the first-order relation

$$(1+z)\tilde d_A(z) = d_A(r(z))[1 - (1 - 1/(\mathcal{H}r))\delta z/(1+z)].$$

Dropping again observer terms, we obtain $\tilde d_A$ as function of the observed redshift,

$$\begin{aligned}\tilde d_A(z) = \bar d_A(z)\Bigg\{1 - \left(1 - \frac{1}{\mathcal{H}r}\right)\left[(\mathbf{v}\cdot\mathbf{n} + \Psi)(r) + \int_0^r dr'(\dot\Phi + \dot\Psi)\right] - \Phi(r)\\ + \int_0^r \frac{dr'}{r}\left[1 - \frac{(r-r')}{2r'}\Delta_\Omega\right](\Psi+\Phi)\Bigg\},\end{aligned} \tag{A10.26}$$

where $\bar d_A(z)$ is the background area (or angular diameter) distance.

Since the Etherington relation $d_L = (1+z)^2 d_A$ is exact, when expressed in terms of the observed redshift, we obtain exactly the same expression for the perturbation of the luminosity distance (we drop the tilde in this final result),

$$\frac{\delta d_L(z,\mathbf{n})}{\bar{d}_L(z)} = -\Phi(r) - \left(1 - \frac{1}{\mathcal{H}r}\right)\left[\Psi + \mathbf{nv} + \int_0^r dr'(\dot{\Psi} + \dot{\Phi})\right]$$
$$+ \int_0^r \frac{dr'}{r}\left[1 - \frac{(r-r')}{2r'}\Delta_\Omega\right](\Psi + \Phi). \qquad \text{(A10.27)}$$

The term on the second line is the Shapiro time delay minus the convergence κ. The first integrated term is the integrated Sachs–Wolfe term, which appears also in the CMB anisotropies. The nonintegrated terms are contributions from the gravitational potential at the source and the Doppler term at the source. Corresponding terms at the observer have been neglected, as they contribute just gauge-dependent monopole and dipole terms.

Comparing this result with Eq. (8.60), we see that it is exactly $-\delta d_L/\bar{d}_L$ which multiplies the factor $5s$ in the expression for the number count perturbation.

In Bonvin *et al.* (2006a) and Sasaki (1987) it is also shown that the expression (A10.27) is gauge invariant, as we expect for a measurable quantity.

References

Abate, A. *et al.* [Large Synoptic Survey Telescope Dark Energy Science Collaboration] (2012). arXiv:1211.0310 [astro-ph.CO].

Abbott, L. F., and Schaefer, R. K. (1986). A general analysis of the cosmic microwave anisotropy. *Astrophys. J.* **308**, 546.

Abbott, T. M. C. *et al.* [DES Collaboration] (2018). Dark Energy Survey year 1 results: Cosmological constraints from galaxy clustering and weak lensing. *Phys. Rev.* **D98**, 043526. DOI:10.1103/PhysRevD.98.043526.

Abraham, R., and Marsden, J. (1982). *Foundations of Mechanics*. New York: Addison and Wesley, chapter 3.

Abramowitz, M., and Stegun, I. A. (1970). *Handbook of Mathematical Functions*, 9th ed. New York: Dover Publications.

Achucarro, A., Atal, V., Hu, B. Ortiz, P., and Torrado, J. (2014). Inflation with moderately sharp features in the speed of sound: Generalized slow roll and in-in formalism for power spectrum and bispectrum. *Phys. Rev.* **D90**, no. 2, 023511. DOI:10.1103/PhysRevD.90.023511.

Achucarro, A., Gong, J. O., Palma, G. A., and Patil, S. P. (2013). Correlating features in the primordial spectra. *Phys. Rev.* **D87**, 121301. DOI:10.1103/PhysRevD.87.121301.

Acquaviva,V., Bartolo, N. Matarrese, S., and Riotto, A. (2003). Second order cosmological perturbations from inflation. *Nucl. Phys. B* **667**, 119. DOI:10.1016/S0550-3213(03)00550-9.

Adams, J. A., Ross, G. G., and Sarkar, S. (1997). Multiple inflation. *Nucl. Phys.* **B503**, 405.

Ade, P. A. R. *et al.* [Planck Collaboration] (2013). Planck intermediate results. X. Physics of the hot gas in the Coma cluster. *Astron. Astrophys.* **554**, A140. DOI:10.1051/0004-6361/201220247.

Ade, P. A. R. *et al.* [Planck Collaboration] (2016). Planck 2015 results. VI. LFI mapmaking. *Astron. Astrophys.* **594**, A6. DOI:10.1051/0004-6361/201525813.

Ade, P. A. R. *et al.* [Planck Collaboration] (2016). Planck 2015 results. XIII. Cosmological parameters. *Astron. Astrophys.* **594**, A13. DOI:10.1051/0004-6361/201525830.

Ade, P. A. R. *et al.* [Planck Collaboration] (2016). Planck 2015 results. XV. Gravitational lensing. *Astron. Astrophys.* **594**, A15. DOI:10.1051/0004-6361/201525941.

Ade, P. A. R. *et al.* [Planck Collaboration] (2016). Planck 2015 results. XVII. Constraints on primordial non-Gaussianity. *Astron. Astrophys.* **594**, A17. DOI:10.1051/0004-6361/201525836.

Ade, P. A. R. *et al.* [Planck Collaboration] (2016). Planck 2015 results. XXIV. Cosmology from Sunyaev-Zeldovich cluster counts. *Astron. Astrophys.* **594**, A24. DOI:10.1051/0004-6361/201525833.

Aghanim, N. *et al.* [Planck Collaboration] (2016). Planck 2015 results. XXII. A map of the thermal Sunyaev-Zeldovich effect. *Astron. Astrophys.* **594**, A22. DOI:10.1051/0004-6361/201525826.

Aghanim, N. *et al.* [Planck Collaboration] (2018). Planck 2018 results. VI. Cosmological parameters. arXiv:1807.06209.

Aghanim, N. *et al.* [Planck Collaboration] (2018). Planck 2018 results. VIII. Gravitational lensing. arXiv:1807.06210.

Akrami, Y. *et al.* [Planck Collaboration] (2018). Planck 2018 results. X. Constraints on inflation. arXiv:1807.06211.

Albrecht, A., Coulson, D., Ferreira, P., and Magueijo, J. (1996). Causality, randomness, and the microwave background. *Phys. Rev. Lett.* **76**, 1413.

Alcock, C., and Paczynsk, B. (1979). An evolution free test for non-zero cosmological constant. *Nature* **281**, 358. DOI:10.1038/281358a0.

Alonso, D., and Ferreira, P. G. (2015). Constraining ultralarge-scale cosmology with multiple tracers in optical and radio surveys. *Phys. Rev. D* **92**, no. 6, 063525. DOI:10.1103/PhysRevD.92.063525.

Amendola, L. *et al.* (2018). Cosmology and fundamental physics with the Euclid satellite. *Living Rev. Rel.* **21**, 2. DOI:10.1007/s41114-017-0010-3.

Anderson L. *et al.* (2012). The clustering of galaxies in the SDSS-III Baryon Oscillation Spectroscopic Survey: Baryon Acoustic Oscillations in the Data Release 9 Spectroscopic Galaxy Sample. *Mon. Not. Roy. Astron. Soc.* **427**, 3435. DOI:10.1111/j.1365-2966.2012.22066.x.

Arfken, G. B., and Weber, H. J. (2001). *Mathematical Methods for Physicists*, 5th ed. San Diego, CA: Academic Press.

Arnol'd, V. I. (1978). *Mathematical Methods of Classical Mechanics*. Berlin: Springer-Verlag.

Bardeen, J. (1980). Gauge-invariant cosmological perturbations. *Phys. Rev.* **D22**, 1882.

Bashinsky, S. (2006). Gravity of cosmological perturbations in the CMB. *Phys. Rev.* **D74**, 043007.

Bautista, J. E. *et al.* (2017). Measurement of baryon acoustic oscillation correlations at $z = 2.3$ with SDSS DR12 Lyα-Forests. *Astron. Astrophys.* **603**, A12. DOI:10.1051/0004-6361/201730533.

Bernardeau, F., Colombi, S., Gaztanaga, E., and Scoccimarro, R. (2002). Large scale structure of the universe and cosmological perturbation theory. *Phys. Rept.* **367**, 1. DOI:10.1016/S0370-1573(02)00135-7.

Bernstein, J., Brown, L., and Feinberg, G. (1989). Cosmological helium production simplified. *Rev. Mod. Phys.* **61**, 25.

Bevis, N., Hindmarsh, M., Kunz, M., and Urrestilla, J. (2007). CMB polarization power spectra contributions from a network of cosmic strings. *Phys. Rev.* **D76**, 043005.

Blas, D., Lesgourgues, J., and Tram, T. (2011). The Cosmic Linear Anisotropy Solving System (CLASS) II: Approximation schemes. *JCAP* **1107**, 034. DOI:10.1088/1475-7516/2011/07/034.

Böhringer, H., Chon G., and Collins, C. A. (2014). The extended ROSAT-ESO Flux Limited X-ray Galaxy Cluster Survey (REFLEX II) IV. X-ray Luminosity Function and First Constraints on Cosmological Parameters. *Astron. Astrophys.* **570**, A31. DOI:10.1051/0004-6361/201323155.

Bonvin, C., and Durrer, R. (2011). What galaxy surveys really measure. *Phys. Rev.* **D84**, 063505.

Bonvin, C., Durrer, R., and Gasparini, M. A. (2006a). Fluctuations of the luminosity distance. *Phys. Rev.* **D73**, 023523.

Bonvin, C., Durrer, R., and Kunz, M. (2006b). The dipole of the luminosity distance: a direct measure of $H(z)$. *Phys. Rev. Lett.* **96**, 191302.

Bowman, J. D., Rogers, A. E. E., Monsalve, R. A. Mozdzen, T. J., and Mahesh, N. (2018). An absorption profile centred at 78 megahertz in the sky-averaged spectrum. *Nature* **555**, 67. DOI:10.1038/nature25792.

Bucher, M., Moodley, K., and Turok, N. (2000). General primordial cosmic perturbations. *Phys. Rev.* **D62,** 083508.

Bucher, M., Moodley, K., and Turok, N. (2001). Constraining iso-curvature perturbations with CMB polarization. *Phys. Rev. Lett.* **87**, 191301.

Burles, S., Nollett, K., and Turner, M. (2001). Big-bang nucleosynthesis predictions for precision cosmology, *Astrophys. J.* **552**, L1.

Cabass, G., Gerbino, M., Giusarma, E., Melchiorri, A., Pagano, L., and Salvati, L. (2015). Constraints on the early and late integrated Sachs-Wolfe effects from the Planck 2015 cosmic microwave background anisotropies in the angular power spectra. *Phys. Rev.* **D92**, 063534. DOI:10.1103/PhysRevD.92.063534.

Caldwell, R. R., Dave, R., and Steinhardt, P. J. (1998). Cosmological imprint of an energy component with general equation of state. *Phys. Rev. Lett.* **80**, 1582.

Caldwell, R. R., Kamionkowski, M., and Weinberg, N. N. (2003). Phantom energy and cosmic doomsday. *Phys. Rev. Lett.* **91**, 071301.

Caprini, C., Durrer, R., and Kahniashvili, T. (2004). The cosmic microwave background and helical magnetic fields: the tensor mode. *Phys. Rev.* **D69**, 063006.

Challinor, A., and Lewis, A. (2005). Lensed CMB power spectra from all-sky correlation functions. *Phys. Rev.* **D71,** 103010.

Challinor, A., and Lewis, A. (2011). The linear power spectrum of observed source number counts. *Phys. Rev.* **D84**, 043516.

Chandrasekhar, S. (1939). *An Introduction to the Study of Stellar Structure*. Chicago: Chicago University Press.

Chluba, J. *et al.* (2019). Spectral distortions of the CMB as a probe of inflation, recombination, structure formation and particle physics: Astro2020 Science White Paper. *Bull. Am. Astron. Soc.* **51**, 184.

Colafrancesco, S. (2007). Beyond the standard lore of the SZ effect. *New Astron. Rev.* **51**, 394.

Conklin, E. K. (1969). Velocity of the Earth with respect to the cosmic background radiation. *Nature* **222**, 971.

Creminelli, P. (2003). On non-Gaussianities in single-field inflation. *JCAP* **0310**, 003. [DOI:10.1088/1475-7516/2003/10/003] [astro-ph/0306122].

Creminelli, P., Noreña, J., and Simonović, M. (2012). Conformal consistency relations for single-field inflation. *JCAP* **1207**, 052. DOI:10.1088/1475-7516/2012/07/052.

Creminelli, P., Senatore, L., and Vasy, A. (2019). Asymptotic behavior of cosmologies with $\Lambda > 0$ in 2+1 dimensions. arXiv:1902.00519 [hep-th].

Dalal, N., Dore, O., Huterer, D., and Shirokov, A. (2008). The imprints of primordial non-gaussianities on large-scale structure: Scale dependent bias and abundance of virialized objects. *Phys. Rev.* **D77**, 123514. DOI:10.1103/PhysRevD.77.123514.

De Petris, M. *et al.* (2002). MITO measurements of the Sunyaev–Zel'dovich effect in the Coma cluster of galaxies. *Astrophys. J.* **574**, L119.

Di Dio, E., Montanari, F., Lesgourgues, J., and Durrer (2013). The CLASSgal code for Relativistic Cosmological Large Scale Structure. *JCAP* **1311**, 044. DOI:10.1088/1475-7516/2013/11/044 [arXiv:1307:1459].

Di Dio, E., Montanari, F., Raccanelli, A., and Durrer, R., Kamionkowski, M., and Lesgourgues, J. (2016). Curvature constraints from Large Scale Structure. *JCAP* 1606(06):013. DOI:10.1088/1475-7516/2016/06/013.

Diu, B., Guthmann, C., Lederer, D., and Roulet, B. (1989). *Physique Statistique*. Paris: Moradinezhad, Hermann, Editeurs des Sciences et des Arts.

Dizgah, A., and Durrer, R. (2016). Lensing corrections to the $E_g(z)$ statistics from large scale structure. *JCAP* **1609**, 035. DOI:10.1088/1475-7516/2016/09/035.

Dodelson, S. (2003). *Modern Cosmology*. New York: Academic Press.

Doran, M. (2005). CMBeasy: An object oriented code for the cosmic microwave background. *JCAP* **0510**, 011.

Durrer, R. (1990). Gauge-invariant cosmological perturbation theory with seeds. *Phys. Rev.* **D42**, 2533.

Durrer, R. (1994). Gauge invariant cosmological perturbation theory. *Fund. Cosmic Phys.* **15**, 209.

Durrer, R., and Straumann, N. (1988). Some applications of the $3+1$ formalism of general relativity. *Helv. Phys. Acta* **61**, 1027.

Durrer, R. *et al.* (1999). Seeds of large-scale anisotropy in string cosmology. *Phys. Rev.* **D59**, 043511.

Durrer, R., Kunz, M., and Melchiorri, A. (2002). Cosmic structure formation with topological defects. *Phys. Rep.* **364**, 1.

Durrer, R., and Tansella, V. (2016). Vector perturbations of galaxy number counts. *JCAP* **1607**(07), 037.

Ehlers, J. (1971). General relativity and kinetic theory. In *Proceedings of the Varenna School 1969, Course XLVII*, ed. R. K. Sachs. New York: Academic Press.

Ellis, G. F. R., and Bruni, M. (1989). Covariant and gauge-invariant approach to cosmological density fluctuations. *Phys. Rev.* **D40**, 1804.

Enqvist, K., and Sloth, M. S. (2002). Adiabatic CMB perturbations in pre - big bang string cosmology. *Nucl. Phys. B* **626**, 395–409. DOI:10.1016/S0550-3213(02)00043-3.

Fan, X. H., Carilli, C. L., and Keating, B. G. (2006). Observational constraints on cosmic reionization. *Ann. Rev. Astron. Astrophys.* **44**, 415. DOI:10.1146/annurev.astro.44.051905.092514.

Fan, X. H. *et al.* (2006). Constraining the evolution of the ionizing background and the epoch of reionization with z 6 quasars. 2. A sample of 19 quasars. *Astron. J.* **132**, 117. DOI:10.1086/504836.

Feng, C., and G. Holder, G. (2019). Searching for patchy reionization from cosmic microwave background with hybrid quadratic estimators. *Phys. Rev.* **D99**, 123502. DOI:10.1103/PhysRevD.99.123502.

Fergusson, J., and Shellard, E. (2009). The shape of primordial non-Gaussianity and the CMB bispectrum. *Phys. Rev.* **D80**, 043510. DOI:10.1103/PhysRevD.80.043510.

Field, G. B. (1959). The spin temperature of intergalactic neutral hydrogen. *Astrophys. J* **129**, 536.

Fields, B. D., and Sarkar, S. (2006). Big-bang nucleosynthesis. *J. Phys.* **G33**, 1.

Fixsen, D. J. *et al.* (1996). The cosmic microwave background spectrum from the full COBE FIRAS data set. *Astrophys. J.* **707**, 916.

Fixsen, D. J. (2009). The temperature of the cosmic microwave background. *Astrophys. J.* **473**, 567.

Friedmann, A. (1922). Über die Krümming des Raumes. *Z. Phys.* **10**, 377.

Friedmann, A. (1924). Über die Möglichkeit einer Welt mit konstanter negativer Krümmung des Raumes. *Z. Phys.* **21**, 326.

Freedman, W. L. *et al.* (2001). Final results from the Hubble Space Telescope key project to measure the Hubble constant. *Astrophys. J.* **553**, 47.

Furlanetto, S., Oh, S. P., and Briggs, F. (2006). Cosmology at low frequencies: The 21 cm transition and the high-redshift universe. *Phys. Rept.* **433**, 181. DOI:10.1016/j.physrep.2006.08.002.

Gamerman, D. (1997). *Markov Chain Monte Carlo: Stochastic Simulations for Bayesian Inference*. London: Chapman and Hall.

Goldberg, J. N. *et al.* (1967). Spin-s Spherical Harmonics and $\eth$. *J. Math. Phys.* **8**, 2155.

Gorski, K. M. (1994). On determining the spectrum of primordial inhomogeneity from the COBE DMR sky maps: I. Method. *Astrophys. J.* **430**, L85.

Gradshteyn, I. S., and Ryzhik, I. M. (2000). *Table of Integrals, Series and Products*, 6th ed. New York: Academic Press.

Gunn, J. E., and Peterson, B. A. (1965). On the density of neutral hydrogen in intergalactic space. *Astrophys. J.* **142**, 1633.

Hajian, A. (2007). Efficient cosmological parameter estimation with Hamiltonian Monte Carlo. *Phys. Rev.* **D75**, 083525.

Hall, A., Bonvin, C., and Challinor, A. (2013). Testing General Relativity with 21-cm intensity mapping. *Phys. Rev.* **D87**, 064026. DOI:10.1103/PhysRevD.87.064026.

Harrison, E. R. (1970). Fluctuations at the threshold of classical cosmology. *Phys. Rev.* **D1**, 2726–2730.

Hawking, S., and Ellis, G. F. R. (1973). *The Large Scale Structure of the Universe*. Cambridge: Cambridge University Press.

Henry, P. S. (1971). Isotropy of the 3 K background. *Nature* **231**, 516.

Hoffman, M., and Turner, M. (2001). Kinematic constraints to the key inflationary observables. *Phys. Rev.* **D64**, 023506.

Hogg, D. *et al.* (2005). Cosmic homogeneity demonstrated with luminous red galaxies. *Astrophys. J.* **624**, 54–58.

Hu, W., and Silk, J. (1993). Thermalization and spectral distortions of the cosmic microwave background radiation. *Phys. Rev.* **D48**, 485.

Hu, W., and Sugiayma, N. (1995). Anisotropies in the cosmic microwave background: an analytic approach. *Astrophys. J.* **444**, 489.

Hu, W., and White, M. (1996). CMB anisotropies in the weak coupling limit. *Astron. Astrophys.* **315**, 33.

Hu, W., and White, M. (1997a). Tensor anisotropies in an open universe. *Astrophys. J.* **486**, L1.

Hu, W., and White, M. (1997b). CMB anisotropies: Total angular momentum method. *Phys. Rev.* **D56**, 596.

Hu, W., Scott, D., Sugiayma, N., and White, M. (1995). The effect of physical assumptions on the calculation of microwave background anisotropies. *Phys. Rev.* **D52**, 5498.

Hu, W., Seljak, U. A., White, M., and Zaldarriaga, M. (1998). Complete treatment of CMB anisotropies in a FRW universe. *Phys. Rev.* **D57**, 3290.

Hubble, E. (1929). A relation between distance and radial velocity among extra-galactic nebulae. *Proc. Natl. Acad. Sci. USA* **15**, 168.

Iršič, V., Di Dio, E., and Viel, M. (2016). Relativistic effects in Lyman-forest. *JCAP* **1602**, 051. DOI:10.1088/1475-7516/2016/02/051.

Israel, W., and Stewart, J. (1980). *Einstein Commemorative Volume*, ed. A. Held. New York: Plenum Press.

Jackson, J. D. (1975). *Classical Electrodynamics*, 2nd ed. New York: John Wiley & Sons.

Jauch, J. M., and Rorlich, F. (1976). *The Theory of Photons and Electrons* (New York, Springer-Verlag).

Kaiser, N. (1987). Clustering in real space and in redshift space. *Mon. Not. Roy. Ast. Soc.* **227**, 1.

Kamionkowski, M., Kosowsky, A., and Stebbins, A. (1997). Statistics of cosmic microwave background polarization. *Phys. Rev. Lett.* **78**, 2058.

Kendall, M. S., and Stuart, A. (1969). *Advanced Theory of Statistics*, vol. II. New York: Van Nostrand.

Kodama, H., and Sasaki, M. (1984). Cosmological perturbation theory. *Prog. Theor. Phys. Suppl.* **78**, 1.

Kogut, A. *et al.* (2006). ARCADE: absolute radiometer for cosmology, astrophysics, and diffuse emission. *New Astron. Rev.* **50**, 925.

Kolb, E., and Turner, M. (1990). *The Early Universe*. Reading, MA: Addison Wesley.

Kovetz, E. D. *et al.* (2017). *Line-Intensity Mapping: 2017 Status Report*. arXiv:1709.09066.

Kunz, M., Trotta, R., and Parkinson, D. (2006). Measuring the effective complexity of cosmological models. *Phys. Rev.* **D74**, 023503.

Lachieze-Rey, M., and Luminet, J. P. (1995). Cosmic topology. *Phys. Rep.* **254**, 135–214.

LaRoque, S. *et al.* (2006). X-ray and Sunyaev–Zel'dovich effect measurements of the gas mass fraction in galaxy clusters. *Astrophys. J.* **652**, 917.

Lazanu, A., Giannantonio, T., Schmittfull, M., and Shellard, E. P. S. (2016). Matter bispectrum of large-scale structure: Three-dimensional comparison between theoretical models and numerical simulations. *Phys. Rev.* **D93**, 083517. DOI:10.1103/PhysRevD.93.083517.

Lemaître, G. (1927). L'univers en expansion. *Ann. Soc. Bruxelles* **47A**, 49.

Lemaître, G. (1931). Expansion of the universe, a homogeneous universe of constant mass and increasing radius accounting for the radial velocity of extra-galactic nebulae. *Mon. Not. R. Ast. Soc.* **91**, 483–490; Expansion of the universe. the expanding universe, *Mon. Not. R. Astron. Soc.* **91**, 490–501.

Lepori, F., Di Dio, E., Viel, M., Baccigalupi, C., and Durrer, R. (2016). The Alcock Paczyski test with Baryon Acoustic Oscillations: Systematic effects for future surveys. *JCAP* **1702** no. 02, 020. DOI:10.1088/1475-7516/2017/02/020.

Lesgourgues, J. (2011). The cosmic linear anisotropy solving system (CLASS) I: Overview. arXiv:1104.2932.

Lewis, A. (2013). Efficient sampling of fast and slow cosmological parameters. *Phys. Rev.* D **87** (2013) no. 10, 103529. DOI:10.1103/PhysRevD.87.103529.

Lewis, A., and Bridle, S. (2002). Cosmological parameters from CMB and other data: A Monte Carlo approach. *Phys. Rev.* **D66**, 103511. (see also http://cosmologist.info/cosmomc/).

Lewis, A., and Challinor, A. (2006). Weak gravitational lensing of the CMB. *Phys. Rep.* **429**, 1.

Lewis, A., Challinor, A., and Lasenby, A. (2000). Efficient computation of CMB anisotropies in closed FRW models. *Astrophys. J.* **538**, 473. (see http://camb.info)

Liddle, A., and Lyth, D. (2000). *Cosmological Inflation and Large Scale Structure*, Cambridge: Cambridge University Press.

Lifshitz, E. (1946). About gravitational stability of expanding worlds *JETP* **10**, 116.

Lifshitz, E., and Pitajewski, L. (1983). *Lehrbuch der Theoretischen Physik*, vol. X. Berlin: Akademie Verlag.

Liguori, M., Sefusatti, E., Fergusson, J. R., and Shellard, E. P. S. (2010). Primordial non-Gaussianity and bispectrum measurements in the cosmic microwave background and large-scale structure. *Adv. Astron.* **2010**, 980523. DOI:10.1155/2010/980523.

Limber, D. N. (1954). The analysis of counts of the extragalactic nebulae in terms of a fluctuating density field. II. *Astrophys. J.* **119**, 655. DOI:10.1086/145870.

Linde, A. (1989). *Inflation and Quantum Cosmology* (New York, Academic Press).

Lizarraga, J., Urrestilla, J., Daverio, D., Hindmarsh, M., and Kunz, M. (2016). New CMB constraints for Abelian Higgs cosmic strings. *JCAP* **1610**, 042. DOI:10.1088/1475-7516/2016/10/042.

Lo Verde, M., and Afshordi, N. (2008). Extended Limber approximation. *Phys. Rev.* **D78**, 123506.

Louis, T. *et al.* [ACTPol Collaboration] (2017). The Atacama Cosmology Telescope: Two-season ACTPol spectra and parameters. *JCAP* **1706**, 031. DOI:10.1088/1475-7516/2017/06/031.

MacKay, D. J. C. (2003). *Information Theory, Inference, and Learning Algorithms*. Cambridge: Cambridge University Press (see also http:www.inference.phy.cam.ac.uk/mackay/itprnn/book.html).

Maggiore, M. (2005). *A Modern Introduction to Quantum Field Theory*. Oxford: Oxford University Press.

Maggiore, M. (2007). *Gravitational Waves*. Oxford: Oxford University Press.

Maldacena, J. (2003). Non-Gaussian features of primordial fluctuations in single field inflationary models. *J. High Energy Phys.* **0305**, 013.

Mao, Q., Berlind, A. A., Scherrer, R. J., Neyrinck, M. C., Scoccimarro, R., Tinker, J. L., McBride, C. K., and Schneider, D. P. (2017). Cosmic voids in the SDSS DR12 BOSS Galaxy Sample: The Alcock Paczynski test. *Astrophys. J.* **835**, 160. DOI:10.3847/1538-4357/835/2/160.

Marozzi, G., Fanizza, G., Di Dio, E., and Durrer, R. (2017). Impact of next-to-leading order contributions to cosmic microwave background lensing. *Phys. Rev. Lett.* **118**, 211301. DOI:10.1103/PhysRevLett.118.211301.

Martin, J., and Schwarz, D. (2003). WKB approximation for inflationary cosmological perturbations. *Phys. Rev.* **D67**, 083512.

Matsubara, T. (2000). The gravitational lensing in redshift-space correlation functions of galaxies and quasars. *Astrophys. J.* **537**, L77. DOI:10.1086/312762.

Mazumdar, A., and Wang, L. F. (2012). Separable and non-separable multi-field inflation and large non-Gaussianity. *JCAP* **1209**, 005. DOI:10.1088/1475-7516/2012/09/005.

McDonald, P. *et al.* (2005). The linear theory power spectrum from the Lyman-alpha forest in the Sloan Digital Sky Survey. *Astrophys. J.* **635**, 761.

Mészáros, P. (1974). The behaviour of point masses in an expanding cosmological substratum. *Astron. Astrophys.* **37**, 225.

Mitsou, E., Yoo, J., Durrer, R., Scaccabarozzi, F., and Tansella, V. (2019). Angular N-point spectra and cosmic variance on the light-cone. arXiv:1905.01293.

Montanari, F., and Durrer, R. (2012). A new method for the Alcock-Paczynski test. *Phys. Rev.* **D86**, 063503. DOI:10.1103/PhysRevD.86.063503.

Montanari, F., and Durrer, R. (2015). Measuring the lensing potential with tomographic galaxy number counts. *JCAP* **1510**, 070.

Moodley, K. *et al.* (2004). Constraints on iso-curvature models from the WMAP first-year data. *Phys. Rev.* **D70**, 103520.

Mukhanov, V. F. (2005). *Physical Foundations of Cosmology*. Cambridge: Cambridge University Press.

Mukhanov, V. F., and Chibisov, G. (1982). The vacuum energy and large scale structure of the Universe. *JETP* **56**, 258.

Mukhanov, V. F., Feldman, H. A., and Brandenberger, R. H. (1992). Theory of cosmological perturbations. *Phys. Rep.* **215**, 203.

Newman, E. T., and Penrose, R. (1966). Note on the Bondi–Metzner–Sachs group. *J. Math. Phys.* **7**, 863.

Newton, I. (1958). Letters from Sir Isaac Newton to Dr. Bentley, Letter I, 203ff quoted by A. Koyré, *From the Classical World to the Infinite Universe*. New York: Harper and Row.

Nussbaumer, H., and Bieri, L. (2009). *Discovering the Expanding Universe*. Cambridge: Cambridge University Press.

Obreschkow, D., Power, C., Bruderer, M., and Bonvin, C. (2013). A robust measure of cosmic structure beyond the power-spectrum: Cosmic filaments and the temperature of dark matter. *Astrophys. J.* **762**, 115. DOI:10.1088/0004-637X/762/2/115.

Okamoto, T., and Hu, W. (2003). CMB lensing reconstruction on the full sky. *Phys. Rev.* **D67**, 083002. DOI:10.1103/PhysRevD.67.083002.

Øksendal, B. K. (2007). *Stochastic Differential Equations*. Berlin: Springer-Verlag.

Olive, K. A., Steigman, G., and Walker, T. P. (2000). Primordial nucleosynthesis: theory and observations. *Phys. Rep.* **333**, 389.

Olive, K. A *et al.* (PDG) (2014). Neutrino mass, mixing, and oscillations. *Chin. Phys.* **C38**, 090001 (http://pdg.lbl.gov).

Padmanabhan, T. (2000). *Theoretical Astrophysics* vol. I. Cambridge: Cambridge University Press.

Padmanabhan, T. (2010). *Gravitation: Foundations and Frontiers*. Cambridge: Cambridge University Press.

Palanque-Delabrouille, N. *et al.* (2015). Constraint on neutrino masses from SDSS-III/BOSS Lyα forest and other cosmological probes. *JCAP* **1502**, 045. DOI:10.1088/1475-7516/2015/02/045.

Park, D. (1974). *Introduction to the Quantum Theory*, 2nd ed. New York: McGraw-Hill.

Particle Data Group (2004). Review of particle physics. *Phys. Lett.* **B592**, 191–227.

Particle Data Group (2006). Review of particle physics, ed. W.-M. Yao *et al.*, *J. Phys.* **G33**, 1.

Peacock, J. A. (1999). *Cosmological Physics*. Cambridge: Cambridge University Press.

Peebles, P. J. E. (1980). *The Large Scale Structure of the Universe*. Princeton, NJ: Princeton University Press.

Peebles, P. J. E. (1993). *Principles of Physical Cosmology*. Princeton, NJ: Princeton University Press.

Perlmutter, S. *et al.* [Supernova Cosmology Project Collaboration] (1999). Measurements of Ω and Λ from 42 high redshift supernovae. *Astrophys. J.* **517**, 565. DOI:10.1086/307221.

Pietrobon, D., Balbi, A., and Marinucci D. (2006). Integrated Sachs–Wolfe effect from the cross correlation of WMAP3 year and the NRAO VLA sky survey data: New results and constraints on dark energy. *Phys. Rev.* **D74**, 043524.

Pratten, G., and Lewis, A. (2016). Impact of post-Born lensing on the CMB. *JCAP* **1608**, 047. DOI:10.1088/1475-7516/2016/08/047.

Press, W., and Schechter, P. (1974). Formation of galaxies and clusters of galaxies by self-similar gravitational condensation. *Astrophys. J.* **187**, 425.

Reed, M., and Simon, B. (1980). *Methods of Modern Mathematical Physics*, vol. 1. *Functional Analysis*. New York: Academic Press.

Regan, D. M., Shellard, E. P. S., and Fergusson, J. R. (2010). General CMB and primordial trispectrum estimation. *Phys. Rev.* **D82** (2010) 023520. DOI:10.1103/PhysRevD.82.023520.

Renaux-Petel, S. (2013). DBI Galileon in the effective field theory of inflation: Orthogonal non-Gaussianities and constraints from the trispectrum. *JCAP* **1308**, 017. DOI:10.1088/1475-7516/2013/08/017.

Riess, A. G. *et al.* [Supernova Search Team] (1998). Observational evidence from supernovae for an accelerating universe and a cosmological constant. *Astron. J.* **116**, 1009. DOI:10.1086/300499.

Riess, A. G. (2019). The expansion of the Universe is faster than expected. *Nature Rev. Phys.* **2**, 10. DOI:10.1038/s42254-019-0137-0.

Robertson, H. P. (1936). Kinematics and world structure I, II, III. *Astrophys. J.* **82**, 284–301; *Astrophys. J.* **83**, 187–201, 257–271.

Rocher, J., and Sakellariadou, M. (2005). Constraints on supersymmetric grand unified theories from cosmology. *J. Cosmol. Astroport. Phys.* **0503**, 004.

Rubino-Martin, J. A., Chluba, J., and Sunyaev, R. A. (2006). Lines in the cosmic microwave background spectrum from the epoch of cosmological hydrogen recombination. *Mon. Not. R. Astron. Soc.* **371**, 1939.

Rybicki, G. B., and Lightman, A. P. (1979). *Radiative Processes in Astrophysics*. New York: John Wiley & Sons.

Sachs, R. K., and Wolfe, A. M. (1967). Perturbations of a cosmological model and angular variations of the microwave background. *Astrophys. J.* **147**, 73.

Sakurai, J. J. (1993). *Modern Quantum Mechanics*. Reading, MA: Addison-Wesley.

Sasaki, M. (1987). The magnitude-redshift relation in a perturbed Friedmann universe. *Mon. Not. Roy. Ast. Soc.* **228**, 653.

Schmalzing, J., and Gorski, K. M. (1998). Minkowski functionals used in the morphological analysis of cosmic microwave background anisotropy maps. *Mon. Not. Roy. Astron. Soc.* **297** 355. DOI:10.1046/j.1365-8711.1998.01467.x.

Schmidt, B. P. *et al.* [Supernova Search Team] (1998 The High Z supernova search: Measuring cosmic deceleration and global curvature of the universe using type Ia supernovae. *Astrophys. J.* **507**, 46. DOI:10.1086/306308.

Schneider, P. (2007). Weak gravitational lensing. In *33rd Saas Fee Lectures*, ed. P. Jetzer and P. North. Berlin: Springer-Verlag.

Schneider, P., Ehlers, J., and Falco, E. E. (1993). Gravitational lenses. Springer-Verlag Berlin.

Schneider, R. (1993). *Convex Bodies: The Brunn-Minkowski Theory*. Cambridge: Cambridge University Press.

Schwarz, D., Terrero-Escalante, C., and Garcia, A. (2001). Higher order corrections to primordial spectra from cosmological inflation. *Phys. Lett.* **B517**, 243–249.

Scodeller, S., Kunz, M., and Durrer, R. (2009). CMB anisotropies from acausal scaling seeds. *Phys. Rev.* **D79**, 083515. DOI:10.1103/PhysRevD.79.083515.

Seager, S., Sasselov, D. D., and Scott, D. (1999). A new calculation of the recombination epoch. *Astrophys. J.* **523**. L1–L5. DOI:10.1086/312250.

Seljak, U. (1996a). Rees–Sciama effect in a CDM universe. *Astrophys. J.* **460**, 549.

Seljak, U. (1996b). Measuring polarization in the cosmic microwave background. *Astrophys. J.* **482**, 6.

Seljak, U., and Zaldarriaga, M. (1996). A line of sight integration approach to cosmic microwave background anisotropies. *Astrophys. J.* **469**, 437.

Sellentin, E., and Durrer, R. (2015). Detecting the cosmological neutrino background in the CMB. *Phys. Rev.* **D92**, 063012. DOI:10.1103/PhysRevD.92.063012.

Senatore, L., Smith, K. M., and Zaldarriaga, M. (2010). Non-Gaussianities in single field inflation and their optimal limits from the WMAP 5-year data. *JCAP* **1001**, 028. DOI:10.1088/1475-7516/2010/01/028.

Shajib, A. J., and Wright, E. L. (2016). Measurement of the integrated Sachs-Wolfe effect using the AllWISE data release. *Astrophys. J.* **827**, 116. DOI:10.3847/0004-637X/827/2/116.

Shaw, J. R., and Chluba, J. (2011). Precise cosmological parameter estimation using CosmoRec. *Mon. Not. Roy. Astron. Soc.* **415**, 1343. DOI:10.1111/j.1365-2966.2011.18782.x.

Silk, J. (1968). Cosmic black body radiation and galaxy formation. *Astrophys. J.* **151**, 459. DOI:10.1086/149449.

Singal, J. *et al.* (2005). Design and performance of sliced-aperture corrugated feed horn antennas. *Rev. Sci. Instrum.* **76**, 124703. (For updates see http://arcade.gsfc.nasa.gov/)

Smoot, G. F. *et al.* (1992). Structure in the COBE differential microwave radiometer first-year maps. *Astrophys. J.* **396**, L1.

Songaila, A., and Cowie, L. L. (1996). Metal enrichment and ionization balance in the Lyman α forest at $z = 3$. *Astron. J.* **112**, 335.

Spergel, D., and Zaldarriaga, M. (1997). CMB polarization as a direct test of inflation. *Phys. Rev. Lett.* **79**, 2180.

Spergel, D. N. *et al.* (2003). First-year Wilkinson Microwave Anisotropy Probe (WMAP) observations: Determinations of cosmological parameters. *Astrophys. J. Suppl.* **148**, 175.

Stewart, J. M. (1971). *Non-equilibrium Relativistic Kinetic Theory*. Springer Lecture Notes. Berlin: Springer-Verlag.

Stewart, J. M., and Walker, M. (1974). Perturbations of space-times in general relativity. *Proc. R. Soc. London* **A341**, 49.

Straumann, N. (1974). Minimal assumptions leading to a Robertson–Walker model of the Universe. *Hel. Phys. Acta* **47**, 379.

Straumann, N. (1984). *General Relativity and Relativistic Astrophysics*. Berlin: Springer-Verlag.

Straumann, N. (2004). *General Relativity with Applications to Astrophysics*. Berlin: Springer-Verlag.

Sylos Labini, F., Montuori, M., and Pietronero, L. (1998). Scale-invariance of galaxy clustering. *Phys. Rep.* **293**, 61–226.

Tanabashi, M. *et al.* (Particle Data Group) (2018 and 2019 update). http://pdg.lbl.gov/2019/reviews/contents~sports.html *Phys. Rev.* **D98**, 030001.

Tansella, V., Bonvin, C., Durrer, R., Ghosh, B., and Sellentin, E. (2018). The full-sky relativistic correlation function and power spectrum of galaxy number counts. Part I: Theoretical aspects. *JCAP* **1803**, 019. DOI:10.1088/1475-7516/2018/03/019.

Tansella, V., Jelic-Cizmek, G., Bonvin, C. and Durrer, R. (2018). Coffe: A code for the full-sky relativistic galaxy correlation function. *JCAP* **1810**, 032. DOI:10.1088/1475-7516/2018/10/032.

Tegmark, M. *et al.* (2004). The 3D power spectrum of galaxies from the SDSS. *Astrophys. J.* **606**, 702.

Thorne, K. (1980). Multipole expansions of gravitational radiation. *Rev. Mod. Phys.* **52**, 299.

Tomita, H. (1986). Curvature invariants of random interface generated by Gaussian fields. *Prog. Theo. Phys.* **76**, 952. DOI:10.1143/PTP.76.952.

Trotta, R. (2017). Bayesian methods in cosmology. arXiv:1701.01467.

Trotta, R., Riazuelo, A., and Durrer, R. (2001). Cosmic microwave background anisotropies with mixed iso-curvature perturbations. *Phys. Rev. Lett.* **87**, 231301. DOI:10.1103/PhysRevLett.87.231301.

Trotta, R., Riazuelo, A. and Durrer, R. (2003). The cosmological constant and general iso-curvature initial conditions, *Phys. Rev.* **D67**, 063520. DOI:10.1103/PhysRevD.67.063520.

Vielva, P., Martnez-Gonzlez, E., Barreiro R. B., Sanz, J. L., and Cayón, L. (2004). Detection of non-Gaussianity in the Wilkinson microwave anisotropy probe first-year *Astrophys. J.* **609**, 22. DOI:10.1086/421007.

Vilenkin, N. Y., and Smorodinskii, Y. A. (1964). Invariant expansion of relativistic amplitudes. *Sov. Phys. JETP*, **19**, 1209.

Wald, R. M. (1984). *General Relativity*. Chicago: University of Chicago Press.

Walker, A. G. (1936). *Proc. Lond. Math. Soc.* **42**, 90.

Weinberg, S. (1995). *The Quantum Theory of Fields I*. Cambridge: Cambridge University Press.

Weinberg, S. (2005). Quantum contributions to cosmological correlations. *Phys. Rev.* **D72**, 043514. DOI:10.1103/PhysRevD.72.043514.

Weinberg, S. (2008). *Cosmology*. Oxford: Oxford University Press.

Wigner, E. P. (1959). *Group Theory*. New York: Academic Press.

Wolf, J. (1974). *Spaces of Constant Curvature*. Boston: American Mathematical Society.

Wong, K. C. *et al.* (2019). H0LiCOW XIII. A 2.4% measurement of H_0 from lensed quasars: 5.3σ tension between early and late-Universe probes. arXiv:1907.04869.

Wong, W. Y., Seager, S., and Scott, D. (2006). Spectral distortions to the cosmic microwave background from the recombination of hydrogen and helium. *Mon. Not. R. Astron. Soc.* **367**, 1666. DOI:10.1111/j.1365-2966.2006.10076.x.

Wouthuysen, S. A. (1952). On the excitation mechanism of the 21-cm (radio-frequency) interstellar hydrogen emission line. *Astron. J.* **57**, 31.

Zaldarriaga, M., and Seljak, U. (1997). An all-sky analysis of polarization in the microwave background. *Phys. Rev.* **D55**, 1830–1840.

Zel'dovich, Y. B. (1972). A hypothesis, unifying the structure and the entropy of the Universe. *Mon. Not. R. Astron. Soc.* **160**, 1.

Zhang, P., Liguori, M., Bean, R., and Dodelson, S. (2007). Probing gravity at cosmological scales by measurements which test the relationship between gravitational lensing and matter overdensity. *Phys. Rev. Lett.* **99**, 141302. DOI:10.1103/PhysRevLett.99.141302.

Index

Printed in the United States
By Bookmasters